SEVENTH EDITION

Modern Control Systems

RICHARD C. DORF

University of California, Davis

ROBERT H. BISHOP

The University of Texas at Austin

ADDISON-WESLEY PUBLISHING COMPANY

Reading, Massachusetts   Menlo Park, California
New York   Don Mills, Ontario   Wokingham, England
Amsterdam   Bonn   Sydney   Singapore   Tokyo   Madrid
San Juan   Milan   Paris

This book is in the Addison-Wesley Series in Electrical and Computer Engineering: Control Engineering

Feedback Control of Dynamic Systems, Third Edition, 52747
Gene F. Franklin and J. David Powell

Digital Control of Dynamic Systems, Second Edition, 11938
Gene F. Franklin, J. David Powell, and Michael L. Workman

Modern Control Systems, Seventh Edition, 50174
Richard C. Dorf and Robert H. Bishop

Adaptive Control, Second Edition, 55866
Karl J. Åstrom and Bjorn Wittenmark

Introduction to Robotics, Second Edition, 09528
John J. Craig

Computer Control of Machines and Processes, 10645
John G. Bollinger and Neil A. Duffie

George J. Horesta, *Sponsoring Editor*
Helen M. Wythe, *Senior Production Editor*
Laurie McGuire, *Development Editor*
Anita M. Devine, *Assistant Editor*
Barbara Gracia, *Woodstock Publishers' Services*
Margaret O. Tsao, *Text Designer*

Scientific Illustrators, *Illustrations*
Alena Konecny, *Art Editor*
Eileen R. Hoff, *Cover Supervisor*
Fahrenheit Design, *Cover Design*
Judith Y. Sullivan, *Senior Manufacturing Coordinator*

Cover: Graph—Family of step responses for the Hubble spacecraft. Photograph of Hubble telescope courtesy of NASA.

Many of the designations used by manufacturers and sellers to distinguish their products are claimed as trademarks. Where those designations appear in this book, and Addison-Wesley was aware of a trademark claim, the designations have been printed in initial caps or all caps.

MATLAB® is a registered trademark of The MathWorks, Inc., 24 Prime Park Way, Natick, MA 01760-1520.
Phone: (508) 653-1415, Fax: (508) 653-2997
Email: info@mathworks.com

Library of Congress Cataloging-in-Publication Data

Dorf, Richard C.
    Modern control systems / Richard C. Dorf, Robert H. Bishop.
      p.   cm.
    Includes bibliographical references and index.
    ISBN 0–201–50174–0
    1. Feedback control systems.   I. Bishop, Robert H., 1957– .
  II. Title.
  TJ216.D67   1995
  629.8'3—dc20   94–5172
                                     CIP

Reprinted with corrections March, 1995

4 5 6 7 8 9 10–DO–98 97

Of the greater teachers—
when they are gone,
their students will say:
we did it ourselves.

Dedicated to:

Lynda Ferrera Bishop
and
Joy MacDonald Dorf

In grateful appreciation

Preface

MODERN CONTROL SYSTEMS—THE BOOK

The Hubble Space Telescope, which appears on the cover, was launched with an optical flaw. In late 1993 after a record five days of extravehicular activity, the Hubble telescope was successfully repaired, including the installation of new solar arrays that eliminate the jitter experienced as the telescope passes from day to night. This allows the Hubble to point within 0.007 arcsec accuracy. This pointing accuracy is equivalent to keeping a laser light located in New York City focused on a six-inch mirror in Los Angeles. Interesting real-world control problems, such as the Hubble telescope, are used as illustrative examples throughout the book. For example, a Hubble pointing control design problem is discussed in the Design Example in Section 5.11, page 251.

Control engineering is an exciting and a challenging field. By its very nature, control engineering is a multidisciplinary subject, and it has taken its place as a core course in the engineering curriculum. It is reasonable to expect different approaches to mastering and practicing the art of control engineering. Since the subject has a strong mathematical foundation, one might approach it from a strictly theoretical point of view, emphasizing theorems and proofs. On the other hand, since the ultimate objective is to implement controllers in real systems, one might take an ad hoc approach relying only on intuition and hands-on experience when designing feedback control systems. Our approach is to present a control engineering methodology that, while based on mathematical fundamentals, stresses physical system modeling and practical control system designs with realistic system specifications.

We believe that the most important and productive approach to learning is for each of us to rediscover and recreate anew the answers and methods of the past. Thus the ideal is to present the student with a series of problems and questions and point to some of the answers that have been obtained over the past decades. The traditional method—to confront the student not with the problem but with the finished solution—is to deprive the student of all excitement, to shut off the creative impulse, to reduce the adventure of humankind to a dusty heap of theorems. The issue, then, is to present some of the unanswered and important problems that we continue to confront. For it may be asserted that what we have truly learned and understood, we discovered ourselves.

The purpose of this book is to present the structure of feedback control theory and to provide a sequence of exciting discoveries as we proceed through the text and problems. If this book is able to assist the student in discovering feedback control system theory and practice, it will have succeeded.

THE AUDIENCE

This text is designed for an introductory undergraduate course in control systems for engineering students. There is very little demarcation between aerospace, chemical, electrical, industrial, and mechanical engineering in control system practice; therefore this text is written without any conscious bias toward one discipline. Thus it is hoped that this book will be equally useful for all engineering disciplines and, perhaps, will assist in illustrating the utility of control engineering. The numerous problems and examples represent all fields, and the examples of the sociological, biological, ecological, and economic control systems are intended to provide the reader with an awareness of the general applicability of control theory to many facets of life. We believe that exposing students of one discipline to examples and problems from other disciplines will provide them with the ability to see beyond their own field of study. Many students pursue careers in related engineering fields other than their own. For example, many electrical and mechanical engineers find themselves in the aerospace industry working alongside aerospace engineers. We hope this introduction to control engineering will give students a broader understanding of control system design and analysis.

In its first six editions, *Modern Control Systems* has been used in senior-level courses for engineering students at more than 400 colleges and universities. It also has been used in courses for engineering graduate students with no previous background in control engineering.

REVISED AND UPDATED!—THE SEVENTH EDITION

With the seventh edition, we have revised and updated the chapters on digital control systems and robust control systems. The coverage of state variable methods has been considerably expanded, with separate chapters on state variable models and the design of state variable feedback systems.

Inexpensive yet powerful computers and software have influenced all aspects of control engineering: system design, analysis, testing, and implementation. In this new edition, we incorporate the idea of computer-aided design and analysis utilizing *Matlab** and the Control System Toolbox. Every chapter but the first contains a separate section covering the relevant *Matlab* topics and establishes a connection between the chapter discussion and the utilization of the computer to assist in the design and analysis of feedback control systems. For the novice, an introduction to *Matlab* basics is presented in Appendix F. Chapter examples are re-solved in the *Matlab* section to demonstrate the utility of the computer. Additionally, *Matlab* problems are included in the problem section for each chapter. Of course, the *Matlab* material can be omitted by simply skipping these sections.

Finally, the format of the book has been enlarged and now includes a two-color scheme to enhance the overall quality and readability. Important concepts and terms are highlighted for emphasis. For example, the discussion of control systems in Section 4.1, page 170, includes the following highlight regarding open-loop systems:

---

* *Matlab* is a registered trademark of The MathWorks, Inc.

An open-loop (direct) system operates without feedback and directly
generates the output in response to an input signal.

MAIN FEATURES

The book is organized around the concepts of control system theory as they have been
developed in the frequency and time domains. A real attempt has been made to make the
selection of topics, as well as the systems discussed in the examples and problems, modern
in the best sense. Therefore this book includes discussions on robust control systems and
system sensitivity, state variable models, controllability and observability, computer control systems, internal model control, robust PID controllers, computer-aided design and
analysis, to name a few. However, the classical topics of control theory that have proved to
be so very useful in practice have been retained and expanded.

Building Basic Principles: From Classical to Modern.   Our goal was to present a
clear exposition of the basic principles of frequency- and time-domain design techniques.
The classical methods of control engineering are thoroughly covered: Laplace transforms
and transfer functions; root locus design; Routh–Hurwitz stability analysis; frequency response methods, including Bode, Nyquist, and Nichols; steady-state error for standard test
signals; second-order system approximations; and phase and gain margin and bandwidth.
In addition, coverage of the state variable method has been significantly expanded. Fundamental notions of controllability and observability for state variable models are discussed
and related to signal-flow graphs. Full state feedback design with Ackermann's formula for
pole placement is presented, along with a discussion on the limitations of state variable
feedback.

Upon this strong foundation of basic principles, the book provides many opportunities
to explore topics beyond the traditional. Recent advances in robust control theory are introduced in Chapter 12. The implementation of digital computer control systems is discussed
in Chapter 13. Each chapter but the first uses a *Matlab* section to introduce the student to
the notion of computer-aided design and analysis. The book concludes with an extensive
References section, divided by chapter, to guide the student to further sources of information on control engineering.

Progressive Development of Problem-Solving Skills.   Reading the chapters, at-
tending lectures and taking notes, and working through the illustrated examples are all part
of the learning process. But the real test comes at the end of the chapter with the problems.
The book takes the issue of problem solving seriously. In each chapter, there are five problem types:

- Exercises
- Problems
- Advanced Problems
- Design Problems
- *Matlab* Problems.

For example, the problem set for The Root Locus Method, Chapter 7 (see page 364) includes 23 exercises, 39 problems, 12 advanced problems, 13 design problems, and 5 *Matlab* problems. The exercises permit the students to utilize readily the concepts and methods introduced in each chapter by solving relatively straightforward exercises before attempting the more complex problems. Answers to selected exercises are provided. The problems require an extension of the concepts of the chapter to new situations. A new edition to the problem set is the advanced problems, which represent problems of increasing complexity. The design problems emphasize the design task; the *Matlab* problems give the student practice with problem solving using computers. In total, the book contains more than 800 problems. The abundance of problems of increasing complexity gives students confidence in their problem-solving ability as they work their way from the exercises to the design and *Matlab* problems. A complete instructor manual, available for all adopters of the text for course use, contains complete solutions to all end-of-chapter problems.

A set of M-files, the *Modern Control Systems Toolbox*, have been developed by the authors to supplement the text. The M-files contain the scripts from each *Matlab* example in the text and Appendix F. You may receive a free diskette (PC or Macintosh format), which includes the functions from The MathWorks, by returning the business reply card bound in the back of this book. Or you may retrieve the M-files from The MathWorks anonymous ftp server at **ftp.mathworks.com** in the directory **pub/books/dorf/**. The *Modern Control Systems Toolbox* is for use with the program, *MATLAB*, available from The MathWorks, Inc. located in Natick, MA, U.S.A. (phone: (508) 653-1415, fax: (508) 653-2997, email: info@mathworks.com, http: www.mathworks.com.)

Design Emphasis Without Compromising Basic Principles.   The all-important topic of design of real-world, complex control systems is a major theme throughout the text. Emphasis on design for real-world applications addresses interest in design by ABET and industry. Each chapter contains at least one design example, including the following:

- turntable speed control (Sec. 1.11, page 19)
- laboratory robot (Sec. 2.11, page 75)
- printer belt drive (Sec. 3.10, page 145)
- Mars rover vehicle (Sec. 4.8, page 189)
- Hubble Space Telescope pointing control (Sec. 5.11, page 251)
- tracked vehicle turning control (Sec. 6.6, page 293)
- laser manipulator control system (Sec. 7.8, page 353)
- engraving machine control system (Sec. 8.7, page 416)
- remotely controlled reconnaissance vehicle (Sec. 9.8, page 480)
- *X-Y* plotter (Sec. 10.13, page 568)
- automatic test system (Sec. 11.9, page 627)
- ultra-precision diamond turning machine (Sec. 12.12, page 678)
- worktable motion control system (Sec. 13.9, page 729).

The *Matlab* sections assist students in utilizing computer-aided design and analysis concepts and rework many of the design examples. For example, in Chapter 6, the tracked vehicle turning control design example is also presented in the *Matlab* section on page 298. A *Matlab* script that can be used to analyze the design is presented in Figure 6.17, p. 300:

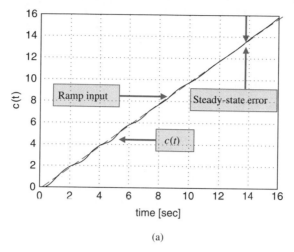

(a)

aKramp.m

```
% This script computes the ramp response
% for the two-track vehicle turning control
% problem with a=0.6 and K=70
%
t=[0:0.01:10]; u=t;          ← u = unit ramp input
numgc=[1 0.6]; dengc=[1 1];
numg=[70]; deng=[1 7 10 0];   ← a = 0.6 and K = 70
[numa,dena]=series(numgc,dengc,numg,deng);
[num,den]=cloop(numa,dena);
[y,x]=lsim(num,den,u,t);      ← Linear simulation
plot(t,y,t,u), grid
xlabel('time [sec]'), ylabel('c(t)')
```

(b)

In general, each script is annotated with comment boxes that highlight important aspects of the script. The accompanying output of the script (generally a graph) also contains comment boxes pointing out significant elements. The scripts can also be utilized with modifications as the foundation for solving other related problems.

Learning Enhancement. Each chapter begins with a chapter Preview describing the topics the student can expect to encounter. The chapters conclude with an end-of-chapter Summary and Terms and Concepts. These sections reinforce the important concepts introduced in the chapter and serve as a reference for later use.

A second color has been introduced to the book to add emphasis when needed and to make the graphs and figures more easily interpreted. Problem 2.47, p. 108, asks the student to model the tire movement for a truck loaded with a force, F, on the support spring. The associated Figure P2.47, p. 109, assists the student with (a) visualizing the problem, and (b) taking the next step to develop the mathematical model:

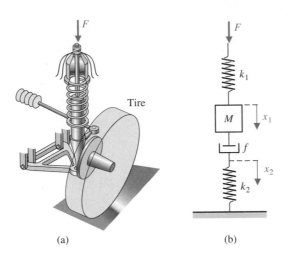

(a)   (b)

THE ORGANIZATION

Chapter 1 Introduction to Control Systems.   Chapter 1 provides an introduction to the basic history of control theory and practice. The purpose of this chapter is to describe the general approach to designing and building a control system.

Chapter 2 Mathematical Models of Systems.   Mathematical models of physical systems in input–output or transfer function form are developed in Chapter 2. A wide range of systems, including mechanical, electrical, and fluid, are considered.

Chapter 3 State Variable Models.   Mathematical models of systems in state variable form are developed in Chapter 3. Using matrix methods, the transient response of control systems and the performance of these systems are examined.

Chapter 4 Feedback Control System Characteristics.   The characteristics of feedback control systems are described in Chapter 4. The advantages of feedback are discussed, and the concept of the system error signal is introduced.

Chapter 5 The Performance of Feedback Control Systems.   In Chapter 5, the performance of control systems is examined. The performance of a control system is correlated with the s-plane location of the poles and zeros of the transfer function of the system.

Chapter 6 The Stability of Linear Feedback Systems.   The stability of feedback systems is investigated in Chapter 6. The relationship of system stability to the characteristic equation of the system transfer function is studied. The Routh–Hurwitz stability criterion is introduced.

Chapter 7 The Root Locus Method.   Chapter 7 deals with the motion of the roots of the characteristic equation in the s-plane as one or two parameters are varied. The locus of roots in the s-plane is determined by a graphical method.

Chapter 8 Frequency Response Methods.   In Chapter 8, a steady-state sinusoidal input signal is utilized to examine the steady-state response of the system as the frequency of the sinusoid is varied. The development of the frequency response plot, called the Bode plot, is considered.

Chapter 9 Stability in the Frequency Domain.   System stability utilizing frequency response methods is investigated in Chapter 9. Relative stability and the Nyquist criterion are discussed.

Chapter 10 The Design of Feedback Control Systems.   Several approaches to designing and compensating a control system are described and developed in Chapter 10. Various candidates for service as compensators are presented and it is shown how they help to achieve improved performance.

Chapter 11 The Design of State Variable Feedback Systems.   The main topic of Chapter 11 is the design of control systems using state variable models. Tests for controllability and observability are presented, and the concept of an internal model design is discussed.

Chapter 12 Robust Control Systems.   Chapter 12 deals with the design of highly accurate control systems in the presence of significant uncertainty. Five methods for robust design are discussed, including root locus, frequency response, ITAE methods for robust PID controllers, internal models, and pseudo-quantitative feedback.

Chapter 13 Digital Control Systems.   Methods for describing and analyzing the performance of computer control systems are described in Chapter 13. The stability and performance of sampled-data systems are discussed.

Appendices.   The six appendices are:
A   Laplace Transform Pairs
B   Symbols, Units, and Conversion Factors
C   An Introduction to Matrix Algebra
D   Decibel Conversion
E   Complex Numbers
F   *Matlab* Basics

ACKNOWLEDGMENTS

We wish to express sincere appreciation to the following individuals who have assisted us with the development of the seventh edition: Ye-Hwa Chen, Georgia Institute of Technology; Robert I. Egbert, The Wichita State University; Kelvin T. Erickson, University of Missouri-Rolla; Robert H. Flake, The University of Texas at Austin; J. W. Howze, Texas

A&M University; David Hullender, The University of Texas at Arlington; Randall S. Janka, SKY Computers; Haniph Latchman, University of Florida; Richard Longman, Columbia University; Peter H. Meckl, Purdue University; D. Subbaram Naidu, Idaho State University; R. Lal Tummala, Michigan State University.

OPEN LINES OF COMMUNICATION

The authors and the staff at Addison-Wesley Publishing Company would like to establish a line of communication with the users of *Modern Control Systems*. We encourage all readers to send Addison-Wesley your e-mail address and pass along comments and suggestions for this and future editions. By doing this, we can keep you informed of any general interest news regarding the textbook and pass along interesting comments of other users.

Keep in touch!

Richard C. Dorf   dorf@ece.ucdavis.edu
Robert H. Bishop   bishop@zeus.ae.utexas.edu
Addison-Wesley Publishing Company   controls@aw.com

Contents

CHAPTER 3   STATE VARIABLE MODELS   114

CHAPTER 4   FEEDBACK CONTROL SYSTEM CHARACTERISTICS   169

CHAPTER 8 FREQUENCY RESPONSE METHODS   387

CHAPTER 9 STABILITY IN THE FREQUENCY DOMAIN   444

CHAPTER 10 **THE DESIGN OF FEEDBACK CONTROL SYSTEMS**   **526**

CHAPTER 11 **THE DESIGN OF STATE VARIABLE
                FEEDBACK SYSTEMS**   **603**

Introduction to Control Systems

PREVIEW

A system, consisting of interconnected components, is built to achieve a desired purpose. The performance of this system can be examined, and methods for controlling its performance can be proposed. It is the purpose of this chapter to describe the general approach to designing and building a control system.

To understand the purpose of a control system, it is useful to examine some examples of control systems through the course of history. Even these early systems incorporated the idea of feedback, which we will discuss throughout this book.

Modern control engineering practice includes the use of control strategies for aircraft, rapid transit, the artificial heart, and steel making, among others. We will examine these very interesting applications of control engineering.

1.1 INTRODUCTION

Engineering is concerned with understanding and controlling the materials and forces of nature for the benefit of humankind. Control system engineers are concerned with understanding and controlling segments of their environment, often called *systems*, to provide useful economic products for society. The twin goals of understanding and control are complementary because effective systems control requires that the systems be understood and modeled. Furthermore, control engineering must often consider the control of poorly understood systems such as chemical process systems. The present challenge to control engineers is the modeling and control of modern, complex, interrelated systems such as traffic control systems, chemical processes, and robotic systems. However, simultaneously, the fortunate engineer has the opportunity to control many very useful and interesting industrial automation systems. Perhaps the most characteristic quality of control engineering is the opportunity to control machines and industrial and economic processes for the benefit of society.

Control engineering is based on the foundations of feedback theory and linear system analysis, and it integrates the concepts of network theory and communication theory. Therefore control engineering is not limited to any engineering discipline but is equally applicable for aeronautical, chemical, mechanical, environmental, civil, and electrical engineering. For example, quite often a control system includes electrical, mechanical, and chemical components. Furthermore, as the understanding of the dynamics of business, social, and political systems increases, the ability to control these systems will increase also.

A *control system* is an interconnection of components forming a system configuration that will provide a desired system response. The basis for analysis of a system is the foundation provided by linear system theory, which assumes a cause–effect relationship for the components of a system. Therefore a component or *process* to be controlled can be represented by a block as shown in Fig. 1.1. The input–output relationship represents the cause-and-effect relationship of the process, which in turn represents a processing of the input signal to provide an output signal variable, often with a power amplification. An *open-loop* control system utilizes a controller or control actuator to obtain the desired response, as shown in Fig. 1.2. An open-loop system is a system without feedback.

An open-loop control system utilizes an actuating device to control the process directly without using feedback.

In contrast to an open-loop control system, a closed-loop control system utilizes an additional measure of the actual output to compare the actual output with the desired output response. The measure of the output is called the *feedback signal*. A simple *closed-loop feedback control system* is shown in Fig. 1.3. A feedback control system is a control system that tends to maintain a prescribed relationship of one system variable to another by comparing functions of these variables and using the difference as a means of control.

A feedback control system often uses a function of a prescribed relationship between the output and reference input to control the process. Often the difference between the output of the process under control and the reference input is amplified and used to control

FIGURE 1.1
Process to be controlled.

FIGURE 1.2
Open-loop control system (without feedback).

FIGURE 1.3
Closed-loop
feedback control
system (with
feedback).

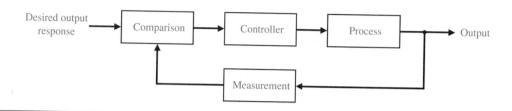

the process so that the difference is continually reduced. The feedback concept has been the foundation for control system analysis and design.

> **A closed-loop control system uses a measurement of the output and feedback of this signal to compare it with the desired input (reference or command).**

Due to the increasing complexity of the system under control and the interest in achieving optimum performance, the importance of control system engineering has grown in the past decade. Furthermore, as the systems become more complex, the interrelationship of many controlled variables must be considered in the control scheme. A block diagram depicting a *multivariable control system* is shown in Fig. 1.4.

A common example of an open-loop control system is an electric toaster in the kitchen. An example of a closed-loop control system is a person steering an automobile (assuming his or her eyes are open) by looking at the auto's location on the road and making the appropriate adjustments.

The introduction of feedback enables us to control a desired output and can improve accuracy, but it requires attention to the issue of stability of response.

FIGURE 1.4
Multivariable
control system.

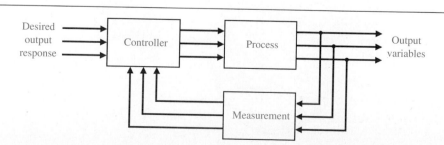

1.2 HISTORY OF AUTOMATIC CONTROL

The use of feedback to control a system has had a fascinating history. The first applications of feedback control rest in the development of float regulator mechanisms in Greece in the period 300 to 1 B.C. [1, 2, 3]. The water clock of Ktesibios used a float regulator (refer to Problem 1.11). An oil lamp devised by Philon in approximately 250 B.C. used a float regulator in an oil lamp for maintaining a constant level of fuel oil. Heron of Alexandria, who lived in the first century A.D., published a book entitled *Pneumatica,* which outlined several forms of water-level mechanisms using float regulators [1].

The first feedback system to be invented in modern Europe was the temperature regulator of Cornelis Drebbel (1572–1633) of Holland [1]. Dennis Papin [1647–1712] invented the first pressure regulator for steam boilers in 1681. Papin's pressure regulator was a form of safety regulator similar to a pressure-cooker valve.

The first automatic feedback controller used in an industrial process is generally agreed to be James Watt's *flyball governor,* developed in 1769 for controlling the speed of a steam engine [1, 2]. The all-mechanical device, shown in Fig. 1.5, measured the speed of the output shaft and utilized the movement of the flyball with speed to control the valve and therefore the amount of steam entering the engine. As the speed increases, the ball weights rise and move away from the shaft axis, thus closing the valve. The flyweights require power from the engine to turn and therefore cause the speed measurement to be less accurate.

The first historical feedback system, claimed by Russia, is the water-level float regulator said to have been invented by I. Polzunov in 1765 [4]. The level regulator system is shown in Fig. 1.6. The float detects the water level and controls the valve that covers the water inlet in the boiler.

The period preceding 1868 was characterized by the development of automatic control systems through intuition and invention. Efforts to increase the accuracy of the control system led to slower attenuation of the transient oscillations and even to unstable systems.

FIGURE 1.5
Watt flyball
governor.

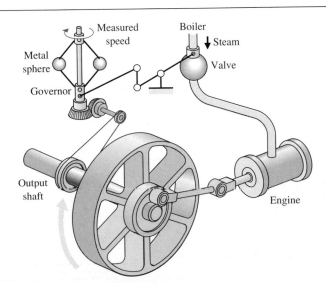

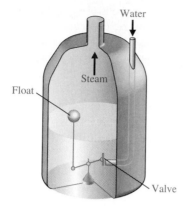

FIGURE 1.6
Water-level float
regulator.

It then became imperative to develop a theory of automatic control. J. C. Maxwell formulated a mathematical theory related to control theory using a differential equation model of a governor [5]. Maxwell's study was concerned with the effect various system parameters had on the system performance. During the same period, I. A. Vyshnegradskii formulated a mathematical theory of regulators [6].

Prior to World War II, control theory and practice developed in the United States and Western Europe in a different manner than in Russia and Eastern Europe. A main impetus for the use of feedback in the United States was the development of the telephone system and electronic feedback amplifiers by Bode, Nyquist, and Black at Bell Telephone Laboratories [7–10, 12]. The frequency domain was used primarily to describe the operation of the feedback amplifiers in terms of bandwidth and other frequency variables. In contrast, the eminent mathematicians and applied mechanicians in the former Soviet Union inspired and dominated the field of control theory. Therefore the Russian theory tended to utilize a time-domain formulation using differential equations.

A large impetus to the theory and practice of automatic control occurred during World War II when it became necessary to design and construct automatic airplane pilots, gun-positioning systems, radar antenna control systems, and other military systems based on the feedback control approach. The complexity and expected performance of these military systems necessitated an extension of the available control techniques and fostered interest in control systems and the development of new insights and methods. Prior to 1940, for most cases, the design of control systems was an art involving a trial-and-error approach. During the 1940s, mathematical and analytical methods increased in number and utility, and control engineering became an engineering discipline in its own right [10–12].

Frequency-domain techniques continued to dominate the field of control following World War II with the increased use of the Laplace transform and the complex frequency plane. During the 1950s, the emphasis in control engineering theory was on the development and use of the s-plane methods and, particularly, the root locus approach. Furthermore, during the 1980s, the utilization of digital computers for control components became routine. The technology of these new control elements to perform accurate and rapid calculations was formerly unavailable to control engineers. There are now over four hundred thousand digital process control computers installed in the United States [14, 27]. These computers are employed especially for process control systems in which many variables are measured and controlled simultaneously by the computer.

With the advent of Sputnik and the space age, another new impetus was imparted to control engineering. It became necessary to design complex, highly accurate control systems for missiles and space probes. Furthermore, the necessity to minimize the weight of satellites and to control them very accurately has spawned the important field of optimal control. Due to these requirements, the time-domain methods developed by Liapunov, Minorsky, and others have met with great interest in the last two decades. Recent theories of optimal control developed by L. S. Pontryagin in the former Soviet Union and R. Bellman in the United States, and recent studies of robust systems, have also contributed to the interest in time-domain methods. It now is clear that control engineering must consider both the time-domain and the frequency-domain approaches simultaneously in the analysis and design of control systems.

A selected history of control system development is summarized in Table 1.1.

TABLE 1.1   Selected Historical Developments of Control Systems

1769	James Watt's steam engine and governor developed. The Watt steam engine is often used to mark the beginning of the Industrial Revolution in Great Britain. During the Industrial Revolution, great strides were made in the development of mechanization, a technology preceding automation.
1800	Eli Whitney's concept of interchangeable parts manufacturing demonstrated in the production of muskets. Whitney's development is often considered as the beginning of mass production.
1868	J. C. Maxwell formulates a mathematical model for a governor control of a steam engine.
1913	Henry Ford's mechanized assembly machine introduced for automobile production.
1927	H. W. Bode analyzes feedback amplifiers.
1932	H. Nyquist develops a method for analyzing the stability of systems.
1952	Numerical control (NC) developed at Massachusetts Institute of Technology for control of machine-tool axes.
1954	George Devol develops "programmed article transfer," considered to be the first industrial robot design.
1960	First Unimate robot introduced, based on Devol's designs. Unimate installed in 1961 for tending die-casting machines.
1970	State variable models and optimal control developed.
1980	Robust control system design widely studied.
1990	Export-oriented manufacturing companies emphasize automation.
1994	Feedback control widely used in automobiles. Reliable, robust systems demanded in manufacturing.

1.3 TWO EXAMPLES OF ENGINEERING CREATIVITY

Harold S. Black graduated from Worcester Polytechnic Institute in 1921 and joined Bell Laboratories of American Telegraph and Telephone (AT&T). In 1921, the major task confronting Bell Labs was the improvement of the telephone system and the design of improved signal amplifiers. Black was assigned the task of linearizing, stabilizing, and improving the amplifiers that were used in tandem to carry conversations over distances of several thousand miles.

Black reports [8]:

> Then came the morning of Tuesday, August 2, 1927, when the concept of the negative feedback amplifier came to me in a flash while I was crossing the Hudson River on the Lackawanna Ferry, on my way to work. For more than 50 years I have pondered how and why the idea came, and I can't say any more today than I could that morning. All I know is that after several years of hard work on the problem, I suddenly realized that if I fed the amplifier output back to the input, in reverse phase, and kept the device from oscillating (singing, as we called it then), I would have exactly what I wanted: a means of canceling out the distortion in the output. I opened my morning newspaper and on a page of *The New York Times* I sketched a simple canonical diagram of a negative feedback amplifier plus the equations for the amplification with feedback. I signed the sketch, and 20 minutes later, when I reached the laboratory at 463 West Street, it was witnessed, understood, and signed by the late Earl C. Blessing.
>
> I envisioned this circuit as leading to extremely linear amplifiers (40 to 50 dB of negative feedback), but an important question is: How did I know I could avoid self-oscillations over very wide frequency bands when many people doubted such circuits would be stable? My confidence stemmed from work that I had done two years earlier on certain novel oscillator circuits and three years earlier in designing the terminal circuits, including the filters, and developing the mathematics for a carrier telephone system for short toll circuits.

Another example of the discovery of an engineering solution to a control system problem was that of the creation of a gun director by David B. Parkinson of Bell Telephone Laboratories. In the spring of 1940, Parkinson was a 29-year-old engineer intent on improving the automatic level recorder, an instrument that used strip-chart paper to plot the record of a voltage. A critical component was a small potentiometer used to control the pen of the recorder through an actuator.

Parkinson had a dream about an antiaircraft gun that was successfully felling airplanes. Parkinson described the situation [13]:

> After three or four shots one of the men in the crew smiled at me and beckoned me to come closer to the gun. When I drew near he pointed to the exposed end of the left trunnion. Mounted there was the control potentiometer of my level recorder!

The next morning Parkinson realized the significance of his dream:

> If my potentiometer could control the pen on the recorder, something similar could, with suitable engineering, control an antiaircraft gun.

After considerable effort, an engineering model was delivered for testing to the U.S. Army on December 1, 1941. Production models were available by early 1943 and eventually 3000 gun controllers were delivered. Input to the controller was provided by radar and the gun was aimed by taking the data of the airplane's present position and calculating the target's future position.

1.4 CONTROL ENGINEERING PRACTICE

Control engineering is concerned with the analysis and design of goal-oriented systems. Therefore the mechanization of goal-oriented policies has grown into a hierarchy of goal-oriented control systems. Modern control theory is concerned with systems that have self-organizing, adaptive, robust, learning, and optimum qualities. This interest has aroused even greater excitement among control engineers.

The control of an industrial process (manufacturing, production, and so on) by automatic rather than manual means is often called *automation*. Automation is prevalent in the chemical, electric power, paper, automobile, and steel industries, among others. The concept of automation is central to our industrial society. Automatic machines are used to increase the production of a plant per worker in order to offset rising wages and inflationary costs. Thus industries are concerned with the productivity per worker of their plant. *Productivity* is defined as the ratio of physical output to physical input [26]. In this case, we are referring to labor productivity, which is real output per hour of work.

Furthermore, industry seeks to provide products that are increasingly precise, reliable, accurate, and robust. For example, precise, reliable control of automobile performance has improved markedly over the past decades.

The transformation of the U.S. labor force in the country's brief history follows the progressive mechanization of work that attended the evolution of the agrarian republic into an industrial world power. In 1820, more than 70% of the labor force worked on the farm. By 1900, fewer than 40% were engaged in agriculture. Today, fewer than 5% work in agriculture [15].

In 1925, some 588,000 people—about 1.3% of the nation's labor force—were needed to mine 520 million tons of bituminous coal and lignite, almost all of it from underground. By 1980, production was up to 774 million tons, but the work force had been reduced to 208,000. Furthermore, only 136,000 of that number were employed in underground mining operations. The highly mechanized and highly productive surface mines, with just 72,000 workers, produced 482 million tons, or 62% of the total [27].

The easing of human labor by technology, a process that began in prehistory, is entering a new stage. The acceleration in the pace of technological innovation inaugurated by the Industrial Revolution has until recently resulted mainly in the displacement of human muscle power from the tasks of production. The current revolution in computer technology is causing an equally momentous social change: the expansion of information gathering and information processing as computers extend the reach of the human brain [16].

Control systems are used to achieve (1) increased productivity and (2) improved performance of a device or system. Automation is used to improve productivity and obtain high-quality products. Automation is the automatic operation or control of a process, device, or system. We utilize automatic control of machines and processes to produce a product within specified tolerances and to achieve high precision [28].

The term *automation* first became popular in the automobile industry. Transfer lines were coupled with automatic machine tools to create long machinery lines that could produce engine parts, such as the cylinder block, virtually without operator intervention. In body parts manufacturing, automatic-feed mechanisms were coupled with high-speed stamping presses to increase productivity in sheet-metal forming. In many other areas where designs were relatively stable, such as radiator production, entire automated lines replaced manual operations.

With the demand for flexible, custom production emerging in the 1990s, a need for flexible automation and robotics is growing [17, 25].

There are about 150,000 control engineers in the United States and also in Japan and in Europe. In the United States alone, the control industry does a business of over $50 billion per year! The theory, practice, and application of automatic control is a large, exciting, and extremely useful engineering discipline. One can readily understand the motivation for a study of modern control systems.

1.5   EXAMPLES OF MODERN CONTROL SYSTEMS

Feedback control is a fundamental fact of modern industry and society. Driving an automobile is a pleasant task when the auto responds rapidly to the driver's commands. Many cars have power steering and brakes, which utilize hydraulic amplifiers for amplification of the force to the brakes or the steering wheel. A simple block diagram of an automobile steering control system is shown in Fig. 1.7(a). The desired course is compared with a measurement of the actual course in order to generate a measure of the error, as shown in Fig. 1.7(b). This measurement is obtained by visual and tactile (body movement) feedback. There is an additional feedback from the feel of the steering wheel by the hand (sensor). This feedback system is a familiar version of the steering control system in an ocean liner or the flight controls in a large airplane.

Control systems operate in a closed-loop sequence, as shown in Fig. 1.8. With an accurate sensor, the measured output is equal to the actual output of the system. The difference between the desired output and the actual output is equal to the error, which is then adjusted by the control device (such as an amplifier). The output of the control device causes the actuator to modulate the process in order to reduce the error. The sequence is such, for instance, that if a ship is heading incorrectly to the right, the rudder is actuated to direct the ship to the left. The system shown in Fig. 1.8 is a *negative feedback* control system, because the output is subtracted from the input and the difference is used as the input signal to the power amplifier.

A basic manually controlled closed-loop system for regulating the level of fluid in a tank is shown in Fig. 1.9. The input is a reference level of fluid that the operator is instructed to maintain. (This reference is memorized by the operator.) The power amplifier is the operator, and the sensor is visual. The operator compares the actual level with the desired level and opens or closes the valve (actuator), adjusting the fluid flow out, to maintain the desired level.

Other familiar control systems have the same basic elements as the system shown in Fig. 1.8. A refrigerator has a temperature setting or desired temperature, a thermostat to measure the actual temperature and the error, and a compressor motor for power amplification. Other examples in the home are the oven, furnace, and water heater. In industry, there are speed controls, process temperature and pressure controls, position, thickness, composition, and quality controls, among many others [14, 17, 18].

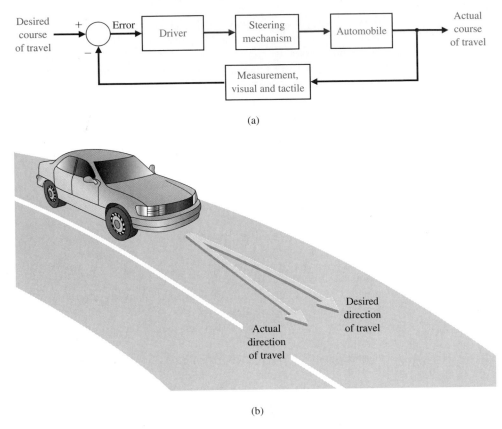

(a)

(b)

FIGURE 1.7
(a) Automobile steering control system. (b) The driver uses the difference between the actual and desired direction of travel to generate a controlled adjustment of the steering wheel.

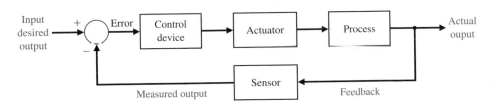

FIGURE 1.8
A negative feedback system block diagram depicting a basic closed-loop control system. The control device is often called a "controller."

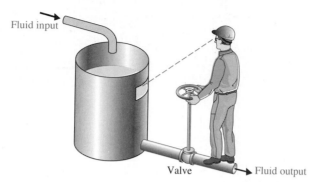

Fluid input

Valve   Fluid output

FIGURE 1.9
A manual control system for regulating the level of fluid in a tank by adjusting the output valve. The operator views the level of fluid through a port in the side of the tank.

In its modern usage, automation can be defined as a technology that uses programmed commands to operate a given process, combined with feedback of information to determine that the commands have been properly executed. Automation is often used for processes that were previously operated by humans. When automated, the process can operate without human assistance or interference. In fact, most automated systems are capable of performing their functions with greater accuracy and precision, and in less time, than humans are able to do. A semiautomated process is one that incorporates both humans and robots. For instance, many automobile assembly line operations require cooperation between a human operator and an intelligent robot.

A *robot* is a computer-controlled machine and is a technology closely associated with automation. Industrial robotics can be defined as a particular field of automation in which the automated machine (i.e., the robot) is designed to substitute for human labor [18, 27, 33]. Thus robots possess certain humanlike characteristics. Today, the most common humanlike characteristic is a mechanical manipulator that is patterned somewhat after the human arm and wrist. We recognize that the automatic machine is well suited to some tasks, as noted in Table 1.2, and that other tasks are best carried out by humans.

Another very important application of control technology is in the control of the modern automobile [19, 20]. Control systems for suspension, steering, and engine control have been introduced. Many new autos have a four-wheel-steering system, as well as an antiskid control system.

A three-axis control system for inspecting individual semiconductor wafers is shown in Fig. 1.10. This system uses a specific motor to drive each axis to the desired position in the x-y-z-axis, respectively. The goal is to achieve smooth, accurate movement in each axis. This control system is an important one for the semiconductor manufacturing industry.

TABLE 1.2   Task Difficulty: Human Versus Automatic Machine

Tasks Difficult for a Machine	Tasks Difficult for a Human
Inspect seedlings in a nursery. Drive a vehicle through rugged terrain. Identify the most expensive jewels on a tray of jewels.	Inspect a system in a hot, toxic environment. Repetitively assemble a clock. Land an airliner at night, in bad weather.

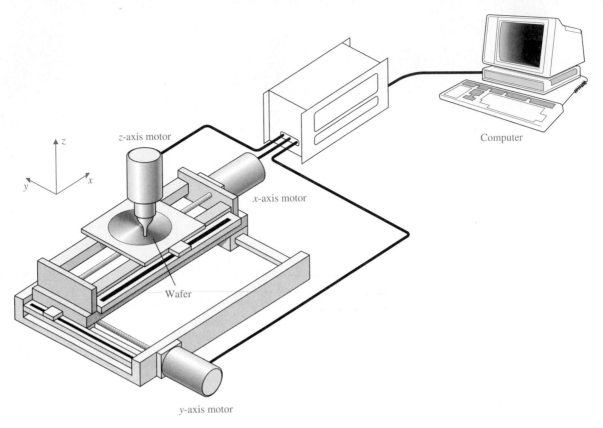

FIGURE 1.10   A three-axis control system for inspecting individual semiconductor wafers with a highly sensitive camera.

There has been considerable discussion recently concerning the gap between practice and theory in control engineering. However, it is natural that theory precedes the applications in many fields of control engineering. Nonetheless, it is interesting to note that in the electric power industry, the largest industry in the United States, the gap is relatively insignificant. The electric power industry is primarily interested in energy conversion, control, and distribution. It is critical that computer control be increasingly applied to the power industry in order to improve the efficient use of energy resources. Also, the control of power *plants* for minimum waste emission has become increasingly important. The modern, large-capacity plants, which exceed several hundred megawatts, require automatic control systems that account for the interrelationship of the process variables and optimum power production. It is common to have as many as 90 or more manipulated variables under coordinated control. A simplified model showing several of the important control variables of a large boiler-generator system is shown in Fig. 1.11. This is an example of the importance of measuring many variables, such as pressure and oxygen, to provide information to the computer for control calculations. It is estimated that more than four hundred thousand computer control systems have been installed in the United States [14, 16, 36, 39]. The diagram of a computer control system is shown in Fig. 1.12; note that the computer is the control device. The electric power industry has utilized the modern aspects of control en-

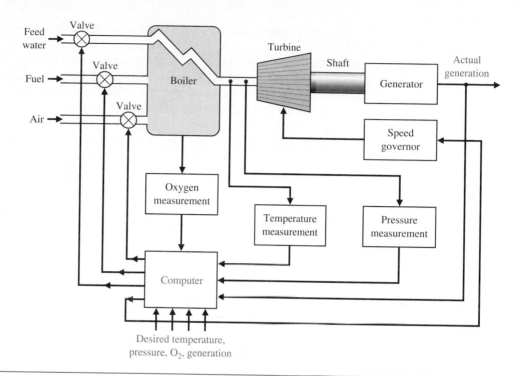

FIGURE 1.11
Coordinated
control system
for a boiler-
generator.

gineering for significant and interesting applications. It appears that in the process industry, the factor that maintains the applications gap is the lack of instrumentation to measure all the important process variables, including the quality and composition of the product. As these instruments become available, the applications of modern control theory to industrial systems should increase measurably.

Another important industry, the metallurgical industry, has had considerable success in automatically controlling its processes. In fact, in many cases, the control applications are beyond the theory. For example, a hot-strip steel mill, which involves a $100-million investment, is controlled for temperature, strip width, thickness, and quality.

Rapidly rising energy costs coupled with threats of energy curtailment are resulting in new efforts for efficient automatic energy management. Computer controls are used to control energy use in industry and to stabilize and connect loads evenly to gain fuel economy.

There has been considerable interest recently in applying the feedback control concepts to automatic warehousing and inventory control. Furthermore, automatic control of agricultural systems (farms) is meeting increased interest. Automatically controlled silos and tractors have been developed and tested. Automatic control of wind turbine generators,

FIGURE 1.12
A computer control
system.

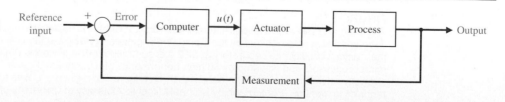

solar heating and cooling, and automobile engine performance are important modern examples [20, 21].

Also, there have been many applications of control system theory to biomedical experimentation, diagnosis, prosthetics, and biological control systems [22, 23, 51]. The control systems under consideration range from the cellular level to the central nervous system and include temperature regulation and neurological, respiratory, and cardiovascular control. Most physiological control systems are closed-loop systems. However, we find not one controller but rather control loop within control loop, forming a hierarchy of systems. The modeling of the structure of biological processes confronts the analyst with a high-order model and a complex structure. Prosthetic devices that aid the 46 million handicapped individuals in the United States are designed to provide automatically controlled aids to the disabled [22, 27, 42]. An artificial hand that uses force feedback signals and is controlled by the amputee's bioelectric control signals, which are called electromyographic signals, is shown in Fig. 1.13.

Finally, it has become of interest and value to attempt to model the feedback processes prevalent in the social, economic, and political spheres. This approach is undeveloped at present but appears to have a reasonable future. Society, of course, is comprised of many

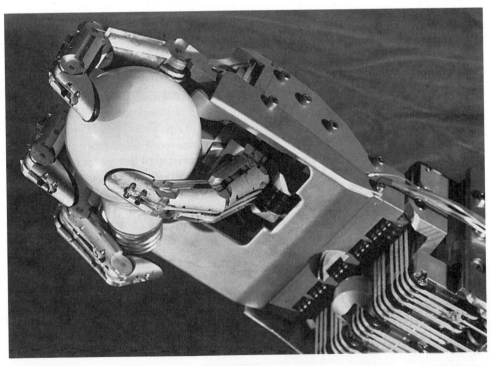

FIGURE 1.13
The Utah/MIT Dextrous Robotic Hand: A dextrous robotic hand having 18 degrees of freedom, developed as a research tool by the Center for Engineering Design at the University of Utah and the Artificial Intelligence Laboratory at MIT. It is controlled by five Motorola 68000 microprocessors and actuated by 36 high-performance electropneumatic actuators via high-strength polymeric tendons. The hand has three fingers and a thumb. It uses touch sensors and tendons for control. (Photograph by Michael Milochik. Courtesy of University of Utah.)

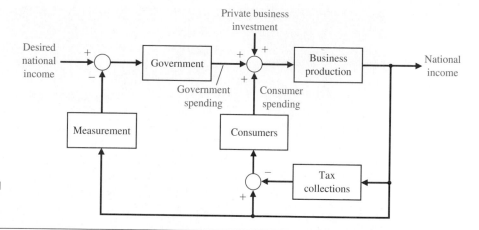

FIGURE 1.14
A feedback control system model of the economy.

feedback systems and regulatory bodies, such as the Interstate Commerce Commission and the Federal Reserve Board, which are controllers exerting the forces on society necessary to maintain a desired output. A simple lumped model of the national income feedback control system is shown in Fig. 1.14. This type of model helps the analyst to understand the effects of government control—granted its existence—and the dynamic effects of government spending. Of course, many other loops not shown also exist, since, theoretically, government spending cannot exceed the tax collected without generating a deficit, which is itself a control loop containing the Internal Revenue Service and the Congress. Of course, in a socialist country, the loop due to consumers is deemphasized and government control is emphasized. In that case, the measurement block must be accurate and must respond rapidly; both are very difficult characteristics to realize from a bureaucratic system. This type of political or social feedback model, while usually nonrigorous, does impart information and understanding.

Feedback control systems are used extensively in industrial applications. A laboratory robot is shown in Fig. 1.15. Thousands of industrial and laboratory robots are currently in use. Manipulators can pick up objects weighing hundreds of pounds and position them with an accuracy of one-tenth of an inch or better [28].

FIGURE 1.15
A laboratory robot used for sample preparation. The robot manipulates small objects, such as test tubes, and probes in and out of tight places at relatively high speeds [41].
(© Copyright 1993 Hewlett-Packard Company. Reproduced with permission.)

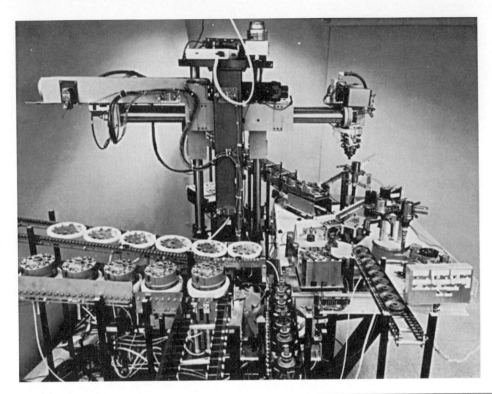

FIGURE 1.16
A programmable robot assembly station can assemble the 17 parts of a commercial automobile alternator in 2 minutes, 42 seconds. At the far right is a control box through which the robot can be taught a sequence of moves that are recorded in the memory of a minicomputer. (Courtesy of *Scientific American.* Photo by Ben Rose.)

1.6 AUTOMATIC ASSEMBLY AND ROBOTS

Automatic handling equipment for home, school, and industry is particularly useful for hazardous, repetitious, dull, or simple tasks. Machines that automatically load and unload, cut, weld, or cast are used by industry to obtain [14, 27, 28] accuracy, safety, economy, and productivity. The use of computers integrated with machines that perform tasks as a human worker does has been foreseen by several authors. In his famous 1923 play, entitled *R.U.R.* [48], Karel Capek called artificial workers *robots,* deriving the word from the Czech noun *robota,* meaning "work."

As stated earlier, robots are programmable computers integrated with machines, and they often substitute for human labor in specific repeated tasks. Some devices even have anthropomorphic mechanisms, including what we might recognize as mechanical arms, wrists, and hands [14, 27, 28]. Robots are used extensively in space exploration and assembly. They also are flexible, accurate aids on assembly lines, as shown in Fig. 1.16.

1.7 THE FUTURE EVOLUTION OF CONTROL SYSTEMS

The continuing goal of control systems is to provide extensive flexibility and a high level of autonomy. Two system concepts are approaching this goal by different evolutionary pathways, as illustrated in Fig. 1.17. Today's industrial robot is perceived as quite autono-

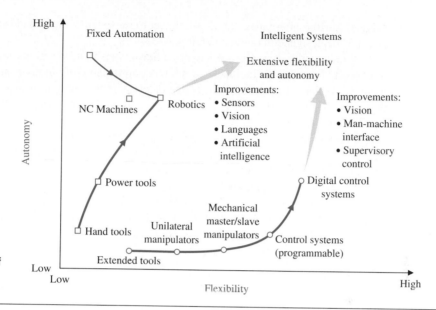

FIGURE 1.17
Future evolution of
control systems
and robotics.

mous—once it is programmed, further intervention is not normally required. Because of sensory limitations, these robotic systems have limited flexibility in adapting to work environment changes, which is the motivation of computer vision research. The control system is very adaptable, but it relies on human supervision. Advanced robotic systems are striving for task adaptability through enhanced sensory feedback. Research areas concentrating on artificial intelligence, sensor integration, computer vision, and off-line CAD/CAM programming will make systems more universal and economical. Control systems are moving toward autonomous operation as an enhancement to human control. Research in supervisory control, human–machine interface methods to reduce operator burden, and computer database management is intended to improve operator efficiency. Many research activities are common to robotics and control systems and are aimed toward reducing implementation cost and expanding the realm of application. These include improved communications methods and advanced programming languages.

1.8 ENGINEERING DESIGN

Engineering design is the central task of the engineer. *Design is the process of conceiving or inventing the forms, parts, and details of a system to achieve a reasoned purpose.* It is a complex process in which both creativity and analysis play a central role.

Design activity can be thought of as planning for the emergence of a particular product or system. Design is a structured, innovative act whereby the engineer creatively uses knowledge and materials to specify the shape, function, and material content of a system. The design steps are (1) to determine a need arising from the values of various groups, covering the spectrum from public policy makers to the consumer; (2) to specify in detail what the solution to that need must be and to embody these values; (3) to develop and evaluate various alternative solutions to meet these specifications; and (4) to decide which one is to be designed in detail and fabricated.

Of course, the design process is not a linear, step-by-step activity, but rather an iterative, reactive process. It includes several test points and continuous examination of intermediate results. Another factor in realistic design is the limitation of time. Design takes place under imposed schedules, and we eventually settle for a design that may be less than ideal but considered "good enough."

A major challenge for the designer is to write the specifications for the technical product. *Specifications* are statements that explicitly state what the device or product is to be and do. The design of technical systems aims to achieve appropriate design specifications and rests on four characteristics: complexity, trade-offs, gaps, and risk.

Complexity of design results from the wide range of tools, issues, and knowledge to be used in the process. The large number of factors to be considered illustrates the complexity of the design specification activity, not only in assigning these factors their relative importance in a particular design, but also in giving them substance either in numerical or written form, or both.

The concept of *trade-off* involves the need to make a judgment about how much of a compromise is made between two conflicting criteria, both of which are desirable. The design process requires an efficient compromise between desirable but conflicting criteria.

In making a technical device, there is frequently a design *gap* or void to the extent that the final product does not appear the same as it had been visualized. For example, our image of the problem we're solving is not what appears in written description and ultimately in the specifications. Such gaps are intrinsic in the progression from an abstract idea to its realization.

This inability to make absolutely sure predictions of the performance of a technological object leads to major uncertainties about the actual effects of the designed devices and products. These uncertainties are embodied in the idea of unintended consequences or *risk*. The result is that designing a system is a risk-taking activity.

Complexity, trade-off, gaps, and risk are inherent in designing new systems and devices. Although they can be minimized by considering all the effects of a given design, they are always present in the design process.

Within engineering design, there is a fundamental difference between the two major types of thinking that must take place: engineering analysis and synthesis. Attention is focused on models of the physical systems that are analyzed to provide insight and that point in directions for improvement. On the other hand, *synthesis* is the process by which these new physical configurations are created.

Design is an iterative process that may proceed in many directions before the desired one is found. It is a deliberate process by which a designer creates something new in response to a recognized need while recognizing realistic constraints.

The main approach to the most effective engineering design is parameter analysis and optimization. Parameter analysis is based on (1) identification of the key parameters, (2) generation of the system configuration, and (3) evaluation of how well the configuration meets the needs. These three steps form an iterative loop. Once the key parameters are identified and the configuration synthesized, the designer can optimize the parameters. Typically, the designer strives to identify a limited set of parameters—hopefully less than five—to be adjusted.

1.9 CONTROL SYSTEM DESIGN

The design of control systems is a specific example of engineering design. Again, the goal of control engineering design is to obtain the configuration, specifications, and identification of the key parameters of a proposed system to meet an actual need.

The first step in the design process consists of establishing the system goals. For example, we may state that our goal is to control the velocity of a motor accurately. The second step is to identify the variables that we desire to control (e.g., the velocity of the motor). The third step is to write the specifications in terms of the accuracy we must attain. This required accuracy of control will then lead to the identification of a sensor to measure the controlled variable.

As designer, we proceed to the first attempt to configure a system that will result in the desired control performance. This system configuration will normally consist of a sensor, the process under control, an actuator, and a controller, as shown in Fig. 1.8. The next step consists of identifying a candidate for the actuator. This will, of course, depend on the process, but the actuation chosen must be capable of effectively adjusting the performance of the process. For example, if we wish to control the speed of a rotating flywheel, we will select a motor as the actuator. The sensor, in this case, will need to be capable of accurately measuring the speed. We then obtain a model for each of these elements.

The next step is the selection of a controller, which will often consist of a summing amplifier that will compare the desired response and the actual response and then forward this error-measurement signal to an amplifier.

The final step in the design process is the adjustment of the parameters of the system in order to achieve the desired performance. If we can achieve the desired performance by adjusting the parameters, we will finalize the design and proceed to document the results. If not, we will need to establish an improved system configuration and perhaps select an enhanced actuator and sensor. Then we will repeat the design steps until we are able to meet the specifications, or until we decide the specifications are too demanding and should be relaxed. The control system design process is summarized in Fig. 1.18.

The performance specifications will describe how the closed-loop system should perform and will include (1) good regulation against disturbances, (2) desirable responses to commands, (3) realistic actuator signals, (4) low sensitivities, and (5) robustness.

In summary, the controller design problem is as follows: Given a model of the system to be controlled (including its sensors and actuators) and a set of design goals, find a suitable controller, or determine that none exists.

1.10 DESIGN EXAMPLE: TURNTABLE SPEED CONTROL

Many modern devices use a turntable to rotate a disk at a constant speed. For example, a CD player, a computer disk drive, and a phonograph record player all require a constant speed of rotation in spite of motor wear and variation and other component changes. Our goal is to design a system for turntable speed control that will ensure that the actual speed of rotation is within a specified percentage of the desired speed [43, 46]. We will consider a system without feedback and a system with feedback.

To obtain disk rotation, we will select a dc motor as the actuator because it provides a

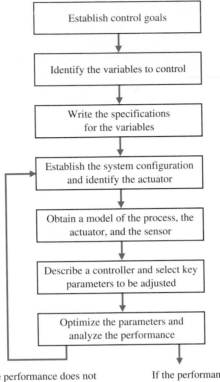

Establish control goals

Identify the variables to control

Write the specifications
for the variables

Establish the system configuration
and identify the actuator

Obtain a model of the process, the
actuator, and the sensor

Describe a controller and select key
parameters to be adjusted

Optimize the parameters and
analyze the performance

If the performance does not
meet the specifications, then
iterate the configuration and
the actuator.

If the performance
meets the specifications,
then finalize the design.

FIGURE 1.18
The control system
design process.

speed proportional to the applied motor voltage. For the input voltage to the motor, we will select an amplifier that can provide the required power.

The open-loop system (without feedback) is shown in Fig. 1.19(a). This system uses a battery source to provide a voltage that is proportional to the desired speed. This voltage is amplified and applied to the motor. The block diagram of the open-loop system identifying control device, actuator, and process is shown in Fig. 1.19(b).

To obtain a feedback system with the general form of Fig. 1.8, we need to select a sensor. One useful sensor is a tachometer that provides an output voltage proportional to the speed of its shaft. Thus the closed-loop feedback system takes the form shown in Fig. 1.20(a). The block diagram model of the feedback system is shown in Fig. 1.20(b). The error voltage is generated by the difference between the input voltage and the tachometer voltage.

We expect the feedback system of Fig. 1.20 to be superior to the open-loop system of Fig. 1.19 because the feedback system will respond to errors and work to reduce them. With precision components, we could expect to reduce the error of the feedback system to one-hundredth of the error of the open-loop system.

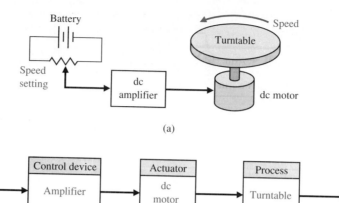

FIGURE 1.19
(a) Open-loop
(without feedback)
control of the
speed of
a turntable.
(b) Block diagram
model.

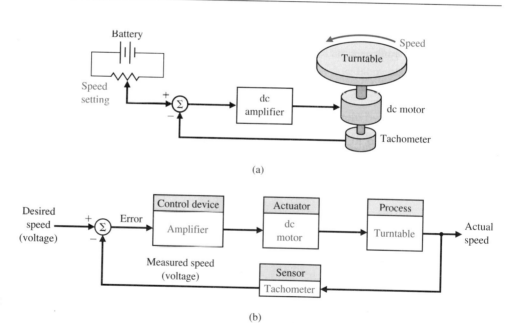

FIGURE 1.20
(a) Closed-loop
control of the
speed of a
turntable. (b) Block
diagram model.

1.11 DESIGN EXAMPLE: INSULIN DELIVERY CONTROL SYSTEM

Control systems have been utilized in the biomedical field to create implanted automatic drug-delivery systems to patients [29–31]. Automatic systems can be used to regulate blood pressure, blood sugar level, and heart rate. A common application of control engineering is in the field of open-loop system drug delivery in which mathematical models of the dose–effect relationship of the drugs are used. A drug-delivery system implanted in the

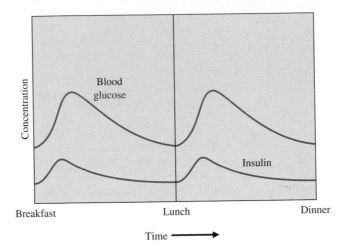

FIGURE 1.21
The blood glucose
and insulin levels
for a healthy
person.

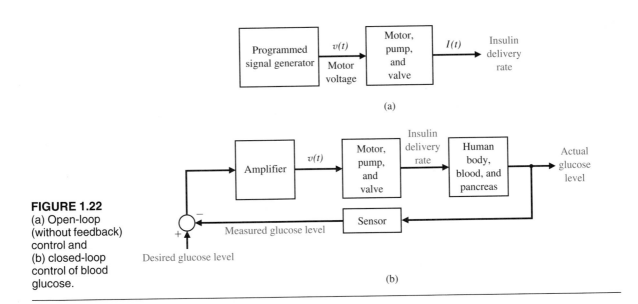

FIGURE 1.22
(a) Open-loop
(without feedback)
control and
(b) closed-loop
control of blood
glucose.

body uses an open-loop system, since miniaturized glucose sensors are not yet available. The best solutions rely on individually programmable pocket-sized insulin pumps that can deliver insulin according to a preset time history. More complicated systems will use closed-loop control for the measured blood glucose levels.

Our goal is (1) to design the block diagram of an open-loop control system and (2) to design a closed-loop control system to regulate the blood sugar concentration of a diabetic. The blood glucose and insulin concentrations for a healthy person are shown in Fig. 1.21. The system must provide the insulin from a reservoir implanted within the person.

An open-loop system would use a preprogrammed signal generator and miniature motor pump to regulate the insulin delivery rate as shown in Fig. 1.22(a). The feedback control system would use a sensor to measure the actual glucose level and compare that level with the desired level, thus turning the motor pump on when it is required, as shown in Fig. 1.22(b).

EXERCISES

(Exercises are straightforward applications of the concepts of the chapter.)

The following systems can be described by a block diagram showing the cause–effect relationship and the feedback (if present). Identify the function of each block and the desired input variable, output variable, and measured variable. Use Fig. 1.8 as a model where appropriate.

E1.1 A precise optical signal source can control the output power level to within 1% [32]. A laser is controlled by an input current to yield the power output. A microprocessor controls the input current to the laser. The microprocessor compares the desired power level with a measured signal proportional to the laser power output obtained from a sensor. Complete the block diagram representing this closed-loop control system shown in Fig. E1.1, identifying the output, input, and measured variables and the control device.

E1.2 An automobile driver uses a control system to maintain the speed of the car at a prescribed level. Draw a block diagram to illustrate this feedback system.

E1.3 The number of robots sold annually for automation of U.S. industry was expected to grow to 20,000 by 1995. Visit a local fast-food restaurant and discuss whether a robot could replace one or more of the workers.

E1.4 An autofocus camera will adjust the distance of the lens from the film by using a beam of infrared or ultrasound to determine the distance to the subject [45]. Draw a block diagram of this open-loop control system, and briefly explain its operation.

E1.5 Because a sailboat can't sail directly into the wind, and traveling straight downwind is usually slow, the shortest sailing distance is rarely a straight line. Thus sailboats tack upwind—the familiar zigzag course—and jibe downwind. A tactician's decision of when to tack and where to go can determine the outcome of a race.

Describe the process of tacking a sailboat as the wind shifts direction. Draw a block diagram depicting this process.

E1.6 Automated highways may be prevalent in the next century. Consider two automated highway lanes merging into a single lane, and describe a control system that ensures that the vehicles merge with a prescribed gap between two vehicles.

E1.7 Describe the block diagram of the speed control system of an automobile with a human driver.

E1.8 Describe the process of human biofeedback used to regulate factors such as pain or body temperature. Biofeedback is a technique whereby a human can, with some success, consciously regulate pulse, reaction to pain, and body temperature.

FIGURE E 1.1
Partial block diagram of an optical source.

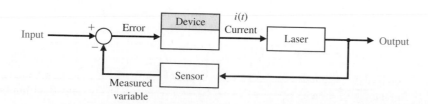

PROBLEMS

(Problems require extending the concepts of this chapter to new situations.)

The following systems may be described by a block diagram showing the cause–effect relationship and the feedback (if present). Each block should describe its function. Use Fig. 1.8 as a model where appropriate.

P1.1   Draw a schematic block diagram of a home heating system. Identify the function of each element of the thermostatically controlled heating system.

P1.2   In the past, control systems used a human operator as part of a closed-loop control system. Draw the block diagram of the valve control system shown in Fig. P1.2.

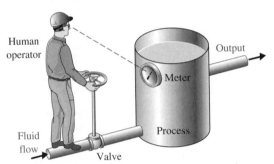

FIGURE P1.2   Fluid-flow control.

P1.3   In a chemical process control system, it is valuable to control the chemical composition of the product. To do so, a measurement of the composition can be obtained by using an infrared stream analyzer, as shown in Fig. P1.3. The valve on the additive stream may be controlled. Complete the control feedback loop, and draw a block diagram describing the operation of the control loop.

P1.4   The accurate control of a nuclear reactor is important for power system generators. Assuming the number of neutrons present is proportional to the power level, an ionization chamber is used to measure the power level. The current, i_o, is proportional to the power level. The position of the graphite control rods moderates the power level. Complete the control system of the nuclear reactor shown in Fig. P1.4 and draw the block diagram describing the operation of the feedback control loop.

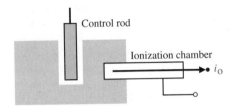

FIGURE P1.4   Nuclear reactor control.

P1.5   A light-seeking control system, used to track the sun, is shown in Fig. P1.5. The output shaft, driven by the motor through a worm reduction gear, has a bracket attached on which are mounted two photocells. Complete the closed-loop system such that the system follows the light source.

P1.6   Feedback systems are not always negative feedback systems in nature. Economic inflation, which is evidenced by continually rising prices, is a *positive feedback* system. A positive feedback control system, as shown in Fig. P1.6, *adds* the feedback signal to the input signal, and the resulting signal is used as the input to the process. A simple model of the price–wage inflationary spiral is shown in Fig. P1.6. Add additional feedback loops, such as legislative control or control of the tax rate, to stabilize the system. It is assumed that an increase in

FIGURE P1.3
Chemical composition control.

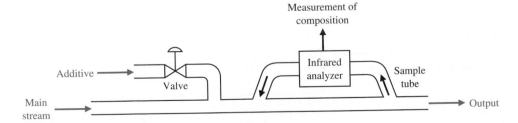

FIGURE P1.5
A photocell is mounted in each tube. The light reaching each cell is the same in both only when the light source is exactly in the middle as shown.

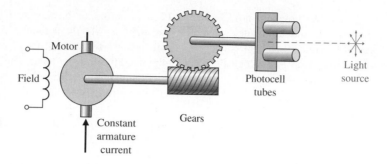

workers' salaries, after some time delay, results in an increase in prices. Under what conditions could prices be stabilized by falsifying or delaying the availability of cost-of-living data? How would a national wage and price economic guideline program affect the feedback system?

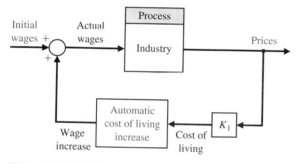

FIGURE P1.6   Positive feedback.

P1.7   The story is told about the sergeant who stopped at the jewelry store every morning at nine o'clock and compared and reset his watch with the chronometer in the window. Finally, one day the sergeant went into the store and complimented the owner on the accuracy of the chronometer.

"Is it set according to time signals from Arlington?" asked the sergeant.

"No," said the owner, "I set it by the five o'clock (P.M.) cannon fired from the fort. Tell me, Sergeant, why do you stop every day and check your watch?"

The sergeant replied, "I'm the gunner at the fort!"

Is the feedback prevalent in this case positive or negative? The jeweler's chronometer loses one minute each 24-hour period and the sergeant's watch loses one minute during each eight hours. What is the net time error of the cannon at the fort after 15 days?

P1.8   The student–teacher learning process is inherently a feedback process intended to reduce the system error to a minimum. The desired output is the knowledge being studied, and the student may be considered the process. With the aid of Fig. 1.3, construct a feedback model of the learning process and identify each block of the system.

P1.9   Models of physiological control systems are valuable aids to the medical profession. A model of the heart-rate control system is shown in Fig. P1.9 [23, 24, 51]. This model includes the processing of the nerve signals by the brain. The heart-rate control system is, in fact, a multivariable system, and the variables x, y, w, v, z, and u are vector variables. In other words, the variable x represents many

FIGURE P1.9
Heart-rate control.

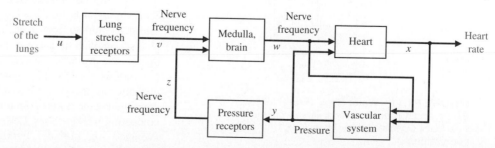

heart variables $x_1, x_2, \ldots, x_n$. Examine the model of the heart-rate control system and add or delete the blocks, if necessary. Determine a control system model of one of the following physiological control systems:

1. Respiratory control system
2. Adrenalin control system
3. Human arm control system
4. Eye control system
5. Pancreas and the blood-sugar-level control system
6. Circulatory system

P1.10   The role of air traffic control systems is increasing as airplane traffic increases at busy airports. Engineers are developing flight control systems, air traffic control systems, and collision avoidance systems [34]. Investigate these and other systems designed to improve air traffic safety; select one and draw a simple block diagram of its operation.

P1.11   Automatic control of water level using a float level was used in the Middle East for a water clock [1, 11]. The water clock (Fig. P1.11) was used from sometime before Christ until the seventeenth century. Discuss the operation of the water clock, and establish how the float provides a feedback control that maintains the accuracy of the clock.

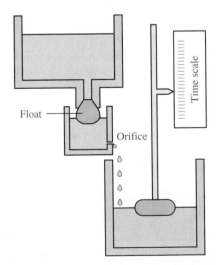

FIGURE P1.11   Water clock. (From Newton, Gould, and Kaiser, *Analytical Design of Linear Feedback Controls.* Wiley, New York, 1957, with permission.)

P1.12   An automatic turning gear for windmills was invented by Meikle in about 1750 [1, 11]. The fantail gear shown in Fig. P1.12 automatically turns the windmill into the wind. The fantail windmill at right angle to the mainsail is used to turn the turret. The gear ratio is of the order of 3000 to 1. Discuss the operation of the windmill, and establish the feedback operation that maintains the main sails into the wind.

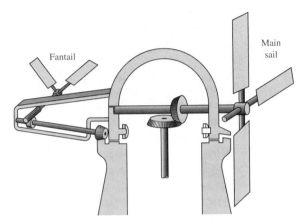

FIGURE P1.12   (From Newton, Gould, and Kaiser, *Analytical Design of Linear Feedback Controls.* Wiley, New York, 1957, with permission.)

P1.13   A common example of a two-input control system is a home shower with separate valves for hot and cold water. The objective is to obtain (1) a desired temperature of the shower water and (2) a desired flow of water. Draw a block diagram of the closed-loop control system. Would you be willing to take a shower under open-loop control by another person?

P1.14   Adam Smith (1723–1790) discussed the issue of free competition between the participants of an economy in his book *Wealth of Nations*. It may be said that Smith employed social feedback mechanisms to explain his theories [44]. Smith suggests that (1) the available workers as a whole compare the various possible employments and enter that one offering the greatest rewards and (2) in any employment the rewards diminish as the number of competing workers rises. Let r = total of rewards averaged over all trades, c = total of rewards in a particular trade, q = influx of workers into the specific trade. Draw a feedback system to represent this system.

P1.15 Small computers are used in automobiles to control emissions and obtain improved gas mileage. A computer-controlled transmission and engine that automatically adjusts itself to the road and driving conditions could improve automobile performance up to 30%. Sketch a block diagram of such a system for an automobile.

P1.16 All humans have experienced a fever associated with an illness. A fever is related to the changing of the control input in your body's thermostat. This thermostat, within the brain, normally regulates your temperature near 98° F in spite of external temperatures ranging from 0 to 100° F or more. For a fever, the input, or desired, temperature is increased. Even to many scientists, it often comes as a surprise to learn that fever does not indicate something wrong with body temperature control but rather well-contrived regulation at an elevated level of desired input. Sketch a block diagram of the temperature control system and explain how aspirin will lower a fever.

P1.17 An outfielder for a baseball team uses feedback to judge a fly ball [35]. Determine the method used by the fielder to judge where the ball will land so he can be in the right spot to catch it.

P1.18 A cutaway view of a commonly used pressure regulator is shown in Fig. P1.18. The desired pressure is set by turning a calibrated screw. This compresses the spring and sets up a force that opposes the upward motion of the diaphragm. The bottom side of the diaphragm is exposed to the water pressure that is to be controlled. Thus the motion of the diaphragm is an indication of the pressure difference between the desired and the actual pressures. It acts like a comparator. The valve is connected to the diaphragm and moves according to the pressure difference until it reaches a position in which the difference is zero. Sketch a block diagram showing the control system with the output pressure as the regulated variable.

P1.19 Ichiro Masaki of General Motors has patented a system that automatically adjusts a car's speed to keep a safe distance from vehicles in front. Using a video camera, the system detects and stores a reference image of the car in front. It then compares this image with a stream of incoming live images as the two cars move down the highway and calculates the distance. Masaki suggests that the system could control steering as well as speed, allowing drivers to lock on to the car ahead and get a "computerized tow." Draw a block diagram for the control system.

P1.20 A high-performance race car with an adjustable wing (airfoil) is shown in Fig. P1.20. Develop a block diagram describing the ability of the airfoil to keep a constant road adhesion between the car's tires and the race track surface. Why is it important to maintain good road adhesion?

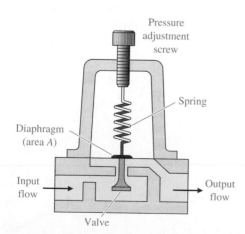

FIGURE P1.18 Pressure regulator.

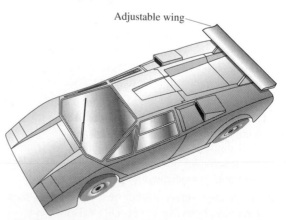

FIGURE P1.20 A high-performance race car with an adjustable wing.

P1.21   The potential of employing two or more helicopters for transporting payloads that are too heavy for a single helicopter is a well-addressed issue in the civil and military rotorcraft design arenas [38].

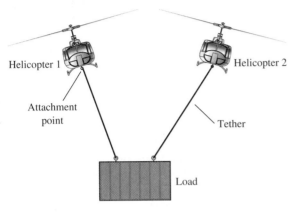

FIGURE P1.21   Two helicopters used to lift and move a large load.

Overall requirements can be satisfied more efficiently with a smaller aircraft by using multilift for infrequent peak demands. Hence the principal motivation for using multilift can be attributed to the promise of obtaining increased productivity without having to manufacture larger and more expensive helicopters. A specific case of a multilift arrangement wherein two helicopters jointly transport payloads has been named *twin lift*. Figure P1.21 shows a typical "two-point pendant" twin lift configuration in the lateral/vertical plane.

Develop the block diagram describing the pilots' action, the position of each helicopter, and the position of the load.

P1.22   Engineers want to design a control system that will allow a building or other structure to react to the force of an earthquake much as a human would. The structure would yield to the force, but only so much, before mustering strength to push back [52]. Develop a block diagram of a control system to reduce the effect of an earthquake force.

DESIGN PROBLEMS

(Design problems emphasize the design task.)

DP1.1   The road and vehicle noise that invades an automobile's cabin hastens occupant fatigue. Design the block diagram of an "antinoise" feedback system that will eliminate the effect of unwanted noises. Indicate the device within each block.

DP1.2   Many cars are fitted with cruise control that, at the press of a button, automatically maintains a set speed. In this way, the driver can cruise at a speed limit or economic speed without continually checking the speedometer. Design a feedback control in block diagram form for a cruise control system.

DP1.3   As part of the automation of a dairy farm, the automation of cow milking is under study [37]. Design a milking machine that can milk cows four or five times a day at the cow's demand. Show a block diagram, and indicate the devices in each block.

DP1.4   A large, braced robot arm for welding large structures is shown in Fig. DP1.4. Draw the block diagram of a closed-loop feedback control system for accurately controlling the location of the weld tip.

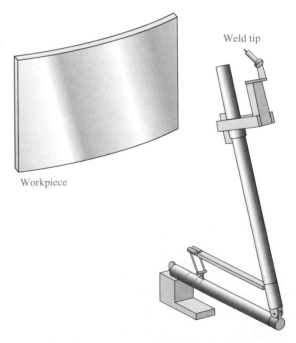

FIGURE DP1.4   Robot welder.

DP1.5 Vehicle traction control, which includes antiskid braking and antispin acceleration, can enhance vehicle performance and handling. The objective of this control is to maximize tire traction by preventing locked brakes as well as tire spinning during acceleration. Wheel slip, the difference between the vehicle speed and the wheel speed, is chosen as the controlled variable because of its strong influence on the tractive force between the tire and the road [19]. The adhesion coefficient, between the wheel and the road, reaches a maximum at a low slip. Develop a block diagram model of one wheel of a traction control system.

DP1.6 The Hubble space telescope was repaired in space in December 1993 [47, 49, 52]. Another challenging problem is damping the jitter that vibrates the spacecraft each time it passes into or out of the earth's shadow. The worst vibration has a period of about 10 seconds, or a frequency of 0.1 hertz.

Design a feedback system that will reduce the vibrations of the Hubble space telescope.

TERMS AND CONCEPTS

Automation The control of a process by automatic means.

Closed-loop feedback control system A system that uses a measurement of the output and compares it with the desired output.

Complexity of design The intricate pattern of interwoven parts and knowledge required.

Control system An interconnection of components forming a system configuration that will provide a desired response.

Design The process of conceiving or inventing the forms, parts, and details of a system to achieve a reasoned purpose.

Engineering design The process of designing a technical system.

Feedback signal A measure of the output of the system used for feedback to control the system.

Flyball governor A mechanical device for controlling the speed of a steam engine.

Gap The void between what is intended (or visualized) as the product or device and the actual, practical form of the final design.

Multivariable control system A system with more than one input variable and more than one output variable.

Negative feedback The output signal is fed back so that it subtracts from the input signal.

Open-loop control system A system that utilizes a device to control the process without using feedback. Thus the output has no effect upon the signal to the process.

Optimization The adjustment of the parameters to achieve the most favorable or advantageous design.

Plant *See* Process.

Positive feedback The output signal is fed back so that it adds to the input signal.

Process The device, plant, or system under control.

Productivity The ratio of physical output to physical input of an industrial process.

Risk Uncertainties embodied in the unintended consequences of a design.

Robot Programmable computers integrated with a manipulator. A reprogrammable, multifunctional manipulator used for a variety of tasks.

Specifications Statements that explicitly state what the device or product is to be and to do. A set of prescribed performance criteria.

Synthesis The process by which new physical configurations are created. The combining of separate elements or devices to form a coherent whole.

System An interconnection of elements and devices for a desired purpose.

Trade-off The need to make a judgment about how much compromise is made between conflicting criteria.

Mathematical Models
of Systems

PREVIEW

To analyze and design control systems, we use quantitative mathematical models of these systems. We will consider a wide range of systems, including mechanical, electrical, and fluid. We will first describe the dynamic behavior of these systems using differential equations. To use the Laplace transform, we will develop the means of obtaining a linear model of components of a system. We then will be able to utilize the differential equations describing the system and obtain the Laplace transform of these equations.

We will then proceed to obtain the input–output relationship for components and subsystems in the form of transfer functions. Then the set of transfer functions representing the interconnected components will be represented by a block diagram model or a signal-flow graph. Using various analytical methods, we will be able to obtain the equations for selected outputs of a control system as they are regulated by selected inputs of the system.

2.1   INTRODUCTION

To understand and control complex systems, one must obtain quantitative *mathematical models* of these systems. It is necessary therefore to analyze the relationships between the system variables and to obtain a mathematical model. Because the systems under consideration are dynamic in nature, the descriptive equations are usually *differential equations.* Furthermore, if these equations can be *linearized,* then the *Laplace transform* can be utilized to simplify the method of solution. In practice, the complexity of systems and the ignorance of all the relevant factors necessitate the introduction of *assumptions* concerning the system operation. Therefore we will often find it useful to consider the physical system, delineate some necessary assumptions, and linearize the system. Then, by using the physical laws describing the linear equivalent system, we can obtain a set of linear differential equations. Finally, utilizing mathematical tools, such as the Laplace transform, we obtain a solution describing the operation of the system. In summary, the approach to dynamic system problems can be listed as follows:

1. Define the system and its components.
2. Formulate the mathematical model and list the necessary assumptions.
3. Write the differential equations describing the model.
4. Solve the equations for the desired output variables.
5. Examine the solutions and the assumptions.
6. If necessary, reanalyze or redesign the system.

2.2 DIFFERENTIAL EQUATIONS OF PHYSICAL SYSTEMS

The differential equations describing the dynamic performance of a physical system are obtained by utilizing the physical laws of the process [1–3]. This approach applies equally well to mechanical [1], electrical [3], fluid, and thermodynamic systems [4]. A summary of the variables of dynamic systems is given in Table 2.1 [5]. We prefer to use the Interna-

TABLE 2.1   Summary of Through- and Across-Variables for Physical Systems

System	Variable Through Element	Integrated Through Variable	Variable Across Element	Integrated Across Variable
Electrical	Current, i	Charge, q	Voltage difference, v_{21}	Flux linkage, λ_{21}
Mechanical translational	Force, F	Translational momentum, P	Velocity difference, v_{21}	Displacement difference, y_{21}
Mechanical rotational	Torque, T	Angular momentum, h	Angular velocity difference, ω_{21}	Angular displacement difference, θ_{21}
Fluid	Fluid volumetric rate of flow, Q	Volume, V	Pressure difference, P_{21}	Pressure momentum, γ_{21}
Thermal	Heat flow rate, q	Heat energy, H	Temperature difference, T_{21}	

TABLE 2.2   The International System of Units (SI)

	Unit	Symbol
Basic Units		
Length	meter	m
Mass	kilogram	kg
Time	second	s
Temperature	kelvin	K
Electric current	ampere	A
Derived Units		
Velocity	meters per second	m/s
Area	square meter	m^2
Force	newton	$N = kgm/s^2$
Torque	kilogram-meter	kgm
Pressure	pascal	Pa
Energy	joule	$J = Nm$
Power	watt	$W = J/s$

tional System of units (SI) in contrast to the British System of units. The International System of units is given in Table 2.2. The conversion of other systems of units to SI units is facilitated by Table 2.3. A summary of the describing equations for lumped, linear, dynamic elements is given in Table 2.4 [5]. The equations in Table 2.4 are idealized descriptions and only approximate the actual conditions (for example, when a linear, lumped approximation is used for a distributed element).

TABLE 2.3   Conversion Factors for Converting to SI Units

From	Multiply by	To Obtain
Length		
inches	25.4	millimeters
feet	30.48	centimeters
Speed		
miles per hour	0.4470	meters per second
Mass		
pounds	0.4536	kilograms
Force		
pounds-force	4.448	newtons
Torque		
foot-pounds	0.1383	kilogram-meters
Power		
horsepower	746	watts
Energy		
British thermal unit	1055	joules
kilowatt-hour	3.6×10^6	joules

TABLE 2.4   Summary of Describing Differential Equations for Ideal Elements

Type of Element	Physical Element	Describing Equation	Energy E or Power $\mathcal{P}$	Symbol
	Electrical inductance	$v_{21} = L \dfrac{di}{dt}$	$E = \dfrac{1}{2} L i^2$	
Inductive storage	Translational spring	$v_{21} = \dfrac{1}{K} \dfrac{dF}{dt}$	$E = \dfrac{1}{2} \dfrac{F^2}{K}$	
	Rotational spring	$\omega_{21} = \dfrac{1}{K} \dfrac{dT}{dt}$	$E = \dfrac{1}{2} \dfrac{T^2}{K}$	
	Fluid inertia	$P_{21} = I \dfrac{dQ}{dt}$	$E = \dfrac{1}{2} I Q^2$	
	Electrical capacitance	$i = C \dfrac{dv_{21}}{dt}$	$E = \dfrac{1}{2} C v_{21}^2$	
	Translational mass	$F = M \dfrac{dv_2}{dt}$	$E = \dfrac{1}{2} M v_2^2$	
Capacitive storage	Rotational mass	$T = J \dfrac{d\omega_2}{dt}$	$E = \dfrac{1}{2} J \omega_2^2$	
	Fluid capacitance	$Q = C_f \dfrac{dP_{21}}{dt}$	$E = \dfrac{1}{2} C_f P_{21}^2$	
	Thermal capacitance	$q = C_t \dfrac{d\tau_2}{dt}$	$E = C_t \tau_2$	
	Electrical resistance	$i = \dfrac{1}{R} v_{21}$	$\mathcal{P} = \dfrac{1}{R} v_{21}^2$	
	Translational damper	$F = f v_{21}$	$\mathcal{P} = f v_{21}^2$	
Energy dissipators	Rotational damper	$T = f \omega_{21}$	$\mathcal{P} = f \omega_{21}^2$	
	Fluid resistance	$Q = \dfrac{1}{R_f} P_{21}$	$\mathcal{P} = \dfrac{1}{R_f} P_{21}^2$	
	Thermal resistance	$q = \dfrac{1}{R_t} \mathcal{T}_{21}$	$\mathcal{P} = \dfrac{1}{R_t} \mathcal{T}_{21}$	

Nomenclature

- *Through-variable:* F = force, T = torque, i = current, Q = fluid volumetric flow rate, q = heat flow rate.
- *Across-variable:* v = translational velocity, ω = angular velocity, v = voltage, P = pressure, T = temperature.
- *Inductive storage:* L = inductance, $1/k$ = reciprocal translational or rotational stiffness, I = fluid inertance.
- *Capacitive storage:* C = capacitance, M = mass, J = moment of inertia, C_f = fluid capacitance, C_t = thermal capacitance.
- *Energy dissipators:* R = resistance, f = viscous friction, R_f = fluid resistance, R_t = thermal resistance.

The symbol $v(t)$ is used for both voltage in electrical circuits and velocity in translational mechanical systems and is distinguished within the context of each differential equation. For mechanical systems, one utilizes Newton's laws, and for electrical systems Kirchhoff's voltage laws. For example, the simple spring-mass-damper mechanical system shown in Fig. 2.1 is described by Newton's second law of motion. (This system could represent, for example, an automobile shock absorber.) Therefore, we obtain

$$M\frac{d^2y(t)}{dt^2} + f\frac{dy(t)}{dt} + Ky(t) = r(t), \tag{2.1}$$

where K is the spring constant of the ideal spring and f is the friction constant. Equation (2.1) is a linear constant coefficient differential equation of second order.

Alternatively, one may describe the electrical *RLC* circuit of Fig. 2.2 by utilizing Kirchhoff's current law. Then we obtain the following integrodifferential equation:

$$\frac{v(t)}{R} + C\frac{dv(t)}{dt} + \frac{1}{L}\int_0^t v(t)\,dt = r(t). \tag{2.2}$$

The solution of the differential equation describing the process may be obtained by classical methods such as the use of integrating factors and the method of undetermined coefficients [1]. For example, when the mass is displaced initially a distance $y(t) = y(0)$

FIGURE 2.1
Spring-mass-damper system.

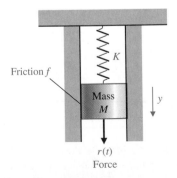

FIGURE 2.2
RLC circuit.

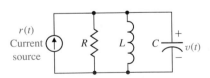

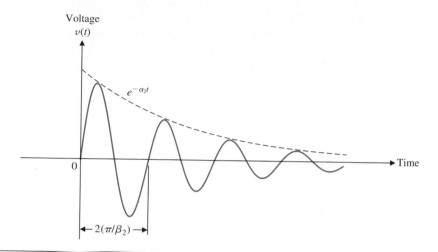

FIGURE 2.3
Typical voltage
response for
underdamped
RLC circuit.

and released, the dynamic response of an *underdamped* system is represented by an equation of the form

$$y(t) = K_1 e^{-\alpha_1 t} \sin(\beta_1 t + \theta_1). \tag{2.3}$$

A similar solution is obtained for the voltage of the *RLC* circuit when the circuit is subjected to a constant current $r(t) = I$. Then the voltage is

$$v(t) = K_2 e^{-\alpha_2 t} \cos(\beta_2 t + \theta_2). \tag{2.4}$$

A voltage curve typical of an underdamped *RLC* circuit is shown in Fig. 2.3.

To reveal further the close similarity between the differential equations for the mechanical and electrical systems, we shall rewrite Eq. (2.1) in terms of velocity:

$$v(t) = \frac{dy(t)}{dt}.$$

Then we have

$$M\frac{dv(t)}{dt} + fv(t) + K\int_0^t v(t)\, dt = r(t). \tag{2.5}$$

One immediately notes the equivalence of Eqs. (2.5) and (2.2) where velocity $v(t)$ and voltage $v(t)$ are equivalent variables, usually called *analogous* variables, and the systems are analogous systems. Therefore the solution for velocity is similar to Eq. (2.4), and the response for an underdamped system is shown in Fig. 2.3. The concept of analogous systems is a very useful and powerful technique for system modeling. The voltage–velocity analogy, often called the force–current analogy, is a natural one because it relates the analogous through- and across-variables of the electrical and mechanical systems. However, another analogy that relates the velocity and current variables is often used and is called the force–voltage analogy [22, 24].

Analogous systems with similar solutions exist for electrical, mechanical, thermal, and fluid systems. The existence of analogous systems and solutions provides the analyst with the ability to extend the solution of one system to all analogous systems with the same describing differential equations. Therefore what one learns about the analysis and design of electrical systems is immediately extended to an understanding of fluid, thermal, and mechanical systems.

2.3 LINEAR APPROXIMATIONS OF PHYSICAL SYSTEMS

A great majority of physical systems are linear within some range of the variables. However, all systems ultimately become nonlinear as the variables are increased without limit. For example, the spring-mass-damper system of Fig. 2.1 is linear and described by Eq. (2.1) as long as the mass is subjected to small deflections $y(t)$. However, if $y(t)$ were continually increased, eventually the spring would be overextended and break. Therefore the question of linearity and the range of applicability must be considered for each system.

A system is defined as linear in terms of the system excitation and response. In the case of the electrical network, the excitation is the input current $r(t)$ and the response is the voltage $v(t)$. In general, a *necessary condition* for a linear system can be determined in terms of an excitation $x(t)$ and a response $y(t)$. When the system at rest is subjected to an excitation $x_1(t)$, it provides a response $y_1(t)$. Furthermore, when the system is subjected to an excitation $x_2(t)$, it provides a corresponding response $y_2(t)$. For a linear system, it is *necessary* that the excitation $x_1(t) + x_2(t)$ result in a response $y_1(t) + y_2(t)$. This is usually called the *principle of superposition.*

Furthermore, it is necessary that the magnitude scale factor be preserved in a linear system. Again, consider a system with an input x that results in an output y. Then it is necessary that the response of a linear system to a constant multiple β of an input x be equal to the response to the input multiplied by the same constant so that the output is equal to βy. This is called the property of *homogeneity.*

A linear system satisfies the properties of superposition and homogeneity.

A system characterized by the relation $y = x^2$ is not linear, because the superposition property is not satisfied. A system represented by the relation $y = mx + b$ is not linear, because it does not satisfy the homogeneity property. However, this device may be considered linear about an operating point x_0, y_0 for small changes Δx and Δy. When $x = x_0 + \Delta x$ and $y = y_0 + \Delta y$, we have

$$y = mx + b$$

or

$$y_0 + \Delta y = mx_0 + m\,\Delta x + b$$

and therefore $\Delta y = m\,\Delta x$, which satisfies the necessary conditions.

The linearity of many mechanical and electrical elements can be assumed over a reasonably large range of the variables [7]. This is not usually the case for thermal and fluid elements, which are more frequently nonlinear in character. Fortunately, however, one can often linearize nonlinear elements assuming small-signal conditions. This is the normal approach used to obtain a linear equivalent circuit for electronic circuits and transistors. Consider a general element with an excitation (through-) variable $x(t)$ and a response (across-) variable $y(t)$. Several examples of dynamic system variables are given in Table 2.1. The relationship of the two variables is written as

$$y(t) = g(x(t)), \tag{2.6}$$

where $g(x(t))$ indicates $y(t)$ is a function of $x(t)$. The normal operating point is designated by x_0. Because the curve (function) is continuous over the range of interest, a *Taylor series* expansion about the operating point may be utilized [7]. Then we have

$$y = g(x) = g(x_0) + \left.\frac{dg}{dx}\right|_{x=x_0} \frac{(x - x_0)}{1!} + \left.\frac{d^2g}{dx^2}\right|_{x=x_0} \frac{(x - x_0)^2}{2!} + \cdots . \tag{2.7}$$

The slope at the operating point,

$$\left.\frac{dg}{dx}\right|_{x=x_0},$$

is a good approximation to the curve over a small range of $(x - x_0)$, the deviation from the operating point. Then, as a reasonable approximation, Eq. (2.7) becomes

$$y = g(x_0) + \left.\frac{dg}{dx}\right|_{x=x_0} (x - x_0) = y_0 + m(x - x_0), \tag{2.8}$$

where m is the slope at the operating point. Finally, Eq. (2.8) can be rewritten as the linear equation

$$(y - y_0) = m(x - x_0)$$

or

$$\Delta y = m \, \Delta x. \tag{2.9}$$

Consider the case of a mass, M, sitting on a nonlinear spring, as shown in Fig. 2.4(a). The normal operating point is the equilibrium position that occurs when the spring force balances the gravitational force Mg, where g is the gravitational constant. Thus $f_0 = Mg$ as shown. For the nonlinear spring with $f = y^2$, the equilibrium position is $y_0 = (Mg)^{1/2}$.

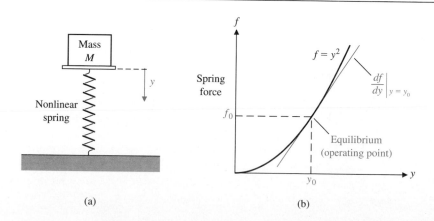

FIGURE 2.4
(a) A mass sitting on a nonlinear spring. (b) The spring force versus y.

The linear model for small deviation is

$$\Delta f = m \, \Delta y,$$

where

$$m = \left. \frac{df}{dy} \right|_{y_0},$$

as shown in Fig. 2.4(b). Thus $m = 2y_0$. A *linear approximation* is as accurate as the assumption of small signals is applicable to the specific problem.

If the dependent variable y depends upon several excitation variables, $x_1, x_2, \ldots, x_n$, then the functional relationship is written as

$$y = g(x_1, x_2, \ldots, x_n). \tag{2.10}$$

The Taylor series expansion about the operating point $x_{1_0}, x_{2_0}, \ldots, x_{n_0}$ is useful for a linear approximation to the nonlinear function. When the higher-order terms are neglected, the linear approximation is written as

$$y = g(x_{1_0}, x_{2_0}, \ldots, x_{n_0}) + \left. \frac{\partial g}{\partial x_1} \right|_{x=x_0} (x_1 - x_{1_0}) + \left. \frac{\partial g}{\partial x_2} \right|_{x=x_0} (x_2 - x_{2_0})$$

$$+ \cdots + \left. \frac{\partial g}{\partial x_n} \right|_{x=x_0} (x_n - x_{n_0}), \tag{2.11}$$

where x_0 is the operating point. An example will clearly illustrate the utility of this method.

EXAMPLE 2.1   Pendulum oscillator model

Consider the pendulum oscillator shown in Fig. 2.5(a). The torque on the mass is

$$T = MgL \sin \theta, \tag{2.12}$$

where g is the gravity constant. The equilibrium condition for the mass is $\theta_0 = 0°$. The nonlinear relation between T and θ is shown graphically in Fig. 2.5(b). The first derivative evaluated at equilibrium provides the linear approximation, which is

$$T - T_0 \cong MgL \left. \frac{\partial \sin \theta}{\partial \theta} \right|_{\theta=\theta_0} (\theta - \theta_0),$$

FIGURE 2.5
Pendulum
oscillator.

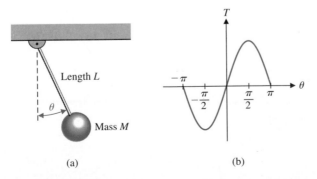

(a)   (b)

where $T_0 = 0$. Then, we have

$$T = MgL(\cos 0°)(\theta - 0°) \tag{2.13}$$
$$= MgL\theta.$$

This approximation is reasonably accurate for $(-\pi/4) \le \theta \le (\pi/4)$. For example, the response of the linear model for the swing through $\pm 30°$ is within 2% of the actual non-linear pendulum response. ■

2.4 THE LAPLACE TRANSFORM

The ability to obtain linear approximations of physical systems allows the analyst to consider the use of the *Laplace transformation*. The Laplace transform method substitutes the relatively easily solved algebraic equations for the more difficult differential equations [1, 3]. The time response solution is obtained by the following operations:

1. Obtain the differential equations.
2. Obtain the Laplace transformation of the differential equations.
3. Solve the resulting algebraic transform of the variable of interest.

The Laplace transform exists for linear differential equations for which the transformation integral converges. Therefore, for $f(t)$ to be transformable, it is sufficient that

$$\int_{0-}^{\infty} |f(t)| e^{-\sigma_1 t} \, dt < \infty$$

for some real, positive σ_1[1]. If the magnitude of $f(t)$ is $|f(t)| < Me^{\alpha t}$ for all positive t, the integral will converge for $\sigma_1 > \alpha$. The region of convergence is therefore given by $\infty > \sigma_1 > \alpha$, and σ_1 is known as the abscissa of absolute convergence. Signals that are physically realizable always have a Laplace transform. The Laplace transformation for a function of time, $f(t)$, is

$$F(s) = \int_{0-}^{\infty} f(t) e^{-st} \, dt = \mathcal{L}\{f(t)\}. \tag{2.14}$$

The *inverse Laplace transform* is written as

$$f(t) = \frac{1}{2\pi j} \int_{\sigma - j\infty}^{\sigma + j\infty} F(s) e^{+st} \, ds. \tag{2.15}$$

The transformation integrals have been used to derive tables of Laplace transforms that are ordinarily used for the great majority of problems. A table of Laplace transforms is provided in Appendix A, and some important Laplace transform pairs are given in Table 2.5.

Alternatively, the Laplace variable s can be considered to be the differential operator so that

$$s \equiv \frac{d}{dt}. \tag{2.16}$$

TABLE 2.5   Important Laplace Transform Pairs

$f(t)$	$F(s)$
Step function, $u(t)$	$\dfrac{1}{s}$
e^{-at}	$\dfrac{1}{s + a}$
$\sin \omega t$	$\dfrac{\omega}{s^2 + \omega^2}$
$\cos \omega t$	$\dfrac{s}{s^2 + \omega^2}$
$e^{-at} f(t)$	$F(s + a)$
t^n	$\dfrac{n!}{s^{n+1}}$
$f^{(k)}(t) = \dfrac{d^k f(t)}{dt^k}$	$s^k F(s) - s^{k-1} f(0^-) - s^{k-2} f'(0^-)$ $- \cdots - f^{(k-1)}(0^-)$
$\displaystyle \int_{-\infty}^{t} f(t) dt$	$\dfrac{F(s)}{s} + \dfrac{1}{s} \displaystyle \int_{-\infty}^{0} f(t) \, dt$
Impulse function $\delta(t)$	1

Then we also have the integral operator

$$\frac{1}{s} \equiv \int_{0^-}^{t} dt. \tag{2.17}$$

The inverse Laplace transformation is usually obtained by using the Heaviside partial fraction expansion. This approach is particularly useful for systems analysis and design because the effect of each characteristic root or eigenvalue can be clearly observed.

To illustrate the usefulness of the Laplace transformation and the steps involved in the system analysis, reconsider the spring-mass-damper system described by Eq. (2.1), which is

$$M\frac{d^2 y}{dt^2} + f\frac{dy}{dt} + Ky = r(t). \tag{2.18}$$

We wish to obtain the response, y, as a function of time. The Laplace transform of Eq. (2.18) is

$$M\left(s^2 Y(s) - s y(0^-) - \frac{dy(0^-)}{dt}\right) + f(sY(s) - y(0^-)) + KY(s) = R(s). \tag{2.19}$$

When $\qquad r(t) = 0, \qquad y(0^-) = y_0, \qquad \left.\dfrac{dy}{dt}\right|_{t=0^-} = 0,$

we have

$$Ms^2 Y(s) - Ms y_0 + fsY(s) - fy_0 + KY(s) = 0. \tag{2.20}$$

Solving for $Y(s)$, we obtain

$$Y(s) = \frac{(Ms + f)y_0}{Ms^2 + fs + K} = \frac{p(s)}{q(s)}. \tag{2.21}$$

The denominator polynomial $q(s)$, when set equal to zero, is called the *characteristic equation* because the roots of this equation determine the character of the time response. The roots of this characteristic equation are also called the *poles* or *singularities* of the system. The roots of the numerator polynomial $p(s)$ are called the *zeros* of the system; for example, $s = -f/M$. Poles and zeros are critical frequencies. At the poles, the function $Y(s)$ becomes infinite, whereas at the zeros, the function becomes zero. The complex frequency *s-plane* plot of the poles and zeros graphically portrays the character of the natural transient response of the system.

For a specific case, consider the system when $K/M = 2$ and $f/M = 3$. Then Eq. (2.21) becomes

$$Y(s) = \frac{(s + 3)y_0}{(s + 1)(s + 2)}. \tag{2.22}$$

The poles and zeros of $Y(s)$ are shown on the *s*-plane in Fig. 2.6.

Expanding Eq. (2.22) in a partial fraction expansion, we obtain

$$Y(s) = \frac{k_1}{s + 1} + \frac{k_2}{s + 2}, \tag{2.23}$$

where k_1 and k_2 are the coefficients of the expansion. The coefficients k_i are called *residues* and are evaluated by multiplying through by the denominator factor of Eq. (2.22) corresponding to k_i and setting s equal to the root. Evaluating k_1 when $y_0 = 1$, we have

$$k_1 = \left. \frac{(s - s_1)p(s)}{q(s)} \right|_{s=s_1}$$

$$= \left. \frac{(s + 1)(s + 3)}{(s + 1)(s + 2)} \right|_{s_1 = -1} = 2 \tag{2.24}$$

and $k_2 = -1$. Alternatively, the residues of $Y(s)$ at the respective poles may be evaluated

FIGURE 2.6
An *s*-plane pole and zero plot.

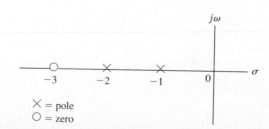

$\times$ = pole
$\bigcirc$ = zero

graphically on the s-plane plot, since Eq. (2.24) may be written as

$$k_1 = \left.\frac{s+3}{s+2}\right|_{s=s_1=-1}$$

(2.25)

$$= \left.\frac{s_1+3}{s_1+2}\right|_{s_1=-1} = 2.$$

The graphical representation of Eq. (2.25) is shown in Fig. 2.7. The graphical method of evaluating the residues is particularly valuable when the order of the characteristic equation is high and several poles are complex conjugate pairs.

The inverse Laplace transform of Eq. (2.22) is then

$$y(t) = \mathcal{L}^{-1}\left\{\frac{2}{s+1}\right\} + \mathcal{L}^{-1}\left\{\frac{-1}{s+2}\right\}.$$

(2.26)

Using Table 2.5, we find that

$$y(t) = 2e^{-t} - 1e^{-2t}.$$

(2.27)

Finally, it is usually desired to determine the *steady-state* or *final value* of the response of $y(t)$. For example, the final or steady-state rest position of the spring-mass-damper system should be calculated. The final value can be determined from the relation

$$\lim_{t\to\infty} y(t) = \lim_{s\to 0} sY(s),$$

(2.28)

where a simple pole of $Y(s)$ at the origin is permitted, but poles on the imaginary axis and in the right half-plane and higher-order poles at the origin are excluded. Therefore, for the specific case of the spring, mass, and damper, we find that

$$\lim_{t\to\infty} y(t) = \lim_{s\to 0} sY(s) = 0.$$

(2.29)

Hence the final position for the mass is the normal equilibrium position $y = 0$.

To illustrate clearly the salient points of the Laplace transform method, let us reconsider the mass-spring-damper system for the underdamped case. The equation for $Y(s)$ may be written as

$$Y(s) = \frac{(s + f/M)(y_0)}{(s^2 + (f/M)s + K/M)} = \frac{(s + 2\zeta\omega_n)(y_0)}{s^2 + 2\zeta\omega_n s + \omega_n^2},$$

(2.30)

where ζ is the dimensionless *damping ratio* and ω_n is the *natural frequency* of the system.

FIGURE 2.7
Graphical
evaluation of the
residues.

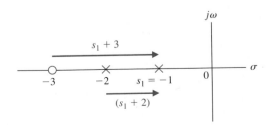

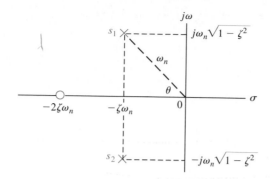

FIGURE 2.8
An s-plane plot of
the poles and
zeros of $Y(s)$.

The roots of the characteristic equation are

$$s_1, s_2 = -\zeta\omega_n \pm \omega_n\sqrt{\zeta^2 - 1}, \tag{2.31}$$

where, in this case, $\omega_n = \sqrt{K/M}$ and $\zeta = f/(2\sqrt{KM})$. When $\zeta > 1$, the roots are real; when $\zeta < 1$, the roots are complex and conjugates. When $\zeta = 1$, the roots are repeated and real and the condition is called *critical damping*.

When $\zeta < 1$, the response is underdamped and

$$s_{1,2} = -\zeta\omega_n \pm j\omega_n\sqrt{1 - \zeta^2}. \tag{2.32}$$

The s-plane plot of the poles and zeros of $Y(s)$ is shown in Fig. 2.8, where $\theta = \cos^{-1}\zeta$. As ζ varies with ω_n constant, the complex conjugate roots follow a circular locus, as shown in Fig. 2.9. The transient response is increasingly oscillatory as the roots approach the imaginary axis when ζ approaches zero.

The inverse Laplace transform can be evaluated using the graphical residue evaluation. The partial fraction expansion of Eq. (2.30) is

$$Y(s) = \frac{k_1}{(s - s_1)} + \frac{k_2}{(s - s_2)}. \tag{2.33}$$

Since s_2 is the complex conjugate of s_1, the residue k_2 is the complex conjugate of k_1 so

FIGURE 2.9
The locus of roots
as ζ varies with
ω_n constant.

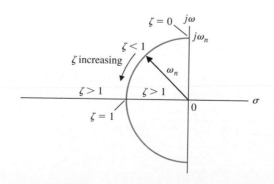

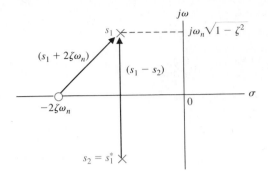

FIGURE 2.10
Evaluation of the
residue k_1.

that we obtain

$$Y(s) = \frac{k_1}{(s - s_1)} + \frac{k_1^*}{(s - s_1^*)},$$

where the asterisk indicates the conjugate relation. The residue k_1 is evaluated from
Fig. 2.10 as

$$k_1 = \frac{(y_0)(s_1 + 2\zeta\omega_n)}{(s_1 - s_1^*)} = \frac{(y_0)M_1 e^{j\theta}}{M_2 e^{j\pi/2}}, \tag{2.34}$$

where M_1 is the magnitude of $(s_1 + 2\zeta\omega_n)$ and M_2 is the magnitude of $(s_1 - s_1^*)$.[†] In this
case, we obtain

$$k_1 = \frac{(y_0)(\omega_n e^{j\theta})}{(2\omega_n\sqrt{1 - \zeta^2}\ e^{j\pi/2})} = \frac{(y_0)}{2\sqrt{1 - \zeta^2}\ e^{j(\pi/2-\theta)}}, \tag{2.35}$$

where $\theta = \cos^{-1}\zeta$. Therefore

$$k_2 = \frac{(y_0)}{2\sqrt{1 - \zeta^2}}\ e^{j(\pi/2-\theta)}. \tag{2.36}$$

Finally, we find that (using $\beta = \sqrt{1 - \zeta^2}$)

$$y(t) = k_1 e^{s_1 t} + k_2 e^{s_2 t}$$

$$= \frac{y_0}{2\sqrt{1 - \zeta^2}}\ (e^{j(\theta - \pi/2)}e^{-\zeta\omega_n t}e^{j\omega_n\beta t} + e^{j(\pi/2-\theta)}e^{-\zeta\omega_n t}e^{-j\omega_n\beta t}) \tag{2.37}$$

$$= \frac{y_0}{\sqrt{1 - \zeta^2}}\ e^{-\zeta\omega_n t}\ \sin\ (\omega_n\sqrt{1 - \zeta^2}t + \theta).$$

The solution, Eq. (2.37), can also be obtained using item 18 of Table A.1 in Appendix A.
The transient responses of the overdamped ($\zeta > 1$) and underdamped ($\zeta < 1$) cases are
shown in Fig. 2.11. The transient response that occurs when $\zeta < 1$ exhibits an oscillation
in which the amplitude decreases with time, and it is called a *damped oscillation*.

[†]A review of complex numbers appears in Appendix E.

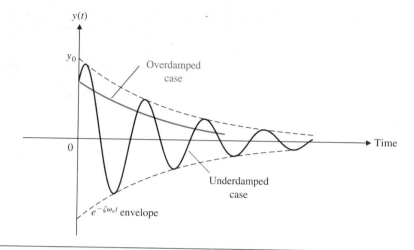

FIGURE 2.11
Response of the
spring-mass-
damper system.

The direct and clear relationship between the s-plane location of the poles and the form of the transient response is readily interpreted from the s-plane pole–zero plots. Furthermore, the magnitude of the response of each root, represented by the residue, is clearly visualized by examining the graphical residues on the s-plane. The Laplace transformation and the s-plane approach is a very useful technique for system analysis and design where emphasis is placed on the transient and steady-state performance. In fact, because the study of control systems is concerned primarily with the transient and steady-state performance of dynamic systems, we have real cause to appreciate the value of the Laplace transform techniques.

2.5 THE TRANSFER FUNCTION OF LINEAR SYSTEMS

The *transfer function* of a linear system is defined as the ratio of the Laplace transform of the output variable to the Laplace transform of the input variable, with all initial conditions assumed to be zero. The transfer function of a system (or element) represents the relationship describing the dynamics of the system under consideration.

A transfer function may be defined only for a linear, stationary (constant parameter) system. A nonstationary system, often called a time-varying system, has one or more time-varying parameters, and the Laplace transformation may not be utilized. Furthermore, a transfer function is an input–output description of the behavior of a system. Thus the transfer function description does not include any information concerning the internal structure of the system and its behavior.

The transfer function of the spring-mass-damper system is obtained from the original describing equation, Eq. (2.19), rewritten with zero initial conditions as follows:

$$Ms^2 Y(s) + fsY(s) + KY(s) = R(s). \tag{2.38}$$

Then the transfer function is

$$\frac{\text{Output}}{\text{Input}} = G(s) = \frac{Y(s)}{R(s)} = \frac{1}{Ms^2 + fs + K}. \tag{2.39}$$

The transfer function of the *RC* network shown in Fig. 2.12 is obtained by writing the Kirchhoff voltage equation, yielding

$$V_1(s) = \left(R + \frac{1}{Cs}\right)I(s). \tag{2.40}$$

The output voltage is

$$V_2(s) = I(s)\left(\frac{1}{Cs}\right). \tag{2.41}$$

Therefore, solving Eq. (2.40) for $I(s)$ and substituting in Eq. (2.41), we have

$$V_2(s) = \frac{(1/Cs)V_1(s)}{R + 1/Cs}.$$

Then the transfer function is obtained as the ratio $V_2(s)/V_1(s)$, which is

$$G(s) = \frac{V_2(s)}{V_1(s)} = \frac{1}{RCs + 1} = \frac{1}{\tau s + 1} = \frac{(1/\tau)}{s + 1/\tau}, \tag{2.42}$$

where $\tau = RC$, the *time constant* of the network. The single pole of $G(s)$ is $s = -1/\tau$. Equation (2.42) could be immediately obtained if one observes that the circuit is a voltage divider, where

$$\frac{V_2(s)}{V_1(s)} = \frac{Z_2(s)}{Z_1(s) + Z_2(s)} \tag{2.43}$$

and $Z_1(s) = R$, $Z_2 = 1/Cs$.

A multiloop electrical circuit or an analogous multiple-mass mechanical system results in a set of simultaneous equations in the Laplace variable. It is usually more convenient to solve the simultaneous equations by using matrices and determinants [1, 3, 16]. An introduction to matrices and determinants is provided in Appendix C.

Let us consider the long-term behavior of a system and determine the response to certain inputs that remain after the transients fade away. Consider the dynamic system represented by the differential equation

$$\frac{d^n y}{dt^n} + q_{n-1}\frac{d^{n-1}y}{dt^{n-1}} + \cdots + q_0 y = p_{n-1}\frac{d^{n-1}r}{dt^{n-1}} + p_{n-2}\frac{d^{n-2}r}{dt^{n-2}} + \cdots + p_0 r, \tag{2.44}$$

where $y(t)$ is the response and $r(t)$ is the input or forcing function. If the initial conditions are all zero, then the transfer function is

$$Y(s) = G(s)R(s) = \frac{p(s)}{q(s)}R(s) = \frac{(p_{n-1}s^{n-1} + p_{n-2}s^{n-2} + \cdots + p_0)}{(s^n + q_{n-1}s^{n-1} + \cdots + q_0)}R(s). \tag{2.45}$$

FIGURE 2.12
An *RC* network.

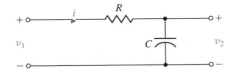

The output response consists of a natural response (determined by the initial conditions) plus a forced response determined by the input. We now have

$$Y(s) = \frac{m(s)}{q(s)} + \frac{p(s)}{q(s)}R(s),$$

where $q(s) = 0$ is the characteristic equation. If the input has the rational form

$$R(s) = \frac{n(s)}{d(s)},$$

then

$$Y(s) = \frac{m(s)}{q(s)} + \frac{p(s)}{q(s)}\frac{n(s)}{d(s)} = Y_1(s) + Y_2(s) + Y_3(s), \tag{2.46}$$

where $Y_1(s)$ is the partial fraction expansion of the natural response, $Y_2(s)$ is the partial fraction expansion of the terms involving factors of $q(s)$, and $Y_3(s)$ is the partial fraction expansion of terms involving factors of $d(s)$.

Taking the inverse Laplace transform yields

$$y(t) = y_1(t) + y_2(t) + y_3(t).$$

The transient response consists of $y_1(t) + y_2(t)$, and the steady-state response is $y_3(t)$.

EXAMPLE 2.2   **Solution of a differential equation**

Consider a system represented by the differential equation

$$\frac{d^2y}{dt^2} + 4\frac{dy}{dt} + 3y = 2r(t),$$

where the initial conditions are $y(0) = 1$, $\frac{dy}{dt}(0) = 0$, and $r(t) - 1$, $t \geq 0$.

The Laplace transform yields

$$[s^2Y(s) - sy(0)] + 4[sY(s) - y(0)] + 3Y(s) = 2R(s).$$

Since $R(s) = 1/s$ and $y(0) = 1$, we obtain

$$Y(s) = \frac{(s+4)}{(s^2 + 4s + 3)} + \frac{2}{s(s^2 + 4s + 3)},$$

where $q(s) = s^2 + 4s + 3 = (s+1)(s+3) = 0$ is the characteristic equation and $d(s) = s$. Then the partial fraction expansion yields

$$Y(s) = \left[\frac{3/2}{(s+1)} + \frac{-1/2}{(s+3)}\right] + \left[\frac{-1}{(s+1)} + \frac{1/3}{(s+3)}\right] + \frac{2/3}{s}$$

$$= Y_1(s) + Y_2(s) + Y_3(s).$$

Hence the response is

$$y(t) = \left[\frac{3}{2}e^{-t} - \frac{1}{2}e^{-3t}\right] + \left[-1e^{-t} + \frac{1}{3}e^{-3t}\right] + \frac{2}{3},$$

and the steady-state response is

$$\lim_{t \to \infty} y(t) = \frac{2}{3}. \ \blacksquare$$

EXAMPLE 2.3   **Transfer function of system**

Consider the mechanical system shown in Fig. 2.13(a) and its electrical circuit analog shown in Fig. 2.13(b). The electrical circuit analog is a force–current analog as outlined in Table 2.1. The velocities, $v_1(t)$ and $v_2(t)$, of the mechanical system are directly analogous to the node voltage $v_1(t)$ and $v_2(t)$ of the electrical circuit. The simultaneous equations, assuming the initial conditions are zero, are

$$M_1 s V_1(s) + (f_1 + f_2)V_1(s) - f_1 V_2(s) = R(s), \tag{2.47}$$

$$M_2 s V_2(s) + f_1(V_2(s) - V_1(s)) + K\frac{V_2(s)}{s} = 0. \tag{2.48}$$

These equations are obtained using the force equations for the mechanical system of Fig. 2.13(a). Rearranging Eqs. (2.47) and (2.48), we obtain

$$(M_1 s + (f_1 + f_2))V_1(s) + (-f_1)V_2(s) = R(s),$$

$$(-f_1)V_1(s) + \left(M_2 s + f_1 + \frac{K}{s}\right)V_2(s) = 0,$$

or, in matrix form, we have

$$\begin{bmatrix} (M_1 s + f_1 + f_2) & (-f_1) \\ (-f_1) & M_2 s + f_1 + \frac{K}{s} \end{bmatrix}\begin{bmatrix} V_1(s) \\ V_2(s) \end{bmatrix} = \begin{bmatrix} R(s) \\ 0 \end{bmatrix}. \tag{2.49}$$

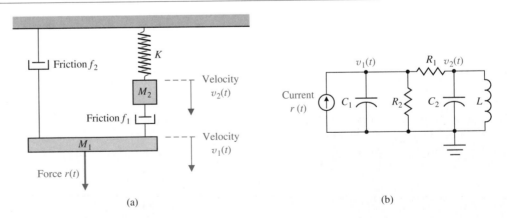

(a)   (b)

FIGURE 2.13
(a) Two-mass mechanical system. (b) Two-node electric circuit analog $C_1 = M_1$, $C_2 = M_2$, $L = 1/K$, $R_1 = 1/f_1$, $R_2 = 1/f_2$.

Assuming the velocity of M_1 is the output variable, we solve for $V_1(s)$ by matrix inversion or Cramer's rule to obtain [1, 3]

$$V_1(s) = \frac{(M_2 s + f_1 + (K/s))R(s)}{(M_1 s + f_1 + f_2)(M_2 s + f_1 + (K/s)) - f_1^2}. \tag{2.50}$$

Then the transfer function of the mechanical (or electrical) system is

$$G(s) = \frac{V_1(s)}{R(s)} = \frac{(M_2 s + f_1 + (K/s))}{(M_1 s + f_1 + f_2)(M_2 s + f_1 + (K/s)) - f_1^2} \tag{2.51}$$

$$= \frac{(M_2 s^2 + f_1 s + K)}{(M_1 s + f_1 + f_2)(M_2 s^2 + f_1 s + K) - f_1^2 s}.$$

If the transfer function in terms of the position $x_1(t)$ is desired, then we have

$$\frac{X_1(s)}{R(s)} = \frac{V_1(s)}{sR(s)} = \frac{G(s)}{s}. \ \blacksquare \tag{2.52}$$

As an example, let us obtain the transfer function of an important electrical control component, the *dc motor* [8]. A dc motor is used to move loads and is called an *actuator*.

An actuator is a device that provides the motive power to the process.

EXAMPLE 2.4   **Transfer function of dc motor**

The *dc motor* is a power actuator device that delivers energy to a load, as shown in Fig. 2.14(a); a sketch of a dc motor is shown in Fig. 2.14(b). A cutaway view of a pancake dc motor is given in Fig. 2.15. The dc motor converts direct current (dc) electrical energy into rotational mechanical energy. A major fraction of the torque generated in the rotor

---

FIGURE 2.14
A dc motor
(a) wiring diagram
and (b) sketch.

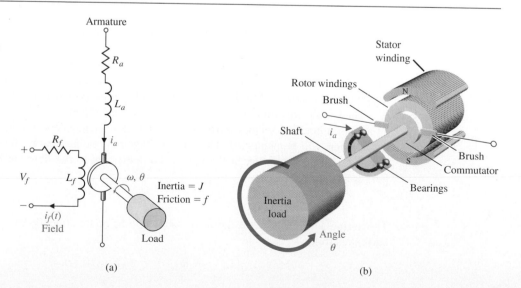

(a)

(b)

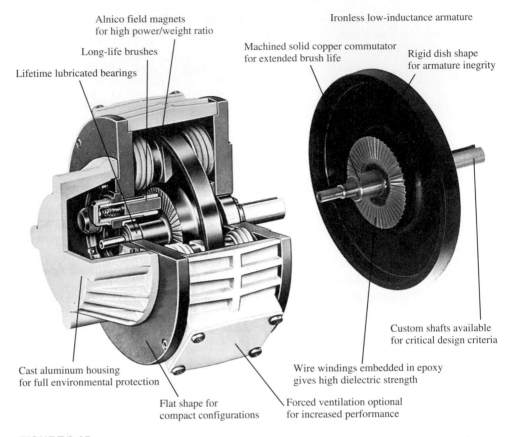

Alnico field magnets
for high power/weight ratio

Ironless low-inductance armature

Long-life brushes

Machined solid copper commutator
for extended brush life

Rigid dish shape
for armature inegrity

Lifetime lubricated bearings

Custom shafts available
for critical design criteria

Cast aluminum housing
for full environmental protection

Wire windings embedded in epoxy
gives high dielectric strength

Flat shape for
compact configurations

Forced ventilation optional
for increased performance

FIGURE 2.15
A pancake dc motor with a flat wound armature and a permanent magnet rotor. These motors are capable of providing high torque with a low rotor inertia. A typical mechanical time constant is in the range of 15 ms. (Courtesy of Mavilor Motors.)

(armature) of the motor is available to drive an external load. Because of features such as high torque, speed controllability over a wide range, portability, well-behaved speed-torque characteristics, and adaptability to various types of control methods, dc motors are still widely used in numerous control applications, including robotic manipulators, tape transport mechanisms, disk drives, machine tools, and servovalve actuators.

The transfer function of the dc motor will be developed for a linear approximation to an actual motor, and second-order effects, such as hysteresis and the voltage drop across the brushes, will be neglected. The input voltage may be applied to the field or armature terminals. The air-gap flux of the motor is proportional to the field current, provided the field is unsaturated, so that

$$\phi = K_f i_f. \tag{2.53}$$

The torque developed by the motor is assumed to be related linearly to ϕ and the armature current as follows:

$$T_m = K_1 \phi i_a(t) = K_1 K_f i_f(t) i_a(t). \tag{2.54}$$

It is clear from Eq. (2.54) that, to have a linear element, one current must be maintained constant while the other current becomes the input current. First, we shall consider the *field current controlled motor,* which provides a substantial power amplification. Then we have in Laplace transform notation

$$T_m(s) = (K_1 K_f I_a) I_f(s) = K_m I_f(s), \tag{2.55}$$

where $i_a = I_a$ is a constant armature current and K_m is defined as the motor constant. The field current is related to the field voltage as

$$V_f(s) = (R_f + L_f s) I_f(s). \tag{2.56}$$

The motor torque $T_m(s)$ is equal to the torque delivered to the load. This relation may be expressed as

$$T_m(s) = T_L(s) + T_d(s), \tag{2.57}$$

where $T_L(s)$ is the load torque and $T_d(s)$ is the disturbance torque, which is often negligible. However, the disturbance torque often must be considered in systems subjected to external forces such as antenna wind-gust forces. The load torque for rotating inertia as shown in Fig. 2.14 is written as

$$T_L(s) = Js^2\theta(s) + fs\theta(s). \tag{2.58}$$

Rearranging Eqs. (2.55)–(2.57), we have

$$T_L(s) = T_m(s) - T_d(s), \tag{2.59}$$

$$T_m(s) = K_m I_f(s), \tag{2.60}$$

$$I_f(s) = \frac{V_f(s)}{R_f + L_f s}. \tag{2.61}$$

Therefore the transfer function of the motor–load combination, with $T_d(s) = 0$, is

$$\frac{\theta(s)}{V_f(s)} = \frac{K_m}{s(Js + f)(L_f s + R_f)} = \frac{K_m/JL_f}{s(s + f/J)(s + R_f/L_f)}. \tag{2.62}$$

The block diagram model of the field-controlled dc motor is shown in Fig. 2.16. Alternatively, the transfer function may be written in terms of the time constants of the motor as

$$\frac{\theta(s)}{V_f(s)} = G(s) = \frac{K_m/fR_f}{s(\tau_f s + 1)(\tau_L s + 1)}, \tag{2.63}$$

where $\tau_f = L_f/R_f$ and $\tau_L = J/f$. Typically, one finds that $\tau_L > \tau_f$ and often the field time constant may be neglected.

FIGURE 2.16
Block diagram
model of field-
controlled
dc motor.

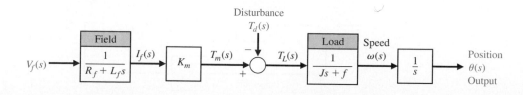

The *armature-controlled dc motor* utilizes a constant field current, and therefore the motor torque is

$$T_m(s) = (K_1 K_f I_f)I_a(s) = K_m I_a(s). \tag{2.64}$$

The armature current is related to the input voltage applied to the armature as

$$V_a(s) = (R_a + L_a s)I_a(s) + V_b(s), \tag{2.65}$$

where $V_b(s)$ is the back electromotive-force voltage proportional to the motor speed. Therefore we have

$$V_b(s) = K_b \omega(s), \tag{2.66}$$

and the armature current is

$$I_a(s) = \frac{V_a(s) - K_b \omega(s)}{(R_a + L_a s)}. \tag{2.67}$$

Equations (2.58) and (2.59) represent the load torque so that

$$T_L(s) = Js^2\theta(s) + fs\theta(s) = T_m(s) - T_d(s). \tag{2.68}$$

The relations for the armature-controlled dc motor are shown schematically in Fig. 2.17. Using Eqs. (2.64), (2.67), and (2.68), or, alternatively, the block diagram, we obtain the transfer function (with $T_d(s) = 0$)

$$G(s) = \frac{\theta(s)}{V_a(s)} = \frac{K_m}{s[(R_a + L_a s)(Js + f) + K_b K_m]} \tag{2.69}$$

$$= \frac{K_m}{s(s^2 + 2\zeta\omega_n s + \omega_n^2)}.$$

However, for many dc motors, the time constant of the armature, $\tau_a = L_a/R_a$, is negligible, and therefore

$$G(s) = \frac{\theta(s)}{V_a(s)} = \frac{K_m}{s[R_a(Js + f) + K_b K_m]} = \frac{[K_m/(R_a f + K_b K_m)]}{s(\tau_1 s + 1)}, \tag{2.70}$$

where the equivalent time constant $\tau_1 = R_a J/(R_a f + K_b K_m)$.

It is of interest to note that K_m is equal to K_b. This equality may be shown by considering the steady-state motor operation and the power balance when the rotor resistance is neglected. The power input to the rotor is $(K_b \omega)i_a$ and the power delivered to the shaft is

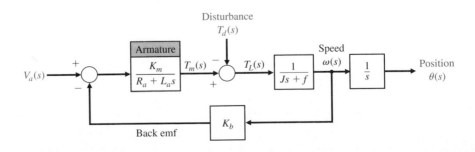

FIGURE 2.17
Armature-controlled dc motor.

TABLE 2.6   Typical Constants for a Fractional Horsepower dc Motor

Motor constant K_m	50×10^{-3} N · m/A
Rotor inertia J_m	1×10^{-3} N · m · s²/rad
Field time constant τ_f	1 ms
Rotor time constant τ	100 ms
Maximum output power	¼ hp, 187 W

$T\omega$. In the steady-state condition, the power input is equal to the power delivered to the shaft so that $(K_b\omega)i_a = T\omega$; since $T = K_m i_a$ (Eq. 2.64), we find that $K_b = K_m$.

Electric motors are used for moving loads when a rapid response is not required and for relatively low power requirements. Typical constants for a fractional horsepower motor are provided in Table 2.6. Actuators that operate as a result of hydraulic pressure are used for large loads. Figure 2.18 shows the usual ranges of use for electromechanical drives as contrasted to electrohydraulic drives. Typical applications are also shown on the figure. ■

EXAMPLE 2.5   **Transfer function of hydraulic actuator**

A useful actuator for the linear positioning of a mass is the hydraulic actuator shown in Table 2.7, entry 9 [9, 10]. The hydraulic actuator is capable of providing a large power amplification. It will be assumed that the hydraulic fluid is available from a constant pres-

FIGURE 2.18
Range of control response time and power to load for electromechanical and electrohydraulic devices.

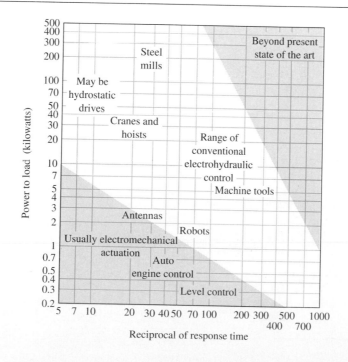

TABLE 2.7   Transfer Functions of Dynamic Elements and Networks

Element or System	$G(s)$

1. Integrating circuit, filter

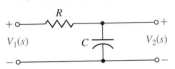

$$\frac{V_2(s)}{V_1(s)} = \frac{1}{RCs + 1}$$

2. Differentiating circuit

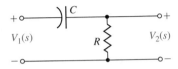

$$\frac{V_2(s)}{V_1(s)} = \frac{RCs}{RCs + 1}$$

3. Differentiating circuit

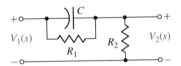

$$\frac{V_2(s)}{V_1(s)} = \frac{s + 1/R_1 C}{s + (R_1 + R_2)/R_1 R_2 C}$$

4. Lead-lag filter circuit

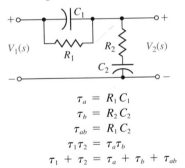

$$\frac{V_2(s)}{V_1(s)} = \frac{(1 + s\tau_a)(1 + s\tau_b)}{\tau_a \tau_b s^2 + (\tau_a + \tau_b + \tau_{ab})s + 1}$$

$$= \frac{(1 + s\tau_a)(1 + s\tau_b)}{(1 + s\tau_1)(1 + s\tau_2)}$$

$$\tau_a = R_1 C_1$$
$$\tau_b = R_2 C_2$$
$$\tau_{ab} = R_1 C_2$$
$$\tau_1 \tau_2 = \tau_a \tau_b$$
$$\tau_1 + \tau_2 = \tau_a + \tau_b + \tau_{ab}$$

5. dc motor, field-controlled, rotational actuator

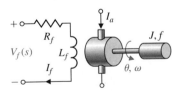

$$\frac{\theta(s)}{V_f(s)} = \frac{K_m}{s(Js + f)(L_f s + R_f)}$$

6. dc motor, armature-controlled, rotational actuator

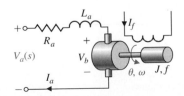

$$\frac{\theta(s)}{V_a(s)} = \frac{K_m}{s[(R_a + L_a s)(Js + f) + K_b K_m]}$$

(continued)

TABLE 2.7   *Continued*

Element or System	$G(s)$

7. ac motor, two-phase control field, rotational
    actuator

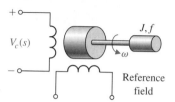

$$\frac{\theta(s)}{V_c(s)} = \frac{K_m}{s(\tau s + 1)}$$

$$\tau = J/(f - m)$$

$m =$ slope of linearized torque-speed
      curve (normally negative)

8. Amplidyne, voltage and power amplifier

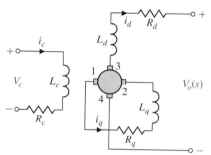

$$\frac{V_o(s)}{V_c(s)} = \frac{(K/R_c R_q)}{(s\tau_c + 1)(s\tau_q + 1)}$$

$$\tau_c = L_c/R_c, \quad \tau_q = L_q/R_q$$

For the unloaded case, $i_d \approx 0$, $\tau_c \approx \tau_q$,
$0.05 \text{ s} < \tau_c < 0.5 \text{ s}$

$$V_{12} = V_q, V_{34} = V_d$$

9. Hydraulic actuator

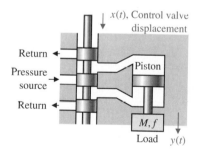

$$\frac{Y(s)}{X(s)} = \frac{K}{s(Ms + B)}$$

$$K = \frac{Ak_x}{k_p}, \quad B = \left(f + \frac{A^2}{k_p}\right)$$

$$k_x = \left.\frac{\partial g}{\partial x}\right|_{x_0}, \quad k_p = \left.\frac{\partial g}{\partial P}\right|_{P_0},$$

$g = g(x, P) =$ flow

$A =$ area of piston

10. Gear train, rotational transformer

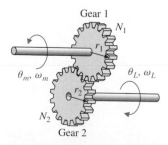

Gear ratio $= n = \dfrac{N_1}{N_2}$

$$N_2\theta_L = N_1\theta_m, \quad \theta_L = n\theta_m$$

$$\omega_L = n\omega_m$$

(*continued*)

TABLE 2.7 *Continued*

Element or System	$G(s)$

11. Potentiometer, voltage control

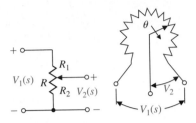

$$\frac{V_2(s)}{V_1(s)} = \frac{R_2}{R} = \frac{R_2}{R_1 + R_2}$$

$$\frac{R_2}{R} = \frac{\theta}{\theta_{max}}$$

12. Potentiometer error detector bridge

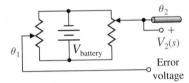

$$V_2(s) = k_s(\theta_1(s) - \theta_2(s))$$

$$V_2(s) = k_s\theta_{error}(s)$$

$$k_s = \frac{V_{battery}}{\theta_{max}}$$

13. Tachometer, velocity sensor

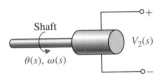

$$V_2(s) = K_t\omega(s) = K_t s\theta(s);$$

$$K_t = \text{constant}$$

14. dc amplifier

$$\frac{V_2(s)}{V_1(s)} = \frac{k_a}{s\tau + 1}$$

$$R_o = \text{output resistance}$$

$$C_o = \text{output capacitance}$$

$$\tau = R_o C_o, \ \tau \ll 1$$

and is often negligible for servomechanism amplifier

15. Accelerometer, acceleration sensor

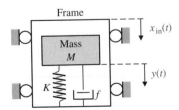

$$x_o(t) = y(t) - x_{in}(t),$$

$$\frac{X_o(s)}{X_{in}(s)} = \frac{-s^2}{s^2 + (f/M)s + K/M}$$

For low-frequency oscillations, where $\omega < \omega_n$,

$$\frac{X_o(j\omega)}{X_{in}(j\omega)} \simeq \frac{\omega^2}{K/M}$$

(continued)

TABLE 2.7 *Continued*

Element or System	$G(s)$

16. Thermal heating system

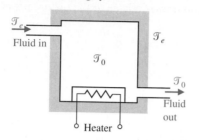

$$\frac{T(s)}{q(s)} = \frac{1}{C_t s + (QS + 1/R)}, \text{ where}$$

$\tau = \tau_o - \tau_e = $ temperature difference
due to thermal process

$C_t = $ thermal capacitance

$Q = $ fluid flow rate $=$ constant

$S = $ specific heat of water

$R_t = $ thermal resistance of insulation

$q(s) = $ rate of heat flow of heating element

17. Rack and pinion

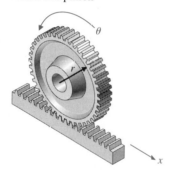

$x = r\theta$
converts radial motion
to linear motion

sure source and that the compressibility of the fluid is negligible. A downward input displacement, x, moves the control valve, and thus fluid passes into the upper part of the cylinder and the piston is forced downward. A small, low-power displacement of $x(t)$ causes a larger, high-power displacement, $y(t)$. The volumetric fluid flow rate Q is related to the input displacement $x(t)$ and the differential pressure across the piston as $Q = g(x, P)$. Using the Taylor series linearization as in Eq. (2.11), we have

$$Q = \left(\frac{\partial g}{\partial x}\right)_{x_0, P_0} x + \left(\frac{\partial g}{\partial P}\right)_{P_0, x_0} P = k_x x - k_P P, \qquad (2.71)$$

where $g = g(x, P)$ and (x_0, P_0) is the operating point. The force developed by the actuator piston is equal to the area of the piston, A, multiplied by the pressure, P. This force is applied to the mass, so we have

$$AP = M \frac{d^2 y}{dt^2} + f \frac{dy}{dt}. \qquad (2.72)$$

Thus, substituting Eq. (2.71) into Eq. (2.72), we obtain

$$\frac{A}{k_P}(k_x x - Q) = M\frac{d^2y}{dt^2} + f\frac{dy}{dt}. \tag{2.73}$$

Furthermore, the volumetric fluid flow is related to the piston movement as

$$Q = A\frac{dy}{dt}. \tag{2.74}$$

Then, substituting Eq. (2.74) into Eq. (2.73) and rearranging, we have

$$\frac{Ak_x}{k_P}x = M\frac{d^2y}{dt^2} + \left(f + \frac{A^2}{k_P}\right)\frac{dy}{dt}. \tag{2.75}$$

Therefore, using the Laplace transformation, we have the transfer function

$$\frac{Y(s)}{X(s)} = \frac{K}{s(Ms + B)}, \tag{2.76}$$

where

$$K = \frac{Ak_x}{k_P} \quad \text{and} \quad B = \left(f + \frac{A^2}{k_P}\right).$$

Note that the transfer function of the hydraulic actuator is similar to that of the electric motor. Also, for an actuator operating at high pressure levels and requiring a rapid response of the load, the effect of the compressibility of the fluid must be accounted for [4, 5].

The SI units of the variables are given in Table B.1 in Appendix B. Also a complete set of conversion factors for the British system of units is given in Table B.2. ■

The transfer function concept and approach is very important because it provides the analyst and designer with a useful mathematical model of the system elements. We shall find the transfer function to be a continually valuable aid in the attempt to model dynamic systems. The approach is particularly useful because the s-plane poles and zeros of the transfer function represent the transient response of the system. The transfer functions of several dynamic elements are given in Table 2.7.

In many situations in engineering, the transmission of rotary motion from one shaft to another is a fundamental requirement. For example, the output power of an automobile engine is transferred to the driving wheels by means of the gearbox and differential. The gearbox allows the driver to select different gear ratios depending on the traffic situation, whereas the differential has a fixed value. The speed of the engine in this case is not constant, since it is under the control of the driver. Another example is a set of gears that transfer the power at the shaft of an electric motor to the shaft of a rotating antenna. Examples of mechanical converters are gears, chain drives, and belt drives. A commonly used electric converter is the electric transformer. An example of a device that converts rotational motion to linear motion is the rack and pinion gear shown as number 17 in Table 2.7.

2.6 BLOCK DIAGRAM MODELS

The dynamic systems that comprise automatic control systems are represented mathematically by a set of simultaneous differential equations. As we have noted in the previous sections, the introduction of the Laplace transformation reduces the problem to the solution

FIGURE 2.19
Block diagram of
dc motor.

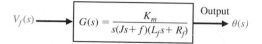

$$V_f(s) \longrightarrow \boxed{G(s) = \dfrac{K_m}{s(Js+f)(L_f s + R_f)}} \xrightarrow{\text{Output}} \theta(s)$$

of a set of linear algebraic equations. Since control systems are concerned with the control of specific variables, the interrelationship of the controlled variables to the controlling variables is required. This relationship is typically represented by the transfer function of the subsystem relating the input and output variables. Therefore one can correctly assume that the transfer function is an important relation for control engineering.

The importance of the cause-and-effect relationship of the transfer function is evidenced by the facility to represent the relationship of system variables by diagrammatic means. The *block diagram* representation of the system relationships is prevalent in control system engineering. Block diagrams consist of *unidirectional,* operational blocks that represent the transfer function of the variables of interest. A block diagram of a field-controlled dc motor and load is shown in Fig. 2.19. The relationship between the displacement $\theta(s)$ and the input voltage $V_f(s)$ is clearly portrayed by the block diagram.

To represent a system with several variables under control, an interconnection of blocks is utilized. For example, the system shown in Fig. 2.20 has two input variables and two output variables [6]. Using transfer function relations, we can write the simultaneous equations for the output variables as

$$C_1(s) = G_{11}(s)R_1(s) + G_{12}(s)R_2(s), \qquad (2.77)$$

$$C_2(s) = G_{21}(s)R_1(s) + G_{22}(s)R_2(s), \qquad (2.78)$$

where $G_{ij}(s)$ is the transfer function relating the ith output variable to the jth input variable. The block diagram representing this set of equations is shown in Fig. 2.21. In general, for J inputs and I outputs, we write the simultaneous equation in matrix form as

$$\begin{bmatrix} C_1(s) \\ C_2(s) \\ \vdots \\ C_I(s) \end{bmatrix} = \begin{bmatrix} G_{11}(s) \cdots G_{1J}(s) \\ G_{21}(s) \cdots G_{2J}(s) \\ \vdots \qquad \vdots \\ G_{I1}(s) \cdots G_{IJ}(s) \end{bmatrix} \begin{bmatrix} R_1(s) \\ R_2(s) \\ \vdots \\ R_J(s) \end{bmatrix} \qquad (2.79)$$

or, simply,

$$\mathbf{C} = \mathbf{GR}. \qquad (2.80)$$

Here the $\mathbf{C}$ and $\mathbf{R}$ matrices are column matrices containing the I output and the J input variables, respectively, and $\mathbf{G}$ is an I by J transfer function matrix. The matrix representation of the interrelationship of many variables is particularly valuable for complex multivariable control systems. An introduction to matrix algebra is provided in Appendix C for those unfamiliar with matrix algebra or who would find a review helpful [2].

FIGURE 2.20
General block
representation of
two-input, two-
output system.

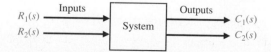

$$\begin{array}{l} R_1(s) \\ R_2(s) \end{array} \xrightarrow{\text{Inputs}} \boxed{\text{System}} \xrightarrow{\text{Outputs}} \begin{array}{l} C_1(s) \\ C_2(s) \end{array}$$

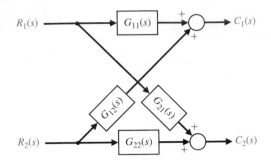

FIGURE 2.21
Block diagram of
interconnected
system.

The block diagram representation of a given system often can be reduced by block diagram reduction techniques to a simplified block diagram with fewer blocks than the original diagram. Since the transfer functions represent linear systems, the multiplication is commutative. Therefore, as in Table 2.8, entry 1, we have

$$X_3(s) = G_2(s)X_2(s) = G_1(s)G_2(s)X_1(s).$$

When two blocks are connected in cascade as in entry 1 of Table 2.8, we assume that

$$X_3(s) = G_2(s)G_1(s)X_1(s)$$

holds true. This assumes that when the first block is connected to the second block, the effect of loading of the first block is negligible. Loading and interaction between interconnected components or systems may occur. If loading of interconnected devices does occur, the engineer must account for this change in the transfer function and use the corrected transfer function in subsequent calculations.

Block diagram transformations and reduction techniques are derived by considering the algebra of the diagram variables. For example, consider the block diagram shown in Fig. 2.22. This negative feedback control system is described by the equation for the actuating signal

$$E_a(s) = R(s) - B(s) = R(s) - H(s)C(s). \tag{2.81}$$

Because the output is related to the actuating signal by $G(s)$, we have

$$C(s) = G(s)E_a(s), \tag{2.82}$$

and therefore

$$C(s) = G(s)[R(s) - H(s)C(s)]. \tag{2.83}$$

Solving for $C(s)$, we obtain

$$C(s)[1 + G(s)H(s)] = G(s)R(s). \tag{2.84}$$

TABLE 2.8   Block Diagram Transformations

Transformation	Original Diagram	Equivalent Diagram
1. Combining blocks in cascade		
2. Moving a summing point behind a block		
3. Moving a pickoff point ahead of a block		
4. Moving a pickoff point behind a block		
5. Moving a summing point ahead of a block		
6. Eliminating a feedback loop		

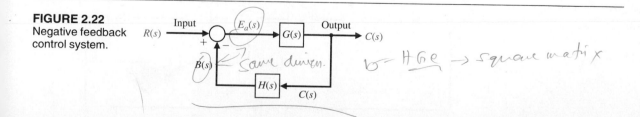

FIGURE 2.22
Negative feedback control system.

Therefore the transfer function relating the output $C(s)$ to the input $R(s)$ is

$$\frac{C(s)}{R(s)} = \frac{G(s)}{1 + G(s)H(s)}. \tag{2.85}$$

This *closed-loop transfer function* is particularly important because it represents many of the existing practical control systems.

The reduction of the block diagram shown in Fig. 2.22 to a single block representation is one example of several useful block diagram reductions. These diagram transformations are given in Table 2.8. All the transformations in Table 2.8 can be derived by simple algebraic manipulation of the equations representing the blocks. System analysis by the method of block diagram reduction affords a better understanding of the contribution of each component element than is possible to obtain by the manipulation of equations. The utility of the block diagram transformations will be illustrated by an example using block diagram reduction.

EXAMPLE 2.6   Block diagram reduction

The block diagram of a multiple-loop feedback control system is shown in Fig. 2.23. It is interesting to note that the feedback signal $H_1(s)C(s)$ is a positive feedback signal and the loop $G_3(s)G_4(s)H_1(s)$ is called a *positive feedback loop*. The block diagram reduction procedure is based on the utilization of rule 6 in Table 2.8, which eliminates feedback loops. Therefore the other transformations are used to transform the diagram to a form ready for eliminating feedback loops. First, to eliminate the loop $G_3G_4H_1$, we move H_2 behind block G_4 by using rule 4, and therefore obtain Fig. 2.24(a). Eliminating the loop $G_3G_4H_1$ by using rule 6, we obtain Fig. 2.24(b). Then, eliminating the inner loop containing H_2/G_4, we obtain Fig. 2.24(c). Finally, by reducing the loop containing H_3 we obtain the closed-loop system transfer function as shown in Fig. 2.24(d). It is worthwhile to examine the form of the numerator and denominator of this closed-loop transfer function. We note that the numerator is composed of the cascade transfer function of the feedforward elements connecting the input $R(s)$ and the output $C(s)$. The denominator is comprised of 1 minus the sum of each loop transfer function. The sign of the loop $G_3G_4H_1$ is plus because it is a positive feedback loop, whereas the loops $G_1G_2G_3G_4H_3$ and $G_2G_3H_2$ are negative feedback loops. To illustrate this point, the denominator can be rewritten as

$$q(s) = 1 - (+G_3G_4H_1 - G_2G_3H_2 - G_1G_2G_3G_4H_3). \tag{2.86}$$

FIGURE 2.23
Multiple-loop
feedback control
system.

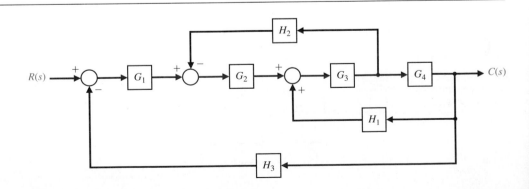

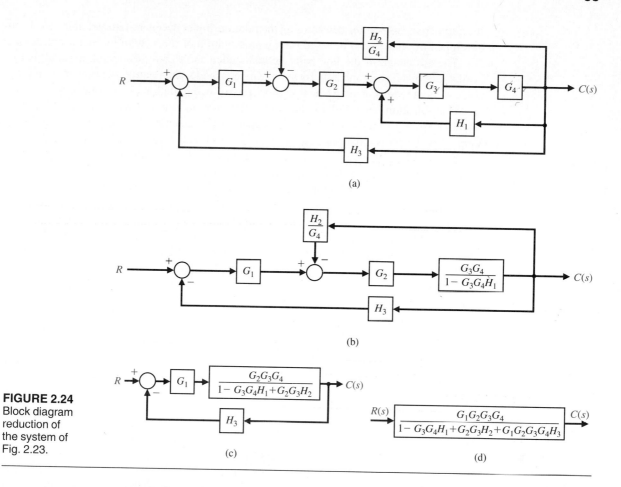

FIGURE 2.24
Block diagram
reduction of
the system of
Fig. 2.23.

This form of the numerator and denominator is quite close to the general form for multiple-loop feedback systems, as we shall find in the following section. ∎

The block diagram representation of feedback control systems is a valuable and widely used approach. The block diagram provides the analyst with a graphical representation of the interrelationships of controlled and input variables. Furthermore, the designer can readily visualize the possibilities for adding blocks to the existing system block diagram to alter and improve the system performance. The transition from the block diagram method to a method utilizing a line path representation instead of a block representation is readily accomplished and is presented in the following section.

2.7 SIGNAL-FLOW GRAPH MODELS

Block diagrams are adequate for the representation of the interrelationships of controlled and input variables. However, for a system with reasonably complex interrelationships, the block diagram reduction procedure is cumbersome and often quite difficult to complete.

An alternative method for determining the relationship between system variables has been developed by Mason and is based on a representation of the system by line segments [4, 25]. The advantage of the line path method, called the signal-flow graph method, is the availability of a flow graph gain formula, which provides the relation between system variables without requiring any reduction procedure or manipulation of the flow graph.

The transition from a block diagram representation to a directed line segment representation is easy to accomplish by reconsidering the systems of the previous section. A *signal-flow graph* is a diagram consisting of nodes that are connected by several directed branches and is a graphical representation of a set of linear relations. Signal-flow graphs are particularly useful for feedback control systems because feedback theory is primarily concerned with the flow and processing of signals in systems. The basic element of a signal-flow graph is a unidirectional path segment called a *branch*, which relates the dependency of an input and an output variable in a manner equivalent to a block of a block diagram. Therefore, the branch relating the output of a dc motor, $\theta(s)$, to the field voltage, $V_f(s)$, is similar to the block diagram of Fig. 2.19 and is shown in Fig. 2.25. The input and output points or junctions are called *nodes*. Similarly, the signal-flow graph representing Eqs. (2.77) and (2.78) and Fig. 2.21 is shown in Fig. 2.26. The relation between each variable is written next to the directional arrow. All branches leaving a node pass the nodal signal to the output node of each branch (unidirectionally). The summation of all signals entering a node is equal to the node variable. A *path* is a branch or a continuous sequence of branches that can be traversed from one signal (node) to another signal (node). A *loop* is a closed path that originates and terminates on the same node, and along the path no node is met twice. Two loops are said to be *nontouching* if they do not have a common node. Two touching loops share one or more common nodes. Therefore, reconsidering Fig. 2.26, we obtain

$$C_1(s) = G_{11}(s)R_1(s) + G_{12}(s)R_2(s), \tag{2.87}$$

$$C_2(s) = G_{21}(s)R_1(s) + G_{22}(s)R_2(s). \tag{2.88}$$

The flow graph is simply a pictorial method of writing a system of algebraic equations so as to indicate the interdependencies of the variables. As another example, consider the following set of simultaneous algebraic equations:

$$a_{11}x_1 + a_{12}x_2 + r_1 = x_1 \tag{2.89}$$

$$a_{21}x_1 + a_{22}x_2 + r_2 = x_2. \tag{2.90}$$

The two input variables are r_1 and r_2, and the output variables are x_1 and x_2. A signal-flow graph representing Eqs. (2.89) and (2.90) is shown in Fig. 2.27. Equations (2.89) and (2.90) may be rewritten as

$$x_1(1 - a_{11}) + x_2(-a_{12}) = r_1, \tag{2.91}$$

$$x_1(-a_{21}) + x_2(1 - a_{22}) = r_2. \tag{2.92}$$

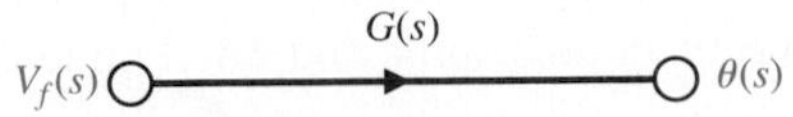

FIGURE 2.25
Signal-flow graph of the dc motor.

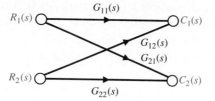

FIGURE 2.26
Signal-flow graph
of interconnected
system.

The simultaneous solution of Eqs. (2.91) and (2.92) using Cramer's rule results in the solutions

$$x_1 = \frac{(1 - a_{22})r_1 + a_{12}r_2}{(1 - a_{11})(1 - a_{22}) - a_{12}a_{21}} = \frac{(1 - a_{22})}{\Delta} r_1 + \frac{a_{12}}{\Delta} r_2, \qquad (2.93)$$

$$x_2 = \frac{(1 - a_{11})r_2 + a_{21}r_1}{(1 - a_{11})(1 - a_{22}) - a_{12}a_{21}} = \frac{(1 - a_{11})}{\Delta} r_2 + \frac{a_{21}}{\Delta} r_1. \qquad (2.94)$$

The denominator of the solution is the determinant Δ of the set of equations and is rewritten as

$$\Delta = (1 - a_{11})(1 - a_{22}) - a_{12}a_{21} = 1 - a_{11} - a_{22} + a_{11}a_{22} - a_{12}a_{21}. \qquad (2.95)$$

In this case, the denominator is equal to 1 minus each self-loop a_{11}, a_{22}, and $a_{12}a_{21}$, plus the product of the two nontouching loops a_{11} and a_{22}. The loops a_{22} and $a_{21}a_{12}$ are touching, as are a_{11} and $a_{21}a_{12}$.

The numerator for x_1 with the input r_1 is 1 times $(1 - a_{22})$, which is the value of Δ not touching the path 1 from r_1 to x_1. Therefore the numerator from r_2 to x_1 is simply a_{12} because the path through a_{12} touches all the loops. The numerator for x_2 is symmetrical to that of x_1.

In general, the linear dependence T_{ij} between the independent variable x_i (often called the input variable) and a dependent variable x_j is given by Mason's *loop rule* [11, 12]:

$$T_{ij} = \frac{\sum_k P_{ijk} \Delta_{ijk}}{\Delta}, \qquad (2.96)$$

where $P_{ijk} = k$th path from variable x_i to variable x_j,

Δ = determinant of the graph,

Δ_{ijk} = cofactor of the path P_{ijk},

FIGURE 2.27
Signal-flow graph
of two algebraic
equations.

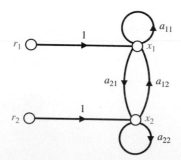

and the summation is taken over all possible k paths from x_i to x_j. The cofactor Δ_{ijk} is the determinant with the loops touching the kth path removed. The determinant Δ is

$$\Delta = 1 - \sum_{n=1}^{N} L_n + \sum_{m=1,q=1}^{M,Q} L_m L_q - \sum L_r L_s L_t + \cdots , \qquad (2.97)$$

where L_q equals the value of the qth loop transmittance. Therefore the rule for evaluating Δ in terms of loops $L_1, L_2, L_3, \ldots, L_N$ is

$\Delta = 1 - $ (sum of all different loop gains)
  $+$ (sum of the gain products of all combinations of 2 nontouching loops)
  $-$ (sum of the gain products of all combinations of 3 nontouching loops)
  $+ \cdots .$

The gain formula is often used to relate the output variable $C(s)$ to the input variable $R(s)$ and is given in somewhat simplified form as

$$T = \frac{\sum_k P_k \Delta_k}{\Delta}, \qquad (2.98)$$

where $T(s) = C(s)/R(s)$. The path gain or transmittance P_k (or P_{ijk}) is defined as the continuous succession of branches that are traversed in the direction of the arrows and with no node encountered more than once. A loop is defined as a closed path in which no node is encountered more than once per traversal.

Several examples will illustrate the utility and ease of this method. Although the gain equation (2.96) appears to be formidable, one must remember that it represents a summation process, not a complicated solution process.

EXAMPLE 2.7   **Transfer function of interacting system**

A two-path signal-flow graph is shown in Fig. 2.28. An example of a control system with multiple signal paths is a multilegged robot. The paths connecting the input $R(s)$ and output $C(s)$ are

path 1: $P_1 = G_1 G_2 G_3 G_4$ and   path 2: $P_2 = G_5 G_6 G_7 G_8$.

There are four self-loops:

$L_1 = G_2 H_2, \qquad L_2 = H_3 G_3, \qquad L_3 = G_6 H_6, \qquad L_4 = G_7 H_7.$

FIGURE 2.28
Two-path
interacting system.

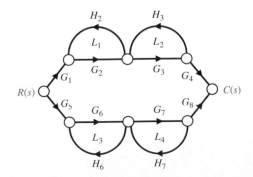

Loops L_1 and L_2 do not touch L_3 and L_4. Therefore the determinant is

$$\Delta = 1 - (L_1 + L_2 + L_3 + L_4) + (L_1L_3 + L_1L_4 + L_2L_3 + L_2L_4). \quad (2.99)$$

The cofactor of the determinant along path 1 is evaluated by removing the loops that touch path 1 from Δ. Therefore we have

$$L_1 = L_2 = 0 \quad \text{and} \quad \Delta_1 = 1 - (L_3 + L_4).$$

Similarly, the cofactor for path 2 is

$$\Delta_2 = 1 - (L_1 + L_2).$$

Therefore the transfer function of the system is

$$\frac{C(s)}{R(s)} = T(s) = \frac{P_1\Delta_1 + P_2\Delta_2}{\Delta} \quad\quad\quad (2.100)$$

$$= \frac{G_1G_2G_3G_4(1 - L_3 - L_4) + G_5G_6G_7G_8(1 - L_1 - L_2)}{1 - L_1 - L_2 - L_3 - L_4 + L_1L_3 + L_1L_4 + L_2L_3 + L_2L_4}. \blacksquare$$

EXAMPLE 2.8   Armature-controlled motor

The block diagram of the armature-controlled dc motor is shown in Fig. 2.17. This diagram was obtained from Eqs. (2.64)–(2.68). The signal-flow diagram can be obtained either from Eqs. (2.64)–(2.68) or from the block diagram and is shown in Fig. 2.29. Using Mason's rule, let us obtain the transfer function for $\theta(s)/V_a(s)$ with $T_d(s) = 0$. The forward path is $P_1(s)$, which touches the one loop, $L_1(s)$, where

$$P_1(s) = \frac{1}{s}G_1(s)G_2(s) \quad\quad \text{and} \quad\quad L_1(s) = -K_bG_1(s)G_2(s).$$

Therefore the transfer function is

$$T(s) = \frac{P_1(s)}{1 - L_1(s)} = \frac{(1/s)G_1(s)G_2(s)}{1 + K_bG_1(s)G_2(s)} = \frac{K_m}{s[(R_a + L_as)(Js + f) + K_aK_m]},$$

which is exactly the same as derived earlier (Eq. 2.69). $\blacksquare$

The signal-flow graph gain formula provides a reasonably straightforward approach for the evaluation of complicated systems. To compare the method with block diagram reduction, which is really not much more difficult, let us reconsider the complex system of Example 2.5.

FIGURE 2.29
The signal-flow graph of the armature-controlled dc motor.

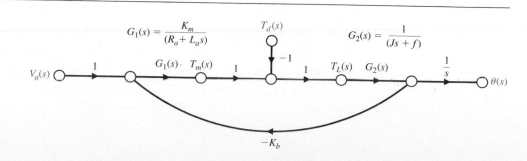

EXAMPLE 2.9   **Transfer function of multiple-loop system**

A multiple-loop feedback system is shown in Fig. 2.23 in block diagram form. There is no reason to redraw the diagram in signal-flow graph form, and so we shall proceed as usual by using the signal-flow gain formula, Eq. (2.98). There is one forward path $P_1 = G_1G_2G_3G_4$. The feedback loops are

$$L_1 = -G_2G_3H_2, \qquad L_2 = G_3G_4H_1, \qquad L_3 = -G_1G_2G_3G_4H_3. \qquad (2.101)$$

All the loops have common nodes and therefore are all touching. Furthermore, the path P_1 touches all the loops, so $\Delta_1 = 1$. Thus the closed-loop transfer function is

$$T(s) = \frac{C(s)}{R(s)} = \frac{P_1\Delta_1}{1 - L_1 - L_2 - L_3} \qquad (2.102)$$

$$= \frac{G_1G_2G_3G_4}{1 + G_2G_3H_2 - G_3G_4H_1 + G_1G_2G_3G_4H_3}. \quad \blacksquare$$

EXAMPLE 2.10   **Transfer function of complex system**

Finally, we shall consider a reasonably complex system that would be difficult to reduce by block diagram techniques. A system with several feedback loops and feedforward paths is shown in Fig. 2.30. The forward paths are

$$P_1 = G_1G_2G_3G_4G_5G_6, \qquad P_2 = G_1G_2G_7G_6, \qquad P_3 = G_1G_2G_3G_4G_8.$$

The feedback loops are

$$L_1 = -G_2G_3G_4G_5H_2, \qquad L_2 = -G_5G_6H_1, \qquad L_3 = -G_8H_1, \qquad L_4 = -G_7H_2G_2,$$
$$L_5 = -G_4H_4, \qquad L_6 = -G_1G_2G_3G_4G_5G_6H_3, \qquad L_7 = -G_1G_2G_7G_6H_3,$$
$$L_8 = -G_1G_2G_3G_4G_8H_3.$$

Loop L_5 does not touch loop L_4 or loop L_7; loop L_3 does not touch loop L_4; and all other loops touch. Therefore the determinant is

$$\Delta = 1 - (L_1 + L_2 + L_3 + L_4 + L_5 + L_6 + L_7 + L_8) \qquad (2.103)$$

$$+ (L_5L_7 + L_5L_4 + L_3L_4).$$

FIGURE 2.30
Multiple-loop
system.

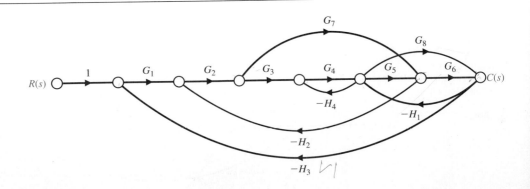

The cofactors are

$$\Delta_1 = \Delta_3 = 1 \quad \text{and} \quad \Delta_2 = 1 - L_5 = 1 + G_4H_4.$$

Finally, the transfer function is

$$T(s) = \frac{C(s)}{R(s)} = \frac{P_1 + P_2\Delta_2 + P_3}{\Delta}. \quad\blacksquare \tag{2.104}$$

Signal-flow graphs and the signal-flow gain formula may be used profitably for the analysis of feedback control systems, analog computer diagrams, electronic amplifier circuits, statistical systems, mechanical systems, among many other examples.

2.8 COMPUTER ANALYSIS OF CONTROL SYSTEMS

A computer model of a system in a mathematical form suitable for demonstrating the system's behavior may be utilized to investigate various designs of a planned system without actually building the system itself. A *computer simulation* uses a model and the actual conditions of the system being modeled and actual input commands to which the system will be subjected.

A system may be simulated using analog or digital computers. An electronic analog computer is used to establish a model of a system, using the analogy between the voltage of the electronic amplifier and the variable of the system being modeled [12–14]. An electronic analog computer usually has available the mathematical functions of integration, multiplication by a constant, multiplication of two variables, and the summation of several variables, among others. These functions are often sufficient to develop a simulation model of a system. The analog simulation model of a second-order system is shown in Fig. 2.31 for the system with negative unity feedback and a plant transfer function

$$G(s) = \frac{C(s)}{E(s)} = \frac{K}{s(s + p)}. \tag{2.105}$$

The differential equation necessary to yield the simulation of the plant is obtained by cross-multiplying in Eq. (2.105) to yield

$$s(s + p)C(s) = KE(s). \tag{2.106}$$

FIGURE 2.31
An analog computer simulation model for a second-order system with negative unity feedback.

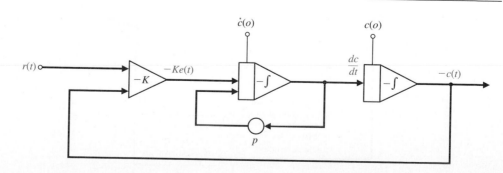

Since $sC(s)$ is the derivative of $c(t)$ in the s-domain we have:

$$\frac{d^2c(t)}{dt^2} = -p\,\frac{dc(t)}{dt} + Ke(t). \tag{2.107}$$

This equation is represented on the analog diagram by the integration in the center of the diagram with the output dc/dt. This analog simulation arrangement can be realized physically on an electronic analog computer to yield an output recording of the simulated response of the system. The parameters K and p can be varied to ascertain the effect of the parameter change.

Simulation models can also be utilized with digital computers. A computer simulation can be developed in common computer language such as Pascal or BASIC or in a language specifically developed for simulation [13, 14]. Two widely used languages for the simulation of systems operating in continuous time are ACSL and *Matlab*. These programs are written to be interactive and are user friendly. They offer user help, error detection, and medium resolution graphics for presentation of system responses.

Assuming that a model and the simulation are reliably accurate, computer simulation has the following advantages [14]:

1. System performance can be observed under all conceivable conditions.
2. Results of field-system performance can be extrapolated with a simulation model for prediction purposes.
3. Decisions concerning future systems presently in a conceptual stage can be examined.
4. Trials of systems under test can be accomplished in a much reduced period of time.
5. Simulation results can be obtained at lower cost than real experimentation.
6. Study of hypothetical situations can be achieved even when the hypothetical situation would be unrealizable in actual life at the present time.
7. Computer modeling and simulation is often the only feasible or safe technique to analyze and evaluate a system.

The analysis and design of a control system is greatly enhanced using simulation as part of the process depicted in Fig. 2.32. In this book, we use *Matlab* as the simulation program, but other, similar programs may be equally useful.

FIGURE 2.32
Analysis and design using a system model.

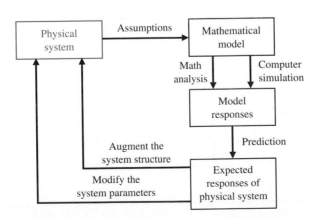

2.9 DESIGN EXAMPLES

EXAMPLE 2.11   Electric traction motor control

A majority of modern trains and local transit vehicles utilize electric traction motors. The electric motor drive for a railway vehicle is shown in block diagram form in Fig. 2.33(a) incorporating the necessary control of the velocity of the vehicle. The goal of the design is to obtain a system model and the closed-loop transfer function of the system, $\omega(s)/\omega_d(s)$, select appropriate resistors R_1, R_2, R_3, and R_4, and then predict the system response.

FIGURE 2.33
Speed control of
an electric traction
motor.

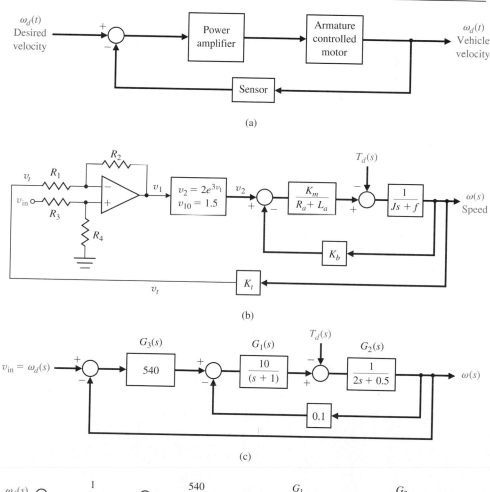

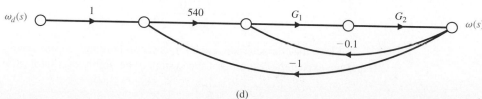

The first step is to describe the transfer function of each block. We propose the use of a tachometer to generate a voltage proportional to velocity and to connect that voltage, v_t, to one input of a difference amplifier, as shown in Fig. 2.33(b). The power amplifier is nonlinear and can be approximately represented by $v_2 = 2e^{3v_1} = 2 \exp(3v_1) = g(v_1)$, an exponential function with a normal operating point, $v_{10} = 1.5V$. Using the technique in Section 2.3, we then obtain a linear model

$$v_2 = \left[\frac{dg(v_1)}{dv_1} \bigg|_{v_{10}} \right] \Delta v_1 = 2[3 \exp(3v_{10})] \Delta v_1 = 2[270] \Delta v_1 = 540 \Delta v_1. \quad (2.108)$$

Then, discarding the delta notation and writing the Laplace transform, we have

$$V_2(s) = 540 V_1(s).$$

The transfer function of the differential amplifier is

$$v_1 = \frac{1 + R_2/R_1}{1 + R_3/R_4} v_{in} - \frac{R_2}{R_1} v_t. \quad (2.109)$$

We wish to obtain an input control that sets $\omega_d(t) = v_{in}$ where the units of ω_d are rad/s and the units of v_{in} are volts. Then, when $v_{in} = 10$ V, the steady-state speed is $\omega = 10$ rad/s. We note that $v_t = K_t \omega_d$ in steady state and we expect, in balance, the steady-state output, v_1, to be

$$v_1 = \frac{1 + R_2/R_1}{1 + R_3/R_4} v_{in} - \left(\frac{R_2}{R_1} \right) K_t(v_{in}). \quad (2.110)$$

When the system is in balance, $v_1 = 0$, and when $K_t = 0.1$, we have

$$\frac{1 + R_2/R_1}{1 + R_3/R_4} = \left(\frac{R_2}{R_1} \right) K_t = 1.$$

This relation can be achieved when

$$R_2/R_1 = 10 \quad \text{and} \quad R_3/R_4 = 10.$$

The parameters of the motor and load are given in Table 2.9. The overall system is shown in Fig. 2.33(b). Using Mason's signal-flow rule with the signal-flow diagram of Fig. 2.33(d), we have

$$\frac{\omega(s)}{\omega_d(s)} = \frac{540 G_1(s) G_2(s)}{1 + 0.1 G_1 G_2 + 540 G_1 G_2} = \frac{540 G_1 G_2}{1 + 540.1 G_1 G_2}$$

$$= \frac{5400}{(s + 1)(2s + 0.5) + 5401} = \frac{5400}{2s^2 + 2.5s + 5401.5} \quad (2.111)$$

$$= \frac{2700}{s^2 + 1.25s + 2700.75}.$$

Since the characteristic equation is second order, we note that $\omega_n = 52$ and $\zeta = 0.012$, and we expect the response of the system to be highly oscillatory (underdamped). ∎

TABLE 2.9   Parameters of a Large dc Motor

$K_m = 10$	$J = 2$
$R_a = 1$	$f = 0.5$
$L_a = 1$	$K_b = 0.1$

EXAMPLE 2.12   Mechanical accelerometer

A mechanical accelerometer is used to measure the acceleration of a levitated test sled, as shown in Fig. 2.34. The test sled is magnetically levitated above a guide rail a small distance δ. The accelerometer provides a measurement of the acceleration $a(t)$ of the sled, since the position y of the mass M, with respect to the accelerometer case, is proportional to the acceleration of the case (and the sled). The goal is to design an accelerometer with an appropriate dynamic responsiveness. We wish to design an accelerometer with an acceptable time for the desired measurement characteristic, $y(t) = qa(t)$, to be attained (q is a constant).

The sum of the forces acting on the mass is

$$-f\frac{dy}{dt} - Ky - M\frac{d^2}{dt^2}(y - x)$$

or

$$M\frac{d^2y}{dt^2} + f\frac{dy}{dt} + Ky = M\frac{d^2x}{dt^2}. \tag{2.112}$$

FIGURE 2.34
An accelerometer mounted on a jet engine test sled.

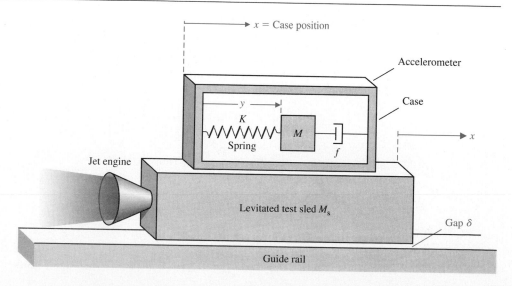

Since

$$M_s \frac{d^2x}{dt^2} = F(t),$$

the engine force, we have

$$M\ddot{y} + f\dot{y} + Ky = \frac{M}{M_s} F(t)$$

or

$$\ddot{y} + \frac{f}{M}\dot{y} + \frac{K}{M}y = \frac{F(t)}{M_s}. \tag{2.113}$$

We select the coefficients where $f/M = 3$, $K/M = 2$, $F(t)/M_s = Q(t)$, and we consider the initial conditions $y(0) = -1$ and $\dot{y}(0) = 2$. We then obtain the Laplace transform equation, when the force and thus $Q(t)$ is a step function, as follows:

$$(s^2Y(s) - sy(0) - \dot{y}(0)) + 3(sY(s) - y(0)) + 2Y(s) = Q(s). \tag{2.114}$$

Since $Q(s) = P/s$, where P is the magnitude of the step function, we obtain

$$(s^2Y(s) + s - 2) + 3(sY(s) + 1) + 2Y(s) = \frac{P}{s}$$

or

$$(s^2 + 3s + 2)Y(s) = \frac{-(s^2 + s - P)}{s}. \tag{2.115}$$

Thus the output transform is

$$Y(s) = \frac{-(s^2 + s - P)}{s(s^2 + 3s + 2)} = \frac{-(s^2 + s - P)}{s(s + 1)(s + 2)}. \tag{2.116}$$

Expanding in partial fraction form,

$$Y(s) = \frac{k_1}{s} + \frac{k_2}{s + 1} + \frac{k_3}{s + 2}. \tag{2.117}$$

We then have

$$k_1 = \frac{-(s^2 + s - P)}{(s + 1)(s + 2)}\bigg|_{s=0} = \frac{P}{2}. \tag{2.118}$$

Similarly, $k_2 = -P$ and $k_3 = \dfrac{P - 2}{2}$. Thus

$$Y(s) = \frac{P}{2s} - \frac{P}{s + 1} + \frac{(P - 2)}{2(s + 2)}. \tag{2.119}$$

Therefore the output measurement is

$$y(t) = \frac{1}{2}[P - 2Pe^{-t} + (P - 2)e^{-2t}], \quad t \geq 0.$$

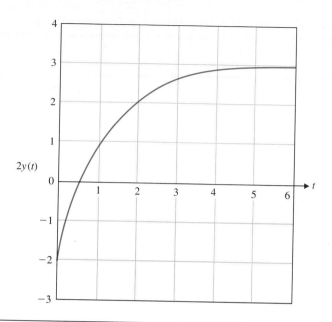

FIGURE 2.35
Accelerometer
response.

A plot of $y(t)$ is shown in Fig. 2.35 for $P = 3$. We can see that $y(t)$ is proportional to the magnitude of the force, after 4 seconds. Thus, in steady-state, after 4 seconds, the response $y(t)$ is proportional to the acceleration, as desired. If this period is excessively long, we must increase the spring constant, K, and the friction, f, while reducing the mass, M. If we are able to select the components so that $f/M = 12$ and $K/M = 32$, the accelerometer will attain the proportional response in one second. (It is left to the reader to show this.) ∎

EXAMPLE 2.13   Design of a laboratory robot

In this example, we endeavor to show the physical design of a laboratory device and demonstrate its complex design. We will also exhibit the many components commonly used in a control system.

A robot for laboratory use is shown in Fig. 1.15. A laboratory robot's work volume must allow the robot to reach the entire bench area and access existing analytical instruments. There must also be sufficient area for a stockroom of supplies for unattended operation.

The laboratory robot can be involved in three types of tasks during an analytical experiment. The first is sample introduction, wherein the robot is trained to accept a number of different sample trays, racks, and containers and to introduce them into the system. The second set of tasks involves the robot's transporting the samples between individual dedicated automated stations for chemical preparation and instrumental analysis. Samples must be scheduled and moved between these stations as necessary to complete the analysis. The third set of tasks for the robot is where flexible automation provides new capability to the analytical laboratory. The robot must be programmed to emulate the human operator or work with various devices. All of these types of operations are required for an effective laboratory robot.

TABLE 2.10   ORCA Robot Arm Hardware Specifications

Arm	Articulated, Rail-Mounted	Teach Pendant	Joy Stick with Emergency Stop
Degrees of freedom	Six	Cycle time	4 s (move 1 in. up, 12 in. across, 1 in. down, and back)
Reach	± 54 cm	Maximum speed	75 cm/s
Height	78 cm	Dwell time	50 ms typical (for moves within a motion)
Rail	1 and 2 m	Payload	0.5 kg continuous, 2.5 kg transient (with restrictions)
Weight	8.0 kg	Vertical deflection	<1.5 mm at continuous payload
Precision	± 0.25 mm	Cross-sectional work envelope	1 m^2
Finger travel (gripper)	40 mm		
Gripper rotation	± 77 revolutions		

Hewlett-Packard has designed the ORCA laboratory robot, which is an anthropomorphic arm, mounted on a rail, designed as the optimum configuration for the analytical laboratory [15]. The rail can be located at the front or back of a workbench, or placed in the middle of a table when access to both sides of the rail is required. Simple software commands permit moving the arm from one side of the rail to the other while maintaining the wrist position (to transfer open containers) or locking the wrist angle (to transfer objects in virtually any orientation). The rectilinear geometry, in contrast to the cylindrical geometry used by many robots, permits more accessories to be placed within the robot workspace and provides an excellent match to the laboratory bench. Movement of all joints is coordinated through software, which simplifies the use of the robot by representing the robot positions and movements in the more familiar Cartesian coordinate space.

The physical and performance specifications of the Hewlett-Packard ORCA system are shown in Table 2.10.

The design for the ORCA laboratory robot progressed to the selection of the component parts required to obtain the total system. The exploded view of the robot is shown in Figure 2.36. This device uses six dc motors, gears, belt drives, and a rail and carriage. The specifications are challenging and require the designer to model the system components and their interconnections accurately. ■

EXAMPLE 2.14   Design of a low-pass filter

Our goal is to design a first-order low-pass filter that passes signals at a frequency below 106.1 Hz and attenuates signals with a frequency above 106 Hz. In addition, the dc gain should be ½.

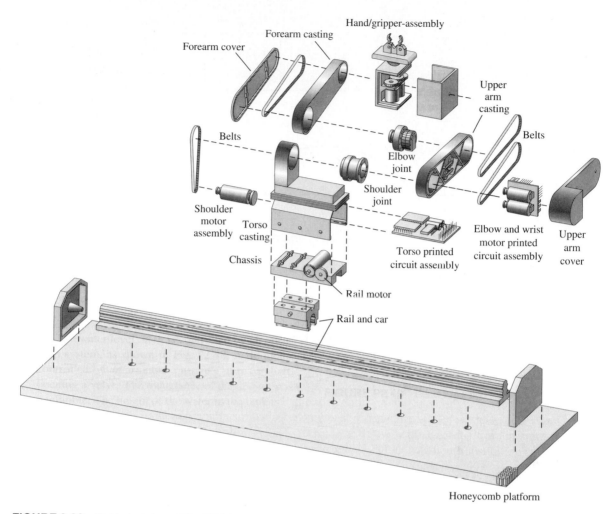

FIGURE 2.36   Exploded view of the ORCA robot showing the components [15]. (*Source:* © Copyright 1993 Hewlett-Packard Company. Reproduced with permission.)

A ladder network with one energy storage element, as shown in Fig. 2.37(a), will act as a first-order low-pass network. Note that the dc gain will be equal to ½ (open-circuit the capacitor). The current and voltage equations are

$$I_1 = (V_1 - V_2)G$$
$$I_2 = (V_2 - V_3)G$$
$$V_2 = (I_1 - I_2)R$$
$$V_3 = I_2 Z,$$

where $G = 1/R$, $Z(s) = 1/Cs$, and $I_1(s) = I_1$ (we omit the (s)). The signal-flow graph constructed for the four equations is shown in Fig. 2.37(b). The three loops are $L_1 = -GR = -1$, $L_2 = -GR = -1$, and $L_3 = -GZ$. All loops touch the forward path. Loops L_1

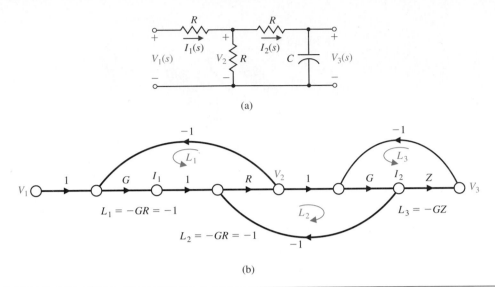

FIGURE 2.37
(a) Ladder network
and (b) its signal-
flow graph.

and L_3 are nontouching. Therefore the transfer function is

$$T(s) = \frac{V_3}{V_1} = \frac{P_1}{1 - (L_1 + L_2 + L_3) + L_1 L_3} = \frac{GZ}{3 + 2GZ}$$

$$= \frac{1}{3RCs + 2} = \frac{(1/3RC)}{(s + 2/3RC)}.$$

Note that the dc gain is $\frac{1}{2}$, as expected. The pole is desired at $p = 2\pi(106.1) = 666.7 = 2000/3$. Therefore we require $RC = 0.001$. Select $R = 1 \text{ k}\Omega$ and $C = 1 \text{ }\mu\text{F}$. Hence we achieve the filter

$$T(s) = \frac{333.35}{(s + 666.7)}. \blacksquare$$

2.10 THE SIMULATION OF SYSTEMS USING *MATLAB*†

Application of the many classical and modern control system design and analysis tools is based on mathematical models. *Matlab* can be used with systems given in the form of transfer function descriptions.

We begin this section by showing how to use *Matlab* to assist in the analysis of a typical spring-mass-damper mathematical model of a mechanical system. Using a *Matlab* script, we will develop an interactive analysis capability to analyze the effects of natural frequency and damping on the unforced response of the mass displacement. This analysis will utilize the fact that we have an analytic solution that describes the unforced time response of the mass displacement.

---

†See Appendix F for an introduction to *Matlab*.

Later, we will discuss transfer functions and block diagrams. In particular, we are interested in how *Matlab* can assist us in manipulating polynomials, computing poles and zeros of transfer functions, computing closed-loop transfer functions, computing block diagram reductions, and computing the response of a system to a unit step input. The section concludes with the electric traction motor control design of Example 2.11.

The functions covered in this section are roots, roots1, series, parallel, feedback, cloop, poly, conv, polyval, printsys, minreal, pzmap, and step.

Spring-Mass-Damper System.   A spring-mass-damper mechanical system is shown in Fig. 2.1. The motion of the mass, denoted by $y(t)$, is described by the differential equation

$$M\ddot{y}(t) + f\dot{x}(t) + Ky(t) = r(t).$$

The unforced dynamic response, $y(t)$, of the spring-mass-damper mechanical system is

$$y(t) = \frac{y(0)}{\sqrt{1 - \zeta^2}} e^{-\zeta\omega_n t} \sin(\omega_n\sqrt{1 - \zeta^2}\, t + \theta),$$

where $\theta = \cos^{-1}\zeta$. The initial displacement is $y(0)$. The transient system response is *underdamped* when $\zeta < 1$, *overdamped* when $\zeta > 1$, and *critically damped* when $\zeta = 1$. We can use *Matlab* to visualize the unforced time response of the mass displacement following an initial displacement of $y(0)$. Consider the overdamped and underdamped cases:

- Case 1: $y(0) = 0.15$ m, $\omega_n = \sqrt{2}\ \frac{\text{rad}}{\text{sec}}$, $\zeta_1 = \frac{3}{2\sqrt{2}}$ $(\frac{K}{M} = 2, \frac{f}{M} = 3)$
- Case 2: $y(0) = 0.15$ m, $\omega_n = \sqrt{2}\ \frac{\text{rad}}{\text{sec}}$, $\zeta_2 = \frac{1}{2\sqrt{2}}$ $(\frac{K}{M} = 2, \frac{f}{M} = 1)$

The *Matlab* commands to generate the plot of the unforced response are shown in Fig. 2.38.

FIGURE 2.38
Script to analyze the spring-mass-damper.

```
>>y0=0.15; wn=sqrt(2);                              ⟵  ωn
>>zeta1=3/(2*sqrt(2)); zeta2=1/(2*sqrt(2));         ⟵
>>t=[0:0.1:10];                                        ζ1 and ζ2
>>unforcedcommands
```

```
unforcedcommands.m
%Compute Unforced Response to an Initial Condition
%
t1=acos(zeta1)*ones(1,length(t));                  ⟵  cos⁻¹ ζ
t2=acos(zeta2)*ones(1,length(t));
c1=(y0/sqrt(1-zeta1^2));c2=(y0/sqrt(1-zeta2^2));    ⟵  y(0)/√(1-ζ²)
y1=c1*exp(-zeta1*wn*t).*sin(wn*sqrt(1-zeta1^2)*t+t1);
y2=c2*exp(-zeta2*wn*t).*sin(wn*sqrt(1-zeta2^2)*t+t2);
%
bu=c2*exp(-zeta2*wn*t);bl=-bu;                      ⟵  e⁻ᶻᵉᵗᵃωn t envelope
%
plot(t,y1,'-',t,y2,'-',t,bu,'--',t,bl,'--'), grid
xlabel('Time[sec]'), ylabel('y(t) Displacement [m]')
text(0.2,0.85,['overdamped zeta1=',num2str(zeta1),...
],'sc')
text(0.2,0.80,['underdamped zeta2=',num2str(zeta2),...
],'sc')
```

In the *Matlab* setup, the variables $y(0)$, ω_n, t, ζ_1, and ζ_2 are input to the workspace at the command level. Then the script unforcedcommands.m is executed to generate the desired plots. This creates an interactive analysis capability to analyze the effects of natural frequency and damping on the unforced response of the mass displacement. One can investigate the effects of the natural frequency and the damping on the time response by simply entering new values of ω_n, ζ_1, or ζ_2 at the command prompt and running the script unforcedcommands.m again. The time-response plot is in Fig. 2.39. Notice that the script automatically labels the plot with the values of the damping coefficients. This avoids confusion when making many interactive simulations. The natural frequency value could also be automatically labeled on the plot. Utilizing scripts is an important aspect of developing an effective interactive design and analysis capability in *Matlab*.

For the spring-mass-damper problem, the unforced solution to the differential equation was readily available. In general, when simulating closed-loop feedback control systems subject to a variety of inputs and initial conditions, it is difficult to attempt to obtain the solution analytically. In these cases, we can use *Matlab* to compute the solutions numerically and to display the solution graphically.

Matlab can be used to analyze systems described by transfer functions. Since the transfer function is a ratio of polynomials, we begin by investigating how *Matlab* handles polynomials, remembering that working with transfer functions means that both a numerator polynomial and a denominator polynomial must be specified.

In *Matlab,* polynomials are represented by row vectors containing the polynomial coefficients in descending order. For example, the polynomial

$$p(s) = s^3 + 3s^2 + 4$$

is entered as shown in Fig. 2.40. Notice that even though the coefficient of the s term is zero, it is included in the input definition of $p(s)$.

FIGURE 2.39
Spring-mass-damper unforced response.

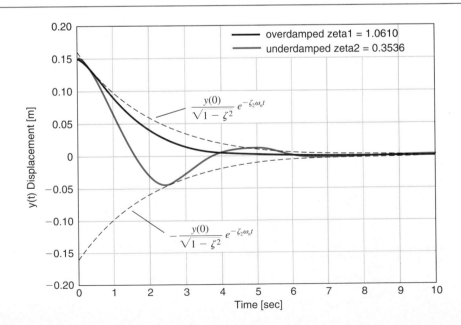

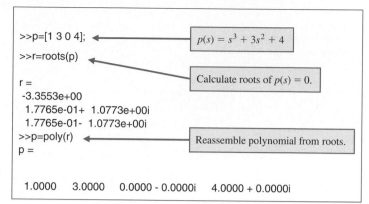

FIGURE 2.40
Entering the
polynomial
$p(s) = s^3 + 3s^2 + 4$
and calculating its
roots.

If **p** is a row vector containing the coefficients of $p(s)$ in descending order, then roots(**p**) is a column vector containing the roots of the polynomial. Conversely, if **r** is a column vector containing the roots of the polynomial, then poly(**r**) is a row vector with the polynomial coefficients in descending order. We can compute the roots of the polynomial $p(s)$, given in the preceding equation, with the roots function as shown in Fig. 2.40. The roots1 function also computes the roots of a polynomial but gives a more accurate result when the polynomial has repeated roots. In Fig. 2.40, we also show how to reassemble the polynomial with the poly function.

Multiplication of polynomials is accomplished with the conv function. Suppose we want to expand the polynomial $n(s)$, where

$$n(s) = (3s^2 + 2s + 1)(s + 4).$$

The associated *Matlab* commands using the conv function are shown in Fig. 2.41. Thus, the expanded polynomial, given by n, is

$$n(s) = 3s^3 + 14s^2 + 9s + 4.$$

The function polyval is used to evaluate the value of a polynomial at the given value of the variable. The polynomial $n(s)$ has the value $n(-5) = -66$, as shown in Fig. 2.41.

In the next example we will obtain a plot of the pole–zero locations in the complex plane. This will be accomplished using the pzmap function, shown in Fig. 2.42. On the pole–zero map, zeros are denoted by an "o" and poles are denoted by an "×". If

FIGURE 2.41
Using **conv**
and **polyval** to
multiply and
evaluate the
polynomials
$(3s^2 + 2s + 1)(s + 4)$.

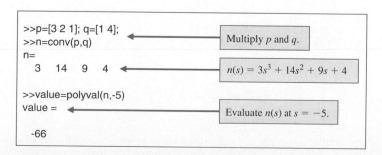

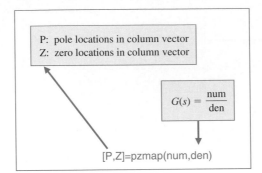

FIGURE 2.42
The **pzmap**
function.

the pzmap function is invoked without left-hand arguments, the plot is automatically generated.

EXAMPLE 2.15   Transfer functions

Consider the transfer functions

$$G(s) = \frac{6s^2 + 1}{s^3 + 3s^2 + 3s + 1} \quad \text{and} \quad H(s) = \frac{(s + 1)(s + 2)}{(s + 2i)(s - 2i)(s + 3)}.$$

Utilizing a *Matlab* script, we can compute the poles and zeros of $G(s)$, the characteristic equation of $H(s)$, and divide $G(s)$ by $H(s)$. We can also obtain a plot of the pole–zero map of $G(s)/H(s)$ in the complex plane.

The pole–zero map of the transfer function $G(s)/H(s)$ is shown in Fig. 2.43, and the associated *Matlab* commands are shown in Fig. 2.44. The pole–zero map shows clearly the five zero locations, but it appears that there are only two poles. This cannot be the case, since we know that the number of poles must be greater than or equal to the number of zeros. Using the roots1 function, we can ascertain that there are in fact four poles at $s = -1$. Hence, multiple poles or multiple zeros at the same location cannot be discerned on the pole–zero map. ∎

FIGURE 2.43
Pole–zero map for
$G(s)/H(s)$.

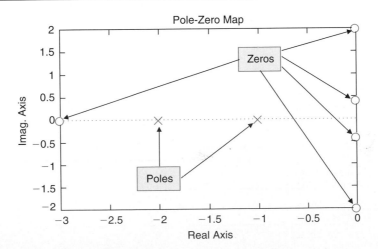

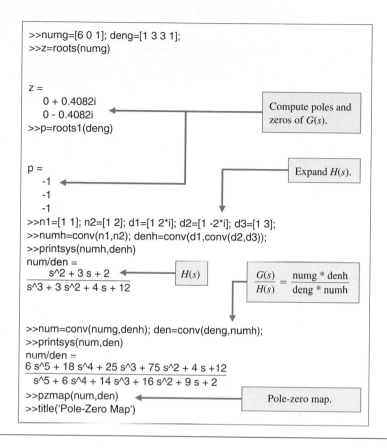

FIGURE 2.44
Transfer function example for $G(s)$ and $H(s)$.

Block Diagram Models.   Suppose we have developed mathematical models in the form of transfer functions for the plant, represented by $G(s)$, and the controller, represented by $G_c(s)$, and possibly many other system components such as sensors and actuators. Our objective is to interconnect these components to form a control system. We will utilize *Matlab* functions to carry out the block diagram transformations.

A simple open-loop control system can be obtained by interconnecting a plant and a controller in series as illustrated in Fig. 2.45. We can use *Matlab* to compute the transfer function from $R(s)$ to $Y(s)$, as follows.

EXAMPLE 2.16   Series connection

Let the process, represented by the transfer function $G(s)$, be

$$G(s) = \frac{1}{500s^2},$$

FIGURE 2.45
Open-loop control system (without feedback).

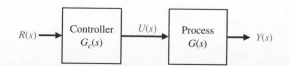

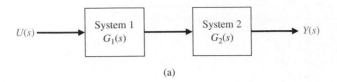

(a)

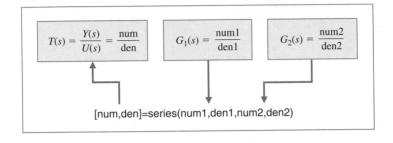

FIGURE 2.46
The **series**
function.

(b)

and let the controller, represented by the transfer function $G_c(s)$, be

$$G_c(s) = \frac{s + 1}{s + 2}.$$

We can use the **series** function to cascade two transfer functions $G_1(s)$ and $G_2(s)$, as shown in Fig. 2.46.

The transfer function $G_c G(s)$ is computed using the **series** function as shown in Fig. 2.47. The resulting transfer function, $G_c G(s)$, is

$$G_c G(s) = \frac{num}{den} = \frac{s + 1}{500s^3 + 1000s^2}. \quad \blacksquare$$

Block diagrams quite often have transfer functions in *parallel*. In such cases, the function **parallel** can be quite useful. The **parallel** function is described in Fig. 2.48.

FIGURE 2.47
Application of the
series function.

$R(s) \rightarrow \boxed{G_c(s) = \dfrac{s + 1}{s + 2}} \xrightarrow{U(s)} \boxed{G(s) = \dfrac{1}{500\,s^2}} \rightarrow Y(s)$

(a)

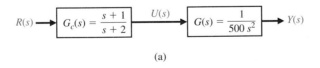

```
>>numg=[1]; deng=[500 0 0];
>>numh=[1 1]; denh=[1 2];
>>[num,den]=series(numg,deng,numh,denh);
>>printsys(num,den)
num/den =
           s + 1
    ─────────────────────
    500 s^3 + 1000 s^2
```

$G_c G(s)$

(b)

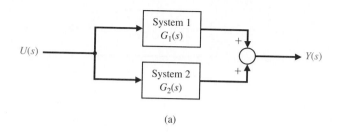

(a)

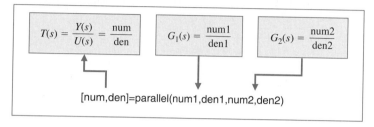

[num,den]=parallel(num1,den1,num2,den2)

FIGURE 2.48
The **parallel**
function.

(b)

We can introduce a *feedback signal* into the control system by closing the loop with *unity* feedback, as shown in Fig. 2.49. The signal $E_a(s)$ is an *error signal;* the signal $R(s)$ is a *reference input.* In this control system, the controller is in the forward path and the closed-loop transfer function is

$$T(s) = \frac{G_c G(s)}{1 \pm G_c G(s)}.$$

There are two functions we can utilize to aid in the block diagram reduction process to compute closed-loop transfer functions for single- and multiple-loop control systems. These functions are cloop and feedback.

The cloop function calculates the closed-loop transfer function, as shown in Fig. 2.50, with the associated system configuration and assumes unity feedback with negative feedback as the default.

The feedback function is shown in Fig. 2.51 with the associated system configuration, which includes $H(s)$ in the feedback path. For both the cloop and feedback functions, if the input "sign" is omitted, then negative feedback is assumed.

FIGURE 2.49
A basic control
system with unity
feedback.

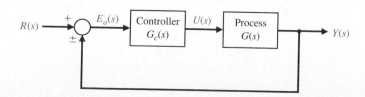

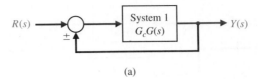

(a)

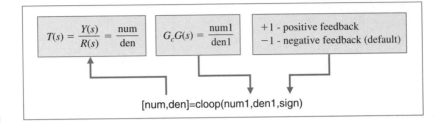

FIGURE 2.50
(a) Block diagram
(b) the **cloop**
function.

(b)

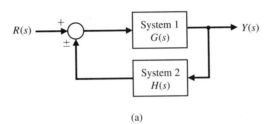

(a)

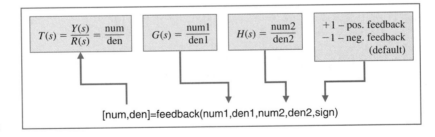

FIGURE 2.51
(a) Block diagram
(b) the **feedback**
function.

(b)

EXAMPLE 2.17 **The cloop function**

Let the process, $G(s)$, and the controller, $G_c(s)$, be as in Fig. 2.47(a). To apply the cloop function we first use the series function to compute $G_cG(s)$, followed by the cloop function to close the loop. The command sequence is shown in Fig. 2.52(b). The closed-loop transfer function, as shown in Fig. 2.52(b), is

$$T(s) = \frac{G_cG(s)}{1 + G_cG(s)} = \frac{num}{den} = \frac{s + 1}{500s^3 + 1000s^2 + s + 1}. \quad \blacksquare$$

Another basic feedback control configuration is shown in Fig. 2.53. In this case, the

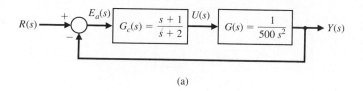

(a)

```
>>numg=[1]; deng=[500 0 0];
>>numc=[1 1]; denc=[1 2];
>>[num1,den1]=series(numg,deng,numc,denc);
>>[num,den]=cloop(num1,den1,-1);
>>printsys(num,den)
num/den =
            s + 1
   ─────────────────────────
   500 s^3 + 1000 s^2 + s + 1
```

$$\frac{Y}{R} = \frac{G_c G}{1 + G_c G}$$

FIGURE 2.52
Application of the
cloop function.

(b)

controller is located in the feedback path. The *error signal, $E_a(s)$*, is also utilized in this control system configuration. The closed-loop transfer function is

$$T(s) = \frac{G(s)}{1 \pm GH(s)}.$$

EXAMPLE 2.18   **The feedback function**

Let the process, $G(s)$, and the controller, $H(s)$, be as in Fig. 2.54(a). To compute the closed-loop transfer function with the controller in the feedback loop, we use the **feedback** function. The command sequence is shown in Fig. 2.54(b). The closed-loop transfer function is

$$T(s) = \frac{num}{den} = \frac{s + 2}{500s^3 + 1000s^2 + s + 1}. \blacksquare$$

The *Matlab* functions **series**, **cloop**, and **feedback** can be used as aids in block diagram manipulations for multiple-loop block diagrams.

EXAMPLE 2.19   **Multiloop reduction**

A multiloop feedback system is shown in Fig. 2.23 on page 62. Our objective is to compute the closed-loop transfer function

$$T(s) = \frac{Y(s)}{R(s)},$$

FIGURE 2.53
A basic control
system with the
controller in the
feedback loop.

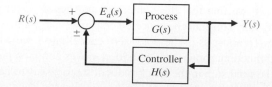

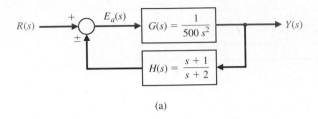

FIGURE 2.54
Application of the
feedback function:
(a) block diagram,
(b) *Matlab* script.

```
>>numg=[1]; deng=[500 0 0];
>>numh=[1 1]; denh=[1 2];
>>[num,den]=feedback(numg,deng,numh,denh,-1);
>>printsys(num,den)
num/den=
```

$$\frac{s + 2}{500 \, s^3 + 1000 \, s^2 + s + 1}$$

$$\frac{Y}{R} = \frac{G}{1 + GH}$$

(b)

when

$$G_1(s) = \frac{1}{s + 10}, \qquad G_2(s) = \frac{1}{s + 1},$$

$$G_3(s) = \frac{s^2 + 1}{s^2 + 4s + 4}, \qquad G_4(s) = \frac{s + 1}{s + 6},$$

and

$$H_1(s) = \frac{s + 1}{s + 2}, \qquad H_2(s) = 2, \qquad H_3(s) = 1.$$

For this example, a five-step procedure is followed:

■ Step 1. Input the system transfer functions into *Matlab*.

■ Step 2. Move H_2 behind G_4.

■ Step 3. Eliminate the $G_3 G_4 H_1$ loop.

■ Step 4. Eliminate the loop containing H_2.

■ Step 5. Eliminate the remaining loop and calculate $T(s)$.

The five steps are utilized in Fig. 2.55, and the corresponding block diagram reduction is shown in Fig. 2.24 on page 63. The result of executing the *Matlab* commands is

$$\frac{num}{den} = \frac{s^5 + 4s^4 + 6s^3 + 6s^2 + 5s + 2}{12s^6 + 205s^5 + 1066s^4 + 2517s^3 + 3128s^2 + 2196s + 712}.$$

We must be careful in calling this the closed-loop transfer function. The transfer function is defined to the input–output relationship *after* pole–zero cancellations. If we compute the poles and zeros of $T(s)$, we find that the numerator and denominator polynomials have $(s + 1)$ as a common factor. This must be canceled before we can claim we have the closed-loop transfer function. To assist us in the pole–zero cancellation, we will use the minreal function. The minreal function, shown in Fig. 2.56, removes common pole–zero factors of a transfer function. The final step in the block reduction process is to cancel out the com-

```
>>ng1=[1]; dg1=[1 10];
>>ng2=[1]; dg2=[1 1];
>>ng3=[1 0 1]; dg3=[1 4 4];                      Step 1
>>ng4=[1 1]; dg4=[1 6];
>>nh1=[1 1]; dh1=[1 2];
>>nh2=[2]; dh2=[1];
>>nh3=[1]; dh3=[1];
>>n1=conv(nh2,dg4); d1=conv(dh2,ng4);            Step 2
>>[n2a,d2a]=series(ng3,dg3,ng4,dg4);
>>[n2,d2]=feedback(n2a,d2a,nh1,dh1,+1);          Step 3
>>[n3a,d3a]=series(ng2,dg2,n2,d2);
>>[n3,d3]=feedback(n3a,d3a,n1,d1);               Step 4
>>[n4,d4]=series(ng1,dg1,n3,d3);
>>[num,den]=cloop(n4,d4,-1);                     Step 5
```

FIGURE 2.55 Multiple-loop block reduction.

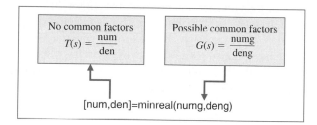

FIGURE 2.56 The **minreal** function.

```
>>numg=[1 4 6 6 5 2];
>>deng=[12 205 1066 2517 3128 2196 712];    <- Input transfer function before pole/zero cancellations.
>>[num,den]=minreal(numg,deng);             <- Cancel common factors.
>>printsys(num,den)

1 pole-zeros cancelled

num/den =    T(s)

      0.08333 s^4 + 0.25 s^3 + 0.25 s^2 + 0.25 s + 0.1667
   ---------------------------------------------------------
   s^5 + 16.08 s^4 + 72.75 s^3 + 137 s^2 + 123.7 s + 59.33
```

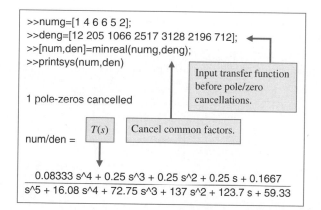

FIGURE 2.57 Application of the **minreal** function.

mon factors, as shown in Fig. 2.57. The closed-loop transfer function is given in Fig. 2.57 as $T(s) = \text{num/den}$. After the application of the minreal function, we find that the order of the denominator polynomial has been reduced from six to five, implying one pole–zero cancellation. ■

EXAMPLE 2.20 Electric traction motor control

Finally, let us reconsider the electric traction motor system of Example 2.11, as described on pages 71–72. The block diagram is shown in Fig. 2.33(c) on page 71. The objective is

```
>>num1=[10]; den1=[1 1]; num2=[1]; den2=[2 0.5]
>>num3=[540]; den3=[1]; num4=[0.1]; den4=[1];
>>[na,da]=series(num1,den1,num2,den2);      Eliminate
                                            inner loop.
>>[nb,db]=feedback(na,da,num4,den4,-1);
>>[nc,dc]=series(num3,den3,nb,db);
>>[num,den]=cloop(nc,dc,-1);           Compute closed-loop
                                       transfer function.
>>printsys(num,den)
num/den=
              5400                      ω(s)
        ──────────────                  ────
        2 s^2 + 2.5 s + 5402            ω_d(s)
```

FIGURE 2.58
Electric traction
motor block
reduction.

to compute the closed-loop transfer function and investigate the response of $\omega(s)$ to a commanded $\omega_d(s)$. The first step, as shown in Fig. 2.58, is to compute the closed-loop transfer
function $\omega/\omega_d = T(s)$. The closed-loop characteristic equation is second-order with $\omega_n =$
52 and $\zeta = 0.012$. Since the damping is low, we expect the response to be highly oscillatory. We can investigate the response $\omega(t)$ to a reference input, $\omega_d(t)$, by utilizing the **step**
function. The **step** function, shown in Fig. 2.59, calculates the unit step response of a linear
system.

The **step** function is very important, since control system performance specifications
are often given in terms of the unit step response. The state response, given by $x(t)$, is an
output of the **step** function and will be discussed in detail in Chapter 3, "State-Variable
Models." Include x in the left-hand argument list (see Fig. 2.59) but do not be concerned
with it for the time being.

If the only objective is to plot the output, $y(t)$, we can use the **step** function without
left-hand arguments and obtain the plot automatically with axis labels. If we need $y(t)$ for

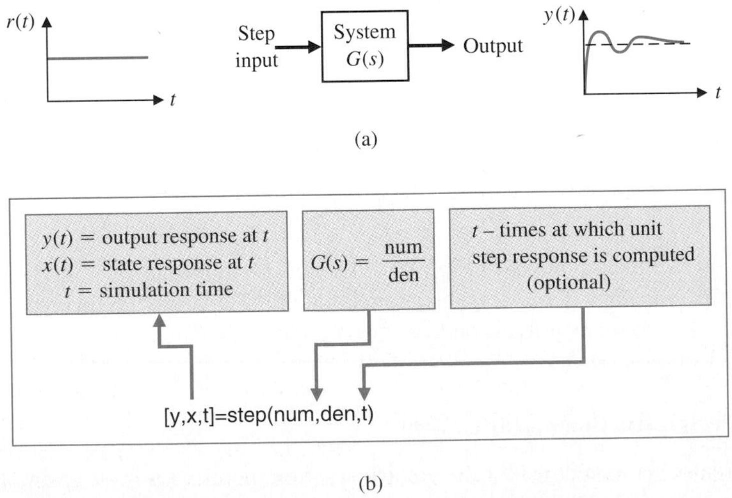

FIGURE 2.59
The **step** function.

(a)

$y(t)$ = output response at t
$x(t)$ = state response at t
t = simulation time

$G(s) = \dfrac{num}{den}$

t – times at which unit
step response is computed
(optional)

[y,x,t]=step(num,den,t)

(b)

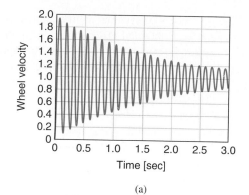

(a)

motoresponse.m

```
% This script computes the step
% response of the Traction Motor
% Wheel Velocity
%
num=[5400]; den=[2 2.5 5402];
t=[0:0.005:3];
[y,x,t]=step(num,den,t);
plot(t,y),grid
xlabel('TIme [sec]')
ylabel('Wheel velocity')
```

FIGURE 2.60
(a) Traction motor
wheel velocity
step response.
(b) *Matlab* script.

(b)

any purpose other than plotting, we must use the **step** function with left-hand arguments, followed by the plot function to plot $y(t)$. We define t as a row vector containing the times at which we wish the value of the output variable $y(t)$.

The step response of the electric traction motor is shown in Fig. 2.60. As expected, the wheel velocity response, given by $y(t)$, is highly oscillatory. Note that the output is $y(t) = \omega(t)$. ∎

2.11 SUMMARY

In this chapter, we have been concerned with quantitative mathematical models of control components and systems. The differential equations describing the dynamic performance of physical systems were utilized to construct a mathematical model. The physical systems under consideration included mechanical, electrical, fluid, and thermodynamic systems. A linear approximation using a Taylor series expansion about the operating point was utilized to obtain a small-signal linear approximation for nonlinear control components. Then, with the approximation of a linear system, one may utilize the Laplace transformation and its related input–output relationship, the transfer function. The transfer function approach to linear systems allows the analyst to determine the response of the system to various input signals in terms of the location of the poles and zeros of the transfer function. Using transfer function notations, block diagram models of systems of interconnected components were developed. The block relationships were obtained. Additionally, an alternative use of transfer function models in signal-flow graph form was investigated. The signal-flow graph gain formula was investigated and was found to be useful for obtaining the relationship between system variables in a complex feedback system. The advantage of the signal-flow graph method was the availability of Mason's flow graph gain formula, which provides the relation between system variables without requiring any reduction or manipulation of the flow graph. Thus, in Chapter 2, we have obtained a useful mathematical model for feedback control systems by developing the concept of a transfer function of a linear system and the relationship among system variables using block diagram and signal-flow graph models. Finally, we considered the utility of the computer simulation of linear and nonlinear systems to determine the response of a system for several conditions of the system parameters and the environment.

EXERCISES

(Exercises are straightforward applications of the concepts of the chapter.)

E2.1   A unity, negative feedback system has a nonlinear function $c = f(e) = e^2$ as shown in Fig. E2.1. For an input r in the range of 0 to 4, calculate and plot the open-loop and closed-loop output versus input and show that the feedback system results in a more linear relationship.

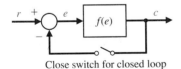

Close switch for closed loop

FIGURE E2.1   Open and closed loop.

E2.2   A thermistor has a response to temperature represented by

$$R = R_o e^{-0.1T},$$

where $R_o = 10,000 \ \Omega$, R = resistance, and T = temperature in degrees Celsius. Find the linear model for the thermistor operating at $T = 20°C$ and for a small range of variation of temperature.

Answer: $\Delta R = -135 \ \Delta T$

E2.3   The force versus displacement for a spring is shown in Fig. E2.3 for the spring-mass-damper system of Fig. 2.1. Graphically find the spring constant for the equilibrium point of $y = 0.5$ cm and a range of operation of ± 1.5 cm.

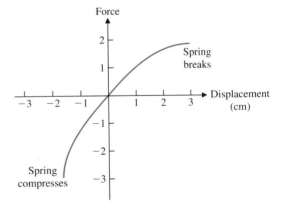

FIGURE E2.3   Spring behavior.

E2.4   A laser jet printer uses a laser beam to print copy rapidly for a computer. The laser is positioned by a control input, $r(t)$, so that we have

$$Y(s) = \frac{500(s + 100)}{s^2 + 60s + 500} R(s).$$

(a) If $r(t)$ is a unit step input, find the output $y(t)$.
(b) What is the final value of $y(t)$?

E2.5   A switching circuit is used to convert one level of dc voltage to an output dc voltage. The filter circuit to filter out the high frequencies is shown in Fig. E2.5. Calculate the transfer function $V_2(s)/V_1(s)$.

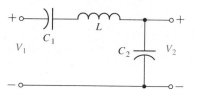

FIGURE E2.5   Filter circuit for 150 W switcher (idealized).

E2.6   A nonlinear device is represented by the function

$$y = f(x) = x^{1/2}$$

where the operating point for the input x is $x_o = 1/2$. Determine the linear approximation in the form of Eq. (2.9).

Answer: $\Delta y = \Delta x / \sqrt{2}$

E2.7   A lamp's intensity stays constant when monitored by an optotransistor-controlled feedback loop. When the voltage drops, the lamp's output

also drops, and optotransistor Q_1 draws less current. As a result, a power transistor conducts more heavily and charges a capacitor more rapidly [25]. The capacitor voltage controls the lamp voltage directly. A flow diagram of the system is shown in Fig. E2.7. Find the closed-loop transfer function, $I(s)/R(s)$ where $I(s)$ is in the lamp intensity and $R(s)$ is the command or desired level of light.

E2.8   A control engineer, N. Minorsky, designed an innovative ship steering system in the 1930s for the U.S. Navy. The system is represented by the signal-flow graph shown in Fig. E2.8 where $C(s)$ is the ship's course, $R(s)$ is the desired course, and $A(s)$ is the rudder angle [17]. Find the transfer function $C(s)/R(s)$.

E2.9   A four-wheel antilock automobile braking system uses electronic feedback to control automatically the brake force on each wheel [16]. A simplified flow graph of a brake control system is shown in Fig. E2.9, where $F_f(s)$ and $F_R(s)$ are the braking force of the front and rear wheels, respectively, and $R(s)$ is the desired automobile response on an icy road. Find $F_f(s)/R(s)$.

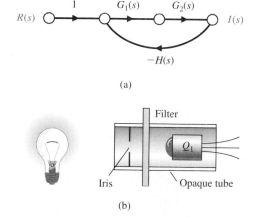

(a)

(b)

FIGURE E2.7   Lamp controller.

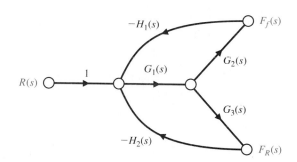

FIGURE E2.9   Brake control system.

FIGURE E2.8
Ship steering
system.

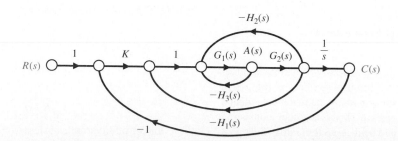

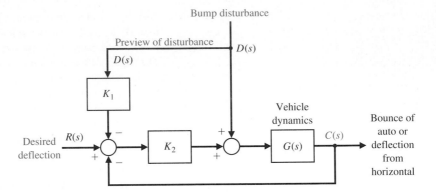

FIGURE E2.10
Active suspension
system.

E2.10 Off-road vehicles experience many disturbance inputs as they traverse over rough roads. An active suspension system can be controlled by a sensor that looks "ahead" at the road conditions. An example of a simple suspension system that can accommodate the bumps is shown in Fig. E2.10. Find the appropriate gain K_1 so that the vehicle does not bounce when the desired deflection is $R(s) = 0$ and the disturbance is $D(s)$.

Answer: $K_1K_2 = 1$

E2.11 A spring exhibits a force versus displacement characteristic as shown in Fig. E2.11. For small deviations from the operating point, find the spring constant when x_0 is (a) -1.4, (b) 0, (c) 3.5.

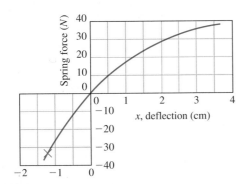

FIGURE E2.11   Spring characteristic.

E2.12 One of the most potentially beneficial applications of automotive control systems is the active control of the suspension system. One feedback control system uses a shock absorber consisting of a cylinder filled with a compressible fluid that provides both spring and damping forces [18]. The cylinder has a plunger activated by a gear motor, a displacement-measuring sensor, and a piston. Spring force is generated by piston displacement, which compresses the fluid. During piston displacement, the pressure imbalance across the piston is used to control damping. The plunger varies the internal volume of the cylinder. This feedback system is shown in Fig. E2.12. Develop a linear model for this device using a block diagram model.

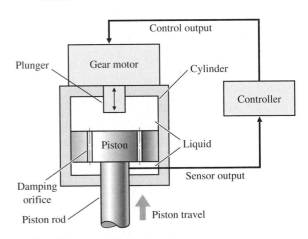

FIGURE E2.12   Shock absorber.

E2.13 Find the transfer function

$$\frac{C_1(s)}{R_2(s)}$$

for the multivariable system in Fig. E2.13.

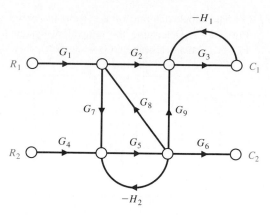

FIGURE E2.13   Multivariable system.

E2.14   Obtain the differential equations in terms of i_1 and i_2 for the circuit in Fig. E2.14.

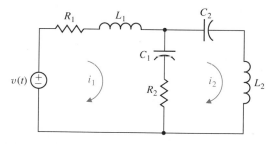

FIGURE E2.14   Electric circuit.

E2.15   The position control system for a spacecraft platform is governed by the following equations:

$$\frac{d^2p}{dt^2} + 2\frac{dp}{dt} + 4p = \theta$$

$$v_1 = r - p$$

$$\frac{d\theta}{dt} = 0.6v_2$$

$$v_2 = 7v_1.$$

The variables involved are as follows:

$r(t) = $ desired platform position

$p(t) = $ actual platform position

$v_1(t) = $ amplifier input voltage

$v_2(t) = $ amplifier output voltage

$\theta(t) = $ motor shaft position

Draw a signal-flow diagram of the system, identifying the component parts and their transmittances, then determine the system transfer function $P(s)/R(s)$.

E2.16   A spring used in an auto shock absorber develops a force, f, represented by the relation

$$f = kx^3,$$

where x is the displacement of the spring. Determine a linear model for the spring when $x_0 = 1$.

E2.17   The output, y, and input, x, of a device are related by

$$y = x + 0.4x^3.$$

(a) Find the values of the output for steady-state operation at the two operating points $x_0 = 1$ and $x_0 = 2$. (b) Obtain a linearized model for both operating points and compare them.

E2.18   The transfer function of a system is

$$\frac{C(s)}{R(s)} = \frac{10(s + 2)}{s^2 + 8s + 15}.$$

Determine $c(t)$ when $r(t)$ is a unit step input.

Answer: $c(t) = 1.33 + 1.67e^{-3t} - 3e^{-5t}, t \geq 0$

E2.19   Determine the transfer function $V_0(s)/V(s)$ of the operational amplifier circuit shown in Fig. E2.19. Assume an ideal operational amplifier. Determine the transfer function when $R_1 - R_2 = 100 \, \text{k}\Omega$, $C_1 = 10 \, \mu\text{f}$, and $C_2 = 5 \, \mu\text{F}$.

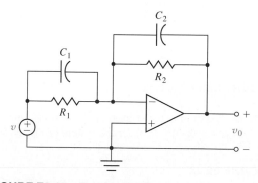

FIGURE E2.19   Op amp circuit.

E2.20   A high-precision positioning slide is shown in Fig. E2.20. Determine the transfer function $X_p(s)/X_{in}(s)$ when the drive shaft friction is $b_d = 1$, the drive shaft spring constant is $k_d = 3$, $m_c = 2/3$, and the sliding friction is $b_s = 1$.

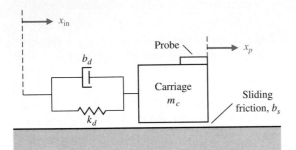

FIGURE E2.20   Precision slide.

E2.21   The rotational velocity ω of the satellite shown in Fig. E2.21 is adjusted by changing the length of the beam L. The transfer function between $\omega(s)$ and the incremental change in beam length $\Delta L(s)$ is

$$\frac{\omega(s)}{\Delta L(s)} = \frac{2.5(s + 2)}{(s + 5)(s + 1)^2}.$$

The beam length change is $\Delta L(s) = 1/(4s)$. Determine the response of the velocity $\omega(t)$.

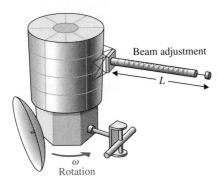

FIGURE E2.21   Satellite.

E2.22   Determine the closed-loop transfer function $T(s) = (C(s))/(R(s))$ for the system of Fig. E2.22.

E2.23   The block diagram of a system is shown in Fig. E2.23. Determine the equivalent system of Fig. 2.22, and determine $G(s)$ when $H(s) = 1$.

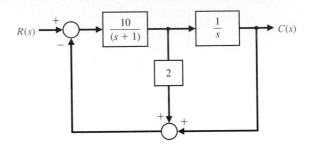

FIGURE E2.23

E2.24   An amplifier may have a region of deadband as shown in Fig. E2.24. Use an approximation that utilizes a cubic equation $y = ax^3$ in the approximately linear region. Select a and determine a linear approximation for the amplifier when the operating point is $x = 0.6$.

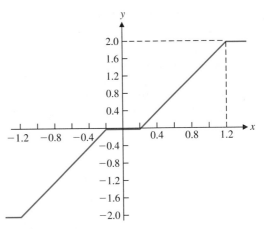

FIGURE E2.24   An amplifier with a deadband region.

FIGURE E2.22

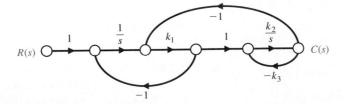

E2.25 Determine the transfer function $X_2(s)/F(s)$ for the system shown in Fig. E2.25. Both masses slide on a frictionless surface, and $k = 1$ N/M.

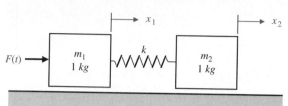

FIGURE E2.25

E2.26 Find the transfer function $C(s)/D(s)$ for the system shown in Fig. E2.26.

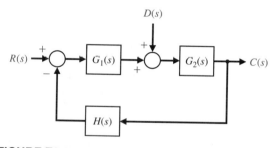

FIGURE E2.26 System with disturbance.

E2.27 Determine the transfer function $V_o(s)/V(s)$ for the op amp circuit shown in Fig. E2.27 [1]. Let $R_1 = 167$ kΩ, $R_2 = 240$ kΩ, $R_3 = 1$ kΩ, $R_4 = 100$ kΩ, and $C = 1$ μF. Assume an ideal op amp.

E2.28 A system is shown in Fig. E2.28(a).

(a) Determine $G(s)$ and $H(s)$ of the block diagram shown in Fig. E2.28(b) that are equivalent to those of the block diagram of Fig. E2.28(a).
(b) Determine $C(s)/R(s)$ for Fig. E2.28(b).

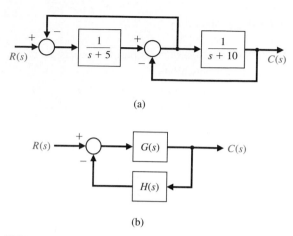

(a)

(b)

FIGURE E2.28

E2.29 A system is shown in Fig. E2.29.

(a) Find the closed-loop transfer function $C(s)/R(s)$, when $G(s) = 24/(s^2 + 30s + 176)$.
(b) Determine $C(s)$ when the input $r(t)$ is a unit step.
(c) Find $c(t)$ and plot it on a chart, also showing the desired step input.

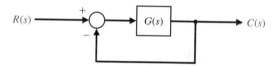

FIGURE E2.29

E2.30 Determine the residues for the partial fraction expansion for $V(s)$ by (a) numerical evaluation and (b) graphical evaluation on the s-plane:

$$V(s) = \frac{540}{s^2 + 8s + 540}.$$

FIGURE E2.27
Op amp
circuit.

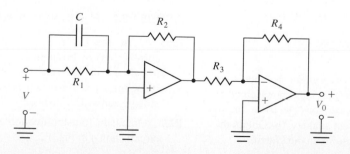

PROBLEMS

(Problems require an extension of the concepts of the chapter to new situations.)

P2.1   An electric circuit is shown in Fig. P2.1. Obtain a set of simultaneous integrodifferential equations representing the network.

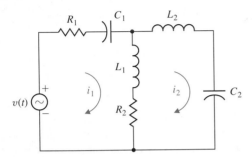

FIGURE P2.1   Electric circuit.

P2.2   A dynamic vibration absorber is shown in Fig. P2.2. This system is representative of many situations involving the vibration of machines containing unbalanced components. The parameters M_2 and k_{12} may be chosen so that the main mass M_1 does not vibrate when $F(t) = a \sin \omega_0 t$. (a) Draw the analogous electrical circuit based on the force–current analogy. (b) Obtain the differential equations describing the system.

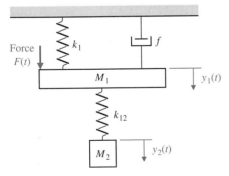

FIGURE P2.2   Vibration absorber.

P2.3   A coupled spring-mass system is shown in Fig. P2.3. The masses and springs are assumed to be equal. (a) Draw an analogous electrical circuit based on the force–current analogy. (b) Obtain the differential equations describing the system.

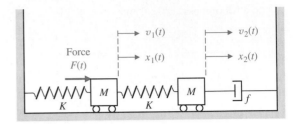

FIGURE P2.3   Two-mass system.

P2.4   A nonlinear amplifier can be described by the following characteristic:

$$v_o(t) = \begin{cases} v_{in}^2 & v_{in} \geq 0 \\ -v_{in}^2 & v_{in} < 0 \end{cases}.$$

The amplifier will be operated over a range for v_{in} of ± 0.5 volts at the operating point. Describe the amplifier by a linear approximation (a) when the operating point is $v_{in} = 0$; and (b) when the operating point is $v_{in} = 1$ volt. Obtain a sketch of the nonlinear function and the approximation for each case.

P2.5   Fluid flowing through an orifice can be represented by the nonlinear equation

$$Q = K(P_1 - P_2)^{1/2},$$

where the variables are shown in Fig. P2.5 and K is a constant [2]. (a) Determine a linear approximation for the fluid-flow equation. (b) What happens to the approximation obtained in part (a) if the operating point is $P_1 - P_2 = 0$?

FIGURE P2.5   Flow through an orifice.

P2.6   Using the Laplace transformation, obtain the current $I_2(s)$ of Problem 2.1. Assume that all the initial currents are zero, the initial voltage across capacitor C_1 is zero, $v(t)$ is zero, and the initial voltage across C_2 is 10 volts.

P2.7   Obtain the transfer function of the differentiating circuit shown in Fig. P2.7.

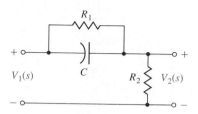

FIGURE P2.7 Differentiating circuit.

P2.8 A bridged-T network is often used in ac control systems as a filter network [8]. The circuit of one bridged-T network is shown in Fig. P2.8. Show that the transfer function of the network is

$$\frac{V_o(s)}{V_{in}(s)} = \frac{1 + 2R_1Cs + R_1R_2C^2s^2}{1 + (2R_1 + R_2)Cs + R_1R_2C^2s^2}.$$

Draw the pole–zero diagram when $R_1 = 0.5$, $R_2 = 1$, and $C = 0.5$.

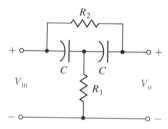

FIGURE P2.8 Bridged-T network.

P2.9 Determine the transfer function $X_1(s)/F(s)$ for the coupled spring-mass system of Problem 2.3. Draw the s-plane pole-zero diagram for low damping when $M = 1$, $f/K = 1$ and

$$\zeta = \frac{1}{2}\frac{f}{\sqrt{KM}} = 0.2.$$

FIGURE P2.11
Amplidyne and armature-controlled motor.

P2.10 Determine the transfer function $Y_1(s)/F(s)$ for the vibration absorber system of Problem 2.2. Determine the necessary parameters M_2 and k_{12} so that the mass M_1 does not vibrate when $F(t) = a \sin \omega_0 t$.

P2.11 For electromechanical systems that require large power amplification, rotary amplifiers are often used [8, 19]. An amplidyne is a power amplifying rotary amplifier. An amplidyne and a servomotor are shown in Fig. P2.11. Obtain the transfer function $\theta(s)/V_c(s)$, and draw the block diagram of the system. Assume $v_d = k_2i_q$ and $v_q = k_1i_c$.

P2.12 A fluid-flow system is shown in Fig. P2.12, where an incompressible fluid is flowing into an open tank. One may assume that the change in outflow ΔQ_2 is proportional to the change in head ΔH. At steady-state $Q_1 = Q_2$ and $Q_2 = kH^{1/2}$. Using a linear approximation, obtain the transfer function of the tank, $\Delta Q_2(s)/\Delta Q_1(s)$ [19].

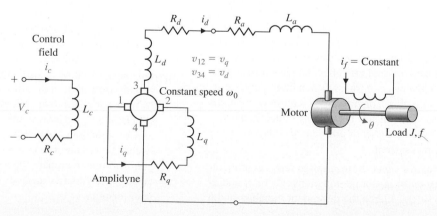

FIGURE P2.12 Fluid-flow system.

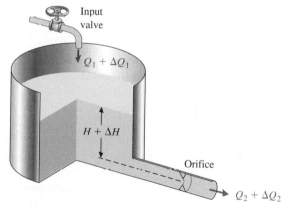

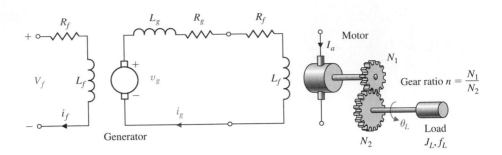

FIGURE P2.13
Motor and
generator.

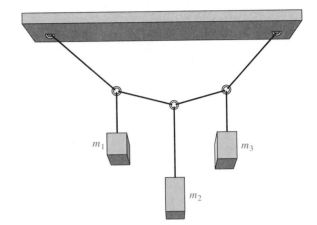

P2.13   An electromechanical open-loop control system is shown in Fig. P2.13. The generator, driven at a constant speed, provides the field voltage for the motor. The motor has an inertia J_m and bearing friction f_m. Obtain the transfer function $\theta_L(s)/V_f(s)$, and draw a block diagram of the system. The generator voltage, v_g, can be assumed to be proportional to the field current, i_f.

P2.14   A rotating load is connected to a field-controlled dc electric motor through a gear system. The motor is assumed to be linear. A test results in the output load reaching a speed of 1 rad/s within ½ s when a constant 80 V is applied to the motor terminals. The output steady-state speed is 2.4 rad/s. Determine the transfer function of the motor, $\theta(s)/V_f(s)$ in rad/V. The inductance of the field may be assumed to be negligible (see Fig. 2.16). Also, note that the application of 80 V to the motor terminals is a step input of 80 V in magnitude.

P2.15   As the complexity of interconnected power systems grows, the potential for interactive dynamic phenomena increases. Mechanical analogies can be used to explain complex interactive dynamics. Consider the weight–rubber band analogy shown in Fig. P2.15. In this weight–rubber band analogy, the weights represent the rotating mass of the turbine generators and the rubber bands are analogous to the inductance of transmission lines. Any disturbance—a "pull" on a weight—causes oscillations to be set up in all synchronous machines in the system. Determine a set of differential equations to describe this system.

P2.16   Obtain a signal-flow graph to represent the following set of algebraic equations where x_1 and x_2 are to be considered the dependent variables and 6

FIGURE P2.15   Three-mass analogy.

and 11 are the inputs:

$$x_1 + 1.5x_2 = 6, \qquad 2x_1 + 4x_2 = 11.$$

Determine the value of each dependent variable by using the gain formula. After solving for x_1 by Mason's formula, verify the solution by using Cramer's rule.

P2.17   A mechanical system is shown in Fig. P2.17, which is subjected to a known displacement $x_3(t)$ with respect to the reference. (a) Determine the two independent equations of motion. (b) Obtain the equations of motion in terms of the Laplace transform, assuming that the initial conditions are zero. (c) Draw a signal-flow graph representing the system of equations. (d) Obtain the relationship between $X_1(s)$ and $X_3(s)$, $T_{13}(s)$, by using Mason's gain formula. Compare the work necessary to obtain $T_{13}(s)$ by matrix methods to that using Mason's gain formula.

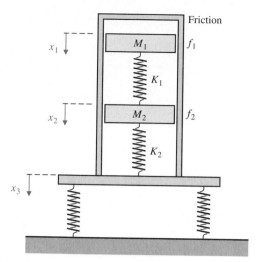

FIGURE P2.17 Mechanical system.

P2.18 An *LC* ladder network is shown in Fig. P2.18. One may write the equations describing the network as follows:

$$I_1 = (V_1 - V_a)Y_1, \qquad V_a = (I_1 - I_a)Z_2,$$
$$I_a = (V_a - V_2)Y_3, \qquad V_2 = I_a Z_4.$$

Construct a flow graph from the equations and determine the transfer function $V_2(s)/V_1(s)$.

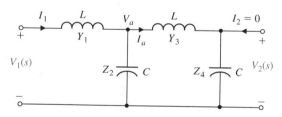

FIGURE P2.18 *LC* Ladder network.

P2.19 The basic noninverting operational amplifier is shown in Fig. P2.19(a), and the signal-flow representation of the equations of the circuit is shown in Fig. P2.19(b) [8]. (a) Write the voltage equations, and verify the representations in the flow graph. (b) Using the signal-flow graph, calculate the gain of the amplifier and verify that $T(s) = (R_1 + R_f)/R_1$ when $A \gg 10^3$.

P2.20 The source follower amplifier provides lower output impedance and essentially unity gain. The circuit diagram is shown in Fig. P2.20(a), and the small signal model is shown in Fig. P2.20(b). This

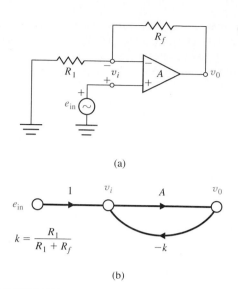

(a)

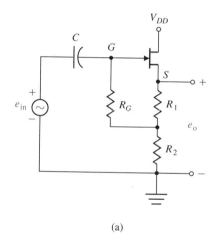

(b)

FIGURE P2.19 Noninverting op amp circuit.

(a)

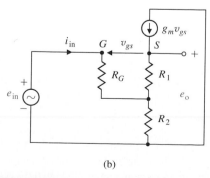

(b)

FIGURE P2.20
The source follower or common drain amplifier using an FET.

circuit uses an FET and provides a gain of approximately unity. Assume that $R_2 \gg R_1$ for biasing purposes and that $R_g \gg R_2$. (a) Solve for the amplifier gain. (b) Solve for the gain when $g_m = 2000$ μmhos and $R_s = 10$ Kohms where $R_s = R_1 + R_2$. (c) Sketch a signal-flow diagram that represents the circuit equations.

P2.21 A hydraulic servomechanism with mechanical feedback is shown in Fig. P2.21 [19]. The power piston has an area equal to A. When the valve is moved a small amount Δz, the oil will flow through to the cylinder at a rate $p \cdot \Delta z$, where p is the port coefficient. The input oil pressure is assumed to be constant. (a) Determine the closed-loop signal-flow graph for this mechanical system. (b) Obtain the closed-loop transfer function $Y(s)/X(s)$.

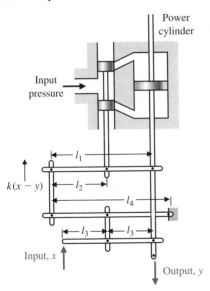

FIGURE P2.21   Hydraulic servomechanism.

P2.22 Figure P2.22 shows two pendulums suspended from frictionless pivots and connected at their midpoints by a spring [1]. Assume that each pendulum can be represented by a mass M at the end of a massless bar of length L. Also assume that the displacement is small and linear approximations can be used for $\sin \theta$ and $\cos \theta$. The spring located in the middle of the bars is unstretched when $\theta_1 = \theta_2$. The input force is represented by $f(t)$, which influences the left-hand bar only. (a) Obtain the equations of motion, and draw a signal-flow diagram for

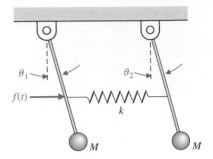

FIGURE P2.22   The bars are each of length L and the spring is located at $L/2$.

them. (b) Determine the transfer function $T(s) = \theta_1(s)/F(s)$. (c) Draw the location of the poles and zeros of $T(s)$ on the s-plane.

P2.23 The small-signal circuit equivalent to a common-emitter transistor amplifier is shown in Fig. P2.23. The transistor amplifier includes a feedback resistor R_f. Obtain a signal-flow graph model of the feedback amplifier, and determine the input–output ratio v_{ce}/v_{in}.

P2.24 A two-transistor series voltage feedback amplifier is shown in Fig. P2.24(a). This ac equivalent circuit neglects the bias resistors and the shunt capacitors. A signal-flow graph representing the circuit is shown in Fig. P2.24(b). This flow graph neglects the effect of h_{re}, which is usually an accurate approximation, and assumes that $(R_2 + R_L) \gg R_1$. (a) Determine the voltage gain, e_o/e_{in}. (b) Determine the current gain, i_{c_2}/i_{b_1}. (c) Determine the input impedance, e_{in}/i_{b_1}.

P2.25 Often overlooked is the fact that H. S. Black, who is noted for developing a negative feedback amplifier in 1927, three years earlier had invented a circuit design technique known as feedforward correction [20]. Recent experiments have shown that this technique offers the potential for yielding excellent amplifier stabilization. Black's amplifier is shown in Fig. P2.25(a) in the form recorded in 1924. The signal-flow graph is shown in Fig. P2.25(b). Determine the transfer function between the output $C(s)$ and the input $R(s)$ and between the output and the disturbance $D(s)$. $G(s)$ is used for the amplifier represented by μ in Fig. P2.25(a).

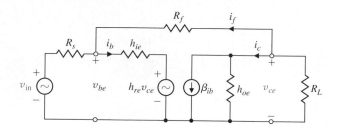

FIGURE P2.23
CE amplifier.

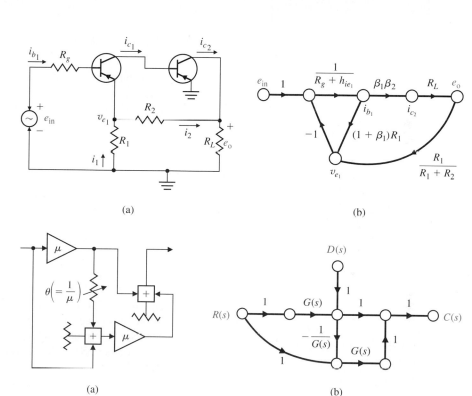

(a)   (b)

FIGURE P2.24
Feedback
amplifier.

FIGURE P2.25
H. S. Black's
amplifier.

(a)   (b)

P2.26 A robot includes significant flexibility in the arm members with a heavy load in the gripper [6, 21]. A two-mass model of the robot is shown in Fig. P2.26. Find the transfer function $Y(s)/F(s)$.

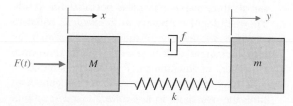

FIGURE P2.26 The spring-mass-damper model of a robot arm.

P2.27 Magnetic levitation trains provide a high speed, very low friction alternative to steel wheels on steel rails. The train floats on an air gap as shown in Fig. P2.27 [27]. The levitation force F_L is controlled by the coil current i in the levitation coils and may be approximated by

$$F_L = k\,\frac{i^2}{z^2},$$

where z is the air gap. This force is opposed by the downward force $F = mg$. Determine the linearized relationship between the air gap z and the controlling current near the equilibrium condition.

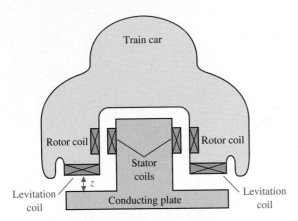

FIGURE P2.27   Cutaway view of train.

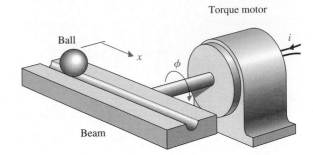

FIGURE P2.29   Tilting beam and ball.

P2.28   A multiple-loop model of an urban ecological system might include the following variables: number of people in the city (P), modernization (M), migration into the city (C), sanitation facilities (S), number of diseases (D), bacteria/area (B), and amount of garbage/area (G), where the symbol for the variable is given in parentheses. The following causal loops are hypothesized:

1. $P \to G \to B \to D \to P$
2. $P \to M \to C \to P$
3. $P \to M \to S \to D \to P$
4. $P \to M \to S \to B \to D \to P$

Draw a signal-flow graph for these causal relationships, using appropriate gain symbols. Indicate whether you believe each gain transmission is positive or negative. For example, the causal link S to B is negative because improved sanitation facilities lead to reduced bacteria/area. Which of the four loops are positive feedback loops and which are negative feedback loops?

P2.29   We desire to balance a rolling ball on a tilting beam as shown in Fig. P2.29. We will assume the motor input current i controls the torque with negligible friction. Assume the beam may be balanced near the horizontal ($\phi = 0$); therefore we have a small deviation of ϕ. Find the transfer function $X(s)/I(s)$, and draw a block diagram illustrating the transfer function showing $\phi(s)$, $X(s)$ and $I(s)$.

P2.30   The measurement or sensor element in a feedback system is important to the accuracy of the system [6]. The dynamic response of the sensor is often important. Most sensor elements possess a transfer function

$$H(s) = \frac{k}{\tau s + 1}.$$

Investigate several sensor elements available today, and determine the accuracy available and the time constant of the sensor. Consider two of the following sensors: (1) linear position, (2) temperature with a thermistor, (3) strain measurement, (4) pressure.

P2.31   A cable reel control system uses a tachometer to measure the speed of the cable as it leaves the reel. The output of the tachometer is used to control the motor speed of the reel as the cable is unwound off the reel. The system is shown in Fig. P2.31. The radius of the reel, R, is 4 meters when full and 2 meters when empty. The moment of inertia of the reel is $I = 18.5R^4 - 221$. The rate of change of the radius is

$$\frac{dR}{dt} = \frac{-D^2\dot{\omega}}{2\pi W},$$

where W = width of the reel and D = diameter of the cable. The actual speed of the cable is $v(t) = R\dot{\omega}$. The desired cable speed is 50 m/sec. Develop a digital computer simulation of this system and obtain the response of the speed over 20 seconds for the three values of gain $K = 0.2, 0.4,$ and 0.6. The reel angular velocity $\dot{\omega} = d\theta/dt$ is equal to $1/I$ times the integral of the torque. Note that the inertia changes with time as the reel is unwound. However, an equation for I within the simulation will account for this change. Select the gain K to limit the overshoot to less than 9% and yet attain the fastest response. Assume $W = 2.0$, $D = 0.1$, and $R = 3.5$ at $t = 0$.

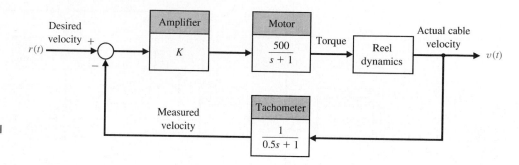

FIGURE P2.31
Cable reel control
system.

P2.32 An interacting control system with two inputs and two outputs is shown in Fig. P2.32. Solve for $C_1(s)/R_1(s)$ and $C_2(s)/R_1(s)$, when $R_2 = 0$.

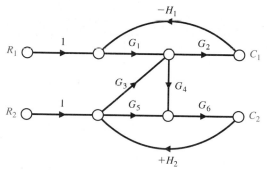

FIGURE P2.32 Interacting system.

P2.33 A system consists of two electric motors that are coupled by a continuous flexible belt. The belt also passes over a swinging arm that is instrumented to allow measurement of the belt speed and tension. The basic control problem is to regulate the belt speed and tension by varying the motor torques.

An example of a practical system similar to that shown occurs in textile fiber manufacturing processes when yarn is wound from one spool to another at high speed. Between the two spools the yarn is processed in a way that may require the yarn speed and tension to be controlled to within defined limits. A model of the system is shown in Fig. P2.33. Find $C_2(s)/R_1(s)$. Determine a relationship for the system that will make C_2 independent of R_1.

P2.34 Find the transfer function for $C(s)/R(s)$ for the idle speed control system for a fuel injected engine as shown in Fig. P2.34.

FIGURE P2.33
A model of the
coupled motor
drives.

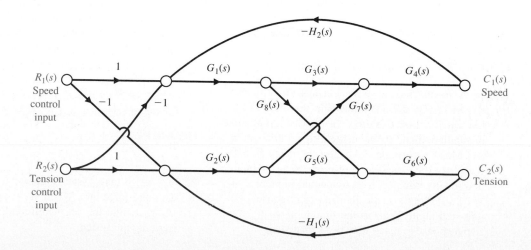

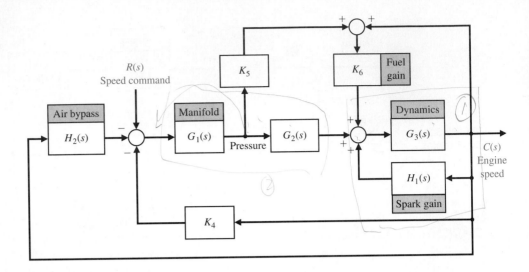

FIGURE P2.34
Idle speed
control system.

P2.35   The suspension system for one wheel of an old-fashioned pickup truck is illustrated in Fig. P2.35. The mass of the vehicle is m_1 and the mass of the wheel is m_2. The suspension spring has a spring constant k_1, and the tire has a spring constant k_2. The damping constant of the shock absorber is f. Obtain the transfer function $Y_1(s)/X(s)$, which represents the vehicle response to bumps in the road.

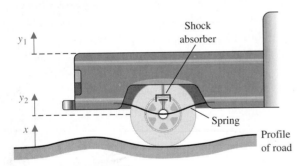

FIGURE P2.35   Pickup truck suspension.

P2.36   A feedback control system has the structure shown in Fig. P2.36. Determine the closed-loop transfer function $C(s)/R(s)$ (a) by block diagram manipulation and (b) by using a signal flow graph and Mason's formula. (c) Select the gains K_1 and K_2 so that the closed-loop response to a step input is critically damped with two equal roots at $s = -10$. (d) Plot the critically damped response for a unit step input. What is the time required for the step response to reach 90% of its final value?

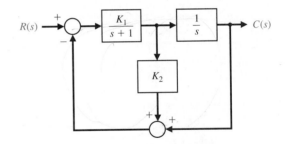

FIGURE P2.36

P2.37   A system is represented by Fig. P2.37. (a) Determine the partial fraction expansion and $c(t)$ for a ramp input, $r(t) = t$, $t \geq 0$. (b) Obtain a plot of $c(t)$ for part (a), and find $c(t)$ for $t = 1.5s$. (c) Determine the impulse response of the system $c(t)$ for $t \geq 0$. (d) Obtain a plot of $c(t)$ for part (c), and find $c(t)$ for $t = 1.5s$.

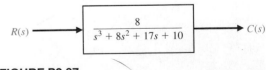

FIGURE P2.37

P2.38   A two-mass system is shown in Fig. P2.38 with an input force $u(t)$. When $m_1 = m_2 = 1$ and $K_1 = K_2 = 1$, find the set of differential equations describing the system.

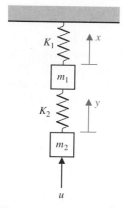

FIGURE P2.38 Two-mass system.

P2.39 A winding oscillator consists of two steel spheres on each end of a long, slender rod, as shown in Fig. P2.39. The rod is hung on a thin wire that can be twisted many revolutions without breaking. The device will be wound up 4000 degrees. How long will it take until the motion decays to a swing of only 10 degrees? Assume that the thin wire has a rotational spring constant of 2×10^{-4} N $\cdot$ m/rad and that the viscous friction coefficient for the sphere in air is 2×10^{-4} N $\cdot$ m/rad. The sphere has a mass of 1 kg.

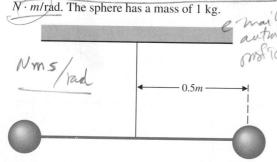

FIGURE P2.39 Winding oscillator.

P2.40 For the circuit of Fig. P2.40, determine the transform of the output voltage $V_0(s)$. Assume that

the circuit is in steady state when $t < 0$. Assume that the switch moves instantaneously from contact 1 to contact 2 at $t = 0$.

P2.41 A damping device is used to reduce the undesired vibrations of machines. A viscous fluid, such as a heavy oil, is placed between the wheels, as shown in Fig. P2.41. When vibration becomes excessive, the relative motion of the two wheels creates damping. When the device is rotating without vibration, there is no relative motion and no damping occurs. Find $\theta_1(s)$ and $\theta_2(s)$. Assume that the shaft has a spring constant K and that f is the damping constant of the fluid. The load torque is T.

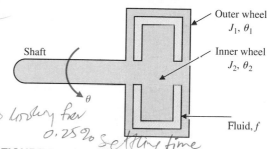

FIGURE P2.41 Cutaway view of damping device.

P2.42 The lateral control of a rocket with a gimbaled engine is shown in Fig. P2.42. The lateral deviation from the desired trajectory is h, and the forward rocket speed is V. The control torque of the engine is T_c, and the disturbance torque is T_d. Derive the describing equations of a linear model of the system, and draw the block diagram with the appropriate transfer functions.

P2.43 In many applications, such as reading product codes in supermarkets and in printing and manufacturing, an optical scanner is utilized to read codes, as shown in Fig. P2.43. As the mirror rotates, a friction force is developed that is propor-

FIGURE P2.40
Model of an electronic circuit.

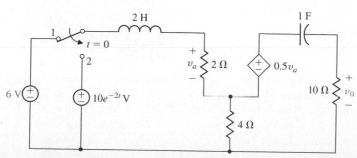

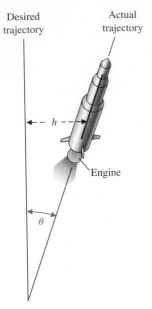

FIGURE P2.42   Rocket with gimbaled engine.

tional to its angular speed. The friction constant is equal to 0.04 $N \cdot s$/rad, and the moment of inertia is equal to 0.1 kg-m^2. The output variable is the velocity, $\omega(t)$. (a) Obtain the differential equation for the motor. (b) Find the response of the system when the input motor torque is a unit step and the initial velocity at $t = 0$ is equal to 0.5.

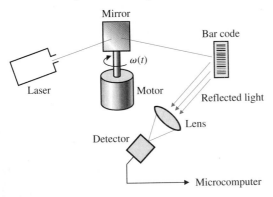

FIGURE P2.43   Optical scanner.

P2.44   An ideal set of gears is shown as item 10 of Table 2.7. Neglect the inertia and friction of the gears and assume that the work done by one gear is equal to that of the other. Derive the relationships given in item 10 of Table 2.7. Also determine the relationship between the torques T_m and T_L.

P2.45   An ideal set of gears is connected to a solid cylinder load as shown in Fig. P2.45. The inertia of the motor shaft and gear G_2 is J_m. Determine (a) the inertia of the load J_L and (b) the torque T at the motor shaft. Assume the friction at the load is f_L and the friction at the motor shaft is f_m. Also assume the density of the load disk is ρ.

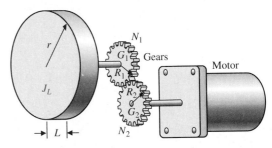

FIGURE P2.45   Motor, gears, and load.

P2.46   To utilize the strength advantage of robot manipulators and the intellectual advantage of humans, a class of manipulators called "extenders" has been examined [23]. The extender is defined as an active manipulator worn by a human to augment the human's strength. The human provides an input $U(s)$ as shown in Fig. P2.46. The endpoint of the extender is $P(s)$. Determine the output $P(s)$ for both $U(s)$ and $F(s)$ in the form

$$P(s) = T_1(s)U(s) + T_2(s)F(s).$$

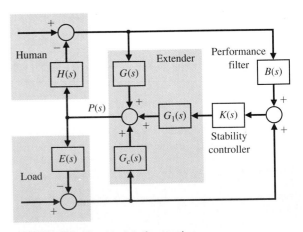

FIGURE P2.46   Model of extender.

P2.47   A load added to a truck results in a force F on the support spring and the tire flexes as shown in Fig. P2.47(a). The model for the tire movement is shown in Fig. P2.47(b). Determine the transfer function $X_1(s)/F(s)$.

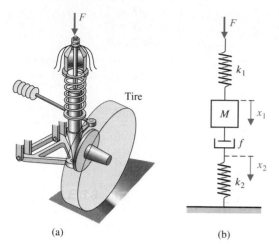

(a)   (b)

FIGURE P2.47   Truck support model.

P2.48   The water level, $h(t)$, is controlled by an open-loop system, as shown in Fig. P2.48. A dc motor controlled by an armature current i_a turns a shaft, opening a valve. The inductance of the dc motor is negligible. The height of the water in the tank is

$$h(t) = \int [1.6\theta(t) - h(t)]dt,$$

the motor constant is $K_m = 10$, and the inertia of the motor shaft and valve is $J = 6 \times 10^{-3}$ kg-m². Determine (a) the differential equation for $h(t)$ and $v(t)$ and (b) the transfer function $H(s)/V(s)$.

P2.49   The circuit shown in Fig. P2.49 is called a lead-lag filter and is used in Chapter 10.

(a) Find the transfer function $V_2(s)/V_1(s)$.
(b) Determine $V_2(s)/V_1(s)$ when $R_1 = 100$ kΩ, $R_2 = 200$ kΩ, $C_1 = 1$ μF, and $C_2 = 0.1$ μF.
(c) Determine the partial fraction expansion for $V_2(s)/V_1(s)$.

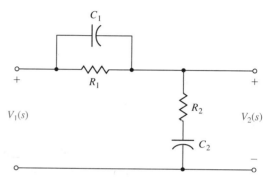

FIGURE P2.49   Lead-lag filter.

FIGURE P2.48
Open-loop control system for the water level of a tank.

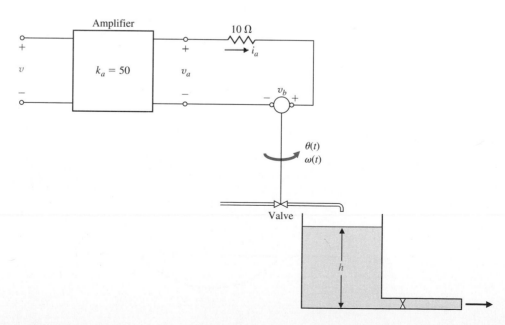

P2.50 A closed-loop control system is shown in Fig. P2.50.

(a) Determine the transfer function

$$T(s) = C(s)/R(s).$$

(b) Determine the poles and zeros of $T(s)$.
(c) Use a unit step input, $R(s) = 1/s$, and obtain the partial fraction expansion for $C(s)$ and the value of the residues.
(d) Plot $c(t)$ and discuss the effect of the real and complex poles of $T(s)$. Do the complex poles or the real poles dominate the response?

P2.51 A closed-loop control system is shown in Fig. P2.51.

(a) Determine the transfer function $T(s) = C(s)/R(s)$.
(b) Determine the poles and zeros of $T(s)$.
(c) Use a unit step input, $R(s) = 1/s$, and obtain the partial fraction expansion for $C(s)$ and the value of the residues.
(d) Plot $c(t)$ and discuss the effect of the real and complex poles of $T(s)$. Do the complex poles or the real poles dominate the response?
(e) Predict the final value of $c(t)$ for the unit step input.

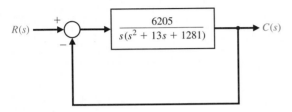

FIGURE P2.50

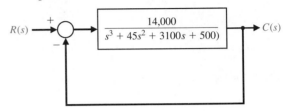

FIGURE P2.51

ADVANCED PROBLEMS

AP2.1 An armature-controlled dc motor is driving a load. The input voltage is 5 V. The speed at $t = 2$ seconds is 30 rad/s, and the steady speed is 70 rad/s when $t \to \infty$. Determine the transfer function $\omega(s)/V(s)$.

AP2.2 A system has a signal-flow diagram as shown in Fig. AP2.2. Determine the transfer function

$$T(s) = \frac{C_2(s)}{R_1(s)}.$$

It is desired to decouple $C_2(s)$ from $R_1(s)$ by obtaining $T(s) = 0$. Select $G_5(s)$ in terms of the other $G_i(s)$ to achieve decoupling.

FIGURE AP2.2

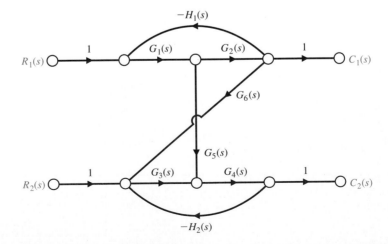

DESIGN PROBLEMS

(Design problems emphasize the design task.)

DP2.1 A control system is shown in Fig. DP2.1. The transfer functions $G_2(s)$ and $H_2(s)$ are fixed. Determine the transfer functions $G_1(s)$ and $H_1(s)$ so that the closed-loop transfer function $C(s)/R(s)$ is exactly equal to 1.

DP2.2 The television beam circuit of a television set is represented by the model in Fig. DP2.2. Select the unknown conductance, G, so that the voltage, v, is 24 V. Each conductance is given in siemens (S).

DP2.3 An input $r(t) = e^{-t}, t \geq 0$ is applied to a black box with a transfer function $G(s)$. The resulting output response, when the initial conditions are zero, is

$$c(t) = 2 - 3e^{-t} + e^{-2t} \cos 2t, \quad t \geq 0.$$

Determine $G(s)$ for this system.

DP2.4 An operational amplifier circuit that can serve as a filter circuit is shown in Fig. DP2.4. (a) Determine the transfer function of the circuit, assuming an ideal op amp. Find $e_0(t)$ when the input is $e_1(t) = At, t \geq 0$.

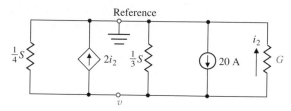

FIGURE DP2.2 Television beam circuit.

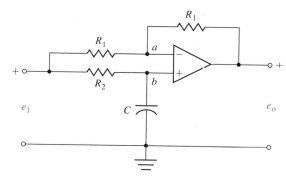

FIGURE DP2.4 Operational amplifier circuit.

FIGURE DP2.1
Selection of
transfer functions.

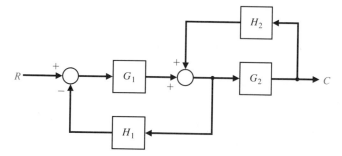

MATLAB PROBLEMS

MP2.1 Consider the feedback system depicted in Fig. MP2.1.

(a) Compute the closed-loop transfer function using the **series** and **cloop** functions, and display the result with the **printsys** function.

(b) Obtain the closed-loop system unit step response with the **step** function, and verify that final value of the output is -0.25.

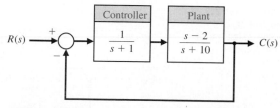

FIGURE MP2.1 A negative feedback control system.

MP2.2 Consider the mechanical system depicted in Fig. MP2.2. The input is given by $f(t)$, and the output is $x(t)$. Determine the transfer function from $f(t)$ to $x(t)$ and, using *Matlab*, plot the system response to a unit step input. Let $m = 10$, $k = 1$, and $b = 0.5$. Show that the peak amplitude of the output is about 1.8.

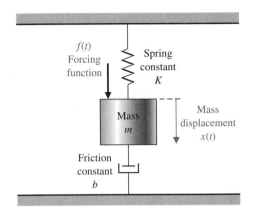

FIGURE MP2.2   A mechanical spring-mass-damper system.

MP2.3 A satellite single-axis attitude control system can be represented by the block diagram in Fig. MP2.3. The variables k, a, and b are controller parameters, and J is the spacecraft moment of inertia. Suppose the nominal moment of inertia is $J = 10.8E+08$, and the controller parameters are $k = 10.8E+08$, $a = 1$, and $b = 8$.

(a) Develop a *Matlab* script to compute the closed-loop transfer function $T(s) = \theta(s)/\theta_d(s)$.

(b) Compute and plot the step response to a $10°$ step input (see hint).

(c) The exact moment of inertia is generally unknown and may change slowly with time. Compare the step response performance of the spacecraft when J is reduced by 20% and 50%. Use the controller parameters $k = 10.8e + 08$, $a = 1$, and $b = 8$ and a $10°$ step input. Discuss your results.

Hint for part (b): Scale the transfer function as follows:

$$\theta(s) = T(s)\theta_d(s) = T(s)\,\frac{a}{s}$$

$$= [aT(s)]\,\frac{1}{s} = T^*(s)\,\frac{1}{s},$$

where $a = 10°$. Use the **step** function with $T^*(s)$ and the output will be $\theta_d(t)$ in degrees.

MP2.4 Consider the block diagram in Fig. MP2.4.

(a) Use *Matlab* to reduce the block diagram in Fig. MP2.4, and compute the closed-loop transfer function.

(b) Generate a pole–zero map of the closed-loop transfer function in graphical form using the **pzmap** function.

(c) Determine explicitly the poles and zeros of the closed-loop transfer function using the **roots** function, and correlate the results with the pole–zero map in part (b).

MP2.5 The relationship between the output $y(t)$ and the input $x(t)$ of a nonlinear system is given by

$$y(x) = x^2 + x \sin x.$$

A linear approximation is given by

$$\tilde{y} = ax,$$

where a is a coefficient to be determined. Compute the parameter a "experimentally" by developing a *Matlab* script that computes and plots the differ-

FIGURE MP2.3
A spacecraft
single-axis attitude
control block
diagram.

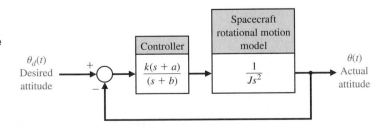

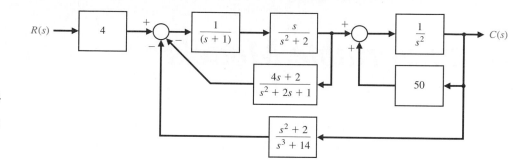

FIGURE MP2.4
A multiple-loop
feedback control
system block
diagram.

ence between y and $\bar{y}$ with the parameter a as an adjustable parameter. Vary the parameter a iteratively so that the maximum difference between y and $\bar{y}$ is less than 20 whenever $0 \leq x \leq 10$. With the final value of a determined, co-plot $y(x)$ versus x and $\bar{y}(x)$ versus x with $0 \leq x \leq 10$.

MP2.6   A system has a transfer function

$$\frac{X(s)}{R(s)} = \frac{(15/z)(s + z)}{s^2 + 35 + 15}.$$

Plot the response of the system when $r(t)$ is a unit step for the parameter $z = 3, 6,$ and 12.

TERMS AND CONCEPTS

Actuator   The device that causes the process to provide the output. The device that provides the motive power to the process.

Block diagrams   Unidirectional, operational blocks that represent the transfer functions of the elements of the system.

Characteristic equation   The relation formed by equating to zero the denominator of a transfer function.

Critical damping   The case where damping is on the boundary between underdamped and overdamped.

Damped oscillation   An oscillation in which the amplitude decreases with time.

Damping ratio   A measure of damping. A dimensionless number for the second-order characteristic equation.

dc motor   An electric actuator that uses an input voltage as a control variable.

Laplace transform   A transformation of a function $f(t)$ from the time domain into the complex frequency domain yielding $F(s)$.

Linear approximation   An approximate model that results in a linear relationship between the output and the input of the device.

Linear system   A system that satisfies the properties of superposition and homogeneity.

Mason loop rule   A rule that enables the user to obtain a transfer function by tracing paths and loops within a system.

Mathematical models   Descriptions of the behavior of a system using mathematics.

Signal-flow graph   A diagram that consists of nodes connected by several directed branches and that is a graphical representation of a set of linear relations.

Simulation   A model of a system that is used to investigate the behavior of a system by utilizing actual input signals.

Transfer function   The ratio of the Laplace transform of the output variable to the Laplace transform of the input variable.

State Variable Models

10/21/97

PREVIEW

In the preceding chapter, we used the Laplace transform to describe the performance of a feedback system. This method is attractive because it provides a practical approach to analysis of a system. In this chapter, we turn to an alternative method of system analysis using time-domain methods. We will reconsider the differential equations describing a control system and select a certain form of differential equations. We will use a set of variables that can be used to establish a set of first-order differential equations.

Using matrix methods, we will be able to determine the transient response of a control system and examine the performance of these systems. These time-domain matrix methods lend themselves readily to computer solution. Furthermore, it is feasible to propose new feedback structures based on the utilization of this new set of time-domain variables, called *state variables*.

3.1 INTRODUCTION

In the preceding chapter, we developed and studied several useful approaches to the analysis and design of feedback systems. The Laplace transform was utilized to transform the differential equations representing the system to an algebraic equation expressed in terms of the complex variable s. Utilizing this algebraic equation, we were able to obtain a transfer function representation of the input–output relationship.

With the ready availability of digital computers, it is convenient to consider the time-domain formulation of the equations representing control systems. The time-domain techniques can be readily utilized for nonlinear, time-varying, and multivariable systems. *A time-varying control system is a system for which one or more of the parameters of the system may vary as a function of time.* For example, the mass of a missile varies as a function of time as the fuel is expended during flight. A multivariable system, as discussed in Section 2.6, is a system with several input and output signals. The solution of a time-domain formulation of a control system problem is facilitated by the availability and ease of use of digital and analog computers. Therefore we are interested in reconsidering the time-domain description of dynamic systems as they are represented by the system differential equation. The *time-domain* is the mathematical domain that incorporates the response and description of a system in terms of time, t.

The time-domain representation of control systems is an essential basis for modern control theory and system optimization. In Chapter 11, we will have an opportunity to design an optimum control system by utilizing time-domain methods. In this chapter, we shall develop the time-domain representation of control systems and illustrate several methods for the solution of the system time response.

3.2   THE STATE VARIABLES OF A DYNAMIC SYSTEM

The time-domain analysis and design of control systems utilize the concept of the state of a system [1–3, 5]. *The state of a system is a set of numbers such that the knowledge of these numbers and the input functions will, with the equations describing the dynamics, provide the future state and output of the system.* For a dynamic system, the state of a system is described in terms of a set of *state variables* $[x_1(t), x_2(t), \dots, x_n(t)]$. The state variables are those variables that determine the future behavior of a system when the present state of the system and the excitation signals are known. Consider the system shown in Fig. 3.1, where $c_1(t)$ and $c_2(t)$ are the output signals and $u_1(t)$ and $u_2(t)$ are the input signals. A set of state variables $(x_1, x_2, \dots, x_n)$ for the system shown in the figure is a set such that knowledge of the initial values of the state variables $[x_1(t_0), x_2(t_0), \dots, x_n(t_0)]$ at the initial time t_0, and of the input signals $u_1(t)$ and $u_2(t)$ for $t \geq t_0$, suffices to determine the future values of the outputs and state variables [2].

The state variables describe the future response of a system, given the present state, the excitation inputs, and the equations describing the dynamics.

The general form of a dynamic system is shown in Fig. 3.2.

A simple example of a state variable is the state of an on–off light switch. The switch

---

FIGURE 3.1
System block diagram.

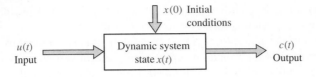

FIGURE 3.2
Dynamic system.

can be in either the on or the off position, and thus the state of the switch can assume one of two possible values. Thus, if we know the present state (position) of the switch at t_0 and if an input is applied, we are able to determine the future value of the state of the element.

The concept of a set of state variables that represent a dynamic system can be illustrated in terms of the spring-mass-damper system shown in Fig. 3.3. The number of state variables chosen to represent this system should be as few as possible in order to avoid redundant state variables. A set of state variables sufficient to describe this system is the position and the velocity of the mass. Therefore we will define a set of state variables as (x_1, x_2), where

$$x_1(t) = y(t) \qquad \text{and} \qquad x_2(t) = \frac{dy(t)}{dt}.$$

The differential equation describes the behavior of the system and is usually written as

$$M\frac{d^2y}{dt^2} + f\frac{dy}{dt} + Ky = u(t). \tag{3.1}$$

To write Eq. (3.1) in terms of the state variables, we substitute the state variables as already defined and obtain

$$M\frac{dx_2}{dt} + fx_2 + Kx_1 = u(t). \tag{3.2}$$

Therefore we can write the differential equations that describe the behavior of the spring-mass-damper system as a set of two first-order differential equations as follows:

$$\frac{dx_1}{dt} = x_2, \tag{3.3}$$

$$\frac{dx_2}{dt} = \frac{-f}{M}x_2 - \frac{K}{M}x_1 + \frac{1}{M}u. \tag{3.4}$$

FIGURE 3.3
A spring-mass-
damper system.

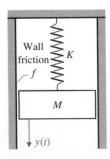

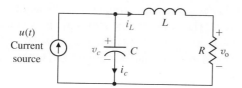

FIGURE 3.4
An *RLC* circuit.

This set of differential equations describes the behavior of the state of the system in terms of the rate of change of each state variable.

As another example of the state variable characterization of a system, let us consider the *RLC* circuit shown in Fig. 3.4. The state of this system can be described in terms of a set of state variables (x_1, x_2), where x_1 is the capacitor voltage $v_c(t)$ and x_2 is equal to the inductor current $i_L(t)$. This choice of state variables is intuitively satisfactory because the stored energy of the network can be described in terms of these variables as

$$\mathcal{E} = (1/2)Li_L^2 + (1/2)Cv_c^2. \tag{3.5}$$

Therefore $x_1(t_0)$ and $x_2(t_0)$ represent the total initial energy of the network and thus the state of the system at $t = t_0$. For a passive *RLC* network, the number of state variables required is equal to the number of independent energy-storage elements. Utilizing Kirchhoff's current law at the junction, we obtain a first-order differential equation by describing the rate of change of capacitor voltage as

$$i_c = C\frac{dv_c}{dt} = +u(t) - i_L. \tag{3.6}$$

Kirchhoff's voltage law for the right-hand loop provides the equation describing the rate of change of inductor current as

$$L\frac{di_L}{dt} = -Ri_L + v_c. \tag{3.7}$$

The output of this system is represented by the linear algebraic equation

$$v_o = Ri_L(t).$$

We can rewrite Eqs. (3.6) and (3.7) as a set of two first-order differential equations in terms of the state variables x_1 and x_2 as follows:

$$\frac{dx_1}{dt} = -\frac{1}{C}x_2 + \frac{1}{C}u(t), \tag{3.8}$$

$$\frac{dx_2}{dt} = +\frac{1}{L}x_1 - \frac{R}{L}x_2. \tag{3.9}$$

The output signal is then

$$c_1(t) = v_o(t) = Rx_2. \tag{3.10}$$

Utilizing Eqs. (3.8) and (3.9) and the initial conditions of the network represented by $[x_1(t_0), x_2(t_0)]$, we can determine the system's future behavior and its output.

The state variables that describe a system are not a unique set, and several alternative sets of state variables can be chosen. For example, for a second-order system, such as the mass-spring-damper or *RLC* circuit, the state variables may be any two independent linear combinations of $x_1(t)$ and $x_2(t)$. Therefore, for the *RLC* circuit, we might choose the set of state variables as the two voltages, $v_c(t)$ and $v_L(t)$, where v_L is the voltage drop across the inductor. Then the new state variables, x_1^* and x_2^*, are related to the old state variables, x_1 and x_2, as

$$x_1^* = v_c = x_1, \tag{3.11}$$

$$x_2^* = v_L = v_c - Ri_L = x_1 - Rx_2. \tag{3.12}$$

Equation (3.12) represents the relation between the inductor voltage and the former state variables v_c and i_L. Thus, in an actual system, there are several choices of a set of state variables that specify the energy stored in a system and therefore adequately describe the dynamics of the system. A widely used choice is a set of state variables that can be readily measured.

An alternative approach to developing a model of a device is the use of the bond graph. Bond graphs can be used for electrical, mechanical, hydraulic, and thermal devices or systems as well as for combinations of various types of elements. Bond graphs produce a set of equations in the state variable form [7].

The state variables of a system characterize the dynamic behavior of a system. The engineer's interest is primarily in physical systems, where the variables are voltages, currents, velocities, positions, pressures, temperatures, and similar physical variables. However, the concept of system state is not limited to the analysis of physical systems and is particularly useful to analyzing biological, social, and economic systems as well as physical systems. For these systems, the concept of state is extended beyond the concept of energy of a physical system to the broader viewpoint of variables that describe the future behavior of the system.

3.3 THE STATE DIFFERENTIAL EQUATION

The state of a system is described by the set of first-order differential equations written in terms of the state variables $(x_1, x_2, \ldots, x_n)$. These first-order differential equations can be written in general form as

$$
\begin{aligned}
\dot{x}_1 &= a_{11}x_1 + a_{12}x_2 + \cdots + a_{1n}x_n + b_{11}u_1 + \cdots + b_{1m}u_m, \\
\dot{x}_2 &= a_{21}x_1 + a_{22}x_2 + \cdots + a_{2n}x_n + b_{21}u_1 + \cdots + b_{2m}u_m, \\
&\vdots \\
\dot{x}_n &= a_{n1}x_1 + a_{n2}x_2 + \cdots + a_{nn}x_n + b_{n1}u_1 + \cdots + b_{nm}u_m,
\end{aligned} \tag{3.13}
$$

where $\dot{x} = dx/dt$. Thus this set of simultaneous differential equations can be written in matrix form as follows [5, 2]:

$$
\frac{d}{dt}
\begin{bmatrix} x_1 \\ x_2 \\ \vdots \\ x_n \end{bmatrix}
=
\begin{bmatrix}
a_{11} & a_{12} & \cdots & a_{1n} \\
a_{21} & a_{22} & \cdots & a_{2n} \\
\vdots & & \cdots & \vdots \\
a_{n1} & a_{n2} & \cdots & a_{nn}
\end{bmatrix}
\begin{bmatrix} x_1 \\ x_2 \\ \vdots \\ x_n \end{bmatrix}
+
\begin{bmatrix}
b_{11} & \cdots & b_{1m} \\
\vdots & & \vdots \\
b_{n1} & \cdots & b_{nm}
\end{bmatrix}
\begin{bmatrix} u_1 \\ \vdots \\ u_m \end{bmatrix}. \tag{3.14}
$$

The column matrix consisting of the state variables is called the *state vector* and is written as

$$\mathbf{x} = \begin{bmatrix} x_1 \\ x_2 \\ \vdots \\ x_n \end{bmatrix}, \tag{3.15}$$

where the boldface indicates a matrix. The matrix of input signals is defined as **u.** Then the system can be represented by the compact notation of the *state differential equation* as

$$\dot{\mathbf{x}} = \mathbf{Ax} + \mathbf{Bu}. \tag{3.16}$$

The differential equation (3.16) is also commonly called the state equation.

The matrix **A** is an $n \times n$ square matrix and **B** is an $n \times m$ matrix.[†] The state differential equation relates the rate of change of the state of the system to the state of the system and the input signals. In general, the outputs of a linear system can be related to the state variables and the input signals by the state equation

$$\mathbf{c} = \mathbf{Dx} + \mathbf{Hu}, \tag{3.17}$$

where **c** is the set of output signals expressed in column vector form.

We use Eqs. (3.8) and (3.9) to obtain the state variable differential equation for the *RLC* of Fig. 3.4 as

$$\dot{\mathbf{x}} = \begin{bmatrix} 0 & \dfrac{-1}{C} \\ \dfrac{1}{L} & \dfrac{-R}{L} \end{bmatrix} \mathbf{x} + \begin{bmatrix} \dfrac{1}{C} \\ 0 \end{bmatrix} u(t) \tag{3.18}$$

and the output as

$$c = \begin{bmatrix} 0 & R \end{bmatrix} \mathbf{x}. \tag{3.19}$$

When $R = 3$, $L = 1$, and $C = 1/2$, we have

$$\dot{\mathbf{x}} = \begin{bmatrix} 0 & -2 \\ 1 & -3 \end{bmatrix} \mathbf{x} + \begin{bmatrix} 2 \\ 0 \end{bmatrix} u$$

and

$$c = \begin{bmatrix} 0 & 3 \end{bmatrix} \mathbf{x}.$$

The solution of the state differential equation (Eq. 3.16) can be obtained in a manner similar to the approach we utilize for solving a first-order differential equation. Consider the first-order differential equation

$$\dot{x} = ax + bu, \tag{3.20}$$

[†]Boldfaced lowercase letters denote vector quantities and boldfaced uppercase letters denote matrices. For an introduction to matrices and elementary matrix operations, refer to Appendix C and references [1] and [2].

where $x(t)$ and $u(t)$ are scalar functions of time. We expect an exponential solution of the form e^{at}. Taking the Laplace transform of Eq. (3.20), we have

$$sX(s) - x(0) = aX(s) + bU(s),$$

and therefore

$$X(s) = \frac{x(0)}{s - a} + \frac{b}{s - a} U(s). \qquad (3.21)$$

The inverse Laplace transform of Eq. (3.21) results in the solution

$$x(t) = e^{at}x(0) + \int_0^t e^{+a(t-\tau)}bu(\tau) \, d\tau. \qquad (3.22)$$

We expect the solution of the state differential equation to be similar to Eq. (3.22) and to be of differential form. The matrix exponential function is defined as

$$e^{\mathbf{A}t} = \exp(\mathbf{A}t) = \mathbf{I} + \mathbf{A}t + \frac{\mathbf{A}^2 t^2}{2!} + \cdots + \frac{\mathbf{A}^k t^k}{k!} + \cdots, \qquad (3.23)$$

which converges for all finite t and any $\mathbf{A}$ [2]. Then the solution of the state differential equation is found to be [2]

$$\mathbf{x}(t) = \exp(\mathbf{A}t)\mathbf{x}(0) + \int_0^t \exp[\mathbf{A}(t - \tau)]\mathbf{B}\mathbf{u}(\tau) \, d\tau. \qquad (3.24)$$

Equation (3.24) may be obtained by taking the Laplace transform of Eq. (3.16) and re-arranging to obtain

$$\mathbf{X}(s) = [s\mathbf{I} - \mathbf{A}]^{-1}\mathbf{x}(0) + [s\mathbf{I} - \mathbf{A}]^{-1}\mathbf{B}\mathbf{U}(s), \qquad (3.25)$$

where we note that $[s\mathbf{I} - \mathbf{A}]^{-1} = \mathbf{\Phi}(s)$, which is the Laplace transform of $\mathbf{\Phi}(t) = \exp(\mathbf{A}t)$. Taking the inverse Laplace transform of Eq. (3.25) and noting that the second term on the right-hand side involves the product $\mathbf{\Phi}(s)\mathbf{B}\mathbf{U}(s)$, we obtain Eq. (3.24). The matrix exponential function describes the unforced response of the system and is called the *fundamental* or *state transition matrix* $\mathbf{\Phi}(t)$. Therefore Eq. (3.24) can be written as

$$\mathbf{x}(t) = \mathbf{\Phi}(t)\mathbf{x}(0) + \int_0^t \mathbf{\Phi}(t - \tau)\mathbf{B}\mathbf{u}(\tau) \, d\tau. \qquad (3.26)$$

The solution to the unforced system (that is, when $\mathbf{u} = 0$) is simply

$$\begin{bmatrix} x_1(t) \\ x_2(t) \\ \vdots \\ x_n(t) \end{bmatrix} = \begin{bmatrix} \phi_{11}(t) & \cdots & \phi_{1n}(t) \\ \phi_{21}(t) & \cdots & \phi_{2n}(t) \\ \vdots & & \vdots \\ \phi_{n1}(t) & \cdots & \phi_{nn}(t) \end{bmatrix} \begin{bmatrix} x_1(0) \\ x_2(0) \\ \vdots \\ x_n(0) \end{bmatrix}. \qquad (3.27)$$

We note therefore that to determine the state transition matrix, all initial conditions are set to 0 except for one state variable, and the output of each state variable is evaluated. That is, the term $\phi_{ij}(t)$ is the response of the ith state variable due to an initial condition on the jth state variable when there are zero initial conditions on all the other states. We shall utilize this relationship between the initial conditions and the state variables to evaluate the coefficients of the transition matrix in a later section. However, first we shall develop several suitable signal-flow state models of systems and investigate the stability of the systems by utilizing these flow graphs.

3.4 SIGNAL-FLOW GRAPH STATE MODELS

The state of a system describes that system's dynamic behavior where the dynamics of the system are represented by a series of first-order differential equations. Alternatively, the dynamics of the system can be represented by a state differential equation as in Eq. (3.16). In either case, it is useful to develop a state flow graph model of the system and use this model to relate the state variable concept to the familiar transfer function representation.

As we have learned in previous chapters, a system can be meaningfully described by an input–output relationship, the transfer function $G(s)$. For example, if we are interested in the relation between the output voltage and the input voltage of the network of Fig. 3.4, we can obtain the transfer function

$$G(s) = \frac{V_0(s)}{U(s)}.$$

The transfer function for the RLC network of Fig. 3.4 is of the form

$$G(s) = \frac{V_0(s)}{U(s)} - \frac{\alpha}{s^2 + \beta s + \gamma}, \tag{3.28}$$

where α, β, and γ are functions of the circuit parameters R, L, and C, respectively. The values of α, β, and γ can be determined from the flow graph representing the differential equations that describe the circuit. For the RLC circuit (see Eqs. 3.8 and 3.9), we have

$$\dot{x}_1 = -\frac{1}{C}x_2 + \frac{1}{C}u(t), \tag{3.29}$$

$$\dot{x}_2 = \frac{1}{L}x_1 - \frac{R}{L}x_2, \tag{3.30}$$

$$v_0 = Rx_2. \tag{3.31}$$

The flow graph representing these simultaneous equations is shown in Fig. 3.5, where $1/s$ indicates an integration. Using Mason's signal-flow gain formula, we obtain the transfer function

$$\frac{V_0(s)}{U(s)} = \frac{+R/LCs^2}{1 + (R/Ls) + (1/LCs^2)} = \frac{+R/LC}{s^2 + (R/L)s + (1/LC)}. \tag{3.32}$$

Unfortunately, many electric circuits, electromechanical systems, and other control systems are not as simple as the RLC circuit of Fig. 3.4, and it is often a difficult task to determine a series of first-order differential equations describing the system. Therefore it is often

FIGURE 3.5
Flow graph for the
RLC network.

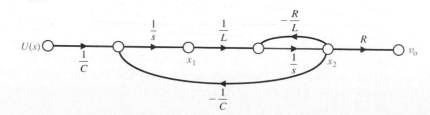

simpler to derive the transfer function of the system by the techniques of Chapter 2 and then derive the state model from the transfer function.

The signal-flow graph state model can be readily derived from the transfer function of a system. However, as we noted in Section 3.3, there is more than one alternative set of state variables, and therefore there is more than one possible form for the signal-flow graph state model. In general, we can represent a transfer function as

$$G(s) = \frac{C(s)}{U(s)} = \frac{s^m + b_{m-1}s^{m-1} + \cdots + b_1 s + b_0}{s^n + a_{n-1}s^{n-1} + \cdots + a_1 s + a_0}, \tag{3.33}$$

where $n \geq m$ and all the a coefficients are real positive numbers. If we multiply the numerator and denominator by s^{-n}, we obtain

$$G(s) = \frac{s^{-(n-m)} + b_{m-1}s^{-(n-m+1)} + \cdots + b_1 s^{-(n-1)} + b_0 s^{-n}}{1 + a_{n-1}s^{-1} + \cdots + a_1 s^{-(n-1)} + a_0 s^{-n}}. \tag{3.34}$$

Our familiarity with Mason's flow graph gain formula causes us to recognize the familiar feedback factors in the denominator and the forward-path factors in the numerator. Mason's flow graph formula was discussed in Section 2.7 and is written as

$$G(s) = \frac{C(s)}{U(s)} = \frac{\sum_k P_k \Delta_k}{\Delta}. \tag{3.35}$$

When all the feedback loops are touching and all the forward paths touch the feedback loops, Eq. (3.35) reduces to

$$G(s) = \frac{\sum_k P_k}{1 - \sum_{q=1}^{N} L_q} = \frac{\text{Sum of the forward-path factors}}{1 - \text{sum of the feedback loop factors}}. \tag{3.36}$$

There are several flow graphs that could represent the transfer function. Two flow graph configurations based on Mason's gain formula are of particular interest, and we will consider these in greater detail. In the next section, we will consider two additional flow graph configurations.

To illustrate the derivation of the signal-flow graph state model, let us initially consider the fourth-order transfer function

$$G(s) = \frac{C(s)}{U(s)} = \frac{b_0}{s^4 + a_3 s^3 + a_2 s^2 + a_1 s + a_0} \tag{3.37}$$

$$= \frac{b_0 s^{-4}}{1 + a_3 s^{-1} + a_2 s^{-2} + a_1 s^{-3} + a_0 s^{-4}}.$$

First we note that the system is fourth order, and hence we identify four state variables (x_1, x_2, x_3, x_4). Recalling Mason's gain formula, we note that the denominator can be considered to be one minus the sum of the loop gains. Furthermore, the numerator of the transfer function is equal to the forward-path factor of the flow graph. The flow graph must utilize a minimum number of integrators equal to the order of the system. Therefore we use four integrators to represent this system. The necessary flow graph nodes and the four integrators are shown in Fig. 3.6. Considering the simplest series interconnection of integrators, we can represent the transfer function by the flow graph of Fig. 3.7. Examining this figure, we note that all the loops are touching and that the transfer function of this flow graph is indeed Eq. (3.37). This can be readily verified by the reader by noting that the

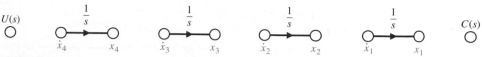

FIGURE 3.6
Flow graph nodes and integrators for fourth-order system.

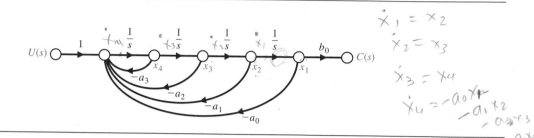

FIGURE 3.7
Flow graph state
model for $G(s)$
of Eq. (3.37).

forward-path factor of the flow graph is b_0/s^4 and the denominator is equal to 1 minus the sum of the loop gains.

Now, consider the fourth-order transfer function when the numerator is a polynomial in s so that we have

$$G(s) = \frac{b_3 s^3 + b_2 s^2 + b_1 s + b_0}{s^4 + a_3 s^3 + a_2 s^2 + a_1 s + a_0}$$

$$= \frac{b_3 s^{-1} + b_2 s^{-2} + b_1 s^{-3} + b_0 s^{-4}}{1 + a_3 s^{-1} + a_2 s^{-2} + a_1 s^{-3} + a_0 s^{-4}}.$$

(3.38)

The numerator terms represent forward-path factors in Mason's gain formula. The forward paths will touch all the loops, and a suitable signal-flow graph realization of Eq. (3.38) is shown in Fig. 3.8. The forward-path factors are b_3/s, b_2/s^2, b_1/s^3, and b_0/s^4 as required to provide the numerator of the transfer function. Recall that Mason's flow graph gain formula indicates that the numerator of the transfer function is simply the sum of the forward-path factors. This general form of a signal-flow graph can represent the general transfer function of Eq. (3.38) by utilizing n feedback loops involving the a_n coefficients and m forward-path factors involving the b_m coefficients. The general form of the flow graph state model shown in Fig. 3.8 is called the *phase variable format*.

The state variables are identified in Fig. 3.8 as the output of each energy storage element, that is, the output of each integrator. To obtain the set of first-order differential equations representing the state model of Fig. 3.8, we will introduce a new set of flow graph nodes immediately preceding each integrator of Fig. 3.8 [5, 6]. The nodes are placed before each integrator, and therefore they represent the derivative of the output of each integrator. The signal-flow graph, including the added nodes, is shown in Fig. 3.9. Using the flow graph of this figure, we are able to obtain the following set of first-order differential equations describing the state of the model:

$$\dot{x}_1 = x_2, \qquad \dot{x}_2 = x_3, \qquad \dot{x}_3 = x_4,$$

$$\dot{x}_4 = -a_0 x_1 - a_1 x_2 - a_2 x_3 - a_3 x_4 + u,$$

(3.39)

where $x_1, x_2, \cdots x_n$ are the n phase variables.

10/21/97

→ generalization of fig 3.7

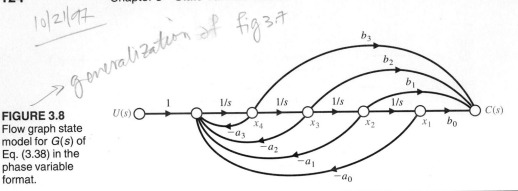

FIGURE 3.8
Flow graph state model for $G(s)$ of Eq. (3.38) in the phase variable format.

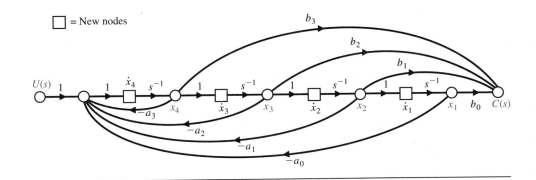

FIGURE 3.9
Flow graph of Fig. 3.8 with nodes inserted.

Furthermore, the output is simply

$$c(t) = b_0 x_1 + b_1 x_2 + b_2 x_3 + b_3 x_4. \tag{3.40}$$

Then, in matrix form, we have

$$\dot{\mathbf{x}} = \mathbf{A}\mathbf{x} + \mathbf{b}u \tag{3.41}$$

or

$$\frac{d}{dt}\begin{bmatrix} x_1 \\ x_2 \\ x_3 \\ x_4 \end{bmatrix} = \begin{bmatrix} 0 & 1 & 0 & 0 \\ 0 & 0 & 1 & 0 \\ 0 & 0 & 0 & 1 \\ -a_0 & -a_1 & -a_2 & -a_3 \end{bmatrix}\begin{bmatrix} x_1 \\ x_2 \\ x_3 \\ x_4 \end{bmatrix} + \begin{bmatrix} 0 \\ 0 \\ 0 \\ 1 \end{bmatrix}u(t), \tag{3.42}$$

and the output is

y = CX + Du

$$c(t) = \mathbf{D}\mathbf{x} = [b_0, \ b_1, \ b_2, \ b_3]\begin{bmatrix} x_1 \\ x_2 \\ x_3 \\ x_4 \end{bmatrix}. \tag{3.43}$$

The flow graph structure of Fig. 3.8 is not a unique representation of Eq. (3.38); another equally useful structure can be obtained. A flow graph that represents Eq. (3.38)

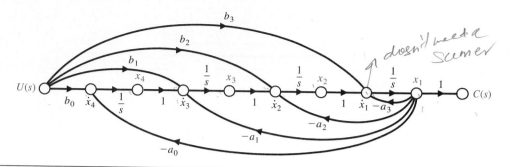

FIGURE 3.10
Alternative flow graph state model for Eq. (3.38). This model is called the input feedforward format.

equally well is shown in Fig. 3.10. In this case, the forward-path factors are obtained by feeding forward the signal $U(s)$. We will call this model the *input feedforward format*.

Then the output signal $c(t)$ is equal to the first state variable $x_1(t)$. This flow graph structure has the forward-path factors b_0/s^4, b_1/s^3, b_2/s^2, b_3/s, and all the forward paths touch the feedback loops. Therefore the resulting transfer function is indeed equal to Eq. (3.38).

Using the flow graph of Fig. 3.10 to obtain the set of first-order differential equations, we obtain

$$\dot{x}_1 = -a_3 x_1 + x_2 + b_3 u, \qquad \dot{x}_2 = -a_2 x_1 + x_3 + b_2 u,$$
$$\dot{x}_3 = -a_1 x_1 + x_4 + b_1 u, \qquad \dot{x}_4 = -a_0 x_1 + b_0 u. \qquad (3.44)$$

Thus in matrix form we have

$$\frac{d\mathbf{x}}{dt} = \begin{bmatrix} -a_3 & 1 & 0 & 0 \\ -a_2 & 0 & 1 & 0 \\ -a_1 & 0 & 0 & 1 \\ -a_0 & 0 & 0 & 0 \end{bmatrix} \mathbf{x} + \begin{bmatrix} b_3 \\ b_2 \\ b_1 \\ b_0 \end{bmatrix} u(t). \qquad (3.45)$$

Although the feedforward flow graph of Fig. 3.10 represents the same transfer function as the phase variable flow graph of Fig. 3.8, the state variables of each graph are not equal because the structure of each flow graph is different. The signal-flow graphs can be recognized as being equivalent to an analog computer diagram. Furthermore, we recognize that the initial conditions of the system can be represented by the initial conditions of the integrators, $x_1(0)$, $x_2(0)$, . . . , $x_n(0)$. Let us consider a control system and determine the state differential equation by utilizing the two forms of flow graph state models.

EXAMPLE 3.1   Two state variable models

A single-loop control system is shown in Fig. 3.11. The closed-loop transfer function of the system is

$$T(s) = \frac{C(s)}{R(s)} = \frac{2s^2 + 8s + 6}{s^3 + 8s^2 + 16s + 6}.$$

FIGURE 3.11
Single-loop control
system.

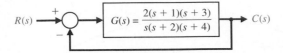

$$R(s) \xrightarrow{\;+\;} \bigcirc \xrightarrow{} \boxed{G(s) = \dfrac{2(s+1)(s+3)}{s(s+2)(s+4)}} \longrightarrow C(s)$$

Multiplying the numerator and denominator by s^{-3}, we have

$$T(s) = \frac{C(s)}{R(s)} = \frac{2s^{-1} + 8s^{-2} + 6s^{-3}}{1 + 8s^{-1} + 16s^{-2} + 6s^{-3}}. \qquad (3.46)$$

The first model is the signal-flow graph phase variable state model using the feedforward of the state variables to provide the output signal, as shown in Fig. 3.12. The state differential equation for this flow graph is

$$\dot{\mathbf{x}} = \begin{bmatrix} 0 & 1 & 0 \\ 0 & 0 & 1 \\ -6 & -16 & -8 \end{bmatrix} \mathbf{x} + \begin{bmatrix} 0 \\ 0 \\ 1 \end{bmatrix} u(t), \qquad (3.47)$$

and the output is

$$c(t) = [6,\ 8,\ 2] \begin{bmatrix} x_1 \\ x_2 \\ x_3 \end{bmatrix}. \qquad (3.48)$$

The second model is the flow graph state model using the feedforward of the input variable, as shown in Fig. 3.13. The vector differential equation for this flow graph is

$$\dot{\mathbf{x}} = \begin{bmatrix} -8 & 1 & 0 \\ -16 & 0 & 1 \\ -6 & 0 & 0 \end{bmatrix} \mathbf{x} + \begin{bmatrix} 2 \\ 8 \\ 6 \end{bmatrix} u(t), \qquad (3.49)$$

and the output is $c(t) = x_1(t)$. ∎

We note that both signal-flow graph representations of the transfer function $T(s)$ are readily obtained. Furthermore, it was not necessary to factor the numerator or denominator polynomial to obtain the state differential equations. Avoiding the factoring of polynomials permits us to avoid the tedious effort involved. Each of the two signal-flow graph state models represents an analog computer simulation of the transfer function. Both models require three integrators because the system is third order. However, it is important to emphasize that the state variables of the state model of Fig. 3.12 are not identical to the state variables of the state model of Fig. 3.13. Of course, one set of state variables is related to

FIGURE 3.12
Phase variable
flow graph state
model for $T(s)$.

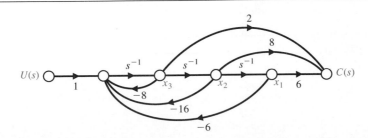

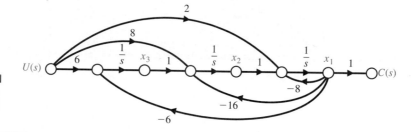

FIGURE 3.13
Alternative flow graph state model for $T(s)$ using the input feedforward format.

the other set of state variables by an appropriate linear transformation of variables. A linear matrix transformation is represented by $\mathbf{y} = \mathbf{Mx}$, which transforms the $\mathbf{x}$-vector into the $\mathbf{y}$-vector by means of the $\mathbf{M}$-matrix (see Appendix C, especially Section C.3, for an introduction to matrix algebra). Finally, we note that the transfer function of Eq. (3.33) represents a single output linear constant coefficient system, and thus the transfer function can represent an nth-order differential equation

$$\frac{d^n c}{dt^n} + a_{n-1}\frac{d^{n-1}c}{dt^{n-1}} + \cdots + a_0 c(t) = \frac{d^m u}{dt^m} + b_{m-1}\frac{d^{m-1}u}{dt^{m-1}} + \cdots + b_0 u(t). \quad (3.50)$$

Thus we can obtain the n first-order equations for the nth-order differential equation by utilizing the signal-flow graph state models of this section.

3.5 ALTERNATIVE SIGNAL-FLOW GRAPH STATE MODELS

Often the control system designer studies an actual control system block diagram that represents physical devices and variables. An example of a model of a dc motor with shaft velocity as the output is shown in Fig. 3.14 [9]. We wish to select the *physical variables* as the state variables. Thus we select $x_1 = c(t)$, the velocity output; $x_2 = i(t)$, the field current; and $x_3 = u(t)$, the field voltage. We may draw a signal-flow graph model for these physical variables, as shown in Fig. 3.15. Note that the state variables x_1, x_2, and x_3 are identified on the flow graph. We will denote this format as the physical state variable flow graph model. This model is particularly useful when we can measure the physical state variables. Note that the signal-flow graph model of each block is separately determined. For example, note that the transfer function for the controller is

$$\frac{U(s)}{R(s)} = G_c(s) = \frac{5(s + 1)}{(s + 5)} = \frac{5 + 5s^{-1}}{1 + 5s^{-1}}$$

and the flow graph between $R(s)$ and $U(s)$ represents $G_c(s)$.

FIGURE 3.14
A block diagram model of an open-loop dc motor control with velocity as the output.

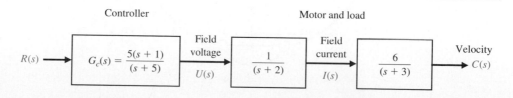

FIGURE 3.15
The physical state
variable signal-flow
graph for the block
diagram of
Fig. 3.14.

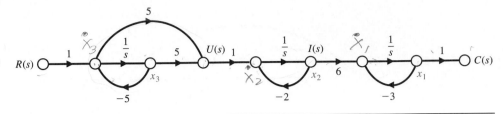

The state variable differential equation is directly obtained from Fig. 3.15 as

$$\dot{\mathbf{x}} = \begin{bmatrix} -3 & 6 & 0 \\ 0 & -2 & -20 \\ 0 & 0 & -5 \end{bmatrix} \mathbf{x} + \begin{bmatrix} 0 \\ 5 \\ 1 \end{bmatrix} r(t) \tag{3.51}$$

and

$$c = [1 \quad 0 \quad 0] \mathbf{x}. \tag{3.52}$$

A second form of flow graph yields the decoupled response modes. The overall input–output transfer function of the block diagram system shown in Fig. 3.14 is

$$\frac{C(s)}{R(s)} = T(s) = \frac{30(s + 1)}{(s + 5)(s + 2)(s + 3)} = \frac{q(s)}{(s - s_1)(s - s_2)(s - s_3)},$$

and the transient response has three modes dictated by s_1, s_2, and s_3. These modes are indicated by the partial fraction expansion as

$$\frac{C(s)}{R(s)} = T(s) = \frac{k_1}{(s + 5)} + \frac{k_2}{(s + 2)} + \frac{k_3}{(s + 3)}. \tag{3.53}$$

Using the procedure described in Chapter 2, we find that $k_1 = -20$, $k_2 = -10$, and $k_3 = 30$. The decoupled state variable flow graph representing Eq. (3.53) is shown in Fig. 3.16. The state variable matrix differential equation is

$$\dot{\mathbf{x}} = \begin{bmatrix} -5 & 0 & 0 \\ 0 & -2 & 0 \\ 0 & 0 & -3 \end{bmatrix} \mathbf{x} + \begin{bmatrix} 1 \\ 1 \\ 1 \end{bmatrix} r(t)$$

and

$$c(t) = [-20 \quad -10 \quad 30] \mathbf{x}. \tag{3.54}$$

FIGURE 3.16
The decoupled
state variable flow
graph model for
the system shown
in block diagram
form in Fig. 3.14.

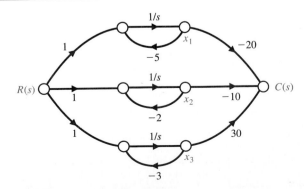

Note that we chose x_1 as the state variable associated with $s_1 = -5$, x_2 associated with $s_2 = -2$, and x_3 associated with $s_3 = -3$, as indicated in Fig. 3.16. This choice of state variables is arbitrary; for example, x_1 could be chosen as associated with the factor $(s + 2)$.

The decoupled form of the state differential matrix equation displays the distinct model poles $-s_1, -s_2, \ldots, -s_n$, and this format is often called the *diagonal form* or the *canonical form*. A system can always be written in diagonal form if it possesses distinct poles; otherwise, it can only be written in a block diagonal form, known as the *Jordan canonical form* [29].

EXAMPLE 3.2   **Spread of an epidemic disease**

The spread of an epidemic disease can be described by a set of differential equations. The population under study is made up of three groups, x_1, x_2, and x_3, such that the group x_1 is susceptible to the epidemic disease, group x_2 is infected with the disease, and group x_3 has been removed from the initial population. The removal of x_3 will be due to immunization, death, or isolation from x_1. The feedback system can be represented by the following equations:

$$\frac{dx_1}{dt} = -\alpha x_1 - \beta x_2 + u_1(t),$$

$$\frac{dx_2}{dt} = \beta x_1 - \gamma x_2 + u_2(t),$$

$$\frac{dx_3}{dt} = \alpha x_1 + \gamma x_2.$$

The rate at which new susceptibles are added to the population is equal to $u_1(t)$ and the rate at which new infectives are added to the population is equal to $u_2(t)$. For a closed population, we have $u_1(t) = u_2(t) = 0$. It is interesting to note that these equations could equally well represent the spread of information of a new idea through a populace.

The physical state variables for this system are x_1, x_2, and x_3. The signal-flow diagram that represents this set of differential equations is shown in Fig. 3.17. The vector differential equation is equal to

$$\frac{d}{dt}\begin{bmatrix} x_1 \\ x_2 \\ x_3 \end{bmatrix} = \begin{bmatrix} -\alpha & -\beta & 0 \\ \beta & -\gamma & 0 \\ \alpha & \gamma & 0 \end{bmatrix}\begin{bmatrix} x_1 \\ x_2 \\ x_3 \end{bmatrix} + \begin{bmatrix} 1 & 0 \\ 0 & 1 \\ 0 & 0 \end{bmatrix}\begin{bmatrix} u_1(t) \\ u_2(t) \end{bmatrix}. \qquad (3.55)$$

FIGURE 3.17
State model flow graph for the spread of an epidemic disease.

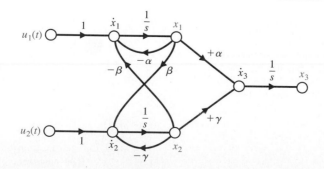

By examining Eq. (3.55) and the signal-flow graph, we find that the state variable x_3 is dependent on x_1 and x_2 and does not affect the variables x_1 and x_2.

Let us consider a closed population so that $u_1(t) = u_2(t) = 0$. The equilibrium point in the state space for this system is obtained by setting $d\mathbf{x}/dt = \mathbf{0}$. The equilibrium point in the state space is the point to which the system settles in the equilibrium, or rest, condition. Examining Eq. (3.55), we find that the equilibrium point for this system is $x_1 = x_2 = 0$. Thus, to determine whether the epidemic disease is eliminated from the population, we must obtain the characteristic equation of the system. From the signal-flow graph shown in Fig. 3.17, we obtain the flow graph determinant

$$\Delta(s) = 1 - (-\alpha s^{-1} - \gamma s^{-1} - \beta^2 s^{-2}) + (\alpha \gamma s^{-2}), \tag{3.56}$$

where there are three loops, two of which are nontouching. Thus the characteristic equation is

$$q(s) = s^2 \Delta(s) = s^2 + (\alpha + \gamma)s + (\alpha\gamma + \beta^2) = 0. \tag{3.57}$$

The roots of this characteristic equation will lie in the left-hand s-plane when $(\alpha + \gamma) > 0$ and $(\alpha\gamma + \beta^2) > 0$. When roots are in the left-hand plane, we expect the unforced response to decay to zero as $t \to \infty$. ∎

EXAMPLE 3.3   Inverted pendulum control

The problem of balancing a broomstick on a person's hand is illustrated in Fig. 3.18. The only equilibrium condition is $\theta(t) = 0$ and $d\theta/dt = 0$. The problem of balancing a broomstick on one's hand is not unlike the problem of controlling the attitude of a missile during the initial stages of launch. This problem is the classic and intriguing problem of the inverted pendulum mounted on a cart, as shown in Fig. 3.19. The cart must be moved so that mass m is always in an upright position. The state variables must be expressed in terms of the angular rotation $\theta(t)$ and the position of the cart $y(t)$. The differential equations describing the motion of the system can be obtained by writing the sum of the forces in the horizontal direction and the sum of the moments about the pivot point [2, 3, 10, 28]. We will assume that $M \gg m$ and the angle of rotation θ is small so that the equations are linear. The sum of the forces in the horizontal direction is

$$M\ddot{y} + ml\ddot{\theta} - u(t) = 0, \tag{3.58}$$

where $u(t)$ equals the force on the cart and l is the distance from the mass m to the pivot point. The sum of the torques about the pivot point is

$$ml\ddot{y} + ml^2\ddot{\theta} - mlg\theta = 0. \tag{3.59}$$

FIGURE 3.18
An inverted pendulum balanced on a person's hand by moving the hand to reduce $\theta(t)$. Assume, for ease, that the pendulum rotates in the x-y plane.

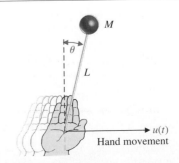

Hand movement

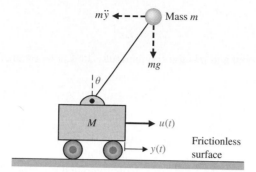

FIGURE 3.19
A cart and
an inverted
pendulum. The
pendulum is
constrained
to pivot in the
vertical plane.

The state variables for the two second-order equations are chosen as $(x_1, x_2, x_3, x_4) = (y, \dot{y}, \theta, \dot{\theta})$. Then Eqs. (3.58) and (3.59) are written in terms of the state variables as

$$M\dot{x}_2 + ml\dot{x}_4 - u(t) = 0 \tag{3.60}$$

and

$$\dot{x}_2 + l\dot{x}_4 - gx_3 = 0. \tag{3.61}$$

To obtain the necessary first-order differential equations, we solve for $l\dot{x}_4$ in Eq. (3.61) and substitute into Eq. (3.60) to obtain

$$M\dot{x}_2 + mgx_3 = u(t), \tag{3.62}$$

since $M \gg m$. Substituting $\dot{x}_2$ from Eq. (3.60) into Eq. (3.61), we have

$$Ml\dot{x}_4 - Mgx_3 + u(t) = 0. \tag{3.63}$$

Therefore the four first-order differential equations can be written as

$$\dot{x}_1 = x_2, \qquad \dot{x}_2 = \frac{-mg}{M} x_3 + \frac{1}{M} u(t),$$

$$\dot{x}_3 = x_4, \qquad \dot{x}_4 = \frac{g}{l} x_3 - \frac{1}{Ml} u(t). \tag{3.64}$$

Thus the system matrices are

$$\mathbf{A} = \begin{bmatrix} 0 & 1 & 0 & 0 \\ 0 & 0 & -(mg/M) & 0 \\ 0 & 0 & 0 & 1 \\ 0 & 0 & g/l & 0 \end{bmatrix}, \qquad \mathbf{b} = \begin{bmatrix} 0 \\ 1/M \\ 0 \\ -1/Ml \end{bmatrix} \blacksquare \tag{3.65}$$

3.6   THE TRANSFER FUNCTION FROM THE STATE EQUATION

Given a transfer function $G(s)$, we can obtain the state variable equations using the signal-flow graph model. Now we turn to the matter of determining the transfer function $G(s)$ of a single-input, single-output (SISO) system. Recalling Eqs. (3.16) and (3.17), we have

$$\dot{\mathbf{x}} = \mathbf{A}\mathbf{x} + \mathbf{B}u \tag{3.66}$$

and

$$c = \mathbf{Dx}, \tag{3.67}$$

where c is the single output and u is the single input. The Laplace transforms of Eqs. (3.66) and (3.67) are

$$s\mathbf{X}(s) = \mathbf{AX}(s) + \mathbf{b}U(s) \tag{3.68}$$

and

$$C(s) = \mathbf{DX}(s), \tag{3.69}$$

where $\mathbf{b}$ is a vector, since u is a single input. Note that we do not include initial conditions, since we seek the transfer function. Reordering Eq. (3.68), we obtain

$$(s\mathbf{I} - \mathbf{A})\mathbf{X}(s) = \mathbf{b}U(s).$$

Since $[s\mathbf{I} - \mathbf{A}]^{-1} = \boldsymbol{\Phi}(s)$, we have

$$\mathbf{X}(s) = \boldsymbol{\Phi}(s)\mathbf{b}U(s).$$

Substituting $\mathbf{X}(s)$ into Eq. (3.67), we obtain

$$C(s) = \mathbf{D}\boldsymbol{\Phi}(s)\mathbf{b}U(s). \tag{3.70}$$

Therefore the transfer function $G(s) = C(s)/U(s)$ is

$$G(s) = \mathbf{D}\boldsymbol{\Phi}(s)\mathbf{b}. \tag{3.71}$$

EXAMPLE 3.4   **Transfer function of *RLC* circuit**

Let us determine the transfer function $G(s) = C(s)/U(s)$ for the *RLC* circuit of Fig. 3.4 as described by the differential equations (see Eqs. 3.18 and 3.19):

$$\dot{\mathbf{x}} = \begin{bmatrix} 0 & \dfrac{-1}{C} \\ \dfrac{1}{L} & \dfrac{-R}{L} \end{bmatrix} \mathbf{x} + \begin{bmatrix} \dfrac{1}{C} \\ 0 \end{bmatrix} u$$

and

$$c = [0 \quad R] \, \mathbf{x}.$$

Then we have

$$[s\mathbf{I} - \mathbf{A}] = \begin{bmatrix} s & \dfrac{1}{C} \\ \dfrac{-1}{L} & \left(s + \dfrac{R}{L} \right) \end{bmatrix}.$$

Therefore we obtain

$$\boldsymbol{\Phi}(s) = [s\mathbf{I} - \mathbf{A}]^{-1} = \frac{1}{\Delta(s)} \begin{bmatrix} \left(s + \dfrac{R}{L} \right) & \dfrac{-1}{C} \\ \dfrac{1}{L} & s \end{bmatrix},$$

where

$$\Delta(s) = s^2 + \frac{R}{L}s + \frac{1}{LC}.$$

Then, the transfer function is

$$G(s) = [0 \quad R] \begin{bmatrix} \dfrac{\left(s + \dfrac{R}{L}\right)}{\Delta(s)} & \dfrac{-1}{C\Delta(s)} \\ \dfrac{1}{L\Delta(s)} & \dfrac{s}{\Delta(s)} \end{bmatrix} \begin{bmatrix} \dfrac{1}{C} \\ 0 \end{bmatrix}$$

$$= \frac{R/LC}{\Delta(s)} = \frac{R/LC}{s^2 + \dfrac{R}{L}s + \dfrac{1}{LC}},$$

which agrees with the result (Eq. 3.32) obtained from the flow graph model using Mason's rule. ∎

3.7 THE TIME RESPONSE AND THE STATE TRANSITION MATRIX

It is often desirable to obtain the time response of the state variables of a control system and thus examine the performance of the system. The transient response of a system can be readily obtained by evaluating the solution to the state vector differential equation. We found in Section 3.3 that the solution for the state differential equation (Eq. 3.24) was

$$\mathbf{x}(t) = \mathbf{\Phi}(t)\mathbf{x}(0) + \int_0^t \mathbf{\Phi}(t - \tau)\mathbf{B}\mathbf{u}(\tau) \; d\tau. \tag{3.72}$$

Clearly, if the initial conditions $\mathbf{x}(0)$, the input $\mathbf{u}(\tau)$, and the state transition matrix $\mathbf{\Phi}(t)$ are known, the time response of $\mathbf{x}(t)$ can be numerically evaluated. Thus the problem focuses on the evaluation of $\mathbf{\Phi}(t)$, the state transition matrix that represents the response of the system. Fortunately, the state transition matrix can be readily evaluated by using the signal-flow graph techniques with which we are already familiar.

However, before proceeding to the evaluation of the state transition matrix using signal-flow graphs, we should note that several other methods exist for evaluating the transition matrix, such as the evaluation of the exponential series

$$\mathbf{\Phi}(t) = \exp(\mathbf{A}t) = \sum_{k=0}^{\infty} \frac{\mathbf{A}^k t^k}{k!} \tag{3.73}$$

in a truncated form [2, 8]. Several efficient methods exist for the evaluation of $\mathbf{\Phi}(t)$ by means of a computer algorithm.

A series of computer programs to assist in the calculation of the state variable response of systems is available. Personal computers can be used to obtain the state variable response of a dynamic system [21]. The personal computer affords the user the advantage of interactive use of the computer for design purposes.

In addition, we found in Eq. (3.25) that $\mathbf{\Phi}(s) = [s\mathbf{I} - \mathbf{A}]^{-1}$. Therefore, if $\mathbf{\Phi}(s)$ is obtained by completing the matrix inversion, we can obtain $\mathbf{\Phi}(t)$ by noting that $\mathbf{\Phi}(t) = \mathscr{L}^{-1}\{\mathbf{\Phi}(s)\}$. However, the matrix inversion process is unwieldy for higher-order systems.

The usefulness of the signal-flow graph state model for obtaining the state transition matrix becomes clear upon consideration of the Laplace transformation version of Eq. (3.72) when the input is zero. Taking the Laplace transformation of Eq. (3.72), when $\mathbf{u}(\tau) = 0$, we have

$$\mathbf{X}(s) = \mathbf{\Phi}(s)\mathbf{x}(0). \tag{3.74}$$

Therefore we can evaluate the Laplace transform of the transition matrix from the signal-flow graph by determining the relation between a state variable $X_i(s)$ and the state initial conditions $[x_1(0), x_2(0), \ldots, x_n(0)]$. Then the state transition matrix is simply the inverse transform of $\mathbf{\Phi}(s)$; that is,

$$\mathbf{\Phi}(t) = \mathscr{L}^{-1}\{\mathbf{\Phi}(s)\}. \tag{3.75}$$

The relationship between a state variable $X_i(s)$ and the initial conditions $\mathbf{x}(0)$ is obtained by using Mason's gain formula. Thus, for a second-order system, we would have

$$X_1(s) = \phi_{11}(s)x_1(0) + \phi_{12}(s)x_2(0), \tag{3.76}$$

$$X_2(s) = \phi_{21}(s)x_1(0) + \phi_{22}(s)x_2(0),$$

and the relation between $X_2(s)$ as an output and $x_1(0)$ as an input can be evaluated by Mason's formula. All the elements of the state transition matrix, $\phi_{ij}(s)$, can be obtained by evaluating the individual relationships between $X_i(s)$ and $x_j(0)$ from the state model flow graph. An example will illustrate this approach to determining the transition matrix.

EXAMPLE 3.5   **Evaluation of the state transition matrix**

We will consider the *RLC* network of Fig. 3.4. We seek to evaluate $\mathbf{\Phi}(s)$ by (1) determining the matrix inversion $\mathbf{\Phi}(s) = [s\mathbf{I} - \mathbf{A}]^{-1}$ and (2) using the signal-flow diagram and Mason's gain formula.

First, we will determine $\mathbf{\Phi}(s)$ by evaluating $\mathbf{\Phi}(s) = [s\mathbf{I} - \mathbf{A}]^{-1}$. We note from Eq. (3.18) that

$$\mathbf{A} = \begin{bmatrix} 0 & -2 \\ 1 & -3 \end{bmatrix}.$$

Then,

$$[s\mathbf{I} - \mathbf{A}] = \begin{bmatrix} s & 2 \\ -1 & (s+3) \end{bmatrix}. \tag{3.77}$$

The inverse matrix is

$$\mathbf{\Phi}(s) = [s\mathbf{I} - \mathbf{A}]^{-1} = \frac{1}{\Delta(s)} \begin{bmatrix} (s+3) & -2 \\ 1 & s \end{bmatrix}, \tag{3.78}$$

where $\Delta(s) = s(s+3) + 2 = s^2 + 3s + 2 = (s+1)(s+2)$.

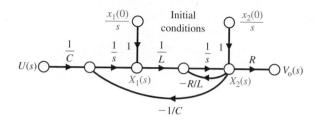

FIGURE 3.20
Flow graph of the
RLC network.

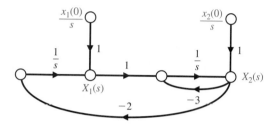

FIGURE 3.21
Flow graph of the
RLC network with
$U(s) = 0$.

The signal-flow graph state model of the *RLC* network of Fig. 3.4 is shown in Fig. 3.5. This *RLC* network, which was discussed in Sections 3.3 and 3.4, can be represented by the state variables $x_1 = v_c$ and $x_2 = i_L$. The initial conditions, $x_1(0)$ and $x_2(0)$, represent the initial capacitor voltage and inductor current, respectively. The flow graph, including the initial conditions of each state variable, is shown in Fig. 3.20. The initial conditions appear as the initial value of the state variable at the output of each integrator.

To obtain $\Phi(s)$, we set $U(s) = 0$. When $R = 3$, $L = 1$, and $C = 1/2$, we obtain the signal-flow graph shown in Fig. 3.21, where the output and input nodes are deleted because they are not involved in the evaluation of $\Phi(s)$. Then, using Mason's gain formula, we obtain $X_1(s)$ in terms of $x_1(0)$ as

$$X_1(s) = \frac{1 \cdot \Delta_1(s) \cdot [x_1(0)/s]}{\Delta(s)}, \qquad (3.79)$$

where $\Delta(s)$ is the graph determinant and $\Delta_1(s)$ is the path cofactor. The graph determinant is

$$\Delta(s) = 1 + 3s^{-1} + 2s^{-2}.$$

The path cofactor is $\Delta_1 = 1 + 3s^{-1}$ because the path between $x_1(0)$ and $X_1(s)$ does not touch the loop with the factor $-3s^{-1}$. Therefore the first element of the transition matrix is

$$\phi_{11}(s) = \frac{(1 + 3s^{-1})(1/s)}{1 + 3s^{-1} + 2s^{-2}} = \frac{(s + 3)}{(s^2 + 3s + 2)}. \qquad (3.80)$$

The element $\phi_{12}(s)$ is obtained by evaluating the relationship between $X_1(s)$ and $x_2(0)$ as

$$X_1(s) = \frac{(-2s^{-1})(x_2(0)/s)}{1 + 3s^{-1} + 2s^{-2}}.$$

Therefore we obtain

$$\phi_{12}(s) = \frac{-2}{s^2 + 3s + 2}.$$ (3.81)

Similarly, for $\phi_{21}(s)$ we have

$$\phi_{21}(s) = \frac{(s^{-1})(1/s)}{1 + 3s^{-1} + 2s^{-2}} = \frac{1}{s^2 + 3s + 2}.$$ (3.82)

Finally, for $\phi_{22}(s)$ we obtain

$$\phi_{22}(s) = \frac{1(1/s)}{1 + 3s^{-1} + 2s^{-2}} = \frac{s}{s^2 + 3s + 2}.$$ (3.83)

Therefore, the state transition matrix in Laplace transformation form is

$$\mathbf{\Phi}(s) = \begin{bmatrix} (s + 3)/(s^2 + 3s + 2) & -2/(s^2 + 3s + 2) \\ 1/(s^2 + 3s + 2) & s/(s^2 + 3s + 2) \end{bmatrix}.$$ (3.84)

The factors of the characteristic equation are $(s + 1)$ and $(s + 2)$ so that

$$(s + 1)(s + 2) = s^2 + 3s + 2.$$

Then the state transition matrix is

$$\mathbf{\Phi}(t) = \mathcal{L}^{-1}\{\mathbf{\Phi}(s)\} = \begin{bmatrix} (2e^{-t} - e^{-2t}) & (-2e^{-t} + 2e^{-2t}) \\ (e^{-t} - e^{-2t}) & (-e^{-t} + 2e^{-2t}) \end{bmatrix}.$$ (3.85)

The evaluation of the time response of the *RLC* network to various initial conditions and input signals can now be evaluated by utilizing Eq. (3.72). For example, when $x_1(0) = x_2(0) = 1$ and $u(t) = 0$, we have

$$\begin{bmatrix} x_1(t) \\ x_2(t) \end{bmatrix} = \mathbf{\Phi}(t) \begin{bmatrix} 1 \\ 1 \end{bmatrix} = \begin{bmatrix} e^{-2t} \\ e^{-2t} \end{bmatrix}.$$ (3.86)

The response of the system for these initial conditions is shown in Fig. 3.22. The trajectory of the state vector $[x_1(t), x_2(t)]$ on the $(x_1$ vs. $x_2)$-plane is shown in Fig. 3.23.

The evaluation of the time response is facilitated by the determination of the state transition matrix. Although this approach is limited to linear systems, it is a powerful method and utilizes the familiar signal-flow graph to evaluate the transition matrix. ∎

FIGURE 3.22
Time response
of the state
variables of the
RLC network for
$x_1(0) = x_2(0) = 1$.

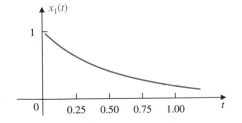
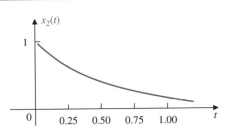

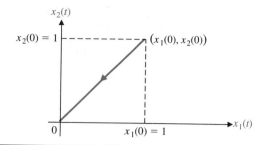

FIGURE 3.23
Trajectory of the
state vector in the
(x_1 vs. x_2)-plane.

3.8 A DISCRETE-TIME EVALUATION OF THE TIME RESPONSE

The response of a system represented by a state vector differential equation can be obtained by utilizing a *discrete-time approximation*. The discrete-time approximation is based on the division of the time axis into sufficiently small time increments. Then the values of the state variables are evaluated at the successive time intervals; that is, $t = 0,\ T,\ 2T,\ 3T, \ldots,$ where T is the increment of time $\Delta t = T$. This approach is a familiar method utilized in numerical analysis and digital computer numerical methods. If the time increment T is sufficiently small compared with the time constants of the system, the response evaluated by discrete-time methods will be reasonably accurate.

The linear state vector differential equation is written as

$$\dot{\mathbf{x}} = \mathbf{A}\mathbf{x} + \mathbf{B}\mathbf{u}. \tag{3.87}$$

The basic definition of a derivative is

$$\dot{\mathbf{x}}(t) = \lim_{\Delta t \to 0} \frac{\mathbf{x}(t + \Delta t) - \mathbf{x}(t)}{\Delta t}. \tag{3.88}$$

Therefore we can utilize this definition of the derivative and determine the value of $\mathbf{x}(t)$ when t is divided in small intervals $\Delta t = T$. Thus, *approximating* the derivative as

$$\dot{\mathbf{x}} = \frac{\mathbf{x}(t + T) - \mathbf{x}(t)}{T}, \tag{3.89}$$

we substitute into Eq. (3.87) to obtain

$$\frac{\mathbf{x}(t + T) - \mathbf{x}(t)}{T} \cong \mathbf{A}\mathbf{x}(t) + \mathbf{B}\mathbf{u}(t). \tag{3.90}$$

Solving for $\mathbf{x}(t + T)$, we have

$$\mathbf{x}(t + T) \cong T\mathbf{A}\mathbf{x}(t) + \mathbf{x}(t) + T\mathbf{B}\mathbf{u}(t) \tag{3.91}$$

$$\cong (T\mathbf{A} + \mathbf{I})\mathbf{x}(t) + T\mathbf{B}\mathbf{u}(t),$$

where t is divided into intervals of width T. Therefore the time t is written as $t = kT$, where k is an integer index so that $k = 0, 1, 2, 3, \ldots$. Then Eq. (3.91) is written as

$$\mathbf{x}[(k + 1)T] \cong (T\mathbf{A} + \mathbf{I})\mathbf{x}(kT) + T\mathbf{B}\mathbf{u}(kT). \tag{3.92}$$

Therefore the value of the state vector at the $(k + 1)$st time instant is evaluated in terms of the values of $\mathbf{x}$ and $\mathbf{u}$ at the kth time instant. Equation (3.92) can be rewritten as

$$\mathbf{x}(k + 1) \cong \psi(t)\mathbf{x}(k) + T\mathbf{B}\mathbf{u}(k), \qquad (3.93)$$

where $\psi(T) = (T\mathbf{A} + \mathbf{I})$ and the symbol T is omitted from the arguments of the variables. Equation (3.93) clearly relates the resulting operation for obtaining $\mathbf{x}(t)$ by evaluating the discrete-time approximation $\mathbf{x}(k + 1)$ in terms of the previous value $\mathbf{x}(k)$. This recurrence operation, known as *Euler's method,* is a sequential series of calculations and is very suitable for digital computer calculation. Other integration approaches, such as the popular Runge–Kutta methods, can also be utilized to evaluate the time response of Eq. (3.87). For example, several Runge–Kutta–Fehlberg integration methods [30] are available in *Matlab.* To illustrate this approximate approach, let us reconsider the evaluation of the response of the *RLC* network of Fig. 3.4.

EXAMPLE 3.6   **Response of *RLC* network**

We shall evaluate the time response of the *RLC* network without determining the transition matrix, by using the discrete-time approximation. Therefore, as in Example 3.5, we will let $R = 3$, $L = 1$, and $C = 1/2$. Then, as we found in Eqs. (3.18) and (3.19), the state vector differential equation is

$$\dot{\mathbf{x}} = \begin{bmatrix} 0 & -(1/C) \\ 1/L & -(R/L) \end{bmatrix} \mathbf{x} + \begin{bmatrix} +(1/C) \\ 0 \end{bmatrix} u(t) \qquad (3.94)$$

$$= \begin{bmatrix} 0 & -2 \\ 1 & -3 \end{bmatrix} \mathbf{x} + \begin{bmatrix} +2 \\ 0 \end{bmatrix} u(t).$$

Now we must choose a sufficiently small time interval T so that the approximation of the derivative (Eq. 3.89) is reasonably accurate and so that the solution to Eq. (3.92) is accurate. Usually we choose T to be less than one-half of the smallest time constant of the system. Therefore, since the shortest time constant of this system is 0.5 s [recalling that the characteristic equation is $[(s + 1)(s + 2)]$], we might choose $T = 0.2$. Alternatively, if a digital computer is used for the calculations and the number of calculations is not important, we would choose $T = 0.05$ in order to obtain greater accuracy. However, we note that as we decrease the increment size, the number of calculations increases proportionally. Using $T = 0.2$ s, Eq. (3.92) is

$$\mathbf{x}(k + 1) \cong (0.2\mathbf{A} + \mathbf{I})\mathbf{x}(k) + 0.2\mathbf{B}\mathbf{u}(k). \qquad (3.95)$$

Therefore

$$\psi(T) = \begin{bmatrix} 1 & -0.4 \\ 0.2 & 0.4 \end{bmatrix} \qquad (3.96)$$

and

$$T\mathbf{B} = \begin{bmatrix} +0.4 \\ 0 \end{bmatrix}. \qquad (3.97)$$

TABLE 3.1

Time t	0	0.2	0.4	0.6	0.8
Exact $x_1(t)$	1	0.67	0.448	0.30	0.20
Approximate $x_1(t)$, $T = 0.1$	1	0.64	0.41	0.262	0.168
Approximate $x_1(t)$, $T = 0.2$	1	0.60	0.36	0.216	0.130

Now let us evaluate the response of the system when $x_1(0) = x_2(0) = 1$ and $u(t) = 0$. The response at the first instant, when $t = T$, or $k = 0$, is

$$\mathbf{x}(1) \cong \begin{bmatrix} 1 & -0.4 \\ 0.2 & 0.4 \end{bmatrix} \mathbf{x}(0) = \begin{bmatrix} 0.6 \\ 0.6 \end{bmatrix}. \tag{3.98}$$

Then the response at the time $t = 2T = 0.4$ second, or $k = 1$, is

$$\mathbf{x}(2) \cong \begin{bmatrix} 1 & -0.4 \\ 0.2 & 0.4 \end{bmatrix} \mathbf{x}(1) = \begin{bmatrix} 0.36 \\ 0.36 \end{bmatrix}. \tag{3.99}$$

The value of the response as $k = 2, 3, 4, \ldots$ is then evaluated in a similar manner.

Now let us compare the actual response of the system evaluated in the previous section using the transition matrix with the approximate response determined by the discrete-time approximation. We found in Example 3.5 that the exact value of the state variables, when $x_1(0) = x_2(0) = 1$, is $x_1(t) = x_2(t) = e^{-2t}$. Therefore the exact values can be readily calculated and compared with the approximate values of the time response in Table 3.1. The approximate time response values for $T = 0.1$ second are also given in Table 3.1. The error, when $T = 0.2$, is approximately a constant equal to 0.07, and thus the percentage error compared to the initial value is 7%. When T is equal to 0.1 second, the percentage error compared to the initial value is approximately 3.5%. If we use $T = 0.05$, the value of the approximation, when time $t = 0.2$ second, is $x_1(t) = 0.655$, and the error has been reduced to 1.5% of the initial value.

Therefore, if one is using a digital computer to evaluate the transient response by evaluating the discrete-time response equations, a value of T equal to one-tenth of the smallest time constant of the system would be selected. The value of this method merits another illustration of the evaluation of the time response of a system. ∎

EXAMPLE 3.7   **Time response of an epidemic**

Let us reconsider the state variable representation of the spread of an epidemic disease presented in Example 3.2. The state vector differential equation was given in Eq. (3.55). When the constants are $\alpha = \beta = \gamma = 1$, we have

$$\dot{\mathbf{x}} = \begin{bmatrix} -1 & -1 & 0 \\ 1 & -1 & 0 \\ 1 & 1 & 0 \end{bmatrix} \mathbf{x} + \begin{bmatrix} 1 & 0 \\ 0 & 1 \\ 0 & 0 \end{bmatrix} \mathbf{u}. \tag{3.100}$$

The characteristic equation of this system, as determined in Eq. (3.57), is $s(s^2 + s + 2) = 0$, and thus the system has complex roots. Let us determine the transient response of the spread of disease when the rate of new susceptibles is zero, that is, when $u_1 = 0$. The rate of adding new infectives is represented by $u_2(0) = 1$ and $u_2(k) = 0$ for $k \geq 1$; that is, one new infective is added at the initial time only (this is equivalent to a pulse input). The time constant of the complex roots is $1/\zeta\omega_n = 2$ seconds, and therefore we will use $T = 0.2$ second. (Note that the actual time units might be months and the units of the input in thousands.)

Then the discrete-time equation is

$$\mathbf{x}(k + 1) = \begin{bmatrix} 0.8 & -0.2 & 0 \\ 0.2 & 0.8 & 0 \\ 0.2 & 0.2 & 1 \end{bmatrix} \mathbf{x}(k) + \begin{bmatrix} 0 \\ 0.2 \\ 0 \end{bmatrix} u_2(k). \tag{3.101}$$

Therefore the response at the first instant, $t = T$, is obtained when $k = 0$ as

$$\mathbf{x}(1) = \begin{bmatrix} 0 \\ 0.2 \\ 0 \end{bmatrix}, \tag{3.102}$$

when $x_1(0) = x_2(0) = x_3(0) = 0$. Then the input $u_2(k)$ is zero for $k \geq 1$ and the response at $t = 2T$ is

$$\mathbf{x}(2) = \begin{bmatrix} 0.8 & -0.2 & 0 \\ 0.2 & 0.8 & 0 \\ 0.2 & 0.2 & 1 \end{bmatrix} \begin{bmatrix} 0 \\ 0.2 \\ 0 \end{bmatrix} = \begin{bmatrix} -0.04 \\ 0.16 \\ 0.04 \end{bmatrix}. \tag{3.103}$$

The response at $t = 3T$ is then

$$\mathbf{x}(3) = \begin{bmatrix} 0.8 & -0.2 & 0 \\ 0.2 & 0.8 & 0 \\ 0.2 & 0.2 & 1 \end{bmatrix} \begin{bmatrix} -0.04 \\ 0.16 \\ 0.04 \end{bmatrix} = \begin{bmatrix} -0.064 \\ 0.120 \\ 0.064 \end{bmatrix},$$

and the ensuing values can then be readily evaluated. Of course, the actual physical value of x_1 cannot become negative. The negative value of x_1 is obtained as a result of an inadequate model.

The discrete-time approximate method is particularly useful for evaluating the time response of nonlinear systems. The transition matrix approach is limited to linear systems, but the discrete-time approximation is not so limited and can be readily applied to nonlinear and time-varying systems. The basic state vector differential equation can be written as

$$\dot{\mathbf{x}} = \mathbf{f}(\mathbf{x}, \mathbf{u}, t), \tag{3.104}$$

where $\mathbf{f}$ is a function, not necessarily linear, of the state vector $\mathbf{x}$ and the input vector $\mathbf{u}$. The column vector $\mathbf{f}$ is the column matrix of functions of $\mathbf{x}$ and $\mathbf{u}$. If the system is a linear function of the control signals, Eq. (3.104) becomes

$$\dot{\mathbf{x}} = \mathbf{f}(\mathbf{x}, t) + \mathbf{B}\mathbf{u}. \tag{3.105}$$

If the system is not time-varying—that is, if the coefficients of the differential equation are constants—Eq. (3.105) is then

$$\dot{\mathbf{x}} = \mathbf{f}(\mathbf{x}) + \mathbf{B}\mathbf{u}. \tag{3.106}$$

Let us consider Eq. (3.106) for a nonlinear system and determine the discrete-time approximation. Using Eq. (3.89) as the approximation to the derivative, we have

$$\frac{\mathbf{x}(t + T) - \mathbf{x}(t)}{T} = \mathbf{f}(\mathbf{x}(t)) + \mathbf{B}\mathbf{u}(t). \tag{3.107}$$

Therefore, solving for $\mathbf{x}(k + 1)$ when $t = kT$, we obtain

$$\mathbf{x}(k + 1) = \mathbf{x}(k) + T[\mathbf{f}(\mathbf{x}(k)) + \mathbf{B}\mathbf{u}(k)]. \tag{3.108}$$

Similarly, the general discrete-time approximation to Eq. (3.104) is

$$\mathbf{x}(k + 1) = \mathbf{x}(k) + T\mathbf{f}(\mathbf{x}(k), \mathbf{u}(k), k). \tag{3.109}$$

Now let us reconsider the previous example when the system is nonlinear. ■

EXAMPLE 3.8   **Improved model of an epidemic**

The spread of an epidemic disease is actually best represented by a set of nonlinear equations as

$$\dot{x}_1 = -\alpha x_1 - \beta x_1 x_2 + u_1(t),$$

$$\dot{x}_2 = \beta x_1 x_2 - \gamma x_2 + u_2(t), \tag{3.110}$$

$$\dot{x}_3 = \alpha x_1 + \gamma x_2,$$

where the interaction between the groups is represented by the nonlinear term $x_1 x_2$. Now, the transition matrix approach and the characteristic equation are not applicable because the system is nonlinear. As in the previous example, we will let $\alpha = \beta = \gamma = 1$ and $u_1(t) = 0$. Also, $u_2(0) = 1$ and $u_2(k) = 0$ for $k \geq 1$. We will select the time increment as $T = 0.2$ second and the initial conditions as $\mathbf{x}^T(0) = [1, 0, 0]$. Then, substituting the $t = kT$ and

$$\dot{x}_i(k) = \frac{x_i(k + 1) - x_i(k)}{T} \tag{3.111}$$

into Eq. (3.110), we obtain

$$\frac{x_1(k + 1) - x_1(k)}{T} = -x_1(k) - x_1(k)x_2(k),$$

$$\frac{x_2(k + 1) - x_2(k)}{T} = +x_1(k)x_2(k) - x_2(k) + u_2(k), \tag{3.112}$$

$$\frac{x_3(k + 1) - x_3(k)}{T} = x_1(k) + x_2(k).$$

Solving these equations for $x_i(k + 1)$ and recalling that $T = 0.2$, we have

$$x_1(k + 1) = 0.8x_1(k) - 0.2x_1(k)x_2(k),$$

$$x_2(k + 1) = 0.8x_2(k) + 0.2x_1(k)x_2(k) + 0.2u_2(k), \tag{3.113}$$

$$x_3(k + 1) = x_3(k) + 0.2x_1(k) + 0.2x_2(k).$$

Then the response of the first instant, $t = T$, is

$$x_1(1) = 0.8x_1(0) = 0.8,$$

$$x_2(1) = 0.2u_2(k) = 0.2,$$

$$x_3(1) = 0.2x_1(0) = 0.2.$$

Again, using Eq. (3.113) and noting that $u_2(1) = 0$, we have

$$x_1(2) = 0.8x_1(1) - 0.2x_1(1)x_2(1) = 0.608,$$

$$x_2(2) = 0.8x_2(1) + 0.2x_1(1)x_2(1) = 0.192, \tag{3.114}$$

$$x_3(2) = x_3(1) + 0.2x_1(1) + 0.2x_2(1) = 0.40.$$

At the third instant, when $t = 3T$, we obtain

$$x_1(3) = 0.463, \qquad x_2(3) = 0.177, \qquad x_3(3) = 0.56.$$

The evaluation of the ensuing values follows in a similar manner. We note that the response of the nonlinear system differs considerably from the response of the linear model considered in the previous example. ∎

The evaluation of the time response of the state variables of linear systems is readily accomplished by using either (1) the transition matrix approach or (2) the discrete-time approximation. The transition matrix of linear systems is readily obtained from the signal-flow graph state model. For a nonlinear system, the discrete-time approximation provides a suitable approach, and the discrete-time approximation method is particularly useful if a digital computer is used for numerical calculations.

3.9 STATE VARIABLE EQUATIONS FOR A SPACECRAFT

In this section, the equations of motion are derived for a rigid spacecraft with control moment gyros (CMGs) in a circular orbit around Earth [12–14]. The nonlinear spacecraft model includes the spacecraft attitude kinematics, rotational dynamics, and CMG momentum. The attitude equations are expressed in terms of Euler's angles relating a body-axis reference frame, fixed to the spacecraft, to a local-vertical local-horizontal reference frame.

We assume that the spacecraft is in a circular orbit with orbital angular rate n. The coordinate systems of interest are the local-vertical, local-horizontal (LVLH) reference frame and the body-axis reference frame. The LVLH reference frame is located at the mass center of the spacecraft and is defined in Fig. 3.24. The X_L-Z_L plane is the instantaneous orbit plane. The Z_L axis lies along the geocentric radius vector to the vehicle and is negative radially outward. The Y_L axis lies along the instantaneous orbital angular momentum vector and is negative in the direction of the angular momentum vector. The X_L axis completes a right-handed system.

When defining the orientation of a spacecraft body with respect to a reference frame, a series of rotations of Euler's angles are performed to determine the orientation uniquely. There are three principal rotations: about the X, Y, and Z axes. The Euler angle sequence associated with our system is pitch, yaw, and roll around Y_B, Z_B, and X_B, respectively (see

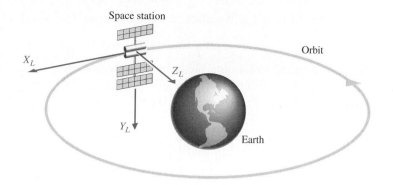

FIGURE 3.24
LVLH reference
frame for a
spacecraft.

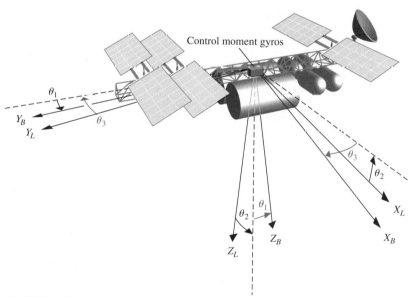

FIGURE 3.25   Spacecraft and control gyros.

Fig. 3.25). In the sequel, $\boldsymbol{\theta} = (\theta_1, \theta_2, \theta_3)^T$ are the roll, pitch, and yaw Euler angles of the body with respect to the LVLH reference frame, and $\boldsymbol{\omega} = (\omega_1, \omega_2, \omega_3)^T$ are the absolute angular velocities in the body axes.

The spacecraft attitude kinematics are given by

$$\dot{\boldsymbol{\theta}} = \mathbf{R}(\boldsymbol{\theta})\boldsymbol{\omega} + \mathbf{n} \tag{3.115}$$

where

$$\mathbf{R}(\boldsymbol{\theta}) = \frac{1}{\cos \theta_3} \begin{bmatrix} \cos \theta_3 & -\cos \theta_1 \sin \theta_3 & \sin \theta_1 \sin \theta_3 \\ 0 & \cos \theta_1 & -\sin \theta_1 \\ 0 & \sin \theta_1 \cos \theta_3 & \cos \theta_1 \cos \theta_3 \end{bmatrix}, \quad \mathbf{n} = \begin{bmatrix} 0 \\ n \\ 0 \end{bmatrix}.$$

The moment equation of the spacecraft in the body-axis frame is [13]

$$\dot{\mathbf{H}}_B + \boldsymbol{\omega} \times \mathbf{H}_B = \boldsymbol{\tau}_{\text{ext}} \tag{3.116}$$

where $\mathbf{H}_B$ denotes the momentum vector in the body frame and $\boldsymbol{\tau}_{\text{ext}}$ is the external torque vector (e.g., gravity gradient torques, aerodynamic torques, and control torques). Since the spacecraft is assumed to be rigid, we have $\mathbf{H}_B = \mathbf{I}\boldsymbol{\omega}$, where $\mathbf{I}$ is the spacecraft inertia matrix in the body frame

$$\mathbf{I} = \begin{bmatrix} I_1 & I_{12} & I_{13} \\ I_{12} & I_2 & I_{23} \\ I_{13} & I_{23} & I_3 \end{bmatrix}.$$

The external torque includes the gravity gradient torque and the CMG control torque. Without external disturbances (such as aerodynamic torques), Eq. (3.116) becomes

$$\mathbf{I}\dot{\boldsymbol{\omega}} + \boldsymbol{\omega} \times \mathbf{I}\boldsymbol{\omega} = \boldsymbol{\tau}_{gg} - \mathbf{u}, \tag{3.117}$$

where $\boldsymbol{\tau}_{gg}$ is the gravity gradient torque vector and $\mathbf{u} = (u_1, u_2, u_3)^T$ is the CMG control torque vector, both in the body frame.

The gravity gradient torque is due to the fact that Earth's gravitational field is not uniform over the body of the spacecraft. With Earth's inverse-square gravitational field, the gravity gradient torque in the body frame can be expressed as

$$\boldsymbol{\tau}_{gg} = 3n^2 \mathbf{c} \times \mathbf{I}\mathbf{c}, \tag{3.118}$$

where

$$\mathbf{c} = \begin{bmatrix} -\sin \theta_2 \cos \theta_3 \\ \sin \theta_1 \cos \theta_2 + \cos \theta_1 \sin \theta_2 \sin \theta_3 \\ \cos \theta_1 \cos \theta_2 - \sin \theta_1 \sin \theta_2 \sin \theta_3 \end{bmatrix}.$$

The rotational dynamics can be expressed in the body axis frame as

$$\mathbf{I}\dot{\boldsymbol{\omega}} = -\boldsymbol{\omega} \times \mathbf{I}\boldsymbol{\omega} + 3n^2 \mathbf{c} \times \mathbf{I}\mathbf{c} - \mathbf{u}. \tag{3.119}$$

The CMG dynamics can be derived similarly and are given by

$$\dot{\mathbf{h}} = -\boldsymbol{\omega} \times \mathbf{h} + \mathbf{u}, \tag{3.120}$$

where $\mathbf{h}$ is the CMG momentum and $\mathbf{u}$ is the CMG control torque, both in the body frame.

The conventional approach to spacecraft attitude control and momentum management design is to develop a linear model representing the spacecraft attitude and CMG momentum by linearizing the nonlinear model in Eqs. (3.115), (3.119), and (3.120). This linearization is accomplished by a standard Taylor series approximation. Linear control design methods can then be readily applied.

Linearizing the spacecraft nonlinear model by assuming small attitude deviations, small rates, small CMG momentum states, and negligible cross products of inertia (i.e., $I_{ij} = 0$, $i \neq j$) results in a decoupling of the pitch axis and the roll and yaw axes. The linearized equations for the pitch axis are

$$\begin{bmatrix} \dot{\theta}_2 \\ \dot{\omega}_2 \\ \dot{h}_2 \end{bmatrix} = \begin{bmatrix} 0 & 1 & 0 \\ 3n^2\Delta_2 & 0 & 0 \\ 0 & 0 & 0 \end{bmatrix} \begin{bmatrix} \theta_2 \\ \omega_2 \\ h_2 \end{bmatrix} + \begin{bmatrix} 0 \\ -1/I_2 \\ 1 \end{bmatrix} u_2,$$

where

$$\Delta_2 = \frac{I_3 - I_1}{I_2}.$$

The linearized equations for the roll and yaw axes are

$$
\begin{bmatrix} \dot{\theta}_1 \\ \dot{\theta}_3 \\ \dot{\omega}_1 \\ \dot{\omega}_3 \\ \dot{h}_1 \\ \dot{h}_3 \end{bmatrix} =
\begin{bmatrix}
0 & n & 1 & 0 & 0 & 0 \\
-n & 0 & 0 & 1 & 0 & 0 \\
-3n^2\Delta_1 & 0 & 0 & -n\Delta_1 & 0 & 0 \\
0 & 0 & -n\Delta_3 & 0 & 0 & 0 \\
0 & 0 & 0 & 0 & 0 & n \\
0 & 0 & 0 & 0 & -n & 0
\end{bmatrix}
\begin{bmatrix} \theta_1 \\ \theta_3 \\ \omega_1 \\ \omega_3 \\ h_1 \\ h_3 \end{bmatrix} +
\begin{bmatrix}
0 & 0 \\
0 & 0 \\
-1/I_1 & 0 \\
0 & -1/I_3 \\
1 & 0 \\
0 & 1
\end{bmatrix}
\begin{bmatrix} u_1 \\ u_3 \end{bmatrix},
$$

where

$$\Delta_1 = \frac{I_2 - I_3}{I_1} \quad \text{and} \quad \Delta_3 = \frac{I_1 - I_2}{I_3}.$$

Notice that the pitch axis control problem is a single-input problem, whereas the roll and yaw control problem is a multiple-input problem. Typical parameter values for the space station are:

$$I_1 = 50.28E+6 \quad (\text{slug-ft}^2),$$
$$I_2 = 10.80E+6,$$
$$I_3 = 58.57E+6,$$

and $n = 0.0011$ rad/s. With these typical parameters, the pitch-axis state-space representation is given by:

$$\dot{x} = Ax + bu, \tag{3.121}$$

where

$$
\mathbf{x} = \begin{bmatrix} \theta_2 \\ \omega_2 \\ h_2 \end{bmatrix}, \qquad
\mathbf{A} = \begin{bmatrix} 0 & 1 & 0 \\ 2.7E-06 & 0 & 0 \\ 0 & 0 & 0 \end{bmatrix}, \qquad
\mathbf{b} = \begin{bmatrix} 0 \\ 9.2E-08 \\ 1 \end{bmatrix}.
$$

3.10 DESIGN EXAMPLE: PRINTER BELT DRIVE

A commonly used low-cost printer for a computer uses a belt drive to move the printing device laterally across the printed page [11]. The printing device may be an ink-jet, a print ball, or thermal. An example of a belt drive printer with a dc motor actuator is shown in Fig. 3.26. In this model, a light sensor is used to measure the position of the printing device and the belt tension adjusts the spring flexibility of the belt. The goal of the design is to determine the effect of the belt spring constant k and select appropriate parameters for the motor, the belt pulley, and the controller. To achieve the analysis, we will determine a model of the belt drive system and select many of its parameters. Using this model, we will obtain the signal-flow graph model and select the state variables. We then will determine an appropriate transfer function for the system and select its other parameters, except for

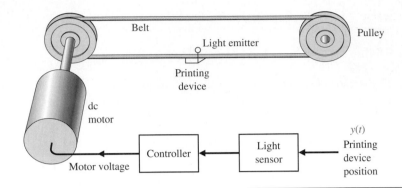

FIGURE 3.26
Printer belt drive
system.

the spring constant. Finally we will examine the effect of varying the spring constant within a realistic range.

We propose the model of the belt drive system shown in Fig. 3.27. This model assumes that the spring constant of the belt is k, the radius of the pulley is r, the angular rotation of the motor shaft is θ, and the angular rotation of the right-hand pulley is θ_p. The mass of the printing device is m and its position is $y(t)$. A light sensor is used to measure y, and the output of the sensor is a voltage v_1, where $v_1 = k_1 y$. The controller provides an output voltage v_2, where v_2 is a function of v_1. The voltage v_2 is connected to the field of the motor. Let us assume that we can use the linear relationship

$$v_2 = -\left[k_2 \frac{dv_1}{dt} + k_3 v_1\right]$$

and elect to use $k_2 = 0.1$ and $k_3 = 0$ (velocity feedback).

The inertia of the motor and pulley is $J = J_{\text{motor}} + J_{\text{pulley}}$. We plan to use a moderate-power dc motor. Selecting a typical 1/8-hp dc motor, we find that $J = 0.05$ kg-m², the field inductance is negligible, the field resistance is $R = 2\ \Omega$, the motor constant is $K_m = 2$ N-m/A, and the motor and pulley friction is $f = 5$ N-ms/rad (assuming low-performance pulley bearings). The radius of the pulley is $r = 0.1$ m. The system parameters are summarized in Table 3.2.

FIGURE 3.27
Printer belt drive
model.

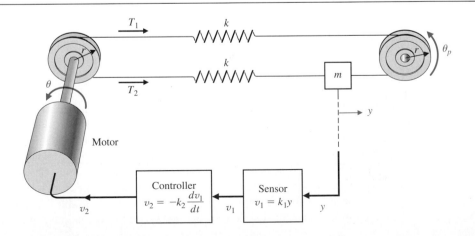

TABLE 3.2   Parameters of Printing Device

Mass	$m = 0.1$ kg
Light sensor	$k_1 = 1$ V/m
Radius	$r = 0.1$ m
Motor	
Inductance	$L \approx 0$
Friction	$f = 5$ N-ms/rad
Resistance	$R = 2\ \Omega$
Constant	$K_m = 2$ N-m/A
Inertia	$J = J_{\text{motor}} + J_{\text{pulley}} : J = 0.05$ kg-m^2

We now proceed to write the equations of the motion for the system; note that $y = r\theta_p$. Then the tension T_1 is

$$T_1 = k(r\theta - r\theta_p) = k(r\theta - y).$$

The tension T_2 is

$$T_2 = k(y - r\theta).$$

The net tension at the mass m is

$$T_1 - T_2 = m\frac{d^2y}{dt^2} \qquad (3.122)$$

and

$$T_1 - T_2 = k(r\theta - y) - k(y - r\theta) = 2k(r\theta - y) = 2kx_1, \qquad (3.123)$$

where the first state variable is $x_1 = (r\theta - y)$. Let the second state variable be $x_2 = dy/dt$, and use Eqs. (3.122) and (3.123) to obtain

$$\frac{dx_2}{dt} = \frac{2k}{m}x_1. \qquad (3.124)$$

The first derivative of x_1 is

$$\frac{dx_1}{dt} = r\frac{d\theta}{dt} - \frac{dy}{dt} = rx_3 - x_2 \qquad (3.125)$$

when we select the third state variable as $x_3 = d\theta/dt$. We now require a differential equation describing the motor rotation. When $L = 0$, we have the field current $i = v_2/R$ and the motor torque $T_m = K_m i$. Therefore

$$T_m = \frac{K_m}{R}v_2,$$

and the motor torque provides the torque to drive the belts plus the disturbance or undesired load torque, so that

$$T_m = T + T_d.$$

The torque T drives the shaft to the pulley so that

$$T = J\frac{d^2\theta}{dt^2} + f\frac{d\theta}{dt} + r(T_1 - T_2).$$

Therefore we note that

$$\frac{dx_3}{dt} = \frac{d^2\theta}{dt^2}$$

and therefore

$$\frac{dx_3}{dt} = \frac{(T_m - T_d)}{J} - \frac{f}{J}x_3 - \frac{2kr}{J}x_1,$$

where

$$T_m = \frac{K_m}{R}v_2 \quad \text{and} \quad v_2 = -k_1k_2\frac{dy}{dt} = -k_1k_2x_2.$$

Therefore we obtain

$$\frac{dx_3}{dt} = \frac{-K_mk_1k_2}{JR}x_2 - \frac{f}{J}x_3 - \frac{2kr}{J}x_1 - \frac{T_d}{J}. \tag{3.126}$$

Equations (3.124)–(3.126) are the three first-order differential equations required to describe this system. The matrix differential equation is

$$\dot{\mathbf{x}} = \begin{bmatrix} 0 & -1 & r \\ \dfrac{2k}{m} & 0 & 0 \\ \dfrac{-2kr}{J} & \dfrac{-K_mk_1k_2}{JR} & \dfrac{-f}{J} \end{bmatrix} \mathbf{x} + \begin{bmatrix} 0 \\ 0 \\ \dfrac{-1}{J} \end{bmatrix} T_d. \tag{3.127}$$

The signal-flow graph representing the matrix differential equation is shown in Fig. 3.28, where we include the identification of the node for the torque disturbance torque T_d.

FIGURE 3.28
Flow graph model for printer belt drive.

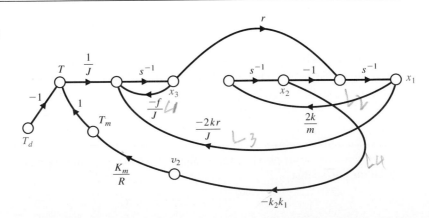

We can use the flow graph to determine the transfer function $X_1(s)/T_d(s)$. The goal is to reduce the effect of the disturbance T_d, and the transfer function will show us how to accomplish this goal. Using Mason's gain formula, we obtain

$$\frac{X_1(s)}{T_d(s)} = \frac{-\dfrac{r}{J}s^{-2}}{1 - (L_1 + L_2 + L_3 + L_4) + L_1 L_2},$$

where

$$L_1 = \frac{-f}{J}s^{-1}, \qquad L_2 = \frac{-2k}{m}s^{-2}, \qquad L_3 = \frac{-2kr^2 s^{-2}}{J}, \qquad L_4 = \frac{-2kK_m k_1 k_2 r s^{-3}}{mJR}.$$

We therefore have

$$\frac{X_1(s)}{T_d(s)} = \frac{-\left(\dfrac{r}{J}\right)s}{s^3 + \left(\dfrac{f}{J}\right)s^2 + \left(\dfrac{2k}{m} + \dfrac{2kr^2}{J}\right)s + \left(\dfrac{2kf}{Jm} + \dfrac{2kK_m k_1 k_2 r}{JmR}\right)}.$$

Substituting the parameter values summarized in Table 3.2, we obtain

$$\frac{X_1(s)}{T_d(s)} = \frac{-s}{s^3 + 100s^2 + 20.4ks + 400k(5 + 0.1k_2)}. \tag{3.128}$$

We wish to select the spring constant k and the gain k_2 so that the state variable x_1 will quickly decline to a low value when a disturbance occurs. For test purposes, consider a step disturbance of unit magnitude so that $T_d(s) = 1/s$. Recalling that $x_1 = r\theta - y$, we thus seek a small magnitude for x_1 so that y is nearly equal to the desired $r\theta$. If we have a perfectly stiff belt with $k \to \infty$, then $y = r\theta$ exactly. With a unit step disturbance, $T_d(s) = 1/s$, we have

$$X_1(s) = \frac{-1}{s^3 + 100s^2 + 20.4ks + 400k(5 + 0.1k_2)}. \tag{3.129}$$

The final value theorem gives

$$\lim_{t \to \infty} x_1(t) = \lim_{s \to 0} sX_1(s) = 0, \tag{3.130}$$

and thus the steady-state value of $x_1(t)$ is zero. We need to use a realistic value for k in the range $1 \le k \le 40$. For an average value of $k = 20$ and $k_2 = 10$, we have

$$X_1(s) = \frac{-1}{s^3 + 100s^2 + 4080s + 48{,}000}$$

$$= \frac{-1}{(s + 18.72)(s^2 + 81.28s + 2558.4)} \tag{3.131}$$

$$= \frac{-1}{(s + 18.72)[(s + 40.6)^2 + (30.1)^2]}.$$

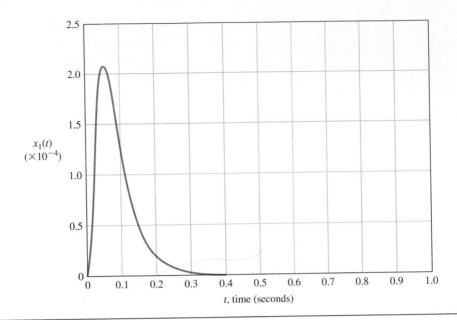

FIGURE 3.29
Response of
$x_1(t)$ to a step
disturbance:
peak value =
2.08×10^{-4}.

The characteristic equation has one real root and two complex roots. The partial fraction expansion yields

$$X_1(s) = \frac{A}{s + 18.72} + \frac{Bs + C}{(s + 40.6)^2 + (30.1)^2},$$ (3.132)

where we find $A = -7.2 \times 10^{-4}$, $B = +7.2 \times 10^{-4}$, and $C = +0.045$. Clearly, with these small residues, the response to the unit disturbance is relatively small. Because A and B are small compared to C, we may approximate $X_1(s)$ as

$$X_1(s) \cong \frac{+0.045}{(s + 40.6)^2 + (30.1)^2}.$$

Using entry 16 of Appendix A, we obtain

$$x_1(t) \cong 0.0015e^{-40.6t} \sin 30.1t.$$ (3.133)

The actual response of x_1 is shown in Fig. 3.29. Clearly, this system will reduce the effect of the unwanted disturbance to a relatively small magnitude. Note that the peak magnitude of $x_1(t)$ is less than one-thousandth of the magnitude of the disturbance. Thus we have achieved our design objective.

3.11 ANALYSIS OF STATE VARIABLE MODELS USING *MATLAB*

The time-domain method utilizes a *state-space* representation of the system model, given by

$$\dot{\mathbf{x}} = \mathbf{Ax} + \mathbf{B}u \quad \text{and} \quad c = \mathbf{Dx} + \mathbf{H}u.$$ (3.134)

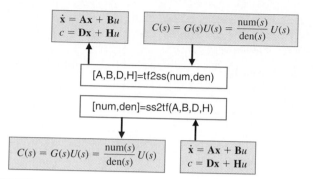

FIGURE 3.30
Linear system
model conversion.

The vector **x** is the *state* of the system, **A** is the constant $n \times n$ system matrix, **B** is the constant $n \times m$ input matrix, **D** is the constant $p \times n$ output matrix, and **H** is a constant $p \times m$ matrix. The number of inputs, m, and the number of outputs, p, are taken to be one, since we are considering only single-input, single-output (SISO) problems. Therefore c and u are not bold (matrix) variables.

The main elements of the state-space representation in Eq. (3.134) are the state vector **x** and the constant matrices (**A, B, D, H**). Since the main computational unit in *Matlab* is the matrix, the state-space representation lends itself well to the *Matlab* environment. In fact, *Matlab* covers so many aspects of state-space methods that we will not be able to discuss them all here. Three new functions covered in this section are tf2ss, ss2tf, and lsim. We also consider the use of the expm function to calculate the state transition matrix.

Given a transfer function, we can obtain an equivalent state-space representation and vice versa. *Matlab* has two functions that convert systems from transfer function to state space and back to transfer function. The function tf2ss converts a transfer-function representation to a state-space representation; the function ss2tf converts a state-space representation to a transfer function. These functions are shown in Fig. 3.30.

For instance, consider the third-order system

$$T(s) = \frac{C(s)}{R(s)} = \frac{2s^2 + 8s + 6}{s^3 + 8s^2 + 16s + 6}. \tag{3.135}$$

We can obtain a state-space phase variable representation using the tf2ss function, as shown in Fig. 3.31. The state-space representation of Eq. (3.135) is given by Eq. (3.134) where

$$\mathbf{A} = \begin{bmatrix} -8 & -16 & -6 \\ 1 & 0 & 0 \\ 0 & 1 & 0 \end{bmatrix}, \quad \mathbf{B} = \begin{bmatrix} 1 \\ 0 \\ 0 \end{bmatrix},$$

and

$$\mathbf{D} = [2 \quad 8 \quad 6], \quad \mathbf{H} = [0].$$

Note the state-space format called for by tf2ss utilizes the state variables defined as shown in Fig. 3.32, which shows x_1 on the leftmost integrator. Notice that the printsys function in convert.m

```
% Convert G(s) = (2s^2+8s+6)/(s^3+8s^2+16s+6)
% to a state-space representation
%
num=[2 8 6]; den=[1 8 16 6];
[A,B,D,H]=tf2ss(num,den);
printsys(A,B,D,H)
```

(a)

```
>>convert
a =
                x1          x2          x3
     x1      -8.00000   -16.00000   -6.00000
     x2       1.00000        0          0
     x3          0        1.00000       0

b =
                u1
     x1       1.00000
     x2          0
     x3          0

c =
                x1          x2          x3
     y1       2.00000     8.00000     6.00000

d =
                u1
     y1          0
```

(b)

FIGURE 3.31
Conversion of
Eq. (3.135) to a
phase variable
state-space
representation.
(a) *Matlab* script,
(b) output printout.

FIGURE 3.32
Signal-flow graph
model with x_1
defined as the
leftmost state
variable.

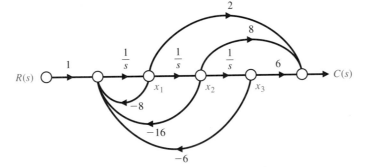

Fig. 3.31 lists the system matrices as a, b, c, d. The conversion to the notation of Eq. (3.134) is as follows:

$$a \mapsto \mathbf{A}, \qquad b \mapsto \mathbf{B}, \qquad c \mapsto \mathbf{D}, \qquad d \mapsto \mathbf{H}.$$

The time response of the system in Eq. (3.134) is given by the solution to the vector differential equation

$$\mathbf{x}(t) = \exp(\mathbf{A}t)\mathbf{x}(0) + \int_0^t \exp[\mathbf{A}(t - \tau)]\mathbf{B}u(\tau)d\tau. \tag{3.136}$$

The matrix exponential function in Eq. (3.136) is the state transition matrix, $\mathbf{\Phi}(t)$, where (Eq. 3.23)

$$\mathbf{\Phi}(t) = \exp(\mathbf{A}t).$$

We can use the function expm to compute the transition matrix for a given time, as illustrated in Fig. 3.33. The expm(A) function computes the matrix exponential. By contrast, the exp(A) function calculates $e^{a_{ij}}$ for each of the elements $a_{ij} \in \mathbf{A}$.

For example, let us consider the *RLC* network of Fig. 3.4 described by the state-space representation of Eq. (3.18) with

$$\mathbf{A} = \begin{bmatrix} 0 & -2 \\ 1 & -3 \end{bmatrix}, \quad \mathbf{B} = \begin{bmatrix} 2 \\ 0 \end{bmatrix}, \quad \mathbf{D} = [1 \quad 0], \quad \mathbf{H} = 0.$$

The initial conditions are $x_1(0) = x_2(0) = 1$ and the input $u(t) = 0$. At $t = 0.2$, the state transition matrix is as given in Fig. 3.33. The state at $t = 0.2$ is predicted by the state transition methods to be

$$\begin{pmatrix} x_1 \\ x_2 \end{pmatrix}_{t-0.2} = \begin{bmatrix} 0.9671 & -0.2968 \\ 0.1484 & 0.5219 \end{bmatrix} \begin{pmatrix} x_1 \\ x_2 \end{pmatrix}_{t=0} = \begin{pmatrix} 0.6703 \\ 0.6703 \end{pmatrix}.$$

The time response of the system of Eq. (3.134) can also be obtained by using the lsim function. The lsim function can accept as input nonzero initial conditions as well as an input function, as shown in Fig. 3.34. Using the lsim function, we can calculate the response for the *RLC* network as shown in Fig. 3.35.

FIGURE 3.33
Computing the state transition matrix for a given time, $\Delta t = dt$.

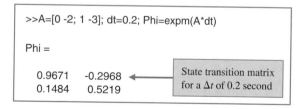

```
>>A=[0 -2; 1 -3]; dt=0.2; Phi=expm(A*dt)

Phi =

    0.9671    -0.2968  ◀——  State transition matrix
    0.1484     0.5219        for a Δt of 0.2 second
```

FIGURE 3.34
The **lsim** function for calculating the output and state response.

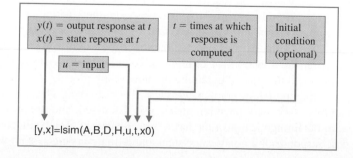

$y(t)$ = output response at t
$x(t)$ = state reponse at t

t = times at which response is computed

Initial condition (optional)

u = input

[y,x]=lsim(A,B,D,H,u,t,x0)

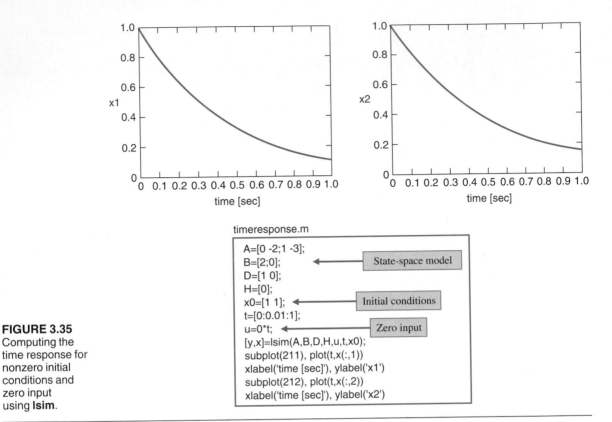

FIGURE 3.35
Computing the
time response for
nonzero initial
conditions and
zero input
using **lsim**.

The state at $t = 0.2$ is predicted with the lsim function to be $x_1(0.2) = x_2(0.2) = 0.6703$. If we can compare the results obtained by the lsim function and by multiplying the initial condition state vector by the state transition matrix, we find exactly the identical results.

3.12 SUMMARY

In this chapter we have considered the description and analysis of systems in the time domain. The concept of the state of a system and the definition of the state variables of a system were discussed. The selection of a set of state variables in terms of the variables that describe the stored energy of a system was examined, and the nonuniqueness of a set of state variables was noted. The state differential equation and the solution for $\mathbf{x}(t)$ were discussed. Two alternative signal-flow graph model structures were considered for representing the transfer function (or differential equation) of a system. Using Mason's gain formula, we noted the ease of obtaining the flow graph model. The state differential equation representing these flow graph models was also examined. The time response of a linear system and its associated transition matrix was discussed, and the utility of Mason's gain formula for obtaining the transition matrix was illustrated. Also, a discrete-time evaluation of the time response of a nonlinear system and a time-varying system were considered. It was noted that the flow graph state model is equivalent to an analog computer diagram where the output of each integrator is a state variable. Also, we found that the discrete-time approximation for a time response, as well as the transition matrix formulation for linear systems, is readily applicable for programming and solution by using a digital computer. Finally, the use of *Matlab* to convert a transfer function to state variable form and calculate the state transition matrix was discussed and illustrated.

EXERCISES

E3.1 For the circuit shown in Fig. E3.1 identify a set of state variables.

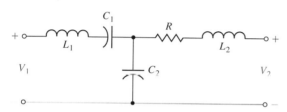

FIGURE E3.1 *RLC* circuit.

E3.2 A robot-arm drive system for one joint can be represented by the differential equation [8]

$$\frac{dv(t)}{dt} = -k_1 v(t) - k_2 y(t) + k_3 i(t),$$

where $v(t)$ = velocity, $y(t)$ = position, and $i(t)$ is the control-motor current. Put the equations in state variable form and set up the matrix form for $k_1 = k_2 = 1$.

E3.3 A system can be represented by the state vector differential equation of Eq. (3.16) where

$$\mathbf{A} = \begin{bmatrix} 0 & 1 \\ -1 & -1 \end{bmatrix}.$$

Find the characteristic roots of the system.

Answer: $s = -1/2 \pm j\sqrt{3}/2$

E3.4 Obtain the state variable matrix in phase variable form for a system with a differential equation

$$2\frac{d^3 y}{dt^3} + 4\frac{d^2 y}{dt^2} + 6\frac{dy}{dt} + 8y = 10u(t).$$

E3.5 A system is represented by a flow graph as shown in Fig. E3.5. Write the state equations for this flow graph in the form of Eqs. (3.16) and (3.17).

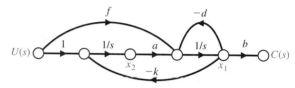

FIGURE E3.5 System flow graph.

E3.6 A system is represented by Eq. (3.16) where

$$\mathbf{A} = \begin{bmatrix} 0 & 1 \\ 0 & 0 \end{bmatrix}.$$

(a) Find the matrix $\mathbf{\Phi}(t)$. (b) For the initial conditions $x_1(0) = x_2(0) = 1$, find $\mathbf{x}(t)$.

Answer: (b) $x_1 = (1 + t), x_2 = 1, t \geq 0$

E3.7 Consider the spring and mass shown in Fig. 3.3 where $M = 1$, $K = 100$, and $f = 20$. (a) Find the state vector differential equation. (b) Find the roots of the characteristic equation for this system.

E3.8 The manual low-altitude hovering task above a moving landing deck of a small ship is very demanding, in particular, in adverse weather and sea conditions. The hovering condition is represented by the **A** matrix

$$\mathbf{A} = \begin{bmatrix} 0 & 1 & 0 \\ 0 & 0 & 1 \\ 0 & -4 & -1 \end{bmatrix}.$$

Find the roots of the characteristic equation.

E3.9 An unforced circuit is shown in Fig. E3.9. The physical state variables are chosen as the charge on each capacitor so that $x_1 = q_1$ and $x_2 = q_2$. (a) Determine the physical state variable flow graph model, and obtain the matrix differential equation. (b) Determine the diagonal (canonical) flow graph model and the matrix differential equation. Note that $i_1 = dq_1/dt = dx_1/dt$ and

$i_2 = dx_2/dt$. Assume that the output variable is the current i_2.

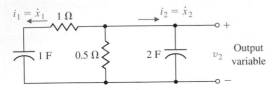

FIGURE E3.9   *RC* circuit.

E3.10   A hovering vehicle (flying saucer) control system is represented by two state variables and [15]

$$A = \begin{bmatrix} 0 & 6 \\ -1 & -5 \end{bmatrix}.$$

(a) Find the roots of the characteristic equation.
(b) Find the state transition matrix $\mathbf{\Phi}(t)$.

E3.11   Determine the state variable description in the phase variable format for the system described by the transfer function

$$T(s) = \frac{C(s)}{R(s)} = \frac{4(s + 3)}{(s + 2)(s + 4)}.$$

E3.12   Use a state variable model to describe the circuit of Fig. E3.12. Obtain the response to an input unit step when the initial current is zero and the initial capacitor voltage is zero.

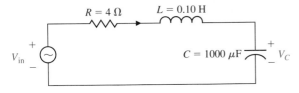

FIGURE E3.12   *RLC* series circuit.

E3.13   A system is described by two differential equations as

$$\frac{dy}{dt} + y - 2u + aw = 0,$$

$$\frac{dw}{dt} - by + 4u = 0,$$

where w and y are functions of time and u is an input $u(t)$. (a) Select a set of state variables. (b) Write the matrix differential equation and specify the elements of the matrices. (c) Find the

characteristic roots of the system in terms of the parameters a and b.

Answer: (c) $s = -1/2 \pm \sqrt{1 - 4ab}/2$

E3.14   Develop the state-space representation of a radioactive material of mass M to which additional radioactive material is added at the rate $r(t) = Ku(t)$, where K is a constant. Identify the state variables.

E3.15   Consider the case of the two masses connected as shown in Fig. E3.15. The sliding friction of each mass has the constant b. Determine the state variable matrix differential equation.

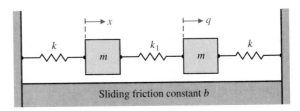

FIGURE E3.15   Two mass system.

E3.16   Two cars with negligible rolling friction are connected as shown in Fig. E3.16. An input force is $u(t)$. Determine the state variable differential matrix equation.

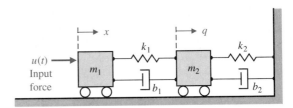

FIGURE E3.16   Two cars with negligible rolling friction.

E3.17   Determine the state variable differential matrix equation for the circuit shown in Fig. E3.17.

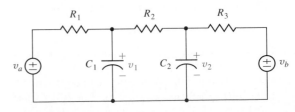

FIGURE E3.17   *RC* circuit.

E3.18 Determine the state variable matrix differential equation for the circuit of Fig. E3.18. Let $x_1 = i_1$, $x_2 = i_2$, and $x_3 = v$.

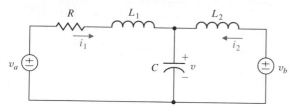

FIGURE E3.18 *RLC* circuit.

E3.19 A single-output, single-output system has the matrix equations

$$\dot{\mathbf{x}} = \begin{bmatrix} 0 & 1 \\ -3 & -4 \end{bmatrix} \mathbf{x} + \begin{bmatrix} 0 \\ 1 \end{bmatrix} u$$

and

$$c = \begin{bmatrix} 10 & 0 \end{bmatrix} \mathbf{x}.$$

Determine the transfer function $G(s) = C(s)/U(s)$ using (a) the signal-flow model and (b) the inverse matrix transform $\mathbf{\Phi}(s)$.

PROBLEMS

P3.1 An *RLC* circuit is shown in Fig. P3.1. (a) Identify a suitable set of state variables. (b) Obtain the set of first-order differential equations in terms of the state variables. (c) Write the state differential equation. (d) Draw the state variable flow graph.

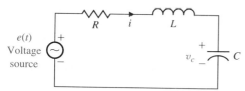

FIGURE P3.1 *RLC* circuit.

P3.2 A *balanced* bridge network is shown in Fig. P3.2. (a) Show that the **A** and **B** matrices for this circuit are

$$\mathbf{A} = \begin{bmatrix} -(2/(R_1 + R_2)C & 0 \\ 0 & -2R_1R_2/(R_1 + R_2)L \end{bmatrix},$$

$$\mathbf{B} = 1/(R_1 + R_2)\begin{bmatrix} 1/C & 1/C \\ R_2/L & -R_2/L \end{bmatrix}.$$

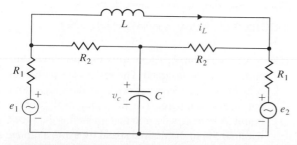

FIGURE P3.2 Balanced bridge network.

(b) Draw the state model flow graph. The state variables are $(x_1, x_2) = (v_c, i_L)$.

P3.3 An *RLC* network is shown in Fig. P3.3. Define the state variables as $x_1 = i_L$ and $x_2 = v_c$. (a) Obtain the state differential equation. (b) Draw the state model flow graph.

Partial answer:

$$\mathbf{A} = \begin{bmatrix} 0 & 1/L \\ -(1/C) & -(1/RC) \end{bmatrix}.$$

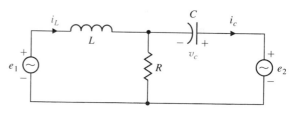

FIGURE P3.3 *RLC* circuit.

P3.4 The transfer function of a system is

$$T(s) = \frac{C(s)}{R(s)} = \frac{s^2 + 2s + 3}{s^3 + 2s^2 + 3s + 4}.$$

Draw the flow graph model, and determine the state variable matrix differential equation for the following formats: (a) phase variables, (b) input feedforward.

P3.5 A closed-loop control system is shown in Fig. P3.5. (a) Determine the closed-loop transfer function $T(s) = C(s)/R(s)$. (b) Draw the state model flow graph for the system, and determine the matrix differential equation for the phase variable

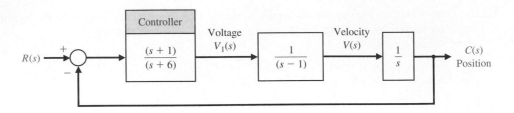

FIGURE P3.5
Closed-loop
system.

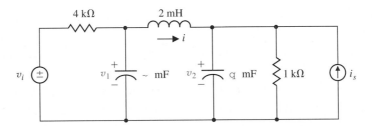

FIGURE P3.6
RLC circuit.

format. (c) Draw the signal-flow graph model, and determine the matrix differential equation for the physical variables shown in Fig. P3.5.

P3.6   Determine the state variable matrix equation for the circuit shown in Fig. P3.6. Let $x_1 = v_1$, $x_2 = v_2$, and $x_3 = i$.

P3.7   An automatic depth-control system for a robot submarine is shown in Fig. P3.7. The depth is measured by a pressure transducer. The gain of the stern plane actuator is $K = 1$ when the velocity is 25 m/s. The submarine has the approximate transfer function

$$G(s) = \frac{(s + 1)^2}{(s^2 + 1)},$$

and the feedback transducer is $H(s) = 1$. (a) Obtain a flow graph state model. (b) Determine the state differential equation for the system. (c) Determine whether the system is stable.

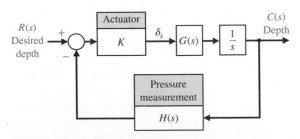

FIGURE P3.7   Submarine depth control.

P3.8   The soft landing of a lunar module descending on the moon can be modeled as shown in Fig. P3.8. Define the state variables as $x_1 = y$, $x_2 = dy/dt$, $x_3 = m$ and the control as $u = dm/dt$. Assume that g is the gravity constant on the moon. Find the state-space equations for this system. Is this a linear model?

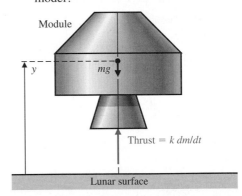

FIGURE P3.8   Lunar module landing control.

P3.9   A speed control system utilizing fluid flow components can be designed. The system is a pure fluid control system because it does not have any moving mechanical parts. The fluid may be a gas or a liquid. A system is desired that maintains the speed within 0.5% of the desired speed by using a tuning fork reference and a valve actuator. Fluid control systems are insensitive and reliable over a wide

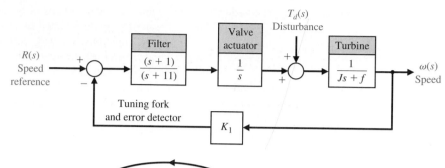

FIGURE P3.9
Steam turbine
control.

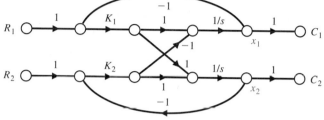

FIGURE P3.10
Two-axis system.

range of temperature, electromagnetic and nuclear radiation, acceleration, and vibration. The amplification within the system is achieved by using a fluid jet deflection amplifier. The system can be designed for a 500-kW steam turbine with a speed of 12,000 rpm. The block diagram of the system is shown in Fig. P3.9. The friction of the large inertia turbine is negligible and thus $f = 0$. The closed-loop gain is $K_1/J = 1$, where $K_1 = J = 10^4$.
(a) Determine the closed-loop transfer function

$$T(s) = \frac{\omega(s)}{R(s)},$$

and draw the state model flow graph for the form of Fig. 3.8, where all the state variables are fed back to the input node. (b) Determine the state vector differential equation. (c) Determine the characteristic equation obtained from the **A** matrix.

P3.10 Many control systems must operate in two dimensions, for example, the x- and the y-axes. A two-axis control system is shown in Fig. P3.10, where a set of state variables is identified. The gain of each axis is K_1 and K_2, respectively. (a) Obtain the state differential equation. (b) Find the characteristic equation from the **A** matrix. (c) Determine the state transition matrix for $K_1 = 1$ and $K_2 = 2$.

P3.11 A system is described by Eq. 3.16 with

$$\mathbf{A} = \begin{bmatrix} 1 & -2 \\ 2 & -3 \end{bmatrix}$$

with $u(t) = 0$ and $x_1(0) = x_2(0) = 10$. Determine $x_1(t)$ and $x_2(t)$.

P3.12 A system is described by its transfer function

$$\frac{C(s)}{R(s)} = T(s) = \frac{8(s + 5)}{s^3 + 12s^2 + 44s + 48}.$$

(a) Determine the phase variable representation.
(b) Determine the canonical variable representation of the state variable matrix equation.
(c) Determine $\mathbf{\Phi}(t)$, the state transition matrix.

P3.13 Reconsider the *RLC* circuit of Problem P3.1 when $R = 2.5$, $L = 1/4$, and $C = 1/6$. (a) Determine whether the system is stable by finding the characteristic equation with the aid of the **A** matrix. (b) Determine the transition matrix of the network. (c) When the initial inductor current is 0.1 amp, $v_c(0) = 0$, and $e(t) = 0$, determine the response of the system. (d) Repeat part (c) when the initial conditions are zero and $e(t) = E$, for $t > 0$, where E is a constant.

P3.14 Determine the signal-flow model and the matrix differential equation using the phase variable format for a system with the transfer function

$$\frac{C(s)}{R(s)} = T(s) = \frac{s^2 + 7s + 2}{s^3 + 9s^2 + 26s + 24}.$$

P3.15 Determine the signal-flow model and the matrix differential equation for (a) the phase variable format and (b) the canonical (diagonal) variable format when

$$\frac{C(s)}{R(s)} = T(s) = \frac{5(s + 6)}{s^3 + 10s^2 + 31s + 30}.$$

P3.16 A system for dispensing radioactive fluid into capsules is shown in Fig. P3.16(a). The horizontal axis moving the tray of capsules is actuated by a linear motor. The x-axis control is shown in Fig. P3.16(b). Assume $K = 500$. Obtain (a) a state variable representation and (b) the unit step-response of the system. (c) Determine the characteristic roots of the system.

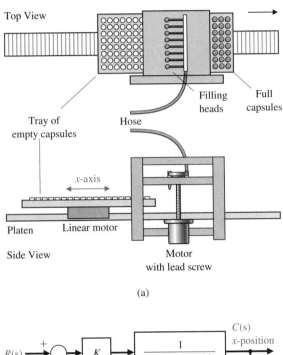

Top View

Tray of empty capsules

Filling heads

Full capsules

Hose

x-axis

Platen   Linear motor

Side View

Motor with lead screw

(a)

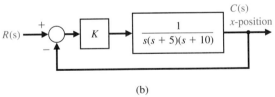

(b)

FIGURE P3.16 Automatic fluid dispenser.

P3.17 The dynamics of a controlled submarine are significantly different from those of an aircraft, missile, or surface ship. This difference results primarily from the moment in the vertical plane due to the buoyancy effect. Therefore it is interesting to consider the control of the depth of a submarine. The equations describing the dynamics of a sub-

marine can be obtained by using Newton's laws and the angles defined in Fig. P3.17. To simplify the equations, we will assume that θ is a small angle and the velocity vc is constant and equal to 25 ft/s. The state variables of the submarine, considering only vertical control, are $x_1 = \theta$, $x_2 = d\theta/dt$ and $x_3 = \alpha$, where α is the angle of attack. Thus the state vector differential equation for this system, when the submarine has an Albacore type hull, is

$$\dot{x} = \begin{bmatrix} 0 & 1 & 0 \\ -0.0071 & -0.111 & 0.12 \\ 0 & 0.07 & -0.3 \end{bmatrix} x + \begin{bmatrix} 0 \\ -0.095 \\ +0.072 \end{bmatrix} u(t),$$

where $u(t) = \delta_s(t)$, the deflection of the stern plane. (a) Determine whether the system is stable. (b) Using the discrete-time approximation, determine the response of the system to a stern plane step command of 0.285° with the initial conditions equal to zero. Use a time increment of T equal to 2 seconds. (c) Using a time increment of $T = 0.5$ second and a digital computer, obtain the transient response for each state for 80 seconds. Compare the response calculated for parts (b) and (c).

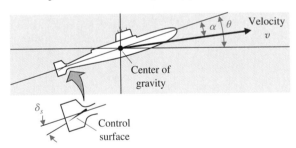

FIGURE P3.17 Submarine depth control.

P3.18 A system is described by the state variable equations

$$\dot{x} = \begin{bmatrix} 0 & 1 & 0 \\ 0 & 0 & 1 \\ -1 & -2 & -3 \end{bmatrix} x + \begin{bmatrix} 0 \\ 0 \\ 1 \end{bmatrix} u,$$

$$c = [20 \quad 30 \quad 10] \; x.$$

Determine $G(s) = C(s)/U(s)$.

P3.19 Consider the control of the robot shown in Fig. P3.19. The motor turning at the elbow moves the wrist through the forearm, which has some flexibility as shown [16]. The spring has a spring constant k and friction damping constant f. Let the state variables be $x_1 = \phi_1 - \phi_2$ and $x_2 = \omega_1/\omega_0$, where

$$\omega_0^2 = \frac{k(J_1 + J_2)}{J_1 J_2}.$$

Write the state variable equation in matrix form when $x_3 = \omega_2/\omega_0$.

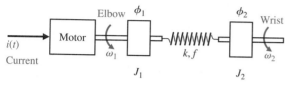

FIGURE P3.19   An industrial robot. (Courtesy of GCA Corporation.)

P3.20   The derivative of a state variable can be approximated by the equation

$$\dot{x}(t) \simeq \frac{1}{2T}[3x(k + 1) - 4x(k) + x(k - 1)].$$

This approximation of the derivative utilizes two past values to estimate the derivative, whereas Eq. (3.89) uses one past value of the state variable. Using this approximation for the derivative, repeat the calculations for Example 3.6. Compare the resulting approximation for $x_1(t)$, $T = 0.2$, with the results given in Table 3.1. Is this approximation more accurate?

P3.21   A nuclear reactor that has been operating in equilibrium at a high thermal-neutron flux level is suddenly shut down. At shutdown, the density of xenon 135(X) and iodine 135(I) are 3×10^{15} and

7×10^{16} atoms per unit volume, respectively. The half lives of I 135 and Xe 135 nucleides are 6.7 and 9.2 hours, respectively. The decay equations are [17, 23]

$$\dot{I} = -\frac{0.693}{6.7}I, \qquad \dot{X} = -\frac{0.693}{9.2}X - I.$$

Determine the concentrations of I 135 and Xe 135 as functions of time following shutdown by determining (a) the transition matrix and the system response and (b) a discrete-time evaluation of the time response. Verify that the response of the system is that shown in Fig. P3.21.

P3.22   There are several forms that are equivalent signal-flow graph state models. Two equivalent state flow graph models for a fourth-order equation (Eq. 3.38) are shown in Figs. 3.8 and 3.10. Another alternative structure for a state flow graph model is shown in Fig. P3.22. In this case, the system is second order and the input–output transfer function is

$$G(s) = \frac{C(s)}{U(s)} = \frac{b_1 s + b_0}{s^2 + a_1 s + a_0}.$$

(a) Verify that the flow graph of Fig. P3.22 is in fact a model of $G(s)$. (b) Show that the vector differ-

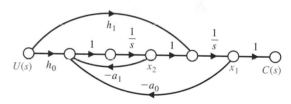

FIGURE P3.22   Model of second-order system.

FIGURE P3.21
Nuclear reactor response.

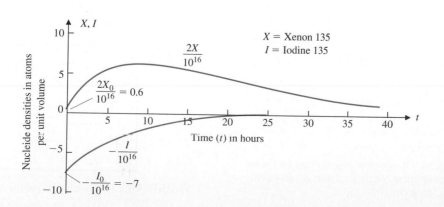

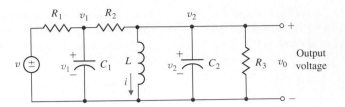

FIGURE P3.23
RLC circuit.

ential equation representing the flow graph model of Fig. P3.22 is

$$\dot{\mathbf{x}} = \begin{bmatrix} 0 & 1 \\ -a_0 & -a_1 \end{bmatrix} \mathbf{x} + \begin{bmatrix} h_1 \\ h_0 \end{bmatrix} u(t),$$

where $h_1 = b_1$ and $h_0 = b_0 - b_1 a_1$.

P3.23 Determine the state variable matrix differential equation for the circuit shown in Fig. P3.23. The state variables are $x_1 = i$, $x_2 = v_1$, and $x_3 = v_2$. The output variable is $v_0(t)$.

P3.24 The two-tank system shown in Fig. P3.24(a) is controlled by a motor adjusting the input valve and ultimately varying the output flow rate. The system has the transfer function

$$\frac{C(s)}{R(s)} = T(s) = \frac{1}{s^3 + 10s^2 + 31s + 30}$$

for the block diagram shown in Fig. P3.24(b). Determine the flow graph model and the matrix differential equation for the following flow graph models: (a) phase variables, (b) input feedforward, (c) physical state variables, and (d) decoupled state variables.

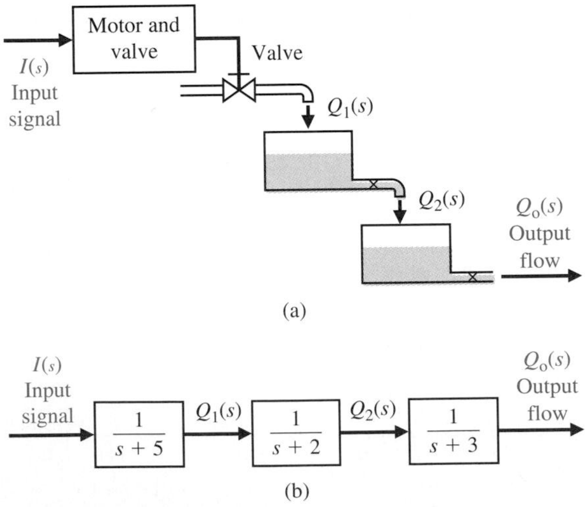

P3.25 It is desirable to use well-designed controllers to maintain building temperature with solar collector space heating systems. One solar heating system can be described by [10]

$$\frac{dx_1}{dt} = 3x_1 + u_1 + u_2,$$

$$\frac{dx_2}{dt} = 2x_2 + u_2 + d,$$

where $x_1 = $ temperature deviation from desired equilibrium and $x_2 = $ temperature of the storage material (such as a water tank). Also, u_1 and u_2 are the respective flow rates of conventional and solar heat, where the transport medium is forced air. A solar disturbance on the storage temperature (such as overcast skies) is represented by d. Write the matrix equations, and solve for the system response when $u_1 = 0$, $u_2 = 1$, and $d = 1$, with zero initial conditions.

P3.26 A system has the following differential equation:

$$\dot{\mathbf{x}} = \begin{bmatrix} -2 & -1 \\ 2 & -5 \end{bmatrix} \mathbf{x} + \begin{bmatrix} 0 \\ 1 \end{bmatrix} r(t).$$

Determine $\mathbf{\Phi}(s)$ and $\mathbf{\Phi}(t)$ for the system.

P3.27 A system has a block diagram as shown in Fig. P3.27. Determine the state variable differential equation and the state transition matrix $\mathbf{\Phi}(s)$.

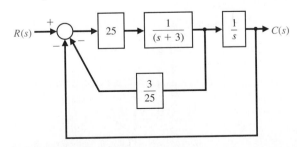

FIGURE P3.24 A two-tank system with the motor current controlling the output flow rate. (a) Physical diagram, (b) block diagram.

FIGURE P3.27 Feedback system.

P3.28 A gyroscope with a single degree of freedom is shown in Fig. P3.28. Gyroscopes sense the angular motion of a system and are used in automatic flight control systems. The gimbal moves about the output axis OB. The input is measured around the input axis OA. The equation of motion about the output axis is obtained by equating the rate of change of angular momentum to the sum of torques. Obtain a state-space representation of the gyro system.

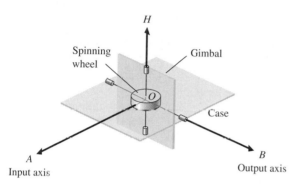

FIGURE P3.28 Gyroscope.

P3.29 A two-mass system is shown in Fig. P3.29. The rolling friction constant is b. Determine the matrix differential equation when the output variable is $y_2(t)$.

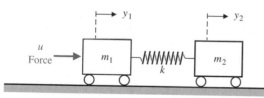

FIGURE P3.29 Two-mass system.

P3.30 There has been considerable engineering effort directed at finding ways to perform manipulative operations in space—for example, assembling a space station and acquiring target satellites. To perform such tasks, space shuttles carry a remote manipulator system (RMS) in the cargo bay [4, 15, 26]. The RMS has proven its effectiveness on recent shuttle missions, but now a new design approach is being considered—a manipulator with inflatable arm segments. Such a design might reduce manipulator weight by a factor of four while

producing a manipulator that, prior to inflation, occupies only one-eighth as much space in the shuttle's cargo bay as the present RMS.

The use of an RMS for constructing a space structure in the shuttle bay is shown in Fig. P3.30(a), and a model of the flexible RMS arm is shown in Fig. P3.30(b), where J is the inertia of the drive motor and L is the distance to the center of gravity of the load component. Derive the state equations for this system.

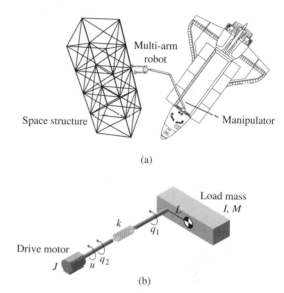

(a)

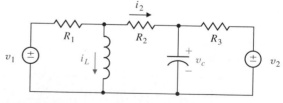

(b)

FIGURE P3.30 Remote manipulator system.

P3.31 Obtain the state equations for the two-input and one-output circuit shown in Fig. P3.31, where the output is i_2.

FIGURE P3.31 Two-input *RLC* circuit.

P3.32 Extenders are robot manipulators that extend (i.e., increase) the strength of the human arm in load maneuvering tasks (Fig. P3.32) [23, 27]. The system is represented by the transfer function

$$\frac{C(s)}{U(s)} = G(s) = \frac{1}{s^2 + 4s + 3}.$$

Using the form in Fig. 3.7, determine the state variable equations and the state transition matrix for the phase variable format.

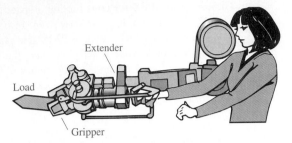

Extender

Load

Gripper

FIGURE P3.32   Extender for increasing the strength of the human arm in load maneuvering tasks.

P3.33   A drug taken orally is ingested at a rate r. The mass of the drug in the gastrointestinal tract is denoted by m_1 and in the bloodstream by m_2. The rate of change of the mass of the drug in the gastrointestinal tract is equal to the rate at which the drug is ingested minus the rate at which the drug enters the bloodstream, a rate that is taken to be proportional to the mass present. The rate of change of the mass in the bloodstream is proportional to the amount coming from the gastrointestinal tract minus the rate at which mass is lost by metabolism, which is proportional to the mass present in the blood. Develop the state space representation of this system.

For the special case where the coefficients of A are equal to 1 (with the appropriate sign), deter-

mine the response when $m_1(0) = 1$ and $m_2(0) = 0$. Plot the state variables versus time and on the $x_1 - x_2$ state plane.

P3.34   The dynamics of a rocket are represented by

$$\frac{C(s)}{U(s)} = G(s) = \frac{1}{s^2},$$

and state variable feedback is used where $x_1 = c(t)$ and $u = -x_2 - 0.5x_1$. Determine the roots of the characteristic equation of this system and the response of the system when the initial conditions are $x_1(0) = 0$ and $x_2(0) = 1$.

P3.35   A system has the transfer function

$$\frac{C(s)}{R(s)} = T(s) = \frac{8}{s^3 + 7s^2 + 14s + 8}.$$

(a) Determine the phase variable form of the state variable matrix differential equation. (b) Determine the element $\phi_{11}(t)$ of the state transition matrix for this system.

P3.36   Determine the state variable flow graph model of the one-tank system shown in Fig. P3.36. The motor inductance is negligible, the motor constant is $K_m = 10$, the back emf constant is $K_b = 0.0706$, the motor friction is negligible. The motor and valve inertia is $J = 0.006$, and the area of the tank is 50 m². Note that the motor is controlled by the armature current i_a. Let $x_1 = h$, $x_2 = \theta$, and $x_3 = d\theta/dt$. Assume that $q_1 = 80\theta$, where θ is the shaft angle. The output flow is $q_0 = 50h(t)$.

FIGURE P3.36
One-tank system.

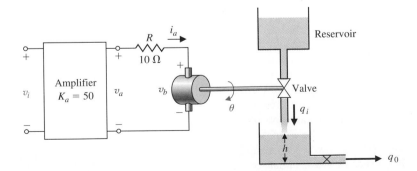

ADVANCED PROBLEMS

AP3.1   Consider the electromagnetic suspension system shown in Fig. AP3.1. An electromagnet is located at the upper part of the experimental system. Utilizing the electromagnetic force f, we desire to suspend the iron ball. Note that this simple electromagnetic suspension system is essentially unworkable. Hence feedback control is indispensable. As a gap sensor, a standard induction probe of eddy current type is placed below the ball [25].

Assume the state variables are $x_1 = x$, $x_2 = dx/dt$, and $x_3 = i$. The electromagnet has an inductance $L = 0.508$ H and a resistance $R = 23.2\ \Omega$. Use a Taylor series approximation for the electromagnetic force. The current is $i_1 = (I_0 + i)$, where $I_0 = 1.06$ A is the operating point and i is the variable. The mass m is equal to 1.75 kg. The gap is $x_g = (X_0 + x)$, where $X_0 = 4.36$ mm is the operating point and x is the variable. The electromagnetic force is $f = k(i_1/x_g)^2$, where $k = 2.9 \times 10^{-4}$ Nm^2/A^2. Determine the matrix differential equation and the equivalent transfer function $X(s)/V(s)$.

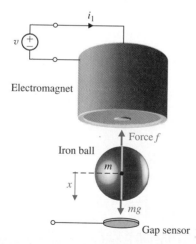

FIGURE AP3.1   Electromagnetic suspension system.

AP3.2   Consider the mass m mounted on the cart, as shown in Fig. AP3.2. Determine the transfer function $Y(s)/U(s)$, and use the transfer function to obtain the matrix differential equation for the phase variable format of Fig. 3.7.

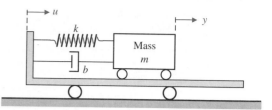

FIGURE AP3.2   Mass on cart.

AP3.3   The control of an autonomous vehicle motion from one point to another point depends on accurate control of the position of the vehicle [18]. The control of the autonomous vehicle position $C(s)$ is obtained by the system shown in Fig. AP3.3. Determine the canonical diagonal form of the matrix differential equation (or as close to the diagonal form as possible).

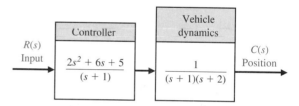

FIGURE AP3.3   Position control.

AP3.4   Front suspensions have become standard equipment on mountain bikes. Replacing the rigid fork that attaches the bicycle's front tire to its frame, such suspensions absorb bump impact energy, shielding both frame and rider from jolts. Commonly used forks, however, use only one spring constant and treat bump impacts at high and low speeds—impacts that vary greatly in severity—essentially the same.

A suspension system with multiple settings that are adjustable while the bike is in motion would be attractive. One air and coil spring with an oil damper is available that permits an adjustment of the damping constant to the terrain as well as to the rider's weight [20]. The suspension system model is shown in Fig. AP3.4, where b is adjustable. Select the appropriate value for b so that the bike accommodates (a) a large bump at high speeds and (b) a small bump at low speeds. Assume that $k_2 = 1$ and $k_1 = 2$.

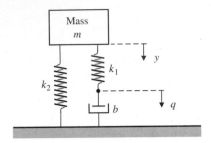

FIGURE AP3.4   Shock absorber.

AP3.5   Figure AP3.5 shows a mass M suspended from another mass m by means of a light rod of length

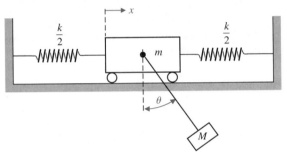

FIGURE AP3.5   Mass suspended from cart.

L. Obtain the state variable differential matrix equation using a linear model assuming a small angle for θ.

AP3.6   Consider a crane moving in the x-direction while the mass m moves in the z-direction, as shown in Fig. AP3.6. The trolley motor and the hoist motor are very powerful with respect to the mass of the trolley, the hoist wire, and the load m. Consider the input control variables as the distances D and R. Also assume that $\theta < 50°$. Determine a linear model, and describe the state variable differential equation.

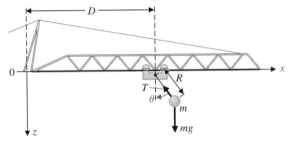

FIGURE AP3.6   A crane moving in the x-direction while the mass moves in the z-direction.

DESIGN PROBLEMS

DP3.1   A spring-mass-damper system, as shown in Fig. 3.3, is used as a shock absorber for a large high-performance motorcycle. The original parameters selected are $m = 1$ kg, $f = 9$ N $\cdot$ s/m, and $k = 20$ N/m. (a) Determine the system matrix, the characteristic roots, and the transition matrix, $\Phi(t)$. The harsh initial conditions are assumed to be $y(0) = 1$ and $dy/dt|_{t=0} = 2$. (b) Plot the response of $y(t)$ and dy/dt for the first two seconds. (c) Redesign the shock absorber by changing the spring constant and the damping constant in order to reduce the effect of a high rate of acceleration force, d^2y/dt^2, on the rider. The mass must remain constant at 1 kg.

DP3.2   A system has the state variable matrix equation in phase variable form

$$\dot{\mathbf{x}} = \begin{bmatrix} 0 & 1 \\ -a & -b \end{bmatrix} \mathbf{x} + \begin{bmatrix} 0 \\ d \end{bmatrix} u(t)$$

and $c = 10x_1$. It is desired that the canonical diagonal form of the differential equation be

$$\dot{\mathbf{y}} = \begin{bmatrix} -3 & 0 \\ 0 & -1 \end{bmatrix} \mathbf{y} + \begin{bmatrix} 1 \\ 1 \end{bmatrix} u,$$

$$c = [-5 \quad 5] \, \mathbf{y}.$$

Determine the parameters a, b, and d to yield the required diagonal matrix differential equation.

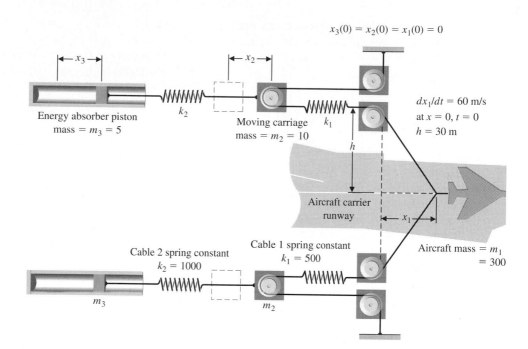

$$x_3(0) = x_2(0) = x_1(0) = 0$$

Energy absorber piston
mass = m_3 = 5

k_2

Moving carriage
mass = m_2 = 10

k_1

dx_1/dt = 60 m/s
at x = 0, t = 0
h = 30 m

h

Aircraft carrier
runway

x_1

Aircraft mass = m_1
= 300

Cable 2 spring constant
k_2 = 1000

Cable 1 spring constant
k_1 = 500

m_3

m_2

FIGURE DP3.3
Aircraft arresting
gear.

DP3.3 An aircraft arresting gear is used on an aircraft carrier as shown in Fig. DP3.3. The linear model of each energy absorber has a drag force $f_D = K_D \dot{x}_3$. It is desired to halt the airplane within 30 m after engaging the arresting cable. The speed of the aircraft on landing is 60 m/s. Select the required constant K_D, and plot the response of the state variables.

MATLAB PROBLEMS

MP3.1 Determine a state variable representation for the following transfer functions using the tf2ss function:

(a) $G(s) = \dfrac{1}{s + 10}$

(b) $G(s) = \dfrac{3s^2 + 10s + 1}{s^2 + 8s + 5}$

(c) $G(s) = \dfrac{s + 14}{s^3 + 3s^2 + 3s + 1}$

MP3.2 Consider the circuit shown in Fig. MP3.2. The transfer function from V_i to V_0 is given by

$$V_0(s)/V_i(s) =$$

$$\frac{1}{R_1 R_2 C_1 C_2 s^2 + (R_1 C_1 + R_2 C_2 + R_1 C_2)s + 1}.$$

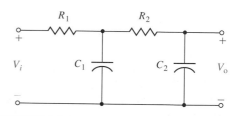

R_1

R_2

V_i

C_1

C_2

V_0

FIGURE MP3.2 An electric circuit.

Let R_1 = 1, R_2 = 10, C_1 = 0.5, and C_2 = 0.1.

(a) Show that a state variable representation is given by

$$\dot{\mathbf{x}} = \begin{bmatrix} -3.2 & -2 \\ 1 & 0 \end{bmatrix} \mathbf{x} + \begin{bmatrix} 1 \\ 0 \end{bmatrix} V_i,$$

(b) Using the state variable representation from part (a), plot the unit step response with the **step** function.

MP3.3   Consider the system

$$\dot{x} = \begin{bmatrix} 0 & 1 & 0 \\ 0 & 0 & 1 \\ -2 & -2 & -4 \end{bmatrix} x + \begin{bmatrix} 0 \\ 0 \\ 1 \end{bmatrix} u,$$

$$c = [1 \quad 0 \quad 0] \; x.$$

(a) Using the **ss2tf** function, determine the transfer function $C(s)/U(s)$.
(b) Plot the response of the system to the initial condition $x(0) = (0 \quad 0 \quad 1)^T$ for $0 \le t \le 10$.
(c) Compute the state transition matrix using the **expm** function, and determine $x(t)$ at $t = 10$ for the initial condition given in part (b). Compare the result with the system response obtained in part (b).

MP3.4   Consider the two systems

$$\dot{x}_1 = \begin{bmatrix} 0 & 1 & 0 \\ 0 & 0 & 1 \\ -4 & -5 & -8 \end{bmatrix} x_1 + \begin{bmatrix} 0 \\ 0 \\ 4 \end{bmatrix} u, \quad (1)$$

$$c = [1 \quad 0 \quad 0] \; x_1$$

and

$$\dot{x}_2 = \begin{bmatrix} 0.5000 & 0.5000 & 0.7071 \\ -0.5000 & -0.5000 & 0.7071 \\ -6.3640 & -0.7071 & -8.0000 \end{bmatrix} x_2 + \begin{bmatrix} 0 \\ 0 \\ 4 \end{bmatrix} u,$$

$$c = [0.7071 \quad -0.7071 \quad 0] \; x_2. \quad (2)$$

(a) Using the **ss2tf** function, determine the transfer function $C(s)/U(s)$ for system (1).
(b) Repeat part (a) for system (2).
(c) Compare the results in parts (a) and (b) and comment.

MP3.5   Consider the closed-loop control system in Fig. MP3.5.

(a) Determine a state variable representation of the controller using *Matlab*.
(b) Repeat part (a) for the plant.
(c) With the controller and plant in state variable form, use the **series** and **cloop** functions to compute a closed-loop system representation in state variable form and plot the closed-loop system impulse response.

FIGURE MP3.5
A closed-loop feedback control system.

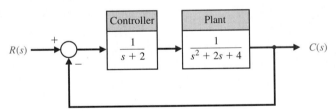

TERMS AND CONCEPTS

Discrete-time approximation   An approximation used to obtain the time response of a system based on the division of the time into small increments Δt.

Fundamental matrix   *See* Transition matrix.

State of a system   A set of numbers such that the knowledge of these numbers and the input function will, with the equations describing the dynamics, provide the future state of the system.

State variable feedback   The control signal for the process is a direct function of all the state variables.

State variables   The set of variables that describe the system.

State vector   The vector matrix containing all n state variables, $x_1, x_2, \ldots, x_n$.

State differential equation   The differential equation for the state vector: $\dot{x} = Ax + Bu$.

Time domain   The mathematical domain that incorporates the time response and the description of a system in terms of time t.

Time varying system   A system for which one or more parameters may vary with time.

Transition matrix, $\Phi(t)$ The matrix exponential function that describes the unforced response of the system.

Feedback Control System Characteristics

P R E V I E W

With the mathematical models obtained in Chapters 2 and 3, we are able to develop analytical tools for describing the characteristics of a feedback control system. In this chapter, we will develop the concepts of the system error signal. This signal is used to control the process, and our ultimate goal is to reduce the error to the smallest feasible amount.

We also will develop the concept of the sensitivity of a system to a parameter change, since it is desirable to minimize the effects of unwanted parameter variation. We then will describe the transient performance of a feedback system and show how this performance can be readily improved.

We wish to reduce the effect of unwanted input signals, called *disturbances,* on the output signal. We will show how we may design a control system to reduce the impact of disturbance signals. Of course, the benefits of a control system come with an attendant cost. We will demonstrate how the cost of using feedback in a control system is associated with the selection of the feedback sensor device.

4.1 OPEN- AND CLOSED-LOOP CONTROL SYSTEMS

Now that we are able to obtain mathematical models of the components of control systems, we will examine the characteristics of control systems. A control system was defined in Section 1.1 as an interconnection of components forming a system configuration that will provide a desired system response. Because a desired system response is known, a signal proportional to the error between the desired and the actual response is generated. The utilization of this signal to control the process results in a closed-loop sequence of operations that is called a feedback system. This closed-loop sequence of operations is shown in Fig. 4.1. The introduction of feedback to improve the control system is often necessary. It is interesting that this is also the case for systems in nature, such as biological and physiological systems, and feedback is inherent in these systems. For example, the human heart-rate control system is a feedback control system.

To illustrate the characteristics and advantages of introducing feedback, we will consider a simple, single-loop feedback system. Although many control systems are not single-loop in character, a single-loop system is illustrative. A thorough comprehension of the benefits of feedback can best be obtained from the single-loop system and then extended to multiloop systems.

A system without feedback, often called a direct system or an open-loop system, is shown in Fig. 4.2.

An open-loop (direct) system operates without feedback and directly generates the output in response to an input signal.

FIGURE 4.1
A closed-loop system.

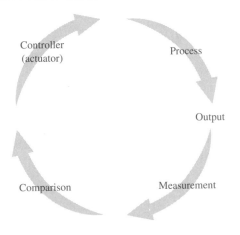

FIGURE 4.2
A direct system. A system without feedback.

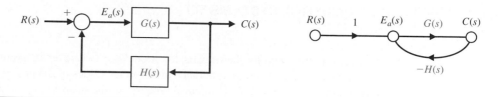

FIGURE 4.3
A closed-loop
control system (a
feedback system).

By contrast, a closed-loop negative feedback control system is shown in Fig. 4.3.

A closed-loop system uses a measurement of the output signal and a comparison with the desired output to generate an error signal that is applied to the actuator.

The two forms of control systems are shown in both block diagram and signal-flow graph form, although signal-flow graphs will be used predominantly for subsequent diagrams.

In many cases, $H(s)$ is equal to 1 or a constant other than 1. The constant accounts for a unit conversion such as radians to volts. First, let us consider the unity-feedback condition with $H(s) = 1$. Then, $E_a(s) = E(s)$ and

$$C(s) = G(s)E(s) = G(s)[R(s) - C(s)].$$

Solving for $C(s)$, we have

$$C(s) = \frac{G(s)}{1 + G(s)} R(s). \tag{4.1}$$

The error signal is

$$E(s) = \frac{1}{1 + G(s)} R(s). \tag{4.2}$$

Thus, to reduce the error, the magnitude of $[1 + G(s)]$ must be greater than one over the range of s under consideration.

Now consider the case where $H(s) \neq 1$. The output of the closed-loop system is

$$C(s) = G(s)E_a(s) = G(s)[R(s) - H(s)C(s)],$$

and therefore

$$C(s) = \frac{G(s)}{1 + GH(s)} R(s).$$

The actuating error signal is

$$E_a(s) = \frac{1}{1 + GH(s)} R(s). \tag{4.3}$$

It is clear that, to reduce the error, the magnitude of $[1 + GH(s)]$ must be greater than one over the range of s under consideration. The signal $E_a(s)$ provides a measure of the error $E(s)$. This measure is increasingly accurate as the dynamics of $H(s)$ become negligible and $H(s) \cong 1$ for the range of s under consideration.

4.2 SENSITIVITY OF CONTROL SYSTEMS TO PARAMETER VARIATIONS

A process, represented by the transfer function $G(s)$, whatever its nature, is subject to a changing environment, aging, ignorance of the exact values of the process parameters, and other natural factors that affect a control process. In the open-loop system, all these errors and changes result in a changing and inaccurate output. However, a closed-loop system senses the change in the output due to the process changes and attempts to correct the output. The *sensitivity* of a control system to parameter variations is of prime importance. A primary advantage of a closed-loop feedback control system is its ability to reduce the system's sensitivity [1–4, 18].

For the closed-loop case, if $GH(s) \gg 1$ for all complex frequencies of interest, then from Eq. (4.2) we obtain

$$C(s) \cong \frac{1}{H(s)} R(s). \tag{4.4}$$

Then, the output is affected only by $H(s)$, which may be a constant. If $H(s) = 1$, we have the desired result; that is, the output is equal to the input. However, before we use this approach for all control systems, we must note that the requirement that $G(s)H(s) \gg 1$ may cause the system response to be highly oscillatory and even unstable. But the fact that as we increase the magnitude of the loop transfer function $G(s)H(s)$, we reduce the effect of $G(s)$ on the output, is an exceedingly useful concept. Therefore the *first advantage* of a feedback system is that the effect of the *variation of the parameters of the process, $G(s)$,* is reduced.

To illustrate the effect of parameter variations, let us consider a change in the process so that the new process is $G(s) + \Delta G(s)$. Then, in the open-loop case, the change in the transform of the output is

$$\Delta C(s) = \Delta G(s)R(s). \tag{4.5}$$

In the closed-loop system, we have

$$C(s) + \Delta C(s) = \frac{G(s) + \Delta G(s)}{1 + (G(s) + \Delta G(s))H(s)} R(s). \tag{4.6}$$

Then the change in the output is

$$\Delta C(s) = \frac{\Delta G(s)}{(1 + GH(s) + \Delta GH(s))(1 + GH(s))} R(s). \tag{4.7}$$

When $GH(s) \gg \Delta GH(s)$, as is often the case, we have

$$\Delta C(s) = \frac{\Delta G(s)}{(1 + GH(s))^2} R(s). \tag{4.8}$$

Examining Eq. (4.8), we note that the change in the output of the closed-loop system is reduced by the factor $[1 + GH(s)]$, which is usually much greater than one over the range of complex frequencies of interest. The factor $1 + GH(s)$ plays a very important role in the characteristics of feedback control systems.

The *system sensitivity* is defined as the ratio of the percentage change in the system transfer function to the percentage change of the process transfer function. The system transfer function is

$$T(s) = \frac{C(s)}{R(s)}, \qquad (4.9)$$

and therefore the sensitivity is defined as

$$S = \frac{\Delta T(s)/T(s)}{\Delta G(s)/G(s)}. \qquad (4.10)$$

In the limit, for small incremental changes, Eq. (4.10) becomes

$$S = \frac{\partial T/T}{\partial G/G} = \frac{\partial \ln T}{\partial \ln G}. \qquad (4.11)$$

System sensitivity is the ratio of the change in the system transfer function to the change of a process transfer function (or parameter) for a small incremental change.

Clearly, from Eq. (4.5), the sensitivity of the open-loop system is equal to one. The sensitivity of the closed-loop is readily obtained by using Eq. (4.11). The system transfer function of the closed-loop system is

$$T(s) = \frac{G}{1 + GH}.$$

Therefore the sensitivity of the feedback system is

$$S_G^T = \frac{\partial T}{\partial G} \cdot \frac{G}{T} = \frac{1}{(1 + GH)^2} \cdot \frac{G}{G/(1 + GH)} = \frac{1}{1 + GH(s)}. \qquad (4.12)$$

Again, we find that the sensitivity of the system may be reduced below that of the open-loop system by increasing $GH(s)$ over the frequency range of interest.

The sensitivity of the feedback system to changes in the feedback element $H(s)$ is

$$S_H^T = \frac{\partial T}{\partial H} \cdot \frac{H}{T} = \left(\frac{G}{1 + GH}\right)^2 \cdot \frac{-H}{G/(1 + GH)} = \frac{-GH}{1 + GH}. \qquad (4.13)$$

When GH is large, the sensitivity approaches unity and the changes in $H(s)$ directly affect the output response. Therefore it is important to use feedback components that will not vary with environmental changes or that can be maintained constant.

Often we seek to determine S_α^T, where α is a parameter within the transfer function of a block G. Using the chain rule, we find

$$S_\alpha^T = S_G^T S_\alpha^G. \qquad (4.14)$$

Very often the transfer function of the system $T(s)$ is a fraction of the form [1]:

$$T(s,\alpha) = \frac{N(s,\alpha)}{D(s,\alpha)}, \qquad (4.15)$$

where α is a parameter that may be subject to variation due to the environment. Then we may obtain the sensitivity to α by rewriting Eq. (4.11) as:

$$S_\alpha^T = \frac{\partial \ln T}{\partial \ln \alpha} = \frac{\partial \ln N}{\partial \ln \alpha}\bigg|_{\alpha_0} - \frac{\partial \ln D}{\partial \ln \alpha}\bigg|_{\alpha_0} = S_\alpha^N - S_\alpha^D, \qquad (4.16)$$

where α_0 is the nominal value of the parameter.

The ability to reduce the effect of the variation of parameters of a control system by adding a feedback loop is an important advantage of feedback control systems. To obtain highly accurate open-loop systems, the components of the open-loop $G(s)$ must be selected carefully in order to meet the exact specifications. However, a closed-loop system allows $G(s)$ to be less accurately specified, because the sensitivity to changes or errors in $G(s)$ is reduced by the loop gain $1 + GH(s)$. This benefit of closed-loop systems is a profound advantage for the electronic amplifiers of the communication industry. A simple example will illustrate the value of feedback for reducing sensitivity.

EXAMPLE 4.1   **Inverting amplifier**

The integrated circuit operational amplifier can be fabricated on a single chip of silicon and sold for less than a dollar. As a result, IC operational amplifiers (op amps) are widely used. The model symbol of an op amp is shown in Fig. 4.4(a) [6]. We can assume that the magnitude of the gain A is greater than 10^4. The basic inverting amplifier circuit is shown in Fig. 4.4(b). Because of the high input impedance of the op amp, the amplifier input current is negligibly small (assume $R_i \rightarrow \infty$). At node n, we may write the current equation as

$$\frac{e_{in} - v_n}{R_1} + \frac{v_o - v_n}{R_f} = 0. \qquad (4.17)$$

Because the gain of the amplifier is $-A$, $v_o = -Av_n$ and therefore

$$v_n = -\frac{v_o}{A}, \qquad (4.18)$$

and we may substitute Eq. (4.18) into Eq. (4.17), obtaining

$$\frac{e_{in}}{R_1} + \frac{v_o}{AR_1} + \frac{v_o}{R_f} + \frac{v_o}{AR_f} = 0. \qquad (4.19)$$

Solving for the output voltage, we have

$$v_o = \frac{-A(R_f/R_1)e_{in}}{1 + (R_f/R_1) + A}. \qquad (4.20)$$

FIGURE 4.4
(a) An operational amplifier model symbol. (b) Inverting amplifier circuit.

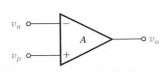

(a)

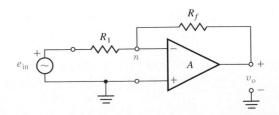

(b)

FIGURE 4.5
Signal-flow graph
of inverting
amplifier.

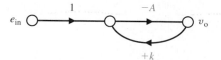

Alternatively, when $A \gg 1$, we may rewrite Eq. (4.20) as follows:

$$\frac{v_o}{e_{in}} = \frac{-A}{1 + A(R_1/R_f)} = \frac{-A}{1 + Ak}, \tag{4.21}$$

where $k = R_1/R_f$. The signal-flow graph representation of the inverting amplifier is shown in Fig. 4.5. Note that when $|A| \gg 1$, we have

$$T = \frac{v_o}{e_{in}} = -\frac{1}{k} = -\frac{R_f}{R_1}. \tag{4.22}$$

The feedback factor in the diagram is $H(s) = k$, and the open-loop transfer function is $G(s) = A$.

The op amp is subject to variations in the amplification A. The sensitivity of the open loop is unity. The sensitivity of the closed-loop amplifier is

$$S_A^T = \frac{\partial T/T}{\partial A/A} = \frac{1}{1 - GH} = \frac{1}{1 + Ak}. \tag{4.23}$$

If $A = 10^4$ and $k = 0.1$, we have

$$S_A^T = \frac{1}{1 + 10^3}, \tag{4.24}$$

or the magnitude of the sensitivity is approximately equal to 0.001, which is one-thousandth of the magnitude of the open-loop sensitivity. The sensitivity due to changes in the feedback resistance R_f (or the factor k) is

$$S_k^T = \frac{GH}{1 - GH} = \frac{-Ak}{1 + Ak}, \tag{4.25}$$

and the magnitude of the sensitivity to k is approximately equal to 1. ∎

We shall return to the concept of sensitivity in subsequent chapters to emphasize the importance of sensitivity in the design and analysis of control systems.

4.3 CONTROL OF THE TRANSIENT RESPONSE OF CONTROL SYSTEMS

One of the most important characteristics of control systems is their transient response. The *transient response* is the response of a system as a function of time. Because the purpose of control systems is to provide a desired response, the transient response of control systems often must be adjusted until it is satisfactory. If an open-loop control system does not provide a satisfactory response, then the process, $G(s)$, must be replaced with a

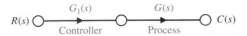

FIGURE 4.6
Cascade controller system (without feedback).

suitable process. By contrast, a closed-loop system can often be adjusted to yield the desired response by adjusting the feedback loop parameters. It should be noted that it is often possible to alter the response of an open-loop system by inserting a suitable cascade controller, $G_1(s)$, preceding the process, $G(s)$, as shown in Fig. 4.6. Then it is necessary to design the cascade transfer function $G_1(s)G(s)$ so that the resulting transfer function provides the desired transient response.

To make this concept more readily comprehensible, let us consider a specific control system, which may be operated in an open- or closed-loop manner. A speed control system, which is shown in Fig. 4.7, is often used in industrial processes to move materials and products. Several important speed control systems are used in steel mills for rolling the steel sheets and moving the steel through the mill [19]. The transfer function of the open-loop system (without feedback) was obtained in Eq. (2.70), and for $\omega(s)/V_a(s)$ we have

$$\frac{\omega(s)}{V_a(s)} = G(s) = \frac{K_1}{(\tau_1 s + 1)}, \tag{4.26}$$

where

$$K_1 = \frac{K_m}{(R_a f + K_b K_m)} \quad \text{and} \quad \tau_1 = \frac{R_a J}{(R_a f + K_b K_m)}.$$

In the case of a steel mill, the inertia of the rolls is quite large and a large armature-controlled motor is required. If the steel rolls are subjected to a step command for a speed change of

$$V_a(s) = \frac{k_2 E}{s}, \tag{4.27}$$

the output response is

$$\omega(s) = G(s)V_a(s). \tag{4.28}$$

The transient speed change is then

$$\omega(t) = K_1(k_2 E)(1 - e^{-t/\tau_1}). \tag{4.29}$$

FIGURE 4.7
Open-loop speed
control system
(without feedback).

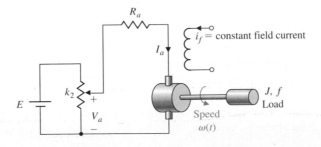

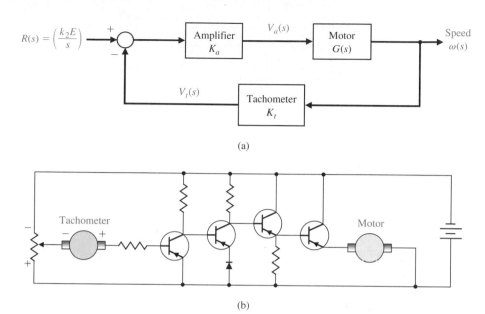

FIGURE 4.8
(a) Closed-loop
speed control
system. (b)
Transistorized
closed-loop speed
control system.

If this transient response is too slow, there is little choice but to choose another motor with a different time constant τ_1, if possible. However, because τ_1 is dominated by the load inertia, J, little hope for much alteration of the transient response remains.

A closed-loop speed control system is easily obtained by utilizing a tachometer to generate a voltage proportional to the speed, as shown in Fig. 4.8(a). This voltage is subtracted from the potentiometer voltage and amplified as shown in Fig. 4.8(a). A practical transistor amplifier circuit for accomplishing this feedback in low-power applications is shown in Fig. 4.8(b) [1, 5, 7]. The closed-loop transfer function is

$$\frac{\omega(s)}{R(s)} = \frac{K_a G(s)}{1 + K_a K_t G(s)}$$

(4.30)

$$= \frac{K_a K_1}{\tau_1 s + 1 + K_a K_t K_1} = \frac{K_a K_1 / \tau_1}{s + [(1 + K_a K_t K_1)/\tau_1]}.$$

The amplifier gain K_a may be adjusted to meet the required transient response specifications. Also, the tachometer gain constant K_t may be varied, if necessary.

The transient response to a step change in the input command is then

$$\omega(t) = \frac{K_a K_1}{(1 + K_a K_t K_1)} (k_2 E)(1 - e^{-pt}),$$

(4.31)

where $p = (1 + K_a K_t K_1)/\tau_1$. Because the load inertia is assumed to be very large, we alter the response by increasing K_a, and we have the approximate response

$$\omega(t) \simeq \frac{1}{K_t} (k_2 E) \left[1 - \exp\left(\frac{-(K_a K_t K_1)t}{\tau_1} \right) \right].$$

(4.32)

For a typical application, the open-loop pole might be $1/\tau_1 = 0.10$, whereas the closed-loop pole could be at least $(K_a K_t K_1)/\tau_1 = 10$, a factor of 100 in the improvement of the

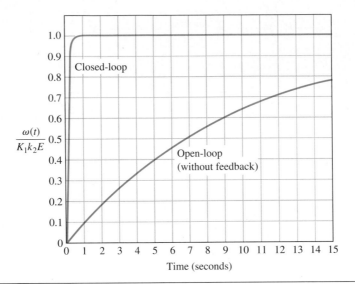

FIGURE 4.9
The response of
the open-loop and
closed-loop speed
control system
when $\tau = 10$ and
$K_t K_a K_t = 100$. The
time to reach 98%
of the final value
for the open-loop
and closed-loop
system is 40
seconds and
0.4 second,
respectively.

speed of response. It should be noted that to attain the gain $K_a K_t K_1$, the amplifier gain K_a must be reasonably large, and the armature voltage signal to the motor and its associated torque signal will be larger for the closed-loop than for the open-loop operation. Therefore a higher-power motor will be required to avoid saturation of the motor. The responses of the closed-loop system and the open-loop system are shown in Fig. 4.9. Note the rapid response of the closed-loop system.

While we are considering this speed control system, it will be worthwhile to determine the sensitivity of the open- and closed-loop systems. As before, the sensitivity of the open-loop system to a variation in the motor constant or the potentiometer constant k_2 is unity. The sensitivity of the closed-loop system to a variation in K_m is

$$S_{K_m}^T = \frac{1}{1 + GH(s)} \tag{4.33}$$

$$= \frac{1}{1 + K_a K_t G(s)} = \frac{[s + (1/\tau_1)]}{[s + (K_a K_t K_1 + 1)/\tau_1]}.$$

Using the typical values given in the previous paragraph, we have

$$S_{K_m}^T = \frac{(s + 0.10)}{(s + 10)}. \tag{4.34}$$

We find that the sensitivity is a function of s and must be evaluated for various values of frequency. This type of frequency analysis is straightforward but will be deferred until a later chapter. However, it is clearly seen that at a specific low frequency—for example, $s = j\omega = j1$—the magnitude of the sensitivity is approximately $|S_{K_m}^T| \cong 0.1$.

4.4 DISTURBANCE SIGNALS IN A FEEDBACK CONTROL SYSTEM

The third most important effect of feedback in a control system is the control and partial elimination of the effect of disturbance signals. A *disturbance signal* is an unwanted input signal that affects the system's output signal. Many control systems are subject to extrane-

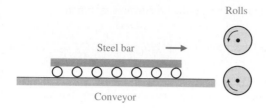

FIGURE 4.10
Steel rolling mill.

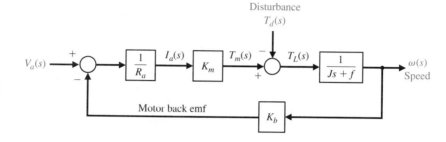

FIGURE 4.11
Open-loop speed
control system
(without external
feedback).

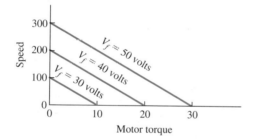

FIGURE 4.12
Motor speed–
torque curves.

ous disturbance signals that cause the system to provide an inaccurate output. Electronic amplifiers have inherent noise generated within the integrated circuits or transistors; radar antennas are subjected to wind gusts; and many systems generate unwanted distortion signals due to nonlinear elements. Feedback systems have the beneficial aspect that the effect of distortion, noise, and unwanted disturbances can be effectively reduced.

As a specific example of a system with an unwanted disturbance, let us reconsider the speed control system for a steel rolling mill. Rolls passing steel through are subject to large load changes or disturbances. As a steel bar approaches the rolls (see Fig. 4.10), the rolls turn unloaded. However, when the bar engages in the rolls, the load on the rolls increases immediately to a large value. This loading effect can be approximated by a step change of disturbance torque as shown in Fig. 4.11. Alternatively, we might examine the speed–torque curves of a typical motor as shown in Fig. 4.12.

The transfer function model of an armature-controlled dc motor with a load torque disturbance was determined in Example 2.3 and is shown in Fig. 4.11, where it is assumed that L_a is negligible. The error for the system shown in Fig. 4.11 is

$$E(s) = R(s) - \omega(s),$$

and $R(s) = \omega_d(s)$, the desired speed. For calculation ease, we let $R(s) = 0$ and examine $E(s) = -\omega(s)$.

The change in speed due to the load disturbance is then

$$E(s) = -\omega(s) = \left(\frac{1}{Js + f + (K_m K_b / R_a)} \right) T_d(s). \tag{4.35}$$

The steady-state error in speed due to the load torque $T_d(s) = D/s$ is found by using the final-value theorem. Therefore, for the open-loop system, we have

$$\lim_{t \to \infty} E(t) = \lim_{s \to 0} sE(s) = \lim_{s \to 0} s \left(\frac{1}{Js + f + (K_m K_b / R_a)} \right) \left(\frac{D}{s} \right) \tag{4.36}$$

$$= \frac{D}{f + (K_m K_b / R_a)} = -\omega_0(\infty).$$

The closed-loop speed control system is shown in block diagram form in Fig. 4.13. The closed-loop system is shown in signal-flow graph form in Fig. 4.14, where $G_1(s) = K_a K_m / R_a$, $G_2(s) = 1/(Js + f)$, and $H(s) = K_t + (K_b / K_a)$. The error, $E(s) = -\omega(s)$, of the closed-loop system of Fig. 4.14 can be obtained by utilizing the signal-flow gain formula:

$$E(s) = -\omega(s) = \frac{G_2(s)}{1 + G_1(s)G_2(s)H(s)} T_d(s). \tag{4.37}$$

Then, if $G_1 G_2 H(s)$ is much greater than one over the range of s, we obtain the approximate result

$$E(s) \simeq \frac{1}{G_1(s)H(s)} T_d(s). \tag{4.38}$$

FIGURE 4.13
Closed-loop speed tachometer control system.

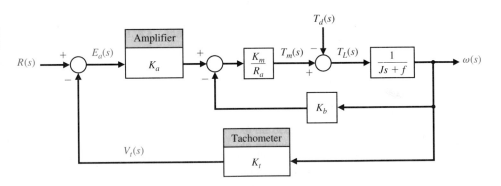

FIGURE 4.14
Signal-flow graph of closed-loop system.

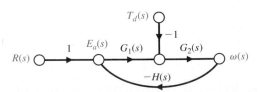

Therefore if $G_1(s)H(s)$ is made sufficiently large, the effect of the disturbance can be decreased by closed-loop feedback. Note that

$$G_1(s)H(s) = \frac{K_a K_m}{R_a}\left(K_t + \frac{K_b}{K_a}\right) \approx \frac{K_a K_m K_t}{R_a},$$

since $K_a \gg K_b$. Thus we strive to obtain a large amplifier gain K_a and keep $R_a < 2\ \Omega$. The error for the system shown in Fig. 4.14 is

$$E(s) = R(s) - \omega(s)$$

and $R(s) = \omega_d(s)$, the desired speed. For calculation ease, we let $R(s) = 0$ and examine $\omega(s)$.

The output for the speed control system of Fig. 4.13 due to the load disturbance, when the input $R(s) = 0$, may be obtained by using Mason's formula as

$$\omega(s) = \frac{-[1/(Js + f)]}{1 + (K_t K_a K_m / R_a)[1/(Js + f)] + (K_m K_b / R_a)[1/(Js + f)]} T_d(s)$$

$$= \frac{-1}{Js + f + (K_m / R_a)(K_t K_a + K_b)} T_d(s). \tag{4.39}$$

Again, the steady-state output is obtained by utilizing the final-value theorem, and we have

$$\lim_{t \to \infty} \omega(t) = \lim_{s \to 0} (s\omega(s)) = \frac{-1}{f + (K_m / R_a)(K_t K_a + K_b)} D; \tag{4.40}$$

when the amplifier gain K_a is sufficiently high, we have

$$\omega(\infty) \simeq \frac{-R_a}{K_a K_m K_t} D = \omega_c(\infty). \tag{4.41}$$

The ratio of closed-loop to open-loop steady-state speed output due to an undesirable disturbance is

$$\frac{\omega_c(\infty)}{\omega_0(\infty)} = \frac{R_a f + K_m K_b}{K_a K_m K_t} \tag{4.42}$$

and is usually less than 0.02.

This advantage of a feedback speed control system can also be illustrated by considering the speed–torque curves for the closed-loop system. The closed-loop system speed–torque curves are shown in Fig. 4.15. Clearly, the improvement of the feedback system is evidenced by the almost horizontal curves, which indicate that the speed is almost independent of the load torque.

In general, a primary reason for introducing feedback is the ability to alleviate the effects of disturbances and noise signals occurring within the feedback loop. A noise signal that is prevalent in many systems is the noise generated by the measurement sensor. This noise, $N(s)$, can be represented as shown in Fig. 4.16. The effect of the noise on the output is

$$C(s) = \frac{-G_1 G_2 H_2(s)}{1 + G_1 G_2 H_1 H_2(s)} N(s), \tag{4.43}$$

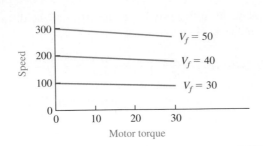

FIGURE 4.15
The closed-loop
system speed–
torque curves.

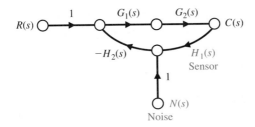

FIGURE 4.16
Closed-loop
control system
with measurement
noise.

which is approximately

$$C(s) \cong - \frac{1}{H_1(s)} N(s). \tag{4.44}$$

Clearly, the designer must obtain a maximum value of $H_1(s)$, which is equivalent to maximizing the signal-to-noise ratio of the measurement sensor. This necessity is equivalent to requiring that the feedback elements $H(s)$ be well designed and operated with minimum noise, drift, and parameter variation. This is equivalent to the requirement determined from the sensitivity function, Eq. (4.13), which showed that $S_H^T \cong 1$. Therefore one must be aware of the necessity of assuring the quality and constancy of the feedback sensors and elements. This is usually possible because the feedback elements operate at low power levels and can be well designed at reasonable cost.

The equivalency of sensitivity S_G^T and the response of the closed-loop system to a disturbance input can be illustrated by considering Fig. 4.14. The sensitivity of the system to G_2 is

$$S_{G_2}^T = \frac{1}{1 + G_1 G_2 H(s)} \cong \frac{1}{G_1 G_2 H(s)}. \tag{4.45}$$

The effect of the disturbance on the output (with $R(s) = 0$) is

$$\frac{C(s)}{T_d(s)} = \frac{-G_2(s)}{1 + G_1 G_2 H(s)} \cong \frac{-1}{G_1 H(s)}. \tag{4.46}$$

In both cases, we found that the undesired effects could be alleviated by increasing $G_1(s) = K_a$, the amplifier gain. The utilization of feedback in control systems is primarily to reduce the sensitivity of the system to parameter variations and the effect of disturbance inputs. It is noteworthy that the effort to reduce the effects of parameter variations or dis-

FIGURE 4.17
Closed-loop
control system
with output noise.

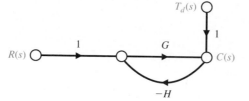

turbances is equivalent and we have the fortunate circumstance that they reduce simulta-neously. As a final illustration of this fact, we note that for the system shown in Fig. 4.17, the effect of the noise or disturbance on the output is

$$\frac{C(s)}{T_d(s)} = \frac{1}{1 + GH(s)}, \tag{4.47}$$

which is identically equal to the sensitivity S_G^T.

Quite often, noise is present at the input to the control system. For example, the signal at the input to the system might be $r(t) + n(t)$, where $r(t)$ is the desired system response and $n(t)$ is the noise signal. The feedback control system, in this case, will simply process the noise as well as the input signal $r(t)$ and will not be able to improve the signal–noise ratio, which is present at the input to the system. However, if the frequency spectrums of the noise and input signals are of a different character, the output signal–noise ratio can be maximized, often by simply designing a closed-loop system transfer function that has a low-pass frequency response.

4.5 STEADY-STATE ERROR

A feedback control system is valuable because it provides the engineer with the ability to adjust the transient response. In addition, as we have seen, the sensitivity of the system and the effect of disturbances can be reduced significantly. However, as a further requirement, one must examine and compare the final steady-state error for an open-loop and a closed-loop system. The *steady-state error* is the error after the transient response has decayed, leaving only the continuous response.

The error of the open-loop system shown in Fig. 4.18 is

$$E_0(s) = R(s) - C(s) = (1 - G(s))R(s). \tag{4.48}$$

The error of the closed-loop system, $E_c(s)$, shown in Fig. 4.19, when $H(s) = 1$, is (Eq. 4.2)[†]

$$E_c(s) = \frac{1}{1 + G(s)} R(s). \tag{4.49}$$

To calculate the steady-state error, we utilize the final-value theorem, which is

$$\lim_{t \to \infty} e(t) = \lim_{s \to 0} sE(s). \tag{4.50}$$

[†] Nonunity feedback ($H(s) \neq 1$) is discussed in Section 5.8.

FIGURE 4.18
Open-loop control system (without feedback).

FIGURE 4.19
Closed-loop
feedback control
system.

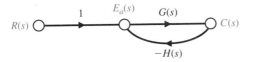

Therefore, using a unit step input as a comparable input, we obtain for the open-loop system

$$e_0(\infty) = \lim_{s \to 0} s(1 - G(s))\left(\frac{1}{s}\right) = \lim_{s \to 0} (1 - G(s)) = 1 - G(0). \qquad (4.51)$$

For the closed-loop system, when $H(s) = 1$, we have

$$e_c(\infty) = \lim_{s \to 0} s\left(\frac{1}{1 + G(s)}\right)\left(\frac{1}{s}\right) = \frac{1}{1 + G(0)}. \qquad (4.52)$$

The value of $G(s)$ when $s = 0$ is often called the dc gain and is normally greater than one. Therefore the open-loop system will usually have a steady-state error of significant magnitude. By contrast, the closed-loop system with a reasonably large dc loop gain $G(0)$ will have a small steady-state error.

Upon examination of Eq. (4.51), one notes that the open-loop control system can possess a zero steady-state error by simply adjusting and calibrating the dc gain, $G(0)$, of the system, so that $G(0) = 1$. Therefore one may logically ask, What is the advantage of the closed-loop system in this case? Again, we return to the concept of the sensitivity of the system to parameter changes as our answer to this question. In the open-loop system, one may calibrate the system so that $G(0) = 1$, but during the operation of the system, it is inevitable that the parameters of $G(s)$ will change due to environmental changes and that the dc gain of the system will no longer be equal to one. However, because it is an open-loop system, the steady-state error will remain other than zero until the system is maintained and recalibrated. By contrast, the closed-loop feedback system continually monitors the steady-state error and provides an actuating signal to reduce the steady-state error. Thus we find that it is the sensitivity of the system to parameter drift, environmental effects, and calibration errors that encourages the introduction of negative feedback. An example of an ingenious feedback control system is shown in Fig. 4.20.

The advantage of the closed-loop system in reducing the steady-state error of the system resulting from parameter changes and calibration errors may be illustrated by an example. Let us consider a system with a process transfer function

$$G(s) = \frac{K}{\tau s + 1}, \qquad (4.53)$$

which would represent a thermal control process, a voltage regulator, or a water-level control process. For a specific setting of the desired input variable, which may be represented

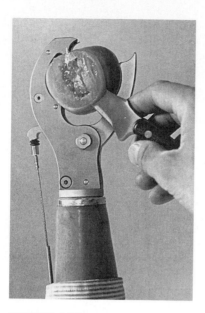

FIGURE 4.20
The Grip II is a prosthetic artificial hand that is cable operated. It can be used to operate an automobile manual shift, drive a nail, slice a tomato, and perform other normal tasks requiring two hands. It is based on a "pull to close" cable action and has a gripping force ranging from 0 to 110 pounds. The hand provides the movement of a thumb and forefinger and closes when effort is exerted on the cable by the person's back muscles. A person's vision system provides feedback, but the person does lack the normal sense of touch most of us use to grasp an object lightly. (Courtesy of Therapeutic Recreation Systems, Inc.)

by the normalized unit step input function, we have $R(s) = 1/s$. Then, the steady-state error of the open-loop system is, as in Eq. (4.51),

$$e_0(\infty) = 1 - G(0) = 1 - K \qquad (4.54)$$

when a consistent set of dimensional units are utilized for $R(s)$ and K. The error for the closed-loop system of Fig. 4.19 is

$$E_c(s) = R(s) - T(s)R(s),$$

where $T(s) = G(s)/(1 + GH(s))$. The steady-state error is

$$e_c(\infty) = \lim_{s \to 0} s\{1 - T(s)\} \frac{1}{s} = 1 - T(0).$$

When $H(s) = 1/(\tau_1 s + 1)$, we obtain $H(0) = 1$ and $G(0) = K$. Then we have

$$e_c(\infty) = 1 - \frac{K}{1 + K} = \frac{1}{1 + K}. \qquad (4.55)$$

For the open-loop system, one would calibrate the system so that $K = 1$ and the steady-state error is zero. For the closed-loop system, one would set a large gain K, for example, $K = 100$. Then the closed-loop system steady-state error is $e_c(\infty) = 1/101$.

If the calibration of the gain setting drifts or changes in some way by $\Delta K/K = 0.1$, a 10% change, the open-loop steady-state error is $\Delta e_0(\infty) = 0.1$ and the percent change from the calibrated setting is

$$\frac{\Delta e_0(\infty)}{|r(t)|} = \frac{0.10}{1}, \tag{4.56}$$

or 10%. By contrast, the steady-state error of the closed-loop system, with $\Delta K/K = 0.1$, is $e_c(\infty) = \tfrac{1}{91}$ if the gain decreases. Thus the change in the

$$\Delta e_c(\infty) = \frac{1}{101} - \frac{1}{91}, \tag{4.57}$$

and the relative change is

$$\frac{\Delta e_c(\infty)}{|r(t)|} = 0.0011, \tag{4.58}$$

or 0.11%. This is indeed a significant improvement.

4.6 THE COST OF FEEDBACK

The addition of feedback to a control system results in the advantages outlined in the previous sections. Naturally, however, these advantages have an attendant cost. The cost of feedback is first manifested in an increased number of *components* and *complexity* in the system. To add the feedback, it is necessary to consider several feedback components, of which the measurement component (sensor) is the key one. The sensor is often the most expensive component in a control system. Furthermore, the sensor introduces noise and inaccuracies into the system.

The second cost of feedback is the *loss of gain*. For example, in a single-loop system, the open-loop gain is $G(s)$ and is reduced to $G(s)/(1 + G(s))$ in a unity negative feedback system. The reduction in closed-loop gain is $1/(1 + G(s))$, which is exactly the factor that reduces the sensitivity of the system to parameter variations and disturbances. Usually, we have open-loop gain to spare, and we are more than willing to trade it for increased control of the system response.

However, we should note that it is the gain of the input–output transmittance that is reduced. The control system does retain the substantial power gain of a power amplifier and actuator, which is fully utilized in the closed-loop system.

Finally, a cost of feedback is the introduction of the possibility of *instability*. Whereas the open-loop system is stable, the closed-loop system may not be always stable. The question of the stability of a closed-loop system is deferred until Chapter 6, where it can be treated more completely.

The addition of feedback to dynamic systems results in several additional problems for the designer. However, for most cases, the advantages far outweigh the disadvantages, and a feedback system is utilized. Therefore it is necessary to consider the additional complexity and the problem of stability when designing a control system.

It has become clear that it is desired that the output of the system $C(s)$ equal the input $R(s)$. However, upon reflection, one might ask, Why not simply set the transfer function $G(s) = C(s)/R(s)$ equal to 1? (See Fig. 4.2.) The answer to this question becomes apparent once we recall that the process (or plant) $G(s)$ was necessary to provide the desired output; that is, the transfer function $G(s)$ represents a real process and possesses dynamics that cannot be neglected. If we set $G(s)$ equal to 1, we imply that the output is directly connected to the input. However, one must recall that a specific output, such as temperature, shaft rotation, or engine speed, is desired, whereas the input might be a potentiometer setting or a voltage. The process $G(s)$ is necessary to provide the physical process between $R(s)$ and $C(s)$. Therefore a transfer function $G(s) = 1$ is unrealizable, and we must settle for a practical transfer function.

4.7 DESIGN EXAMPLE: ENGLISH CHANNEL BORING MACHINES

The construction of the tunnel under the English Channel from France to Great Britain began in December 1987. The first connection of the boring tunnels from each country was achieved in November 1990. The tunnel is 23.5 miles long and is bored 200 feet below sea level. The tunnel, completed in 1992 at a total cost of $14 billion, accommodates 500 train trips daily. This construction is a critical link between Europe and Great Britain, making it possible for a train to reach Paris from London in three hours.

The machines operating from both ends of the Channel bored toward the middle. To link up in the middle of the Channel with the necessary accuracy, a laser guidance system kept the machines precisely aligned. A model of the boring machine control is shown in Fig. 4.21, where $C(s)$ is the actual angle of direction of travel of the boring machine and $R(s)$ is the desired angle. The effect of load on the machine is represented by the disturbance $D(s)$.

The design objective is to select the gain K so that the response to input angle changes is desirable while maintaining minimal error due to the disturbance. Using Mason's rule, the output due to the two inputs is

$$C(s) = T(s)R(s) + T_d(s)D(s)$$

$$= \frac{KG(s)}{1 + KG(s)} R(s) + \frac{G(s)}{1 + KG(s)} D(s) \qquad (4.59)$$

$$= \frac{K}{s^2 + 12s + K} R(s) + \frac{1}{s^2 + 12s + K} D(s).$$

Thus, to reduce the effect of the disturbance, we wish to set the gain greater than 10. When we select $K = 100$ and let $d(t) = 0$, we have the step response for a unit step input $r(t)$, as shown in Fig. 4.22(a). Letting the input $r(t) = 0$ and determining the response to the unit step disturbance, we obtain $c(t)$, as shown in Fig. 4.22(b). Clearly, the effect of the disturbance is quite small. If we set the gain K equal to 50, the responses of $c(t)$ due to a unit step input $r(t)$ and $d(t)$ are displayed together in Fig. 4.23. Since the overshoot of the response is small (less than 2%) and the steady state is attained in one second, we would prefer $K = 50$. The results are summarized in Table 4.1.

FIGURE 4.21
A block diagram model of a boring machine control system.

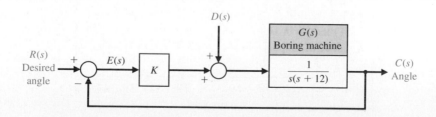

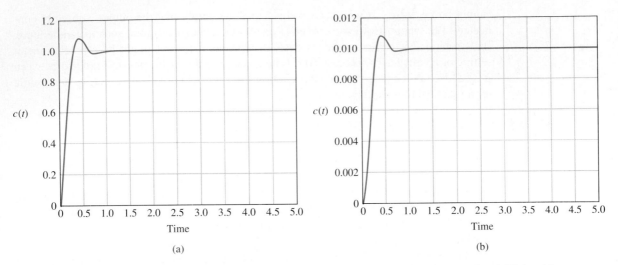

FIGURE 4.22   The response $c(t)$ to (a) a unit input step $r(t)$ and (b) a unit disturbance step input $D(s) = 1/s$ for $K = 100$.

The steady-state error of the system to a unit step input $R(s) = 1/s$ is

$$\lim_{t \to \infty} e(t) = \lim_{s \to 0} s \, \frac{1}{1 + KG(s)} \left(\frac{1}{s}\right) = 0. \tag{4.60}$$

The steady-state value of $c(t)$ when the disturbance is a unit step, $D(s) = 1/s$, and the desired value is $r(t) = 0$ is

$$\lim_{t \to \infty} c(t) = \lim_{s \to 0} s \left[\frac{G(s)}{1 + KG(s)}\right] D(s)$$

$$= \lim_{s \to 0} \left[\frac{1}{s(s + 12) + K}\right] = \frac{1}{K}. \tag{4.61}$$

Thus the steady-state value is 0.01 and 0.02 for $K = 100$ and 50, respectively.

FIGURE 4.23
The response $c(t)$
for a unit step
input (solid line)
and for a unit step
disturbance
(dashed line) for
$K = 50$.

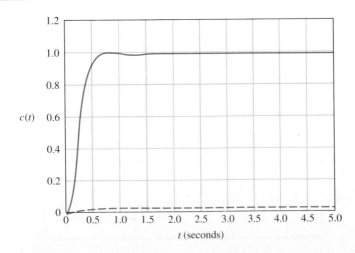

TABLE 4.1   Response of the Boring System for Two Gains

Gain K	Overshoot of Response to $r(t)$ = Step	Time for Response to $r(t)$ = Step to Reach Steady State	Steady-State Response $c(t)$ for $d(t)$ = Step with $r(t)$ = 0	Steady-State Error of Response to $r(t)$ = Step with $d(t)$ = 0
100	10%	1.5 s	0.01	0
50	1%	1.0 s	0.02	0

Finally, we examine the sensitivity of the system to a change in the process $G(s)$ using Eq. (4.12). Then

$$S_G^T = \frac{1}{1 + KG(s)} = \frac{s(s + 12)}{s(s + 12) + K}. \tag{4.62}$$

For low frequencies ($|s| < 4$), the sensitivity can be approximated by

$$S_G^T \simeq \frac{12s}{K}, \tag{4.63}$$

where $K \geq 50$. Thus the sensitivity of the system is reduced by increasing the gain K. In this case, we choose $K = 50$ for a reasonable design compromise.

4.8 DESIGN EXAMPLE: MARS ROVER VEHICLE

A solar-powered Mars rover vehicle is shown in Fig. 4.24. The vehicle can be controlled from Earth by sending it path commands, $r(t)$. We expect to use a ramp input signal, $r(t) = t, t > 0$. The system may be operated without feedback, as shown in Fig. 4.25(a), or with feedback, as shown in Fig. 4.25(b). The goal is to operate the rover with modest effects from disturbances (such as rocks) and with low sensitivity to changes in the gain K.

The transfer function for the open-loop system is

$$T_o(s) = \frac{C(s)}{R(s)} = \frac{K}{s^2 + 4s + 5} \tag{4.64}$$

and the transfer function for the closed-loop system is

$$T_c(s) = \frac{C(s)}{R(s)} = \frac{K}{s^2 + 4s + 3 + K}. \tag{4.65}$$

Then, for $K = 2$,

$$T(s) = T_o(s) = T_c(s) = \frac{2}{s^2 + 4s + 5}.$$

Hence we can compare the sensitivity of the open-loop and closed-loop systems for the same transfer function.

The sensitivity for the open-loop system is

$$S_K^{T_o} = \frac{dT_o}{dK} \frac{K}{T_o} = 1, \tag{4.66}$$

FIGURE 4.24
Solar-powered
Mars rover
controlled by an
operator on Earth
using cameras on
the rover.

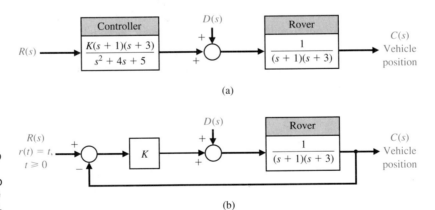

FIGURE 4.25
Control system for
rover (a) open-loop
(without feedback)
and (b) closed-loop
with feedback. The
input is $R(s) = 1/s$.

and the sensitivity for the closed-loop system is

$$S_K^{T_c} = \frac{dT_c}{dK}\frac{K}{T_c} = \frac{s^2 + 4s + 3}{s^2 + 4s + 3 + K}. \tag{4.67}$$

To examine the effect of the sensitivity at low frequencies, we let $s = j\omega$ to obtain

$$S_K^{T_c} = \frac{(3 - \omega^2) + j4\omega}{(3 + K - \omega^2) + j4\omega}. \tag{4.68}$$

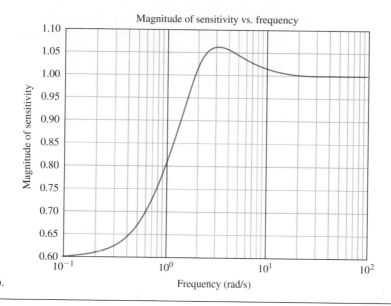

FIGURE 4.26
The magnitude of
the sensitivity of
the closed-loop
system for the
Mars rover vehicle.

For $K = 2$, the sensitivity at low frequencies, $\omega < 0.1$, is $|S| \doteq 0.6$.

A frequency plot of the magnitude of the sensitivity is shown in Fig. 4.26. Note the sensitivity for low frequencies is

$$|S| < 0.8, \qquad \omega \leq 0.8.$$

The effect of the disturbance can be determined by setting $R(s) = 0$ and letting $D(s) = 1/s$. Then, for the open-loop system, we have the steady-state value

$$c(\infty) = \lim_{s \to 0} s \left\{ \frac{1}{(s + 1)(s + 3)} \right\} \frac{1}{s} = \frac{1}{3}. \tag{4.69}$$

The output of the closed-loop system with a unit step disturbance, $D(s) = 1/s$, is (see Section 4.4)

$$c(\infty) = \lim_{s \to 0} s \left\{ \frac{1}{(s^2 + 4s + 3 + K)} \right\} \frac{1}{s} = \frac{1}{3 + K}. \tag{4.70}$$

When $K = 2$, $c(\infty) = 1/5$. Because we seek to minimize the effect of the disturbance, it is clear that a larger value of K would be desirable. An increased value of K, such as $K = 50$, will further reduce the effect of the disturbance as well as reduce the magnitude of the sensitivity (Eq. 4.68). However, as we increase K beyond $K = 50$, we begin to deteriorate the transient performance of the system for the ramp input $r(t)$.

4.9 CONTROL SYSTEM CHARACTERISTICS USING *MATLAB*

In this section, the advantages of feedback will be illustrated with two examples. Our objective is to illustrate the use of *Matlab* in the control system analysis. In the first example, we will introduce feedback control to a speed tachometer system in an effort to reject disturbances. The tachometer speed control system example can be found on page 177. The reduction in system sensitivity to plant variations, adjustment of the transient response, and reduction in steady-state error will be demonstrated in a second example. This is the English Channel boring machine example in Section 4.7.

EXAMPLE 4.2   **Speed control system**

The open-loop block diagram description of the armature-controlled dc motor with a load torque disturbance, $T_d(s)$, is shown in Fig. 4.13 on page 180. The values for the various parameters, taken from Fig. 4.11 on page 179, are given in Table 4.2. We have two inputs to our system, $V_a(s)$ and $T_d(s)$. Relying on the *principle of superposition*, which applies to our linear system, we consider each input separately. To investigate the effects of disturbances on the system, we let $V_a(s) = 0$ and consider only the disturbance $T_d(s)$. Conversely, to investigate the response of the system to a reference input, we let $T_d(s) = 0$ and consider only the input $V_a(s)$.

The closed-loop speed tachometer control system block diagram is shown in Fig. 4.13 on page 180. The values for K_a and K_t are given in Table 4.2.

If our system displays good disturbance rejection, then we expect the disturbance $T_d(s)$ to have a small effect on the output $\omega(s)$. Consider the open-loop system in Fig. 4.11 first. We can use *Matlab* to compute the transfer function from $T_d(s)$ to $\omega(s)$ and evaluate the output response to a unit step disturbance (i.e., $T_d(s) = 1/s$). The time response to a unit step disturbance is shown in Fig. 4.27(a). The script opentach.m, shown in Fig. 4.27(b), is used to analyze the open-loop speed tachometer system.

The open-loop transfer function is

$$\frac{\omega(s)}{T_d(s)} = \frac{\text{num}}{\text{den}} = \frac{-1}{2s + 1.5}.$$

Since our desired value of $\omega(t)$ is zero (remember that $V_a(s) = 0$), the steady-state error is just the final value of $\omega(t)$, which we denote by $\omega_o(t)$ to indicate open-loop. The steady-state error, shown on the plot in Fig. 4.27(a), is approximately the value of the speed when $t = 7$ seconds. We can obtain an approximate value of the steady-state error by looking at the last value in the output vector $\mathbf{y}_o$, which we computed in the process of generating the plot in Fig. 4.27(a). The approximate steady-state value of ω_o is

$$\omega_o(\infty) \approx \omega_o(7) = -0.663 \text{ rad/s}.$$

The plot verifies that we have in fact reached steady state.

In a similar fashion, we begin the closed-loop system analysis by computing the closed-loop transfer function from $T_d(s)$ to $\omega(s)$, and then generating the time-response of

TABLE 4.2   Tachometer Control System Parameters

R_a	K_m	J	f	K_b	K_a	K_t
1	10	2	0.5	0.1	54	1

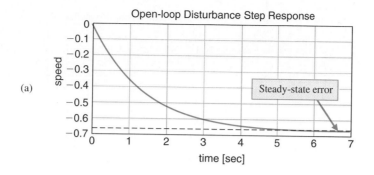

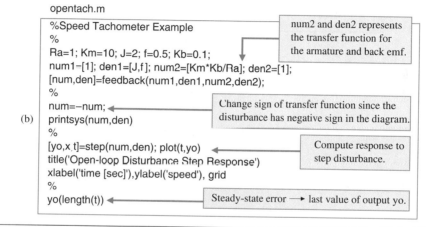

FIGURE 4.27
Analysis of the open-loop speed control system. (a) Response and (b) *Matlab* script.

$\omega(t)$ to a unit step disturbance input. The output response and the script closedtach.m are shown in Fig. 4.28. The closed-loop transfer function from the disturbance input is

$$\frac{\omega(s)}{T_d(s)} = \frac{num}{den} = \frac{-1}{2s + 541.5}.$$

As before, the steady-state error is just the final value of $\omega(t)$, which we denote by $\omega_c(t)$ to indicate closed-loop. The steady-state error is shown on the plot in Fig. 4.28(a). We can obtain an approximate value of the steady-state error by looking at the last value in the output vector $\mathbf{y}_c$, which we computed in the process of generating the plot in Fig. 4.28(a). The approximate steady-state value of ω is

$$\omega_c(\infty) \approx \omega_c(0.02) = -0.002 \text{ rad/s}.$$

We generally expect that $\omega_c(\infty)/\omega_o(\infty) < 0.02$. The ratio of closed-loop to open-loop steady-state speed output due to a unit step disturbance input, in this example, is

$$\frac{\omega_c(\infty)}{\omega_o(\infty)} = 0.003.$$

We have achieved a remarkable improvement in disturbance rejection. It is clear that the addition of the negative feedback loop reduced the effect of the disturbance on the output. This demonstrates the *disturbance rejection* property of closed-loop feedback systems. ∎

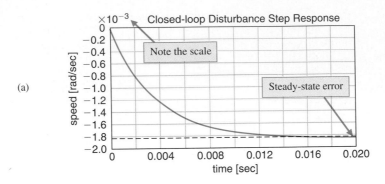

(a)

closedtach.m

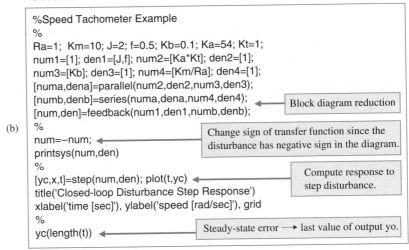

(b)

FIGURE 4.28
Analysis of the
closed-loop speed
control system.
(a) Response and
(b) *Matlab* script.

EXAMPLE 4.3   **English Channel boring machines**

The block diagram description of the English Channel boring machines is shown in Fig. 4.21 on page 187. The transfer function of the output due to the two inputs is

$$C(s) = \frac{K}{s^2 + 12s + K} R(s) + \frac{1}{s^2 + 12s + K} D(s).$$

The effect of the control gain K on the transient response is shown in Fig. 4.29 along with the script english1.m used to generate the plots. Comparing the two plots in parts (a) and (b), it can be seen that decreasing K decreases the *overshoot*. Although it is not as obvious from the plots in Fig. 4.29, it is also true that decreasing K decreases the *settling time*. This can be verified by taking a closer look (at the *Matlab* command level) at the data used to generate the plots. This example demonstrates how the transient response can be altered by feedback control gain K. Based on our analysis thus far, we would prefer to use $K = 50$. Other considerations must be taken into account, however, before we can establish the final design.

Before making the final choice of K, it is important to consider the system response to a unit step disturbance, as shown in Fig. 4.30. We see that increasing K reduces the steady-state response of $c(t)$ to the step disturbance. The steady-state value of $c(t)$ is 0.02 and 0.01

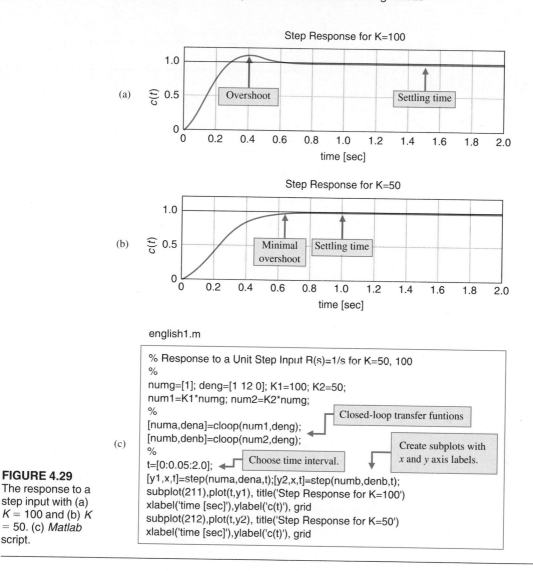

FIGURE 4.29
The response to a step input with (a) $K = 100$ and (b) $K = 50$. (c) *Matlab* script.

for $K = 50$ and 100, respectively. The steady-state errors, percent overshoot, and settling times are summarized in Table 4.3. The steady-state values are predicted from the final-value theorem for a disturbance input as follows:

$$\lim_{t \to \infty} c(t) = \lim_{s \to 0} s \left\{ \frac{1}{s(s + 12) + K} \right\} \frac{1}{s} = \frac{1}{K}.$$

If our only design consideration is disturbance rejection, we would prefer to use $K = 100$.

We have just experienced a very common trade-off situation in control system design. In this particular example, increasing K leads to better disturbance rejection, whereas decreasing K leads to better performance (via less overshoot and quicker settling time). The final decision on how to choose K rests with the designer. Although *Matlab* can certainly assist in the control system design, it cannot replace the engineer's decision-making capability and intuition.

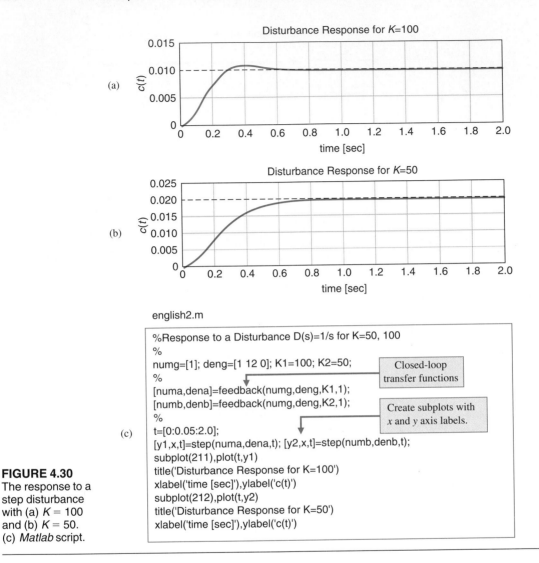

FIGURE 4.30
The response to a step disturbance with (a) $K = 100$ and (b) $K = 50$. (c) *Matlab* script.

english2.m

```
%Response to a Disturbance D(s)=1/s for K=50, 100
%
numg=[1]; deng=[1 12 0]; K1=100; K2=50;
%
[numa,dena]=feedback(numg,deng,K1,1);
[numb,denb]=feedback(numg,deng,K2,1);
%
t=[0:0.05:2.0];
[y1,x,t]=step(numa,dena,t); [y2,x,t]=step(numb,denb,t);
subplot(211),plot(t,y1)
title('Disturbance Response for K=100')
xlabel('time [sec]'),ylabel('c(t)')
subplot(212),plot(t,y2)
title('Disturbance Response for K=50')
xlabel('time [sec]'),ylabel('c(t)')
```

Closed-loop transfer functions

Create subplots with *x* and *y* axis labels.

TABLE 4.3   Response of the Boring Machine Control System for $K = 50$ and $K = 100$

	$K = 50$	$K = 100$
P.O.	1%	10%
T_s	1.0s	1.5s
e_{ss}	2%	1%

The final step in the analysis is to look at the system sensitivity to changes in the plant. The system sensitivity is given by (Eq. 4.62)

$$S_G^T = \frac{1}{1 + KG(s)} = \frac{s(s + 12)}{s(s + 12) + K}.$$

We can compute the values of $S(s)$ for different values of s and generate a plot of the system sensitivity. For low frequencies, we can approximate the system sensitivity by

$$S_G^T \approx \frac{12s}{K}.$$

Increasing the gain K reduces the system sensitivity. The system sensitivity plots when $s = j\omega$ are shown in Fig. 4.31 for $K = 50$. ∎

FIGURE 4.31
(a) System sensitivity to plant variations ($s = j\omega$).
(b) *Matlab* script.

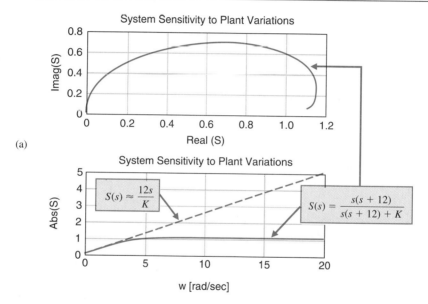

(a)

(b)

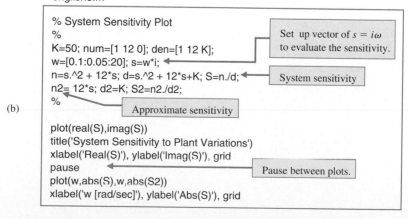

english3.m

```
% System Sensitivity Plot
%
K=50; num=[1 12 0]; den=[1 12 K];
w=[0.1:0.05:20]; s=w*i;
n=s.^2 + 12*s; d=s.^2 + 12*s+K; S=n./d;
n2= 12*s; d2=K; S2=n2./d2;
%

plot(real(S),imag(S))
title('System Sensitivity to Plant Variations')
xlabel('Real(S)'), ylabel('Imag(S)'), grid
pause
plot(w,abs(S),w,abs(S2))
xlabel('w [rad/sec]'), ylabel('Abs(S)'), grid
```

Set up vector of $s = i\omega$ to evaluate the sensitivity.

System sensitivity

Approximate sensitivity

Pause between plots.

4.10 SUMMARY

The fundamental reasons for using feedback, despite its cost and additional complexity, are as follows:

1. Decrease in the sensitivity of the system to variations in the parameters of the process $G(s)$
2. Ease of control and adjustment of the transient response of the system
3. Improvement in the rejection of the disturbance and noise signals within the system
4. Improvement in the reduction of the steady-state error of the system

The benefits of feedback can be illustrated by considering the system shown in Fig. 4.32(a). This system can be considered for several values of gain, K. Table 4.4 summarizes the results of the system operated as an open-loop system (disconnect the feedback path) and for several values of gain, K, with the feedback connected. It is clear that the rise time and sensitivity of the system are reduced as the gain is increased. Also, the feedback system demonstrates excellent reduction of the steady-state error as the gain is increased. Finally, Fig. 4.32(b) shows the response for a unit step disturbance (when $R(s) = 0$) and shows how a larger gain will reduce the effect of the disturbance.

FIGURE 4.32
(a) A single-loop feedback control system. (b) The error response for a unit step disturbance when $R(s) = 0$.

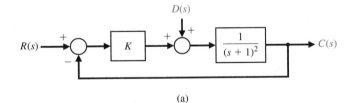

(a)

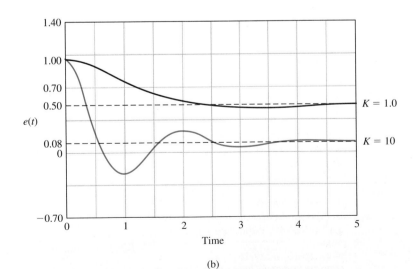

(b)

TABLE 4.4 System Response

	Open Loop*	Closed Loop		
	K = 1	K = 1	K = 8	K = 10
Rise time (seconds) (10% to 90% of final value)	3.35	1.52	0.45	0.38
Percent overshoot (%)	0	4.31	33	40
Final value of $c(t)$ due to a disturbance, $D(s) = 1/s$	1.0	0.50	0.11	0.09
Percent steady-state error for unit step input	0	50%	11%	9%
Percent change in steady-state error due to 10% decrease in K	10%	5.3%	1.2%	0.9%

*Response only when K = 1 exactly.

> Feedback control systems possess many beneficial characteristics, and it is not surprising that one finds a multitude of feedback control systems in industry, government, and nature.

EXERCISES

E4.1 A closed-loop system is used to track the sun to obtain maximum power from a photovoltaic array. The tracking system may be represented by Fig. 4.3 with $H(s) = 1$ and

$$G(s) = \frac{100}{\tau s + 1}.$$

where $\tau = 3$ nominally. (a) Calculate the sensitivity of this system for a small change in τ. (b) Calculate the time constant of the closed-loop system response.

Answers: $S = -3s/(3s + 101)$; $\tau_c = 3/101$

E4.2 A digital audio system is designed to minimize the effect of disturbances and noise as shown in

Fig. E4.2. As an approximation, we may represent $G(s) = K_2$. (a) Calculate the sensitivity of the system due to K_2. (b) Calculate the effect of the disturbance on V_o. (c) What value would you select for K_1 to minimize the effect of the disturbance?

E4.3 A robot arm and camera could be used to pick fruit, as shown in Fig. E4.3(a). The camera is used to close the feedback loop to a microcomputer, which controls the arm [8, 9]. The process is

$$G(s) = \frac{K}{(s + 3)^2}.$$

(a) Calculate the expected steady-state error of the gripper for a step command A as a function of K.

FIGURE E4.2
Digital audio system.

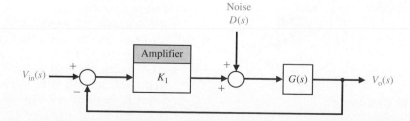

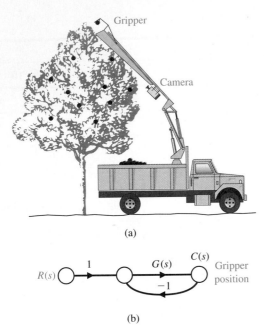

Gripper

Camera

(a)

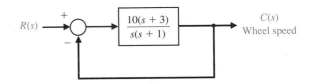

(b)

FIGURE E4.3   Robot fruit picker.

(b) Name a possible disturbance signal for this system.

E4.4   A magnetic disk drive requires a motor to position a read/write head over tracks of data on a spinning disk, as shown in Fig. E4.4. The motor and head may be represented by

$$G(s) = \frac{10}{s(\tau s + 1)},$$

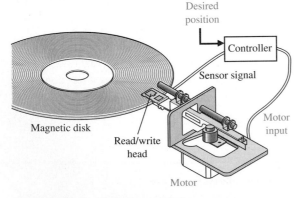

FIGURE E4.4   Disk drive control.

where $\tau = 0.001$ second. The controller takes the difference of the actual and desired positions and generates an error. This error is multiplied by an amplifier K. (a) What is the steady-state position error for a step change in the desired input? (b) Calculate the required K in order to yield a steady-state error of 0.1 mm for a ramp input of 10 cm/s.

Answers: $e_{ss} = 0$; $K = 100$

E4.5   Most people have experienced an out-of-focus slide projector. A projector with an automatic focus adjusts for variations in slide position and temperature disturbances [11]. Draw the block diagram of an auto-focus system, and describe how the system works. An unfocused slide projection is a visual example of steady-state error.

E4.6   Four-wheel drive autos are popular in regions where winter road conditions are often slippery due to snow and ice. A four-wheel drive with anti-lock brakes uses a sensor to keep each wheel rotating to maintain traction. One system is shown in Fig. E4.6. Find the closed-loop response of this system as it attempts to maintain a constant speed of the wheel. Use a computer program to determine the response when $R(s) = A/s$.

FIGURE E4.6   Four-wheel drive auto.

E4.7   Submersibles with clear plastic hulls have the potential to revolutionize underwater leisure. One small submersible vehicle has a depth-control system as illustrated in Fig. E4.7.

(a) Determine the closed-loop transfer function $T(s) = C(s)/R(s)$.
(b) Determine the sensitivity $S_{K_1}^T$ and S_K^T.
(c) Determine the steady-state error due to a disturbance $D(s) = 1/s$.
(d) Calculate the response $c(t)$ for a step input $R(s) = 1/s$ when $K = K_2 = 1$ and $1 < K_1 < 10$. Select K_1 for the fastest response.

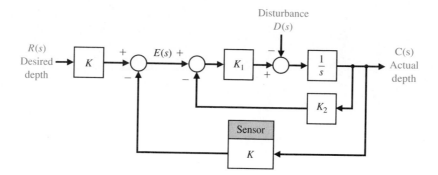

FIGURE E4.7
Depth control
system.

E4.8 Consider the feedback control system shown in Fig. E4.8. (a) Determine the steady-state error for a step input in terms of the gain K. (b) Determine the overshoot for the step response for $40 \leq K \leq 400$. (c) Plot the overshoot and the steady-state error versus K.

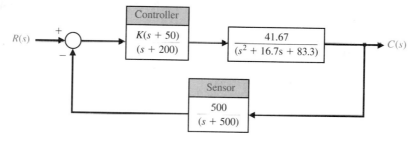

FIGURE E4.8

PROBLEMS

P4.1 The open-loop transfer function of a fluid-flow system obtained in Problem 2.12 can be written as

$$G(s) = \frac{\Delta Q_2(s)}{\Delta Q_1(s)} = \frac{1}{\tau s + 1},$$

where $\tau = RC$, R is a constant equivalent to the resistance offered by the orifice so that $1/R = \frac{1}{2}kH_0^{-1/2}$, and $C =$ the cross-sectional area of the tank. Since $\Delta H = R\Delta Q_2$, we have for the transfer function relating the head to the input change:

$$G_1(s) = \frac{\Delta H(s)}{\Delta Q_1(s)} = \frac{R}{RCs + 1}.$$

For a closed-loop feedback system, a float level sensor and valve may be used as shown in Fig. P4.1. Assuming the float is a negligible mass, the valve is controlled so that a reduction in the flow rate, ΔQ_1, is proportional to an increase in

head, ΔH, or $\Delta Q_1 = -K\Delta H$. Draw a closed-loop flow graph or block diagram. Determine and compare the open-loop and closed-loop systems for (a) sensitivity to changes in the equivalent coefficient R and the feedback coefficient K; (b) ability to reduce the effects of a disturbance in the level $\delta H(s)$; and (c) steady-state error of the level (head) for a step change of the input $\Delta Q_1(s)$.

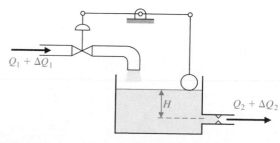

FIGURE P4.1 Tank level control.

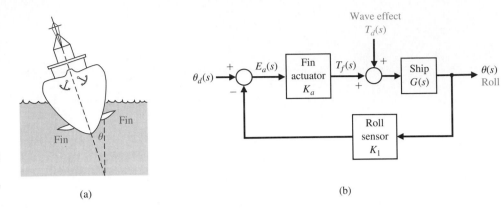

FIGURE P4.2
Ship stabilization system. The effect of the waves is a torque on the ship, $T_d(s)$.

(a)

(b)

P4.2 It is important to ensure passenger comfort on ships by stabilizing the ship's oscillations due to waves [13]. Most ship stabilization systems use fins or hydrofoils projecting into the water to generate a stabilization torque on the ship. A simple diagram of a ship stabilization system is shown in Fig. P4.2. The rolling motion of a ship can be regarded as an oscillating pendulum with a deviation from the vertical of θ degrees and a typical period of 3 seconds. The transfer function of a typical ship is

$$G(s) = \frac{\omega_n^2}{s^2 + 2\zeta\omega_n s + \omega_n^2},$$

where $\omega_n = 2\pi/T = 2$, $T = 3.14$ seconds, and $\zeta = 0.10$. With this low damping factor ζ, the oscillations continue for several cycles and the rolling amplitude can reach $18°$ for the expected amplitude of waves in a normal sea. Determine and compare the open-loop and closed-loop system for (a) sensitivity to changes in the actuator constant K_a and the roll sensor K_1; and (b) the ability to re-

duce the effects of the disturbance of the waves. Note that the desired roll $\theta_d(s)$ is zero degrees.

P4.3 One of the most important variables that must be controlled in industrial and chemical systems is temperature. A simple representation of a thermal control system is shown in Fig. P4.3 [14]. The temperature $\mathcal{T}$ of the process is controlled by the heater with a resistance R. An approximate representation of the dynamics linearly relates the heat loss from the process to the temperature difference $(\mathcal{T} - \mathcal{T}_e)$. This relation holds if the temperature difference is relatively small and the energy storage of the heater and the vessel walls is negligible. Also, it is assumed that the voltage connected to the heater, e_h, is proportional to $e_{desired}$ or $e_h = kE_b = k_a E_b e(t)$, where k_a is the constant of the actuator. Then the linearized open-loop response of the system is

$$\mathcal{T}(s) = \frac{(k_1 k_a E_b)}{\tau s + 1} E(s) + \frac{\mathcal{T}_e(s)}{\tau s + 1},$$

FIGURE P4.3
Temperature control system.

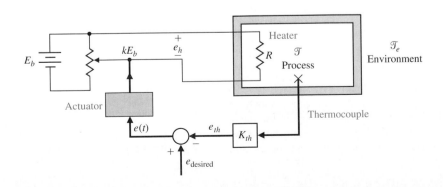

where

$$\tau = MC/\rho A,$$
M = mass in tank,
A = surface area of tank,
ρ = heat transfer constant,
C = specific heat constant,
k_1 = a dimensionality constant,
e_{th} = output voltage of thermocouple.

Determine and compare the open-loop and closed-loop systems for (a) sensitivity to changes in the constant $K = k_1 k_a E_b$; (b) ability to reduce the effects of a step disturbance in the environmental temperature $\Delta \mathcal{T}_e(s)$; and (c) the steady-state error of the temperature controller for a step change in the input, $e_{desired}$.

P4.4 A control system has two forward signal paths as shown in Fig. P4.4. (a) Determine the overall transfer function $T(s) = C(s)/R(s)$. (b) Calculate the sensitivity S_G^T using Eq. (4.16). (c) Does the sensitivity depend on $U(s)$ or $M(s)$?

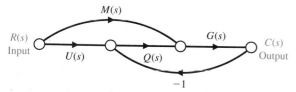

FIGURE P4.4 Two-path system.

P4.5 Large microwave antennas have become increasingly important for radio astronomy and satellite tracking. A large antenna, for example, with a diameter of 60 ft, is subject to large wind gust torques. A proposed antenna is required to have an error of less than 0.10° in a 35 mph wind. Experiments show that this wind force exerts a maximum disturbance at the antenna of 200,000 ft-lb at 35 mph, or equivalent to 10 volts at the input, $T_d(s)$,

to the amplidyne. Also, one problem of driving large antennas is the form of the system transfer function that possesses a structural resonance. The antenna servosystem is shown in Fig. P4.5. The transfer function of the antenna, drive motor, and amplidyne is approximated by

$$G(s) = \frac{\omega_n^2}{s^2 + 2\zeta\omega_n s + \omega_n^2},$$

where $\zeta = 0.707$ and $\omega_n = 10$. The transfer function of the power amplifier is approximately

$$G_1(s) = \frac{k_a}{\tau s + 1},$$

where $\tau = 0.10$ second. (a) Determine the sensitivity of the system to a change of the parameter k_a. (b) The system is subjected to a disturbance $T_d(s) = 10/s$. Determine the required magnitude of k_a in order to maintain the steady-state error of the system less than 0.10° when the input $R(s)$ is zero. (c) Determine the error of the system when subjected to a disturbance $T_d(s) = 10/s$ when it is operating as an open-loop system ($k_s = 0$) with $R(s) = 0$.

P4.6 An automatic speed control system will be necessary for passenger cars traveling on the automatic highways of the future. A model of a feedback speed control system for a standard vehicle is shown in Fig. P4.6. The load disturbance due to a percent grade, $\Delta D(s)$, is shown also. The engine gain K_e varies within the range of 10 to 1000 for various models of automobiles. The engine time constant, τ_e, is 20 seconds. (a) Determine the sensitivity of the system to changes in the engine gain K_e. (b) Determine the effect of the load torque on the speed. (c) Determine the constant percent grade $\Delta D(s) = \Delta d/s$ for which the vehicle stalls (velocity $V(s) = 0$) in terms of the gain factors. Note that since the grade is constant, the

FIGURE P4.5
Antenna control
system.

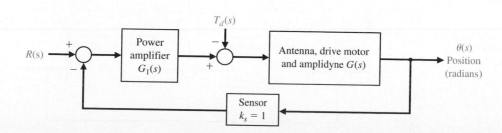

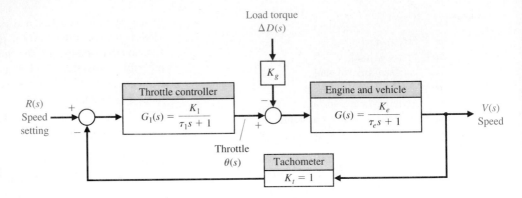

FIGURE P4.6
Automobile speed
control.

steady-state solution is sufficient. Assume that $R(s) = 30/s$ km/hr and that $K_eK_1 \gg 1$. When $(K_g/K_1) = 2$, what percent grade Δd would cause the automobile to stall?

P4.7   A robot uses feedback to control the orientation of each joint axis. Also, the load effect varies due to varying load objects and the extended position of the arm. The system will be deflected by the load carried in the gripper. Thus the system may be represented by Fig. P4.7 where the load torque is D/s. Assume $R(s) = 0$ at the index position. (a) What is the effect of $T_L(s)$ on $C(s)$? (b) Determine the sensitivity of the closed loop to k_2. (c) What is the steady-state error when $R(s) = 1/s$ and $T_L(s) = 0$?

P4.8   Extreme temperature changes result in many failures of electronic circuits [1]. Temperature control feedback systems reduce the change of tem-

perature by using a heater to overcome outdoor low temperatures. A block diagram of one system is shown in Fig. P4.8. The effect of a drop in environmental temperature is a step decrease in $D(s)$. The actual temperature of the electronic circuit is $C(s)$. The dynamics of the electronic circuit temperature change are represented by

$$G(s) = \frac{324}{s^2 + 20s + 324}.$$

(a) Determine the sensitivity of the system to K.
(b) Obtain the effect of the disturbance $D(s)$ on the output $C(s)$.

P4.9   A useful unidirectional sensing device is the photoemitter sensor [15]. A light source is sensitive to the emitter current flowing and alters the resistance of the photosensor. Both the light source and

FIGURE P4.7
Robot control
system.

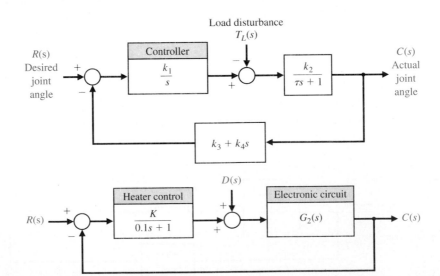

FIGURE P4.8
Temperature
control system.

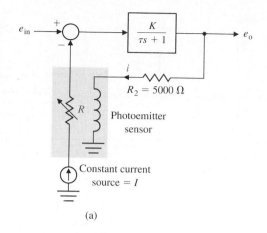

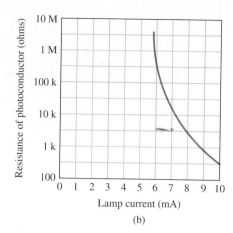

FIGURE P4.9
Photosensor
system.

(a)

(b)

the photoconductor are packaged in a single four-terminal device. This device provides a large gain and total isolation. A feedback circuit utilizing this device is shown in Fig. P4.9(a), and the non-linear resistance–current characteristic is shown in Fig. 4.9(b) for the Raytheon CK1116. The resistance curve can be represented by the equation

$$\log_{10} R = \frac{0.175}{(i - 0.005)^{1/2}},$$

where i is the lamp current. The normal operating point is obtained when $e_o = 35$ V, and $e_{in} = 2.0$ V. (a) Determine the closed-loop transfer function of the system. (b) Determine the sensitivity of the system to changes in the gain K.

P4.10 For a paper processing plant, it is important to maintain a constant tension on the continuous sheet of paper between the windoff and windup rolls. The tension varies as the widths of the rolls change, and an adjustment in the take-up motor speed is necessary, as shown in Fig. P4.10. If the windup motor speed is uncontrolled, as the paper transfers from the windoff roll to the windup roll, the ve-

locity v_0 decreases and the tension of the paper drops [10, 14]. The three-roller and spring combination provides a measure of the tension of the paper. The spring force is equal to $k_1 y$ and linear differential transformer, rectifier, and amplifier may be represented by $e_0 = -k_2 y$. Therefore the measure of the tension is described by the relation $2T(s) = k_1 y$, where y is the deviation from the equilibrium condition and $T(s)$ is the vertical component of the deviation in tension from the equilibrium condition. The time constant of the motor is $\tau = L_a/R_a$, and the linear velocity of the windup roll is twice the angular velocity of the motor; that is, $v_0(t) = 2\omega_0(t)$. The equation of the motor is then

$$E_0(s) = \frac{1}{K_m} [\tau s \omega_0(s) + \omega_0(s)] + k_3 \Delta T(s),$$

where $\Delta T = $ a tension disturbance. (a) Draw the closed-loop block diagram for the system, including the disturbance $\Delta T(s)$. (b) Add the effect of a disturbance in the windoff roll velocity $\Delta V_1(s)$ to the block diagram. (c) Determine the sensitivity of

FIGURE P4.10
Paper tension
control.

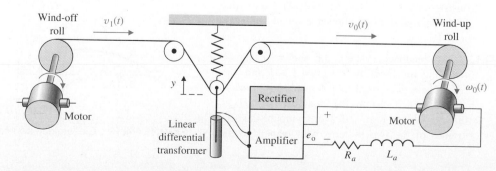

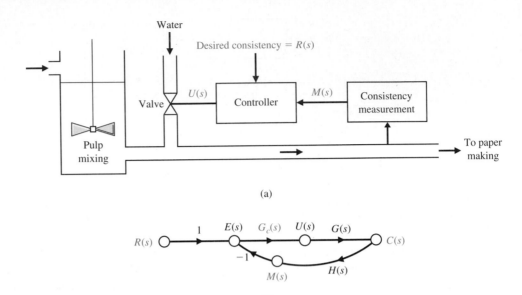

Water

Desired consistency = $R(s)$

$U(s)$

Valve

Controller

$M(s)$

Consistency measurement

Pulp mixing

To paper making

(a)

$R(s)$ $\xrightarrow{1}$ $E(s)$ $G_c(s)$ $U(s)$ $G(s)$ $C(s)$

-1

$M(s)$ $H(s)$

(b)

FIGURE P4.11
Paper-making
control.

the system to the motor constant K_m. (d) Determine
the steady-state error in the tension when a step disturbance in the input velocity $\Delta V_1(s) = A/s$ occurs.

P4.11   One important objective of the paper-making
process is to maintain uniform consistency of the
stock output as it progresses to drying and rolling.
A diagram of the thick stock consistency dilution
control system is shown in Fig. P4.11(a). The
amount of water added determines the consistency.
The signal-flow diagram of the system is shown in
Fig. P4.11(b). Let $H(s) = 1$ and

$$G_c(s) = \frac{K}{(10s + 1)}, \qquad G(s) = \frac{1}{(2s + 1)}.$$

Determine (a) the closed-loop transfer function
$T(s) = C(s)/R(s)$, (b) the sensitivity S_K^T, and (c) the
steady-state error for a step change in the desired
consistency $R(s) = A/s$. (d) Calculate the value
of K required for an allowable steady-state error of 1%.

P4.12   Two feedback systems are shown in signal-
flow diagram form in Figs. P4.12(a) and (b).
(a) Evaluate the closed-loop transfer functions T_1
and T_2 for each system. (b) Show that $T_1 = T_2 =$
100 when $K_1 = K_2 = 100$. (c) Compare the sensitivities of the two systems with respect to the

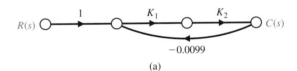

$R(s)$ $\xrightarrow{1}$ $\xrightarrow{K_1}$ $\xrightarrow{K_2}$ $C(s)$

-0.0099

(a)

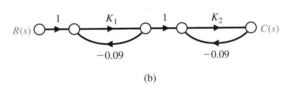

$R(s)$ $\xrightarrow{1}$ $\xrightarrow{K_1}$ $\xrightarrow{1}$ $\xrightarrow{K_2}$ $C(s)$

-0.09 -0.09

(b)

FIGURE P4.12   Two feedback systems.

parameter K_1 for the nominal values of $K_1 = K_2$
$= 100$.

P4.13   One form of closed-loop transfer function is

$$T(s) = \frac{G_1(s) + kG_2(s)}{G_3(s) + kG_4(s)}.$$

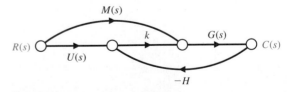

$M(s)$

$R(s)$ $\longrightarrow$ $\xrightarrow{k}$ $\xrightarrow{G(s)}$ $C(s)$

$U(s)$

$-H$

FIGURE P4.13   Closed-loop system.

(a) Use Eq. (4.16) to show that [1]

$$S_k^T = \frac{k(G_2G_3 - G_1G_4)}{(G_3 + kG_4)(G_1 + kG_2)}.$$

(b) Determine the sensitivity of the system shown in Fig. P4.13, using the equation verified in part (a).

P4.14 A proposed hypersonic plane would climb to 100,000 feet and fly 3800 miles per hour, and cross the Pacific in 2 hours. Control of speed of the aircraft could be represented by the model of Fig. P4.14. Find the sensitivity of the closed-loop transfer function, $T(s)$, to a small change in the parameter, a.

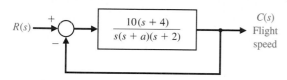

$R(s)$ $\dfrac{10(s+4)}{s(s+a)(s+2)}$ $C(s)$ Flight speed

FIGURE P4.14 Hypersonic airplane speed control.

P4.15 The steering control of a modern ship may be represented by the system shown in Fig. P4.15 [16, 20]. Find the steady-state effect of a constant wind force represented by $D(s) = 1/s$ for $K = 5$ and $K = 30$. (a) Assume that the rudder input $R(s)$ is zero, without any disturbance, and has not been

adjusted. (b) Show that the rudder can then be used to bring the ship deviation back to zero.

P4.16 A two-tank system containing a heated liquid has the model shown in Fig. P4.16, where T_0 is the temperature of the fluid flowing into the first tank and T_2 is the temperature of the liquid flowing out of the second tank. The system of two tanks has a heater in tank 1 with a controllable heat input Q. The time constants are $\tau_1 = 10s$ and $\tau_2 = 50s$. (a) Determine $T_2(s)$ in terms of $T_0(s)$ and $T_{2d}(s)$. (b) If $T_{2d}(s)$, the desired output temperature, is changed instantaneously from $T_{2d}(s) = A/s$ to $T_{2d}(s) = 2A/s$ where $T_0(s) = A/s$, determine the transient response of $T_2(t)$ when $G_c(s) = K = 500$. (c) Find the steady-state error, e_{ss}, for the system of part (b), where $E(s) = T_{2d}(s) - T_2(s)$.

P4.17 A robot gripper, shown in part (a) of Fig. P4.17, is to be controlled so that it closes to an angle θ by using a dc motor control system, as shown in part (b). The model of the control system is shown in part (c), where $K_m = 30$, $R_f = 1\ \Omega$, $K_f = K_i = 1$, $J = 0.1$, and $f = 1$. (a) Determine the response, $\theta(t)$, of the system to a step change in $\theta_d(t)$ when $K = 20$. (b) Assuming $\theta_d(t) = 0$, find the effect of a load disturbance $T_d(s) = A/s$. (c) Determine the steady-state error, e_{ss}, when the input is $r(t) = t$, $t > 0$. (Assume that $T_d = 0$.)

FIGURE P4.15
Ship steering control.

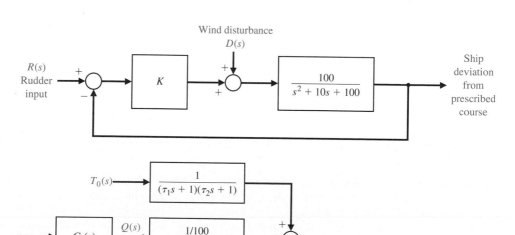

Wind disturbance
$D(s)$

$R(s)$
Rudder input

K

$\dfrac{100}{s^2 + 10s + 100}$

Ship deviation from prescribed course

FIGURE P4.16
Two-tank temperature control.

$T_0(s)$ $\dfrac{1}{(\tau_1 s + 1)(\tau_2 s + 1)}$

$G_c(s)$ $Q(s)$ $\dfrac{1/100}{(\tau_1 s + 1)(\tau_2 s + 1)}$ $T_2(s)$

$E(s)$

$T_{2d}(s)$

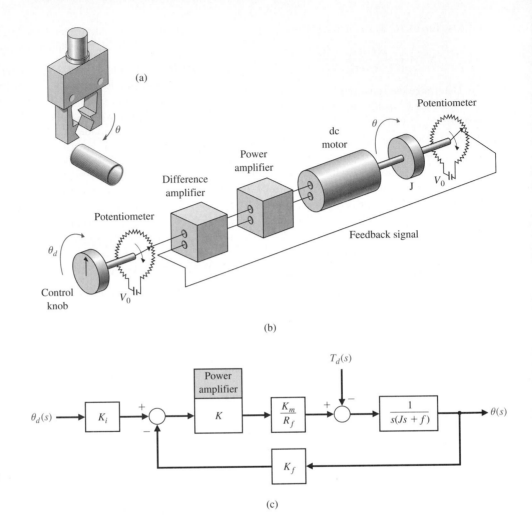

FIGURE P4.17
Robot gripper
control.

ADVANCED PROBLEMS

AP4.1   A tank level regulator control is shown in Fig. AP4.1(a). It is desired to regulate the level h in response to a disturbance change q_3. The block diagram is shown for small variable changes about the equilibrium conditions so that the desired $h_d(t) = 0$. Determine the equation for the error $E(s)$, and determine the steady-state error for a unit step disturbance when (a) $G(s) = K$ and (b) $G(s) = K/s$.

AP4.2   The shoulder joint of a robot arm uses a dc motor with armature control and a set of gears on the output shaft. The model of the system is shown in Fig. AP4.2 with a disturbance torque $D(s)$, which represents the effect of the load. Determine the steady-state error when the desired angle input is a step so that $\theta_d(s) = A/s$, $G_c(s) = K$, and the disturbance input is zero. When $\theta_d(s) = 0$ and the load

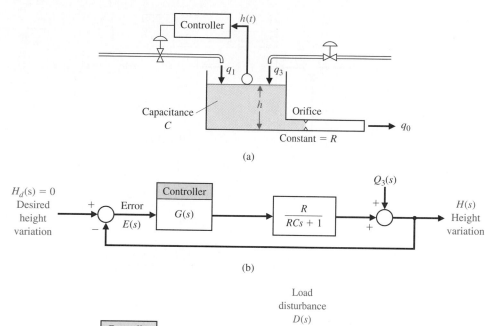

FIGURE AP4.1
A tank level
regulator.

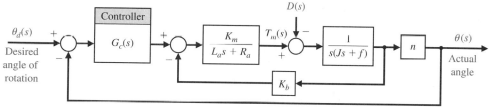

FIGURE AP4.2
Robot joint control.

effect is $D(s) = M/s$, determine the steady-state error when (a) $G_c(s) = K$ and (b) $G_c(s) = K/s$.

AP4.3 A machine tool is designed to follow a desired path so that

$$r(t) = (2 - t + 0.5t^2)u(t),$$

where $u(t)$ is the unit step function. The machine tool control system is shown in Fig. AP4.3.

(a) Determine the steady-state error when $r(t)$ is the desired path as given and $D(s) = 0$.
(b) Plot the error $e(t)$ for the desired path for part (a) for $0 < t \le 10$ seconds.
(c) If the desired input is $r(t) = 0$, find the steady-state error when $D(s) = 1/s$.
(d) Plot the error $e(t)$ for part (c) for $0 < t \le 10$ seconds.

FIGURE AP4.3

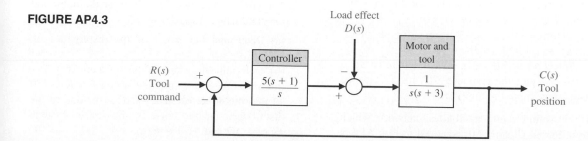

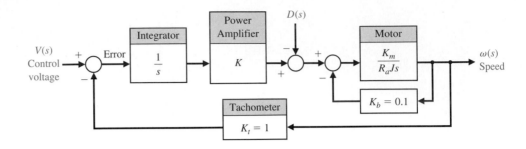

FIGURE AP4.4
dc motor with
feedback.

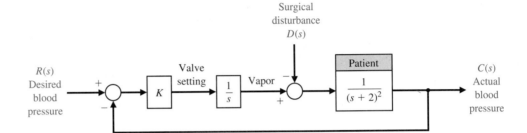

FIGURE AP4.5
Blood pressure
control.

AP4.4   An armature-controlled dc motor with tachometer feedback is shown in Fig. AP4.4. Assume that $K_m = 10$, $J = 1$, and $R = 1$.

(a) Determine the required gain K to restrict the
steady-state error to a ramp input $v(t) = t, t > 0$ to
0.1 (assume that $D(s) = 0$).
(b) For the gain selected in part (a), determine and
plot the error $e(t)$ due to a ramp disturbance so that
$d(t) = t, 0 \leq t \leq 5$ seconds.

AP4.5   A system that controls the mean arterial pressure during anesthesia has been designed and
tested [12]. The level of arterial pressure is postu-
lated to be a proxy for depth of anesthesia during
surgery. A block diagram of the system is shown in
Fig. AP4.5, where the impact of surgery is represented by the disturbance $D(s)$.

(a) Determine the steady-state error due to a disturbance $D(s) = 1/s$ (let $R(s) = 0$).
(b) Determine the steady-state error for a ramp input $r(t) = t, t > 0$ (let $D(s) = 0$).
(c) Select a suitable value of K less than or equal
to 10, and plot the response $C(t)$ for a unit step
disturbance input (assume $r(t) = 0$).

AP4.6   A useful circuit, called a lead network, which
we utilize in Chapter 10, is shown in Fig. AP4.6.

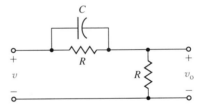

FIGURE AP4.6   A lead network.

(a) Determine the transfer function $G(s) = V_0(s)/V(s)$.
(b) Determine the sensitivity of $G(s)$ with respect
to the capacitance C.
(c) Determine and plot the transient response $v_0(t)$
for a step input, $V(s) = 1/s$.

AP4.7   A feedback control system with sensor noise
and a disturbance input is shown in Fig. AP4.7.
The goal is to reduce the effects of the noise and
the disturbance. Let $R(s) = 0$.

(a) Determine the effect of the disturbance on
$C(s)$.
(b) Determine the effect of the noise on $C(s)$.
(c) Select the best value for K when $1 \leq K \leq 100$
so that the effect of steady-state error due to the
disturbance and the noise is minimized. Assume
$D(s) = A/s$ and $N(s) = B/s$.

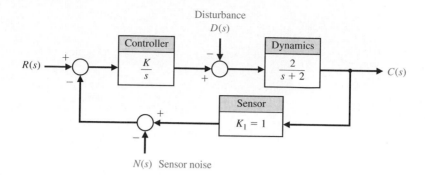

FIGURE AP4.7

AP4.8 The block diagram of a machine tool control system is shown in Fig. AP4.8.

(a) Determine the transfer function $T(s) = C(s)/R(s)$.

(b) Determine the sensitivity S_b^T.

(c) Select K when $1 \leq K \leq 50$ so that the effects of the disturbance and S_b^T are minimizcd.

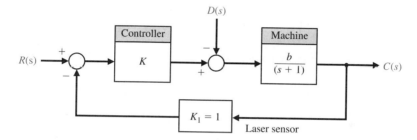

FIGURE AP4.8
Machine tool
control.

DESIGN PROBLEMS

DP4.1 A closed-loop speed control system is subjected to a disturbance due to a load, as shown in Fig. DP4.1. The desired speed is $\omega_d(t) = 100$ rad/s, and the load disturbance is a unit step input $D(s) = 1/s$. Assume that the speed has attained the no-load speed of 100 rad/s and is in steady state. (a) Determine the steady-state effect of the load disturbance, and (b) plot $c(t)$ for the step disturbance for selected values of gain so that $10 \leq K \leq 25$. Determine a suitable value for the gain K.

FIGURE DP4.1
Speed control
system.

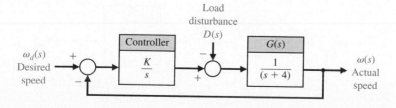

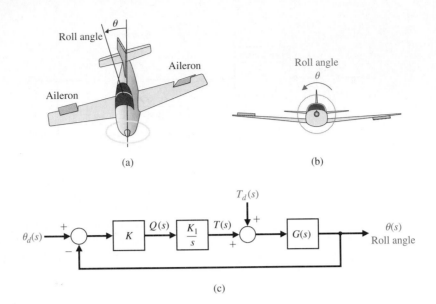

(a)   (b)

FIGURE DP4.2
Control of the
roll angle of an
airplane.

$\theta_d(s)$ → + $\ominus$ − → K → $Q(s)$ → $\dfrac{K_1}{s}$ → $T(s)$ → + $\oplus$ + ← $T_d(s)$ → $G(s)$ → $\theta(s)$ Roll angle

(c)

DP4.2   The control of the roll angle of an airplane is achieved by using the torque developed by the ailerons, as shown in Figs. DP4.2(a) and (b). A linear model of the roll control system for a small experimental aircraft is shown in Fig. DP4.2(c), where $q(t)$ is the flow of fluid into a hydraulic cylinder and

$$G(s) = \frac{1}{s^2 + 4s + 9}.$$

The goal is to maintain a small roll angle θ due to disturbances. Select an appropriate gain, KK_1, that will reduce the effect of the disturbance while attaining a desirable transient response to a step disturbance, with $\theta_d(t) = 0$. To require a desirable transient response, let $KK_1 < 35$.

DP4.3   The speed control system of Fig. DP4.1 is altered so that $G(s) = 1/(s + 3)$ and the feedback is K_1, as shown in Fig. DP4.3.

(a) Determine the range of K_1 allowable so that the steady state is $e_{ss} \leq 1\%$.

(b) Determine a suitable value for K_1 and K so that the magnitude of the steady-state error to a wind disturbance $d(t) = 2t$ mrad/s, $0 \leq t < 5$ s, is less than 0.1 mrad.

DP4.4   Lasers have been used in eye surgery for more than 25 years. They can cut tissue or aid in coagulation [17]. The laser allows the ophthalmologist to apply heat to a location in the eye in a controlled manner. Many procedures use the retina as a laser target. The retina is the thin sensory tissue that rests on the inner surface of the back of the eye and is the actual transducer of the eye, converting light energy into electrical pulses. On occasion, this layer will detach from the wall, resulting in death of the detached area from lack of blood and leading to partial if not total blindness in that eye. A laser

FIGURE DP4.3
Speed control
system.

$\omega_d(s)$ → + $\ominus$ − → K/s → $\ominus$ ← $D(s)$ → $G(s)$ $\dfrac{1}{(s+3)}$ → $\omega(s)$ Speed ← K_1 Tachometer

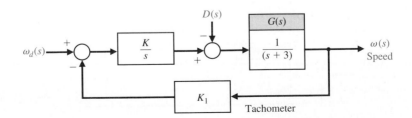

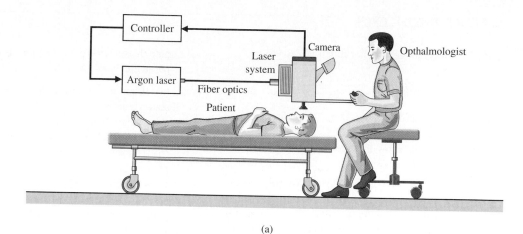

(a)

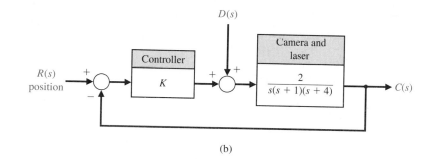

FIGURE DP4.4
Laser eye surgery
system.

(b)

can be used to "weld" the retina into its proper place on the inner wall.

Automated control of position enables the ophthalmologist to indicate to the controller where lesions should be inserted. The controller then monitors the retina and controls the laser's position such that each lesion is placed at the proper location. A wide-angle video-camera system is required to monitor the movement of the retina, as shown in Fig. DP4.4(a). If the eye moves during the irradiation, the laser must be either redirected or turned off. The position-control system is shown in Fig. DP4.4(b). Select an appropriate gain for the controller so that the transient response to a step change in $r(t)$ is satisfactory and the effect of the disturbance due to noise in the system is minimized. Also ensure that the steady-state error for a step input command is zero. To ensure acceptable transient response, require that $K < 10$.

DP4.5 An op amp circuit can be used to generate a short pulse. The circuit shown in Fig. DP4.5 can generate a pulse $v_0(t) = 5e^{-100t}$, $t > 0$, when the input $v(t)$ is a unit step [6]. Select appropriate values for the resistors and capacitors. Assume an ideal op amp.

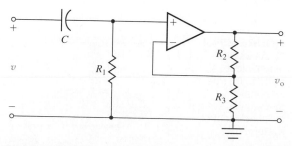

FIGURE DP4.5 Op amp circuit.

MATLAB PROBLEMS

MP4.1   Consider the transfer function

$$G(s) = \frac{5}{s^2 + 2s + 25}.$$

When the input is a unit step, the desired steady-state value of the output is 1. Using the *Matlab* step function, show that the steady-state error to a unit step input is 0.8.

MP4.2   Consider the closed-loop control system shown in Fig. MP4.2. Develop a *Matlab* script to assist in the search for a value of k so that the percent overshoot to a unit step input is greater than 1% but less than 10%. The script should compute the closed-loop transfer function $T(s) = C(s)/R(s)$ and generate the step response. Verify graphically that the steady-state error to a unit step input is zero.

MP4.3   Consider the closed-loop control system shown in Fig. MP4.3. The controller gain $K = 2$. The nominal value of the plant parameter is $a = 1$. The nominal value is used for design purposes only, since in reality the value is not precisely known. The objective of our analysis is to investi-

gate the sensitivity of the closed-loop system to the parameter a.

(a) When $a = 1$, show analytically that the steady-state value of $c(t)$ is equal to 2 when $r(t)$ is a unit step. Verify that the unit step response is within 2% of the final value after 4 seconds.

(b) The sensitivity of the system to changes in the parameter a can be investigated by studying the effects of parameter changes on the transient response. Plot the unit step response for $a = 0.5, 2,$ and 5 and discuss the results.

MP4.4   Consider the torsional mechanical system in Fig. MP4.4(a). The torque due to the twisting of the shaft is $-k\theta$; the damping torque due to the braking device is $-c\dot{\theta}$; the disturbance torque is $d(t)$; the input torque is $r(t)$; and the moment of inertia of the mechanical system is J. The transfer function of the torsional mechanical system is

$$G(s) = \frac{1/J}{s^2 + c/Js + k/J}.$$

A closed-loop control system for the system is

FIGURE MP4.2
A closed-loop negative feedback control system.

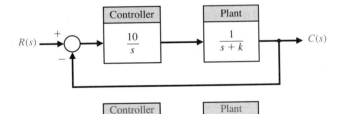

FIGURE MP4.3
A closed-loop control system with uncertain parameter a.

FIGURE MP4.4
(a) A torsional mechanical system. (b) The torsional mechanical system feedback control system.

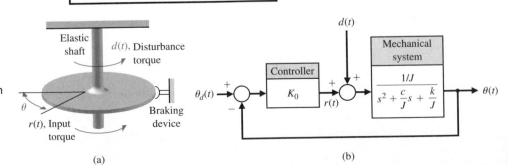

shown in Fig. MP4.4(b). Suppose the desired angle $\theta_d = 0°$, $k = 5$, $c = 0.9$, and $J = 1$.

(a) Determine the open-loop response $\theta(t)$ of the system for a unit step disturbance $d(t)$ using *Matlab* (set $r(t) = 0$).

(b) With the controller gain $K_0 = 50$, determine the closed-loop response, $\theta(t)$, to a unit step disturbance, $d(t)$, using *Matlab*.

(c) Co-plot the open-loop versus the closed-loop response to the disturbance input. Discuss your results and make an argument for using closed-loop feedback control to improve the disturbance rejection properties of the system.

MP4.5 A negative feedback control system is depicted in Fig. MP4.5. Suppose that our design objective is to find a controller, $G_c(s)$, of minimal complexity such that our closed-loop system can track a unit step input with zero steady-state error.

(a) As a first try, consider a simple proportional controller

$$G_c(s) = K,$$

where K is a fixed gain. Let $K = 2$. Using *Matlab*, plot the unit step response and determine the steady-state error from the plot.

(b) Now consider a more complex controller

$$G_c(s) = K_0 + \frac{K_1}{s},$$

where $K_0 = 2$ and $K_1 = 20$. This controller is known as a proportional, integral (PI) controller. Plot the unit step response, and determine the steady-state error from the plot.

(c) Compare the results from parts (a) and (b), and discuss the trade-off between controller complexity and steady-state tracking error performance.

FIGURE MP4.5
A simple single-loop feedback control system.

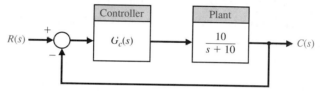

TERMS AND CONCEPTS

Closed-loop system A system with a measurement of the output signal and a comparison with the desired output to generate an error signal that is applied to the actuator.

Direct system *See* Open-loop system.

Disturbance signal An unwanted input signal that affects the system's output signal.

Error signal The difference between the desired output, $R(s)$, and the actual output, $C(s)$. Therefore $E(s) = R(s) - C(s)$.

Open-loop system A system without feedback that directly generates the output in response to an input signal.

Steady-state error The error when the time period is large and the transient response has decayed, leaving the continuous response.

System sensitivity The proportional change of the transfer function of a system to a proportional change in a system parameter.

Transient response The response of a system as a function of time.

The Performance of Feedback Control Systems

PREVIEW

The ability to adjust the transient and steady-state response of a control system is a beneficial outcome of the design of a feedback system. We wish to adjust one or more parameters to provide a desirable system response. Thus we must define the desired response in terms of specifications for the system.

We will use selected input signals to test the response of a control system. This response will be characterized by a selected set of response measures such as the overshoot of a response to a step input. We will then show how the performance of a system is correlated with the *s*-plane location of the poles and zeros of the transfer function of the system.

We will see that one of the most important measures of performance is the steady-state error. The concept of a performance index that adequately represents the system's performance by a single number (or index) will be considered. In this chapter, we will strive to delineate a set of quantitative performance measures that adequately represent the performance of the control system. This approach will enable us to adjust the system for the best achievable performance.

5.1 INTRODUCTION

The ability to adjust the transient and steady-state performance is a distinct advantage of feedback control systems. To analyze and design a control system, we must define and measure its performance. Then, based on the desired performance of the control system, the system parameters may be adjusted to provide the desired response. Because control systems are inherently dynamic, their performance is usually specified in terms of both the transient response and the steady-state response. The *transient response* is the response that disappears with time. The *steady-state response* is that which exists a long time following any input signal initiation.

The design *specifications* for control systems normally include several time response indices for a specified input command as well as a desired steady-state accuracy. However, often in the course of any design, the specifications are revised to effect a compromise. Therefore specifications are seldom a rigid set of requirements, but rather a first attempt at listing a desired performance. The effective compromise and adjustment of specifications can be graphically illustrated by examining Fig. 5.1. Clearly, the parameter p may minimize the performance measure M_2 by selecting p as a very small value. However, this results in large measure M_1, an undesirable situation. Obviously, if the performance measures are equally important, the crossover point at p_{min} provides the best compromise. This type of compromise is normally encountered in control system design. It is clear that if the original specifications called for both M_1 and M_2 to be zero, the specifications could not be simultaneously met and would have to be altered to allow for the compromise resulting with p_{min} [1, 12, 17, 23].

The specifications stated in terms of the measures of performance indicate to the designer the quality of the system. In other words, the performance measures help to answer the question, How well does the system perform the task for which it was designed?

FIGURE 5.1
Two performance measures versus parameter p.

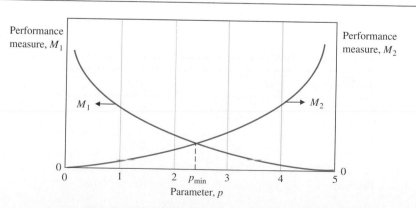

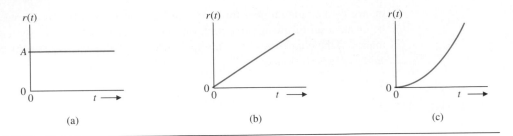

FIGURE 5.2
Test input signals:
(a) step, (b) ramp,
(c) parabolic.

5.2 TEST INPUT SIGNALS

The time-domain performance specifications are important indices because control systems are inherently time-domain systems. That is, the system transient or time performance is the response of prime interest for control systems. It is necessary to determine initially whether the system is stable, by utilizing the techniques of ensuing chapters. If the system is stable, then the response to a specific input signal will provide several measures of the performance. However, because the actual input signal of the system is usually unknown, a standard *test input signal* is normally chosen. This approach is quite useful because there is a reasonable correlation between the response of a system to a standard test input and the system's ability to perform under normal operating conditions. Furthermore, using a standard input allows the designer to compare several competing designs. Also, many control systems experience input signals very similar to the standard test signals.

The standard test input signals commonly used are the step input, the ramp input, and the parabolic input. These inputs are shown in Fig. 5.2. The equations representing these test signals are given in Table 5.1, where the Laplace transform can be obtained by using Table 2.5. The ramp signal is the integral of the step input, and the parabola is simply the integral of the ramp input. A *unit impulse* function is also useful for test signal purposes. The unit impulse is based on a rectangular function $f_\epsilon(t)$ such that

$$f_\epsilon(t) = \begin{cases} 1/\epsilon, & 0 \le t \le \epsilon, \\ 0, & t > \epsilon, \end{cases}$$

TABLE 5.1   Test Signal Inputs

Test Signal	$r(t)$	$R(s)$
Step	$r(t) = A, t > 0$ $= 0, t < 0$	$R(s) = A/s$
Ramp	$r(t) = At, t > 0$ $= 0, t < 0$	$R(s) = A/s^2$
Parabolic	$r(t) = At^2, t > 0$ $= 0, t < 0$	$R(s) = 2A/s^3$

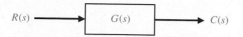

FIGURE 5.3
Direct (open-loop) control system (without feedback).

where $\epsilon > 0$. As ϵ approaches zero, the function $f_\epsilon(t)$ approaches the impulse function $\delta(t)$, which has the following properties:

$$\int_0^\infty \delta(t)dt = 1, \qquad \int_0^\infty \delta(t - a)g(t) \, dt = g(a). \tag{5.1}$$

The impulse input is useful when one considers the convolution integral for an output $c(t)$ in terms of an input $r(t)$, which is written as

$$c(t) = \int_0^t g(t - \tau)r(\tau)d\tau = \mathcal{L}^{-1}\{G(s)R(s)\}. \tag{5.2}$$

[handwritten: $\mathcal{L}[c(t)] = C(s) = G(s)R(s)$]

This relationship is shown in block diagram form in Fig. 5.3. Clearly, if the input is an impulse function of unit amplitude, we have

$$c(t) = \int_0^t g(t - \tau)\delta(\tau)d\tau. \tag{5.3}$$

The integral has a value only at $\tau = 0$, and therefore

$$c(t) = g(t),$$

the impulse response of the system $G(s)$. The impulse response test signal can often be used for a dynamic system by subjecting the system to a large amplitude, narrow width pulse of area A.

The standard test signals are of the general form

$$r(t) = t^n, \tag{5.4}$$

and the Laplace transform is

$$R(s) = \frac{n!}{s^{n+1}}. \tag{5.5}$$

Hence one may relate the response to one test signal to the response of another test signal of the form of Eq. (5.4). The step input signal is the easiest to generate and evaluate and is usually chosen for performance tests.

Consider the response of the system shown in Fig. 5.3 for a unit step input when

$$G(s) = \frac{9}{(s + 10)}.$$

[handwritten: $r = A, \; R(s) = \dfrac{A}{s}$]

Then the output is

[handwritten: $C(s) = R(s)\,G(s)$]

$$C(s) = \frac{9}{s(s + 10)}, \quad = \frac{A}{s} + \frac{B}{s+10}$$

[handwritten: $\dfrac{9}{10}\left(\dfrac{1}{s} - \dfrac{9}{s+10}\right)$]

[handwritten: $A(s+10) + Bs = 9$]
[handwritten: $s = -10, \; B = \dfrac{-9}{10}, \quad s = 0, \; A = \dfrac{9}{10}$]

the transient response is

$$c(t) = 0.9(1 - e^{-10t}),$$

and the steady-state response is

$$c(\infty) = 0.9.$$

If the error is $E(s) = R(s) - C(s)$, then the steady-state error is

$$e_{ss} = \lim_{s \to 0} sE(s) = 0.1.$$

5.3 PERFORMANCE OF A SECOND-ORDER SYSTEM

Let us consider a single-loop second-order system and determine its response to a unit step input. A closed-loop feedback control system is shown in Fig. 5.4. The closed-loop output is

$$C(s) = \frac{G(s)}{1 + G(s)} R(s) = \frac{K}{s^2 + ps + K} R(s). \tag{5.6}$$

Utilizing the generalized notation of Section 2.4, we may rewrite Eq. (5.6) as

$$C(s) = \frac{\omega_n^2}{s^2 + 2\zeta\omega_n s + \omega_n^2} R(s). \tag{5.7}$$

With a unit step input, we obtain

$$C(s) = \frac{\omega_n^2}{s(s^2 + 2\zeta\omega_n s + \omega_n^2)}, \tag{5.8}$$

for which the transient output, as obtained from the Laplace transform table in Appendix A, is

$$c(t) = 1 - \frac{1}{\beta} e^{-\zeta\omega_n t} \sin(\omega_n \beta t + \theta), \tag{5.9}$$

where $\beta = \sqrt{1 - \zeta^2}$ and $\theta = \tan^{-1}\beta/\zeta$. The transient response of this second-order system for various values of the damping ratio ζ is shown in Fig. 5.5. As ζ decreases, the closed-loop roots approach the imaginary axis and the response becomes increasingly oscillatory. The response as a function of ζ and time is also shown in Fig. 5.5(b) for a step input.

FIGURE 5.4
Closed-loop
control system
(with feedback).

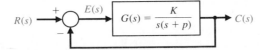

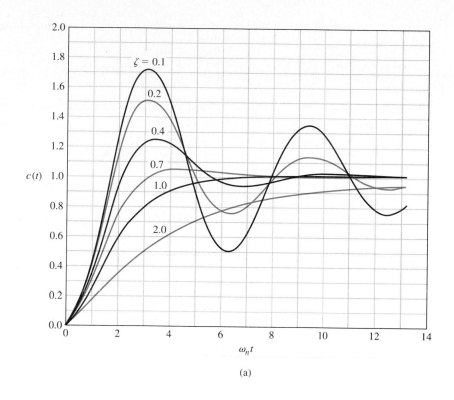

(a)

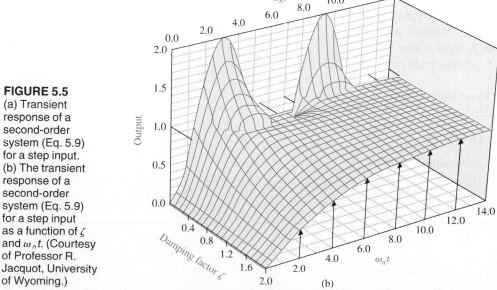

(b)

FIGURE 5.5
(a) Transient response of a second-order system (Eq. 5.9) for a step input. (b) The transient response of a second-order system (Eq. 5.9) for a step input as a function of ζ and $\omega_n t$. (Courtesy of Professor R. Jacquot, University of Wyoming.)

The Laplace transform of the unit impulse is $R(s) = 1$, and therefore the output for an impulse is

$$C(s) = \frac{\omega_n^2}{s^2 + 2\zeta\omega_n s + \omega_n^2}, \tag{5.10}$$

which is $T(s) = C(s)/R(s)$, the transfer function of the closed-loop system. The transient response for an impulse function input is then

$$c(t) = \frac{\omega_n}{\beta} e^{-\zeta\omega_n t} \sin \omega_n \beta t, \tag{5.11}$$

which is simply the derivative of the response to a step input. The impulse response of the second-order system is shown in Fig. 5.6 for several values of the damping ratio, ζ. The designer is able to select several alternative performance measures from the transient response of the system for either a step or impulse input.

Standard performance measures are usually defined in terms of the step response of a system as shown in Fig. 5.7. The swiftness of the response is measured by the *rise time, T_r,* and the *peak time, T_p.* For underdamped systems with an overshoot, the 0 to 100% rise time is a useful index. If the system is overdamped, then the peak time is not defined and the 10–90% rise time, T_{r_1}, is normally used. The similarity with which the actual response matches the step input is measured by the percent overshoot and settling time T_s. The *percent overshoot,* P.O., is defined as

$$\text{P.O.} = \frac{M_{p_t} - fv}{fv} \times 100\% \tag{5.12}$$

for a unit step input, where M_{p_t} is the peak value of the time response and fv is the final value of the response. Normally fv is the magnitude of the input, but many systems have a final value significantly different from the desired input magnitude. For the system with a unit step represented by Eq. (5.8), we have $fv = 1$.

FIGURE 5.6
Response of a second-order system for an impulse function input.

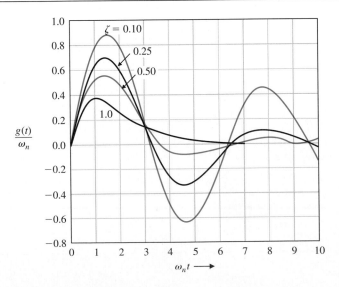

[handwritten: Smallest real part decays the slowest]

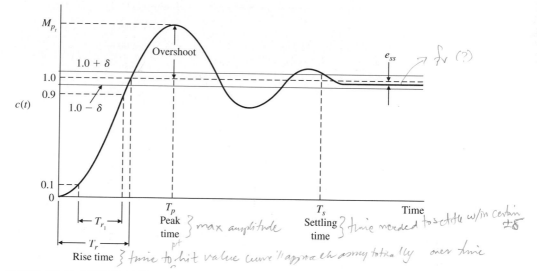

FIGURE 5.7
Step response of a
control system
(Eq. 5.9).

[handwritten annotations on figure: Peak time { max amplitude ; Settling time { time needed to settle w/in certain ±δ ; Rise time { time to hit value curve'll approach asymptotically over time ; $e_{ss} \to$ f(v (?)]

The *settling time*, T_s, is defined as the time required for the system to settle within a certain percentage δ of the input amplitude. This band of $\pm \delta$ is shown in Fig. 5.7. For the second-order system with closed-loop damping constant $\zeta \omega_n$, with a response described by Eq. (5.9), we seek to determine the time, T_s, for which the response remains within 2% of the final value. This occurs approximately when

$$e^{-\zeta \omega_n T_s} < 0.02$$

or

$$\zeta \omega_n T_s \cong 4.$$

Therefore we have

$$T_s = 4\tau = \frac{4}{\zeta \omega_n}. \tag{5.13}$$

Hence we will define the settling time as four time constants of the dominant roots of the characteristic equation. Finally, the steady-state error of the system may be measured on the step response of the system as shown in Fig. 5.7.

Consequently, the transient response of the system may be described in terms of two factors:

1. The swiftness of response, as represented by the rise time and the peak time.
2. The closeness of the response to the desired response as represented by the overshoot and settling time.

As nature would have it, these are contradictory requirements and a compromise must be obtained. To obtain an explicit relation for M_{p_t} and T_p as a function of ζ, one can differentiate Eq. (5.9) and set it equal to zero. Alternatively, one can utilize the differentiation property of the Laplace transform, which may be written as

$$\mathcal{L}\left\{\frac{dc(t)}{dt}\right\} = sC(s)$$

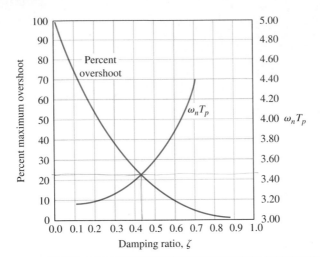

FIGURE 5.8
Percent overshoot and peak time versus damping ratio ζ for a second-order system (Eq. 5.8).

when the initial value of $c(t)$ is zero. Therefore we may acquire the derivative of $c(t)$ by multiplying Eq. (5.8) by s and thus obtaining the right side of Eq. (5.10). Taking the inverse transform of the right side of Eq. (5.10), we obtain Eq. (5.11), which is equal to zero when $\omega_n \beta t = \pi$. Therefore we find that the peak time relationship for this second-order system is

$$T_p = \frac{\pi}{\omega_n \sqrt{1 - \zeta^2}}, \tag{5.14}$$

and the peak response is

$$M_{p_t} = 1 + e^{-\zeta\pi/\sqrt{1-\zeta^2}}. \tag{5.15}$$

Therefore the percent overshoot is

$$\text{P.O.} = 100 e^{-\zeta\pi/\sqrt{1-\zeta^2}}. \tag{5.16}$$

The percent overshoot versus the damping ratio ζ is shown in Fig. 5.8. Also, the normalized peak time, $\omega_n T_p$, is shown versus the damping ratio ζ in Fig. 5.8. The percent overshoot versus the damping ratio is listed in Table 5.2 for selected values of the damping ratio. Again, we are confronted with a necessary compromise between the swiftness of response and the allowable overshoot.

The swiftness of step response can be measured as the time it takes to rise from 10% to 90% of the magnitude of the step input. This is the definition of the rise time T_{r_1} shown in Fig. 5.7. The normalized rise time $\omega_n T_{r_1}$ versus ζ ($0.05 \leq \zeta \leq 0.95$) is shown in Fig. 5.9.

TABLE 5.2   Percent Peak Overshoot Versus Damping Ratio for a Second-Order System

Damping ratio	0.9	0.8	0.7	0.6	0.5	0.4	0.3
Percent overshoot	0.2	1.5	4.6	9.5	16.3	25.4	37.2

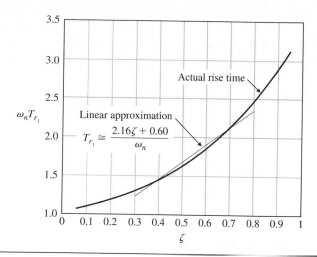

FIGURE 5.9
Normalized rise time T_{r_1} versus ζ for a second-order system.

Although it is difficult to obtain exact analytic expressions for T_{r_1}, we can utilize the linear approximation:

$$T_{r_1} = \frac{2.16\zeta + 0.60}{\omega_n}, \qquad (5.17)$$

which is accurate for $0.3 \le \zeta \le 0.8$. This linear approximation is shown in Fig. 5.9.

The swiftness of a response to a step input as described by Eq. (5.17) is dependent on ζ and ω_n. For a given ζ, the response is faster for larger ω_n, as shown in Fig. 5.10. Note that the overshoot is independent of ω_n.

For a given ω_n, the response is faster for lower ζ, as shown in Fig. 5.11. The swiftness of the response, however, will be limited by the overshoot that can be accepted.

FIGURE 5.10
The step response for $\zeta = 0.2$ for $\omega_n = 1$ and $\omega_n = 10$.

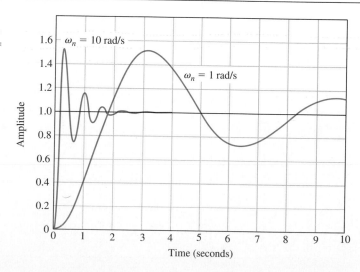

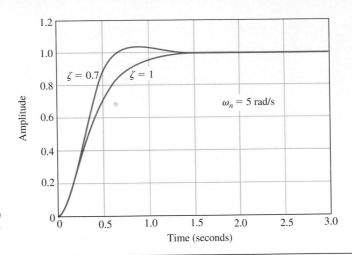

FIGURE 5.11
The step response
for $\omega_n = 5$ with $\zeta = 0.7$ and $\zeta = 1$.

5.4 EFFECTS OF A THIRD POLE AND A ZERO ON THE SECOND-ORDER SYSTEM RESPONSE

The curves presented in Fig. 5.8 are exact only for the second-order system of Eq. (5.8). However, they provide a remarkably good source of data because many systems possess a dominant pair of roots and the step response can be estimated by utilizing Fig. 5.8. This approach, although an approximation, avoids the evaluation of the inverse Laplace transformation in order to determine the percent overshoot and other performance measures. For example, for a third-order system with a closed-loop transfer function

$$T(s) = \frac{1}{(s^2 + 2\zeta s + 1)(\gamma s + 1)}, \tag{5.18}$$

the s-plane diagram is shown in Fig. 5.12. This third-order system is normalized with $\omega_n = 1$. It was ascertained experimentally that the performance as indicated by the percent

FIGURE 5.12
An s-plane
diagram of a third-
order system.

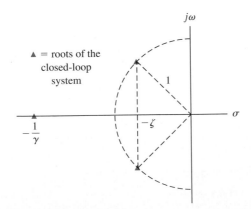

overshoot, $P.O._{p_t}$, and the settling time, T_s, was represented by the second-order system curves when [4]

$$|1/\gamma| \geq 10|\zeta\omega_n|.$$

In other words, the response of a third-order system can be approximated by the *dominant roots* of the second-order system as long as the real part of the dominant roots is less than $\frac{1}{10}$ of the real part of the third root [17, 23].

Using a computer simulation, when $\zeta = 0.45$, one can determine the response of a system to a unit step input. When $\gamma = 2.25$, we find that the response is overdamped because the real part of the complex poles is -0.45, whereas the real pole is equal to -0.444. The settling time is found via the simulation to be 9.6 seconds. If $\gamma = 0.90$ or $1/\gamma = 1.11$ is compared to $\zeta\omega_n = 0.45$ of the complex poles, we find that the overshoot is 12% and the settling time is 8.8 seconds. If the complex roots were dominant, we would expect the overshoot to be 20% and the settling time to be $4/\zeta\omega_n = 8.9$ seconds. The results are summarized in Table 5.3.

Also, we must note that the performance measures of Fig. 5.8 are correct only for a transfer function without finite zeros. If the transfer function of a system possesses finite zeros and they are located relatively near the dominant complex poles, then the zeros will materially affect the transient response of the system [5].

The transient response of a system with one zero and two poles may be affected by the location of the zero [5]. The percent overshoot for a step input as a function of $a/\zeta\omega_n$, when $\zeta \leq 1$, is given in Fig. 5.13(a) for the system transfer function

$$T(s) = \frac{(\omega_n^2/a)(s + a)}{s^2 + 2\zeta\omega_n s + \omega_n^2}.$$

The actual transient response for a step input is shown in Fig. 5.13(b) for selected values of $a/\zeta\omega_n$. The actual response for these selected values is summarized in Table 5.4 (p. 229) when $\zeta = 0.45$.

The correlation of the time-domain response of a system with the s-plane location of the poles of the closed-loop transfer function is very useful for selecting the specifications of a system. To illustrate clearly the utility of the s-plane, let us consider a simple example.

TABLE 5.3   Effect of a Third Pole (Eq. 5.18) for $\zeta = 0.45$

γ	$\dfrac{1}{\gamma}$	Percent Overshoot	Settling Time*
2.25	0.444	0	9.63
1.5	0.666	3.9	6.3
0.9	1.111	12.3	8.81
0.4	2.50	18.6	8.67
0.05	20.0	20.5	8.37
0	∞	20.5	8.24

* *Note:* Settling time is normalized time, $\omega_n T_s$.

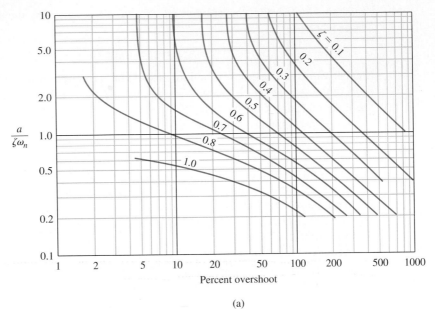

(a)

FIGURE 5.13
(a) Percent overshoot as a function of ζ and ω_n when a second-order transfer function contains a zero. (From R. N. Clark, *Introduction to Automatic Control Systems,* New York, Wiley, 1962, redrawn with permission.) (b) The response for the second-order transfer function with a zero for four values of the ratio $(a/\zeta\omega_n)$: $A = 5$, $B = 2$, $C = 1$, and $D = 0.5$ when $\zeta = 0.45$.

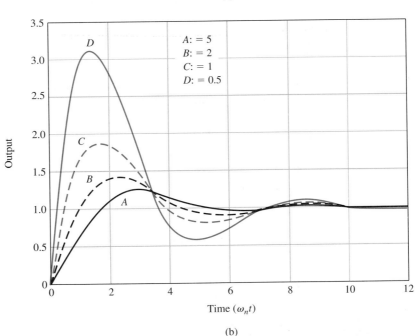

(b)

EXAMPLE 5.1 Parameter selection

A single-loop feedback control system is shown in Fig. 5.14. We desire to select the gain K and the parameter p so that the time-domain specifications will be satisfied. The transient

TABLE 5.4   The Response of a Second-Order System with a Zero and $\zeta = 0.45$

$a/\zeta\,\omega_n$	Percent Overshoot	Settling Time	Peak Time
5	23.1	8.0	3.0
2	39.7	7.6	2.2
1	89.9	10.1	1.8
0.5	210.0	10.3	1.5

Note: Time is normalized as $\omega_n t$.

response to a step should be as fast in responding as is attainable while retaining an overshoot of less than 5%. Furthermore, the settling time should be less than 4 seconds. The damping ratio ζ for an overshoot of 4.3% is 0.707. This damping ratio is shown graphically as a line in Fig. 5.15. Because the settling time is

$$T_s = \frac{4}{\zeta\omega_n} \leq 4 \text{ s},$$

we require that the real part of the complex poles of $T(s)$ be

$$\zeta\omega_n \geq 1.$$

This region is also shown in Fig. 5.15. The region that will satisfy both time-domain requirements is shown cross-hatched on the s-plane of Fig. 5.15.

FIGURE 5.14
Single-loop
feedback control
system.

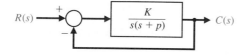

FIGURE 5.15
Specifications and
root locations on
the s-plane.

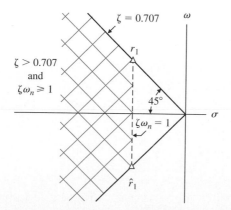

When the closed-loop roots are chosen as $r_1 = -1 + j1$ and $\hat{r}_1 = -1 - j1$, then we have $T_s = 4s$ and an overshoot of 4.3%. Therefore $\zeta = 1/\sqrt{2}$ and $\omega_n = 1/\zeta = \sqrt{2}$. The closed-loop transfer function is

$$T(s) = \frac{G(s)}{1 + G(s)} = \frac{K}{s^2 + ps + K} = \frac{\omega_n^2}{s^2 + 2\zeta\omega_n s + \omega_n^2}.$$

Hence we require that $K = \omega_n^2 = 2$ and $p = 2\zeta\omega_n = 2$. A full comprehension of the correlation between the closed-loop root location and the system transient response is important to the system analyst and designer. Therefore we shall consider the matter more fully in the following sections. ∎

EXAMPLE 5.2 **Dominant poles of $T(s)$**

Consider a system with a closed-loop transfer function

$$\frac{C(s)}{R(s)} = T(s) = \frac{\dfrac{\omega_n^2}{a}(s + a)}{(s^2 + 2\zeta\omega_n s + \omega_n^2)(1 + \tau s)}.$$

Both the zero and the real pole may affect the transient response. If $a \gg \zeta\omega_n$ and $\tau \ll \zeta\omega_n$, then the pole and zero will have little effect on the step response.

Assume that we have

$$T(s) = \frac{62.5(s + 2.5)}{(s^2 + 6s + 25)(s + 6.25)}.$$

Note that the dc gain is equal to 1 ($T(0) = 1$), and we expect zero steady-state error for a step input. We have $\zeta\omega_n = 3$, $\tau = 0.16$, and $a = 2.5$. The poles and the zero are shown on the s-plane in Fig. 5.16. As a first approximation, we neglect the real pole and obtain

$$T(s) \approx \frac{10(s + 2.5)}{(s^2 + 6s + 25)}.$$

We now have $\zeta = 0.6$ and $\omega_n = 5$ for dominant poles with one accompanying zero for

---

FIGURE 5.16
The poles and zeros on the s-plane for a third-order system.

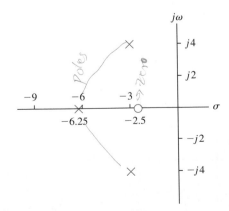

which $a/(\zeta\omega_n) = 0.833$. Using Fig. 5.13(a), we find that the percent overshoot is 55%. We expect a settling time

$$T_s = \frac{4}{\zeta\omega_n} = \frac{4}{0.6(5)} = 1.33 \text{ s.}$$

Using a computer simulation for the actual third-order system, we find that the percent overshoot is equal to 38% and the settling time is 1.6 seconds. Thus the effect of the damping of the third pole of $T(s)$ is to dampen the overshoot and increase the settling time (hence the real pole cannot be neglected). ∎

5.5 ESTIMATION OF THE DAMPING RATIO

The damping ratio can be estimated from a system's response to a step input [14]. The step response of a second-order system for a unit step input is given in Eq. (5.9) and repeated here:

$$c(t) = 1 - \frac{1}{\beta}e^{-\zeta\omega_n t}\sin(\omega_n\beta t + \theta),$$

where $\beta = \sqrt{1 - \zeta^2}$ and $\theta = \tan^{-1}(\beta/\zeta)$. Hence the frequency of the damped sinusoidal term for $\zeta < 1$ is

$$\omega = \omega_n(1 - \zeta^2)^{1/2} = \omega_n\beta,$$

and the number of cycles in one second is $\omega/2\pi$.

The time constant for the exponential decay is $\tau = 1/(\zeta\omega_n)$ in seconds. The number of cycles of the damped sinusoid during one time constant is

$$\text{(cycles) } \tau = \frac{\omega}{2\pi\zeta\omega_n} = \frac{\omega_n\beta}{2\pi\zeta\omega_n} = \frac{\beta}{2\pi\zeta}.$$

Assuming that the response decays in n visible time constants, we have

$$\text{cycles visible} = \frac{n\beta}{2\pi\zeta}. \tag{5.19}$$

For the second-order system, the response remains within 2% of the steady-state value after four time constants (4τ). Hence $n = 4$ and

$$\text{cycles visible} = \frac{4\beta}{2\pi\zeta} = \frac{4(1 - \zeta^2)^{1/2}}{2\pi\zeta} \approx \frac{0.55}{\zeta} \tag{5.20}$$

for $0.2 \leq \zeta \leq 0.6$.

As an example, examine the response shown in Fig. 5.5(a) for $\zeta = 0.4$. Start at $c(t) = 0$ as the first minimum point and count 1.4 cycles visible (until the response settles with 2% of the final value). Then we estimate ζ as

$$\zeta = \frac{0.55}{\text{cycles}} = \frac{0.55}{1.4} = 0.39.$$

We can use this approximation for systems with dominant complex poles so that

$$T(s) \approx \frac{\omega_n^2}{s^2 + 2\zeta\omega_n s + \omega_n^2}.$$

Then we are able to estimate the damping ratio ζ from the actual system response of a physical system.

An alternative method of estimating ζ is to determine the percent overshoot for the step response and use Fig. 5.8 to estimate ζ. For example, we determine an overshoot of 25% for $\zeta = 0.4$ from the response of Fig. 5.5(a). Using Fig. 5.8, we estimate that $\zeta = 0.4$, as expected.

5.6 THE s-PLANE ROOT LOCATION AND THE TRANSIENT RESPONSE

The transient response of a closed-loop feedback control system can be described in terms of the location of the poles of the transfer function. The closed-loop transfer function is written in general as

$$T(s) = \frac{C(s)}{R(s)} = \frac{\Sigma P_i(s)\Delta_i(s)}{\Delta(s)},$$

where $\Delta(s) = 0$ is the characteristic equation of the system. For the single-loop system of Fig. 5.14, the characteristic equation reduces to $1 + G(s) = 0$. It is the poles and zeros of $T(s)$ that determine the transient response. However, for a closed-loop system, the poles of $T(s)$ are the roots of the characteristic equation $\Delta(s) = 0$ and the poles of $\Sigma P_i(s)\Delta_i(s)$. The output of a system without repeated roots and a unit step input can be formulated as a partial fraction expansion as

$$C(s) = \frac{1}{s} + \sum_{i=1}^{M} \frac{A_i}{s + \sigma_i} + \sum_{k=1}^{N} \frac{B_k}{s^2 + 2\alpha_k s + (\alpha_k^2 + \omega_k^2)}, \tag{5.21}$$

where the A_i and B_k are constants. The roots of the system must be either $s = -\sigma_i$ or complex conjugate pairs as $s = -\alpha_k \pm j\omega_k$. Then the inverse transform results in the transient response as a sum of terms:

$$c(t) = 1 + \sum_{i=1}^{M} A_i e^{-\sigma_i t} + \sum_{k=1}^{N} B_k \left(\frac{1}{\omega_k}\right) e^{-\alpha_k t} \sin \omega_k t. \tag{5.22}$$

The transient response is composed of the steady-state output, exponential terms, and damped sinusoidal terms. Obviously, for the response to be stable—that is, bounded for a step input—one must require that the real part of the roots, σ_i or α_k, be in the left-hand portion of the s-plane. The impulse response for various root locations is shown in Fig. 5.17. The information imparted by the location of the roots is graphic, indeed, and usually well worth the effort of determining the location of the roots in the s-plane.

It is important for the control system analyst to understand the relationship between the complex-frequency representation of a linear system, through the poles and zeros of its transfer function, and its time-domain response to step and other inputs. Many of the analysis and design calculations in such areas as signal processing and control are done in the

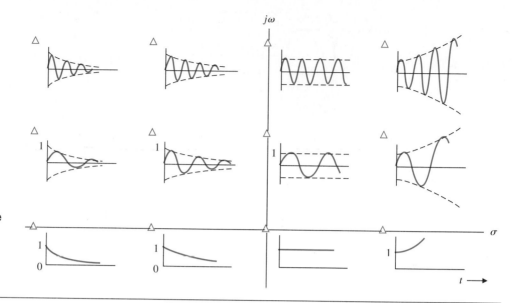

FIGURE 5.17
Impulse response for various root locations in the s-plane. (The conjugate root is not shown.)

complex-frequency plane, where a system model is represented in terms of the poles and zeros of its transfer function $T(s)$. On the other hand, system performance is often analyzed by examining time-domain responses, particularly when dealing with control systems.

The capable system designer will be able to envision the effects on the step and impulse responses of adding, deleting, or moving poles and zeros of $T(s)$ in the s-plane. Likewise, the designer should be able to visualize what changes should be made in the poles and zeros of $T(s)$ in order to effect desired changes in the model's step and impulse responses.

An experienced designer is aware of the effects of zero locations on system response. The poles of $T(s)$ determine the particular response modes that will be present and the zeros of $T(s)$ establish the relative weightings of the individual mode functions. For example, moving a zero closer to a specific pole will reduce the relative contribution of the mode function corresponding to the pole.

A computer program can be developed that allows a person to specify arbitrary sets of poles and zeros for the transfer function of a linear system. Then the computer will evaluate and plot the system's impulse and step responses individually. It will also display them in reduced form along with the pole–zero plot.

Once the program has been run for a set of poles and zeros, the user can modify the locations of one or more of them. Plots are then presented showing the old and new poles and zeros in the complex plane and the old and new impulse and step responses.

5.7 THE STEADY-STATE ERROR OF FEEDBACK CONTROL SYSTEMS

One of the fundamental reasons for using feedback, despite its cost and increased complexity, is the attendant improvement in the reduction of the steady-state error of the system. As was illustrated in Section 4.5, the steady-state error of a stable closed-loop system is

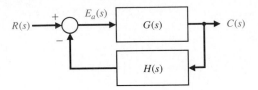

FIGURE 5.18
Closed-loop
control system.

usually several orders of magnitude smaller than the error of an open-loop system. The system actuating signal, which is a measure of the system error, is denoted as $E_a(s)$. However, the actual system error is $E(s) = R(s) - C(s)$. Considering the closed-loop feedback system of Fig. 5.18, we have

$$E(s) = R(s) - \frac{G(s)}{1 + GH(s)} R(s) = \frac{[1 + GH(s) - G(s)]}{1 + GH(s)} R(s).$$

The system error is equal to the signal, $E_a(s)$, when $H(s) = 1$. Then

$$E(s) = \frac{1}{1 + G(s)} R(s).$$

The steady-state error, when $H(s) = 1$, is then

$$\lim_{t \to \infty} e(t) = e_{ss} = \lim_{s \to 0} \frac{sR(s)}{1 + G(s)}. \tag{5.23}$$

It is useful to determine the steady-state error of the system for the three standard test inputs for a unity feedback system ($H(s) = 1$).

Step Input.   The steady-state error for a step input of magnitude A is therefore

$$e_{ss} = \lim_{s \to 0} \frac{s(A/s)}{1 + G(s)} = \frac{A}{1 + G(0)}.$$

Clearly, it is the form of the loop transfer function $G(s)$ that determines the steady-state error. The loop transfer function is written in general form as

$$G(s) = \frac{K \prod_{i=1}^{M} (s + z_i)}{s^N \prod_{k=1}^{Q} (s + p_k)}, \tag{5.24}$$

where $\prod$ denotes the product of the factors. Therefore the loop transfer function as s approaches zero depends on the number of integrations N. If N is greater than zero, then $G(0)$ approaches infinity and the steady-state error approaches zero. The number of integrations is often indicated by labeling a system with a *type number* that simply is equal to N.

Therefore, for a type-zero system, $N = 0$, the steady-state error is

$$e_{ss} = \frac{A}{1 + G(0)} = \frac{A}{1 + \left(K \prod_{i=1}^{M} z_i \Big/ \prod_{k=1}^{Q} p_k \right)}. \tag{5.25}$$

The constant $G(0)$ is denoted by K_p, the position error constant, so that

$$e_{ss} = \frac{A}{1 + K_p}.$$ (5.26)

Hence the steady-state error for a unit step input with one integration or more, $N \geq 1$, is zero because

$$e_{ss} = \lim_{s \to 0} \frac{A}{1 + (K \prod z_i / s^N \prod p_k)}$$ (5.27)

$$= \lim_{s \to 0} \frac{A s^N}{s^N + (K \prod z_i / \prod p_k)} = 0.$$

Ramp Input.   The steady-state error for a ramp (velocity) input with a slope A is

$$e_{ss} = \lim_{s \to 0} \frac{s(A/s^2)}{1 + G(s)} = \lim_{s \to 0} \frac{A}{s + sG(s)} = \lim_{s \to 0} \frac{A}{sG(s)}.$$ (5.28)

Again, the steady-state error depends upon the number of integrations N. For a type-zero system, $N - 0$, the steady-state error is infinite. For a type-one system, $N = 1$, the error is

$$e_{ss} = \lim_{s \to 0} \frac{A}{s\{[K \prod (s + z_i)] / [s \prod (s + p_k)]\}}$$ (5.29)

$$= \frac{A}{(K \prod z_i / \prod p_k)} = \frac{A}{K_v},$$

where K_v is designated the *velocity error constant*. When the transfer function possesses two or more integrations, $N \geq 2$, we obtain a steady-state error of zero. When $N = 1$, a steady-state error exists. However, the steady-state velocity of the output is equal to the velocity of input (see Fig. 5.20).

Acceleration Input.   When the system input is $r(t) = At^2/2$, the steady-state error is then

$$e_{ss} = \lim_{s \to 0} \frac{s(A/s^3)}{1 + G(s)} = \lim_{s \to 0} \frac{A}{s^2 G(s)}.$$ (5.30)

The steady-state error is infinite for one integration; for two integrations, $N = 2$, we obtain

$$e_{ss} = \frac{A}{K \prod z_i / \prod p_k} = \frac{A}{K_a},$$ (5.31)

where K_a is designated the acceleration constant. When the number of integrations equals or exceeds three, then the steady-state error of the system is zero.

Control systems are often described in terms of their type number and the error constants, K_p, K_v, and K_a. Definitions for the error constants and the steady-state error for the three inputs are summarized in Table 5.5. The usefulness of the error constants can be illustrated by considering a simple example.

TABLE 5.5   Summary of Steady-State Errors

Number of Integrations in $G(s)$, Type Number	Input		
	Step, $r(t) = A$, $R(s) = A/s$	Ramp, At, A/s^2	Parabola, $At^2/2$, A/s^3
0	$e_{ss} = \dfrac{A}{1 + K_p}$	Infinite	Infinite
1	$e_{ss} = 0$	$\dfrac{A}{K_v}$	Infinite
2	$e_{ss} = 0$	0	$\dfrac{A}{K_a}$

EXAMPLE 5.3   Mobile robot steering control

A severely disabled person could use a mobile robot to serve as an assisting device or servant [8]. The steering control system for such a robot can be represented by the block diagram shown in Fig. 5.19. The steering controller, $G_1(s)$, is

$$G_1(s) = K_1 + K_2/s. \qquad (5.32)$$

The steady-state error of the system for a step input when $K_2 = 0$ and $G_1(s) = K_1$ is therefore

$$e_{ss} = \frac{A}{1 + K_p}, \qquad (5.33)$$

where $K_p = KK_1$. When K_2 is greater than zero, we have a type-one system,

$$G_1(s) = \frac{K_1 s + K_2}{s},$$

and the steady-state error is zero for a step input.

If the steering command is a ramp input, the steady-state error is then

$$e_{ss} = \frac{A}{K_v}, \qquad (5.34)$$

where

$$K_v = \lim_{s \to 0} sG_1(s)G(s) = K_2 K.$$

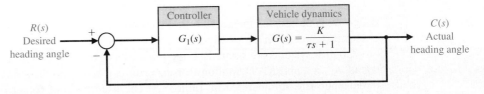

FIGURE 5.19
Block diagram of steering control system for a mobile robot.

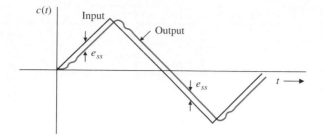

FIGURE 5.20
Triangular wave
response.

The transient response of the vehicle to a triangular wave input when $G_1(s) = (K_1s + K_2)/s$ is shown in Fig. 5.20. The transient response clearly shows the effect of the steady-state error, which may not be objectionable if K_v is sufficiently large. Note that the output attains the desired velocity as required by the input, but it exhibits a steady-state error. ■

The error constants, K_p, K_v, and K_a, of a control system describe the ability of a system to reduce or eliminate the steady-state error. They are therefore utilized as numerical measures of the steady-state performance. The designer determines the error constants for a given system and attempts to determine methods of increasing the error constants while maintaining an acceptable transient response. In the case of the steering control system, it is desirable to increase the gain factor KK_2 in order to increase K_v and reduce the steady-state error. However, an increase in KK_2 results in an attendant decrease in the damping ratio, ζ, of the system and therefore a more oscillatory response to a step input. Again, a compromise would be determined that would provide the largest K_v based on the smallest ζ allowable.

5.8 THE STEADY-STATE ERROR OF NONUNITY FEEDBACK SYSTEMS

A general feedback system with nonunity feedback is shown in Fig. 5.21 with $H(s)$ not equal to unity. For a system in which the feedback is not unity, the units of the output, $C(s)$, are usually different from the output of the sensor. For example, a speed control system is shown in Fig. 5.22, where $H(s) = K_2$. The constants K_1 and K_2 account for the conversion of one set of units to another set of units (here we convert rad/s to volts). We can select K_1 and thus we set $K_1 = K_2$ and move the block for K_1 and K_2 past the summing node. Then we obtain the equivalent block diagram shown in Fig. 5.21. Thus we obtain a unity feedback system as desired.

Let us return to the system of Fig. 5.21 with $H(s)$. The case where

$$H(s) = \frac{K_2}{(\tau s + 1)}$$

FIGURE 5.21
A nonunity
feedback system.

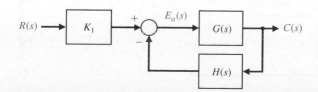

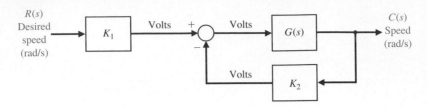

FIGURE 5.22
A speed control
system.

has a dc gain for $H(s)$ of

$$\lim_{s \to 0} H(s) = K_2.$$

Again, the factor K_2 is a conversion of units factor. If we set $K_2 = K_1$, then the system is transformed to that of Fig. 5.23 (for the dc gain or steady-state calculation). Then the error of the system shown in Fig. 5.23 is $E(s)$, where

$$E(s) = R(s) - C(s) = [1 - T(s)]R(s), \tag{5.35}$$

since $C(s) = T(s)R(s)$. Note that

$$T(s) = \frac{K_1 G(s)}{1 + K_1 G(s)}$$

and therefore

$$E(s) = \frac{1}{1 + K_1 G(s)} R(s).$$

Then the steady-state error for a unit step input is

$$e_{ss} = \lim_{s \to 0} E(s) = \frac{1}{1 + K_1 G(0)}. \tag{5.36}$$

In general, we can always determine the actual system error by using Eq. (5.23).

EXAMPLE 5.4   Steady-state error

Let us determine the appropriate value of K_1 and calculate the steady-state error for a unit step input for the system shown in Fig. 5.21 when

$$G(s) = \frac{40}{(s + 5)}$$

and

$$H(s) = \frac{20}{(s + 10)} = \frac{K_2}{(0.1s + 1)}.$$

FIGURE 5.23
The speed control
system of Fig. 5.22
with $K_1 = K_2$.

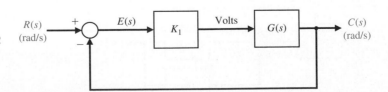

We can rewrite $H(s)$ as

$$H(s) = \frac{2}{(0.1s + 1)}.$$

Selecting $K_1 = K_2 = 2$, we can use Eq. (5.36) to determine

$$e_{ss} = \frac{1}{1 + K_1 G(0)} = \frac{1}{1 + 2(8)} = \frac{1}{17},$$

or 5.9% of the magnitude of the step input. ∎

EXAMPLE 5.5   **Feedback system**

Let us consider the system of Fig. 5.24, where we assume we cannot insert a gain K_1 following $R(s)$ as we did for the system of Fig. 5.21. Then the actual error is Eq. (5.35)

$$E(s) = [1 - T(s)]R(s).$$

Let us determine an appropriate gain K so that the steady-state error to a step input is minimized. The steady-state error is

$$e_{ss} = \lim_{s \to 0} s[1 - T(s)]\frac{1}{s},$$

where

$$T(s) = \frac{G(s)}{1 + G(s)H(s)} = \frac{K(s + 4)}{(s + 2)(s + 4) + 2K}.$$

Then we have

$$T(0) = \frac{4K}{8 + 2K}.$$

The steady-state error for a unit step input is

$$e_{ss} = [1 - T(0)].$$

Thus, to achieve zero steady-state error, we require

$$T(0) = \frac{4K}{8 + 2K} = 1,$$

or $8 + 2K = 4K$. Thus $K = 4$ will yield a zero steady-state error. ∎

FIGURE 5.24
A system with a
feedback $H(s)$.

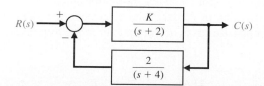

5.9 PERFORMANCE INDICES

Increasing emphasis on the mathematical formulation and measurement of control system performance can be found in the recent literature on automatic control. Modern control theory assumes that the systems engineer can specify quantitatively the required system performance. Then a performance index can be calculated or measured and used to evaluate the system's performance. A quantitative measure of the performance of a system is necessary for the operation of modern adaptive control systems, for automatic parameter optimization of a control system, and for the design of optimum systems.

Whether the aim is to improve the design of a system or to design a control system, a performance index must be chosen and measured.

A performance index is a quantitative measure of the performance of a system and is chosen so that emphasis is given to the important system specifications.

A system is considered an *optimum control system* when the system parameters are adjusted so that the index reaches an extremum value, commonly a minimum value. A performance index, to be useful, must be a number that is always positive or zero. Then the best system is defined as the system that minimizes this index.

A suitable performance index is the integral of the square of the error, ISE, which is defined as

$$\text{ISE} = \int_0^T e^2(t)dt. \tag{5.37}$$

The upper limit T is a finite time chosen somewhat arbitrarily so that the integral approaches a steady-state value. It is usually convenient to choose T as the settling time, T_s. The step response for a specific feedback control system is shown in Fig. 5.25(b), and the error in Fig. 5.25(c). The error squared is shown in Fig. 5.25(d), and the integral of the error squared in Fig. 5.25(e). This criterion will discriminate between excessively overdamped and excessively underdamped systems. The minimum value of the integral occurs for a compromise value of the damping. The performance index of Eq. (5.37) is easily adapted for practical measurements because a squaring circuit is readily obtained. Furthermore, the squared error is mathematically convenient for analytical and computational purposes.

Another readily instrumented performance criterion is the integral of the absolute magnitude of the error, IAE, which is written as

$$\text{IAE} = \int_0^T |e(t)|dt. \tag{5.38}$$

This index is particularly useful for computer simulation studies.

To reduce the contribution of the large initial error to the value of the performance integral, as well as to emphasize errors occurring later in the response, the following index has been proposed [6]:

$$\text{ITAE} = \int_0^T t|e(t)|dt. \tag{5.39}$$

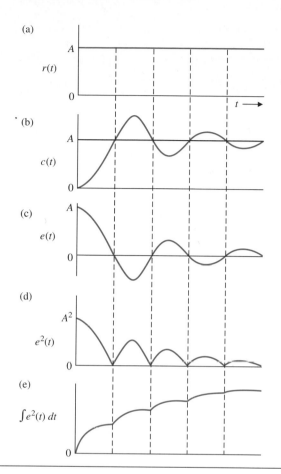

FIGURE 5.25
The calculation of
the integral
squared error.

This performance index is designated the integral of time multiplied by absolute error, ITAE. Another similar index is the integral of time multiplied by the squared error, ITSE:

$$\text{ITSE} = \int_0^T te^2(t)dt. \tag{5.40}$$

The performance index ITAE provides the best selectivity of the performance indices; that is, the minimum value of the integral is readily discernible as the system parameters are varied. The general form of the performance integral is

$$I = \int_0^T f(e(t),\ r(t),\ c(t),\ t)dt, \tag{5.41}$$

where f is a function of the error, input, output, and time. Clearly, one can obtain numerous indices based on various combinations of the system variables and time. It is worth noting that the minimization of IAE or ISE is often of practical significance. For example, the minimization of a performance index can be directly related to the minimization of fuel consumption for aircraft and space vehicles.

Performance indices are useful for the analysis and design of control systems. Two examples will illustrate the utility of this approach.

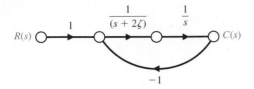

FIGURE 5.26
Single-loop
feedback control
system.

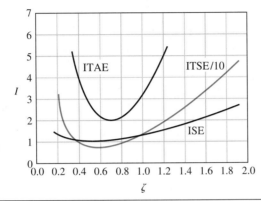

FIGURE 5.27
Three performance
criteria for a
second-order
system.

EXAMPLE 5.6   **Performance criteria**

A single-loop feedback control system is shown in Fig. 5.26, where the natural frequency is the normalized value, $\omega_n = 1$. The closed-loop transfer function is then

$$T(s) = \frac{1}{s^2 + 2\zeta s + 1}. \tag{5.42}$$

Three performance indices—ISE, ITSE, and ITAE—calculated for various values of the damping ratio ζ and for a step input are shown in Fig. 5.27. These curves show the selectivity of the ITAE index in comparison with the ISE index. The value of the damping ratio ζ selected on the basis of ITAE is 0.7, which, for a second-order system, results in a swift response to a step with a 4.6% overshoot. ∎

EXAMPLE 5.7   **Space telescope control system**

The signal-flow graph of a space telescope pointing control system is shown in Fig. 5.28 [11]. We desire to select the magnitude of the gain K_3 to minimize the effect of the disturbance $U(s)$. The disturbance in this case is equivalent to an initial attitude error. The closed-loop transfer function for the disturbance is obtained by using the signal-flow gain formula as

$$
\begin{aligned}
\frac{C(s)}{D(s)} &= \frac{P_1(s)\Delta_1(s)}{\Delta(s)} \\[2mm]
&= \frac{1 \cdot (1 + K_1 K_3 s^{-1})}{1 + K_1 K_3 s^{-1} + K_1 K_2 K_p s^{-2}} \\[2mm]
&= \frac{s(s + K_1 K_3)}{s^2 + K_1 K_3 s + K_1 K_2 K_p}.
\end{aligned}
\tag{5.43}
$$

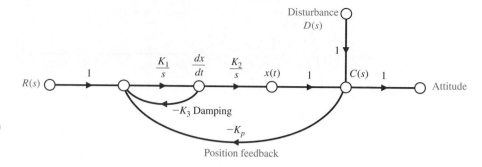

FIGURE 5.28
A space telescope pointing control system.

Typical values for the constants are $K_1 = 0.5$ and $K_1 K_2 K_p = 2.5$. Then the natural frequency of the vehicle is $f_n = \sqrt{2.5}/2\pi = 0.25$ cycles/s. For a unit step disturbance, the minimum ISE can be analytically calculated. The attitude $c(t)$ is

$$c(t) = \frac{\sqrt{10}}{\beta}\left[e^{-0.25K_3t}\,\sin\left(\frac{\beta}{2}t + \psi\right)\right], \tag{5.44}$$

where $\beta = K_3\sqrt{(K_3^2/8) - 5}$. Squaring $c(t)$ and integrating the result, we have

$$
\begin{aligned}
I &= \int_0^\infty \frac{10}{\beta^2}e^{-0.5K_3t}\,\sin^2\left(\frac{\beta}{2}t + \psi\right)dt \\
&= \int_0^\infty \frac{10}{\beta^2}e^{-0.5K_3t}\left(\frac{1}{2} - \frac{1}{2}\cos(\beta t + 2\psi)\right)dt \tag{5.45} \\
&= \left(\frac{1}{K_3} + 0.1K_3\right).
\end{aligned}
$$

Differentiating I and equating the result to zero, we obtain

$$\frac{dI}{dK_3} = -K_3^{-2} + 0.1 = 0. \tag{5.46}$$

Therefore the minimum ISE is obtained when $K_3 = \sqrt{10} = 3.2$. This value of K_3 corresponds to a damping ratio ζ of 0.50. The values of ISE and IAE for this system are plotted in Fig. 5.29. The minimum for the IAE performance index is obtained when $K_3 = 4.2$ and $\zeta = 0.665$. While the ISE criterion is not as selective as the IAE criterion, it is clear that

FIGURE 5.29
The performance indices of the telescope control system versus K_3.

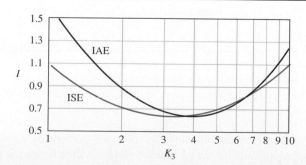

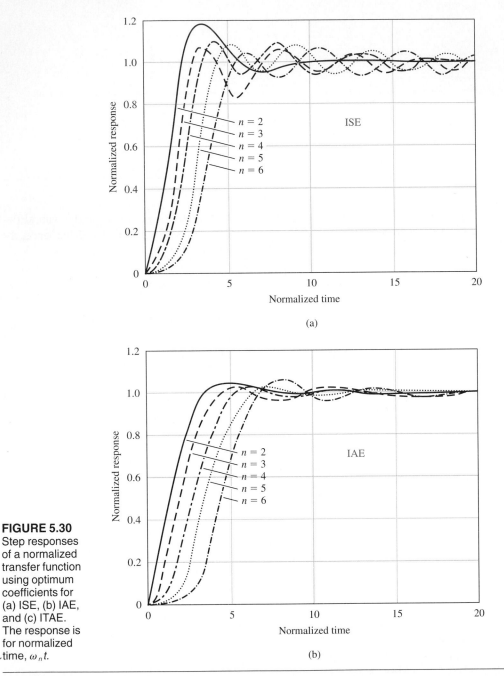

FIGURE 5.30
Step responses
of a normalized
transfer function
using optimum
coefficients for
(a) ISE, (b) IAE,
and (c) ITAE.
The response is
for normalized
.time, $\omega_n t$.

it is possible to solve analytically for the minimum value of ISE. The minimum of IAE is obtained by measuring the actual value of IAE for several values of the parameter of interest. ∎

A control system is optimum when the selected performance index is minimized. However, the optimum value of the parameters depends directly on the definition of opti-

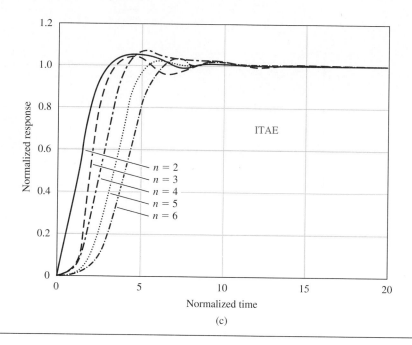

FIGURE 5.30
Continued

(c)

mum, that is, the performance index. Therefore, in Examples 5.6 and 5.7 we found that the optimum setting varied for different performance indices.

The coefficients that will minimize the ITAE performance criterion for a step input have been determined for the general closed-loop transfer function [6]:

$$T(s) = \frac{C(s)}{R(s)} = \frac{b_0}{s^n + b_{n-1}s^{n-1} + \cdots + b_1 s + b_0}. \quad (5.47)$$

This transfer function has a steady-state error equal to zero for a step input. Note that the transfer function has n poles and no zeros. The optimum coefficients for the ITAE criterion are given in Table 5.6. The responses using optimum coefficients for a step input are given in Fig. 5.30 for ISE, IAE, and ITAE. The responses are provided for normalized time, $\omega_n t$. Other standard forms based on different performance indices are available and can be useful in aiding the designer to determine the range of coefficients for a specific problem. A final example will illustrate the utility of the standard forms for ITAE.

EXAMPLE 5.8   Two-camera control

A very accurate and rapidly responding control system is required for a system that allows live actors seemingly to perform inside of complex miniature sets. The two-camera system is shown in Fig. 5.31(a), where one camera is trained on the actor and the other on the miniature set. The challenge is to obtain rapid and accurate coordination of the two cameras by using sensor information from the foreground camera to control the movement of the background camera. The block diagram of the background camera system is shown in Fig. 5.31(b) for one axis of movement of the background camera. The closed-loop transfer function is

$$T(s) = \frac{K_a K_m \omega_0^2}{s^3 + 2\zeta\omega_0 s^2 + \omega_0^2 s + K_a K_m \omega_0^2}. \quad (5.48)$$

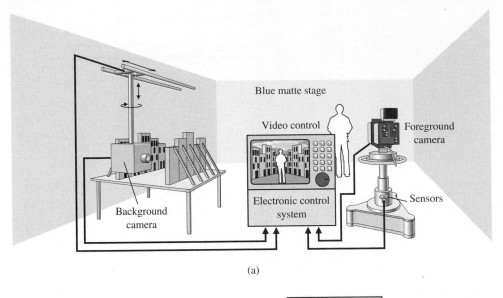

(a)

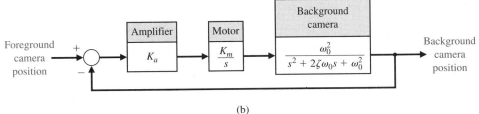

(b)

FIGURE 5.31
The foreground camera, which may be either a film or video camera, is trained on the blue cyclorama stage. The electronic servocontrol installation permits the slaving, by means of electronic servodevices, of the two cameras. The background camera reaches into the miniature set with a periscope lens and instantaneously reproduces all movements of the foreground camera in the scale of the miniature. The video control installation allows the composite image to be monitored and recorded live. (Part (a) reprinted with permission from *Electronic Design* 24, 11, May 24, 1976. Copyright © Hayden Publishing Co., Inc., 1976.)

The standard form for a third-order system given in Table 5.6 requires that

$$2\zeta\omega_0 = 1.75\omega_n, \qquad \omega_0^2 = 2.15\omega_n^2, \qquad K_a K_m \omega_0^2 = \omega_n^3.$$

Examining Fig. 5.30(c) for $n = 3$, we estimate that the settling time is approximately 14 seconds (normalized time). Therefore we estimate

$$\omega_n T_s \doteq 14.$$

Because a rapid response is required, a large ω_n will be selected so that the settling time will be less than one second. Thus, ω_n will be set equal to 50 rad/s. Then, for an ITAE system, it is necessary that the parameters of the camera dynamics be

$$\omega_0 = 73 \text{ rad/s}$$

TABLE 5.6   The Optimum Coefficients of $T(s)$ Based on the ITAE Criterion for a Step Input

$$s + \omega_n$$

$$s^2 + 1.4\omega_n s + \omega_n^2$$

$$s^3 + 1.75\omega_n s^2 + 2.15\omega_n^2 s + \omega_n^3$$

$$s^4 + 2.1\omega_n s^3 + 3.4\omega_n^2 s^2 + 2.7\omega_n^3 s + \omega_n^4$$

$$s^5 + 2.8\omega_n s^4 + 5.0\omega_n^2 s^3 + 5.5\omega_n^3 s^2 + 3.4\omega_n^4 s + \omega_n^5$$

$$s^6 + 3.25\omega_n s^5 + 6.60\omega_n^2 s^4 + 8.60\omega_n^3 s^3 + 7.45\omega_n^4 s^2 + 3.95\omega_n^5 s + \omega_n^6$$

and

$$\zeta = 0.60.$$

The amplifier and motor gain are required to be

$$K_a K_m = \frac{\omega_n^3}{\omega_0^2} = \frac{\omega_n^3}{2.15\omega_n^2} = \frac{\omega_n}{2.15} = 23.2.$$

Then, the closed-loop transfer function is

$$T(s) = \frac{125{,}000}{s^3 + 87.5s^2 + 537.5s + 125{,}000}$$

$$= \frac{125{,}000}{(s + 35.5)(s + 26 + j53.4)(s + 26 - j53.4)}.$$

(5.49)

The locations of the closed-loop roots dictated by the ITAE system are shown in Fig. 5.32. The damping ratio of the complex roots is $\zeta = 0.49$. However, the complex roots

FIGURE 5.32
The closed-loop roots of a minimum ITAE system.

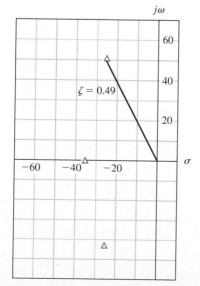

TABLE 5.7   The Optimum Coefficients of $T(s)$ Based on the ITAE Criterion for a Ramp Input

$$s^2 + 3.2\omega_n s + \omega_n^2$$

$$s^3 + 1.75\omega_n s^2 + 3.25\omega_n^2 s + \omega_n^3$$

$$s^4 + 2.41\omega_n s^3 + 4.93\omega_n^2 s^2 + 5.14\omega_n^3 s + \omega_n^4$$

$$s^5 + 2.19\omega_n s^4 + 6.50\omega_n^2 s^3 + 6.30\omega_n^3 s^2 + 5.24\omega_n^4 s + \omega_n^5$$

do not dominate. The actual response to a step input using a computer simulation showed the overshoot to be only 2% and the settling time equal to 0.08 second. This illustrates the damping effect of the real root (see Section 5.5).

For a ramp input, the coefficients have been determined that minimize the ITAE criterion for the general closed-loop transfer function [6]:

$$T(s) = \frac{b_1 s + b_0}{s^n + b_{n-1}s^{n-1} + \cdots + b_1 s + b_0}. \tag{5.50}$$

This transfer function has a steady-state error equal to zero for a ramp input. The optimum coefficients for this transfer function are given in Table 5.7. The transfer function, Eq. (5.50), implies that the plant $G(s)$ has two or more pure integrations, as required to provide zero steady-state error. ∎

5.10 THE SIMPLIFICATION OF LINEAR SYSTEMS

It is quite useful to study complex systems with high-order transfer functions by using lower-order approximate models. Thus, for example, a fourth-order system could be approximated by a second-order system leading to a use of the performance indices in Fig. 5.8. Several methods are available for reducing the order of a systems transfer function.

One relatively simple way to delete a certain insignificant pole of a transfer function is to note a pole that has a negative real part that is much larger than the other poles. Thus that pole is expected to affect the transient response insignificantly.

For example, if we have a system plant where

$$G(s) = \frac{K}{s(s + 2)(s + 30)},$$

we can safely neglect the impact of the pole at $s = -30$. However, we must retain the steady-state response of the system, and thus we reduce the system to

$$G(s) = \frac{(K/30)}{s(s + 2)}.$$

A second, more complicated system reduction scheme uses an algebraic method. We will let the high-order system be described by the transfer function

$$H(s) = K \frac{a_m s^m + a_{m-1}s^{m-1} + \cdots + a_1 s + 1}{b_n s^n + b_{n-1}s^{n-1} + \cdots + b_1 s + 1}, \tag{5.51}$$

in which the poles are in the left-hand s-plane and $m \leq n$. The lower-order approximate transfer function is

$$L(s) = K \frac{c_p s^p + \cdots + c_1 s + 1}{d_g s^g + \cdots + d_1 s + 1}, \tag{5.52}$$

where $p \leq g < n$. Notice that the gain constant K is the same for the original and approximate system so as to ensure the same steady-state response. The method outlined in Example 5.9 is based on selecting c_i and d_i in such a way that $L(s)$ has a frequency response (see Chapter 8) very close to that of $H(s)$. This is equivalent to stating that $H(j\omega)/L(j\omega)$ is required to deviate the least amount from unity for various frequencies. The c and d coefficients are obtained by utilizing the following equation:

$$M^{(k)}(s) = \frac{d^k}{ds^k} M(s) \tag{5.53}$$

and

$$\Delta^{(k)}(s) = \frac{d^k}{ds^k} \Delta(s), \tag{5.54}$$

where $M(s)$ and $\Delta(s)$ are the numerator and denominator polynomials of $H(s)/L(s)$, respectively. We also define

$$M_{2q} = \sum_{k=0}^{2q} \frac{(-1)^{k+q} M^{(k)}(0) M^{(2q-k)}(0)}{k!(2q - k)!}, \qquad q = 0, 1, 2 \ldots \tag{5.55}$$

and a completely identical equation of Δ_{2q}. The solutions for the c and d coefficients are obtained by equating

$$M_{2q} = \Delta_{2q} \tag{5.56}$$

for $q = 1, 2, \ldots$ up to the number required to solve for the unknown coefficients.

Let us consider an example to clarify the use of these equations.

EXAMPLE 5.9   A simplified model

Consider the third-order system

$$H(s) = \frac{6}{s^3 + 6s^2 + 11s + 6} = \frac{1}{1 + (^{11}\!/_6)s + s^2 + (^1\!/_6)s^3}. \tag{5.57}$$

Using the second-order model

$$L(s) = \frac{1}{1 + d_1 s + d_2 s^2}, \tag{5.58}$$

$M(s) = 1 + d_1 s + d_2 s^2$ and $\Delta(s) = 1 + (^{11}\!/_6)s + s^2 + (^1\!/_6)s^3$. Then

$$M^0(s) = 1 + d_1 s + d_2 s^2 \tag{5.59}$$

and $M^0(0) = 1$. Similarly,

$$M^1 = \frac{d}{ds}(1 + d_1 s + d_2 s^2) = d_1 + 2d_2 s. \tag{5.60}$$

Therefore $M^1(0) = d_1$. Continuing this process, we find that

$$M^0(0) = 1 \qquad \Delta^0(0) = 1,$$
$$M^1(0) = d_1 \qquad \Delta^1(0) = {}^{11}\!/_6, \tag{5.61}$$
$$M^2(0) = 2d_2 \qquad \Delta^2(0) = 2,$$
$$M^3(0) = 0 \qquad \Delta^3(0) = 1.$$

We now equate $M_{2q} = \Delta_{2q}$ for $q = 1$ and 2. We find that for $q = 1$,

$$M_2 = (-1)\frac{M^0(0)M^2(0)}{2} + \frac{M^1(0)M^1(0)}{1} + (-1)\frac{M^2(0)M^0(0)}{2} \tag{5.62}$$
$$= -d_2 + d_1^2 - d_2 = -2d_2 + d_1^2.$$

Then, because the equation for Δ_2 is identical, we have

$$\Delta_2 = \frac{\Delta^0(0)\Delta^2(0)}{2} + \frac{\Delta^1(0)\Delta^1(0)}{1} + (-1)\frac{\Delta^2(0)\Delta^0(0)}{2} \tag{5.63}$$
$$= -1 + \frac{121}{36} - 1 = \frac{49}{36}.$$

Therefore, because $M_2 = \Delta_2$, we have

$$-2d_2 + d_1^2 = {}^{49}\!/_{36}. \tag{5.64}$$

Completing the process for $M_4 = \Delta_4$, when $q = 2$, we obtain

$$d_2^2 = {}^7\!/_{18}. \tag{5.65}$$

Then the solution for $L(s)$ is $d_1 = 1.615$ and $d_2 = 0.625$. (The other sets of solutions are rejected because they lead to unstable poles.) It is interesting to see that the poles of $H(s)$ are $s = -1, -2, -3$, whereas the poles of $L(s)$ are at $s = -1.029$ and -1.555. The lower-order system transfer function is

$$L(s) = \frac{1}{1 + 1.615s + 0.625s^2} = \frac{1.60}{s^2 + 2.584s + 1.60}. \tag{5.66}$$

Because the lower-order model has two poles, we can estimate that we would obtain a slightly overdamped step response with a settling time of approximately four seconds. ∎

It is sometimes desirable to retain the dominant poles of the original system, $H(s)$, in the low-order model. This can be accomplished by specifying the denominator of $L(s)$ to be the dominant poles of $H(s)$ and allow the numerator of $L(s)$ to be subject to approximation. A complex system such as the robot hand shown in Fig. 5.33 is an example of a high-order system that can be favorably represented by a low-order model.

Another novel and useful method for reducing the order is the Routh approximation method, based on the idea of truncating the Routh table used to determine stability. The Routh approximants can be computed by a finite recursive algorithm that is suited for programming on a digital computer [22].

The Routh approximation algorithm starts from the polynomial coefficients of the original transfer function, as given in Eq. (5.51), and determines a lower-order model,

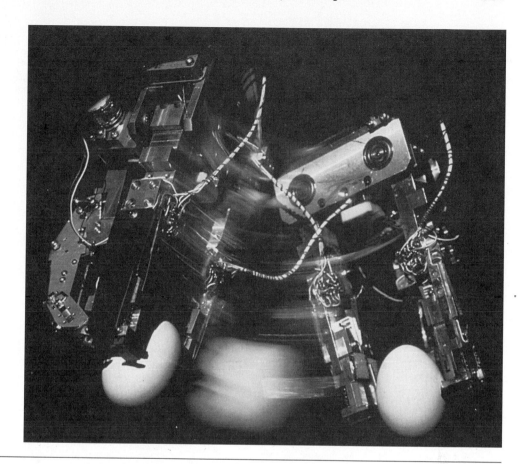

FIGURE 5.33
A high-performance robot hand with a delicate touch illustrates the challenge of modern high-performance control systems. (Photo courtesy of Hitachi America Ltd.)

Eq. (5.52). The Routh approximation method results in a fast and efficient algorithm for computer computation of the reduced-order transfer functions.

5.11 DESIGN EXAMPLE: HUBBLE TELESCOPE POINTING CONTROL

The Hubble space telescope, the most complex and expensive scientific instrument that has ever been built, is orbiting the earth. Launched to 380 miles above the earth on April 24, 1990, the telescope has pushed technology to new limits. The telescope's 2.4 meter (94.5-inch) mirror has the smoothest surface of any mirror made, and its pointing system can center it on a dime 400 miles away [9, 10, 21]. The mirror had a spherical aberration that was largely corrected during a space mission in December 1993 [24]. Furthermore, the Hubble telescope can point accurately. Consider the model of the telescope-pointing system shown in Fig. 5.34.

The goal of the design is to choose K_1 and K so that (1) the percent overshoot of the output to a step command, $r(t)$, is less than or equal to 10%, (2) the steady-state error to a ramp command is minimized, and (3) the effect of a step disturbance is reduced. Since the system has an inner loop, block diagram reduction can be used to obtain the simplified system of Fig. 5.34(b).

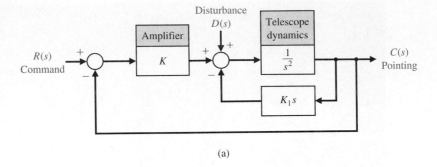

(a)

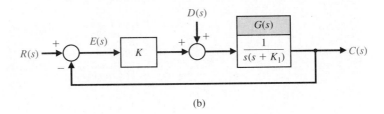

(b)

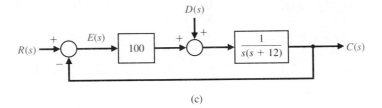

(c)

FIGURE 5.34
(a) The Hubble
telescope
pointing system,
(b) reduced
block diagram,
(c) system design,
and (d) system
response to a
unit step input
command and
a unit step
disturbance input.

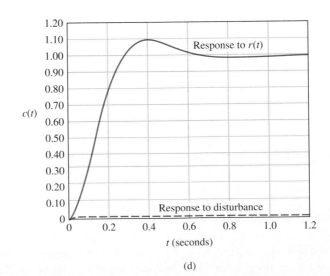

(d)

Mason's formula can be used to obtain the output due to the two inputs of the system of Fig. 5.34(b):

$$C(s) = T(s)R(s) + [T(s)/K]D(s), \tag{5.67}$$

where

$$T(s) = \frac{KG(s)}{1 + KG(s)} = \frac{KG(s)}{1 + L(s)}.$$

The error $E(s)$ is

$$E(s) = \frac{1}{1 + L(s)}R(s) - \frac{G(s)}{1 + L(s)}D(s). \tag{5.68}$$

First, let us select K and K_1 to meet the percent overshoot requirement for a step input, $R(s) = A/s$. Setting $D(s) = 0$, we have

$$C(s) = \frac{KG(s)}{1 + KG(s)}R(s)$$

$$= \frac{K}{s(s + K_1) + K}\left(\frac{A}{s}\right) = \frac{K}{s^2 + K_1 s + K}\left(\frac{A}{s}\right). \tag{5.69}$$

To set the overshoot less than 10%, we select $\zeta = 0.6$ by examining Fig. 5.8 or using Eq. (5.16) to determine that the overshoot will be 9.5% for $\zeta = 0.6$. We next examine the steady-state error for a ramp, $r(t) = Bt, t \geq 0$, using (Eq. 5.28):

$$e_{ss} = \lim_{s \to 0} \left\{\frac{B}{sKG(s)}\right\} = \frac{B}{(K/K_1)}. \tag{5.70}$$

The steady-state error due to a step disturbance is equal to zero. (Can you show this?) The transient response of the error due to the step disturbance input can be reduced by increasing K (see Eq. 5.68). Thus, in summary, we seek a large K and a large value of (K/K_1) to obtain a low steady-state error for the ramp input (see Eq. 5.70). However, we also require $\zeta = 0.6$ to limit the overshoot.

For our design, we need to select K. The characteristic equation of the system is (with $\zeta = 0.6$)

$$(s^2 + 2\zeta\omega_n s + \omega_n^2) = (s^2 + 2(0.6)\omega_n s + K). \tag{5.71}$$

Therefore $\omega_n = \sqrt{K}$ and the second term of the denominator of Eq. (5.69) requires $K_1 = 2(0.6)\omega_n$. Then $K_1 = 1.2\sqrt{K}$ or the ratio K/K_1 becomes

$$\frac{K}{K_1} = \frac{K}{1.2\sqrt{K}} = \frac{\sqrt{K}}{1.2}.$$

Selecting $K = 25$, we have $K_1 = 6$ and $K/K_1 = 4.17$. If we select $K = 100$, we have $K_1 = 12$ and $K/K_1 = 8.33$. Realistically, we must limit K so that the system's operation remains linear. Using $K = 100$, we obtain the system shown in Fig. 5.34(c). The responses of the system to a unit step input command and a unit step disturbance input are shown in Fig. 5.34(d). Note how the effect of the disturbance is relatively insignificant.

Finally, we note that the steady-state error for a ramp input is

$$e_{ss} = \frac{B}{8.33} = 0.12B.$$

This design, using $K = 100$, is an excellent system.

5.12 SYSTEM PERFORMANCE USING *MATLAB*

In this section, we will investigate time-domain performance specifications given in terms of transient response to a given input signal and the resulting steady-state tracking errors. We conclude with a discussion of the simplification of linear systems. The *Matlab* functions introduced in this section are impulse and lsim. These functions are used to simulate linear systems.

Time-Domain Specifications.   Time-domain performance specifications are generally given in terms of the transient response of a system to a given input signal. Because the actual input signals are generally unknown, a standard test input signal is used. Consider the second-order system shown in Fig. 5.35. The closed-loop output is

$$C(s) = \frac{\omega_n^2}{s^2 + 2\zeta\omega_n s + \omega_n^2} R(s). \tag{5.72}$$

We have already discussed the use of the step function to compute the step response of a system. Now we address another important test signal: the impulse. The impulse response is the time derivative of the step response. We compute the impulse response with the impulse function shown in Fig. 5.36.

FIGURE 5.35
Single-loop
second-order
feedback system.

FIGURE 5.36
The **impulse**
function.

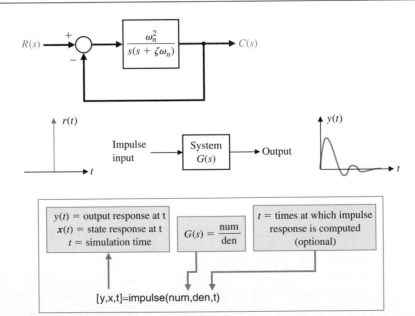

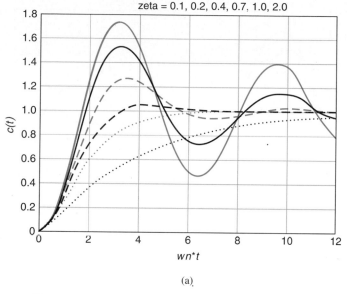

(a)

stepresponse.m

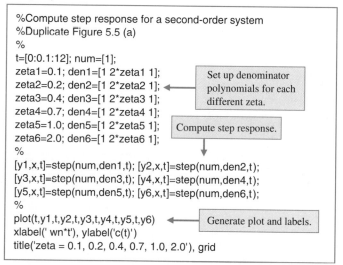

```
%Compute step response for a second-order system
%Duplicate Figure 5.5 (a)
%
t=[0:0.1:12]; num=[1];
zeta1=0.1; den1=[1 2*zeta1 1];
zeta2=0.2; den2=[1 2*zeta2 1];
zeta3=0.4; den3=[1 2*zeta3 1];
zeta4=0.7; den4=[1 2*zeta4 1];
zeta5=1.0; den5=[1 2*zeta5 1];
zeta6=2.0; den6=[1 2*zeta6 1];
%
[y1,x,t]=step(num,den1,t); [y2,x,t]=step(num,den2,t);
[y3,x,t]=step(num,dcn3,t); [y4,x,t]=step(num,den4,t);
[y5,x,t]=step(num,den5,t); [y6,x,t]=step(num,den6,t);
%
plot(t,y1,t,y2,t,y3,t,y4,t,y5,t,y6)
xlabel(' wn*t'), ylabel('c(t)')
title('zeta = 0.1, 0.2, 0.4, 0.7, 1.0, 2.0'), grid
```

Set up denominator polynomials for each different zeta.

Compute step response.

Generate plot and labels.

FIGURE 5.37
(a) Response of a second-order system to a step input. (b) *Matlab* script.

(b)

We can obtain a plot similar to that of Fig. 5.5(a) with the step function, as shown in Fig. 5.37. Using the impulse function, we can obtain a plot similar to that of Fig. 5.6. The response of a second-order system for an impulse function input is shown in Fig. 5.38. In the script, we set $\omega_n = 1$, which is equivalent to computing the step response versus $\omega_n t$. This gives us a more general plot valid for any $\omega_n > 0$.

In many cases, it may be necessary to simulate the system response to an arbitrary but known input. In these cases, use the lsim function. The lsim function is shown in Fig. 5.39. An example of the use of lsim is given in Example 5.10.

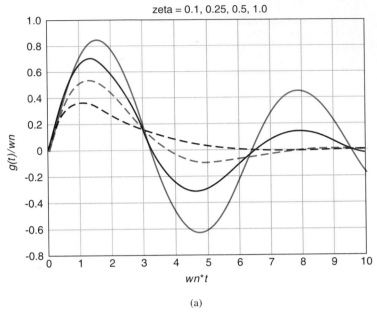

zeta = 0.1, 0.25, 0.5, 1.0

(a)

impulseresponse.m

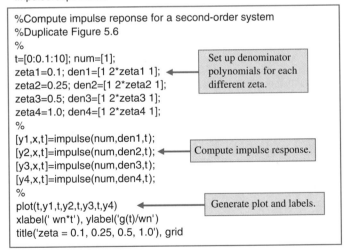

```
%Compute impulse reponse for a second-order system
%Duplicate Figure 5.6
%
t=[0:0.1:10]; num=[1];
zeta1=0.1; den1=[1 2*zeta1 1];        Set up denominator
zeta2=0.25; den2=[1 2*zeta2 1];       polynomials for each
zeta3=0.5; den3=[1 2*zeta3 1];        different zeta.
zeta4=1.0; den4=[1 2*zeta4 1];
%
[y1,x,t]=impulse(num,den1,t);
[y2,x,t]=impulse(num,den2,t);         Compute impulse response.
[y3,x,t]=impulse(num,den3,t);
[y4,x,t]=impulse(num,den4,t);
%
plot(t,y1,t,y2,t,y3,t,y4)             Generate plot and labels.
xlabel(' wn*t'), ylabel('g(t)/wn')
title('zeta = 0.1, 0.25, 0.5, 1.0'), grid
```

(b)

FIGURE 5.38
(a) Response of
a second-order
system to an
impulse function
input. (b) *Matlab*
script.

EXAMPLE 5.10   Mobile robot steering control

The block diagram for a steering control system for a mobile robot is shown in Fig. 5.23 on page 238 (see Example 5.3).

Suppose the steering controller, $G_1(s)$, is

$$G_1(s) = K_1 + \frac{K_2}{s}.$$

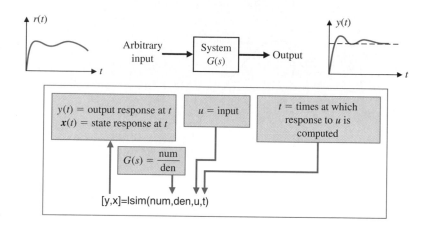

FIGURE 5.39
The **lsim** function.

When the input is a ramp, the steady-state error is

$$e_{ss} = \frac{A}{K_v},\tag{5.73}$$

where

$$K_v = K_2 K.$$

The effect of the controller constant, K_2, on the steady-state error is evident from Eq. (5.73). Whenever K_2 is large, the steady-state error is small.

We can simulate the closed-loop system response to a ramp input using the lsim function. The controller gains K_1, K_2 and the system gain K can be represented symbolically in the script so that various values can be selected and simulated. The results are shown in Fig. 5.40 for $K_1 = K = 1$, $K_2 = 2$, and $\tau = 1/10$. ∎

Simplification of Linear Systems.   It may be possible to develop a lower-order approximate model that closely matches the input–output response of a high-order model. A procedure for approximating transfer functions is given in Section 5.10. We can use *Matlab* to compare the approximate model to the actual model, as illustrated in the following example.

EXAMPLE 5.11   A simplified model

Consider the third-order system

$$H(s) = \frac{6}{s^3 + 6s^2 + 11s + 6}.$$

A second-order approximation (see Example 5.9) is

$$L(s) = \frac{1.60}{s^2 + 2.584s + 1.60}.$$

A comparison of their respective step responses is given in Fig. 5.41. ∎

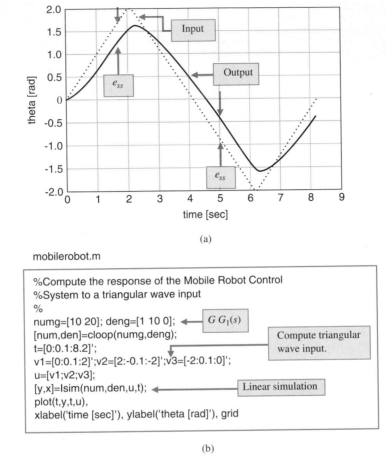

(a)

mobilerobot.m

```
%Compute the response of the Mobile Robot Control
%System to a triangular wave input
%
numg=[10 20]; deng=[1 10 0];        ← G G₁(s)
[num,den]=cloop(numg,deng);                    Compute triangular
t=[0:0.1:8.2]';                                wave input.
v1=[0:0.1:2]';v2=[2:-0.1:-2]';v3=[-2:0.1:0]';
u=[v1;v2;v3];
[y,x]=lsim(num,den,u,t);        ← Linear simulation
plot(t,y,t,u),
xlabel('time [sec]'), ylabel('theta [rad]'), grid
```

FIGURE 5.40
(a) Transient response of the mobile robot steering control system to a ramp input. (b) *Matlab* script.

(b)

5.13 SUMMARY

In this chapter we have considered the definition and measurement of the performance of a feedback control system. The concept of a performance measure or index was discussed, and the usefulness of standard test signals was outlined. Then several performance measures for a standard step input test signal were delineated. For example, the overshoot, peak time, and settling time of the response of the system under test for a step input signal were considered. The fact that often the specifications on the desired response are contradictory was noted, and the concept of a design compromise was proposed. The relationship between the location of the *s*-plane root of the system transfer function and the system response was discussed. A most important measure of system performance is the steady-state error for specific test input signals. Thus the relationship of the steady-state error of a system in terms of the system parameters was developed by utilizing the final-value theorem. The capability of a feedback control system is demonstrated in Fig. 5.42. Finally, the utility of an integral performance index was outlined, and several examples of design that mini-

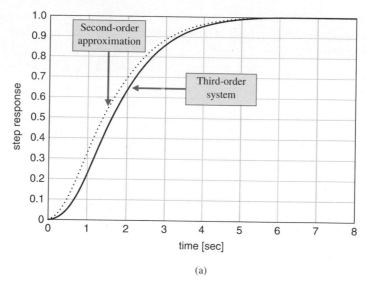

(a)

stepcompare.m

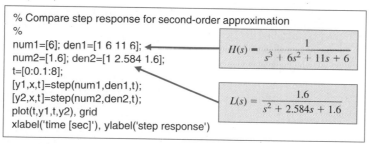

FIGURE 5.41
(a) Step response comparison for an approximate transfer function versus the actual transfer function.
(b) *Matlab* script.

(b)

FIGURE 5.42
The response of a feedback system to a ramp input with $K = 1, 2,$ and 8 when $G(s) = K/[s(s + 1)(s + 3)]$. Note the steady-state error is reduced as K is increased, but the response becomes oscillatory at $K = 8$.

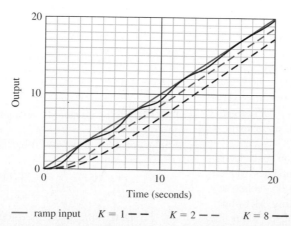

mized a system's performance index were completed. Thus we have been concerned with the definition and usefulness of quantitative measures of the performance of feedback control systems.

EXERCISES

E5.1 A motor control system for a computer disk drive must reduce the effect of disturbances and parameter variations, as well as reduce the steady-state error. We desire to have no steady-state error for the head-positioning control system, which is of the form shown in Fig. 5.18 where $H(s) = 1$. (a) What type number is required? (How many integrations?) (b) If the input is a ramp signal, then, to achieve a zero steady-state error, what type number is required?

E5.2 The engine, body, and tires of a racing vehicle affect the acceleration and speed attainable [11]. The speed control of the car is represented by the model shown in Fig. E5.2. (a) Calculate the steady-state error of the car to a step command in speed. (b) Calculate overshoot of the speed to a step command.

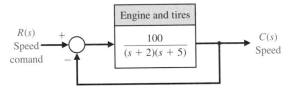

FIGURE E5.2   Racing car speed control.

E5.3 For years, Amtrak has struggled to attract passengers on its routes in the Midwest, using technology developed decades ago. During the same time, foreign railroads were developing new passenger rail systems that could profitably compete with air travel. Two of these systems, the French TGV and the Japanese Shinkansen, reach speeds of 160 mph [20]. The Transrapid-06, a U.S. experimental magnetic levitation train, is shown in Fig. E5.3(a).

The use of magnetic levitation and electromagnetic propulsion to provide contactless vehicle movement makes the Transrapid-06 technology radically different from the existing Metroliner. The underside of the TR-06 carriage (where the wheel trucks would be on a conventional car)

wraps around a guideway. Magnets on the bottom of the guideway attract electromagnets on the "wraparound," pulling it up toward the guideway. This suspends the vehicles about one centimeter above the guideway. (See Problem 2.27.)

The levitation control is represented by Fig. E5.3(b). (a) Using Table 5.6 for a step input, select K so that the system provides an optimum ITAE response. (b) Using Fig. 5.8, determine the expected overshoot to a step input of $I(s)$.

Answers: $K = 100; 4.6\%$

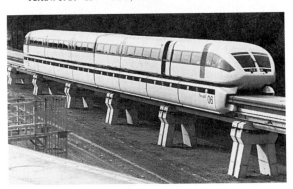

(a)

FIGURE E5.3   Levitated train control.

E5.4 A feedback system with negative unity feedback has a plant

$$G(s) = \frac{2(s + 8)}{s(s + 4)}.$$

(a) Determine the closed-loop transfer function $T(s) = C(s)/R(s)$. (b) Find the time response $c(t)$

for a step input $r(t) = A$ for $t > 0$. (c) Using Fig. 5.13(a), determine the overshoot of the response. (d) Using the final-value theorem, determine the steady-state value of $c(t)$.

E5.5 A low-inertia plotter is shown in Fig. E5.5(a). This system may be represented by the block dia-

(a)

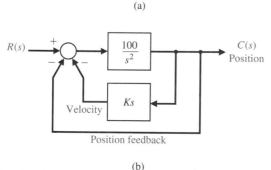

(b)

FIGURE E5.5 (a) The Hewlett-Packard *x-y* plotter. (Courtesy of Hewlett-Packard Co.) (b) Block diagram of plotter.

FIGURE E5.6
Blood-sugar level
control.

gram shown in Fig. E5.5(b) [18]. (a) Calculate the steady-state error for a ramp input. (b) Select a value of K that will result in zero overshoot to a step input but as rapid response as is attainable.

Plot the poles and zeros of this system and discuss the dominance of the complex poles. What overshoot for a step input do you expect?

E5.6 Effective control of insulin injections can result in better lives for diabetic persons. Automatically controlled insulin injection by means of a pump and a sensor that measures blood sugar can be very effective. A pump and injection system has a feedback control as shown in Fig. E5.6. Calculate the suitable gain K so that the overshoot of the step response due to the drug injection is approximately 7%. $R(s)$ is the desired blood-sugar level and $C(s)$ is the actual blood-sugar level. (*Hint:* Use Fig. 5.13a.)

Answer: $K = 1.67$

E5.7 A control system for positioning the head of a floppy disk drive has the closed-loop transfer function

$$T(s) = \frac{0.313(s + 0.8)}{(s + 0.25)(s^2 + 0.3s + 1)}.$$

Plot the poles and zeros of this system and discuss the dominance of the complex poles. What overshoot for a step input do you expect?

E5.8 A prototype for a microwave-powered aircraft was recently tested. The energy to drive the small airplane's electric engine is powered by microwave power beamed up from an earth station. A commercial version will have a 36-m wingspan and a 24-m long fuselage and will be used for unmanned reconnaissance and radio relay. If the aircraft flies at 50,000 ft, determine the accuracy of the beam transmission required if it is cruising at 24 mph. Assume the movement is a ramp input, and determine the maximum steady-state error allowable.

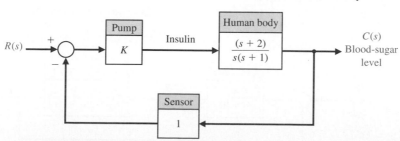

·E5.9   A second-order control system has the closed-loop transfer function $T(s) = C(s)/R(s)$. The system specifications for a step input follow:

(1) Percent overshoot $\leq 5\%$.
(2) Settling time < 4 seconds.
(3) Peak time $T_p < 1$ second.

Show the permissible area for the poles of $T(s)$ in order to achieve the desired response.

E5.10   A system with unity feedback is shown in Fig. E5.10. Determine the percent steady-state error for a step and a ramp input when

$$G(s) = \frac{10(s + 4)}{s(s + 1)(s + 2)(s + 5)}.$$

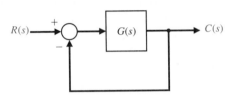

FIGURE E5.10   Unity-feedback system.

E5.11   We are all familiar with the Ferris wheel featured at state fairs and carnivals. George Ferris, born in Galesburg, Illinois, in 1859, later moved to Nevada and then graduated from Rensselaer Polytechnic Institute in 1881. By 1891, Ferris had considerable experience with iron, steel, and bridge construction. He conceived and constructed his famous wheel for the 1893 Columbian Exposition in Chicago [9]. To avoid upsetting passengers, let us set a requirement that the steady-state speed be controlled to within 5% of the desired speed for the system shown in Fig. E5.11.

(a) Determine the required gain K to achieve the steady-state requirement.
(b) For the gain of part (a), determine and plot the error $e(t)$ for a disturbance $D(s) = 1/s$. Does the speed change more than 5%? (Set $R(s) = 0$ for ease of computation.)

E5.12   For the system with unity feedback shown in Fig. E5.10, determine the percent steady-state error for a step and a ramp input when

$$G(s) = \frac{10}{s^2 + 14s + 50}.$$

E5.13   A feedback system is shown in Fig. E5.13.

(a) Determine the steady-state error for a unit step when $K = 0.4$ and $G_p(s) = 1$.
(b) Select an appropriate value for $G_p(s)$ so that the steady-state error is equal to zero for the unit step input.

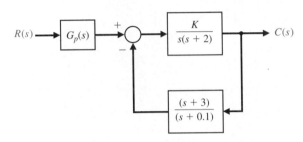

FIGURE E5.13   Feedback system.

E5.14   A closed-loop control system has a transfer function $T(s)$ as

$$\frac{C(s)}{R(s)} = T(s) = \frac{500}{(s + 10)(s^2 + 10s + 50)}.$$

Plot $c(t)$ for a step input $R(s)$ when (a) the actual $T(s)$ is used, and (b) using the relatively dominant complex poles. Compare the results.

E5.15   A second-order system is

$$\frac{C(s)}{R(s)} = T(s) = \frac{(10/z)(s + z)}{(s + 1)(s + 8)}$$

Consider the case where $1 < z < 8$. Obtain the partial fraction expansion, and plot $c(t)$ for a step input $r(t)$ for z equal to 2, 4, and 6.

E5.16   A closed-loop control system transfer function $T(s)$ has two dominant complex conjugate poles.

FIGURE E5.11
Speed control of a
Ferris wheel.

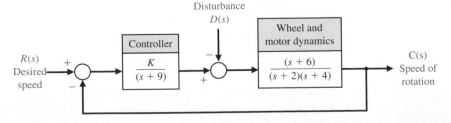

Sketch the region in the left-hand s-plane where the complex poles should be located to meet the given specifications.

(a) $0.6 \leq \zeta \leq 0.8, \quad \omega_n \leq 10$
(b) $0.5 \leq \zeta \leq 0.707, \quad \omega_n \geq 10$
(c) $\zeta \geq 0.5, \quad 5 \leq \omega_n \leq 10$
(d) $\zeta \leq 0.707, \quad 5 \leq \omega_n \leq 10$
(e) $\zeta \geq 0.6, \quad \omega_n \leq 6$

E5.17 A system is shown in Fig. E5.17(a). The response to a unit step, when $K = 1$, is shown in Fig. E5.17(b). Determine the value of K so that the steady-state error is equal to zero.

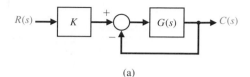

(a)

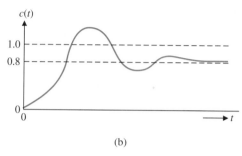

(b)

FIGURE E5.17

PROBLEMS

P5.1 An important problem for television systems is the jumping or wobbling of the picture due to the movement of the camera. This effect occurs when the camera is mounted in a moving truck or airplane. The Dynalens system has been designed to reduce the effect of rapid scanning motion; see Fig. P5.1. A maximum scanning motion of 25°/s is

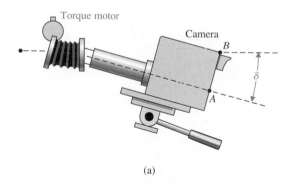

(a)

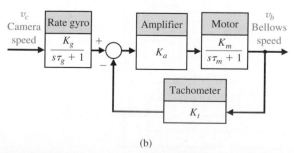

(b)

FIGURE P5.1 Camera wobble control.

expected. Let $K_g = K_t = 1$ and assume that τ_g is negligible. (a) Determine the error of the system $E(s)$. (b) Determine the necessary loop gain, $K_a K_m K_t$, when a 1°/s steady-state error is allowable. (c) The motor time constant is 0.40 s. Determine the necessary loop gain so that the settling time of v_b is less than or equal to 0.03 s.

P5.2 A specific closed-loop control system is to be designed for an underdamped response to a step input. The specifications for the system are as follows:

$$20\% > \text{percent overshoot} > 10\%,$$

$$\text{Settling time} < 0.6 \text{ s.}$$

(a) Identify the desired area for the dominant roots of the system. (b) Determine the smallest value of a third root, r_3, if the complex conjugate roots are to represent the dominant response. (c) The closed-loop system transfer function $T(s)$ is third order, and the feedback has a unity gain. Determine the forward transfer function $G(s) = C(s)/E(s)$ when the settling time is 0.6 s and the percent overshoot is 20%.

P5.3 A laser beam can be used to weld, drill, etch, cut, and mark metals, as shown in Fig. P5.3(a), on the next page [16]. Assume we have a work requirement for an accurate laser to mark a parabolic path with a closed-loop control system as shown in Fig. P5.3(b). Calculate the necessary gain to result in a steady-state error of 1 mm for $r(t) = t^2$, cm/s².

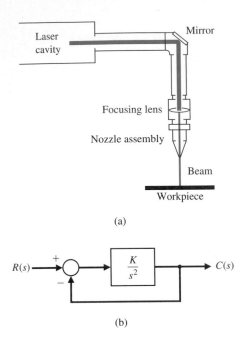

(a)

(b)

FIGURE P5.3   Laser beam control.

P5.4   The open-loop transfer function of a unity negative feedback system is

$$G(s) = \frac{K}{s(s + 2)}.$$

A system response to a step input is specified as follows:

$$\text{peak time } T_p = 1.1 \text{ s,}$$

$$\text{percent overshoot} = 5\%.$$

(a) Determine whether both specifications can be met simultaneously. (b) If the specifications cannot be met simultaneously, determine a compromise value for K so that the peak time and percent overshoot specifications are relaxed the same percentage.

P5.5   A space telescope is to be launched to carry out astronomical experiments [9]. The pointing control system is desired to achieve 0.01 minute of arc and track solar objects with apparent motion up to 0.21 minute per second. The system is illustrated

FIGURE P5.5
(a) The space telescope. (b) The space telescope pointing control system.

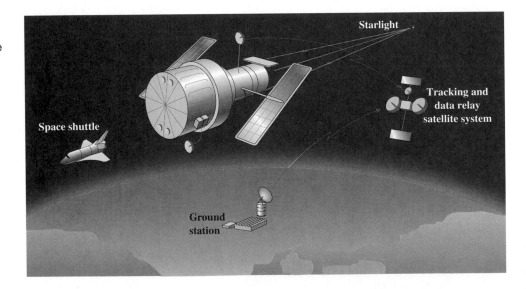

(a)

(b)

in Fig. P5.5(a). The control system is shown in Fig. P5.5(b). Assume that $\tau_1 = 1$ second and $\tau_2 = 0$ (an approximation). (a) Determine the gain $K = K_1 K_2$ required so that the response to a step command is as rapid as reasonable with an overshoot of less than 5%. (b) Determine the steady-state error of the system for a step and a ramp input. (c) Determine the value of $K_1 K_2$ for an ITAE optimal system for (1) a step input and (2) a ramp input.

P5.6 A robot is programmed to have a tool or welding torch follow a prescribed path [8, 13]. Consider a robot tool that is to follow a sawtooth path as shown in Fig. P5.6(a). The transfer function of the plant is

$$G(s) = \frac{800(s + 2)}{s(s + 10)(s + 14)}$$

for the closed-loop system shown in Fig. 5.6(b). Calculate the steady-state error.

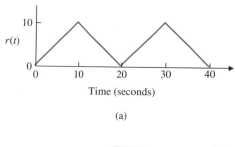

(a)

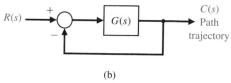

(b)

FIGURE P5.6   Robot path control.

P5.7 Astronaut Bruce McCandless II took the first untethered walk in space on February 7, 1984, using the gas-jet propulsion device illustrated in

FIGURE P5.7
(a) Astronaut Bruce McCandless II is shown a few meters away from the earth-orbiting space shuttle *Challenger*. He used a nitrogen-propelled hand-controlled device called the manned maneuvering unit. (Courtesy of National Aeronautics and Space Administration.)
(b) Block diagram of controller.

(a)

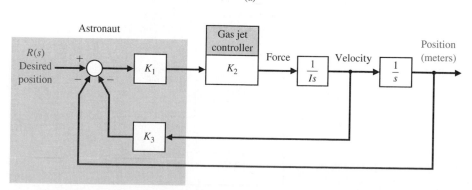

(b)

Fig. P5.7(a). The controller can be represented by a gain K_2 as shown in Fig. P5.7(b). The inertia of the man and equipment with his arms at his sides is 25 Kg-m². (a) Determine the necessary gain K_3 to maintain a steady-state error equal to 1 cm when the input is a ramp $r(t) = t$ (meters). (b) With this gain K_3, determine the necessary gain $K_1 K_2$ in order to restrict the percent overshoot to 10%. (c) Determine analytically the gain $K_1 K_2$ in order to minimize the ISE performance index for a step input.

P5.8 Photovoltaic arrays (solar cells) generate a dc voltage that can be used to drive dc motors or that can be converted to ac power and added to the distribution network. It is desirable to maintain the power out of the array at its maximum available as the solar incidence changes during the day. One such closed-loop system is shown in Fig. P5.8. The transfer function for the process is

$$G(s) = \frac{K}{s + 6},$$

where $K = 12$. Find (a) the time constant of the closed-loop system and (b) the settling time of the system when disturbances such as clouds occur.

P5.9 The antenna that receives and transmits signals to the *Telstar* communication satellite is the largest horn antenna ever built. The microwave antenna is 177 ft long, weighs 340 tons, and rolls on a circular track. A photo of the antenna is shown in Fig. P5.9. The *Telstar* satellite is 34 inches in diameter and moves about 16,000 mph at an altitude of 2500 miles. The antenna must be positioned accurately to 1/10 of a degree, because the microwave beam is 0.2° wide and highly attenuated by the large distance. If the antenna is following the moving satellite, determine the K_v necessary for the system.

FIGURE P5.9   A model of the antenna for the Telstar System at Andover, Maine. (Photo courtesy of Bell Telephone Laboratories, Inc.)

P5.10 A speed control system of an armature-controlled dc motor uses the back emf voltage of the motor as a feedback signal. (a) Draw the block diagram of this system (see Eq. 2.66). (b) Calculate the steady-state error of this system to a step input command setting the speed to a new level. Assume that $R_a = L_a = J = f = 1$, the motor constant is $K_m = 1$, and $K_b = 1$. (c) Select a feedback gain for the back emf signal to yield a step response with an overshoot of 15%.

P5.11 A simple unity feedback control system has a process transfer function

$$\frac{C(s)}{E(s)} = G(s) = \frac{K}{s}.$$

The system input is a step function with an amplitude A. The initial condition of the system at time t_0 is $c(t_0) = Q$, where $c(t)$ is the output of the system. The performance index is defined as

FIGURE P5.8
Solar cell control.

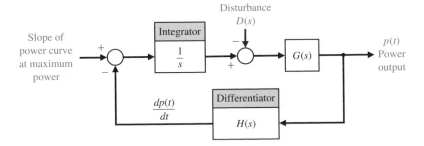

$$I = \int_0^\infty e^2(t)dt.$$

(a) Show that $I = (A - Q)^2/2K$. (b) Determine the gain K that will minimize the performance index I. Is this gain a practical value? (c) Select a practical value of gain and determine the resulting value of the performance index.

P5.12 Train travel between cities will increase as trains are developed that travel at high speeds, making the train travel time from city center to city center equivalent to airline travel time. The Japanese National Railway has a train called the Bullet Express that travels between Tokyo and Osaka on the Tokaido line. This train travels the 320 miles in 3 hours and 10 minutes, an average speed of 101 mph [20]. This speed will be increased as new systems are used, such as magnetically levitated systems to float vehicles above an aluminum guideway. To maintain a desired speed, a speed control system is proposed that yields a zero steady-state error to a ramp input. A third-order system is sufficient. Determine the optimum system transfer function, $T(s)$, for an ITAE performance criterion. Estimate the settling time and overshoot for a step input when $\omega_n = 5$.

P5.13 It is desired to approximate a fourth-order system by a lower-order model. The transfer function of the original system is

$$H(s) = \frac{s^3 + 7s^2 + 24s + 24}{s^4 + 10s^3 + 35s^2 + 50s + 24}$$

$$= \frac{s^3 + 7s^2 + 24s + 24}{(s + 1)(s + 2)(s + 3)(s + 4)}.$$

Show that if we obtain a second-order model by the method of Section 5.10, and we do not specify the poles and the zero of $L(s)$, we have

$$L(s) = \frac{0.2917s + 1}{0.399s^2 + 1.375s + 1}$$

$$= \frac{0.731(s + 3.428)}{(s + 1.043)(s + 2.4)}.$$

P5.14 For the original system of Problem 5.13, it is desired to find the lower-order model when the poles of the second-order model are specified as -1 and -2 and the model has one unspecified zero. Show that this low-order model is

$$L(s) = \frac{0.986s + 2}{s^2 + 3s + 2} = \frac{0.986(s + 2.028)}{(s + 1)(s + 2)}.$$

P5.15 A magnetic amplifier with a low-output impedance is shown in Fig. P5.15 in cascade with a low-pass filter and a preamplifier. The amplifier has a high-input impedance and a gain of one and is used for adding the signals as shown. Select a value for the capacitance C so that the transfer function $V_0(s)/V_{in}(s)$ has a damping ratio of $1/\sqrt{2}$. The time constant of the magnetic amplifier is equal to one second and the gain is $K = 10$. Calculate the settling time of the resulting system.

P5.16 Electronic pacemakers for human hearts regulate the speed of the heart pump. A proposed closed-loop system that includes a pacemaker and the measurement of the heart rate is shown in Fig. P5.16 [2, 3]. The transfer function of the heart pump and the pacemaker is found to be

$$G(s) = \frac{K}{s(s/12 + 1)}.$$

Design the amplifier gain to yield a system with a settling time to a step disturbance of less than one second. The overshoot to a step in desired heart rate should be less than 10%. (a) Find a suitable range of K. (b) If the nominal value of K is $K = 10$, find the sensitivity of the system to small changes in K. (c) Evaluate the sensitivity of part (b)

FIGURE P5.15
Feedback
amplifier.

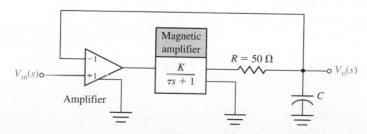

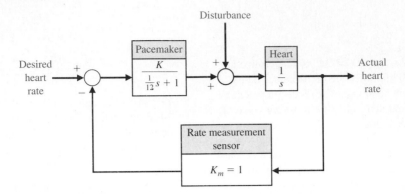

FIGURE P5.16
Heart pacemaker.

at DC (set $s = 0$). (d) Evaluate the magnitude of the sensitivity at the normal heart rate of 60 beats/minute.

P5.17 Consider the original third-order system given in Example 5.9. Determine a first-order model with one pole unspecified and no zeros that will represent the third-order system.

P5.18 A closed-loop control system with negative unity feedback has a plant with a transfer function

$$G(s) = \frac{8}{s(s^2 + 6s + 12)}.$$

(a) Determine the closed-loop transfer function $T(s)$. (b) Determine a second-order approximation for $T(s)$ using the method of Section 5.10.

(c) Using *Matlab* or a suitable computer program, plot the response of $T(s)$ and the second-order approximation to a unit step input and compare the results.

P5.19 A system is shown in Fig. P5.19.

(a) Determine the steady-state error for a unit step input in terms of K and K_1.

(b) Select K_1 so that the steady-state error is zero.

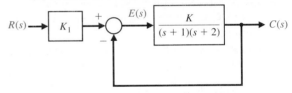

FIGURE P5.19   System with pregain, K_1.

ADVANCED PROBLEMS

AP5.1 A closed-loop transfer function is

$$T(s) = \frac{C(s)}{R(s)} = \frac{96(s + 3)}{(s + 8)(s^2 + 8s + 36)}.$$

(a) Determine the steady-state error for a unit step input $R(s) = 1/s$.

(b) Assume that the complex poles dominate, and determine the overshoot and settling time.

(c) Plot the actual system response, and compare it with the estimates of part (b).

AP5.2 A closed-loop system is shown in Fig. AP5.2. Plot the response to a unit step input for the system for $\tau_z = 0, 0.05, 0.1,$ and 0.5. Record the percent overshoot, rise time, and settling time as τ_z varies. Describe the effect of varying τ_z. Compare the lo-

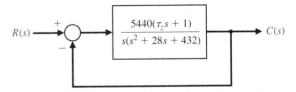

FIGURE AP5.2   System with a variable zero.

cation of the zero, $1/\tau_z$, with the location of the closed-loop roots.

AP5.3 A closed-loop system is shown in Fig. AP5.3. Plot the response to a unit step input for the system with $\tau_p = 0, 0.5, 2,$ and 5. Record the percent overshoot, rise time, and settling time as τ_p varies. Describe the effect of varying τ_p. Compare the location of the open-loop pole $1/\tau_p$ with the location of the closed-loop roots.

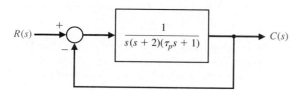

FIGURE AP5.3 System with a variable pole in the plant.

AP5.4 The speed control of a high-speed train is represented by the system shown in Fig. AP5.4 [20]. Determine the equation for steady-state error for K for a unit step input $r(t)$. Consider the three values for K equal to 1, 10, and 100.

(a) Determine the steady-state error.
(b) Determine and plot the response $c(t)$ for (i) a unit step input $r(t)$ and (ii) a unit step input $d(t)$.
(c) Create a table showing overshoot, settling time, e_{ss} for $r(t)$, and $|c/d|_{max}$ for the three values of K. Select the best compromise value.

AP5.5 A system with a controller is shown in Fig. AP5.5. The zero of the controller may be varied. Let $\alpha = 0, 10, 100$.

(a) Determine the steady-state error for a step input $r(t)$ for $\alpha = 0$ and $\alpha \neq 0$.
(b) Plot the response of the system to a step input disturbance for the three values of α. Compare the results and select the best value of the three values of α.

AP5.6 The block diagram model of an armature-current-controlled dc motor is shown in Fig. AP5.6.

(a) Determine the steady-state error to a ramp input, $r(t) = t, t \geq 0$, in terms of K, K_b, and K_m.
(b) Let $K_m = 10$ and $K_b = 0.05$, and select K so that steady-state error is equal to 1.
(c) Plot the response to a unit step input and a unit ramp input for 20 seconds. Are the responses acceptable?

FIGURE AP5.4
Speed control.

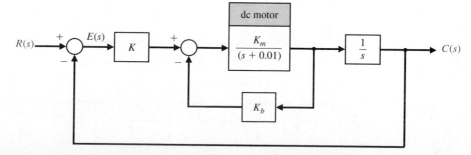

FIGURE AP5.5
System with control parameter α.

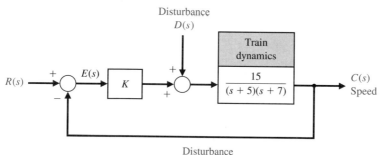

FIGURE AP5.6
dc motor control.

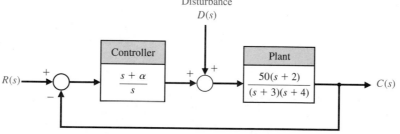

DESIGN PROBLEMS

DP5.1 The roll control autopilot of a jet fighter is shown in Fig. DP5.1. The goal is to select a suitable K so that the response to a unit step command $\phi_d(t) = A$, $t \geq 0$, will provide a response $\phi(t)$ with a fast response and an overshoot of less than 20%. (a) Determine the closed-loop transfer function $\phi(s)/\phi_d(s)$. (b) Determine the roots of the characteristic equation for $K = 0.7$, 3, and 6. (c) Using the concept of dominant roots, find the expected overshoot and peak time for the approximate second-order system. (d) Plot the actual response and compare with the approximate results of part (c). (e) Select the gain K so that the percentage overshoot is equal to 16%. What is the resulting peak time?

DP5.2 The design of the control for a welding arm with a long reach requires the careful selection of the parameters [13]. The system is shown in Fig. DP5.2, where $\zeta = 0.2$ and the gain K and the natural frequency ω_n can be selected. (a) Determine K and ω_n so that the response a unit step input achieves a peak time for the first overshoot (above the desired level of one) is less than or equal

to 1 second and the overshoot is less than 5%. (*Hint:* Try $0.1 < K/\omega_n < 0.3$.) (b) Plot the response of the system designed in part (a) to a step input.

DP5.3 Active suspension systems for modern automobiles provide a quality, firm ride. The design of an active suspension system adjusts the valves of the shock absorber so that the ride fits the conditions. A small electric motor, as shown in Fig. DP5.3, changes the valve settings [15]. Select a design value for K and the parameter q in order to satisfy the ITAE performance for a step command, $R(s)$, and a settling time for the step response of less than or equal to 0.5 second. Upon completion of your design, predict the resulting overshoot for a step input.

DP5.4 The space satellite shown in Fig. DP5.4(a) uses a control system to readjust its orientation, as shown in Fig. DP5.4(b).

(a) Determine a second-order model for the closed-loop system.

(b) Using the second-order model, select a gain K

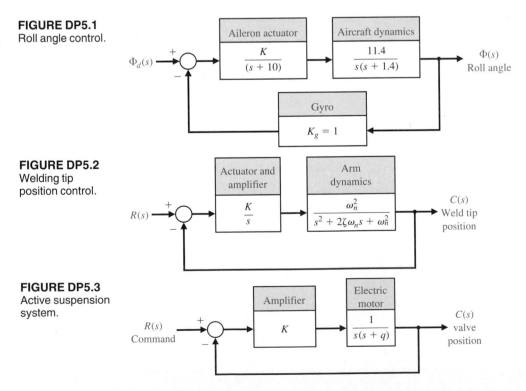

FIGURE DP5.1
Roll angle control.

Aileron actuator: $\dfrac{K}{(s + 10)}$

Aircraft dynamics: $\dfrac{11.4}{s(s + 1.4)}$

$\Phi_d(s)$ — $\Phi(s)$ Roll angle

Gyro: $K_g = 1$

FIGURE DP5.2
Welding tip
position control.

Actuator and amplifier: $\dfrac{K}{s}$

Arm dynamics: $\dfrac{\omega_n^2}{s^2 + 2\zeta\omega_n s + \omega_n^2}$

$R(s)$ — $C(s)$ Weld tip position

FIGURE DP5.3
Active suspension
system.

$R(s)$ Command

Amplifier: K

Electric motor: $\dfrac{1}{s(s + q)}$

$C(s)$ valve position

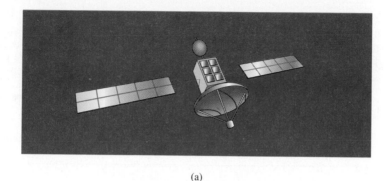

(a)

$R(s)$

$G_c(s)$
$\dfrac{K}{(s + 90)}$

$G(s)$
$\dfrac{10}{(s + 1)(s + 9)}$

$C(s)$
Angle of orientation

FIGURE DP5.4
Control of a space
satellite.

(b)

so that the percent overshoot is less than 15% and the steady-state error to a step is less than 12%.
(c) Verify your design by determining the actual performance of the third-order system.

DP5.5 A deburring robot can be used to smooth off machined parts by following a preplanned path (input command signal). In practice, errors occur due to robot inaccuracy, machining errors, large tolerances, and tool wear. These errors can be eliminated using force feedback to modify the path on-line [9, 13].

While force control has been able to address the problem of accuracy, it has been more difficult to solve the contact stability problem. In fact, by closing the force loop and introducing a compliant wrist force sensor (the most common type of force control), one can add to the stability problem.

A model of a robot deburring system is shown in Fig. DP5.5. Determine the region of stability for the system for K_1 and K_2. Assume both adjustable gains are greater than zero.

FIGURE DP5.5
Deburring robot.

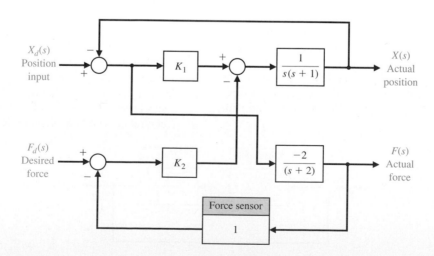

$X_d(s)$
Position
input

K_1

$\dfrac{1}{s(s + 1)}$

$X(s)$
Actual
position

$F_d(s)$
Desired
force

K_2

$\dfrac{-2}{(s + 2)}$

$F(s)$
Actual
force

Force sensor
1

DP5.6   The model for a position control system using a dc motor is shown in Fig. DP5.6. The goal is to select K_1 and K_2 so that the peak time is 0.8 second and the overshoot (P.O.) for a step input is negligible ($0.5\% < $ P.O. $< 2\%$).

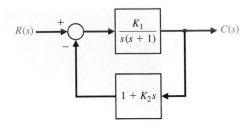

FIGURE DP5.6   Position control robot.

MATLAB PROBLEMS

MP5.1   A working knowledge of the relationship between the pole locations of the second-order system shown in Fig. MP5.1 and the transient response is important in control design. With that in mind, consider the four cases

(1) $\omega_n = 2$, $\zeta = 0$
(2) $\omega_n = 2$, $\zeta = 0.1$
(3) $\omega_n = 1$, $\zeta = 0$
(4) $\omega_n = 1$, $\zeta = 0.2$.

Using the impulse and subplot functions, create a plot containing four subplots, with each subplot depicting the impulse response of one of the four cases listed above. Compare the plot with Fig. 5.17 in Section 5.3, and discuss the results.

$$R(s) \longrightarrow \boxed{\dfrac{\omega_n^2}{s^2 + 2\zeta\omega_n s + \omega_n^2}} \longrightarrow C(s)$$

FIGURE MP5.1   A simple second-order system.

MP5.2   Consider the control system shown in Fig. MP5.2.

(a) Show analytically that the expected percent overshoot of the closed-loop system response to a unit step input is about 50%.
(b) Using *Matlab,* plot the unit step response of the closed-loop system and estimate the percent overshoot from the plot. Compare the result with part (a).

MP5.3   The open-loop transfer function of a unity negative feedback system is

$$G(s) = \frac{50}{s(s + 5)}.$$

Plot the unit step response with *Matlab,* and determine the approximate values of peak overshoot, M_p, time to peak, T_p, and settling time, T_s (2% criterion) from the plot. Label the plot (by hand) with the approximate values of M_p, T_p, and T_s.

MP5.4   An autopilot designed to hold an aircraft in straight and level flight is shown in Fig. MP5.4.

(a) Suppose the controller is a constant gain controller given by $G_c(s) = 2$. Using the lsim function, compute and plot the ramp response for $\theta_d(t) = at$, where $a = 0.5°/s$. Determine the attitude error after 10 seconds.
(b) If we increase the complexity of the controller, we can reduce the steady-state tracking error. With this objective in mind, suppose we replace the constant gain controller with the more sophisticated controller

$$G_c(s) = K_1 + \frac{K_2}{s} = 2 + \frac{1}{s}.$$

This type of controller is known as a proportional, integral (PI) controller. Repeat the simulation of part (a) with the PI controller, and compare the steady-state tracking errors of the constant gain controller versus the PI controller.

FIGURE MP5.2
A negative
feedback control
system.

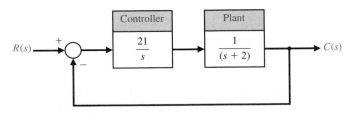

FIGURE MP5.4
An aircraft
autopilot block
diagram.

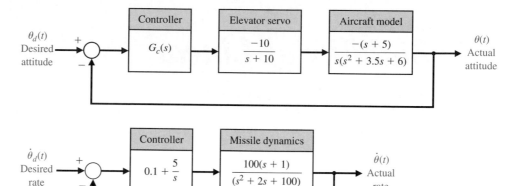

FIGURE MP5.5
A missile rate loop
autopilot.

MP5.5 The block diagram of a rate loop for a missile autopilot is shown in Fig. MP5.5. Using the analytic formulas for second-order systems, predict M_p, T_p, and T_s for the closed-loop system due to a unit step input. Compare the predicted results with the actual unit step response obtained with *Matlab*. Explain any differences.

TERMS AND CONCEPTS

Dominant roots The roots of the characteristic equation that cause the dominant transient response of the system.

Optimum control system A system whose parameters are adjusted so that the performance index reaches an extremum value.

Overshoot The amount the system output response proceeds beyond the desired response.

Peak time The time for a system to respond to a step input and rise to a peak response.

Performance index A quantitative measure of the performance of a system.

Rise time The time for a system to respond to a step input and attain a response equal to a percentage of the magnitude of the input. The 0–100% rise time, T_r, measures the time to 100% of the magnitude of the input. Alternatively, T_{r_1} measures the time from 10% to 90% of the response to the step input.

Settling time The time required for the system output to settle within a certain percentage of the input amplitude.

Specifications A set of prescribed performance criteria.

Test input signal An input signal used as a standard test of a system's ability to respond adequately.

Type number The number N of poles of the transfer function, $G(s)$, at the origin. $G(s)$ is the forward path transfer function.

Velocity error constant, K_v The constant evaluated as $\lim_{s \to 0}[sG(s)]$ for a type-one system. The steady-state error for a ramp input for a system is equal to A/K_v.

The Stability of Linear Feedback Systems

P R E V I E W

The idea of a stable system is familiar to us. Knowing that an unstable device will exhibit an erratic and destructive response, we seek to ensure that a system is stable and exhibits a bounded transient response.

The stability of a feedback system is related to the location of the roots of the characteristic equation of the system transfer function. Thus we wish to develop a few methods for determining whether a system is stable, and if so, how stable it is.

In this chapter we will consider the characteristic equation and examine the determination of the location of its roots. We also will consider a method of the determination of a system's stability that does not require the determination of the roots, but uses only the polynomial coefficients of the characteristic equation.

6.1 THE CONCEPT OF STABILITY

The transient response of a feedback control system is of primary interest and must be investigated. A very important factor regarding the transient performance of a system is the *stability* of the system. A *stable system* is defined as a system with a bounded (limited) system response. That is, if the system is subjected to a bounded input or disturbance and the response is bounded in magnitude, the system is said to be stable.

**A stable system is a dynamic system with a bounded response
to a bounded input.**

The concept of stability can be illustrated by considering a right circular cone placed on a plane horizontal surface. If the cone is resting on its base and is tipped slightly, it returns to its original equilibrium position. This position and response is said to be stable. If the cone rests on its side and is displaced slightly, it rolls with no tendency to leave the position on its side. This position is designated as the neutral stability. On the other hand, if the cone is placed on its tip and released, it falls onto its side. This position is said to be unstable. These three positions are illustrated in Fig. 6.1.

The stability of a dynamic system is defined in a similar manner. The response to a displacement, or initial condition, will result in either a decreasing, neutral, or increasing response. Specifically, it follows from the definition of stability that a linear system is stable if and only if the absolute value of its impulse response, $g(t)$, integrated over an infinite range, is finite. That is, in terms of the convolution integral Eq. (5.1) for a bounded input, one requires that $\int_0^\infty |g(t)|\, dt$ be finite. The location in the s-plane of the poles of a system indicates the resulting transient response. The poles in the left-hand portion of the s-plane result in a decreasing response for disturbance inputs. Similarly, poles on the $j\omega$-axis and in the right-hand plane result in a neutral and an increasing response, respectively, for a disturbance input. This division of the s-plane is shown in Fig. 6.2. Clearly the poles of desirable dynamic systems must lie in the left-hand portion of the s-plane [1–3].

A common example of the potential destabilizing effect of feedback is that of feedback in audio amplifier and speaker systems used for public address in auditoriums. In this case, a loudspeaker produces an audio signal that is an amplified version of the sounds picked up by a microphone. In addition to other audio inputs, the sound coming from the speaker itself may be sensed by the microphone. How strong this particular signal is depends upon the distance between the loudspeaker and the microphone. Because of the attenuating

---

FIGURE 6.1
The stability of
a cone.

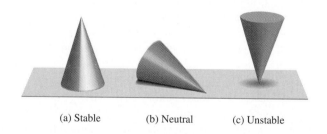

(a) Stable   (b) Neutral   (c) Unstable

FIGURE 6.2
Stability in the
s-plane.

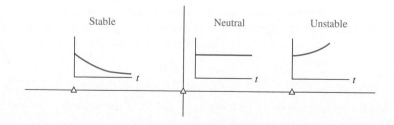

properties of air, the larger this distance is, the weaker the signal is that reaches the microphone. In addition, due to the finite propagation speed of sound waves, there is a time delay between the signal produced by the loudspeaker and that sensed by the microphone. In this case, the output from the feedback path is added to the external input. This is an example of positive feedback.

As the distance between the loudspeaker and the microphone decreases, we find that if the microphone is placed too close to the speaker, then the system will be unstable. The result of this instability is an excessive amplification and distortion of audio signals and an oscillatory squeal.

Another example of an unstable system is shown in Fig. 6.3. The first bridge across the Tacoma Narrows at Puget Sound, Washington, was opened to traffic on July 1, 1940. The bridge was found to oscillate whenever the wind blew. After four months, on November 7, 1940, a wind produced an oscillation that grew in amplitude until the bridge broke apart. Figure 6.3(a) shows the condition of beginning oscillation; Fig. 6.3(b) shows the catastrophic failure [5].

In terms of linear systems, we recognize that the stability requirement may be defined in terms of the location of the poles of the closed-loop transfer function. The closed-loop system transfer function is written as

$$T(s) = \frac{p(s)}{q(s)} = \frac{K\prod_{i=1}^{M} (s + z_i)}{s^N \prod_{k=1}^{Q} (s + \sigma_k) \prod_{m=1}^{R} [s^2 + 2\alpha_m s + (\alpha_m^2 + \omega_m^2)]}, \qquad (6.1)$$

where $q(s) = \Delta(s)$ is the characteristic equation whose roots are the poles of the closed-loop system. The output response for an impulse function input is then

$$c(t) = \sum_{k=1}^{Q} A_k e^{-\sigma_k t} + \sum_{m=1}^{R} B_m \left(\frac{1}{\omega_m}\right) e^{-\alpha_m t} \sin \omega_m t \qquad (6.2)$$

when $N = 0$. Clearly, to obtain a bounded response, the poles of the closed-loop system must be in the left-hand portion of the s-plane. Thus *a necessary and sufficient condition that a feedback system be stable is that all the poles of the system transfer function have negative real parts.* A system is stable if all the poles of the transfer function are in the left-hand s-plane. We will call a system not stable if not all the roots are in the left-hand plane. If the characteristic equation has simple roots on the imaginary axis ($j\omega$-axis) with all other roots in the left half-plane, the steady-state output will be sustained oscillations for a bounded input, unless the input is a sinusoid (which is bounded) whose frequency is equal to the magnitude of the $j\omega$-axis roots. For this case, the output becomes unbounded. Such a system is called *marginally stable,* since only certain bounded inputs (sinusoids of the frequency of the poles) will cause the output to become unbounded. For an unstable system, the characteristic equation has at least one root in the right half of the s-plane or repeated $j\omega$ roots; for this case, the output will become unbounded for any input.

For example, if the characteristic equation of a closed-loop system is

$$(s + 10)(s^2 + 16) = 0,$$

then the system is said to be marginally stable. If this system is excited by a sinusoid of frequency, $\omega = 4$, the output becomes unbounded.

(a)

FIGURE 6.3
Tacoma Narrows
Bridge (a) as
oscillation begins
and (b) at
catastrophic
failure.

(b)

To ascertain the stability of a feedback control system, one could determine the roots of the characteristic equation $q(s)$. However, we are first interested in determining the answer to the question, Is the system stable? If we calculate the roots of the characteristic equation in order to answer this question, we have determined much more information than is necessary. Therefore, several methods have been developed that provide the required yes or no answer to the stability question. The three approaches to the question of stability are (1) the s-plane approach, (2) the frequency plane ($j\omega$) approach, and (3) the time-domain approach. The real frequency ($j\omega$) approach is outlined in Chapter 9, and the discussion of the time-domain approach is considered in Section 6.4.

6.2 THE ROUTH–HURWITZ STABILITY CRITERION

The discussion and determination of stability has occupied the interest of many engineers. Maxwell and Vishnegradsky first considered the question of stability of dynamic systems. In the late 1800s, A. Hurwitz and E. J. Routh published independently a method of investigating the stability of a linear system [6, 7]. The Routh–Hurwitz stability method provides an answer to the question of stability by considering the characteristic equation of the system. The characteristic equation in the Laplace variable is written as

$$\Delta(s) = q(s) = a_n s^n + a_{n-1} s^{n-1} + \cdots + a_1 s + a_0 = 0. \tag{6.3}$$

To ascertain the stability of the system, it is necessary to determine whether any one of the roots of $q(s)$ lies in the right half of the s-plane. If Eq. (6.3) is written in factored form, we have

$$a_n(s - r_1)(s - r_2) \cdots (s - r_n) = 0, \tag{6.4}$$

where $r_i = i$th root of the characteristic equation. Multiplying the factors together, we find that

$$\begin{aligned} q(s) = a_n s^n &- a_n(r_1 + r_2 + \cdots + r_n)s^{n-1} \\ &+ a_n(r_1 r_2 + r_2 r_3 + r_1 r_3 + \cdots)s^{n-2} \\ &- a_n(r_1 r_2 r_3 + r_1 r_2 r_4 \cdots)s^{n-3} + \cdots \\ &+ a_n(-1)^n r_1 r_2 r_3 \cdots r_n = 0. \end{aligned} \tag{6.5}$$

In other words, for an nth-degree equation, we obtain

$$\begin{aligned} q(s) = a_n s^n &- a_n(\text{sum of all the roots})s^{n-1} \\ &+ a_n(\text{sum of the products of the roots taken 2 at a time})s^{n-2} \\ &- a_n(\text{sum of the products of the roots taken 3 at a time})s^{n-3} \\ &+ \cdots + a_n(-1)^n(\text{product of all } n \text{ roots}) = 0. \end{aligned} \tag{6.6}$$

Examining Eq. (6.5), we note that all the coefficients of the polynomial must have the same sign if all the roots are in the left-hand plane. Also, it is necessary that all the coefficients for a stable system be nonzero. However, although necessary, these requirements are not sufficient. That is, we immediately know the system is unstable if they are not satisfied;

yet if they are satisfied, we must proceed to ascertain the stability of the system. For example, when the characteristic equation is

$$q(s) = (s + 2)(s^2 - s + 4) = (s^3 + s^2 + 2s + 8), \tag{6.7}$$

the system is unstable and yet the polynomial possesses all positive coefficients.

The *Routh–Hurwitz criterion* is a necessary and sufficient criterion for the stability of linear systems. The method was originally developed in terms of determinants, but we shall utilize the more convenient array formulation.

The Routh–Hurwitz criterion is based on ordering the coefficients of the characteristic equation

$$a_n s^n + a_{n-1} s^{n-1} + a_{n-2} s^{n-2} + \cdots + a_1 s + a_0 = 0 \tag{6.8}$$

into an array or schedule as follows [4]:

$$
\begin{array}{c|cccc}
s^n & a_n & a_{n-2} & a_{n-4} & \cdots \\
s^{n-1} & a_{n-1} & a_{n-3} & a_{n-5} & \cdots
\end{array}
$$

Further rows of the schedule are then completed as follows:

$$
\begin{array}{c|cccc}
s^n & a_n & a_{n-2} & a_{n-4} \\
s^{n-1} & a_{n-1} & a_{n-3} & a_{n-5} \\
s^{n-2} & b_{n-1} & b_{n-3} & b_{n-5} \\
s^{n-3} & c_{n-1} & c_{n-3} & c_{n-5} \\
\cdot & \cdot & \cdot & \cdot \\
\cdot & \cdot & \cdot & \cdot \\
\cdot & \cdot & \cdot & \cdot \\
s^0 & h_{n-1} &
\end{array}
$$

where

$$b_{n-1} = \frac{(a_{n-1})(a_{n-2}) - a_n(a_{n-3})}{a_{n-1}} = \frac{-1}{a_{n-1}} \begin{vmatrix} a_n & a_{n-2} \\ a_{n-1} & a_{n-3} \end{vmatrix},$$

$$b_{n-3} = -\frac{1}{a_{n-1}} \begin{vmatrix} a_n & a_{n-4} \\ a_{n-1} & a_{n-5} \end{vmatrix},$$

and

$$c_{n-1} = \frac{-1}{b_{n-1}} \begin{vmatrix} a_{n-1} & a_{n-3} \\ b_{n-1} & b_{n-3} \end{vmatrix},$$

and so on. The algorithm for calculating the entries in the array can be followed on a determinant basis or by using the form of the equation for b_{n-1}.

The Routh–Hurwitz criterion states that the number of roots of $q(s)$ with positive real parts is equal to the number of changes in sign of the first column of the array. This criterion requires that there be no changes in sign in the first column for a stable system. This requirement is both necessary and sufficient.

Four distinct cases or configurations of the first column array must be considered, and each must be treated separately and requires suitable modifications of the array calculation procedure: (1) No element in the first column is zero; (2) there is a zero in the first column, but some other elements of the row containing the zero in the first column are nonzero; (3) there is a zero in the first column, and the other elements of the row containing the zero are also zero; and (4) as in (3) with *repeated* roots on the $j\omega$-axis.

To illustrate this method clearly, several examples will be presented for each case.

Case 1.   *No element in the first column is zero.*

EXAMPLE 6.1   **Second-order system**

The characteristic equation of a second-order system is

$$q(s) = a_2 s^2 + a_1 s + a_0.$$

The array is written as

$$
\begin{array}{c|cc}
s^2 & a_2 & a_0 \\
s & a_1 & 0 \\
s^0 & b_1 & 0 \;,
\end{array}
$$

where

$$b_1 = \frac{a_1 a_0 - (0) a_2}{a_1} = \frac{-1}{a_1} \begin{vmatrix} a_2 & a_0 \\ a_1 & 0 \end{vmatrix} = a_0.$$

Therefore the requirement for a stable second-order system is simply that all the coefficients be positive. ∎

EXAMPLE 6.2   **Third-order system**

The characteristic equation of a third-order system is

$$q(s) = a_3 s^3 + a_2 s^2 + a_1 s + a_0.$$

The array is

$$
\begin{array}{c|cc}
s^3 & a_3 & a_1 \\
s^2 & a_2 & a_0 \\
s^1 & b_1 & 0 \\
s^0 & c_1 & 0 \;,
\end{array}
$$

where

$$b_1 = \frac{a_2 a_1 - a_0 a_3}{a_2} \qquad \text{and} \qquad c_1 = \frac{b_1 a_0}{b_1} = a_0.$$

For the third-order system to be stable, it is necessary and sufficient that the coefficients be positive and $a_2 a_1 \geq a_0 a_3$. The condition when $a_2 a_1 = a_0 a_3$ results in a marginal stability case, and one pair of roots lies on the imaginary axis in the s-plane. This marginal case is recognized as Case 3 because there is a zero in the first column when $a_2 a_1 = a_0 a_3$, and it will be discussed under Case 3.

As a final example of characteristic equations that result in no zero elements in the first row, let us consider a polynomial

$$q(s) = (s - 1 + j\sqrt{7})(s - 1 - j\sqrt{7})(s + 3) = s^3 + s^2 + 2s + 24. \qquad (6.9)$$

The polynomial satisfies all the necessary conditions because all the coefficients exist and are positive. Therefore, utilizing the Routh–Hurwitz array, we have

$$
\begin{array}{c|cc}
s^3 & 1 & 2 \\
s^2 & 1 & 24 \\
s^1 & -22 & 0 \\
s^0 & 24 & 0.
\end{array}
$$

Because two changes in sign appear in the first column, we find that two roots of $q(s)$ lie in the right-hand plane, and our prior knowledge is confirmed. ■

Case 2.   *Zeros in the first column while some other elements of the row containing a zero in the first column are nonzero.* If only one element in the array is zero, it may be replaced with a small positive number ϵ that is allowed to approach zero after completing the array. For example, consider the following characteristic equation:

$$
q(s) = s^5 + 2s^4 + 2s^3 + 4s^2 + 11s + 10. \tag{6.10}
$$

The Routh–Hurwitz array is then

$$
\begin{array}{c|ccc}
s^5 & 1 & 2 & 11 \\
s^4 & 2 & 4 & 10 \\
s^3 & \epsilon & 6 & 0 \\
s^2 & c_1 & 10 & 0 \\
s^1 & d_1 & 0 & 0 \\
s^0 & 10 & 0 & 0,
\end{array}
$$

where

$$
c_1 = \frac{4\epsilon - 12}{\epsilon} = \frac{-12}{\epsilon} \quad \text{and} \quad d_1 = \frac{6c_1 - 10\epsilon}{c_1} \rightarrow 6.
$$

There are two sign changes due to the large negative number in the first column, $c_1 = -12/\epsilon$. Therefore the system is unstable and two roots lie in the right half of the plane.

EXAMPLE 6.3   **Unstable system**

As a final example of the type of Case 2, consider the characteristic equation

$$
q(s) = s^4 + s^3 + s^2 + s + K, \tag{6.11}
$$

where it is desired to determine the gain K that results in marginal stability. The Routh–Hurwitz array is then

$$
\begin{array}{c|ccc}
s^4 & 1 & 1 & K \\
s^3 & 1 & 1 & 0 \\
s^2 & \epsilon & K & 0 \\
s^1 & c_1 & 0 & 0 \\
s^0 & K & 0 & 0,
\end{array}
$$

where

$$
c_1 = \frac{\epsilon - K}{\epsilon} \rightarrow \frac{-K}{\epsilon}.
$$

Therefore, for any value of K greater than zero, the system is unstable. Also, because the last term in the first column is equal to K, a negative value of K will result in an unstable system. Therefore the system is unstable for all values of gain K. ∎

Case 3. *Zeros in the first column, and the other elements of the row containing the zero are also zero.* Case 3 occurs when all the elements in one row are zero or when the row consists of a single element that is zero. This condition occurs when the polynomial contains singularities that are symmetrically located about the origin of the s-plane. Therefore Case 3 occurs when factors such as $(s + \sigma)(s - \sigma)$ or $(s + j\omega)(s - j\omega)$ occur. This problem is circumvented by utilizing the *auxiliary* polynomial, $U(s)$, which immediately precedes the zero entry in the Routh array. The order of the auxiliary polynomial is always even and indicates the number of symmetrical root pairs.

To illustrate this approach, let us consider a third-order system with a characteristic equation:

$$q(s) = s^3 + 2s^2 + 4s + K, \tag{6.12}$$

where K is an adjustable loop gain. The Routh array is then

$$
\begin{array}{c|cc}
s^3 & 1 & 4 \\
s^2 & 2 & K \\
s^1 & \dfrac{8 - K}{2} & 0 \\
s^0 & K & 0.
\end{array}
$$

Therefore, for a stable system, we require that

$$0 \le K \le 8.$$

When $K = 8$, we have two roots on the $j\omega$-axis and a marginal stability case. Note that we obtain a row of zeros (Case 3) when $K = 8$. The auxiliary polynomial, $U(s)$, is the equation of the row preceding the row of zeros. The equation of the row preceding the row of zeros is, in this case, obtained from the s^2-row. We recall that this row contains the coefficients of the even powers of s and therefore in this case, we have

$$U(s) = 2s^2 + Ks^0 = 2s^2 + 8 = 2(s^2 + 4) = 2(s + j2)(s - j2). \tag{6.13}$$

To show that the auxiliary polynomial, $U(s)$, is indeed a factor of the characteristic equation, we divide $q(s)$ by $U(s)$ to obtain

$$
\begin{array}{r}
\tfrac{1}{2}s \; + 1 \\
2s^2 + 8 \overline{) \, s^3 + 2s^2 + 4s + 8} \\
\underline{s^3 \qquad\quad + 4s} \\
2s^2 \qquad + 8 \\
\underline{2s^2 \qquad + 8.}
\end{array}
$$

Therefore, when $K = 8$, the factors of the characteristic equation are

$$q(s) = (s + 2)(s + j2)(s - j2). \tag{6.14}$$

Strictly, the marginal case response is an unacceptable oscillation.

Case 4.   *Repeated roots of the characteristic equation on the jω-axis.* If the roots of the characteristic equation are simple, the system is neither stable nor unstable; it is instead called marginally stable, since it has an undamped sinusoidal mode. If the *jω*-axis roots are repeated, the system response will be unstable, with a form $t[\sin(\omega t + \phi)]$. The Routh–Hurwitz criteria will not reveal this form of instability [21].

Consider the system with a characteristic equation

$$q(s) = (s + 1)(s + j)(s - j)(s + j)(s - j) = s^5 + s^4 + 2s^3 + 2s^2 + s + 1.$$

The Routh array is

$$
\begin{array}{c|ccc}
s^5 & 1 & 2 & 1 \\
s^4 & 1 & 2 & 1 \\
s^3 & \epsilon & \epsilon & 0 \\
s^2 & 1 & 1 & \\
s^1 & \epsilon & 0 & \\
s^0 & 1 & &
\end{array}
\quad,
$$

where $\epsilon \to 0$. Note the absence of sign changes, a condition that falsely indicates that the system is marginally stable. The impulse response of the system increases with time as $t \sin(t + \phi)$. The auxiliary equation at the s^2 line is $(s^2 + 1)$ and the auxiliary equation at the s^4 line is $(s^4 + 2s^2 + 1) = (s^2 + 1)^2$, indicating the repeated roots on the *jω*-axis.

EXAMPLE 6.4   **Robot control**

Let us consider the control of a robot arm as shown in Fig. 6.4. It is predicted that there will be about 100,000 robots in service throughout the world by 2000 [10]. The robot shown in Fig. 6.4 is a six-legged micro robot system using highly flexible legs with high-gain controllers that may become unstable and oscillate. Under this condition, we have the characteristic polynomial

$$q(s) = s^5 + s^4 + 4s^3 + 24s^2 + 3s + 63. \tag{6.15}$$

The Routh–Hurwitz array is

$$
\begin{array}{c|ccc}
s^5 & 1 & 4 & 3 \\
s^4 & 1 & 24 & 63 \\
s^3 & -20 & -60 & 0 \\
s^2 & 21 & 63 & 0 \\
s^1 & 0 & 0 & 0.
\end{array}
$$

Therefore the auxiliary polynomial is

$$U(s) = 21s^2 + 63 = 21(s^2 + 3) = 21(s + j\sqrt{3})(s - j\sqrt{3}), \tag{6.16}$$

which indicates that two roots are on the imaginary axis. To examine the remaining roots, we divide by the auxiliary polynomial to obtain

$$\frac{q(s)}{s^2 + 3} = s^3 + s^2 + s + 21.$$

FIGURE 6.4
A completely integrated, six-legged, micro robot system. The legged design provides maximum dexterity. Legs also provide a unique sensory system for environmental interaction. It is equipped with a sensor network that includes 150 sensors of 12 different types. The legs are instrumented so that the robot can determine the lay of the terrain, the surface texture, hardness, and even color. The gyro-stabilized camera and range finder can be used for gathering data beyond the robot's immediate reach. This high-performance system is able to walk quickly, climb over obstacles, and perform dynamic motions. (Courtesy of IS Robotics Corporation.)

Establishing a Routh–Hurwitz array for this equation, we have

$$
\begin{array}{c|cc}
s^3 & 1 & 1 \\
s^2 & 1 & 21 \\
s^1 & -20 & 0 \\
s^0 & 21 & 0.
\end{array}
$$

The two changes in sign in the first column indicate the presence of two roots in the right-hand plane, and the system is unstable. The roots in the right-hand plane are $s = +1 \pm j\sqrt{6}$. ∎

EXAMPLE 6.5   **Disk-drive control**

Large disk-storage devices are used with today's computers [1]. The read head is moved to different positions on the spinning disk, and rapid, accurate response is required. A block diagram of a disk-storage read-head positioning system is shown in Fig. 6.5. It is desired to determine the range of K and a for which the system is stable. The characteristic equation is

$$
1 + G(s) = 1 + \frac{K(s + a)}{s(s + 1)(s + 2)(s + 3)} = 0.
$$

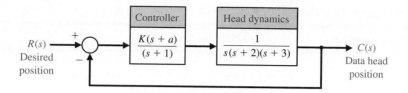

FIGURE 6.5
Disk drive head position control.

Therefore $q(s) = s^4 + 6s^3 + 11s^2 + (K + 6)s + Ka = 0$. Establishing the Routh array, we have

$$
\begin{array}{c|ccc}
s^4 & 1 & 11 & Ka \\
s^3 & 6 & (K + 6) & \\
s^2 & b_3 & Ka & \\
s^1 & c_3 & & \\
s^0 & Ka & &
\end{array}
$$

where

$$b_3 = \frac{60 - K}{6} \quad \text{and} \quad c_3 = \frac{b_3(K + 6) - 6Ka}{b_3}.$$

The coefficient c_3 sets the acceptable range of K and a, while b_3 requires that K be less than 60. Setting $c_3 = 0$, we obtain

$$(K - 60)(K + 6) + 36Ka = 0.$$

The required relationship between K and a is then

$$a \leq \frac{(60 - K)(K + 6)}{36K}$$

when a is positive. Therefore, if $K = 40$, we require $a \leq 0.639$. ■

The general form of the characteristic equation of an nth-order system is

$$s^n + a_{n-1}s^{n-1} + a_{n-2}s^{n-2} + \cdots + a_1 s + \omega_n^n = 0.$$

We divide through by ω_n^n and use $\overset{*}{s} = s/\omega_n$ to obtain the normalized form of the characteristic equation

$$\overset{*}{s}{}^n + b\overset{*}{s}{}^{n-1} + c\overset{*}{s}{}^{n-2} + \cdots + 1 = 0.$$

For example, we normalize

$$s^3 + 5s^2 + 2s + 8 = 0$$

by dividing through by $8 = \omega_n^3$, obtaining

$$\frac{s^3}{\omega_n^3} + \frac{5}{2}\frac{s^2}{\omega_n^2} + \frac{2}{4}\left(\frac{s}{\omega_n}\right) + 1 = 0,$$

or

$$\overset{*}{s}{}^3 + 2.5\overset{*}{s}{}^2 + 0.5\overset{*}{s} + 1 = 0,$$

where $\overset{*}{s} = s/\omega_n$. In this case, $b = 2.5$ and $c = 0.5$. Using this normalized form of the

TABLE 6.1   The Routh–Hurwitz Stability Criterion

n	Characteristic Equation	Criterion
2	$s^2 + bs + 1 = 0$	$b > 0$
3	$s^3 + bs^2 + cs + 1 = 0$	$bc - 1 > 0$
4	$s^4 + bs^3 + cs^2 + ds + 1 = 0$	$bcd - d^2 - b^2 > 0$
5	$s^5 + bs^4 + cs^3 + ds^2 + es + 1 = 0$	$bcd + b - d^2 - b^2e > 0$
6	$s^6 + bs^5 + cs^4 + ds^3 + es^2 + fs + 1 = 0$	$(bcd + bf - d^2 - b^2e)e + b^2c - bd - bc^2f - f^2 + bfe + cdf > 0$

Note: The equations are normalized by $(\omega_n)^n$.

characteristic equation, we summarize the stability criterion for up to a sixth-order characteristic equation, as provided in Table 6.1. Note that $bc = 1.25$ and that the system is stable.

6.3 THE RELATIVE STABILITY OF FEEDBACK CONTROL SYSTEMS

The verification of stability using the Routh–Hurwitz criterion provides only a partial answer to the question of stability. The Routh–Hurwitz criterion ascertains the absolute stability of a system by determining whether any of the roots of the characteristic equation lies in the right half of the s-plane. However, if the system satisfies the Routh–Hurwitz criterion and is absolutely stable, it is desirable to determine the *relative stability;* that is, it is necessary to investigate the relative damping of each root of the characteristic equation. The relative stability of a system can be defined as the property that is measured by the relative real part of each root or pair of roots. Thus root r_2 is relatively more stable than the roots r_1, $\hat{r}_1$, as shown in Fig. 6.6. The relative stability of a system can also be defined in terms of the relative damping coefficients ζ of each complex root pair and therefore in terms of the speed of response and overshoot instead of settling time.

Hence the investigation of the relative stability of each root is clearly necessary because, as we found in Chapter 5, the location of the closed-loop poles in the s-plane determines the performance of the system. Thus it is imperative that we reexamine the characteristic equation $q(s)$ and consider several methods for the determination of relative stability.

Because the relative stability of a system is dictated by the location of the roots of the characteristic equation, a first approach using an s-plane formulation is to extend the Routh–Hurwitz criterion to ascertain relative stability. This can be simply accomplished by utilizing a change of variable, which shifts the s-plane axis in order to utilize the Routh–Hurwitz criterion. Examining Fig. 6.6, we notice that a shift of the vertical axis in the s-plane to $-\sigma_1$ will result in the roots r_1, $\hat{r}_1$ appearing on the shifted axis. The correct magnitude to shift the vertical axis must be obtained on a trial-and-error basis. Then, without solving the fifth-order polynomial $q(s)$, one may determine the real part of the dominant roots r_1, $\hat{r}_1$.

EXAMPLE 6.6   Axis shift

Consider the simple third-order characteristic equation

$$q(s) = s^3 + 4s^2 + 6s + 4. \tag{6.17}$$

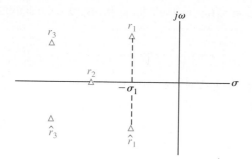

FIGURE 6.6
Root locations in
the s-plane.

As a first try, let $s_n = s + 2$ and note that we obtain a Routh–Hurwitz array without a zero occurring in the first column. However, upon setting the shifted variable s_n equal to $s + 1$, we obtain

$$(s_n - 1)^3 + 4(s_n - 1)^2 + 6(s_n - 1) + 4 = s_n^3 + s_n^2 + s_n + 1. \quad (6.18)$$

Then the Routh array is established as

s_n^3	1	1
s_n^2	1	1
s_n^1	0	0
s_n^0	1	0.

Clearly, there are roots on the shifted imaginary axis that can be obtained from the auxiliary polynomial

$$U(s_n) = s_n^2 + 1 = (s_n + j)(s_n - j) = (s + 1 + j)(s + 1 - j). \ \blacksquare \quad (6.19)$$

The shifting of the s-plane axis to ascertain the relative stability of a system is a very useful approach, particularly for higher-order systems with several pairs of closed-loop complex conjugate roots.

6.4 THE STABILITY OF STATE VARIABLE SYSTEMS

The stability of a system modeled by a state variable flow graph model can be readily ascertained. The stability of a system with an input–output transfer function $T(s)$ can be determined by examining the denominator polynomial of $T(s)$. Therefore if the transfer function is written as

$$T(s) = \frac{p(s)}{q(s)},$$

where $p(s)$ and $q(s)$ are polynomials in s, then the stability of the system is represented by the roots of $q(s)$. The polynomial $q(s)$, when set equal to zero, is called the characteristic equation. The roots of the characteristic equation must lie in the left-hand s-plane for the system to exhibit a stable time response. Therefore, to ascertain the stability of a system represented by a transfer function, we investigate the characteristic equation and utilize the Routh–Hurwitz criterion. If the system we are investigating is represented by a signal-flow graph state model, we obtain the characteristic equation by evaluating the flow graph determinant. As an illustration of this method, let us investigate the stability of the system of Example 3.1.

EXAMPLE 6.7   Stability of a system

The transfer function $T(s)$ examined in Example 3.1 is

$$T(s) = \frac{2s^2 + 8s + 6}{s^3 + 8s^2 + 16s + 6}. \tag{6.20}$$

Clearly, the characteristic polynomial for this system is

$$q(s) = s^3 + 8s^2 + 16s + 6. \tag{6.21}$$

Of course, this characteristic polynomial is also readily obtained from either the flow graph model shown in Fig. 3.8 or the one shown in Fig. 3.10. Using the Routh–Hurwitz criterion, we find that the system is stable and that all the roots of $q(s)$ lie in the left-hand s-plane. ■

We often determine the flow graph state model directly from a set of state differential equations. In this case, we can use the flow graph directly to determine the stability of the system by obtaining the characteristic equation from the flow graph determinant $\Delta(s)$. An illustration of this approach will aid in comprehending this method.

EXAMPLE 6.8   Stability of a second-order system

A second-order system is described by the two first-order differential equations

$$\dot{x}_1 = -3x_1 + x_2, \qquad \dot{x}_2 = +1x_2 - Kx_1 + Ku,$$

where the dot notation implies the first derivative and $u(t)$ is the input. The flow graph model of this set of differential equations is shown in Fig. 6.7.

Using Mason's flow graph rule, we note three loops:

$$L_1 = s^{-1}, \qquad L_2 = -3s^{-1}, \qquad \text{and} \qquad L_3 = -Ks^{-2},$$

where L_1 and L_2 do not share a common node. Therefore the determinant is

$$\Delta = 1 - (L_1 + L_2 + L_3) + L_1 L_2 = 1 - (s^{-1} - 3s^{-1} - Ks^{-2}) + (-3s^{-2}).$$

We multiply by s^2 to obtain the characteristic equation

$$s^2 + 2s + (K - 3) = 0.$$

Since all coefficients must be positive, we require $K > 3$ for stability. ■

FIGURE 6.7
Flow graph model for state variable equations of Example 6.8.

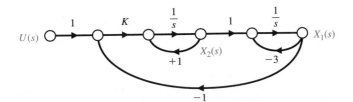

A method of obtaining the characteristic equation directly from the vector differential equation is based on the fact that the solution to the unforced system is an exponential function. The vector differential equation without input signals is

$$\dot{\mathbf{x}} = \mathbf{A}\mathbf{x}, \tag{6.22}$$

where $\mathbf{x}$ is the state vector. The solution is of exponential form, and we can define a constant λ such that the solution of the system for one state might be $x_i(t) = k_i e^{\lambda_i t}$. The λ_i are called the characteristic roots of the system, which are simply the roots of the characteristic equation. If we let $\mathbf{x} = \mathbf{k} e^{\lambda t}$ and substitute into Eq. (6.22), we have

$$\lambda \mathbf{k} e^{\lambda t} = \mathbf{A} \mathbf{k} e^{\lambda t} \tag{6.23}$$

or

$$\lambda \mathbf{x} = \mathbf{A}\mathbf{x}. \tag{6.24}$$

Equation (6.24) can be rewritten as

$$(\lambda \mathbf{I} - \mathbf{A})\mathbf{x} = \mathbf{0}, \tag{6.25}$$

where $\mathbf{I}$ equals the identity matrix and $\mathbf{0}$ equals the null matrix. The solution of this set of simultaneous equations has a nontrivial solution if and only if the determinant vanishes, that is, only if

$$\det(\lambda \mathbf{I} - \mathbf{A}) = 0. \tag{6.26}$$

The nth-order equation in λ resulting from the evaluation of this determinant is the characteristic equation, and the stability of the system can be readily ascertained. Let us reconsider the third-order system described in Example 3.2 to illustrate this approach.

EXAMPLE 6.9   **Closed epidemic system**

The vector differential equation of the epidemic system is given in Eq. (3.55) and repeated here as

$$\frac{d\mathbf{x}}{dt} = \begin{bmatrix} -\alpha & -\beta & 0 \\ \beta & -\gamma & 0 \\ \alpha & \gamma & 0 \end{bmatrix} \mathbf{x} + \begin{bmatrix} 1 & 0 \\ 0 & 1 \\ 0 & 0 \end{bmatrix} \begin{bmatrix} u_1 \\ u_2 \end{bmatrix}.$$

The characteristic equation is then

$$
\begin{aligned}
\det(\lambda \mathbf{I} - \mathbf{A}) &= \det\left\{ \begin{bmatrix} \lambda & 0 & 0 \\ 0 & \lambda & 0 \\ 0 & 0 & \lambda \end{bmatrix} - \begin{bmatrix} -\alpha & -\beta & 0 \\ \beta & -\gamma & 0 \\ \alpha & \gamma & 0 \end{bmatrix} \right\} \\
&= \det\begin{bmatrix} (\lambda + \alpha) & \beta & 0 \\ -\beta & (\lambda + \gamma) & 0 \\ -\alpha & -\gamma & \lambda \end{bmatrix} \\
&= \lambda[(\lambda + \alpha)(\lambda + \gamma) + \beta^2] \\
&= \lambda[\lambda^2 + (\alpha + \gamma)\lambda + (\alpha\gamma + \beta^2)] = 0.
\end{aligned}
$$

Thus we obtain the characteristic equation of the system, and it is similar to that obtained in Eq. (3.57) by flow graph methods. The additional root $\lambda = 0$ results from the definition of x_3 as the integral of $(\alpha x_1 + \gamma x_2)$, and x_3 does not affect the other state variables. Thus the root $\lambda = 0$ indicates the integration connected with x_3. The characteristic equation indicates that the system is marginally stable when $(\alpha + \gamma) > 0$ and $(\alpha\gamma + \beta^2) > 0$. ■

As another example, reconsider the inverted pendulum described in Example 3.4. The system matrix is

$$\mathbf{A} = \begin{bmatrix} 0 & 1 & 0 & 0 \\ 0 & 0 & -(mg/M) & 0 \\ 0 & 0 & 0 & 1 \\ 0 & 0 & g/l & 0 \end{bmatrix}.$$

The characteristic equation can be obtained from the determinant of $(\lambda\mathbf{I} - \mathbf{A})$ as follows:

$$\det\begin{bmatrix} \lambda & -1 & 0 & 0 \\ 0 & \lambda & mg/M & 0 \\ 0 & 0 & \lambda & -1 \\ 0 & 0 & -(g/l) & \lambda \end{bmatrix} = \lambda\left[\lambda\left(\lambda^2 - \frac{g}{l}\right)\right] = \lambda^2\left(\lambda^2 - \frac{g}{l}\right) = 0.$$

The characteristic equation indicates that there are two roots at $\lambda = 0$, a root at $\lambda = +\sqrt{g/l}$, and a root at $\lambda = -\sqrt{g/l}$. Hence the system is unstable, because there is a root in the right-hand plane at $\lambda = +\sqrt{g/l}$. The two roots at $\lambda = 0$ will also result in an unbounded response.

6.5 THE DETERMINATION OF ROOT LOCATIONS IN THE s-PLANE

The relative stability of a feedback control system is directly related to the location of the closed-loop roots of the characteristic equation in the s-plane. Therefore it is often necessary and easiest simply to determine the values of the roots of the characteristic equation. This approach has become particularly attractive today due to the availability of digital computer programs for determining the roots of polynomials. Moreover, this approach may even be the most logical when using manual calculations if the order of the system is relatively low. For a third-order system or lower, it is usually simpler to utilize manual calculation methods.

The determination of the roots of a polynomial can be obtained by utilizing *synthetic division*, which is based on the remainder theorem; that is, upon dividing the polynomial by a factor, the remainder is zero when the factor is a root of the polynomial. Synthetic division is commonly used to carry out the division process. The relations for the roots of a polynomial as obtained in Eq. (6.6) are utilized to aid in the choice of a first estimate of a root.

EXAMPLE 6.10   Synthetic division

Let us determine the roots of the polynomial

$$q(s) = s^3 + 4s^2 + 6s + 4. \tag{6.27}$$

Establishing a table of synthetic division, we have

1	4	6	4	$\underline{-1}$ = trial root
	-1	-3	-3	
1	3	3	1	= remainder

for a trial root of $s = -1$. In this table, we multiply by the trial root and successively add in each column. With a remainder of one, we might try $s = -2$, which results in the form

$$
\begin{array}{cccc|c}
1 & 4 & 6 & 4 & \underline{-2} \\
 & -2 & -4 & -4 & \\
\hline
1 & 2 & 2 & 0 &
\end{array}
$$

Because the remainder is zero, one root is equal to -2 and the remaining roots may be obtained from the remaining polynomial $(s^2 + 2s + 2)$ by using the quadratic root formula. ∎

The search for a root of the polynomial can be aided considerably by utilizing the rate of change of the polynomial at the estimated root in order to obtain a new estimate. The *Newton–Raphson method* is a rapid method utilizing synthetic division to obtain the value of

$$
\left. \frac{dq(s)}{ds} \right|_{s=s_1},
$$

where s_1 is a first estimate of the root. The Newton–Raphson method is an iteration approach utilized in many digital computer root-solving programs. A new estimate s_{n+1} of the root is based on the last estimate as [4, 12]

$$
s_{n+1} = s_n - \frac{q(s_n)}{q'(s_n)}, \tag{6.28}
$$

where

$$
q'(s_n) = \left. \frac{dq(s)}{ds} \right|_{s=s_n}.
$$

The synthetic division process may be utilized to obtain $q(s_n)$ and $q'(s_n)$. The synthetic division process for a trial root may be written for the polynomial

$$
q(s) = a_m s^m + a_{m-1} s^{m-1} + \cdots + a_1 s + a_0
$$

as

$$
\begin{array}{ccccc|c}
a_m & a_{m-1} & \cdots & a_1 & a_0 & \underline{s_n} \\
 & b_m s_n & \cdots & & b_1 s_n & \\
\hline
b_m & b_{m-1} & \cdots & b_1 & b_0 &
\end{array}
$$

where $b_0 = q(s_n)$, the remainder of the division process. When s_n is a root of $q(s)$, the remainder is equal to zero and the remaining polynomial

$$
b_m s^{m-1} + b_{m-1} s^{m-2} + \cdots + b_1
$$

may itself be factored. The derivative evaluated at the nth estimate of the root, $q'(s_n)$, may also be obtained by repeating the synthetic division process on the $b_m, b_{m-1}, \ldots, b_1$ coefficients. The value of $q'(s_n)$ is the remainder of this repeated synthetic division process. This process converges as the square of the absolute error. The Newton–Raphson method, using synthetic division, is readily illustrated, as can be seen by repeating Example 6.10.

EXAMPLE 6.11   **Newton–Raphson method**

For the polynomial $q(s) = s^3 + 4s^2 + 6s + 4$, we establish a table of synthetic division for a first estimate as follows:

$$
\begin{array}{cccc|l}
1 & 4 & 6 & 4 & \underline{-1} \\
  & -1 & -3 & -3 & \\
\hline
1 & 3 & 3 & 1 & = q(s_1) \\
  & -1 & -2 & & \\
\hline
1 & 2 & 1 & & = q'(s_1)
\end{array}
$$

The derivative of $q(s)$ evaluated at s_1 is determined by continuing the synthetic division as shown. Then the second estimate becomes

$$
s_2 = s_1 - \frac{q(s_1)}{q'(s_1)} = -1 - \left(\frac{1}{1}\right) = -2.
$$

As we found in Example 6.10, s_2 is, in fact, a root of the polynomial and results in a zero remainder. ■

EXAMPLE 6.12   **Third-order system**

Let us consider the polynomial

$$
q(s) = s^3 + 3.5s^2 + 6.5s + 10. \tag{6.29}
$$

From Eq. (6.6), we note that the sum of all the roots is equal to -3.5 and that the product of all the roots is -10. Therefore, as a first estimate, we try $s_1 = -1$ and obtain the following table:

$$
\begin{array}{cccc|l}
1 & 3.5 & 6.5 & 10 & \underline{-1} \\
  & -1 & -2.5 & -4 & \\
\hline
1 & 2.5 & 4 & 6 & = q(s_1) \\
  & -1 & -1.5 & & \\
\hline
1 & 1.5 & 2.5 & & = q'(s_1)
\end{array}
$$

Therefore a second estimate is

$$
s_2 = -1 - \left(\frac{6}{2.5}\right) = -3.40.
$$

Now let us use a second estimate that is convenient for these manual calculations. Therefore, on the basis of the calculation of $s_2 = -3.40$, we will choose a second estimate $s_2 = -3.00$. Establishing a table for $s_2 = -3.00$ and completing the synthetic division, we find that

$$
s_3 = -s_2 - \frac{q(s_2)}{q'(s_2)} = -3.00 - \frac{(-5)}{12.5} = -2.60.
$$

Finally, completing a table for $s_4 = -2.50$, we find that the remainder is zero and that the polynomial factors are $q(s) = (s + 2.5)(s^2 + s + 4)$. ∎

The availability of a digital computer or a programmable calculator enables one to readily determine the roots of a polynomial by using root determination programs, which usually use the Newton–Raphson algorithm, Eq. (6.28). The availability of a personal computer is particularly advantageous for a control engineer, because the ability to perform on-line calculations aids in the iterative analysis and design process.

6.6 DESIGN EXAMPLE: TRACKED VEHICLE TURNING CONTROL

The design of a turning control for a tracked vehicle involves the selection of two parameters [8]. In Fig. 6.8 the system shown in part (a) has the model shown in part (b). The two tracks are operated at different speeds in order to turn the vehicle. Select K and a so that the system is stable and the steady-state error for a ramp command is less than or equal to 24% of the magnitude of the command.

The characteristic equation of the feedback system is

$$1 + G_c G(s) = 0$$

or

$$1 + \frac{K(s + a)}{s(s + 1)(s + 2)(s + 5)} = 0. \tag{6.30}$$

Therefore we have

$$s(s + 1)(s + 2)(s + 5) + K(s + a) = 0$$

FIGURE 6.8
(a) Turning control system for a two-track vehicle.
(b) Block diagram.

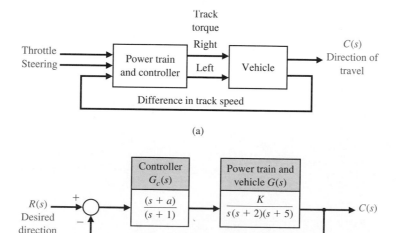

(a)

(b)

or

$$s^4 + 8s^3 + 17s^2 + (K + 10)s + Ka = 0. \tag{6.31}$$

To determine the stable region for K and a, we establish the Routh array as

$$
\begin{array}{c|ccc}
s^4 & 1 & 17 & Ka \\
s^3 & 8 & (K + 10) & 0 \\
s^2 & b_3 & Ka \\
s^1 & c_3 \\
s^0 & Ka
\end{array}
$$

where

$$b_3 = \frac{126 - K}{8} \quad \text{and} \quad c_3 = \frac{b_3(K + 10) - 8Ka}{b_3}.$$

For the elements of the first column to be positive, we require that Ka, b_3, and c_3 be positive. We therefore require

$$K < 126$$

$$Ka > 0 \tag{6.32}$$

$$(K + 10)(126 - K) - 64Ka > 0.$$

The region of stability for $K > 0$ is shown in Fig. 6.9. The steady-state error to a ramp input $r(t) = At, t > 0$ is

$$e_{ss} = A/K_v,$$

where

$$K_v = \lim_{s \to 0} sG_c G = Ka/10.$$

FIGURE 6.9
The stable region.

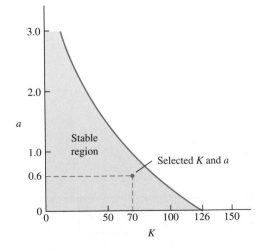

Therefore we have

$$e_{ss} = \frac{10A}{Ka}.$$ (6.33)

When e_{ss} is equal to 23.8% of A, we require that $Ka = 42$. This can be satisfied by the selected point in the stable region when $K = 70$ and $a = 0.6$, as shown in Fig. 6.9. Of course, another acceptable design would be attained when $K = 50$ and $a = 0.84$. We can calculate a series of possible combinations of K and a that can satisfy $Ka = 42$ and that lie within the stable region, and all will be acceptable design solutions. However, not all selected values of K and a will lie within the stable region. Note that K cannot exceed 126.

6.7 SYSTEM STABILITY USING *MATLAB*

This section begins with a discussion of the Routh–Hurwitz stability method. We will see how *Matlab* can assist us in the stability analysis by providing an easy and accurate method for computing the poles of the characteristic equation. For the case of the characteristic equation as a function of a single parameter, it will be possible to generate a plot displaying the *movement* of the poles as the parameter varies. The section concludes with an example.

The function introduced in this section is the function for, which is used to repeat a number of statements a specific number of times.

Routh–Hurwitz Stability.   As stated earlier the Routh–Hurwitz criterion is a necessary and sufficient criterion for stability. Given a characteristic equation with fixed coefficients, we can use Routh–Hurwitz to determine the number of roots in the right half-plane. For example, consider the characteristic equation

$$q(s) = s^3 + s^2 + 2s + 24 = 0$$

associated with the closed-loop control system shown in Fig. 6.10. The corresponding Routh–Hurwitz array is shown in Fig. 6.11. The two sign changes in the first column indicate that there are two roots of the characteristic polynomial in the right half-plane; hence the closed-loop system is unstable. Using *Matlab*, we can verify the Routh–Hurwitz result by directly computing the roots of the characteristic equation, as shown in Fig. 6.12, using the roots function. Recall that the roots function computes the roots of a polynomial.

Whenever the characteristic equation is a function of a single parameter, the Routh–

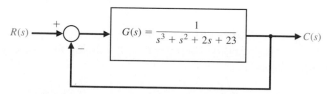

FIGURE 6.10
Closed-loop control system with
$T(s) = C(s)/R(s) = 1/(s^3 + s^2 + 2s + 24)$.

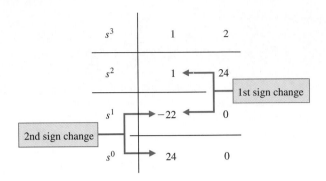

FIGURE 6.11
Routh–Hurwitz array for the closed-loop control system
with $T(s) = C(s)/R(s) = 1/(s^3 + s^2 + 2s + 24)$.

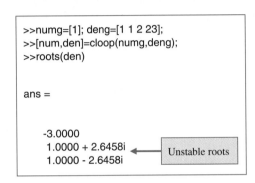

FIGURE 6.12
Using the **roots**
function to
compute the
closed-loop control
system poles of
the system shown
in Fig. 6.10.

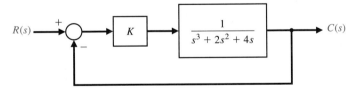

FIGURE 6.13
Closed-loop control system with
$T(s) = C(s)/R(s) = K/(s^3 + 2s^2 + 4s + K)$.

Hurwitz method can be utilized to determine the range of values that the parameter may take while maintaining stability. Consider the closed-loop feedback system in Fig. 6.13. The characteristic equation is

$$q(s) = s^3 + 2s^2 + 4s + K = 0.$$

Using a Routh–Hurwitz approach, we find that we require $0 \le K \le 8$ for stability (see Eq. 6.12). We can use *Matlab* to verify this result graphically. As shown in Fig. 6.14(a), we establish a vector of values for K at which we wish to compute the roots of the charac-

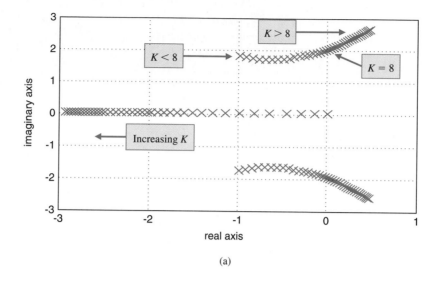

(a)

```
% This script computes the roots of the characteristic
% equation q(s) = s^3 + 2 s^2 + 4 s + K for 0<K<20
%
K=[0:0.5:20];
for i=1:length(K)
q=[1 2 4 K(i)];
p(:,i)=roots(q);
end
plot(real(p),imag(p),'x'), grid
xlabel('real axis'), ylabel('imaginary axis')
```

Loop for roots as
a function of *K*

(b)

FIGURE 6.14
(a) Plot of root locations of
$q(s) = s^3 + 2s^2 + 4s + K$ for $0 \le K \le 20$.
(b) *Matlab* script.

teristic equation. Then using the roots function, we calculate and plot the roots of the characteristic equation, as shown in Fig. 6.14(b). It can be seen that as K increases, the roots of the characteristic equation move toward the right half-plane as the gain tends toward $K = 8$, and eventually into the right half-plane when $K > 8$.

The script in Fig. 6.14 contains the for function. This function provides a mechanism for repeatedly executing a series of statements a given number of times. The for function connected to an end statement sets up a repeating calculation loop. Figure 6.15 describes the for function format and provides an illustrative example of its usefulness. The example sets up a loop that repeats ten times. During the ith iteration, where $1 \le i \le 10$, the ith element of the vector a is set equal to 20 and the scalar b is recomputed.

The Routh–Hurwitz method allows us to make definitive statements regarding absolute stability of a linear system. The method does not address the issue of *relative*

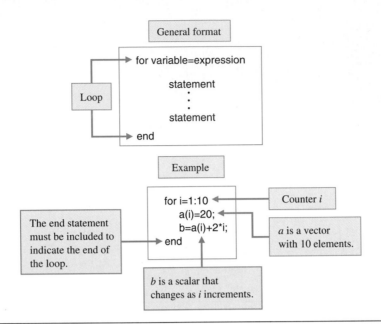

FIGURE 6.15
The **for** function
and an illustrative
example.

stability, which is directly related to the location of the roots of the characteristic equation. Routh–Hurwitz tells us how many poles lie in the right half-plane, but not the specific location of the poles. With *Matlab,* we can easily calculate the poles explicitly, thus allowing us to comment on the system relative stability.

EXAMPLE 6.13   Tracked vehicle control

The block diagram of the control system for the two-track vehicle is shown in Fig. 6.8 on page 293. The design objective is to find a and K such that the system is stable and the steady-state error for a ramp input is less than or equal to 24% of the command. We can use the Routh–Hurwitz method to aid in the search for appropriate values of a and K. The closed-loop characteristic equation is

$$q(s) = s^4 + 8s^3 + 17s^2 + (K + 10)s + aK = 0.$$

Using the Routh–Hurwitz array, we find that for stability we require

$$K < 126, \qquad aK > 0.$$

For positive K it follows that we can restrict our search to $0 < K < 126$ and $a > 0$. Our approach will be to use *Matlab* to find a parameterized a versus K region in which stability is assured. Then we can find a set of (a, K) belonging to the stable region such that the steady-state error specification is met. This procedure, shown in Fig. 6.16, involves selecting a range of values for a and K and computing the roots of the characteristic for specific values of a and K. For each value of K, we find the first value of a that results in at least one root of the characteristic equation in the right half-plane. The process is repeated until the entire selected range of a and K is exhausted. Then, the plot of the (a, K) pairs defines the separation between the stable and unstable regions.

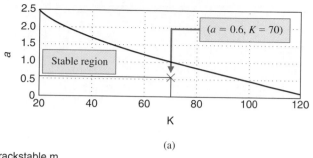

(a)

twotrackstable.m

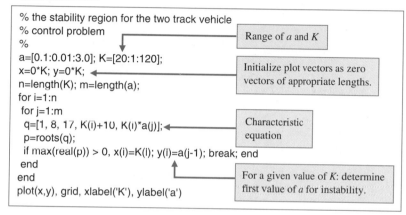

(b)

FIGURE 6.16
(a) Stability region
for *a* and *K* for two-
track vehicle
turning control.
(b) *Matlab* script.

The region to the left of the plot of *a* versus *K* in Fig. 6.16 is the stable region, since that corresponds to $K < 126$.

If we assume that $r(t) = At,\ t > 0$, then the steady-state error is

$$e_{ss} = \lim_{s \to 0} s \cdot \frac{s(s + 1)(s + 2)(s + 5)}{s(s + 1)(s + 2)(s + 5) + K(s + a)} \cdot \frac{A}{s^2} = \frac{10A}{aK},$$

where we have used the fact that

$$E(s) = \frac{1}{1 + G_c G(s)} R(s) = \frac{s(s + 1)(s + 2)(s + 5)}{s(s + 1)(s + 2)(s + 5) + K(s + a)} R(s).$$

Given the steady-state specification, $e_{ss} < 0.24A$, we find that the specification is satisfied when

$$\frac{10A}{aK} < 0.24A,$$

or

$$aK > 41.67. \tag{6.34}$$

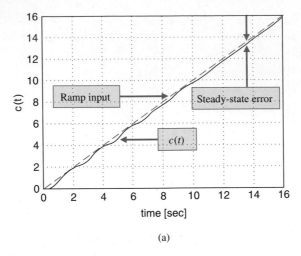

(a)

aKramp.m

```
% This script computes the ramp response
% for the two-track vehicle turning control
% problem with a=0.6 and K=70
%
t=[0:0.01:10]; u=t;          ← u = unit ramp input
numgc=[1 0.6]; dengc=[1 1];
numg=[70]; deng=[1 7 10 0];  ← a = 0.6 and K = 70
[numa,dena]=series(numgc,dengc,numg,deng);
[num,den]=cloop(numa,dena);
[y,x]=lsim(num,den,u,t);     ← Linear simulation
plot(t,y,t,u), grid
xlabel('time [sec]'), ylabel('c(t)')
```

FIGURE 6.17
(a) Ramp response
for $a = 0.6$ and
$K = 70$ for two-
track vehicle
turning control.
(b) *Matlab* script.

(b)

Any values of a and K that lie in the stable region in Fig. 6.16 and satisfy Eq. (6.34) will lead to an acceptable design. For example, $K = 70$ and $a = 0.6$ will satisfy all the design requirements. The closed-loop transfer function (with $a = 0.6$ and $K = 70$) is

$$T(s) = \frac{70s + 42}{s^4 + 8s^3 + 17s^2 + 80s + 42}.$$

The associated closed-loop poles are

$$s = -7.0767,$$
$$s = -0.5781,$$
$$s = -0.1726 + 3.1995i, \text{ and}$$
$$s = -0.1726 - 3.1995i.$$

The corresponding unit ramp input response is shown in Fig. 6.17. The steady-state error is less than 0.24, as desired. ∎

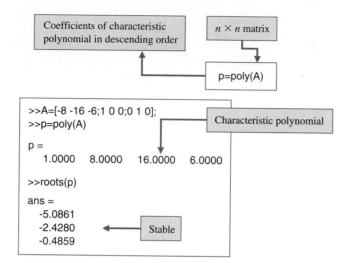

FIGURE 6.18
Computing the characteristic equation of **A** with the poly function.

The Stability of State Variable Systems. Now let us turn to determining the stability of systems described in state variable form. Suppose we have a system in state-space form as in Eq. (6.22). The stability of the system can be evaluated with the *characteristic equation* associated with the system matrix **A**. The characteristic equation is

$$\det(s\mathbf{I} - \mathbf{A}) = 0. \tag{6.35}$$

The characteristic equation is a polynomial in *s*. If all of the roots of the characteristic equation have negative real parts (i.e., $\text{Re}(s_i) < 0$), then the system is stable.

When the system model is given in the state-space form, we must calculate the characteristic polynomial associated with the **A** matrix. In this regard, we have several options. We can calculate the characteristic equation directly from Eq. (6.35) by manually computing the determinant of $(s\mathbf{I} - \mathbf{A})$. Then we can compute the roots using the roots function to check for stability, or alternatively, we can utilize the Routh–Hurwitz method to detect any unstable roots. Unfortunately, the manual computations can become lengthy, especially if the dimension of **A** is large. We would like to avoid this manual computation if possible. As it turns out, *Matlab* can assist in this endeavor.

The poly function described in Section 2.10 can be used to compute the characteristic equation associated with **A**. Recall that poly is used to form a polynomial from a vector of roots. It can also be used to compute the characteristic equation of **A**, as illustrated in Fig. 6.18, wherein input matrix **A** is

$$\mathbf{A} = \begin{bmatrix} -8 & -16 & -6 \\ 1 & 0 & 0 \\ 0 & 1 & 0 \end{bmatrix}$$

and the associated characteristic polynomial is

$$s^3 + 8s^2 + 16s + 6.$$

If **A** is an $n \times n$ matrix, **poly(A)** is the characteristic equation represented by the $n + 1$ element row vector whose elements are the coefficients of the characteristic equation.

EXAMPLE 6.14   **Stability region for an unstable plant**

A jump-jet aircraft has a control system as shown in Fig. 6.19 [17]. Assume that $z > 0$ and $p > 0$. The system is open-loop unstable (without feedback), since the characteristic equation of the plant and controller is

$$s(s - 1)(s + p) = s[s^2 + (p - 1)s - p] = 0.$$

Note that since one term within the bracket has a negative coefficient, the characteristic equation has at least one root in the right-hand s-plane. The characteristic equation of the closed-loop system is

$$s^3 + (p - 1)s^2 + (K - p)s + Kz = 0.$$

The goal is to determine the region of stability for K, p, and z. The Routh array is

s^3	1	$(K - p)$
s^2	$(p - 1)$	Kz
s	b_2	
s^0	Kz	

where

$$b_2 = \frac{(p - 1)(K - p) - Kz}{(p - 1)}.$$

From the Routh criterion, we find that we require $Kz > 0$ and $p > 1$. Setting $b_2 = 0$, we have

$$(p - 1)(K - p) - Kz = K[(p - 1) - z] - p(p - 1) = 0.$$

Therefore we require that

$$K > \frac{p(p - 1)}{(p - 1) - z}. \tag{6.36}$$

Consider three cases:

1. $z > p - 1$: Since $p > 1$, any $K > 0$ will satisfy the stability conditions.
2. $z = p - 1$: There is no $0 < K < \infty$ that leads to stability.
3. $z < p - 1$: Any $0 < K < \infty$ satisfying stability condition Eq. (6.36) for a given p and z will result in stability.

FIGURE 6.19
Control system for
jump-jet aircraft.
Assume that $z > 0$
and $p > 0$.

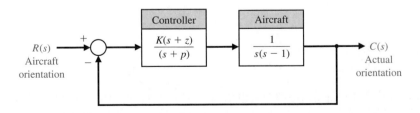

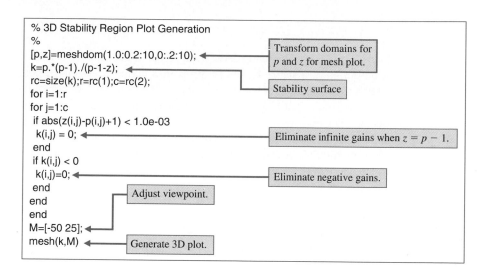

```
% 3D Stability Region Plot Generation
%
[p,z]=meshdom(1.0:0.2:10,0:.2:10);   ◄——  Transform domains for
k=p.*(p-1)./(p-1-z);                         p and z for mesh plot.
rc=size(k);r=rc(1);c=rc(2);          ◄——  Stability surface
for i=1:r
for j=1:c
 if abs(z(i,j)-p(i,j)+1) < 1.0e-03
  k(i,j) = 0;                        ◄——  Eliminate infinite gains when z = p − 1.
 end
 if k(i,j) < 0
  k(i,j)=0;                          ◄——  Eliminate negative gains.
 end
end
end                    Adjust viewpoint.
M=[-50 25];            ◄——
mesh(k,M)              ◄——  Generate 3D plot.
```

FIGURE 6.20
Matlab script for
stability region.

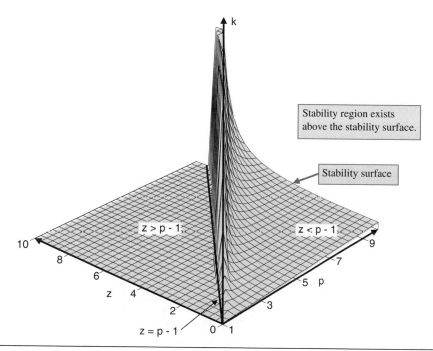

Stability region exists
above the stability surface.

Stability surface

z > p - 1

z < p - 1

z = p - 1

FIGURE 6.21
The three-
dimensional region
of stability lies
above the surface
shown.

The stability conditions can be depicted graphically. The *Matlab* script used to generate a three-dimensional stability surface is shown in Fig. 6.20. This script uses **mesh** to create the three-dimensional surface and **meshdom** to generate arrays for use with the mesh surface.

The three-dimensional plot of the stability region for K, p, and z is shown in Fig. 6.21. One acceptable stability point is $z = 1$, $p = 10$, and $K = 15$. ∎

6.8 SUMMARY

In this chapter, we have considered the concept of the stability of a feedback control system. A definition of a stable system in terms of a bounded system response was outlined and related to the location of the poles of the system transfer function in the s-plane.

The Routh–Hurwitz stability criterion was introduced, and several examples were considered. The relative stability of a feedback control system was also considered in terms of the location of the poles and zeros of the system transfer function in the s-plane. The determination of the roots of the characteristic equation was considered, and the Newton–Raphson method was illustrated. The stability of state variable systems was considered.

EXERCISES

E6.1   A system has a characteristic equation $s^3 + 3Ks^2 + (2 + K)s + 4 = 0$. Determine the range of K for a stable system.

   Answer: $K > 0.53$

E6.2   A system has a characteristic equation $s^3 + 9s^2 + 26s + 24 = 0$. (a) Using the Routh criterion, show that the system is stable. (b) Using the Newton–Raphson method, find the three roots.

E6.3   Find the roots of the characteristic equation $s^4 + 9.5s^3 + 30.5s^2 + 37s + 12 = 0$.

E6.4   A control system has the structure shown in Fig. E6.4. Determine the gain at which the system will become unstable.

E6.5   A feedback system has a loop transfer function

$$GH(s) = \frac{K}{(s + 1)(s + 3)(s + 6)},$$

where $K = 10$. Find the roots of this system's characteristic equation.

E6.6   For the feedback system of Exercise E6.5, find the value of K when two roots lie on the imaginary axis. Determine the value of the three roots.

   Answer: $s = -10, \pm j5.2$

E6.7   A negative feedback system has a loop transfer function

$$GH(s) = \frac{K(s + 2)}{s(s - 1)}.$$

(a) Find the value of the gain when the ζ of the closed-loop roots is equal to 0.707. (b) Find the value of the gain when the closed-loop system has two roots on the imaginary axis.

E6.8   Designers have developed small, fast, vertical-takeoff fighter aircraft that are invisible to radar (Stealth aircraft). This aircraft concept uses quickly turning jet nozzles to steer the airplane [22]. The control system for the heading or direction control is shown in Fig. E6.8. Determine the maximum gain of the system for stable operation.

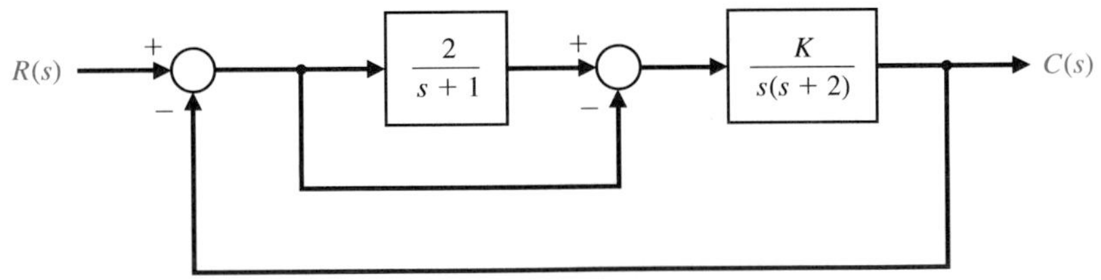

FIGURE E6.4
Feedforward system.

FIGURE E6.8
Aircraft heading control.

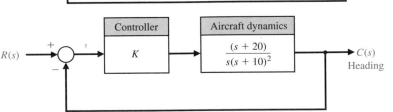

E6.9 A system has a characteristic equation

$$s^3 + 3s^2 + (K + 1)s + 4 = 0.$$

Find the range of K for a stable system.

Answer: $K > \frac{1}{3}$

E6.10 We all use our eyes and ears to achieve balance. Our orientation system allows us to sit or stand in a desired position even while in motion. This orientation system is primarily run by the information received in the inner ear, where the semicircular canals sense angular acceleration and the otoliths measure linear acceleration. But these acceleration measurements need to be supplemented by visual signals. Try the following experiment: (a) Stand with one foot in front of another and with your hands resting on your hips and your elbows bowed outward. (b) Close your eyes. Did you find that you experienced a low-frequency oscillation that grew until you lost balance? Is this orientation position stable with and without the use of your eyes?

E6.11 A system with a transfer function $C(s)/R(s)$ is

$$\frac{C(s)}{R(s)} = \frac{23(s + 1)}{s^4 + 6s^3 + 2s^2 + s + 3}.$$

Determine the steady-state error to a unit step input. Is the system stable?

E6.12 By using magnetic bearings, a rotor is supported contactless. The technique of contactless support for rotors becomes more important in light and heavy industrial applications [14]. The matrix differential equation for a magnetic bearing system is

$$\dot{\mathbf{x}} = \begin{bmatrix} 0 & 1 & 0 \\ -3 & -1 & 0 \\ -2 & -1 & -2 \end{bmatrix} \mathbf{x},$$

where $\mathbf{x}^T = [y, dy/dt, i]$, $y = $ bearing gap, and i is the electromagnetic current. Determine whether the system is stable.

E6.13 A system has a characteristic equation

$$q(s) = s^6 + 9s^5 + 31.25s^4 + 61.25s^3 + 67.75s^2 + 14.75s + 15.$$

(a) Determine whether the system is stable, using the Routh–Hurwitz criterion. (b) Determine the roots of the characteristic equation.

E6.14 A system has a characteristic equation

$$q(s) = s^4 + 9s^3 + 45s^2 + 87s + 50.$$

(a) Determine whether the system is stable, using the Routh–Hurwitz criterion. (b) Determine the roots of the characteristic equation.

E6.15 The matrix differential equation of a state variable model of a system has

$$\mathbf{A} = \begin{bmatrix} 0 & 1 & -1 \\ -6 & -11 & 6 \\ -6 & -11 & 5 \end{bmatrix}.$$

(a) Determine the characteristic equation. (b) Determine whether the system is stable. (c) Determine the roots of the characteristic equation.

E6.16 A system has a characteristic equation

$$q(s) = s^3 + 10s^2 + 16s + 160.$$

(a) Determine whether the system is stable, using the Routh–Hurwitz criterion. (b) Determine the roots of the characteristic equation.

E6.17 Determine whether the systems with the following characteristic equations are stable or unstable:

(a) $s^3 - 4s^2 + 6s + 100$
(b) $s^4 - 6s^3 - s^2 - 17s - 6$
(c) $s^2 + 6s + 3$

E6.18 Using the Newton–Raphson method, find the roots of the following polynomials:

(a) $s^3 + 10.4s^2 + 36.2s + 40.8$
(b) $s^3 + 9s^2 + 27s + 27$

E6.19 A system has the characteristic equation

$$q(s) = s^3 + 10s^2 + 29s + K.$$

Shift the vertical axis to the right by 2 by using $s = s_n - 2$, and determine the value of gain K so that the complex roots are $s = -2 \pm j$.

E6.20 A system has a transfer function $C(s)/R(s) = T(s) = 1/s$. (a) Is this system stable? (b) If $r(t)$ is a unit step input, determine the response $c(t)$.

E6.21 A system is represented by Eq. (6.22) where

$$\mathbf{A} = \begin{bmatrix} 0 & 1 & 0 \\ 0 & 0 & 1 \\ -1 & -k & -2 \end{bmatrix}.$$

Find the range of k where the system is stable.

PROBLEMS

P6.1 Utilizing the Routh–Hurwitz criterion, determine the stability of the following polynomials:

(a) $s^2 + 4s + 1$
(b) $s^3 + 4s^2 + 6s + 6$
(c) $s^3 + 3s^2 - 6s + 10$
(d) $s^4 + s^3 + 2s^2 + 5s + 8$
(e) $s^4 + s^3 + 3s^2 + 2s + K$
(f) $s^5 + s^4 + 2s^3 + s + 5$
(g) $s^5 + s^4 + 2s^3 + s^2 + s + K$

Determine the number of roots, if any, in the right-hand plane. Also, when it is adjustable, determine the range of K that results in a stable system.

P6.2 An antenna control system was analyzed in Problem 4.5, and it was determined that to reduce the effect of wind disturbances, the gain of the magnetic amplifier k_a should be as large as possible. (a) Determine the limiting value of gain for maintaining a stable system. (b) It is desired to have a system settling time equal to 1.5 seconds. Using a shifted axis and the Routh–Hurwitz criterion, determine the value of gain that satisfies this requirement. Assume that the complex roots of the closed-loop system dominate the transient response. (Is this a valid approximation in this case?)

P6.3 Arc welding is one of the most important areas of application for industrial robots [11]. In most manufacturing welding situations, uncertainties in dimensions of the part, geometry of the joint, and the welding process itself require the use of sensors for maintaining weld quality. Several systems use a vision system to measure the geometry of the puddle of melted metal as shown in Fig. P6.3. This system uses a constant rate of feeding the wire to be melted. (a) Calculate the maximum value for K for the system that will result in a stable system. (b) For ½ of the maximum value of K found in

part (a), determine the roots of the characteristic equation. (c) Estimate the overshoot of the system of part (b) when it is subjected to a step input.

P6.4 A feedback control system is shown in Fig. P6.4. The process transfer function is

$$G(s) = \frac{K(s + 40)}{s(s + 10)},$$

and the feedback transfer function is $H(s) = 1/(s + 20)$. (a) Determine the limiting value of gain K for a stable system. (b) For the gain that results in marginal stability, determine the magnitude of the imaginary roots. (c) Reduce the gain to ½ the magnitude of the marginal value and determine the relative stability of the system (1) by shifting the axis and using the Routh–Hurwitz criterion and (2) by determining the root locations. Show the roots are between -1 and -2.

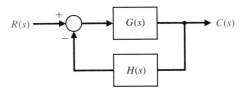

FIGURE P6.4   Feedback system.

P6.5 Determine the relative stability of the systems with the following characteristic equations (1) by shifting the axis in the s-plane and using the Routh–Hurwitz criterion and (2) by determining the location of the complex roots in the s-plane:

(a) $s^3 + 3s^2 + 4s + 2 = 0$
(b) $s^4 + 9s^3 + 30s^2 + 42s + 20 = 0$
(c) $s^3 + 19s^2 + 110s + 200 = 0$

P6.6 A unity-feedback control system is shown in Fig. P6.6. Determine the relative stability of the

FIGURE P6.3
Welder control.

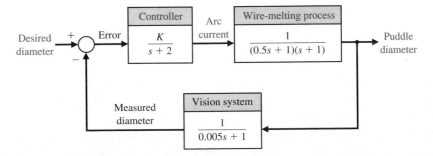

system with the following transfer functions by lo-cating the complex roots in the s-plane:

(a) $G(s) = \dfrac{65 + 33s}{s^2(s + 9)}$

(b) $G(s) = \dfrac{24}{s(s^3 + 10s^2 + 35s + 50)}$

(c) $G(s) = \dfrac{3(s + 4)(s + 8)}{s(s + 5)^2}$

FIGURE P6.6 Unity feedback system.

P6.7 The linear model of a phase detector (phase-lock loop) can be represented by Fig. P6.7 [9]. The phase-lock systems are designed to maintain zero difference in phase between the input carrier signal and a local voltage-controlled oscillator. Phase-lock loops find application in color television, missile tracking, and space telemetry. The filter for a particular application is chosen as

$$F(s) = \frac{10(s + 10)}{(s + 1)(s + 100)}.$$

It is desired to minimize the steady-state error of the system for a ramp change in the phase information signal. (a) Determine the limiting value of the gain $K_a K = K_v$ in order to maintain a stable system. (b) It is decided that a steady-state error

equal to $1°$ is acceptable for a ramp signal of 100 rad/s. For that value of gain K_v, determine the location of the roots of the system.

P6.8 A very interesting and useful velocity control system has been designed for a wheelchair control system. It is desirable to enable people paralyzed from the neck down to drive themselves about in motorized wheelchairs. A proposed system utilizing velocity sensors mounted in a headgear is shown in Fig. P6.8. The headgear sensor provides an output proportional to the magnitude of the head movement. There is a sensor mounted at $90°$ intervals so that forward, left, right, or reverse can be commanded. Typical values for the time constants are $\tau_1 = 0.5$ s, $\tau_3 = 1$ s, and $\tau_4 = \frac{1}{4}$ s. (a) Determine the limiting gain $K = K_1 K_2 K_3$ for a stable system. (b) When the gain K is set equal to $\frac{1}{3}$ of the limiting value, determine if the settling time of the system is less than 4 s. (c) Determine the value of gain that results in a system with a settling time of 4 s. Also, obtain the value of the roots of the characteristic equation when the settling time is equal to 4 s.

P6.9 A cassette tape storage device has been designed for mass-storage [1]. It is necessary to control accurately the velocity of the tape. The speed control of the tape drive is represented by the system shown in Fig. P6.9. (a) Determine the limiting gain for a stable system. (b) Determine a suitable gain so that the overshoot to a step command is approximately 5%.

FIGURE P6.7
Phase-lock loop
system.

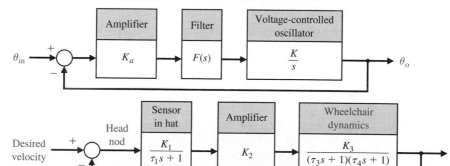

FIGURE P6.8
Wheelchair control
system.

FIGURE P6.9
Tape drive control.

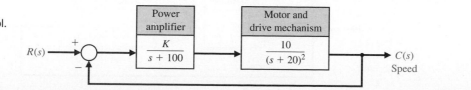

P6.10   Robots can be used in manufacturing and assembly operations that require accurate, fast, and versatile manipulation [10, 11]. The open-loop transfer function of a direct-drive arm may be approximated by

$$GH(s) = \frac{K(s + 2)}{s(s + 5)(s^2 + 2s + 5)}.$$

(a) Determine the value of gain K when the system oscillates. (b) Calculate the roots of the closed-loop system for the K determined in part (a).

P6.11   A feedback control system has a characteristic equation

$$s^3 + (5 + K)s^2 + 7s + 18 + 9K = 0.$$

The parameter K must be positive. What is the maximum value K can assume before the system becomes unstable? When K is equal to the maximum value, the system oscillates. Determine the frequency of oscillation.

P6.12   A feedback control system has a characteristic equation

$$s^6 + 2s^5 + 5s^4 + 8s^3 + 8s^2 + 8s + 4 = 0.$$

Determine whether the system is stable, and determine the values of the roots.

P6.13   The stability of a motorcycle and rider is an important area for study because many motorcycle designs result in vehicles that are difficult to control [12, 13]. The handling characteristics of a motorcycle must include a model of the rider as well as one of the vehicle. The dynamics of one motorcycle and rider can be represented by an open-loop transfer function (Fig. P6.4)

$$GH(s) = \frac{K(s^2 + 30s + 1125)}{s(s + 20)(s^2 + 10s + 125)(s^2 + 60s + 3400)}.$$

(a) As an approximation, calculate the acceptable range of K for a stable system when the numerator polynomial (zeros) and the denominator poly-

nomial $(s^2 + 60s + 3400)$ are neglected. (b) Calculate the actual range of acceptable K accounting for all zeros and poles.

P6.14   A system has a transfer function

$$T(s) = \frac{1}{s^3 + 1.3s^2 + 2.0s + 1}.$$

(a) Determine whether the system is stable. (b) Determine the roots of the characteristic equation. (c) Plot the response of the system to a unit step input.

P6.15   On July 16, 1993, the elevator in Yokohama's 70-story Landmark Tower, operating at a peak speed of 45 km/hr (28 mph), was inaugurated as the fastest super-fast elevator. To reach such a speed without leaving passengers' stomachs on the ground floor, the lift accelerates for longer periods, rather than more precipitously. Going up, it reaches full speed only at the 27th floor; it begins decelerating 15 floors later. The result is a peak acceleration similar to that of other skyscraper elevators—a bit less than a tenth of the force of gravity.

Admirable ingenuity has gone into making this safe and comfortable. Special ceramic brakes had to be developed; iron ones would melt. Computer-controlled systems damp out vibrations. The lift has been streamlined to reduce the wind noise as it hurtles up and down [20]. One proposed control system for the elevator's vertical position is shown in Fig. P6.15. Determine the range of K for a stable system.

P6.16   Consider the case of rabbits and foxes in Australia. The number of rabbits is x_1 and if left alone would grow indefinitely (until the food supply was exhausted) so that

$$\dot{x}_1 = kx_1.$$

However, with foxes present on the continent, we have

$$\dot{x}_1 = kx_1 - ax_2,$$

FIGURE P6.15
Elevator control
system.

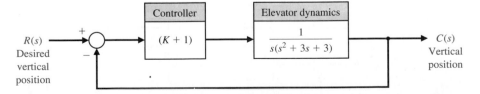

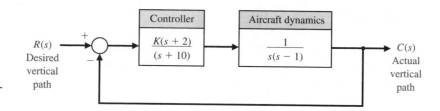

FIGURE P6.17
Control of a jump-
jet aircraft.

where x_2 is the number of foxes. Now, if the foxes must have rabbits to exist, we have

$$\dot{x}_2 = -hx_2 + bx_1.$$

Determine whether this system is stable and thus decays to the condition $x_1(t) = x_2(t) = 0$ at $t = \infty$. What are the requirements on a, b, h, and k for a stable system? What is the result when k is greater than h?

P6.17 The goal of vertical takeoff and landing (VTOL) aircraft is to achieve operation from relatively small airports and yet operate as a normal aircraft in level flight [17]. An aircraft taking off in a form similar to a missile (on end) is inherently

unstable (see Example 3.4 for a discussion of the inverted pendulum). A control system using adjustable jets can control the vehicle as shown in Fig. P6.17. (a) Determine the range of gain for which the system is stable. (b) Determine the gain K for which the system is marginally stable and the roots of the characteristic equation for this value of K.

P6.18 A vertical-liftoff vehicle is shown in Fig. P6.18(a). The four engines swivel for liftoff. The control system for aircraft altitude is shown in Fig. P6.18(b). (a) For $K = 1$, determine whether the system is stable. (b) Determine a range of stability, if any, for $K > 0$.

FIGURE P6.18
(a) Vertical-takeoff
aircraft (courtesy
of Moller
International).
(b) Control system.

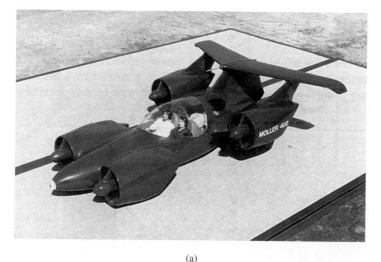

(a)

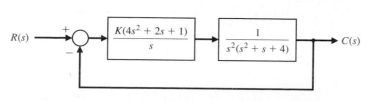

(b)

ADVANCED PROBLEMS

AP6.1 A teleoperated control system incorporates both a person (operator) and a remote machine. The normal teleoperation system is based on a one-way link to the machine and limited feedback to the operator. Two-way coupling using bilateral information exchange enables better operation [19]. In the case of remote control of a robot, force feedback plus position feedback is useful. The characteristic equation for a teleoperated system as shown in Fig. AP6.1 is

$$s^4 + 10s^3 + K_1 s^2 + 2s + K_2 = 0,$$

where K_1 and K_2 are feedback gain factors. Determine and plot the region of stability for this system for K_1 and K_2.

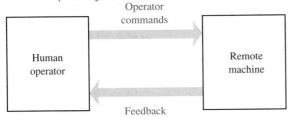

FIGURE AP6.1   Model of a teleoperated machine.

AP6.2 Consider the case of a navy pilot landing an aircraft on an aircraft carrier. The pilot has three basic tasks. First, the pilot must guide the aircraft's

approach to the ship along the extended centerline of the runway. The second task is maintaining the aircraft on the correct glideslope, the third task that of maintaining the correct speed. A model of a lateral position control system is shown in Fig. AP6.2. Determine the range of stability for $K \geq 0$.

AP6.3 A control system is shown in Fig. AP6.3. It is desired that the system be stable and the steady-state error for a unit step input be less than or equal to 0.05 (5%). (a) Determine the range of α that satisfies the error requirement. (b) Determine the range of α that satisfies the stability required. (c) Select an α that meets both requirements.

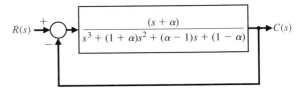

FIGURE AP6.3

AP6.4 A bottle-filling line uses a feeder screw mechanism as shown in Fig. AP6.4. The tachometer feedback is used to maintain accurate speed control. Determine and plot the range of K and p that permits stable operation.

FIGURE AP6.2
Lateral position
control for landing
on an aircraft
carrier.

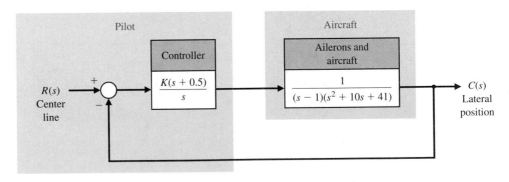

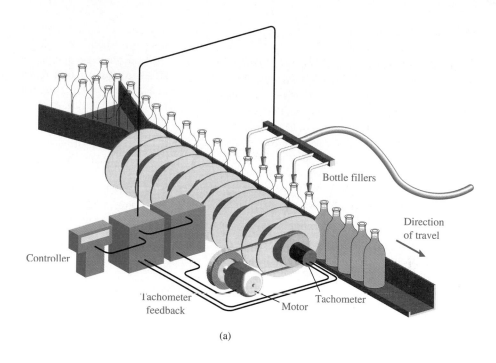

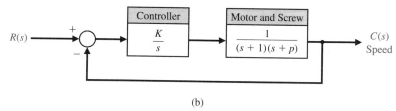

FIGURE AP6.4
Speed control of a
bottle-filling line:
(a) system layout,
(b) block diagram.

(b)

DESIGN PROBLEMS

DP6.1 The control of the spark ignition of an automotive engine requires constant performance over
a wide range of parameters [15]. The control system is shown in Fig. DP6.1, with a controller gain
K to be selected. The parameter p is equal to 2 for
many autos but can equal zero for those with high

FIGURE DP6.1
Automobile engine
control.

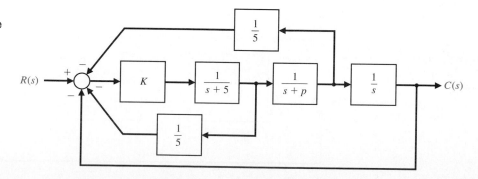

performance. Select a gain K that will result in a stable system for both values of p.

DP6.2 An automatically guided vehicle on Mars is represented by the system in Fig. DP6.2. The system has a steerable wheel in both the front and back of the vehicle, and the design requires that $H(s) = Ks + 1$. Determine (a) the value of K required for stability, (b) the value of K when one root of the characteristic equation is equal to $s = -\frac{1}{2}$, and (c) the value of the two remaining roots for the gain selected in part (b). (d) Find the response of the system to a step command for the gain selected in part (b).

DP6.3 A unity negative feedback system with

$$G(s) = \frac{K(s + 2)}{s(1 + \tau s)(1 + 2s)}$$

has two parameters to be selected. (a) Determine and plot the regions of stability for this system. (b) Select τ and K so that the steady-state error to a ramp input is less than or equal to 25% of the input magnitude. (c) Determine the percent overshoot for a step input for the design selected in part (b).

DP6.4 The attitude control system of a space shuttle rocket is shown in Fig. DP6.4 [18]. (a) Determine the range of gain K and parameter m so that the system is stable, and plot the region of stability. (b) Select the gain and parameter values so that the steady-state error to a ramp input is less than or equal to 10% of the input magnitude. (c) Determine the percent overshoot for a step input for the design selected in part (b).

DP6.5 A traffic control system is designed to control the distance between vehicles as shown in Fig. DP6.5 [15]. (a) Determine the range of gain K for which the system is stable. (b) If K_m is the maximum value of K so that the characteristic roots are on the $j\omega$-axis, then let $K = K_m/N$ where $6 < N < 7$. We desire that the peak time be less than 2 seconds and the percent overshoot be less than 18%. Determine an appropriate value for N.

DP6.6 Consider the potential for a robot steering a motorcycle as shown in Fig. DP6.6(a) [22]. The block diagram of the system model is shown in Fig. DP6.6(b). Determine the range of K for stable operation of the cycle when $\alpha_1 = g/h = 9$, $\alpha_2 = V^2/hc = 2.7$, and $\alpha_3 = V/hc = 1.35$. We assume the motorcycle is moving with a constant velocity $V = 2$ m/s. The time constant of the controller is $\tau = 0.2$ s.

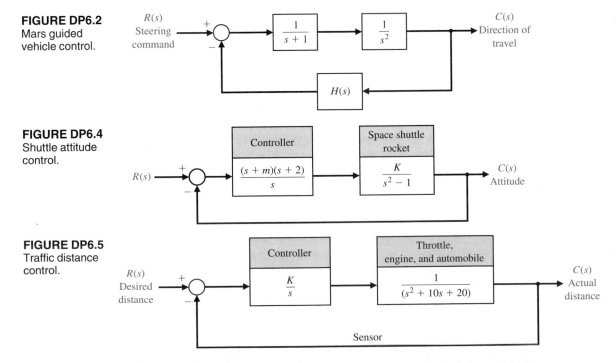

FIGURE DP6.2
Mars guided vehicle control.

FIGURE DP6.4
Shuttle attitude control.

FIGURE DP6.5
Traffic distance control.

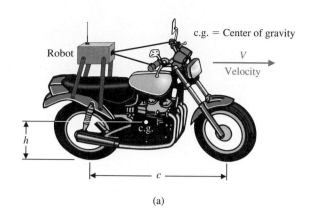

c.g. = Center of gravity

Robot

V
Velocity

c.g.

h

c

(a)

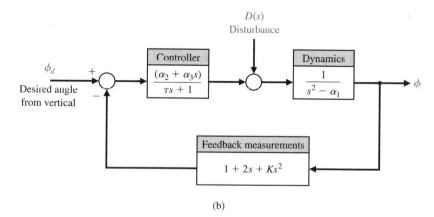

$D(s)$
Disturbance

ϕ_d
Desired angle
from vertical

Controller
$$\frac{(\alpha_2 + \alpha_3 s)}{\tau s + 1}$$

Dynamics
$$\frac{1}{s^2 - \alpha_1}$$

ϕ

Feedback measurements
$$1 + 2s + Ks^2$$

(b)

FIGURE DP6.6
(a) Robot
controlled
motorcycle.
(b) System block
diagram.

MATLAB PROBLEMS

MP6.1 A unity negative feedback system has the open-loop transfer function

$$G(s) = \frac{s + 1}{s^3 + 4s^2 + 6s + 10}.$$

Using *Matlab,* determine the closed-loop transfer function and show that the roots of the characteristic equation are $s_1 = -2.89$, $s_{2,3} = -0.55 \pm 1.87i$.

MP6.2 Consider the transfer function

$$T(s) = \frac{1}{s^5 + 2s^4 + 2s^3 + 4s^2 + s + 2}.$$

(a) Using Routh–Hurwitz methods, determine whether the system is stable. If not, how many poles are in the right half-plane? (b) Using *Matlab,* compute the poles of $T(s)$ and verify the result in part (a). (c) Plot the unit step response, and discuss the results.

MP6.3 A "paper-pilot" model is sometimes utilized in aircraft control design and analysis to represent the pilot in the loop. A block diagram of an aircraft with a pilot "in-the-loop" is shown in Fig. MP6.3. The variable τ represents the pilot's time delay. We can represent a slower pilot with $\tau = 0.5$ and a faster pilot with $\tau = 0.25$. The remaining variables in the pilot model are assumed to be $K = 1$, $\tau_1 = 2$, and $\tau_2 = 0.5$. Using *Matlab,* compute the closed-loop system poles for the fast and slow pilots. Comment on the results. What is the maximum pilot time delay allowable for stability?

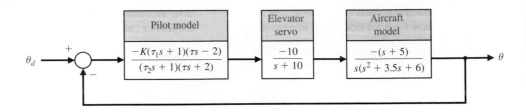

FIGURE MP6.3
An aircraft with a
pilot in the loop.

MP6.4   Consider the feedback control system in Fig. MP6.4. Using the for function, develop a *Matlab* script to compute the closed-loop transfer function poles for $0 \le K \le 5$ and plot the results denoting the poles with the "$\times$" symbol. Determine the maximum range of K for stability with the Routh–Hurwitz method. Compute the roots of the characteristic equation when K is the minimum value allowed for stability.

MP6.5   Consider the system in state variable form

$$\dot{x} = \begin{bmatrix} 0 & 1 & 0 \\ 0 & 0 & 1 \\ -10 & -15 & -10 \end{bmatrix} x + \begin{bmatrix} 0 \\ 0 \\ 10 \end{bmatrix} u,$$

$$c = \begin{bmatrix} 1 & 1 & 0 \end{bmatrix} x.$$

(a) Compute the characteristic equation using the poly function. (b) Compute the roots of the characteristic equation, and determine whether the system is stable. (c) Obtain the response plot of $c(t)$ when $u(t)$ is a unit step and when the system has zero initial conditions.

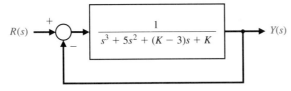

FIGURE MP6.4   A single-loop feedback control
system with parameter K.

TERMS AND CONCEPTS

Auxiliary polynomial   The equation that immediately precedes the zero entry in the Routh array.

Newton–Raphson method   An iterative approach to solving for roots of a polynomial equation.

Relative stability   The property that is measured by the relative real part of each root or pair of roots of the characteristic equation.

Routh–Hurwitz criterion   A criterion for determining the stability of a system by examining the characteristic equation of the transfer function. The criterion states that the number of roots of the char-

acteristic equation with positive real parts is equal to the number of changes of sign of the coefficients in the first column of the Routh array.

Stability   A performance measure of a system. A system is stable if all the poles of the transfer function have negative real parts.

Stable system   A dynamic system with a bounded system response to a bounded input.

Synthetic division   A method of determining the roots of the characteristic equation based on the remainder theorem of mathematics.

The Root Locus Method

PREVIEW

Since the performance of a feedback system can be adjusted by changing one or more parameters, we described that performance in terms of the s-plane location of the roots of the characteristic equation in the preceding chapters. Thus it is very useful to determine how the roots of the characteristic equations move around the s-plane as we change a parameter.

The locus of roots in the s-plane can be determined by a graphical method. Once you understand this graphical method, then we will proceed to demonstrate the utility of computer generation of the locus of roots as a parameter is varied.

Since we can determine how the roots migrate as one parameter varies, it is possible to show they will vary as two parameters vary. This provides us with the opportunity to design a system with two adjustable parameters so as to achieve a very desirable performance.

Using the concept of the root locus as a parameter varies, we also will be able to define a measure of the sensitivity of a specified root to a small incremental change in the parameter.

7.1 INTRODUCTION

The relative stability and the transient performance of a closed-loop control system are directly related to the location of the closed-loop roots of the characteristic equation in the s-plane. Also, it is frequently necessary to adjust one or more system parameters in order to obtain suitable root locations. Therefore it is worthwhile to determine how the roots of the characteristic equation of a given system migrate about the s-plane as the parameters are varied; that is, it is useful to determine the *locus* of roots in the s-plane as a parameter is varied. The *root locus method* was introduced by Evans in 1948 and has been developed and utilized extensively in control engineering practice [1–3]. The root locus technique is a graphical method for drawing the locus of roots in the s-plane as a parameter is varied. In fact, the root locus method provides the engineer with a measure of the sensitivity of the roots of the system to a variation in the parameter being considered. The root locus technique may be used to great advantage in conjunction with the Routh–Hurwitz criterion and the Newton–Raphson method.

The root locus method provides graphical information, and therefore an approximate sketch can be used to obtain qualitative information concerning the stability and performance of the system. Furthermore, the locus of roots of the characteristic equation of a multiloop system may be investigated as readily as for a single-loop system. If the root locations are not satisfactory, the necessary parameter adjustments often can be readily ascertained from the root locus [4].

7.2 THE ROOT LOCUS CONCEPT

The dynamic performance of a closed-loop control system is described by the closed-loop transfer function

$$T(s) = \frac{C(s)}{R(s)} = \frac{p(s)}{q(s)}, \qquad (7.1)$$

where $p(s)$ and $q(s)$ are polynomials in s. The roots of the characteristic equation $q(s)$ determine the modes of response of the system. In the case of the simple single-loop system, as shown in Fig. 7.1, we have the characteristic equation

$$1 + KG(s) = 0, \qquad (7.2)$$

where K is a variable parameter. The characteristic roots of the system must satisfy Eq. (7.2), where the roots lie in the s-plane. Because s is a complex variable, Eq. (7.2) may be rewritten in polar form as

$$|KG(s)|\ \underline{/KG(s)} = -1 + j0, \qquad (7.3)$$

FIGURE 7.1
Closed-loop control system with a variable parameter K.

and therefore it is necessary that

$$|KG(s)| = 1$$

and

$$\underline{/KG(s)} = 180° \pm k360°, \tag{7.4}$$

where $k = 0, \pm 1, \pm 2, \pm 3, \ldots$. The graphical computation required for Eq. (7.4) is readily accomplished by using a protractor for estimating angles.

The root locus is the path of the roots of the characteristic equation traced out in the s-plane as a system parameter is changed.

The simple second-order system considered in the previous chapters is shown in Fig. 7.2. The characteristic equation representing this system is

$$\Delta(s) = 1 + KG(s) = 1 + \frac{K}{s(s + 2)} = 0$$

or, alternatively,

$$q(s) = s^2 + 2s + K = s^2 + 2\zeta\omega_n s + \omega_n^2 = 0. \tag{7.5}$$

The locus of the roots as the gain K is varied is found by requiring that

$$|G(s)| = \left| \frac{K}{s(s + 2)} \right| = 1 \tag{7.6}$$

and

$$\underline{/G(s)} = \pm 180°, \pm 540°, \ldots \tag{7.7}$$

The gain K may be varied from zero to an infinitely large positive value. For a second-order system, the roots are

$$s_1, s_2 = -\zeta\omega_n \pm \omega_n\sqrt{\zeta^2 - 1}, \tag{7.8}$$

and for $\zeta < 1$, we know that $\theta = \cos^{-1}\zeta$. Graphically, for two open-loop poles as shown in Fig. 7.3, the locus of roots is a vertical line for $\zeta \leq 1$ in order to satisfy the angle requirement, Eq. (7.7). For example, as shown in Fig. 7.4, at a root s_1, the angles are

$$\underline{\Big/ \frac{K}{s(s + 2)}}\Bigg|_{s=s_1} = -\underline{/s_1} - \underline{/(s_1 + 2)} = -[(180° - \theta) + \theta] = -180°. \tag{7.9}$$

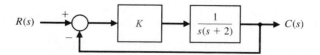

FIGURE 7.2
Unity feedback control system. The gain K is a variable parameter.

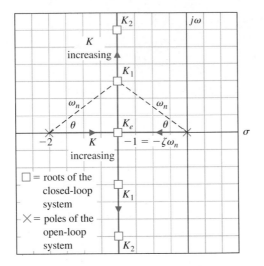

FIGURE 7.3
Root locus for a second-order system when K_e $< K_1 < K_2$. The locus is shown in color.

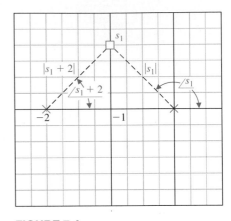

FIGURE 7.4
Evaluation of the angle and gain at s_1 for gain $K = K_1$.

This angle requirement is satisfied at any point on the vertical line that is a perpendicular bisector of the line 0 to -2. Furthermore, the gain K at the particular point s_1 is found by using Eq. (7.6) as

$$\left|\frac{K}{s(s+2)}\right|_{s=s_1} = \frac{K}{|s_1|\,|s_1+2|} = 1 \tag{7.10}$$

and thus

$$K = |s_1|\,|s_1+2|, \tag{7.11}$$

where $|s_1|$ is the magnitude of the vector from the origin to s_1, and $|s_1 + 2|$ is the magnitude of the vector from -2 to s_1.

For a multiloop closed-loop system, we found in Section 2.7 that by using Mason's signal-flow gain formula, we had

$$\Delta(s) = 1 - \sum_{n=1}^{N} L_n + \sum_{m,q}^{M,N} L_m L_q - \sum L_r L_s L_t + \cdots, \tag{7.12}$$

where L_q equals the value of the qth self-loop transmittance. Hence we have a characteristic equation, which may be written as

$$q(s) = \Delta(s) = 1 + F(s). \tag{7.13}$$

To find the roots of the characteristic equation, we set Eq. (7.13) equal to zero and obtain

$$1 + F(s) = 0. \tag{7.14}$$

Of course, Eq. (7.14) may be rewritten as

$$F(s) = -1 + j0, \tag{7.15}$$

and the roots of the characteristic equation must also satisfy this relation.

In general, the function $F(s)$ may be written as

$$F(s) = \frac{K(s + z_1)(s + z_2)(s + z_3) \cdots (s + z_m)}{(s + p_1)(s + p_2)(s + p_3) \cdots (s + p_n)}.$$

Then the magnitude and angle requirement for the root locus are

$$|F(s)| = \frac{K|s + z_1| \, |s + z_2| \cdots}{|s + p_1| \, |s + p_2| \cdots} = 1 \qquad (7.16)$$

and

$$\underline{/F(s)} = \underline{/s + z_1} + \underline{/s + z_2} + \cdots \qquad (7.17)$$

$$- (\underline{/s + p_1} + \underline{/s + p_2} + \cdots) = 180° \pm q360°$$

where q is an integer. The magnitude requirement, Eq. (7.16), enables one to determine the value of K for a given root location s_1. A test point in the s-plane, s_1, is verified as a root location when Eq. (7.17) is satisfied. The angles are all measured in a counterclockwise direction from a horizontal line.

To further illustrate the root locus procedure, let us reconsider the second-order system of Fig. 7.5(a). The effect of varying the parameter, a, can be effectively portrayed by rewriting the characteristic equation for the root locus form with a as the multiplying factor in the numerator. Then the characteristic equation is

$$1 + KG(s) = 1 + \frac{K}{s(s + a)} = 0$$

or, alternatively,

$$s^2 + as + K = 0.$$

Dividing by the factor $(s^2 + K)$, we obtain

$$1 + \frac{as}{s^2 + K} = 0. \qquad (7.18)$$

FIGURE 7.5
(a) Single-loop system. (b) Root locus as a function of the parameter a.

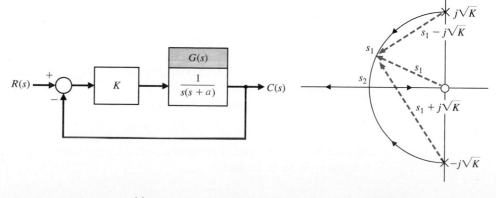

(a)   (b)

Then the magnitude criterion is satisfied when

$$\frac{a|s_1|}{|s_1^2 + K|} = 1 \qquad (7.19)$$

at the root s_1. The angle criterion is

$$\underline{/s_1} - (\underline{/s_1 + j\sqrt{K}} + \underline{/s_1 - j\sqrt{K}}) = \pm 180°, \ \pm 540°, \ \ldots.$$

To construct the root locus, we find the points in the s-plane that satisfy the angle criterion. These points are located on a trial-and-error basis by searching in an orderly manner for a point with a total angle of $\pm 180°$, $\pm 540°$, or in general,

$$\frac{\pm (2q + 1)180°}{n_p - n_z}$$

where n_p = number of poles, n_z = number of zeros, and q is an integer. The algebraic sum of the angles from the poles and zeros is measured with a protractor, and we find the locus of roots as shown in Fig. 7.5(b). Specifically, at the root s_1, the magnitude of the parameter, a, is found from Eq. (7.19) as

$$a = \frac{|s_1 - j\sqrt{K}| \ |s_1 + j\sqrt{K}|}{|s_1|}. \qquad (7.20)$$

The roots of the system merge on the real axis at the point s_2 and provide a critically damped response to a step input. The parameter, a, has a magnitude at the critically damped roots $s_2 = \sigma_2$ equal to

$$a = \frac{|\sigma_2 - j\sqrt{K}| \ |\sigma_2 + j\sqrt{K}|}{\sigma_2} = \frac{1}{\sigma_2} (\sigma_2^2 + K), \qquad (7.21)$$

where σ_2 is evaluated from the s-plane vector lengths as $\sigma_2 = \sqrt{K}$. As a increases beyond the critical value, the roots are both real and distinct; one root is larger than σ_2 and one is smaller.

In general, an orderly process for locating the locus of roots as a parameter varies is desirable. In the following section, we will develop such an orderly approach to obtaining a root locus diagram.

7.3 THE ROOT LOCUS PROCEDURE

The roots of the characteristic equation of a system provide a valuable insight concerning the response of the system. To locate the roots of the characteristic equation in a graphical manner on the s-plane, we will develop an orderly procedure of 12 steps that facilitates the rapid sketching of the locus.

Step 1: Write the characteristic equation as

$$1 + F(s) = 0 \qquad (7.22)$$

and rearrange the equation, if necessary, so that the parameter of interest, k, appears as the multiplying factor in the form,

$$1 + kP(s) = 0. \qquad (7.23)$$

Step 2: Factor $P(s)$, if necessary, and write the polynomial in the form of poles and zeros as follows:

$$1 + k \frac{\displaystyle\prod_{i=1}^{M} (s + z_i)}{\displaystyle\prod_{j=1}^{n} (s + p_j)} = 0. \tag{7.24}$$

Step 3: Locate the poles and zeros on the s-plane with selected symbols. We are usually interested in determining the locus of roots as k varies as

$$0 \le k \le \infty.$$

Rewriting Eq. (7.24), we have

$$\prod_{j=1}^{n} (s + p_j) + k \prod_{i=1}^{M} (s + z_i) = 0. \tag{7.25}$$

Step 4: Locate the segments of the real axis that are root loci.

(a) When $k = 0$, the roots of the characteristic equation are simply the poles of $P(s)$. Furthermore, when k approaches infinity, the roots of the characteristic equation are simply the zeros of $P(s)$. Therefore we note that *the locus of the roots of the characteristic equation $1 + kP(s) = 0$ begins at the poles of $P(s)$ and ends at the zeros of $P(s)$ as k increases from 0 to infinity*. For most functions, $P(s)$, that we will encounter, several of the zeros of $P(s)$ lie at infinity in the s-plane.

(b) *The root locus on the real axis always lies in a section of the real axis to the left of an odd number of poles and zeros.* This fact is clearly ascertained by examining the angle criterion of Eq. (7.17). These four useful steps in plotting a root locus will be illustrated by a suitable example.

EXAMPLE 7.1   **Second-order system**

A single-loop feedback control system possesses the following characteristic equation (step 1):

$$1 + GH(s) = 1 + \frac{K(\frac{1}{2}s + 1)}{s(\frac{1}{4}s + 1)} = 0. \tag{7.26}$$

Step 2: The transfer function $GH(s)$ is rewritten in terms of poles and zeros as follows:

$$1 + \frac{2K(s + 2)}{s(s + 4)} = 0, \tag{7.27}$$

and the multiplicative gain parameter is $k = 2K$. To determine the locus of roots for the gain $0 \le K \le \infty$ (step 3), we locate the poles and zeros on the real axis as shown in Fig. 7.6(a). Finally (step 4), the angle criterion is satisfied on the real axis between the points 0 and -2, because the angle from pole p_1 at the origin is 180° and the angle from the zero and pole p_2 at $s = -4$ is zero degrees. The locus begins at the pole and ends at the zeros, and therefore the locus of roots appears as shown in Fig. 7.6(b), where the direction of the locus as K is increasing $(K\uparrow)$ is shown by an arrow. We note that because the system has two real poles and one real zero, the second locus segment ends at a zero at

FIGURE 7.6
(a) The zero and poles of a second-order system,
(b) the root locus segments, and
(c) the magnitude of each vector at s_1.

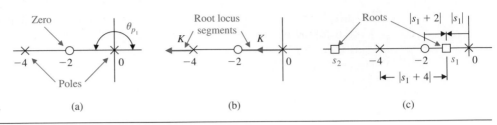

(a)   (b)   (c)

negative infinity. To evaluate the gain K at a specific root location on the locus, we utilize the magnitude criterion, Eq. (7.16). For example, the gain K at the root $s = s_1 = -1$ is found from (7.16) as

$$\frac{(2K) \, |s_1 + 2|}{|s_1| \, |s_1 + 4|} = 1$$

or

$$K = \frac{|-1| \, |-1 + 4|}{2|-1 + 2|} = \frac{3}{2}. \tag{7.28}$$

This magnitude can also be evaluated graphically as is shown in Fig. 7.6(c). Finally, for the gain of $K = \frac{3}{2}$, one other root exists, located on the locus to the left of the pole at -4. The location of the second root is found graphically to be located at $s = -6$, as shown in Fig. 7.6(c). ∎

We now return to developing a general list of root locus steps.

Step 5: Determine the number of separate loci, *SL*. Because the loci begin at the poles and end at the zeros, the *number of separate loci* is equal to the number of poles if the number of poles is greater than the number of zeros. In the unusual case when the number of zeros is greater than the number of poles, the number of separate loci would be the number of zeros. Therefore, as we found in Fig. 7.6, the number of separate loci is equal to two because there are two poles and one zero.

Step 6: *The root loci must be symmetrical with respect to the horizontal real axis* because the complex roots must appear as pairs of complex conjugate roots.

Step 7: The loci proceed to the zeros at infinity along asymptotes centered at σ_A and with angles ϕ_A. When the number of finite zeros of $P(s)$, n_z, is less than the number of poles, n_p, by the number $N = n_p - n_z$, then N sections of loci must end at zeros at infinity. These sections of loci proceed to the zeros at infinity along *asymptotes* as k approaches infinity. These linear *asymptotes are centered* at a point on the real axis given by

$$\sigma_A = \frac{\sum \text{poles of } P(s) - \sum \text{zeros of } P(s)}{n_p - n_z} = \frac{\sum_{j=1}^{n} (-p_j) - \sum_{i=1}^{M} (-z_i)}{n_p - n_z}. \tag{7.29}$$

The angle of the asymptotes with respect to the real axis is

$$\phi_A = \frac{(2q + 1)}{n_p - n_z} 180°, \qquad q = 0, 1, 2, \ldots, (n_p - n_z - 1), \tag{7.30}$$

where q is an integer index [3]. The usefulness of this rule is obvious for sketching the approximate form of a root locus. Equation (7.30) can be readily derived by considering a point on a root locus segment at a remote distance from the finite poles and zeros in the s-plane. The net phase angle at this remote point is 180°, because it is a point on a root locus segment. The finite poles and zeros of $P(s)$ are a great distance from the remote point, and so the angle from each pole and zero, ϕ, is essentially equal and therefore the net angle is simply $\phi(n_p - n_z)$, where n_p and n_z are the number of finite poles and zeros, respectively. Thus we have

$$\phi(n_p - n_z) = 180°,$$

or, alternatively,

$$\phi = \frac{180°}{n_p - n_z}.$$

Accounting for all possible root locus segments at remote locations in the s-plane, we obtain Eq. (7.30).

The center of the linear asymptotes, often called the *asymptote centroid,* is determined by considering the characteristic equation $1 + GH(s) = 0$ (Eq. 7.24). For large values of s, only the higher-order terms need be considered so that the characteristic equation reduces to

$$1 + \frac{k s^M}{s^n} = 0.$$

However, this relation, which is an approximation, indicates that the centroid of $(n - M)$ asymptotes is at the origin, $s = 0$. A better approximation is obtained if we consider a characteristic equation of the form

$$1 + \frac{k}{(s - \sigma_A)^{n-M}} = 0$$

with a centroid at σ_A.

The centroid is determined by considering the first two terms of Eq. (7.24), which may be found from the relation

$$1 + \frac{k \prod\limits_{i=1}^{M} (s + z_i)}{\prod\limits_{j=1}^{n} (s + p_j)} = 1 + k \frac{(s^M + b_{M-1}s^{M-1} + \cdots + b_0)}{(s^n + a_{n-1}s^{n-1} + \cdots + a_0)}.$$

From Chapter 6, especially Eq. (6.5), we note that

$$b_{M-1} = \sum_{i=1}^{M} z_i \quad \text{and} \quad a_{n-1} = \sum_{j=1}^{n} p_j.$$

Considering only the first two terms of this expansion, we have

$$1 + \frac{k}{s^{n-M} + (a_{n-1} - b_{M-1})s^{n-M-1}} = 0.$$

The first two terms of

$$1 + \frac{k}{(s - \sigma_A)^{n-M}} = 0$$

are

$$1 + \frac{k}{s^{n-M} - (n - M)\sigma_A s^{n-M-1}} = 0.$$

Equating the term for s^{n-M-1}, we obtain

$$(a_{n-1} - b_{M-1}) = -(n - M)\sigma_A,$$

which is equivalent to Eq. (7.29).

    For example, reexamine the system shown in Fig. 7.2 and discussed in Section 7.2. The characteristic equation is written as

$$1 + \frac{K}{s(s + 2)} = 0.$$

Because $n_p - n_z = 2$, we expect two loci to end at zeros at infinity. The asymptotes of the loci are located at a center

$$\sigma_A = \frac{-2}{2} = -1,$$

and at angles of

$$\phi_A = 90°, \qquad q = 0, \qquad \text{and} \qquad \phi_A = 270°, \qquad q = 1.$$

Therefore the root locus is readily sketched and the locus as shown in Fig. 7.3 is obtained. An example will further illustrate the process of utilizing the asymptotes.

EXAMPLE 7.2   **Fourth-order system**

A single-loop feedback control system has a characteristic equation as follows:

$$1 + GH(s) = 1 + \frac{K(s + 1)}{s(s + 2)(s + 4)^2}. \tag{7.31}$$

We wish to sketch the root locus in order to determine the effect of the gain K. The poles and zeros are located in the s-plane as shown in Fig. 7.7(a). The root loci on the real axis must be located to the left of an odd number of poles and zeros and are therefore located as shown in Fig. 7.7(a) as heavy lines. The intersection of the asymptotes is

$$\sigma_A = \frac{(-2) + 2(-4) - (-1)}{4 - 1} = \frac{-9}{3} = -3. \tag{7.32}$$

The angles of the asymptotes are

$$\phi_A = +60°, \qquad q = 0,$$
$$\phi_A = 180°, \qquad q = 1,$$
$$\phi_A = 300°, \qquad q = 2,$$

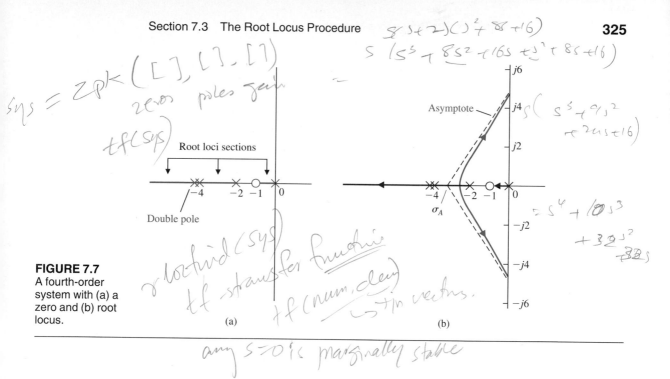

FIGURE 7.7
A fourth-order system with (a) a zero and (b) root locus.

(a)   (b)

where there are three asymptotes, since $n_p - n_z = 3$. Also, we note that the root loci must begin at the poles, and therefore two loci must leave the double pole at $s = -4$. Then, with the asymptotes drawn in Fig. 7.7(b), we may sketch the form of the root locus as shown in Fig. 7.7(b). The actual shape of the locus in the area near σ_A would be graphically evaluated, if necessary. ■

We now proceed to develop more steps for the process of determining the root loci.

Step 8: Determine the point at which the locus crosses the imaginary axis (if it does so), using the Routh criterion. *The actual point at which the root locus crosses the imaginary axis* is readily evaluated by utilizing the Routh–Hurwitz criterion and other methods of Chapter 6.

Step 9: Determine the breakaway point on the real axis (if any). The root locus in Example 7.2 left the real axis at a *breakaway point.* The locus breakaway from the real axis occurs where the net change in angle caused by a small displacement is zero. The locus leaves the real axis where there are a multiplicity of roots, typically two. The breakaway point for a simple second-order system is shown in Fig. 7.8(a) and, for a special case of a fourth-order system, in Fig. 7.8(b). In general, due to the phase criterion, *the tangents to the loci at the breakaway point are equally spaced over 360°.* Therefore, in Fig. 7.8(a), we find that the two loci at the breakaway point are spaced 180° apart, whereas in Fig. 7.8(b), the four loci are spaced 90° apart.

The breakaway point on the real axis can be evaluated graphically or analytically. The most straightforward method of evaluating the breakaway point involves the rearranging of the characteristic equation to isolate the multiplying factor K. Then the characteristic equation is written as

$$p(s) = K. \qquad (7.33)$$

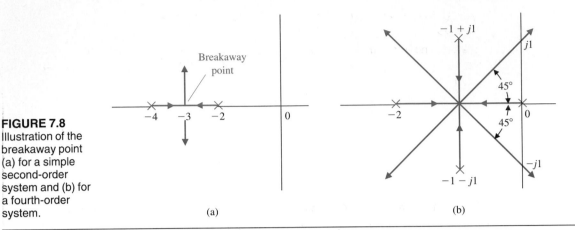

For example, consider a unity feedback closed-loop system with an open-loop transfer function

$$G(s) = \frac{K}{(s + 2)(s + 4)},$$

which has a characteristic equation as follows:

$$1 + G(s) = 1 + \frac{K}{(s + 2)(s + 4)} = 0. \tag{7.34}$$

Alternatively, the equation may be written as

$$K = p(s) = -(s + 2)(s + 4). \tag{7.35}$$

The root loci for this system are shown in Fig. 7.8(a). We expect the breakaway point to be near $s = \sigma = -3$ and plot $p(s)|_{s=\sigma}$ near that point, as shown in Fig. 7.9. In this case, $p(s)$ equals zero at the poles $s = -2$ and $s = -4$. The plot of $p(s)$ versus σ is symmetrical, and the maximum point occurs at $s = \sigma = -3$, the breakaway point.

Analytically, the very same result may be obtained by determining the maximum of $K = p(s)$. To find the maximum analytically, we differentiate, set the differentiated polynomial equal to zero, and determine the roots of the polynomial. Therefore we may evaluate

$$\frac{dK}{ds} = \frac{dp(s)}{ds} = 0 \tag{7.36}$$

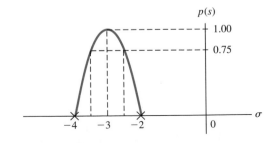

in order to find the breakaway point. Clearly, Eq. (7.36) is an analytical expression of the graphical procedure outlined in Fig. 7.9 and will result in an equation of only one order less than the total number of poles and zeros ($n_p + n_z - 1$). In almost all cases, we will prefer to use the graphical method of locating the breakaway point when it is necessary to do so.

The proof of Eq. (7.36) is obtained from a consideration of the characteristic equation

$$1 + F(s) = 1 + \frac{KY(s)}{X(s)} = 0,$$

which may be written as

$$X(s) + KY(s) = 0. \tag{7.37}$$

For a small increment in K, we have

$$X(s) + (K + \Delta K)Y(s) = 0$$

dividing by $X(s) + KY(s)$, we have

$$1 + \frac{\Delta KY(s)}{X(s) + KY(s)} = 0. \tag{7.38}$$

Because the denominator is the original characteristic equation, at a breakaway point a multiplicity of roots exists and

$$\frac{Y(s)}{X(s) + KY(s)} = \frac{C_i}{(s - s_i)^n} = \frac{C_i}{(\Delta s_i)^n}. \tag{7.39}$$

Then we may write Eq. (7.38) as

$$1 + \frac{\Delta K C_i}{(\Delta s_i)^n} = 0 \tag{7.40}$$

or, alternatively,

$$\frac{\Delta K}{\Delta s} = \frac{-(\Delta s)^{n-1}}{C_i}. \tag{7.41}$$

Therefore, as we let Δs approach zero, we obtain

$$\frac{dK}{ds} = 0 \tag{7.42}$$

at the breakaway points.

Now, reconsidering the specific case where

$$G(s) = \frac{K}{(s + 2)(s + 4)},$$

we obtain $p(s)$ as

$$K = p(s) = -(s + 2)(s + 4) = -(s^2 + 6s + 8). \tag{7.43}$$

Then, differentiating, we have

$$\frac{dp(s)}{ds} = -(2s + 6) = 0 \tag{7.44}$$

or the breakaway point occurs at $s = -3$. A more complicated example will illustrate the approach and exemplify the advantage of the graphical technique.

EXAMPLE 7.3   Third-order system

A feedback control system is shown in Fig. 7.10. The characteristic equation is

$$1 + G(s)H(s) = 1 + \frac{K(s + 1)}{s(s + 2)(s + 3)} = 0. \tag{7.45}$$

The number of poles, n_p, minus the number of zeros, n_z, is equal to two, and so we have two asymptotes at $\pm 90°$ with a center at $\sigma_A = -2$. The asymptotes and the sections of loci on the real axis are shown in Fig. 7.11(a). A breakaway point occurs between $s = -2$ and $s = -3$. To evaluate the breakaway point, we rewrite the characteristic equation so that K is separated:

$$s(s + 2)(s + 3) + K(s + 1) = 0$$

or

$$p(s) = \frac{-s(s + 2)(s + 3)}{(s + 1)} = K. \tag{7.46}$$

FIGURE 7.10
Closed-loop
system.

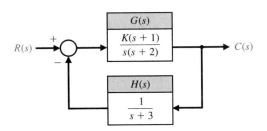

FIGURE 7.11
Evaluation of the
(a) asymptotes and
(b) breakaway
point.

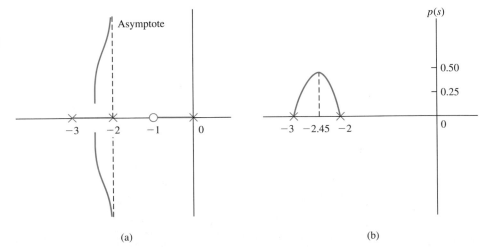

(a)   (b)

TABLE 7.1

$p(s)$	0	+0.412	+0.420	+0.417	+0.390	0
s	-2.00	-2.40	-2.45	-2.50	-2.60	-3.0

Then, evaluating $p(s)$ at various values of s between $s = -2$ and $s = -3$, we obtain the results of Table 7.1 as shown in Fig. 7.11(b). Alternatively, we differentiate Eq. (7.46) and set equal to zero to obtain

$$\frac{d}{ds}\left(\frac{-s(s + 2)(s + 3)}{(s + 1)}\right) = \frac{(s^3 + 5s^2 + 6s) - (s + 1)(3s^2 + 10s + 6)}{(s + 1)^2} = 0$$

$$= 2s^3 + 8s^2 + 10s + 6 = 0. \tag{7.47}$$

Now, to locate the maximum of $p(s)$, we locate the roots of Eq. (7.47) by synthetic division or by the Newton–Raphson method to obtain $s = -2.45$. It is evident from this one example that the numerical evaluation of $p(s)$ near the expected breakaway point will result in the simplest method of evaluating the breakaway point. As the order of the characteristic equation increases, the usefulness of the graphical (or tabular) evaluation of the breakaway point will increase in contrast to the analytical approach. ∎

Step 10: Determine *the angle of departure of the locus from a pole* and *the angle of arrival at the locus at a zero,* using the phase angle criterion. The angle of locus departure from a pole is the difference between the net angle due to all other poles and zeros and the criterion angle of $\pm 180° (2q + 1)$, and similarly for the locus angle of arrival at zero. The angle of departure (or arrival) is particularly of interest for complex poles (and zeros) because the information is helpful in completing the root locus. For example, consider the third-order open-loop transfer function

$$F(s) = G(s)H(s) = \frac{K}{(s + p_3)(s^2 + 2\zeta\omega_n s + \omega_n^2)}. \tag{7.48}$$

The pole locations and the vector angles at one complex pole p_1 are shown in Fig. 7.12(a). The angles at a test point s_1, an infinitesimal distance from p_1, must meet the angle criterion. Therefore, since $\theta_2 = 90°$, we have

$$\theta_1 + \theta_2 + \theta_3 = \theta_1 + 90° + \theta_3 = +180°,$$

or the angle of departure at pole p_1 is

$$\theta_1 = 90° - \theta_3$$

as shown in Fig. 7.12(b). The departure at pole p_2 is the negative of that at p_1 because p_1 and p_2 are complex conjugates. Another example of a departure angle is shown in Fig. 7.13. In this case, the departure angle is found from

$$\theta_2 - (\theta_1 + \theta_3 + 90°) = 180°.$$

Since $(\theta_2 - \theta_3) = \gamma$, we find that the departure angle is $\theta_1 = 90° + \gamma$.

The final two steps of the root locus procedure are used to determine a root location, s_q, and the value of the parameter k at that root location.

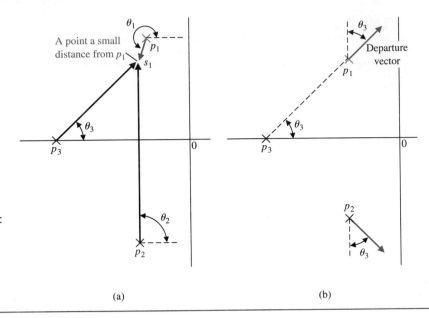

FIGURE 7.12
Illustration of the
angle of departure:
(a) Test point
infinitesimal
distance from p_1;
(b) actual
departure vector
at p_1.

(a)   (b)

Step 11: Determine the root locations that satisfy the phase criterion at the root s_x, $x = 1, 2, \ldots, n_p$. The phase criterion (Eq. 7.17) is

$$\underline{/P(s)} = 180° \pm q360°, \quad q = 1, 2, \ldots$$

Step 12: Determine the parameter value k_x at a specific root s_x using the magnitude requirement (Eq. 7.16). The magnitude requirement at s_x is

$$k_x = \left. \frac{\displaystyle\prod_{j=1}^{n} |(s + p_j)|}{\displaystyle\prod_{i=1}^{m} |(s + z_i)|} \right|_{s=s_x}$$

It is worthwhile at this point to summarize the 12 steps utilized in the root locus method (Table 7.2) and then illustrate their use in a complete example.

FIGURE 7.13
Evaluation of the
angle of departure.

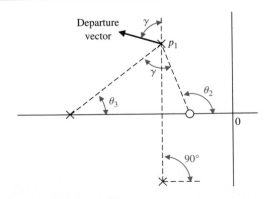

TABLE 7.2   The 12 Steps of the Root Locus Procedure

Step	Related Equation or Rule
1. Write the characteristic equation so that the parameter of interest k appears as a multiplier.	$1 + kP(s) = 0$
2. Factor $P(s)$ in terms of n_p poles and n_z zeros.	$1 + k \dfrac{\prod\limits_{i=1}^{n_z} (s + z_i)}{\prod\limits_{j=1}^{n_p} (s + p_j)} = 0$
3. Locate the open-loop poles and zeros of $F(s)$ in the s-plane with selected symbols.	$\times$ = poles, $\bigcirc$ = zeros, $\square$ or $\triangle$ = roots of characteristic equation
4. Locate the segments of the real axis that are root loci.	(a) Locus begins at a pole and ends at a zero (b) Locus lies to the left of an odd number of poles and zeros
5. Determine the number of separate loci, SL.	$SL = n_p$ when $n_p \geq n_z$; n_p = number of finite poles, n_z = number of finite zeros
6. The root loci are symmetrical with respect to the horizontal real axis.	
7. The loci proceed to the zeros at infinity along asymptotes centered at σ_A and with angles ϕ_A.	$\sigma_A = \dfrac{\sum(-p_j) - \sum(-z_i)}{n_p - n_z}$ $\phi_A = \dfrac{(2q + 1)}{n_p - n_z} 180°$, $q = 0, 1, 2, \ldots (n_p - n_z - 1)$
8. By utilizing the Routh–Hurwitz criterion, determine the point at which the locus crosses the imaginary axis (if it does so).	See Section 6.2.
9. Determine the breakaway point on the real axis (if any).	a) Set $K = p(s)$. b) Obtain $\dfrac{dp(s)}{ds} = 0$. c) Determine roots of (b) *or* use graphical method to find maximum of $p(s)$
10. Determine the angle of locus departure from complex poles and the angle of locus arrival at complex zeros, using the phase criterion.	$\underline{/P(s)} = 180° \pm q360°$ at $s = p_j$ or z_i
11. Determine the root locations that satisfy the phase criterion.	$\underline{/P(s)} = 180° \pm q360°$ at a root location s_x
12. Determine the parameter value k_x at a specific root s_x.	$k_x = \dfrac{\prod\limits_{j=1}^{n_p}|(s + p_j)|}{\prod\limits_{i=1}^{n_z}|(s + z_i)|}\Bigg|_{s=s_A}$

EXAMPLE 7.4   **Fourth-order system**

1. We desire to plot the root locus for the characteristic equation of a system as K varies for $K > 0$, when

$$1 + \frac{K}{s^4 + 12s^3 + 64s^2 + 128s} = 0.$$

2. Determining the poles, we have

$$1 + \frac{K}{s(s + 4)(s + 4 + j4)(s + 4 - j4)} = 0 \qquad (7.49)$$

as K varies from zero to infinity. This system has no finite zeros.

3. The poles are located on the s-plane as shown in Fig. 7.14(a).

4. A segment of the root locus exists on the real axis between $s = 0$ and $s = -4$.

5. Because the number of poles n_p is equal to 4, we have four separate loci.

6. The root loci are symmetrical with respect to the real axis.

7. The angles of the asymptotes are

$$\phi_A = \frac{(2q + 1)}{4} 180°, \qquad q = 0, 1, 2, 3$$

$$\phi_A = +45°, \ 135°, \ 225°, \ 315°.$$

The center of the asymptotes is

$$\sigma_A = \frac{-4 - 4 - 4}{4} = -3.$$

Then the asymptotes are drawn as shown in Fig. 7.14(a).

8. The characteristic equation is rewritten as

$$s(s + 4)(s^2 + 8s + 32) + K = s^4 + 12s^3 + 64s^2 + 128s + K = 0. \qquad (7.50)$$

Therefore, the Routh–Hurwitz array is

s^4	1	64	K
s^3	12	128	
s^2	b_1	K	
s	c_1		
s^0	K		

where

$$b_1 = \frac{12(64) - 128}{12} = 53.33 \quad \text{and} \quad c_1 = \frac{53.33(128) - 12K}{53.33}.$$

Hence the limiting value of gain for stability is $K = 568.89$ and the roots of the auxiliary equation are

$$53.33s^2 + 570 = 53.33(s^2 + 10.6) = 53.33(s + j3.266)(s - j3.266). \qquad (7.51)$$

The points where the locus crosses the imaginary axis are shown in Fig. 7.14(a).

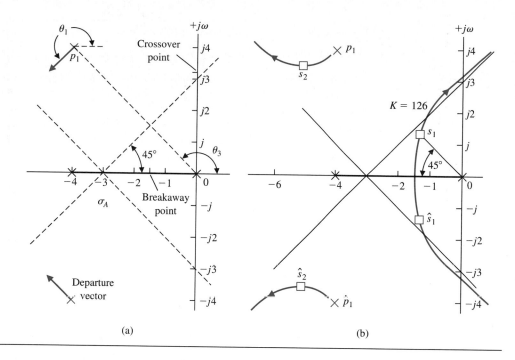

FIGURE 7.14
The root locus for
Example 7.4:
locating (a) the
poles and (b) the
asymptotes.

9. The breakaway point is estimated by evaluating

$$K = p(s) = -s(s + 4)(s + 4 + j4)(s + 4 - j4)$$

between $s = -4$ and $s = 0$. We expect the breakaway point to lie between $s = -3$ and $s = -1$ and so we search for a maximum value of $p(s)$ in that region. The resulting values of $p(s)$ for several values of s are given in Table 7.3. The maximum of $p(s)$ is found to lie at approximately $s = -1.5$, as indicated in the table. A more accurate estimate of the breakaway point is normally not necessary or worthwhile. The breakaway point is then indicated on Fig. 7.14(a).

10. The angle of departure at the complex pole p_1 can be estimated by utilizing the angle criterion as follows:

$$\theta_1 + 90° + 90° + \theta_3 = 180°,$$

where θ_3 is the angle subtended by the vector from pole p_3. The angles from the pole at $s = -4$ and $s = -4 - j4$ are each equal to 90°. Since $\theta_3 = 135°$, we find that

$$\theta_1 = -135° = +225°$$

as shown in Fig. 7.14(a).

TABLE 7.3

$p(s)$	0	51	68.5	80	85	75	0
s	-4.0	-3.0	-2.5	-2.0	-1.5	-1.0	0

11. Determine the root locations that satisfy the phase criterion as shown in Fig. 7.14(b).

12. Determine the value of K at $s = s_1$.

Utilizing all the information obtained from the 12 steps of the root locus method, the complete root locus is plotted by using a protractor to locate points that satisfy the angle criterion. The root locus for this system is shown in Fig. 7.14(b). When the complex roots near the origin have a damping ratio of $\zeta = 0.707$, the gain K can be determined graphically as shown in Fig. 7.14(b). The vector lengths to the root location s_1 from the open-loop poles are evaluated and result in a gain at s_1 of

$$K = |s_1| \, |s_1 + 4| \, |s_1 - p_1| \, |s_1 - \hat{p}_1| = (1.9)(2.9)(3.8)(6.0) = 126. \quad (7.52)$$

The remaining pair of complex roots occurs at s_2 and $\hat{s}_2$ when $K = 126$. The effect of the complex roots at s_2 and $\hat{s}_2$ on the transient response will be negligible compared to the roots s_1 and $\hat{s}_1$. This fact can be ascertained by considering the damping of the response due to each pair of roots. The damping due to s_1 and $\hat{s}_1$ is

$$e^{-\zeta_1 \omega_n t} = e^{-\sigma_1 t},$$

and the damping factor due to s_2 and $\hat{s}_2$ is

$$e^{-\zeta_2 \omega_{n2} t} = e^{-\sigma_2 t},$$

where σ_2 is approximately five times as large as σ_1. Therefore, the transient response term due to s_2 will decay much more rapidly than the transient response term due to s_1. Thus the response to a unit step input may be written as

$$
\begin{aligned}
c(t) &= 1 + c_1 e^{-\sigma_1 t} \sin(\omega_1 t + \theta_1) + c_2 e^{-\sigma_2 t} \sin(\omega_2 t + \theta_2) \qquad (7.53) \\
&\cong 1 + c_1 e^{-\sigma_1 t} \sin(\omega_1 t + \theta_1).
\end{aligned}
$$

The complex conjugate roots near the origin of the s-plane relative to the other roots of the closed-loop system are labeled the *dominant roots* of the system because they represent or dominate the transient response. The relative dominance of the complex roots is determined by the ratio of the real root to the real part of the complex roots and will result in approximate dominance for ratios exceeding five.

Of course, the dominance of the second term of Eq. (7.53) also depends upon the relative magnitudes of the coefficients c_1 and c_2. These coefficients, which are the residues evaluated at the complex roots, in turn depend upon the location of the zeros in the s-plane. Therefore, the concept of dominant roots is useful for estimating the response of a system but must be used with caution and with a comprehension of the underlying assumptions. ∎

Using *Matlab* or a similar program, one can readily determine a breakaway point. For example, refer back to Example 7.2. After several tries, you should be able to determine that when $K = 1.925$, the breakaway occurs from the real axis with two real roots at $s = -2.6$. In the same way, you will find that when $K = 200$, the two roots on the imaginary axis are at $s = \pm j4.82$.

Similarly, for Example 7.4, using *Matlab,* one can determine that breakaway from the real axis occurs at $K = 84$ when $s = -1.57$. In the same way when $K = 600$, we find that two roots are on the imaginary axis at $s = \pm j3.33$.

7.4 AN EXAMPLE OF A CONTROL SYSTEM ANALYSIS AND DESIGN UTILIZING THE ROOT LOCUS METHOD

The analysis and design of a control system can be accomplished by utilizing the Laplace transform, a signal-flow diagram, the s-plane, and the root locus method. It will be worthwhile at this point to examine a control system and select suitable parameter values based on the root locus method.

An automatic self-balancing scale in which the weighing operation is controlled by the physical balance function through an electrical feedback loop is shown in Fig. 7.15 [5]. The balance is shown in the equilibrium condition, and x is the travel of the counterweight W_c from an unloaded equilibrium condition. The weight to be measured, W, is applied 5 cm from the pivot, and the length of the beam to the viscous damper, l_i, is 20 cm. It is desired to accomplish the following items:

1. Select the parameters and the specifications of the feedback system.
2. Obtain a model and signal-flow diagram representing the system.
3. Select the gain K based on a root locus diagram.
4. Determine the dominant mode of response.

An inertia of the beam equal to 0.05 kg-m² will be chosen. We must select a battery voltage that is large enough to provide a reasonable position sensor gain, so let us choose $E_{bb} = 24$ volts. We will utilize a lead screw of 20 turns/cm and a potentiometer for x equal to 6 cm in length. Accurate balances are required, and therefore an input potentiometer 0.5 cm in length for y will be chosen. A reasonable viscous damper will be chosen with a damping constant $f = 10\sqrt{3}$ kg/m/s. Finally, a counterweight W_c is chosen so that the expected range of weights W can be balanced. Therefore, in summary, the *parameters* of the system are selected as listed in Table 7.4.

Specifications.   A rapid and accurate response resulting in a small steady-state weight measurement error is desired. Therefore we will require that the system be at least a type one so that a zero measurement error is obtained. An underdamped response to a step change in the measured weight W is satisfactory, and therefore a dominant response with

FIGURE 7.15
An automatic self-balancing scale. (From J. H. Goldberg, *Automatic Controls*, Allyn and Bacon, Boston, 1964, with permission.)

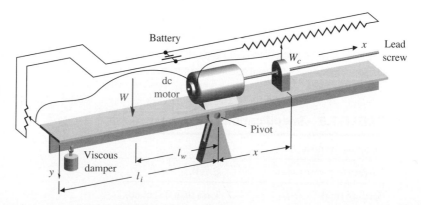

TABLE 7.4

$W_c = 2 \text{ N}$	Lead screw gain $K_s = \dfrac{1}{4000\pi}$ m/rad
$I = 0.05$ kg-m^2	
$l_w = 5$ cm	Input potentiometer gain $K_i = 4800$ V/m
$l_i = 20$ cm	
$f = 10\sqrt{3}$ kg/m/s	Feedback potentiometer gain $K_f = 400$ V/m

$\zeta = 0.5$ will be specified. The settling time of the balance following the introduction of a weight to be measured should be less than 2 seconds in order to provide a rapid weight-measuring device. The specifications are summarized in Table 7.5.

The derivation of a model of the electromechanical system may be accomplished by obtaining the equations of motion of the balance. For small deviations from balance, the deviation angle θ is

$$\theta \cong \frac{y}{l_i}. \tag{7.54}$$

The motion of the beam about the pivot is represented by the torque equation:

$$I\frac{d^2\theta}{dt^2} = \Sigma \text{ torques.}$$

Therefore, in terms of the deviation angle, the motion is represented by

$$I\frac{d^2\theta}{dt^2} = l_w W - xW_c - l_i^2 f\frac{d\theta}{dt}. \tag{7.55}$$

The input voltage to the motor is

$$v_m(t) = K_i y - K_f x. \tag{7.56}$$

The transfer function of the motor is

$$\frac{\theta_m(s)}{V_m(s)} = \frac{K_m}{s(\tau s + 1)}, \tag{7.57}$$

where τ will be considered to be negligible with respect to the time constants of the overall system, and θ_m is the output shaft rotation. A signal-flow graph representing Eqs. (7.55)

TABLE 7.5   Specifications

Steady-state error	$K_p = \infty$
Underdamped response	$\zeta = 0.5$
Settling time	Less than 2 seconds

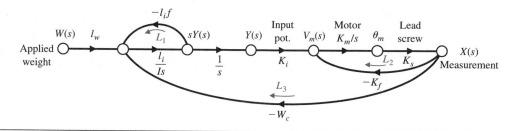

FIGURE 7.16
Signal-flow graph
model of the
automatic self-
balancing scale.

through (7.57) is shown in Fig. 7.16. Examining the forward path from W to $X(s)$, we find that the system is a type one due to the integration preceding $Y(s)$. Therefore the steady-state error of the system is zero.

The closed-loop transfer function of the system is obtained by utilizing Mason's flow graph formula and is found to be

$$\frac{X(s)}{W(s)} = \frac{(l_w l_i K_i K_m K_s / Is^3)}{1 + (l_i^2 f / Is) + (K_m K_s K_f / s) + (l_i K_i K_m K_s W_c / Is^3) + (l_i^2 f K_m K_s K_f / Is^2)}, \quad (7.58)$$

where the numerator is the path factor from W to X, the second term in the denominator is the loop L_1, the third term is the loop factor L_2, the fourth term is the loop L_3, and the fifth term is the two nontouching loops $L_1 L_2$. Therefore the closed-loop transfer function is

$$\frac{X(s)}{W(s)} = \frac{l_w l_i K_i K_m K_s}{s(Is + l_i^2 f)(s + K_m K_s K_f) + W_c K_m K_s K_i l_i}. \quad (7.59)$$

The steady-state gain of the system is then

$$\lim_{t \to \infty} \left(\frac{x(t)}{|W|} \right) = \lim_{s \to 0} \left(\frac{X(s)}{W(s)} \right) = \frac{l_w}{W_c} = 2.5 \text{ cm/kg} \quad (7.60)$$

when $W(s) = |W|/s$. To obtain the root locus as a function of the motor constant K_m, we substitute the selected parameters into the characteristic equation, which is the denominator of Eq. (7.59). Therefore we obtain the following characteristic equation:

$$s(s + 8\sqrt{3}) \left(s + \frac{K_m}{10\pi} \right) + \frac{96 K_m}{10\pi} = 0. \quad (7.61)$$

Rewriting the characteristic equation in root locus form, we first isolate K_m as follows:

$$s^2(s + 8\sqrt{3}) + s(s + 8\sqrt{3}) \frac{K_m}{10\pi} + \frac{96 K_m}{10\pi} = 0. \quad (7.62)$$

Then, rewriting Eq. (7.62) in root locus form, we have

$$1 + KP(s) = 1 + \frac{(K_m/10\pi)[s(s + 8\sqrt{3}) + 96]}{s^2(s + 8\sqrt{3})} = 0$$

$$= 1 + \frac{(K_m/10\pi)(s + 6.93 + j6.93)(s + 6.93 - j6.93)}{s^2(s + 8\sqrt{3})}. \quad (7.63)$$

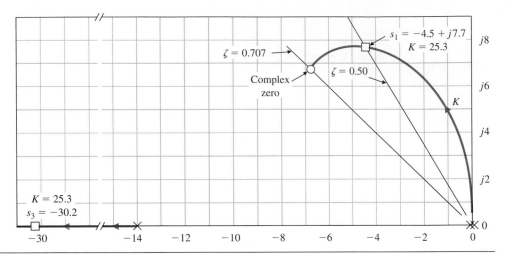

FIGURE 7.17
Root locus as K_m varies. One locus leaves the two poles at the origin and goes to the two complex zeros as K increases. The other locus is to the left of the pole at $s = -14$.

The root locus as K_m varies is shown in Fig. 7.17. The dominant roots can be placed at $\zeta = 0.5$ when $K = 25.3 = K_m/10\pi$. To achieve this gain,

$$K_m = 795 \frac{\text{rad/s}}{\text{volt}} = 7600 \frac{\text{rpm}}{\text{volt}}, \tag{7.64}$$

an amplifier would be required to provide a portion of the required gain. The real part of the dominant roots is greater than four and therefore the settling time, $4/\sigma$, is less than 1 second, and the settling time requirement is satisfied. The third root of the characteristic equation is a real root at $s = -30.2$, and the underdamped roots clearly dominate the response. Therefore the system has been analyzed by the root locus method and a suitable design for the parameter K_m has been achieved. The efficiency of the s-plane and root locus methods is clearly demonstrated by this example.

7.5 PARAMETER DESIGN BY THE ROOT LOCUS METHOD

The original development of the root locus method was concerned with the determination of the locus of roots of the characteristic equation as the system gain, K, is varied from zero to infinity. However, as we have seen, the effect of other system parameters may be readily investigated by using the root locus method. Fundamentally, the root locus method is concerned with a characteristic equation (Eq. 7.22), which may be written as

$$1 + F(s) = 0. \tag{7.65}$$

Then the standard root locus method we have studied may be applied. The question arises: How do we investigate the effect of two parameters, α and β? It appears that the root locus method is a single-parameter method; fortunately, however, it can be readily extended to the investigation of two or more parameters. This method of *parameter design* uses the root locus approach to select the values of the parameters.

The characteristic equation of a dynamic system may be written as

$$a_n s^n + a_{n-1} s^{n-1} + \cdots + a_1 s + a_0 = 0. \tag{7.66}$$

Hence the effect of the coefficient a_1 may be ascertained from the root locus equation

$$1 + \frac{a_1 s}{a_n s^n + a_{n-1} s^{n-1} + \cdots + a_2 s^2 + a_0} = 0. \tag{7.67}$$

If the parameter of interest, α, does not appear solely as a coefficient, the parameter is isolated as

$$a_n s^n + a_{n-1} s^{n-1} + \cdots + (a_{n-q} - \alpha) s^{n-q}$$

$$+ \alpha s^{n-q} + \cdots + a_1 s + a_0 = 0. \tag{7.68}$$

Then, for example, a third-order equation of interest might be

$$s^3 + (3 + \alpha) s^2 + 3s + 6 = 0. \tag{7.69}$$

To ascertain the effect of the parameter α, we isolate the parameter and rewrite the equation in root locus form as shown in the following steps:

$$s^3 + 3s^2 + \alpha s^2 + 3s + 6 = 0, \tag{7.70}$$

$$1 + \frac{\alpha s^2}{s^3 + 3s^2 + 3s + 6} = 0. \tag{7.71}$$

Then, to determine the effect of two parameters, we must repeat the root locus approach twice. Thus, for a characteristic equation with two variable parameters, α and β, we have

$$a_n s^n + a_{n-1} s^{n-1} + \cdots + (a_{n-q} - \alpha) s^{n-q} + \alpha s^{n-q} + \cdots$$

$$+ (a_{n-r} - \beta) s^{n-r} + \beta s^{n-r} + \cdots + a_1 s + a_0 = 0. \tag{7.72}$$

The two variable parameters have been isolated and the effect of α will be determined, followed by the determination of the effect of β. For example, for a certain third-order characteristic equation with α and β as parameters, we obtain

$$s^3 + s^2 + \beta s + \alpha = 0. \tag{7.73}$$

In this particular case, the parameters appear as the coefficients of the characteristic equation. The effect of varying β from zero to infinity is determined from the root locus equation

$$1 + \frac{\beta s}{s^3 + s^2 + \alpha} = 0. \tag{7.74}$$

One notes that the denominator of Eq. (7.74) is the characteristic equation of the system with $\beta = 0$. Therefore one first evaluates the effect of varying α from zero to infinity by utilizing the equation

$$s^3 + s^2 + \alpha = 0,$$

rewritten as

$$1 + \frac{\alpha}{s^2(s + 1)} = 0, \tag{7.75}$$

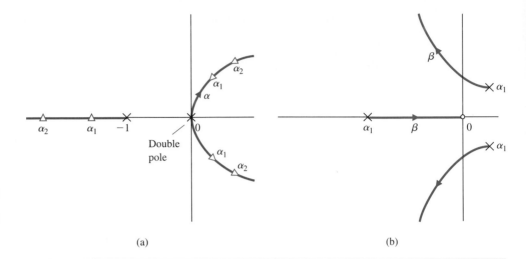

FIGURE 7.18
Root loci as a function of α and β: (a) loci as α varies, (b) loci as β varies for one value of $\alpha = \alpha_1$.

where β has been set equal to zero in Eq. (7.73). Then, upon evaluating the effect of α, a value of α is selected and used with Eq. (7.74) to evaluate the effect of β. This two-step method of evaluating the effect of α and then β may be carried out as a two-root locus procedure. First we obtain a locus of roots as α varies, and we select a suitable value of α; the results are satisfactory root locations. Then we obtain the root locus for β by noting that the poles of Eq. (7.74) are the roots evaluated by the root locus of Eq. (7.75). A limitation of this approach is that one will not always be able to obtain a characteristic equation that is linear in the parameter under consideration, for example, α.

To illustrate this approach effectively, let us obtain the root locus for α and then β for Eq. (7.73). A sketch of the root locus as α varies for Eq. (7.75) is shown in Fig. 7.18(a), where the roots for two values of gain α are shown. If the gain α is selected as α_1, then the resultant roots of Eq. (7.75) become the poles of Eq. (7.74). The root locus of Eq. (7.74) as β varies is shown in Fig. 7.18(b), and a suitable β can be selected on the basis of the desired root locations.

Using the root locus method, we will further illustrate this parameter design approach by a specific design example.

EXAMPLE 7.5   Disk drive control

A disk drive for a computer requires an accurate control system for positioning the read-write head [4]. The feedback control system is to be designed to satisfy the following specifications:

1. Steady-state error for a ramp input ≤ 35% of input slope

2. Damping ratio of dominant roots ≥ 0.707

3. Settling time of the system ≤ 3 seconds

The structure of the feedback control system is shown in Fig. 7.19, where the amplifier gain K_1 and the derivative feedback gain K_2 are to be selected. The steady-state error specifica-

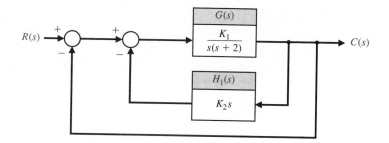

FIGURE 7.19
Block diagram of
disk drive control
system.

tion can be written as follows:

$$e_{ss} = \lim_{t \to \infty} e(t) = \lim_{s \to 0} sE(s) = \lim_{s \to 0} \frac{s(|R|/s^2)}{1 + G_2(s)}, \tag{7.76}$$

where $G_2(s) = G(s)/(1 + G(s)H_1(s))$. Therefore the steady-state error requirement is

$$\frac{e_{ss}}{|R|} = \frac{2 + K_1 K_2}{K_1} \leq 0.35. \tag{7.77}$$

Thus we will select a small value of K_2 to achieve a low value of steady-state error. The damping ratio specification requires that the roots of the closed-loop system be below the line at 45° in the left-hand s-plane. The settling time specification can be rewritten in terms of the real part of the dominant roots as

$$T_s = \frac{4}{\sigma} \leq 3 \text{ seconds.} \tag{7.78}$$

Therefore it is necessary that $\sigma \geq \frac{4}{3}$; this area in the left-hand s-plane is indicated along with the ζ-requirement in Fig. 7.20. To satisfy the specifications, all the roots must lie within the shaded area of the left-hand plane.

FIGURE 7.20
A region in the
s-plane for desired
root location.

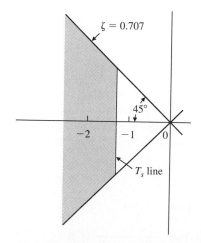

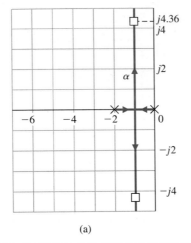 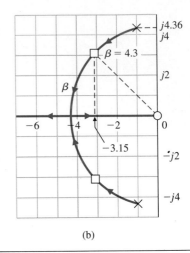

FIGURE 7.21
Root loci as a function of (a) α and (b) β.

(a)　　　　　　　　　　　　　　　(b)

The parameters to be selected are $\alpha = K_1$ and $\beta = K_2 K_1$. The characteristic equation is

$$1 + GH(s) = s^2 + 2s + \beta s + \alpha = 0. \tag{7.79}$$

The locus of roots as $\alpha = K_1$ varies (set $\beta = 0$) is determined from the following equation:

$$1 + \frac{\alpha}{s(s + 2)} = 0. \tag{7.80}$$

(See Fig. 7.21a.) For a gain of $K_1 = \alpha = 20$, the roots are indicated on the locus. Then the effect of varying $\beta = 20K_2$ is determined from the locus equation

$$1 + \frac{\beta s}{s^2 + 2s + \alpha} = 0, \tag{7.81}$$

where the poles of this root locus are the roots of the locus of Fig. 7.21(a). The root locus for Eq. (7.81) is shown in Fig. 7.21(b) and roots with $\zeta = 0.707$ are obtained when $\beta = 4.3 = 20K_2$ or when $K_2 = 0.215$. The real part of these roots is $\sigma = 3.15$, and therefore the settling time is equal to 1.27 seconds, which is considerably less than the specification of 3 seconds. ∎

The root locus method may be extended to more than two parameters by extending the number of steps in the method outlined in this section. Furthermore, a family of root loci can be generated for two parameters in order to determine the total effect of varying two parameters. For example, let us determine the effect of varying α and β of the following characteristic equation:

$$s^3 + 3s^2 + 2s + \beta s + \alpha = 0. \tag{7.82}$$

The root locus equation as a function of α is (set $\beta = 0$)

$$1 + \frac{\alpha}{s(s + 1)(s + 2)} = 0. \tag{7.83}$$

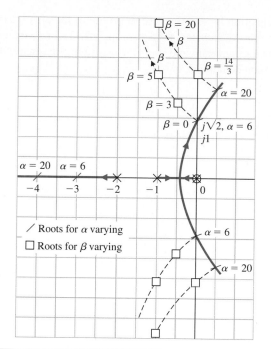

FIGURE 7.22
Two-parameter
root locus. The loci
for α varying are
in color.

The root locus as a function of β is

$$1 + \frac{\beta s}{s^3 + 3s^2 + 2s + \alpha} = 0. \tag{7.84}$$

The root locus for Eq. (7.83) as a function of α is shown in Fig. 7.22 (unbroken lines). The roots of this locus, indicated by a cross, become the poles for the locus of Eq. (7.84). Then the locus of Eq. (7.84) is continued on Fig. 7.22 (dotted lines), where the locus for β is shown for several selected values of α. This family of loci, often called *root contours,* illustrates the effect of α and β on the roots of the characteristic equation of a system [3].

7.6 SENSITIVITY AND THE ROOT LOCUS

One of the prime reasons for the utilization of negative feedback in control systems is to reduce the effect of parameter variations. The effect of parameter variations, as we found in Section 4.2, can be described by a measure of the *sensitivity* of the system performance to specific parameter changes. In Section 4.2, we defined the *logarithmic sensitivity* originally suggested by Bode as

$$S_k^T = \frac{d \ln T}{d \ln k} = \frac{\partial T/T}{\partial k/k}, \tag{7.85}$$

where the system transfer function is $T(s)$ and the parameter of interest is k.

In recent years, with the increased utilization of the pole–zero (s-plane) approach, it has become useful to define a measure of sensitivity in terms of the positions of the roots of the characteristic equation [7–9]. Because the roots of the characteristic equation represent the dominant modes of transient response, the effect of parameter variations on the position of the roots is an important and useful measure of the sensitivity. The *root sensitivity* of a system $T(s)$ can be defined as

$$S_k^{r_i} = \frac{\partial r_i}{\partial \ln k} = \frac{\partial r_i}{\partial k/k}, \tag{7.86}$$

where r_i equals the ith root of the system so that

$$T(s) = \frac{K_1 \prod_{j=1}^{m} (s + Z_j)}{\prod_{i=1}^{n} (s + r_i)}, \tag{7.87}$$

and k is the parameter. The root sensitivity relates the changes in the location of the root in the s-plane to the change in the parameter. The root sensitivity is related to the logarithmic sensitivity by the relation

$$S_k^T = \frac{\partial \ln K_1}{\partial \ln k} - \sum_{i=1}^{n} \frac{\partial r_i}{\partial \ln k} \cdot \frac{1}{(s + r_i)} \tag{7.88}$$

when the zeros of $T(s)$ are independent of the parameter k so that

$$\frac{\partial Z_j}{\partial \ln k} = 0.$$

This logarithmic sensitivity can be readily obtained by determining the derivative of $T(s)$, Eq. (7.87), with respect to k. For the particular case, when the gain of the system is independent of the parameter k, we have

$$S_k^T = -\sum_{i=1}^{n} S_k^{r_i} \cdot \frac{1}{(s + r_i)}, \tag{7.89}$$

and the two sensitivity measures are directly related.

The evaluation of the root sensitivity for a control system can be readily accomplished by utilizing the root locus methods of the preceding section. The root sensitivity $S_k^{r_i}$ may be evaluated at root r_i by examining the root contours for the parameter k. We can change k by a small, finite amount Δk and evaluate the new, modified root $(r_i + \Delta r_i)$ at $k + \Delta k$. Then, using Eq. (7.86), we have

$$S_k^{r_i} \cong \frac{\Delta r_i}{\Delta k/k}. \tag{7.90}$$

Equation (7.90) is an approximation that approaches the actual value of the sensitivity as $\Delta k \to 0$. An example will illustrate the process of evaluating the root sensitivity.

EXAMPLE 7.6   Root sensitivity of a control system

The characteristic equation of the feedback control system shown in Fig. 7.23 is

$$1 + \frac{K}{s(s + \beta)} = 0$$

FIGURE 7.23
A feedback control
system.

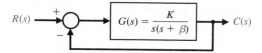

or, alternatively,

$$s^2 + \beta s + K = 0. \tag{7.91}$$

The gain K will be considered to be the parameter α. Then the effect of a change in each parameter can be determined by utilizing the relations

$$\alpha = \alpha_0 \pm \Delta\alpha, \qquad \beta = \beta_0 \pm \Delta\beta,$$

where α_0 and β_0 are the nominal or desired values for the parameters α and β, respectively. We shall consider the case when the nominal pole value is $\beta_0 = 1$ and the desired gain is $\alpha_0 = K = 0.5$. Then the root locus as a function of $\alpha = K$ can be obtained by utilizing the root locus equation

$$1 + \frac{K}{s(s + \beta_0)} = 1 + \frac{K}{s(s + 1)} = 0 \tag{7.92}$$

as shown in Fig. 7.24. The nominal value of gain $K = \alpha_0 = 0.5$ results in two complex roots, $r_1 = -0.5 + j0.5$ and $r_2 = \hat{r}_1$, as shown in Fig. 7.24. To evaluate the effect of unavoidable changes in the gain, the characteristic equation with $\alpha = \alpha_0 \pm \Delta\alpha$ becomes

$$s^2 + s + \alpha_0 \pm \Delta\alpha = s^2 + s + 0.5 \pm \Delta\alpha$$

or

$$1 + \frac{\pm\Delta\alpha}{s^2 + s + 0.5} = 1 + \frac{\pm\Delta\alpha}{(s + r_1)(s + \hat{r}_1)} = 0. \tag{7.93}$$

Therefore the effect of changes in the gain can be evaluated from the root locus of Fig. 7.24. For a 20% change in α, we have $\pm\Delta\alpha = \pm 0.1$. The root locations for a gain $\alpha = 0.4$ and $\alpha = 0.6$ are readily determined by root locus methods, and the root locations for $\pm\Delta\alpha = \pm 0.1$ are shown in Fig. 7.24. When $\alpha = K = 0.6$, the root in the second quadrant of the s-plane is

$$r_1 + \Delta r_1 = -0.5 + j0.59,$$

FIGURE 7.24
The root locus
for K.

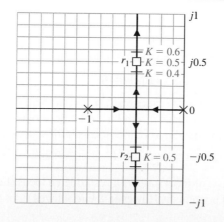

and the change in the root is $\Delta r_1 = +j0.09$. When $\alpha = K = 0.4$, the root in the second quadrant is

$$r_1 + \Delta r_1 = -0.5 + j0.387,$$

and the change in the root is $\Delta r = -j0.11$. Thus the root sensitivity for r_1 is

$$S^{r_1}_{+\Delta K} = S^{r_1}_{K+} = \frac{\Delta r_1}{\Delta K/K} = \frac{+j0.09}{+0.2} = j0.45 = 0.45\ \underline{/+90°} \tag{7.94}$$

for positive changes of gain. For negative increments of gain, the sensitivity is

$$S^{r_1}_{-\Delta K} = S^{r_1}_{K-} = \frac{\Delta r_1}{\Delta K/K} = \frac{-j0.11}{+0.2} = -j0.55 = 0.55\ \underline{/-90°}.$$

Of course, for infinitesimally small changes in the parameter ∂K, the sensitivity will be equal for negative or positive increments in K. The angle of the root sensitivity indicates the direction the root would move as the parameter varies. The angle of movement for $+\Delta\alpha$ is always 180° minus the angle of movement for $-\Delta\alpha$ at the point $\alpha = \alpha_0$.

The pole β varies due to environmental changes, and it may be represented by $\beta = \beta_0 + \Delta\beta$, where $\beta_0 = 1$. Then the effect of variation of the poles is represented by the characteristic equation

$$s^2 + s + \Delta\beta s + K = 0,$$

or, in root locus form, we have

$$1 + \frac{\Delta\beta s}{s^2 + s + K} = 0. \tag{7.95}$$

Again, the denominator of the second term is the unchanged characteristic equation when $\Delta\beta = 0$. The root locus for the unchanged system ($\Delta\beta = 0$) is shown in Fig. 7.24 as a function of K. For a design specification requiring $\zeta = 0.707$, the complex roots lie at

$$r_1 = -0.5 + j0.5 \qquad \text{and} \qquad r_2 = \hat{r}_1 = -0.5 - j0.5.$$

Then, because the roots are complex conjugates, the root sensitivity for r_1 is the conjugate of the root sensitivity for $\hat{r}_1 = r_2$. Using the parameter root locus techniques discussed in the preceding section, we obtain the root locus for $\Delta\beta$ as shown in Fig. 7.25. We are normally interested in the effect of a variation for the parameter so that $\beta = \beta_0 \pm \Delta\beta$, for which the locus as $\Delta\beta$ decreases is obtained from the root locus equation

$$1 + \frac{-(\Delta\beta)s}{s^2 + s + K} = 0. \tag{7.96}$$

Examining Eq. (7.96), we note that the equation is of the form

$$1 - kP(s) = 0.$$

Comparing this equation with Eq. (7.23) (Section 7.3), we find that the sign preceding the gain k is negative in this case. In a manner similar to the development of the root locus method in Section 7.3, we require that the root locus satisfy the equations

$$|kP(s)| = 1, \quad \underline{/P(s)} = 0° \pm q360°, \tag{7.97}$$

where $q = 0, 1, 2, \ldots$. The locus of roots follows a zero-degree locus (Eq. 7.97) in contrast with the 180° locus considered previously. However, the root locus rules of Section 7.3 may

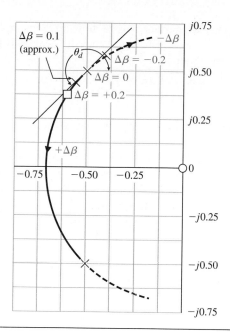

FIGURE 7.25
The root locus for
the parameter β.

be altered to account for the zero-degree phase angle requirement, and then the root locus may be obtained as in the preceding sections. Therefore, to obtain the effect of reducing β, one determines the zero-degree locus in contrast to the 180° locus as shown by a dotted locus in Fig. 7.25. Therefore, to find the effect of a 20% change of the parameter β, we evaluate the new roots for $\pm\Delta\beta = \pm0.20$ as shown in Fig. 7.25. The root sensitivity is readily evaluated graphically and, for a positive change in β, is

$$S^{r_1}_{\beta+} = \frac{\Delta r_1}{(\Delta\beta/\beta)} = \frac{0.16\,\underline{/-131°}}{0.20} = 0.80\,\underline{/-131°}. \qquad (7.98)$$

The root sensitivity for a negative change in β is

$$S^{r_1}_{\beta-} = \frac{\Delta r_1}{(\Delta\beta/\beta)} = \frac{0.125\,\underline{/38°}}{0.20} = 0.625\,\underline{/+38°}. \qquad (7.99)$$

As the percentage change $(\Delta\beta/\beta)$ decreases, the sensitivity measures, $S^{r_1}_{\beta+}$ and $S^{r_1}_{\beta-}$, will approach equality in magnitude and a difference in angle of 180°. Thus, for small changes when $\Delta\beta/\beta \leq 0.10$, the sensitivity measures are related as

$$|S^{r_1}_{\beta+}| = |S^{r_1}_{\beta-}| \qquad (7.100)$$

and

$$\underline{/S^{r_1}_{\beta+}} = 180° + \underline{/S^{r_1}_{\beta-}}. \qquad (7.101)$$

Often the desired root sensitivity measure is for small changes in the parameter. When the relative change in the parameter is such that $\Delta\beta/\beta = 0.10$, a root locus approximation is satisfactory. The root locus for Eq. (7.97) when $\Delta\beta$ is varying leaves the pole at $\Delta\beta = 0$ at an angle of departure θ_d. Since θ_d is readily evaluated, one can estimate the increment in the root change by approximating the root locus with the line at θ_d. This approximation is shown in Fig. 7.25 and is accurate for only relatively small changes in $\Delta\beta$. However, the use of this approximation allows the analyst to avoid drawing the complete root locus diagram. Therefore, for Fig. 7.25, the root sensitivity may be evaluated for $\Delta\beta/\beta = 0.10$ along the departure line, and one obtains

$$S_{\beta+}^{r_1} = \frac{0.74 \, \underline{/-135°}}{0.10} = 0.74 \, \underline{/-135°}. \tag{7.102}$$

The root sensitivity measure for a parameter variation is useful for comparing the sensitivity for various design parameters and at different root locations. Comparing Eq. (7.102) for β with Eq. (7.93) for α, we find that the sensitivity for β is greater in magnitude by approximately 50% and that the angle for $S_{\beta-}^{r_1}$ indicates that the approach of the root toward the $j\omega$-axis is more sensitive for changes in β. Therefore the tolerance requirements for β would be more stringent than for α. This information provides the designer with a comparative measure of the required tolerances for each parameter. ■

EXAMPLE 7.7   **Root sensitivity to a parameter**

A unity feedback control system has a forward transfer function

$$G(s) = \frac{20.7(s + 3)}{s(s + 2)(s + \beta)}, \tag{7.103}$$

where $\beta = \beta_0 + \Delta\beta$ and $\beta_0 = 8$. The characteristic equation as a function of $\Delta\beta$ is

$$s(s + 2)(s + 8 + \Delta\beta) + 20.7(s + 3) = 0$$

or

$$s(s + 2)(s + 8) + \Delta\beta s(s + 2) + 20.7(s + 3) = 0. \tag{7.104}$$

When $\Delta\beta = 0$, the roots may be determined by the root locus method or the Newton–Raphson method, and thus we evaluate the roots as

$$r_1 = -2.36 + j2.48, \qquad r_2 = \hat{r}_1, \qquad r_3 = -5.27.$$

The root locus for $\Delta\beta$ is determined by using the root locus equation

$$1 + \frac{\Delta\beta s(s + 2)}{(s + r_1)(s + \hat{r}_1)(s + r_3)} = 0. \tag{7.105}$$

The poles and zeros of Eq. (7.105) are shown in Fig. 7.26. The angle of departure at r_1 is evaluated from the angles as follows:

$$180° = -(\theta_d + 90° + \theta_{p3}) + (\theta_{z_1} + \theta_{z_2}) \tag{7.106}$$
$$= -(\theta_d + 90° + 40°) + (133° + 98°).$$

Therefore $\theta_d = -80°$ and the locus is approximated near r_1 by the line at an angle of θ_d. For a change of $\Delta r_1 = 0.2 \, \underline{/-80°}$ along the departure line, the $+\Delta\beta$ is evaluated by determining the vector lengths from the poles and zeros. Then we have

$$+\Delta\beta = \frac{4.8(3.75)(0.2)}{(3.25)(2.3)} = 0.48. \tag{7.107}$$

Therefore the sensitivity at r_1 is

$$S_\beta^{r_1} = \frac{\Delta r_1}{\Delta\beta/\beta} = \frac{0.2 \, \underline{/-80°}}{0.48/8} = 3.34 \, \underline{/-80°}, \tag{7.108}$$

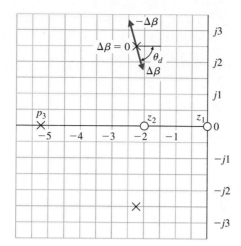

FIGURE 7.26
Pole and zero
diagram for the
parameter β.

which indicates that the root is quite sensitive to this 6% change in the parameter β. For comparison, it is worthwhile to determine the sensitivity of the root, r_1, to a change in the zero, $s = -3$. Then the characteristic equation is

$$s(s + 2)(s + 8) + 20.7(s + 3 + \Delta\gamma) = 0$$

or

$$1 + \frac{20.7\,\Delta\gamma}{(s + r_1)(s + \hat{r}_1)(s + r_3)} = 0. \tag{7.109}$$

The pole–zero diagram for Eq. (7.109) is shown in Fig. 7.27. The angle of departure at root r_1 is $180° = -(\theta_d + 90° + 40°)$ or

$$\theta_d = +50°. \tag{7.110}$$

FIGURE 7.27
Pole–zero diagram
for the parameter γ.

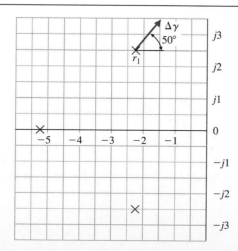

For a change of $r_1 = 0.2 \underline{/+50°}$, the $\Delta\gamma$ is positive, and obtaining the vector lengths, we find

$$|\Delta\gamma| = \frac{5.22(4.18)(0.2)}{20.7} = 0.21. \tag{7.111}$$

Therefore the sensitivity at r_1 for $+\Delta\gamma$ is

$$S_\gamma^{r_1} = \frac{\Delta r_1}{\Delta\gamma/\gamma} = \frac{0.2 \underline{/+50°}}{0.21/3} = 2.84 \underline{/+50°}. \tag{7.112}$$

Thus we find that the magnitude of the root sensitivity for the pole β and the zero γ is approximately equal. However, the sensitivity of the system to the pole can be considered to be less than the sensitivity to the zero because the angle of the sensitivity $S_\gamma^{r_1}$ is equal to $+50°$ and the direction of the root change is toward the $j\omega$-axis.

Evaluating the root sensitivity in the manner of the preceding paragraphs, we find that for the pole $s = -\delta_0 = -2$ the sensitivity is

$$S_{\delta-}^{r_1} = 2.1 \underline{/+27°}. \tag{7.113}$$

Thus, for the parameter δ, the magnitude of the sensitivity is less than for the other parameters, but the direction of the change of the root is more important than for β and γ. ■

To utilize the root sensitivity measure for the analysis and design of control systems, a series of calculations must be performed for various selections of possible root configurations and the zeros and poles of the open-loop transfer function. Therefore the use of the root sensitivity measure as a design technique is somewhat limited by the relatively large number of calculations required and by the lack of an obvious direction for adjusting the parameters in order to provide a minimized or reduced sensitivity. However, the root sensitivity measure can be utilized as an analysis measure, which permits the designer to compare the sensitivity for several system designs based on a suitable method of design. The root sensitivity measure is a useful index of sensitivity of a system to parameter variations expressed in the s-plane. The weakness of the sensitivity measure is that it relies on the ability of the root locations to represent the performance of the system. As we have seen in the preceding chapters, the root locations represent the performance quite adequately for many systems, but due consideration must be given to the location of the zeros of the closed-loop transfer function and the dominancy of the pertinent roots. The root sensitivity measure is a suitable measure of system performance sensitivity and can be used reliably for system analysis and design.

7.7 GAIN PLOTS

The root locus depicts the location of the characteristic roots (the eigenvalues) in the complex s-plane as the proportional gain (or any real-valued parameter) is changed. The open-loop transfer function is $G(s)$, and the closed-loop transfer function (see Fig. 7.28) is

$$T(s) = \frac{KG(s)}{1 + KG(s)}$$

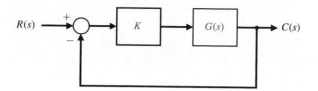

FIGURE 7.28
A closed-loop
system.

for the gain K. A root locus displays the migration of the roots of $1 + KG(s)$ for $0 \le K < \infty$.

An alternative visualization of the root locus plot can be obtained by graphing (1) the magnitude of the roots versus gain and (2) the angle of the roots versus gain. The magnitude gain plot employs a log-log scale, whereas the angle gain plot uses a semilog scale. In these plots, the solid colored lines track the loci of the poles that start at $s = -1$ and $s = -2$ [8, 9].

For example, consider the case where

$$G(s) = \frac{(s + 3)}{(s + 1)(s + 2)}. \tag{7.114}$$

The root locus for $KG(s)$ is shown in Fig. 7.29. It is possible to show the gain graduation on the locus with tick marks denoting equal values of K. However, even if the root locus is scaled, it is not convenient for determining the gain associated with a given point on the locus. For example, from Fig. 7.29, the gain $K = 5.8$ generating the break-in point at $s \approx -4.4$ cannot be determined by inspection.

The plots of the magnitude of the characteristic roots versus gain K and the angle of the characteristic roots versus gain K are shown in Fig. 7.30 for $KG(s)$ for $G(s)$ of Eq. (7.114). The roots start at $s = -1$ and $s = -2$ for small K and come together and then break away when $K = 0.17$. The roots are complex for $0.17 < K < 5.83$.

Performance measures are available directly from the gain plots. In particular, for complex conjugate eigenvalues, the natural frequency, ω_n(rad/s), is the magnitude presented in the magnitude plot, and the damping ratio, $\zeta = -\cos(\theta)$, where θ is the angle in the angle

FIGURE 7.29
The root locus for
$KG(s)$.

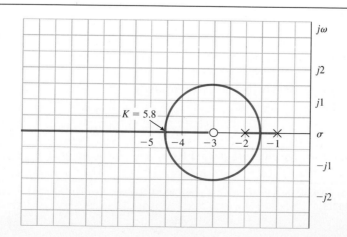

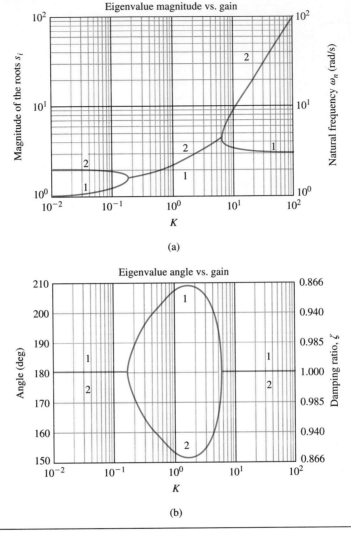

FIGURE 7.30
(a) Magnitude of the roots versus K.
(b) Angle of the roots versus K.

plot. In Fig. 7.30, additional axes have been added to display ω_n and ζ. *Matlab* software is available for the calculation and plotting of the gain plots.[†]

It is also possible to plot the root sensitivity

$$S_K^{r_i} = \frac{d \ln r_i}{d \ln K} = \frac{dr_i/r_i}{dK/K}.$$

The root sensitivity plots are plotted for $|S|$ and the angle of S versus K. The sensitivity

---

[†] Software is available from Professors T. Kurfess and M. Nagurka, Carnegie-Mellon University, Pittsburgh, Pennsylvania 15213.

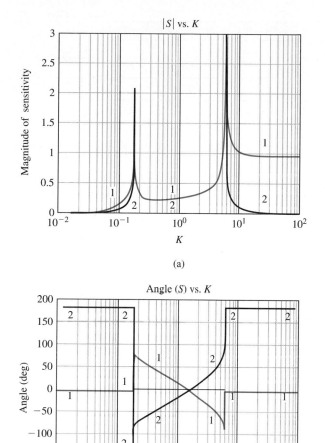

FIGURE 7.31
(a) Magnitude of the sensitivity of the roots s_i to K (the numbers 1 and 2 indicate the two roots).
(b) Angle of the sensitivity of the roots s_i to changes in K.

plots for the $G(s)$ of Eq. (7.114) are shown in Fig. 7.31. They clearly highlight the infinite sensitivities at the breakaway and entry points on the real s-axis.

7.8 DESIGN EXAMPLE: LASER MANIPULATOR CONTROL SYSTEM

Lasers can be used to drill the hip socket for appropriate insertion of an artificial hip joint. The use of lasers for surgery requires high accuracy for position and velocity response. Let us consider the system shown in Fig. 7.32, which uses a dc motor manipulator for the laser. The amplifier gain K must be adjusted so that the steady-state error for a ramp input,

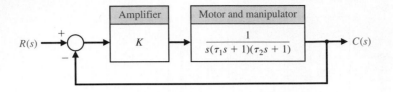

$R(s)$ ⟶ + ⊙ − ⟶ Amplifier K ⟶ Motor and manipulator $\dfrac{1}{s(\tau_1 s + 1)(\tau_2 s + 1)}$ ⟶ $C(s)$

FIGURE 7.32
Laser manipulator
control system.

$r(t) = At$ (where $A = 1$ mm/s), is less than or equal to 0.1 mm while a stable response is maintained.

To obtain the steady-state error required and a good response, we select a motor with a field time constant $\tau_1 = 0.1s$ and a motor plus load time constant of $\tau_2 = 0.2s$. We then have

$$T(s) = \frac{KG(s)}{1 + KG(s)} = \frac{K}{s(\tau_1 s + 1)(\tau_2 s + 1) + K}$$

$$= \frac{K}{0.02s^3 + 0.3s^2 + s + K} = \frac{50K}{s^3 + 15s^2 + 50s + 50K}.$$ (7.115)

The steady-state error for a ramp, $R(s) = A/s^2$, from Eq. (5.29), is

$$e_{ss} = \frac{A}{K_v} = \frac{A}{K}.$$

Since we desire $e_{ss} = 0.1$ mm (or less) and $A = 1$ mm, we require $K = 10$ (or greater).

To ensure a stable system, we obtain the characteristic equation from Eq. (7.115) as

$$s^3 + 15s^2 + 50s + 50K = 0.$$

Establishing the Routh–Hurwitz array, we have

$$
\begin{array}{c|cc}
s^3 & 1 & 50 \\
s^2 & 15 & 50K \\
s^1 & b_1 & 0 \\
s_0 & 50K
\end{array}
$$

where

$$b_1 = \frac{750 - 50K}{15}.$$

Therefore the system is stable for

$$0 \le K \le 15.$$

Then using $K = 10$, where the system is stable, we examine the root locus for $K > 0$. Since there are three loci and the centroid $\sigma = -5$, we obtain the root locus shown in Fig. 7.33. The breakaway point is $s = -2.11$, and the roots at $K = 10$ are $r_2 = -13.98$, $r_1 = -0.51 + j5.96$, and $\hat{r}_1$. The ζ of the complex roots is 0.085 and $\zeta\omega_n = 0.51$. Thus, assuming that the complex roots are dominant, for a step input we expect (using Eq. 5.16) an overshoot of 76% and a settling time of

$$T_s = \frac{4}{\zeta\omega_n} = \frac{4}{0.51} = 7.8s.$$

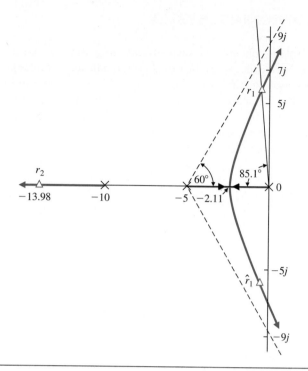

FIGURE 7.33
Root locus for a
laser control
system.

Plotting the actual system response, we find that the overshoot is 72% and that the settling time is 7.9 seconds. Thus the complex roots are essentially dominant. The system response to a step input is highly oscillatory and cannot be tolerated for laser surgery. The command signal must be limited to a low-velocity ramp signal. The response to a ramp signal is shown in Fig. 7.34.

FIGURE 7.34
The response to a
ramp input for a
laser control
system.

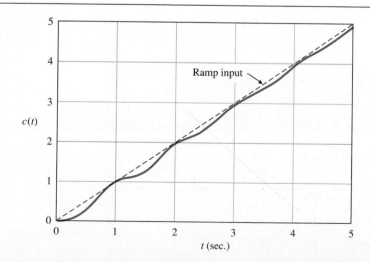

7.9 THE DESIGN OF A ROBOT CONTROL SYSTEM

The concept of robot replication is relatively easy to grasp. The central idea is that robots replicate themselves and develop a factory that automatically produces robots. An example of a robot replication facility is shown in Fig. 7.35. To achieve the rapid and accurate control of a robot, it is important to keep the robot arm stiff and yet lightweight [6].

The specifications for controlling the motion of a lightweight, flexible arm are (1) a settling time of less than 2 seconds, (2) a percent overshoot of less than 10% for a step input, and (3) a steady-state error of zero for a step input.

The block diagram of the proposed system with a controller is shown in Fig. 7.36. The configuration proposes the use of velocity feedback as well as the use of a controller $G_c(s)$. Since the robot is quite light and flexible, the transfer function of the arm is

$$\frac{C(s)}{U(s)} = \left(\frac{1}{s^2}\right) G(s)$$

and

$$G(s) = \frac{(s^2 + 4s + 10,004)(s^2 + 12s + 90,036)}{(s + 10)(s^2 + 2s + 2501)(s^2 + 6s + 22,509)}. \tag{7.116}$$

Therefore the complex zeros are located at

$$s = -2 \pm j100 \quad \text{and} \quad s = -6 \pm j300.$$

The complex poles are located at

$$s = -1 \pm j50 \quad \text{and} \quad s = -3 \pm j150.$$

FIGURE 7.35
A robot replication facility.

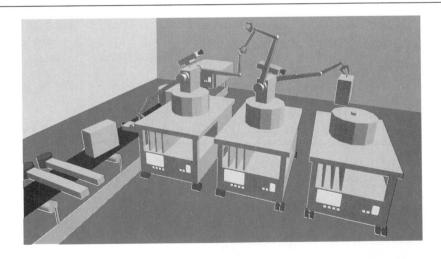

FIGURE 7.36
Proposed configuration for control of the lightweight robot arm.

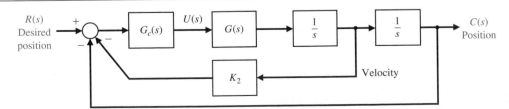

A sketch of the root locus when $K_2 = 0$ and the controller is an adjustable gain, $G_c(s) = K_1$, is shown in Fig. 7.37. Clearly, the system is unstable, since two roots of the characteristic equation appear in the right-hand s-plane for $K_1 > 0$.

It is clear that we need to introduce the use of velocity feedback by setting K_2 to a positive magnitude. Then we have $H(s) = (1 + K_2 s)$ and therefore the loop transfer function is

$$\left(\frac{1}{s^2}\right) G(s)H(s) = \frac{K_1 K_2 \left(s + \dfrac{1}{K_2}\right)(s^2 + 4s + 10{,}004)(s^2 + 12s + 90{,}036)}{s^2(s + 10)(s^2 + 2s + 2501)(s^2 + 6s + 22{,}509)},$$

where K_1 is the gain of $G_c(s)$. We now have available two parameters, K_1 and K_2, that we may adjust. We select $5 < K_2 < 10$ in order to place the adjustable zero near the origin.

When $K_2 = 5$ and K_1 is varied, we obtain the root locus sketched in Fig. 7.38. When $K_1 = 0.8$ and $K_2 = 5$, we obtain a step response with a percent overshoot of 12% and a settling time of 1.8 seconds. This is the optimum achievable response. If one tries $K_2 = 7$ or $K_2 = 4$, the overshoot will be larger than desired. Therefore we have achieved the best performance with this system. If we desired to continue the design process, we would use a controller $G_c(s)$ with a pole and zero in addition to retaining the velocity feedback with $K_2 = 5$.

One possible selection of a controller is

$$G_c(s) = \frac{K_1(s + z)}{(s + p)}. \tag{7.117}$$

If one selects $z = 1$ and $p = 5$, then when $K_1 = 5$ we obtain a step response with an overshoot of 8% and a settling time of 1.6 seconds.

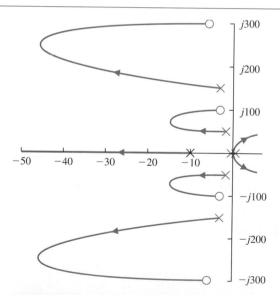

FIGURE 7.37   Root locus of the system if K_2 = 0 and K_1 is varied from $K_1 = 0$ to $K_1 = \infty$ and $G_c(s) = K_1$.

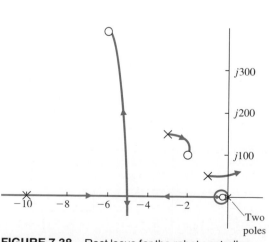

FIGURE 7.38   Root locus for the robot controller with a zero inserted at $s = -0.2$ with $G_c(s) = K_1$.

7.10 THE ROOT LOCUS USING *MATLAB*

An approximate root locus sketch can be obtained by applying the orderly procedure summarized in Table 7.2. Alternatively, we can use *Matlab* to obtain an accurate root locus plot. However, do not be tempted to rely solely on *Matlab* for obtaining root locus plots while neglecting the manual steps in developing an approximate root locus. The fundamental concepts behind the root locus method are embedded in the manual steps and it is essential to fully understand their application.

The section begins with a discussion on obtaining a root locus plot with *Matlab*. This is followed by a discussion of the connections between the partial fraction expansion, dominant poles, and the closed-loop system response. Root sensitivity is covered in the final paragraphs.

The functions covered in this section are rlocus, rlocfind, and residue. The functions rlocus and rlocfind are used to obtain root locus plots, and the residue function is utilized for partial fraction expansions of rational functions.

Obtaining a Root Locus Plot. Consider the closed-loop control system in Fig. 7.10 on page 328. The closed-loop transfer function is

$$T(s) = \frac{C(s)}{R(s)} = \frac{K(s + 1)(s + 3)}{s(s + 2)(s + 3) + K(s + 1)}.$$

The characteristic equation can be written as

$$1 + K\frac{(s + 1)}{s(s + 2)(s + 3)} = 0. \tag{7.118}$$

The form of the characteristic equation in Eq. (7.118) is necessary to use the rlocus function for generating root locus plots. The general form of the characteristic equation necessary for application of the rlocus function is

$$1 + k\frac{p(s)}{q(s)} = 0, \tag{7.119}$$

where k is the parameter of interest to be varied from $0 \leq k \leq \infty$. The rlocus function is shown in Fig. 7.39. The steps to obtaining the root locus plot associated with Eq. (7.118), along with the associated root locus plot, are shown in Fig. 7.40. Invoking the rlocus function without left-hand arguments results in an automatic generation of the root locus plot. When invoked with left-hand arguments, the rlocus function returns a matrix of root locations and the associated gain vector.

FIGURE 7.39
The **rlocus**
function.

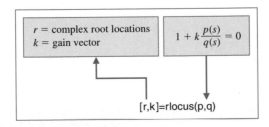

r = complex root locations
k = gain vector

$$1 + k\frac{p(s)}{q(s)} = 0$$

[r,k]=rlocus(p,q)

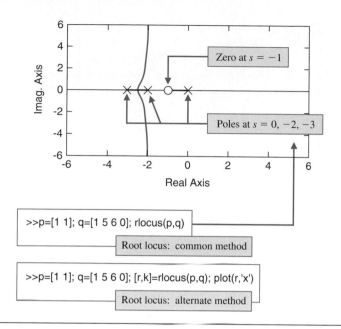

FIGURE 7.40
The root locus for
the characteristic
equation (7.118).

The steps to obtain a root locus plot with *Matlab* are as follows:

1. Obtain the characteristic equation in the form given in Eq. (7.119) where k is the parameter of interest.

2. Use the rlocus function to generate the plots.

Referring to Fig. 7.40, we can see that as K increases, two branches of the root locus break away from the real axis. This means that for some values of K, the closed-loop system characteristic equation will have two complex roots. Suppose we want to find the value of K corresponding to a pair of complex roots. We can use the rlocfind function to do this, but only after a root locus has been obtained with the rlocus function. Executing the rlocfind function will result in a cross-hair marker appearing on the root locus plot. You move the cross-hair marker to the location on the locus of interest and hit the enter key. The value of the parameter K and the value of the selected point will then be displayed in the command display. The use of the rlocfind function is illustrated in Fig. 7.41.

Continuing our third-order root locus example, we find that when $K = 20.5775$, the closed-loop transfer function has three poles and two zeros at

$$\text{poles: } s = \begin{pmatrix} -2.0505 + 4.3227i \\ -2.0505 - 4.3227i \\ -0.8989 \end{pmatrix}, \qquad \text{zeros: } s = \begin{pmatrix} -1 \\ -3 \end{pmatrix}.$$

Considering the closed-loop pole locations only, we would expect that the real pole at $s = -.8989$ would be the *dominant* pole. To verify this, we can study the closed-loop system response to a step input, $R(s) = 1/s$. For a step input we have

$$C(s) = \frac{20.5775(s + 1)(s + 3)}{s(s + 2)(s + 3) + 20.5775(s + 1)} \cdot \frac{1}{s}. \tag{7.120}$$

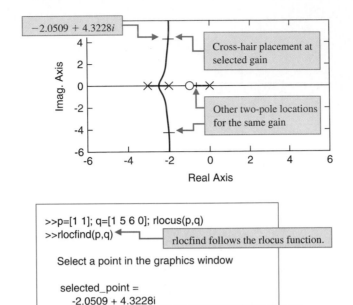

FIGURE 7.41
Using the **rlocfind** function.

Generally, the first step in computing $c(t)$ is to expand Eq. (7.120) in a partial fraction expansion. The **residue** function can be used to expand Eq. (7.120), as shown in Fig. 7.42. The **residue** function is described in Fig. 7.43.

The partial fraction expansion of Eq. (7.120) is

$$C(s) = \frac{-1.3786 + 1.7010i}{s + 2.0505 + 4.3228i} + \frac{-1.3786 - 1.7010i}{s + 2.0505 - 4.3228i} + \frac{-0.2429}{s + 0.8989} + \frac{3}{s}.$$

Comparing the residues, we see that the coefficient of the term corresponding to the pole at $s = -0.8989$ is considerably smaller than the coefficient of the terms corresponding to the complex-conjugate poles at $s = -2.0505 \pm 4.3227i$. From this we expect that the influence of the pole at $s = -0.8989$ on the output response $c(t)$ is not *dominant*. The settling time is then predicted by considering the complex-conjugate poles. The poles at $s = -2.0505 \pm 4.3227i$ correspond to a damping of $\zeta = 0.4286$ and a natural frequency of $\omega_n = 4.7844$. Thus the settling time is predicted to be

$$T_s \approx \frac{4}{\zeta \omega_n} = 1.95 \text{ seconds.}$$

Using the **step** function, as shown in Fig. 7.44, we find that $T_s \approx 1.6$ seconds. Hence our approximation of settling time $T_s \approx 1.95$ is a fairly good approximation. The percent overshoot can be predicted using Fig. 5.12, since the zero of $T(s)$, $s = -3$, will impact the system response. Using Fig. 5.12, we predict an overshoot of 60%. As can be seen in Fig. 7.44, the actual overshoot is 50%.

In this example, the role of the system zeros on the transient response is illustrated. The proximity of the zero at $s = -1$ to the pole at $s = -0.8989$ reduces the impact of that

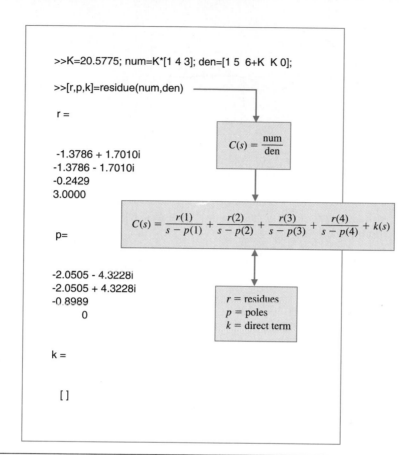

FIGURE 7.42
Partial fraction
expansion of
Eq. (7.120).

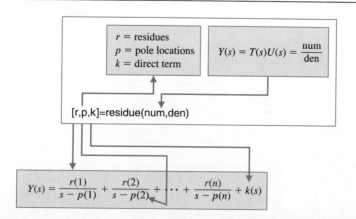

FIGURE 7.43
The residue
function.

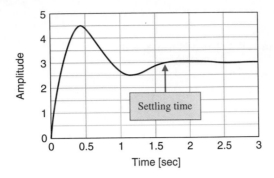

FIGURE 7.44
Step response for
the closed-loop
system in Fig. 7.10
with K = 20.5775.

```
>>K=20.5775;num=K*[1 4 3]; den=[1 5 6+K K];
>>step(num,den), grid
```

pole on the transient response. The main contributors to the transient response are the complex-conjugate poles at $s = -2.0505 \pm -4.3228i$ and the zero at $s = -3$.

One final point regarding the residue function: You can convert the partial fraction expansion back to the polynomials num/den, given the residues (r), the pole locations (p), and the direct terms (k), with the command shown in Fig. 7.45.

Sensitivity and the Root Locus.   The roots of the characteristic equation play an important role in defining the closed-loop system transient response. The effect of parameter variations on the roots of the characteristic equation is a useful measure of sensitivity. The root sensitivity can be defined to be

$$\frac{\partial r_i}{\partial k / k}. \tag{7.121}$$

We can utilize Eq. (7.121) to investigate the sensitivity of the roots of the characteristic equation to variations in the parameter k. If we change k by a small finite amount Δk, and evaluate the modified root $r_i + \Delta r_i$, it follows that

$$S_k^{r_i} = \frac{\Delta r_i}{\Delta k / k}. \tag{7.122}$$

The quantity $S_k^{r_i}$ is a complex number. Referring back to the third-order example of Fig. 7.10 (Eq. 7.118), if we change K a factor of 5%, we find that the dominant complex-

FIGURE 7.45
Converting a
partial fraction
expansion back to
a rational function.

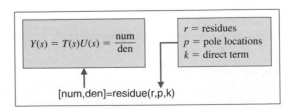

pfsensitivity.m

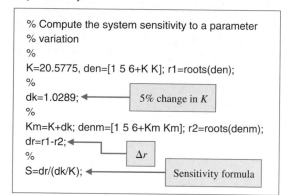

```
% Compute the system sensitivity to a parameter
% variation
%
K=20.5775, den=[1 5 6+K K]; r1=roots(den);
%
dk=1.0289;         ◄—————  5% change in K
%
Km=K+dk; denm=[1 5 6+Km Km]; r2=roots(denm);
dr=r1-r2; ◄—
%                   Δr
S=dr/(dk/K); ◄—————  Sensitivity formula
```

FIGURE 7.46
Sensitivity
calculations for
the root locus for
a 5% change in
$K = 20.5775$.

conjugate pole at $s = -2.0505 + 4.3228i$ changes by

$$\Delta r_i = -0.0025 - 0.1168i$$

when K changes from $K = 20.5775$ to $K = 21.6064$. From Eq. (7.122), it follows that

$$S_k^{r_i} = \frac{-0.0025 - 0.1168i}{1.0289/20.5775} = -0.0494 - 2.3355i.$$

The sensitivity $S_k^{r_i}$ can also be written in the form

$$S_k^{r_i} = 2.34 \ \underline{/268.79°}.$$

The magnitude and direction of $S_k^{r_i}$ provides a measure of the root sensitivity. The script used to perform these sensitivity calculations is shown in Fig. 7.46.

The root sensitivity measure may be useful for comparing the sensitivity for various system parameters at different root locations.

7.11 SUMMARY

The relative stability and the transient response performance of a closed-loop control system are directly related to the location of the closed-loop roots of the characteristic equation. Therefore we have investigated the movement of the characteristic roots on the s-plane as the system parameters are varied by utilizing the root locus method. The root locus method, a graphical technique, can be used to obtain an approximate sketch in order to analyze the initial design of a system and determine suitable alterations of the system structure and the parameter values. A computer is commonly used to calculate several accurate roots at important points on the locus. A summary of 15 typical root locus diagrams is shown in Table 9.6.

Furthermore, we extended the root locus method for the design of several parameters for a closed-loop control system. Then the sensitivity of the characteristic roots was investigated for undesired parameter variations by defining a root sensitivity measure. It is clear that the root locus method is a powerful and useful approach for the analysis and design of modern control systems and will continue to be one of the most important procedures of control engineering.

EXERCISES

E7.1 Let us consider a device that consists of a ball rolling on the inside rim of a hoop [11]. This model is similar to the problem of liquid fuel sloshing in a rocket. The hoop is free to rotate about its horizontal principal axis as shown in Fig. E7.1. The angular position of the hoop may be controlled via the torque T applied to the hoop from a torque motor attached to the hoop drive shaft. If negative feedback is used, the system characteristic equation is

$$1 + \frac{Ks(s + 4)}{s^2 + 2s + 2} = 0.$$

(a) Sketch the root locus. (b) Find the gain when the roots are both equal. (c) Find these two equal roots. (d) Find the settling time of the system when the roots are equal.

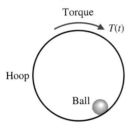

Torque

$T(t)$

Hoop

Ball

FIGURE E7.1 Hoop rotated by motor.

E7.2 A tape recorder has a speed control system so that $H(s) = 1$ with negative feedback and

$$G(s) = \frac{K}{s(s + 2)(s^2 + 4s + 5)}.$$

(a) Draw a root locus for K, and show that the dominant roots are $s = -0.35 \pm j0.80$ when $K = 6.5$. (b) For the dominant roots of part (a), calculate the settling time and overshoot for a step input.

E7.3 A control system for an automobile suspension tester has negative unity feedback and a process [12]

$$G(s) = \frac{K(s^2 + 4s + 8)}{s^2(s + 4)}.$$

It is desired that the dominant roots have a ζ equal to 0.5. Using the root locus, show that $K = 7.35$ is required and the dominant roots are $s = -1.3 \pm j2.2$.

E7.4 Consider a unity feedback system with

$$G(s) = \frac{K(s + 1)}{s^2 + 4s + 5}.$$

(a) Find the angle of departure of the root locus from the complex poles. (b) Find the entry point for the root locus as it enters the real axis.

Answers: $\pm 225°$; $- 2.4$

E7.5 Consider a feedback system with a loop transfer function

$$GH(s) = \frac{K}{(s + 1)(s + 3)(s + 6)}.$$

(a) Find the breakaway point on the real axis. (b) Find the asymptote centroid. (c) Find the value of K at the breakaway point.

E7.6 The United States is planning to have an operating space station in orbit by the late-1990s. One version of the space station is shown in Fig. E7.6. It is critical to keep this station in the proper orientation toward the sun and the earth for generating power and communications. The orientation controller may be represented by a unity feedback system with an actuator and controller:

$$G(s) = \frac{K(s + 20)}{s(s^2 + 24s + 144)}.$$

Sketch the root locus of the system as K increases. Find the value of K that results in an oscillatory response.

E7.7 The elevator in a modern office building travels at a top speed of 25 feet per second and is still able to stop within one-eighth of an inch of the floor outside. The transfer function of the unity feedback elevator position control is

$$G(s) = \frac{K(s + 10)}{s(s + 1)(s + 20)(s + 50)}.$$

Determine the gain K when the complex roots have a ζ equal to 0.8.

E7.8 Draw the root locus for a unity feedback system (Fig. 7.1) with

$$G(s) = \frac{K(s + 1)}{s^2(s + 9)}.$$

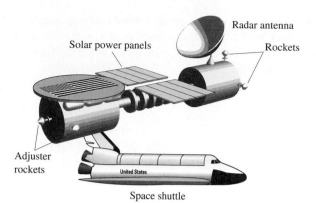

Radar antenna

Solar power panels

Rockets

Adjuster
rockets

United States

FIGURE E7.6
Space station.

Space shuttle

(a) Find the gain when all three roots are real and equal. (b) Find the roots when all the roots are equal as in part (a).

Answers: $K = 27$; $s = -3$

E7.9 The world's largest telescope, completed in 1990, is located in Hawaii. The primary mirror has a diameter of 10 m and consists of a mosaic of 36 hexagonal segments with the orientation of each segment actively controlled. This unity feedback system for the mirror segments (Fig. 7.1) has

$$G(s) = \frac{K}{s(s^2 + 2s + 5)}.$$

(a) Find the asymptotes and draw them in the s-plane. (b) Find the angle of departure from the complex poles. (c) Determine the gain when two roots lie on the imaginary axis. (d) Sketch the root locus.

E7.10 A unity feedback system (Fig. 7.1) has

$$KG(s) = \frac{K(s + 2)}{s(s + 1)}.$$

(a) Find the breakaway and entry points on the real axis. (b) Find the gain and the roots when the real part of the complex roots is located at -2. (c) Sketch the locus.

Answers: (a) -0.59, -3.41; (b) $K = 3$, $s = -2 \pm j\sqrt{2}$

E7.11 A robot force control system with unity feedback (Fig. 7.1) has a plant [6]

$$KG(s) = \frac{K(s + 2.5)}{(s^2 + 2s + 2)(s^2 + 4s + 5)}.$$

(a) Find the gain K that results in dominant roots with a damping ratio of 0.707. Sketch the locus. (b) Find the actual percent overshoot and peak time for the gain K of part (a).

E7.12 A unity feedback system (Fig. 7.1) has a plant

$$KG(s) = \frac{K(s + 1)}{s(s^2 + 4s + 8)}.$$

Find (a) the root locus for $K > 0$; (b) the roots when $K = 10$ and 20; (c) the 0–100% rise time, percent overshoot, and settling time of the system for a unit step input when $K = 10$ and 20.

E7.13 A unity feedback system has a process

$$G(s) = \frac{4(s + z)}{s(s + 1)(s + 3)}.$$

(a) Draw the root locus as z varies from 0 to 100. (b) Using the locus, estimate the percent overshoot and settling time of the system at $z = 0.6$, 2, and 4 for a step input. (c) Determine the actual overshoot and settling time at $z = 0.6$, 2, and 4.

E7.14 A unity feedback system has the process

$$G(s) = \frac{K(s + 10)}{s(s + 5)}.$$

(a) Determine the breakaway and entry points of the root locus, and sketch the root locus for $K > 0$.
(b) Determine the gain K when the two characteristic roots have a ζ of $1/\sqrt{2}$. (c) Calculate the roots.

E7.15   (a) Plot the root locus for

$$GH(s) = \frac{K(s + 1)(s + 3)}{s^3}.$$

(b) Calculate the range of K for which the system is stable. (c) Predict the steady-state error of the system for a ramp input.

E7.16   A negative unity feedback system has a plant transfer function

$$G(s) = \frac{Ke^{-sT}}{s + 1},$$

where $T = 0.1$ second. Show that an approximation for the time delay is

$$e^{-sT} \cong \frac{\left(\dfrac{2}{T} - s\right)}{\left(\dfrac{2}{T} + s\right)}.$$

Using

$$e^{-0.1s} = \frac{20 - s}{20 + s},$$

obtain the root locus for the system for $K > 0$. Determine the range of K for which the system is stable.

E7.17   A control system as shown in Fig. E7.17 has a plant

$$G(s) = \frac{1}{s(s - 1)}.$$

(a) When $G_c(s) = K$, show that the system is always unstable by sketching the root locus.
(b) When

$$G_c(s) = \frac{K(s + 2)}{(s + 20)},$$

sketch the root locus and determine the range of K for which the system is stable. Determine the value of K and the complex roots when two roots lie on the $j\omega$-axis.

FIGURE E7.17   Feedback system.

E7.18   A closed-loop negative feedback system is used to control the yaw of the A-6 Intruder attack jet, which was widely used in the Persian Gulf war. When $H(s) = 1$ and

$$G(s) = \frac{K}{s(s + 3)(s^2 + 2s + 2)},$$

determine (a) the root locus breakaway point and (b) the value of the roots on the $j\omega$-axis and the gain required for those roots. Sketch the root locus.

Answer: breakaway: $s = -2.29$
$j\omega$-axis: $s = \pm j1.09$, $K = 8$

E7.19   A unity feedback system has a plant

$$G(s) = \frac{K}{s(s + 3)(s^2 + 6s + 64)}.$$

(a) Determine the angle of departure of the root locus at the complex poles. (b) Sketch the root locus. (c) Determine the gain K when the roots are on the $j\omega$-axis, and determine the location of these roots.

E7.20   A unity feedback system has a plant

$$G(s) = \frac{K(s + 1)}{s(s - 1)(s + 4)}.$$

(a) Determine the range of K for stability. (b) Draw the root locus. (c) Determine the maximum ζ of the stable complex roots.

E7.21   A unity feedback system has a plant

$$G(s) = \frac{Ks}{s^3 + 5s^2 + 10}.$$

Plot the root locus. Also determine the gain K when the complex roots of the characteristic equation have a ζ approximately equal to 0.66.

E7.22   A high-performance missile for launching a satellite has a unity feedback system with a plant transfer function

$$G(s) = \frac{K(s^2 + 10)(s + 2)}{(s^2 - 2)(s + 10)}.$$

Sketch the root locus as K varies greater than 1.

E7.23   A unity feedback system has a plant

$$G(s) = \frac{4(s^2 + 1)}{s(s + a)}.$$

Plot the root locus as a varies when $a \geq 0$.

PROBLEMS

P7.1 Draw the root locus for the following open-loop transfer functions of the system shown in Fig. P7.1 when $0 < K < \infty$:

(a) $GH(s) = \dfrac{K}{s(s + 1)^2}$

(b) $GH(s) = \dfrac{K}{(s^2 + 2s + 2)(s + 2)}$

(c) $GH(s) = \dfrac{K(s + 1)}{s(s + 2)(s + 3)}$

(d) $GH(s) = \dfrac{K(s^2 + 4s + 8)}{s^2(s + 4)}$

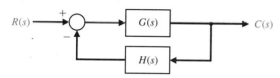

FIGURE P7.1

P7.2 The linear model of a phase detector was presented in Problem 6.7. Draw the root locus as a function of the gain $K_v = K_a K$. Determine the value of K_v attained if the complex roots have a damping ratio equal to 0.60 [13].

P7.3 A unity feedback system has

$$G(s) = \frac{K}{s(s + 1)(s + 4)}.$$

Find (a) the breakaway point in the real axis and the gain K for this point, (b) the gain and the roots when two roots lie on the imaginary axis, and (c) the roots when $K = 2.5$. (d) Sketch the root locus.

P7.4 The analysis of a large antenna was presented in Problem 4.5. Plot the root locus of the system as

$0 < k_a < \infty$. Determine the maximum allowable gain of the amplifier for a stable system.

P7.5 Automatic control of helicopters is necessary because, unlike fixed-wing aircraft, which possess a fair degree of inherent stability, the helicopter is quite unstable. A helicopter control system that utilizes an automatic control loop plus a pilot stick control is shown in Fig. P7.5. When the pilot is not using the control stick, the switch may be considered to be open. The dynamics of the helicopter are represented by the transfer function

$$G_2(s) = \frac{25(s + 0.03)}{(s + 0.4)(s^2 - 0.36s + 0.16)}.$$

(a) With the pilot control loop open (hands-off control), plot the root locus for the automatic stabilization loop. Determine the gain K_2 that results in a damping for the complex roots equal to $\zeta = 0.707$. (b) For the gain K_2 obtained in part (a), determine the steady-state error due to a wind gust $T_d(s) = 1/s$. (c) With the pilot loop added, draw the root locus as K_1 varies from zero to ∞ when K_2 is set at the value calculated in part (a). (d) Recalculate the steady-state error of part (b) when K_1 is equal to a suitable value based on the root locus.

P7.6 An attitude control system for a satellite vehicle within the earth's atmosphere is shown in Fig. P7.6. The transfer functions of the system are

$$G(s) = \frac{K(s + 0.20)}{(s + 0.90)(s - 0.60)(s - 0.10)},$$

$$G_c(s) = \frac{(s + 2 + j1.5)(s + 2 - j1.5)}{(s + 4.0)}.$$

(a) Draw the root locus of the system as K varies from 0 to ∞. (b) Determine the gain K that results in a system with a 2.5% settling time less than 12

FIGURE P7.5
Helicopter control.

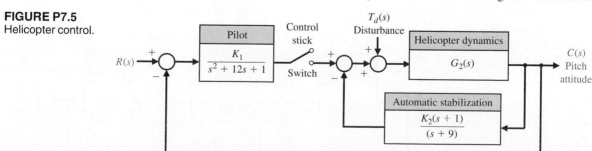

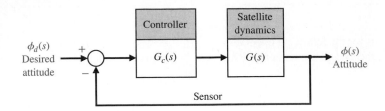

FIGURE P7.6
Satellite attitude
control.

seconds and a damping ratio for the complex roots greater than 0.50.

P7.7 The speed control system for an isolated power system is shown in Fig. P7.7. The valve controls the steam flow input to the turbine in order to account for load changes, $\Delta L(s)$, within the power distribution network. The equilibrium speed desired results in a generator frequency equal to 60 cps. The effective rotary inertia, J, is equal to 4000 and the friction constant, f, is equal to 0.75. The steady-state speed regulation factor, R, is represented by the equation $R \cong (\omega_0 - \omega_r)/\Delta L$, where ω_r equals the speed at rated load and ω_0 equals the speed at no load. Clearly, it is desired to obtain a very small R, usually less than 0.10. (a) Using root locus techniques, determine the regulation, R, attainable when the damping ratio of the roots of the system must be greater than 0.60. (b) Verify that the steady-state speed deviation for a load torque change, $\Delta L(s) = \Delta L/s$, is, in fact, approximately equal to $R\Delta L$ when $R \leq 0.1$.

P7.8 Reconsider the power control system of Problem 7.7 when the steam turbine is replaced by a hydroturbine. For hydroturbines, the large inertia of the water used as a source of energy causes a considerably larger time constant. The transfer function of a hydroturbine may be approximated by

$$G_t(s) = \frac{-\tau s + 1}{(\tau/2)s + 1},$$

where $\tau = 1$ second. With the rest of the system

remaining as given in Problem 7.7, repeat parts (a) and (b) of Problem 7.7.

P7.9 The achievement of safe, efficient control of the spacing of automatically controlled guided vehicles is an important part of future use of the vehicles in a manufacturing plant [14, 15]. It is important that the system eliminate the effects of disturbances such as oil on the floor as well as maintain accurate spacing between vehicles on a guideway. The system can be represented by the block diagram of Fig. P7.9. The vehicle dynamics can be represented by

$$G(s) = \frac{(s + 0.1)(s^2 + 2s + 289)}{s(s - 0.4)(s + 0.8)(s^2 + 1.45s + 361)}.$$

(a) Neglect the pole of the feedback sensor and draw the root locus of the system. (b) Determine all the roots when the loop gain $K = K_1K_2K_4/250$ is equal to 4000.

P7.10 Unlike the present day *Concorde*, a turn-of-the-century supersonic passenger jet would have the range to cross the Pacific in a single hop and the efficiency to make it economical [16]. This new aircraft shown in Fig. P7.10(a) will require the use of temperature-resistant, lightweight materials and advanced computer control systems.

The plane would carry 300 passengers at three times the speed of sound for up to 7500 miles. The flight control system requires good quality handling and comfortable flying conditions. An auto-

FIGURE P7.7
Power system
control.

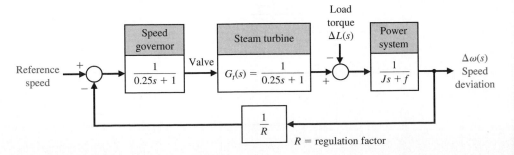

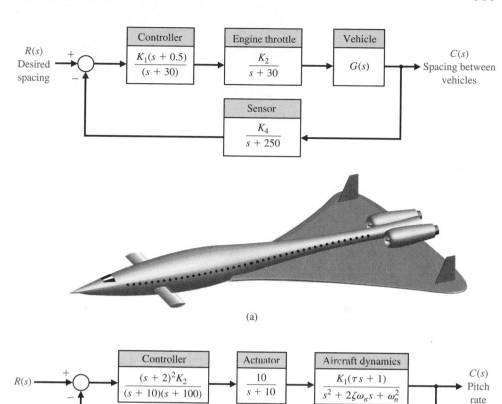

FIGURE P7.9
Guided vehicle
control.

FIGURE P7.10
(a) A supersonic
jet aircraft of the
future. (b) Control
system.

matic flight control system can be designed for SST vehicles. The desired characteristics of the dominant roots of the control system shown in Fig. P7.10(b) have a $\zeta = 0.707$. The characteristics of the aircraft are $\omega_n = 2.5$, $\zeta = 0.30$, and $\tau = 0.1$. The gain factor K_1, however, will vary over the range 0.02 at medium-weight cruise conditions to 0.20 at lightweight descent conditions. (a) Draw the root locus as a function of the loop gain K_1K_2. (b) Determine the gain K_2 necessary to yield roots with $\zeta = 0.707$ when the aircraft is in the medium-cruise condition. (c) With the gain K_2 as found in part (b), determine the ζ of the roots when the gain K_1 results from the condition of light-descent.

P7.11   A computer system requires a high-performance magnetic tape transport system [17]. However, the environmental conditions imposed on the system result in a severe test of control engineering

design. A direct-drive dc motor system for the magnetic tape reel system is shown in Fig. P7.11, where r equals the reel radius and J equals the reel and rotor inertia. A complete reversal of the tape reel direction is required in 6 ms, and the tape reel must follow a step command in 3 ms or less. The tape is normally operating at a speed of 100 in./s. The motor and components selected for this system possess the following characteristics:

$K_b = 0.40$	$r = 0.2$
$K_p = 1$	$K_1 = 2.0$
$\tau_1 = \tau_a = 1$ ms	K_2 is adjustable.
$K_T/LJ = 2.0$	

The inertia of the reel and motor rotor is 2.5×10^{-3} when the reel is empty, and 5.0×10^{-3} when the reel is full. A series of photocells is used for an

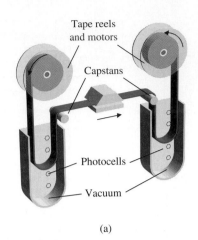

(a)

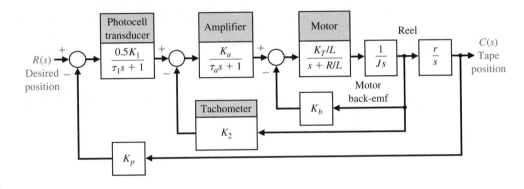

(b)

FIGURE P7.11
(a) Tape control
system. (b) Block
diagram.

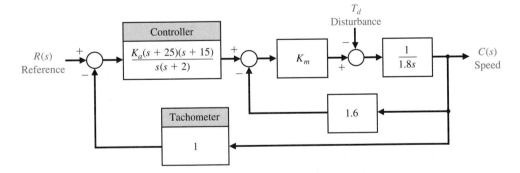

FIGURE P7.12
Speed control.

error sensing device. The time constant of the motor is $L/R = 0.5$ ms. (a) Draw the root locus for the
system when $K_2 = 10$ and $J = 5.0 \times 10^{-3}$, $0 <$
$K_a < \infty$. (b) Determine the gain K_a that results in a
well-damped system so that the ζ of all the roots is
greater than or equal to 0.60. (c) With the K_a determined from part (b), draw a root locus for $0 <$
$K_2 < \infty$.

P7.12 A precision speed control system (Fig. P7.12)
is required for a platform used in gyroscope and
inertial system testing where a variety of closely
controlled speeds is necessary. A direct-drive dc
torque motor system was utilized to provide (1) a
speed range of 0.01°/s to 600°/s, and (2) 0.1%
steady-state error maximum for a step input. The
direct-drive dc torque motor avoids the use of a

gear train with its attendant backlash and friction. Also the direct-drive motor has a high-torque capability, high efficiency, and low motor time constants. The motor gain constant is nominally $K_m = 1.8$ but is subject to variations up to 50%. The amplifier gain K_a is normally greater than 10 and subject to a variation of 10%. (a) Determine the minimum loop gain necessary to satisfy the steady-state error requirement. (b) Determine the limiting value of gain for stability. (c) Draw the root locus as K_a varies from 0 to ∞. (d) Determine the roots when $K_a = 40$, and estimate the response to a step input.

P7.13 A unity feedback system (Fig. 7.1) has

$$G(s) = \frac{K}{s(s + 3)(s^2 + 4s + 7.84)}.$$

(a) Find the breakaway point on the real axis and the gain for this point. (b) Find the gain to provide two complex roots nearest the $j\omega$-axis with a damping ratio of 0.707. (c) Are the two roots of part (b) dominant? (d) Determine the settling time of the system when the gain of part (b) is used.

P7.14 The open-loop transfer function of a single-loop negative feedback system is

$$GH(s) = \frac{K(s + 2)(s + 3)}{s^2(s + 1)(s + 24)(s + 30)}.$$

This system is called *conditionally stable* because it is stable for only a range of the gain K as follows:

$k_1 < K < k_2$. Using the Routh–Hurwitz criteria and the root locus method, determine the range of the gain for which the system is stable. Sketch the root locus for $0 < K < ∞$.

P7.15 Let us again consider the stability and ride of a rider and high performance motorcycle as outlined in Problem 6.13. The dynamics of the motorcycle and rider can be represented by the open-loop transfer function

$$GH(s) = \frac{K(s^2 + 30s + 625)}{s(s + 20)(s^2 + 20s + 200)(s^2 + 60s + 3400)}.$$

Draw the root locus for the system; determine the ζ of the dominant roots when $K = 3 \times 10^4$.

P7.16 Control systems for maintaining constant tension on strip steel in a hot strip finishing mill are called "loopers." A typical system is shown in Fig. P7.16. The looper is an arm 2 to 3 ft long with a roller on the end and is raised and pressed against the strip by a motor [18]. The typical speed of the strip passing the looper is 2000 ft/min. A voltage proportional to the looper position is compared with a reference voltage and integrated where it is assumed that a change in looper position is proportional to a change in the steel strip tension. The time constant of the filter, τ, is negligible relative to the other time constants in the system. (a) Draw the root locus of the control system for $0 < K_a < ∞$. (b) Determine the gain K_a that results in a sys-

FIGURE P7.16
Steel mill control system.

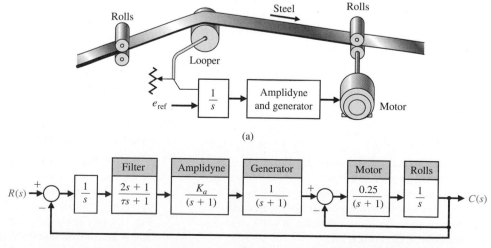

(a)

(b)

tem whose roots have a damping ratio of $\zeta = 0.707$ or greater. (c) Determine the effect of τ as τ increases from a negligible quantity.

P7.17   Reconsider the vibration absorber discussed in Problems 2.2 and 2.10 as a design problem. Using the root locus method, determine the effect of the parameters M_2 and k_{12}. Determine the specific values of the parameters M_2 and k_{12} so that the mass M_1 does not vibrate when $F(t) = a \sin \omega_0 t$. Assume that $M_1 = 1$, $k_1 = 1$, and $f = 1$. Also assume that $k_{12} < 1$ and the term k_{12}^2 may be neglected.

P7.18   A feedback control system is shown in Fig. P7.18. The filter $G_c(s)$ is often called a compensator, and the design problem is that of selecting the parameters α and β. Using the root locus method, determine the effect of varying the parameters. Select a suitable filter so that the settling time is less than 4 seconds and the damping ratio of the dominant roots is greater than 0.60.

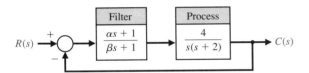

FIGURE P7.18   Filter design.

P7.19   In recent years, many automatic control systems for guided vehicles in factories have been utilized. One such system uses a guidance cable embedded in the floor to guide the vehicle along the desired lane [10, 15]. An error detector, composed of two coils mounted on the front of the cart, senses a magnetic field produced by the current in the guidance cable. An example of a guided vehicle in a factory is shown in Fig. P7.19(a). We have

$$G(s) = \frac{K_a(s^2 + 3.6s + 81)}{s(s + 1)(s + 5)}$$

when K_a equals the amplifier gain. (a) Draw a root locus and determine a suitable gain K_a so that the damping ratio of the complex roots is 0.707. (b) Determine the root sensitivity of the system for the complex root r_1 as a function of (1) K_a and (2) the pole of $G(s)$ at $s = -1$.

(a)

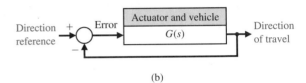

(b)

FIGURE P7.19   (a) An automatically guided vehicle. (Photo courtesy of Control Engineering Corp.) (b) Block diagram.

P7.20   Determine the root sensitivity for the dominant roots of the design for Problem 7.18 for the gain $K = 4\alpha/\beta$ and the pole $s = -2$.

P7.21   Determine the root sensitivity of the dominant roots of the power system of Problem 7.7. Evaluate the sensitivity for variations of (a) the poles at $s = -4$, and (b) the feedback gain, $1/R$.

P7.22   Determine the root sensitivity of the dominant roots of Problem 7.1(a) when K is set so that the damping ratio of the unperturbed roots is 0.707. Evaluate and compare the sensitivity as a function of the poles and zeros of $GH(s)$.

P7.23   Repeat Problem 7.22 for the open-loop transfer function $GH(s)$ of Problem 7.1(c).

P7.24   For systems of relatively high degree, the form of the root locus can often assume an unexpected pattern. The root loci of four different feedback systems of third order or higher are shown in Fig. P7.24. The open-loop poles and zeros of $KF(s)$ are shown, and the form of the root loci as

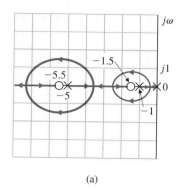

(a)

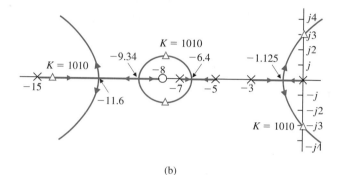

(b)

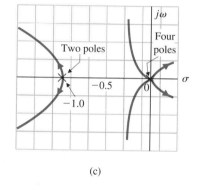

(c)

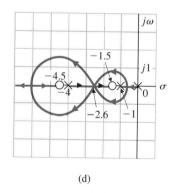

(d)

FIGURE P7.24
Root loci of four
systems.

K varies from zero to infinity is presented. Verify the diagrams of Fig. P7.24 by constructing the root loci.

P7.25 Solid-state integrated electronic circuits are comprised of distributed R and C elements. Therefore feedback electronic circuits in integrated circuit form must be investigated by obtaining the transfer function of the distributed RC networks. It

has been shown that the slope of the attenuation curve of a distributed RC network is n 3 dB/octave, where n is the order of the RC filter [13]. This attenuation is in contrast with the normal n 6 dB/octave for the lumped parameter circuits. (The concept of the slope of an attenuation curve is considered in Chapter 8. If the reader is unfamiliar with this concept, this problem may be reexamined following the study of Chapter 8.) An interesting case arises when the distributed RC network occurs in a series-to-shunt feedback path of a transistor amplifier. Then the loop transfer function may be written as

$$GH(s) = \frac{K(s - 1)(s + 3)^{1/2}}{(s + 1)(s + 2)^{1/2}}.$$

(a) Using the root locus method, determine the locus of roots as K varies from zero to infinity. (b) Calculate the gain at borderline stability and the frequency of oscillation for this gain.

P7.26 A single-loop negative feedback system has a loop transfer function

$$GH(s) = \frac{K(s + 1)^2}{s(s^2 + 1)(s + 4)}.$$

(a) Sketch the root locus for $0 \leq K \leq \infty$ to indicate the significant features of the locus. (b) Determine the range of the gain K for which the system is stable. (c) For what value of K in the range $K \geq 0$ do purely imaginary roots exist? What are the values of these roots? (d) Would the use of the dominant roots approximation for an estimate of settling time be justified in this case for a large magnitude of gain $(K > 10)$?

P7.27 A unity negative feedback system has a transfer function

$$G(s) = \frac{K(s^2 + 0.105625)}{s(s^2 + 1)}$$

$$= \frac{K(s + j0.325)(s - j0.325)}{s(s^2 + 1)}.$$

Draw the root locus as a function of K. Carefully calculate where the segments of the locus enter and leave the real axis.

P7.28 To meet current U.S. emissions standards for automobiles, hydrocarbon (HC) and carbon monoxide (CO) emissions are usually controlled by a catalytic converter in the automobile exhaust. Federal standards for nitrogen oxides (NO$_x$) emissions are met mainly by exhaust-gas recirculation (EGR)

techniques. However, as NO$_x$ emissions standards were tightened from the current limit of 2.0 grams per mile to 1.0 gram per mile, these techniques alone were no longer sufficient.

Although many schemes are under investigation for meeting the emissions standards for all three emissions, one of the most promising employs a three-way catalyst—for HC, CO, and NO$_x$ emissions—in conjunction with a closed-loop engine-control system. The approach is to use a closed-loop engine control as shown in Fig. P7.28 [19, 23]. The exhaust gas sensor gives an indication of a rich or lean exhaust and compares it to a reference. The difference signal is processed by the controller, and the output of the controller modulates the vacuum level in the carburetor to achieve the best air–fuel ratio for proper operation of the catalytic converter. The open-loop transfer function is represented by

$$GH(s) = \frac{K(s + 2)(s + 7)}{s(s + 5)(s + 3)}.$$

Calculate the root locus as a function of K. Carefully calculate where the segments of the locus enter and leave the real axis. Determine the roots when $K = 4$. Predict the step response of the system when $K = 4$.

P7.29 A unity feedback control system has a transfer function

$$G(s) = \frac{K(s^2 + 4s + 8)}{s^2(s + 4)}.$$

It is desired that the dominant roots have a damping ratio equal to 0.5. Find the gain K when this condition is satisfied. Show that at this gain the imaginary roots are $s = -1.3 \pm j2.2$.

P7.30 An RLC network is shown in Fig. P7.30. The nominal values (normalized) of the network elements are $L = C = 1$ and $R = 2.5$. Show that the root sensitivity of the two roots of the input impedance $Z(s)$ to a change in R is different by a factor of 4.

FIGURE P7.28
Auto engine
control.

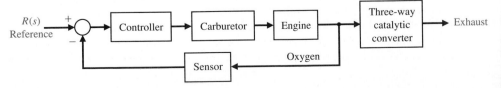

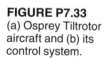

FIGURE P7.30   *RLC* network.

P7.31   The development of high-speed aircraft and missiles requires information about aerodynamic parameters prevailing at very high speeds. Wind tunnels are used to test these parameters. These wind tunnels are constructed by compressing air to very high pressures and releasing it through a valve to create a wind. Since the air pressure drops as the air escapes, it is necessary to open the valve wider to maintain a constant wind speed. Thus a control system is needed to adjust the valve to maintain a constant wind speed. The open-loop transfer function for a unity feedback system is

$$GH(s) = \frac{K(s + 4)}{s(s + 0.16)(s + p)(s + p^*)}.$$

where $p = -7.3 + 9.7831j$. Draw the root locus and show the location of the roots for $K = 326$ and $K = 1350$.

P7.32   A mobile robot suitable for nighttime guard duty is available. This guard never sleeps and

can tirelessly patrol large warehouses and outdoor yards. The steering control system for the mobile robot has a unity feedback with

$$G(s) = \frac{K(s + 1)(s + 5)}{s(s + 1.5)(s + 2)}.$$

(a) Find K for all breakaway and entry points on the real axis. (b) Find K when the damping ratio of the complex roots is 0.707. (c) Find the minimum value of the damping ratio for the complex roots and the associated gain K. (d) Find the overshoot and settling time for a unit step input for the gain, K, determined in parts (b) and (c).

P7.33   The Bell-Boeing V-22 Osprey Tiltrotor is both an airplane and a helicopter. Its advantage is the ability to rotate its engines to 90° from a vertical position, as shown in Fig. P7.33(a), for takeoffs and landings and then to switch the engines to a horizontal position for cruising as an airplane [20]. The altitude control system in the helicopter mode is shown in Fig. P7.33(b). (a) Determine the root locus as K varies and determine the range of K for a stable system. (b) For $K = 280$, find the actual $c(t)$ for a unit step input $r(t)$ and the percentage overshoot and settling time. (c) When $K = 280$ and $r(t) = 0$, find $c(t)$ for a unit step disturbance,

FIGURE P7.33
(a) Osprey Tiltrotor aircraft and (b) its control system.

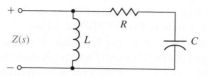

(a)

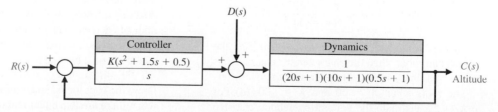

(b)

$D(s) = 1/s$. (d) Add a prefilter between $R(s)$ and the summing node so that

$$G_p(s) = \frac{0.5}{s^2 + 1.5s + 0.5}$$

and repeat part (b).

P7.34 The fuel control for an automobile uses a diesel pump that is subject to parameter variations. A unity negative feedback has a plant

$$G(s) = \frac{K(s + 1.5)}{(s + 1)(s + 2)(s + 4)(s + 10)}.$$

(a) Sketch the root locus as K varies from 0 to 2000. (b) Find the roots for K equal to 400, 500, and 600. (c) Predict how the percent overshoot to a step will vary for the gain K, assuming dominant roots. (d) Find the actual time response for a step input for all three gains and compare the actual overshoot with the predicted overshoot.

P7.35 A powerful electrohydraulic forklift can be used to lift pallets weighing several tons on top of 35-ft scaffolds at a construction site. The negative unity feedback system has a plant transfer function

$$G(s) = \frac{K(s + 1)^2}{s(s^2 + 1)}.$$

(a) Sketch the root locus for $K > 0$. (b) Find the gain K when two complex roots have a ζ of 0.707, and calculate all three roots. (c) Find the entry point of the root locus at the real axis. (d) Estimate the expected overshoot to a step input, and compare it with the actual overshoot determined from a computer program.

P7.36 A microrobot with a high-performance manipulator has been designed for testing very small particles such as simple living cells [6]. The single-loop unity negative feedback system has a plant transfer function

$$G(s) = \frac{K(s + 1)(s + 2)(s + 3)}{s^3(s - 1)}.$$

(a) Sketch the root locus for $K > 0$. (b) Find the gain and roots when the characteristic equation has two imaginary roots. (c) Determine the characteristic roots when $K = 20$ and $K = 100$. (d) For $K = 20$, estimate the percent overshoot to a step input, and compare the estimate to the actual overshoot determined from a computer program.

P7.37 Identify the parameters K, a, and b of the system shown in Fig. P7.37. The system is subject to a unit step input, and the output response has an overshoot but ultimately attains the final value of 1. When the closed-loop system is subjected to a ramp input, the output response follows the ramp input with a finite steady-state error. When the gain is doubled to $2K$, the output response to an impulse input is a pure sinusoid with a period 0.314 second. Determine K, a, and b.

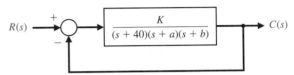

FIGURE P7.37   Feedback system.

P7.38 A unity feedback system, as shown in Fig. 7.1, has

$$G(s) = \frac{K(s + 1)}{s(s - 3)}.$$

This system is open-loop unstable. (a) Determine the range of K so that the system is stable. (b) Plot the root locus. (c) Determine the roots for $K = 10$. (d) For $K = 10$, predict the percent overshoot for a step input using Fig. 5.13. (e) Determine the actual overshoot by plotting the response.

P7.39 High-speed trains for U.S. railroad track must traverse twists and turns. In conventional trains, the axles are fixed in steel frames called trucks. The trucks pivot as the train goes into a curve, but the fixed axles stay parallel to each other, even though the front axle wants to go in a different direction from the real axle [24]. If the train is going fast, it may jump the tracks. One solution uses axles that pivot independently. To counterbalance the strong centrifugal forces in a curve, the train also has a computerized hydraulic system that tilts each car as it rounds a turn. On-board sensors calculate the train's speed and the sharpness of the curve and feed this information to hydraulic pumps under the floor of each car. The pumps tilt the car as much as eight degrees, causing it to lean into the curve like a race car on a banked track.

The tilt control system is shown in Fig. P7.39. Draw the root locus, and determine the value of K when there are two equal real roots and a third real root with a smaller magnitude. Predict the response of this system to a step input $R(s)$.

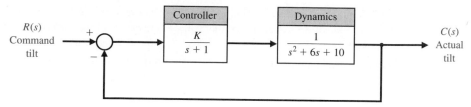

FIGURE P7.39
Tilt control for a
high-speed train.

ADVANCED PROBLEMS

AP7.1 The top view of a high-performance jet aircraft is shown in Fig. AP7.1(a) [20]. Plot the root locus and determine the gain K so that the ζ of the complex poles near the $j\omega$-axis is the maximum achievable. Evaluate the roots at this K, and predict the response to a step input. Determine the actual response, and compare it to the predicted response.

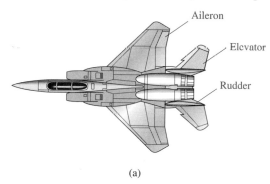

(a)

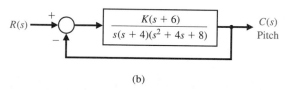

(b)

FIGURE AP7.1 (a) High-performance aircraft.
(b) Pitch control system.

AP7.2 A magnetically levitated high-speed train "flies" on an air gap above its rail system [24]. The air gap control system has a unity feedback system with a plant

$$G(s) = \frac{K(s+1)(s+3)}{s(s-1)(s+4)(s+8)}.$$

The goal is to select K so that the response for a unit step input is reasonably damped and the set-

tling time is less than 3 seconds. Draw the root locus, and select K so that all of the complex roots have a ζ greater than 0.6. Determine the actual response for the selected K and the percent overshoot and settling time.

AP7.3 A compact disc player for portable use requires good rejection of disturbances and accurate position of the optical reader sensor. The position control system uses unity feedback and a plant transfer function

$$G(s) = \frac{10}{s(s+1)(s+p)}.$$

The parameter p can be chosen by selecting the appropriate dc motor. Plot the root locus as a function of p. Select p so that the ζ of the complex roots of the characteristic equation is approximately $1/\sqrt{2}$.

AP7.4 A remote manipulator control system has unity feedback and a plant

$$G(s) =$$
$$\frac{(s+\alpha)}{s^3 + (1+\alpha)s^2 + (\alpha-1)s + 1 - \alpha}.$$

It is desired that the steady-state position error for a step input be less than or equal to 10% of the magnitude of the input. Plot the root locus as a function of the parameter α. Determine the range of α required for the desired steady-state error. Locate the roots for the allowable value of α to achieve the required steady-state error, and estimate the step response of the system.

AP7.5 A unity feedback system has a plant

$$G(s) = \frac{K}{s^3 + 10s^2 + 7s - 18}.$$

(a) Plot the root locus and determine K for a stable system with complex roots with ζ equal to $1/\sqrt{2}$. (b) Determine the root sensitivity of the complex roots of part (a). (c) Determine the percent change in K (increase or decrease) so that the roots lie on the $j\omega$-axis.

AP7.6   A unity feedback system has a plant

$$G(s) = \frac{K(s^2 + 3s + 3.25)}{s^2(s + 1)(s + 10)(s + 20)}.$$

Draw the root locus for $K > 0$, and select a value for K that will provide an acceptable step response.

AP7.7   A feedback system with positive feedback is shown in Fig. AP7.7. The root locus for $K > 0$ must meet the condition

$$KG(s) = 1 \; \underline{/\pm k360°}$$

for $k = 0, 1, 2, \ldots$.

Sketch the root locus for $0 < K < \infty$.

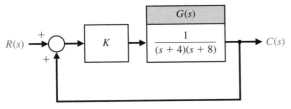

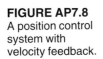

FIGURE AP7.7   A closed-loop system with positive feedback.

AP7.8   A position control system for a dc motor is shown in Fig. AP7.8. Obtain the root locus for the velocity feedback constant k, and select the gain k so that all the roots of the characteristic equation

are real (two are equal and real). Estimate the step response of the system for the k selected. Compare the estimate with the actual response.

AP7.9   A control system is shown in Fig. AP7.9. Sketch the root loci for the following transfer functions $G_c(s)$:

a) $G_c(s) = K$

b) $G_c(s) = K(s + 1)$

c) $G_c(s) = \dfrac{K(s + 1)}{(s + 10)}$

d) $G_c(s) = \dfrac{K(s + 1)(s + 3)}{(s + 10)}$

AP7.10   A feedback system is shown in Fig. AP7.10. Draw the root locus as K varies when $K \geqslant 0$. Determine a value for K that will provide a step response with an overshoot less than 5% and a settling time less than 2.5 seconds.

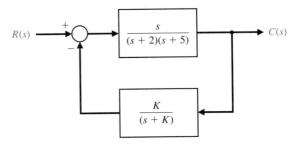

FIGURE AP7.10

AP7.11   Reconsider the design of the robot motorcycle described in Design Problem 6.6. Plot the root

FIGURE AP7.8
A position control system with velocity feedback.

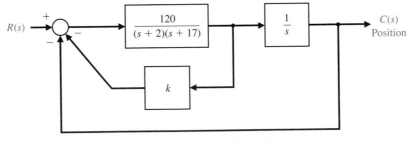

FIGURE AP7.9

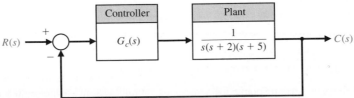

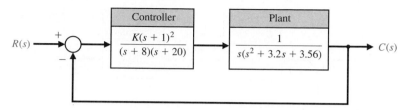

FIGURE AP7.12

locus for $K > 0$, and select a value for K that will result in steady performance as disturbances occur.

AP7.12 A control system is shown in Fig. AP7.12.

Plot the root locus, and select a gain K so that the step response of the system has an overshoot of less than 20% and the settling time is less than 5 seconds.

DESIGN PROBLEMS

DP7.1 A high-performance aircraft, shown in Fig. DP7.1(a), uses the ailerons, rudder, and elevator to steer through a three-dimensional flight path [20]. The pitch rate control system for a fighter aircraft at 10,000 m and Mach 0.9 can be represented by the system in Fig. DP7.1(b), where

$$G(s) = \frac{-18(s + 0.015)(s + 0.45)}{(s^2 + 1.2s + 12)(s^2 + 0.01s + 0.0025)}.$$

(a) Plot the root locus when the controller is a gain, so that $G_c(s) = K$, and determine K when ζ for the roots with $\omega_n > 2$ is larger than 0.15 (seek a maximum ζ). (b) Plot the response, $q(t)$, for a step input $r(t)$. (c) A designer suggests an anticipatory controller so that $G_c(s) = K_1 + K_2s = K(s + 2)$. Plot the root locus for this system as K varies and determine K so that the ζ of all the closed-loop roots is

$0.8 < \zeta < 0.6$. (d) Plot the response, $q(t)$, for a step input $r(t)$.

DP7.2 A large helicopter uses two tandem rotors rotating in opposite directions, as shown in Fig. P7.33(a). The controller adjusts the tilt angle of the main rotor and thus the forward motion as shown in Fig. DP7.2. The helicopter dynamics are represented by

$$G(s) = \frac{10}{s^2 + 4.5s + 9},$$

and the controller is selected as

$$G_c(s) = K_1 + \frac{K_2}{s} = \frac{K(s + 1)}{s}.$$

(a) Plot the root locus of the system and determine K when ζ of the complex roots is equal to 0.6.

FIGURE DP7.1
(a) High-performance aircraft. (b) Pitch rate control system.

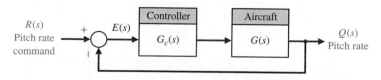

(a)

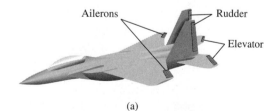

(b)

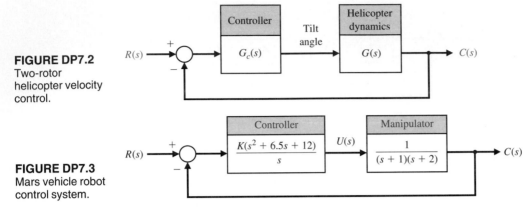

FIGURE DP7.2
Two-rotor
helicopter velocity
control.

FIGURE DP7.3
Mars vehicle robot
control system.

(b) Plot the response of the system to a step input $r(t)$ and find the settling time and overshoot for the system of part (a). What is the steady-state error for a step input? (c) Repeat parts (a) and (b) when the ζ of the complex roots is 0.41. Compare the results with those obtained in parts (a) and (b).

DP7.3   The vehicle Rover has been designed for maneuvering at 0.25 mph over Martian terrain. Because Mars is 189 million miles from Earth and it would take up to 40 minutes each way to communicate with Earth [22, 27], Rover must act independently and reliably. Resembling a cross between a small flatbed truck and a jacked-up jeep, Rover will be constructed of three articulated sections, each with its own two independent axle-bearing one-meter conical wheels. A pair of sampling arms—one for chipping and drilling, the other for manipulating fine objects—jut from its front end like pincers. The control of the arms can be represented by the system shown in Fig. DP7.3. (a) Plot the root locus for K and identify the roots for $K = 4.1$ and 41. (b) Determine the gain K that results in an overshoot to a step of approximately 1%. (c) Determine the gain that minimizes the set-

tling time while maintaining an overshoot of less than 1%.

DP7.4   A welding torch is remotely controlled to achieve high accuracy while operating in changing and hazardous environments [21]. A model of the welding arm position control is shown in Fig. DP7.4, with the disturbance representing the environmental changes. (a) With $D(s) = 0$, select K_1 and K to provide high-quality performance of the position control system. Select a set of performance criteria, and examine the results of your design. (b) For the system in part (a), let $R(s) = 0$ and determine the effect of a unit step $D(s) = 1/s$ by obtaining $c(t)$.

DP7.5   A high-performance jet aircraft with an autopilot control system has a unity feedback and control system as shown in Fig. DP7.5. Draw the root locus, and predict the step response of the system. Determine the actual response of the system, and compare it to the predicted response.

DP7.6   A system to aid and control the walk of a partially disabled person could use automatic control of the walking motion [25]. One model of a system

FIGURE DP7.4
Remotely
controlled welder.

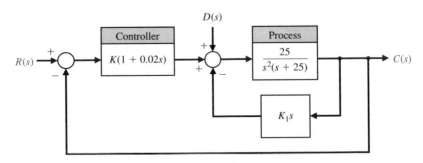

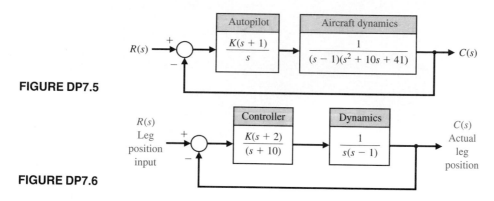

FIGURE DP7.5

FIGURE DP7.6

that is open-loop unstable is shown in Fig. DP7.6. Using a root locus, select K for the maximum achievable ζ of the complex roots. Predict the step response of the system, and compare it with the actual step response.

DP7.7 Most commercial op amps are designed to be unity gain stable [26]. That is, they are stable when used in a unity gain configuration. To achieve higher bandwidth, some op amps relax the requirement to be unity gain stable. One such amplifier has a dc gain of 10^5 and bandwidth of 10 kHz. The amplifier, $G(s)$, is connected in the feedback circuit shown in Fig. DP7.7(a). The amplifier is represented by the model shown in Fig. DP7.7(b),

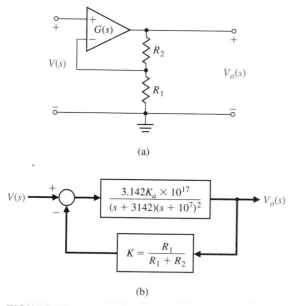

(a)

(b)

FIGURE DP7.7 (a) Op amp circuit and (b) control system.

where $K_a = 10^5$. Plot the root locus of the system for K. Determine the minimum value of the dc gain of the closed-loop amplifier for stability. Select a dc gain and the resistors R_1 and R_2.

DP7.8 A robot arm actuated at the elbow joint is shown in Fig. DP7.8(a), and the control system for the actuator is shown in Fig. DP7.8(b). Plot the root locus for $K \geq 0$. Select $G_p(s)$ so that the steady-state error for a step input is equal to zero. Using the $G_p(s)$ selected, plot $c(t)$ for K equal to 1, 1.5, and 2.85. Record the rise time, settling time, and percent overshoot for the three gains. We wish to limit the overshoot to less than 6% while achieving the shortest rise time possible. Select the best system for $1 \leq K \leq 2.85$.

DP7.9 The four-wheel-steering automobile has several benefits. The system gives the driver a greater degree of control over the automobile. The driver gets a more forgiving vehicle over a wide variety of conditions. The system enables the driver to make sharp, smooth lane transitions. It also prevents yaw, which is the swaying of the rear end during sudden movements. Furthermore, the four-wheel-steering system gives a car increased maneuverability. This enables the driver to park the car in extremely tight quarters. Finally, with additional closed-loop computer operating systems, a car could be prevented from sliding out of control in abnormal icy or wet road conditions.

The system works by moving the rear wheels relative to the front-wheel-steering angle. The control system takes information about the front wheels' steering angle and passes it to the actuator in the back. This actuator then moves the rear wheels appropriately.

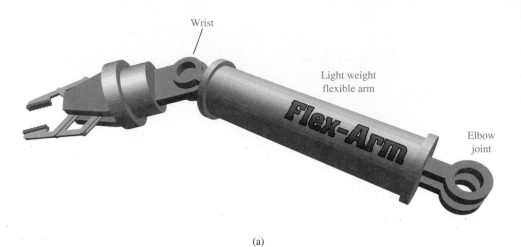

(a)

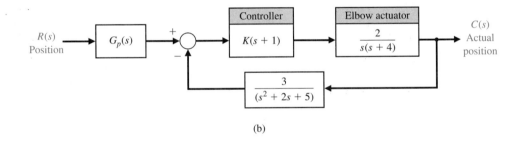

FIGURE DP7.8
(a) A robot arm
actuated at the
joint elbow, and
(b) its control
system.

(b)

FIGURE DP7.10
(a) Pilot crane
control system;
(b) block diagram.

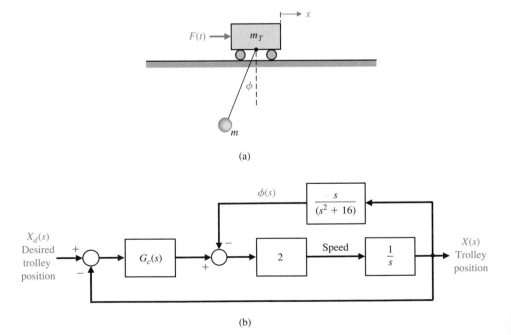

(a)

(b)

(a)

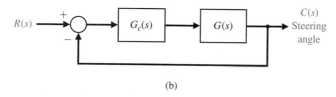

FIGURE DP7.11
(a) Planetary
rover vehicle.
(b) Steering control
system.

(b)

When the rear wheels are given a steering angle relative to the front ones, the vehicle can vary its lateral acceleration response according to the transfer function

$$G(s) = K \frac{1 + (1 + \lambda)T_1 s + (1 + \lambda)T_2 s^2}{s[1 + (2\zeta/\omega_n)s + (1/\omega_n^2)s^2]},$$

where $\lambda = 2q/(1 - q)$ and q is the ratio of rear wheel angle to front wheel steering angle [14]. We will assume that $T_1 = T_2 = 1$ second and $\omega_n = 4$. Design a unity feedback system for $G(s)$, selecting an appropriate set of parameters (λ, K, ζ) so that the steering control response is rapid and yet will yield modest overshoot characteristics. In addition, q must be between 0 and 1.

DP7.10 A pilot crane control is shown in Fig. DP7.10(a). The trolley is moved by an input $F(t)$ in order to control $x(t)$ and $\phi(t)$ [13]. The model of the pilot crane control is shown in Fig. DP 7.10(b). Design a controller that will achieve control of the desired variables when $G_c(s) = K$.

DP7.11 A rover vehicle designed for use on other planets and moons is shown in Fig. DP7.11(a) [21]. The block diagram of the steering control is shown in Fig. DP7.11(b), where

$$G(s) = \frac{(s + 1.5)}{(s + 1)(s + 2)(s + 4)(s + 10)}.$$

(a) When $G_c(s) = K$, draw the root locus as K varies from 0 to 1000. Find the roots for K equal to

100, 300, and 600. (b) Predict the overshoot, settling time, and steady-state error for a step input, assuming dominant roots. (c) Determine the actual time response for a step input for the three values of the gain K, and compare the actual results with the predicted results.

DP7.12 Electronic systems currently make up about 6% of a car's value. That figure will climb to 20% by the year 2000 as antilock brakes, active suspensions, and other computer-dependent technologies move into full production. Much of the added computing power will be used for new technology for smart cars and smart roads, or IVHS (intelligent vehicle/highway systems) [14]. The term refers to a varied assortment of electronics that provide real-time information on accidents, congestion, routing, and roadside services to drivers and traffic controllers. IVHS also encompasses devices that would make vehicles more autonomous: collision-avoidance systems and lane-tracking technology that alert drivers to impending disaster or allow a car to drive itself.

An example of an automated highway system is shown in Fig. DP7.12(a). A position control system for maintaining the distance between vehicles is shown in Fig. DP7.12(b). Select K_a and K_t so that the steady-state error for a ramp input is less than 25% of the input magnitude, A, of the ramp $R(s) = A/s^2$. The response to a step command should have an overshoot of less than 3% and a settling time of less than 1.5 seconds.

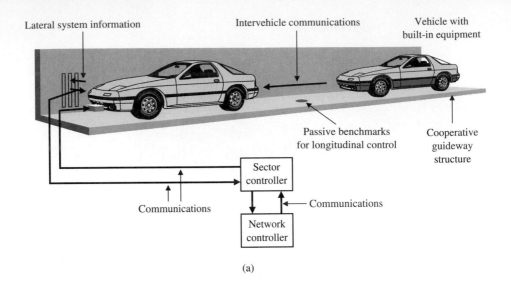

FIGURE DP7.12
(a) Automated
highway system.
(b) Vehicle
distance control
system.

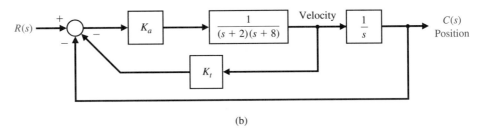

(b)

FIGURE DP7.13
(a) An airplane with
a set of ailerons.
(b) The block
diagram for
controlling the roll
rate of the airplane.

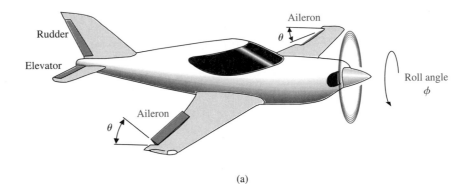

(a)

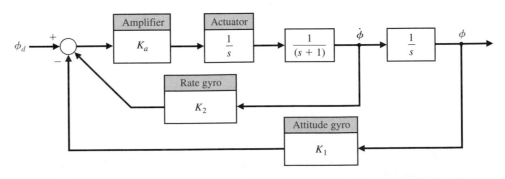

(b)

DP7.13 The automatic control of an airplane is one example where multiple-variable feedback methods are required. In this system, the attitude of an aircraft is controlled by three sets of surfaces: elevators, rudder, and ailerons, as shown in Fig. DP7.13(a). By manipulating these surfaces, a pilot can set the aircraft on a desired flight path [20].

An autopilot, which will be considered here, is an automatic control system that controls the roll angle ϕ by adjusting aileron surfaces. The deflection of the aileron surfaces by an angle θ generates a torque due to air pressure on these surfaces. This causes a rolling motion of the aircraft. The aileron surfaces are controlled by a hydraulic actuator with a transfer function b/s.

The actual roll angle ϕ is measured and compared with the input. The difference between the desired roll angle ϕ_d and the actual angle ϕ will drive the hydraulic actuator, which in turn adjusts the deflection of the aileron surface.

A simplified model where the rolling motion can be considered independent of other motions is assumed, and its block diagram is shown in Fig. DP7.13(b). Assume that $K_1 = 1$ and that the roll rate ϕ is fed back using a rate gyro. The step response desired has an overshoot less than 10% and a settling time less than 9 seconds. Select the parameters K_a and K_2.

MATLAB PROBLEMS

MP7.1 Using the rlocus function, obtain the root locus for the following transfer functions of the system shown in Fig. MP7.1 when $0 \le k \le \infty$.

(a) $G(s) = \dfrac{1}{s^3 + 4s^2 + 6s + 1}$

(b) $G(s) = \dfrac{s + 2}{s^2 + 2s + 1}$

(c) $G(s) = \dfrac{s^2 + s + 1}{s(s^2 + 4s + 6)}$

(d) $G(s) =$
$$\dfrac{s^5 + 4s^4 + 6s^3 + 8s^2 + 6s + 4}{s^6 + 2s^5 + 2s^4 + s^3 + s^2 + 10s + 1}$$

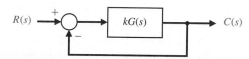

FIGURE MP7.1 A single-loop feedback system with parameter k.

MP7.2 A unity negative feedback system has the open-loop transfer function

$$kG(s) = k\,\dfrac{(s^2 - 2s + 2)}{s(s^2 + 3s + 2)}.$$

Using *Matlab*, plot the root locus, and show with the rlocfind function that the maximum value of k for a stable system is $k = 0.79$.

MP7.3 Compute the partial fraction expansion of

$$C(s) = \dfrac{s + 2}{s(s^2 + 4s + 3)}$$

and verify the result with *Matlab* using the residue function.

MP7.4 Consider the feedback control system in Fig. MP7.4. We have three potential controllers for our system:

(1) $G_c(s) = K$ ⟨proportional controller⟩
(2) $G_c(s) = K/s$ ⟨integral controller⟩
(3) $G_c(s) = K(1 + 1/s)$ ⟨proportional, integral (PI) controller⟩

The design specifications are $T_s \le 10$ seconds and $M_p \le 10\%$ for a unit step input.

(a) For the proportional controller, draw the root locus using *Matlab* for $0 \le K \le \infty$, and determine the value of K such that the design specifications are satisfied.
(b) Repeat part (a) for the integral controller.
(c) Repeat part (a) for the PI controller.
(d) Co-plot the unit step responses for the closed-loop systems with each controller designed in parts (a)–(c).
(e) Compare and contrast the three controllers obtained in parts (a)–(c), concentrating your discussion on the steady-state errors and transient performance.

FIGURE MP7.4
A single-loop
feedback control
system with
controller $G_c(s)$.

FIGURE MP7.5
A spacecraft
attitude control
system with a
proportional-
derivative
controller.

MP7.5   Consider the spacecraft single-axis attitude control system shown in Fig. MP7.5. The controller is known as a proportional-derivative (PD) controller. Suppose that we require the ratio of K_1/K_2 = 5. Then, using root locus methods on *Matlab*, find the values of K_2/J and K_1/J such that the settling time, T_s, is less than or equal to 4 seconds, and the peak overshoot, M_p, is less than or equal to 10% for a unit step input.

TERMS AND CONCEPTS

Angle of departure   The angle at which a locus leaves a complex pole in the *s*-plane.

Asymptote   The path the root locus follows as the parameter becomes very large and approaches infinity. The number of asymptotes is equal to the number of poles minus the number of zeros.

Asymptote centroid   The center of the linear asymptotes, σ_A.

Breakaway point   The point on the real axis where the locus departs from the real axis of the *s*-plane.

Locus   A path or trajectory that is traced out as a parameter is changed.

Number of separate loci   Equal to the number of poles of the transfer function, assuming that the number of poles is greater than the number of zeros of the transfer function.

Parameter design   A method of selecting one or two parameters using the root locus method.

Root locus   The locus or path of the roots traced out on the *s*-plane as a parameter is changed.

Root locus method   The method for determining the locus of roots of the characteristic equation $1 + KP(s) = 0$ as K varies from 0 to infinity.

Root locus segments on the real axis   The root locus lying in a section of the real axis to the left of an odd number of poles and zeros.

Root sensitivity   The sensitivity of the roots as a parameter changes from its normal value. The root sensitivity is the incremental change in the root divided by the proportional change of the parameter.

CHAPTER 8

Frequency Response Methods

PREVIEW

We have examined the use of test input signals such as a step and a ramp signal. In this chapter, we will use a steady-state sinusoidal input signal and consider the steady-state response of the system as the frequency of the sinusoid is varied. Thus we will look at the response of the system to a changing frequency, ω.

We will examine the transfer function $G(s)$ when $s = j\omega$ and develop several forms of plotting the complex number for $G(j\omega)$ when ω is varied. These plots provide insight regarding the performance of a system. We are able to develop several time-domain performance measures in terms of the frequency response of a system. The measures can be used as system specifications and the parameters adjusted to meet the time-domain specifications.

We will consider the graphical development of one or more forms for the frequency response plot. We can then proceed to use computer-generated data to readily obtain these plots.

8.1 INTRODUCTION

In the preceding chapters the response and performance of a system have been described in terms of the complex frequency variable s and the location of the poles and zeros on the s-plane. A very practical and important alternative approach to the analysis and design of a system is the *frequency response* method.

> The frequency response of a system is defined as the steady-state response of the system to a sinusoidal input signal. The sinusoid is a unique input signal, and the resulting output signal for a linear system, as well as signals throughout the system, is sinusoidal in the steady state; it differs from the input waveform only in amplitude and phase angle.

For example, consider the system $C(s) = T(s)R(s)$ with $r(t) = A \sin \omega t$. We have

$$R(s) = \frac{A\omega}{s^2 + \omega^2}$$

and

$$T(s) = \frac{m(s)}{q(s)} = \frac{m(s)}{\prod\limits_{i=1}^{n} (s + p_i)}.$$

Then, in partial fraction form,

$$C(s) = \frac{k_1}{s + p_1} + \cdots + \frac{k_n}{s + p_n} + \frac{\alpha s + \beta}{s^2 + \omega^2}.$$

If the system is stable, then all p_i have negative nonzero real parts and

$$\lim_{t \to \infty} \mathcal{L}^{-1} \left[\frac{k_i}{s + p_i} \right] = 0.$$

In the limit, for $c(t)$, we obtain for $t \to \infty$ (the steady state)

$$c(t) = \mathcal{L}^{-1} \left[\frac{\alpha s + \beta}{s^2 + \omega^2} \right]$$

$$= \frac{1}{\omega} \left| A\omega T(j\omega) \right| \sin(\omega t + \phi) \qquad (8.1)$$

$$= A \left| T(j\omega) \right| \sin(\omega t + \phi),$$

where $\phi = \underline{/T(j\omega)}$.

Thus the steady-state output signal depends only on the magnitude and phase of $T(j\omega)$ at a specific frequency ω. Notice that the steady-state response as described in Eq. (8.1) is true only for stable systems, $T(s)$.

One advantage of the frequency response method is the ready availability of sinusoid test signals for various ranges of frequencies and amplitudes. Thus the experimental determination of the frequency response of a system is easily accomplished and is the most reliable and uncomplicated method for the experimental analysis of a system. Often, as we shall find in Section 8.4, the unknown transfer function of a system can be deduced from the experimentally determined frequency response of a system [1, 2]. Furthermore, the design of a system in the frequency domain provides the designer with control of the bandwidth of a system and some measure of the response of the system to undesired noise and disturbances.

A second advantage of the frequency response method is that the transfer function describing the sinusoidal steady-state behavior of a system can be obtained by replacing s with $j\omega$ in the system transfer function $T(s)$. The transfer function representing the sinusoidal steady-state behavior of a system is then a function of the complex variable $j\omega$ and is itself a complex function $T(j\omega)$ that possesses a magnitude and phase angle. The magnitude and phase angle of $T(j\omega)$ are readily represented by graphical plots that provide a significant insight for the analysis and design of control systems.

The basic disadvantage of the frequency response method for analysis and design is the indirect link between the frequency and the time domain. Direct correlations between the frequency response and the corresponding transient response characteristics are somewhat tenuous, and in practice the frequency response characteristic is adjusted by using various design criteria that will normally result in a satisfactory transient response.

The Laplace transform pair was given in Section 2.4 and is written as

$$F(s) = \mathcal{L}\{f(t)\} = \int_0^\infty f(t)e^{-st}\,dt \tag{8.2}$$

and

$$f(t) = \mathcal{L}^{-1}\{F(s)\} = \frac{1}{2\pi j}\int_{\sigma-j\infty}^{\sigma+j\infty} F(s)e^{st}\,ds, \tag{8.3}$$

where the complex variable $s = \sigma + j\omega$. Similarly, the *Fourier transform* pair is written as

$$F(j\omega) = \mathcal{F}\{f(t)\} = \int_{-\infty}^\infty f(t)e^{-j\omega t}\,dt \tag{8.4}$$

and

$$f(t) = \mathcal{F}^{-1}\{F(j\omega)\} = \frac{1}{2\pi}\int_{-\infty}^\infty F(j\omega)e^{j\omega t}\,d\omega. \tag{8.5}$$

The Fourier transform exists for $f(t)$ when

$$\int_{-\infty}^\infty |f(t)|dt < \infty.$$

The Fourier and Laplace transforms are closely related, as we can see by examining Eqs. (8.2) and (8.4). When the function $f(t)$ is defined only for $t \geq 0$, as is often the case, the lower limits on the integrals are the same. Then we note that the two equations differ only in the complex variable. Thus, if the Laplace transform of a function $f_1(t)$ is known to be $F_1(s)$, we can obtain the Fourier transform of this same time function $F_1(j\omega)$ by setting $s = j\omega$ in $F_1(s)$ [3].

Again, we might ask, because the Fourier and Laplace transforms are so closely related, why not always use the Laplace transform? Why use the Fourier transform at all? The Laplace transform permits us to investigate the s-plane location of the poles and zeros of a transfer $T(s)$ as in Chapter 7. However, the frequency response method allows us to consider the transfer function $T(j\omega)$ and to concern ourselves with the amplitude and phase characteristics of the system. This ability to investigate and represent the character of a system by amplitude and phase equations and curves is an advantage for the analysis and design of control systems.

If we consider the frequency response of the closed-loop system, we might have an input $r(t)$ that has a Fourier transform, in the frequency domain, as follows:

$$R(j\omega) = \int_{-\infty}^{\infty} r(t)e^{-j\omega t} \, dt.$$

Then the output frequency response of a single-loop control system can be obtained by substituting $s = j\omega$ in the closed-loop system relationship, $C(s) = T(s)R(s)$, so that we have

$$C(j\omega) = T(j\omega)R(j\omega) = \frac{G(j\omega)}{1 + G(j\omega)H(j\omega)}R(j\omega). \tag{8.6}$$

Utilizing the inverse Fourier transform, the output transient response would be

$$c(t) = \mathscr{F}^{-1}\{C(j\omega)\} = \frac{1}{2\pi} \int_{-\infty}^{\infty} C(j\omega)e^{j\omega t} \, d\omega. \tag{8.7}$$

However, it is usually quite difficult to evaluate this inverse transform integral for any but the simplest systems, and a graphical integration may be used. Alternatively, as we will note in succeeding sections, several measures of the transient response can be related to the frequency characteristics and utilized for design purposes.

8.2 FREQUENCY RESPONSE PLOTS

The transfer function of a system $G(s)$ can be described in the frequency domain by the relation[†]

$$G(j\omega) = G(s)\big|_{s=j\omega} = R(\omega) + jX(\omega), \tag{8.8}$$

where

$$R(\omega) = \text{Re}[G(j\omega)] \quad \text{and} \quad X(\omega) = \text{Im}[G(j\omega)].$$

Alternatively, the transfer function can be represented by a magnitude $|G(j\omega)|$ and a phase $\phi(j\omega)$ as

$$G(j\omega) = |G(j\omega)|e^{j\phi(j\omega)} = |G(\omega)| \underline{/\phi(\omega)}, \tag{8.9}$$

where

$$\phi(\omega) = \tan^{-1}\frac{X(\omega)}{R(\omega)} \quad \text{and} \quad |G(\omega)|^2 = [R(\omega)]^2 + [X(\omega)]^2.$$

The graphical representation of the frequency response of the system $G(j\omega)$ can utilize either Eq. (8.8) or Eq. (8.9). The *polar plot* representation of the frequency response is obtained by using Eq. (8.8). The coordinates of the polar plot are the real and imaginary parts of $G(j\omega)$, as shown in Fig. 8.1. An example of a polar plot will illustrate this approach.

[†] See Appendix E for a review of complex numbers.

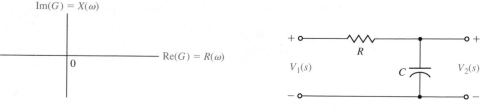

FIGURE 8.1
The polar plane.

FIGURE 8.2
An *RC* filter.

EXAMPLE 8.1   **Frequency response of an *RC* filter**

A simple *RC* filter is shown in Fig. 8.2. The transfer function of this filter is

$$G(s) = \frac{V_2(s)}{V_1(s)} = \frac{1}{RCs + 1},\tag{8.10}$$

and the sinusoidal steady-state transfer function is

$$G(j\omega) = \frac{1}{j\omega(RC) + 1} = \frac{1}{j(\omega/\omega_1) + 1},\tag{8.11}$$

where

$$\omega_1 = \frac{1}{RC}.$$

Then the polar plot is obtained from the relation

$$G(j\omega) = R(\omega) + jX(\omega)$$

$$= \frac{1 - j(\omega/\omega_1)}{(\omega/\omega_1)^2 + 1}\tag{8.12}$$

$$= \frac{1}{1 + (\omega/\omega_1)^2} - \frac{j(\omega/\omega_1)}{1 + (\omega/\omega_1)^2}.$$

The first step is to determine $R(\omega)$ and $X(\omega)$ at the two frequencies, $\omega = 0$ and $\omega = \infty$. At $\omega = 0$, we have $R(\omega) = 1$ and $X(\omega) = 0$. At $\omega = \infty$, we have $R(\omega) = 0$ and $X(\omega) = 0$. These two points are shown on Fig. 8.3. The locus of the real and imaginary parts is also shown in Fig. 8.3 and is easily shown to be a circle with the center at $(\frac{1}{2}, 0)$. When $\omega = \omega_1$, the real and imaginary parts are equal, and the angle $\phi(\omega) = 45°$. The polar plot can also be readily obtained from Eq. (8.9) as

$$G(j\omega) = |G(\omega)| \underline{/\phi(\omega)},\tag{8.13}$$

where

$$|G(\omega)| = \frac{1}{[1 + (\omega/\omega_1)^2]^{1/2}} \quad \text{and} \quad \phi(\omega) = -\tan^{-1}(\omega/\omega_1).$$

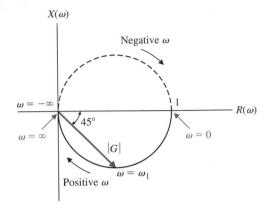

FIGURE 8.3
Polar plot for
RC filter.

Hence when $\omega = \omega_1$, magnitude is $|G(\omega_1)| = 1/\sqrt{2}$ and phase $\phi(\omega_1) = -45°$. Also, when ω approaches $+\infty$, we have $|G(\omega)| \to 0$ and $\phi(\omega) = -90°$. Similarly, when $\omega = 0$, we have $|G(\omega)| = 1$ and $\phi(\omega) = 0$. ∎

EXAMPLE 8.2   **Polar plot of a transfer function**

The polar plot of a transfer function will be useful for investigating system stability and will be utilized in Chapter 9. Therefore it is worthwhile to complete another example at this point. Consider a transfer function

$$|G(s)|_{s=j\omega} = G(j\omega) = \frac{K}{j\omega(j\omega\tau + 1)} = \frac{K}{j\omega - \omega^2\tau}. \tag{8.14}$$

Then the magnitude and phase angle are written as

$$|G(\omega)| = \frac{K}{(\omega^2 + \omega^4\tau^2)^{1/2}} \quad \text{and} \quad \phi(\omega) = -\tan^{-1}\left(\frac{1}{-\omega\tau}\right).$$

The phase angle and the magnitude are readily calculated at the frequencies $\omega = 0$, $\omega = 1/\tau$, and $\omega = +\infty$. The values of $|G(\omega)|$ and $\phi(\omega)$ are given in Table 8.1, and the polar plot of $G(j\omega)$ is shown in Fig. 8.4.

An alternative solution uses the real and imaginary parts of $G(j\omega)$ as

$$G(j\omega) = \frac{K}{j\omega - \omega^2\tau} = \frac{K(-j\omega - \omega^2\tau)}{\omega^2 + \omega^4\tau^2} = R(\omega) + jX(\omega), \tag{8.15}$$

where $R(\omega) = -K\omega^2\tau/M(\omega)$ and $X(\omega) = -j\omega K/M(\omega)$ and where $M(\omega) = \omega^2 + \omega^4$. Then, when $\omega = \infty$, we have $R(\omega) = 0$ and $X(\omega) = 0$. When $\omega = 0$, we have $R(\omega) = -K\tau$ and $X(\omega) = -j\omega$. When $\omega = 1/\tau$, we have $R(\omega) = -K\tau/2$ and $X(\omega) = -jK\tau/2$ as shown in Fig. 8.4.

TABLE 8.1

ω	0	½τ	1/τ	∞
$|G(\omega)|$	∞	$4K\tau/\sqrt{5}$	$K\tau/\sqrt{2}$	0
$\phi(\omega)$	$-90°$	$-117°$	$-135°$	$-180°$

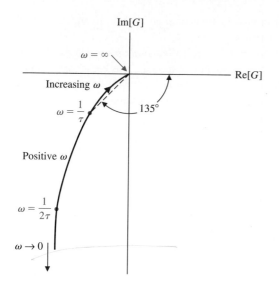

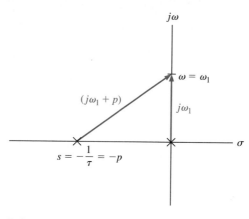

FIGURE 8.4
Polar plot for $G(j\omega) = K/j\omega(j\omega\tau + 1)$.
Note that $\omega = \infty$ at the origin.

FIGURE 8.5
Two vectors on the s-plane to evaluate $G(j\omega_1)$.

Another method of obtaining the polar plot is to evaluate graphically the vector $G(j\omega)$ at specific frequencies, ω, along the $s = j\omega$ axis on the s-plane. We consider

$$G(s) = \frac{K\tau}{s(s + 1/\tau)}$$

with the two poles shown on the s-plane in Fig. 8.5.
When $s = j\omega$, we have

$$G(j\omega) = \frac{K\tau}{j\omega(j\omega + p)},$$

where $p = 1/\tau$. The magnitude and phase of $G(j\omega)$ can be evaluated at a specific frequency, ω_1, on the $j\omega$-axis, as shown in Fig. 8.5. The magnitude and the phase $\phi(\omega)$ is

$$|G(\omega_1)| = \frac{K\tau}{|j\omega_1|\,|j\omega_1 + p|}$$

and

$$\phi(\omega) = -\underline{/(j\omega_1)} - \underline{/(j\omega_1 + p)} = -90° - \tan^{-1}(\omega_1/p). \quad \blacksquare$$

There are several possibilities for coordinates of a graph portraying the frequency response of a system. As we have seen, we may choose to utilize a polar plot to represent the frequency response (Eq. 8.8) of a system. However, the limitations of polar plots are readily apparent. The addition of poles or zeros to an existing system requires the recalculation of the frequency response as outlined in Examples 8.1 and 8.2. (See Table 8.1.) Furthermore, the calculation of the frequency response in this manner is tedious and does not indicate the effect of the individual poles or zeros.

Therefore the introduction of *logarithmic plots,* often called *Bode plots,* simplifies the determination of the graphical portrayal of the frequency response. The logarithmic plots are called Bode plots in honor of H. W. Bode, who used them extensively in his studies of feedback amplifiers [4, 5]. The transfer function in the frequency domain is

$$G(j\omega) = |G(\omega)|e^{j\phi(\omega)}. \tag{8.16}$$

The natural logarithm of Eq. (8.16) is

$$\ln G(j\omega) = \ln|G(\omega)| + j\phi(\omega), \tag{8.17}$$

where $\ln|G|$ is the magnitude in nepers. The logarithm of the magnitude is normally expressed in terms of the logarithm to the base 10, so that we use

$$\text{Logarithmic gain} = 20 \log_{10}|G(\omega)|,$$

where the units are decibels (dB). A decibel conversion table is given in Appendix D. The logarithmic gain in dB and the angle $\phi(\omega)$ can be plotted versus the frequency ω by utilizing several different arrangements. For a Bode diagram, the plot of logarithmic gain in dB versus ω is normally plotted on one set of axes, and the phase $\phi(\omega)$ versus ω on another set of axes, as shown in Fig. 8.6. For example, the Bode diagram of the transfer function of Example 8.1 can be readily obtained, as we will find in the following example.

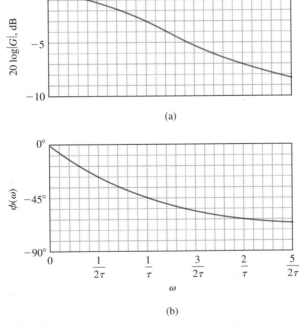

(a)

(b)

FIGURE 8.6
Bode diagram for $G(j\omega) = 1/(j\omega\tau + 1)$: (a) magnitude plot and (b) phase plot.

EXAMPLE 8.3   **Bode diagram of an *RC* filter**

The transfer function of Example 8.1 is

$$\frac{1}{\frac{s}{a}+1}$$

$$= \frac{a}{s-a}$$

$$G(j\omega) = \frac{1}{j\omega(RC) + 1} = \frac{1}{j\omega\tau + 1}, \qquad (8.18)$$

where

$$\tau = RC,$$

the time constant of the network. The logarithmic gain is

$$20 \log |G| = 20 \log \left(\frac{1}{1 + (\omega\tau)^2}\right)^{1/2} = -10 \log(1 + (\omega\tau)^2). \qquad (8.19)$$

For small frequencies—that is, $\omega \ll 1/\tau$—the logarithmic gain is

$$20 \log |G| = -10 \log(1) = 0 \text{ dB}, \qquad \omega \ll 1/\tau. \qquad (8.20)$$

For large frequencies—that is, $\omega \gg 1/\tau$—the logarithmic gain is

$$20 \log |G| = -20 \log \omega\tau \qquad \omega \gg 1/\tau, \qquad (8.21)$$

and at $\omega = 1/\tau$, we have

$$20 \log |G| = -10 \log 2 = -3.01 \text{ dB}.$$

The magnitude plot for this network is shown in Fig. 8.6(a). The phase angle of this network is

$$\phi(j\omega) = -\tan^{-1} \omega\tau. \qquad (8.22)$$

The phase plot is shown in Fig. 8.6(b). The frequency $\omega = 1/\tau$ is often called the *break frequency* or *corner frequency*. ∎

Examining the Bode diagram of Fig. 8.6, we find that a linear scale of frequency is not the most convenient or judicious choice and we should consider the use of a logarithmic scale of frequency. The convenience of a logarithmic scale of frequency can be seen by considering Eq. (8.21) for large frequencies $\omega \gg 1/\tau$, as follows:

$$20 \log |G| = -20 \log \omega\tau = -20 \log \tau - 20 \log \omega. \qquad (8.23)$$

Then, on a set of axes where the horizontal axis is log ω, the asymptotic curve for $\omega \gg 1/\tau$ is a straight line, as shown in Fig. 8.7. The slope of the straight line can be ascertained

FIGURE 8.7
Asymptotic curve
for $(j\omega\tau + 1)^{-1}$.

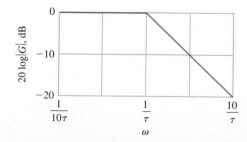

from Eq. (8.21). An interval of two frequencies with a ratio equal to 10 is called a decade, so that the range of frequencies from ω_1 to ω_2, where $\omega_2 = 10\omega_1$, is called a decade. Then, the difference between the logarithmic gains, for $\omega \gg 1/\tau$, over a decade of frequency is

$$20 \log |G(\omega_1)| - 20 \log |G(\omega_2)| = -20 \log \omega_1\tau - (-20 \log \omega_2\tau)$$

$$= -20 \log \frac{\omega_1\tau}{\omega_2\tau} \tag{8.24}$$

$$= -20 \log (\tfrac{1}{10}) = +20 \text{ dB}.$$

That is, the slope of the asymptotic line for this first-order transfer function is -20 dB/decade, and the slope is shown for this transfer function in Fig. 8.7. Instead of using a horizontal axis of log ω and linear rectangular coordinates, it is simpler to use semilog paper with a linear rectangular coordinate for dB and a logarithmic coordinate for ω. Alternatively, we could use a logarithmic coordinate for the magnitude as well as for frequency and avoid the necessity of calculating the logarithm of the magnitude.

The frequency interval $\omega_2 = 2\omega_1$ is often used and is called an octave of frequencies. The difference between the logarithmic gains for $\omega \gg 1/\tau$, for an octave, is

$$20 \log |G(\omega_1)| - 20 \log |G(\omega_2)| = -20 \log \frac{\omega_1\tau}{\omega_2\tau} \tag{8.25}$$

$$= -20 \log (\tfrac{1}{2}) = 6.02 \text{ dB}.$$

Therefore the slope of the asymptotic line is -6 dB/octave or -20 dB/decade.

The primary advantage of the logarithmic plot is the conversion of multiplicative factors such as $(j\omega\tau + 1)$ into additive factors $20 \log (j\omega\tau + 1)$ by virtue of the definition of logarithmic gain. This can be readily ascertained by considering a generalized transfer function as

$$G(j\omega) = \frac{K_b \displaystyle\prod_{i=1}^{Q} (1 + j\omega\tau_i)}{(j\omega)^N \displaystyle\prod_{m=1}^{M} (1 + j\omega\tau_m) \displaystyle\prod_{k=1}^{R} [(1 + (2\zeta_k/\omega_{n_k})j\omega + (j\omega/\omega_{n_k})^2)]}. \tag{8.26}$$

This transfer function includes Q zeros, N poles at the origin, M poles on the real axis, and R pairs of complex conjugate poles. Clearly, obtaining the polar plot of such a function would be a formidable task indeed. However, the logarithmic magnitude of $G(j\omega)$ is

$$20 \log |G(\omega)| = 20 \log K_b + 20 \sum_{i=1}^{Q} \log |1 + j\omega\tau_i|$$

$$- 20 \log |(j\omega)^N| - 20 \sum_{m=1}^{M} \log |1 + j\omega\tau_m| \tag{8.27}$$

$$- 20 \sum_{k=1}^{R} \log \left|1 + \left(\frac{2\zeta_k}{\omega_{n_k}}\right)j\omega + \left(\frac{j\omega}{\omega_{n_k}}\right)^2\right|,$$

and the Bode diagram can be obtained by adding the plot due to each individual factor. Furthermore, the separate phase angle plot is obtained as

$$\phi(\omega) = + \sum_{i=1}^{Q} \tan^{-1} \omega\tau_i - N(90°) - \sum_{m=1}^{M} \tan^{-1} \omega\tau_m$$

$$- \sum_{k=1}^{R} \tan^{-1}\left(\frac{2\zeta_k \omega_{n_k}\omega}{\omega_{n_k}^2 - \omega^2}\right), \tag{8.28}$$

which is simply the summation of the phase angles due to each individual factor of the transfer function.

Therefore the four different kinds of factors that may occur in a transfer function are as follows:

1. Constant gain K_b
2. Poles (or zeros) at the origin $(j\omega)$
3. Poles (or zeros) on the real axis $(j\omega\tau + 1)$
4. Complex conjugate poles (or zeros) $[1 + (2\zeta/\omega_n)j\omega + (j\omega/\omega_n)^2]$

We can determine the logarithmic magnitude plot and phase angle for these four factors and then utilize them to obtain a Bode diagram for any general form of a transfer function. Typically, the curves for each factor are obtained and then added together graphically to obtain the curves for the complete transfer function. Furthermore, this procedure can be simplified by using the asymptotic approximations to these curves and obtaining the actual curves only at specific important frequencies.

Constant Gain K_b.   The logarithmic gain is

$$20 \log K_b = \text{constant in dB},$$

and the phase angle is zero. The gain curve is simply a horizontal line on the Bode diagram.

If the gain is a negative value, $-K_b$, the logarithmic gain remains $20 \log K_b$. The negative sign is accounted for by the phase angle, $-180°$.

Poles (or Zeros) at the Origin $(j\omega)$.   A pole at the origin has a logarithmic magnitude

$$20 \log \left|\frac{1}{j\omega}\right| = -20 \log \omega \text{ dB} \tag{8.29}$$

and a phase angle $\phi(\omega) = -90°$. The slope of the magnitude curve is -20 dB/decade for a pole. Similarly for a multiple pole at the origin, we have

$$20 \log \left|\frac{1}{(j\omega)^N}\right| = -20 N \log \omega, \tag{8.30}$$

and the phase is $\phi(\omega) = -90°N$. In this case the slope due to the multiple pole is $-20N$ dB/decade. For a zero at the origin, we have a logarithmic magnitude

$$20 \log |j\omega| = +20 \log \omega, \tag{8.31}$$

where the slope is $+20$ dB/decade and the phase angle is $+90°$. The Bode diagram of the magnitude and phase angle of $(j\omega)^{\pm N}$ is shown in Fig. 8.8 for $N = 1$ and $N = 2$.

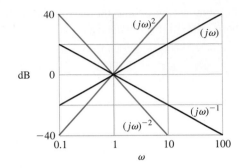
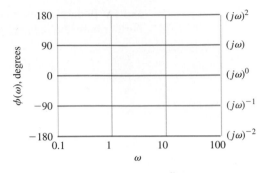

FIGURE 8.8
Bode diagram for
$(j\omega)^{\pm N}$.

Poles or Zeros on the Real Axis.   The pole factor $(1 + j\omega\tau)^{-1}$ has been considered previously and we found that

$$20 \log \left| \frac{1}{1 + j\omega\tau} \right| = -10 \log (1 + \omega^2\tau^2). \qquad (8.32)$$

The asymptotic curve for $\omega \ll 1/\tau$ is $20 \log 1 = 0$ dB, and the asymptotic curve for $\omega \gg 1/\tau$ is $-20 \log \omega\tau$, which has a slope of -20 dB/decade. The intersection of the two asymptotes occurs when

$$20 \log 1 = 0 \text{ dB} = -20 \log \omega\tau$$

or when $\omega = 1/\tau$, the *break frequency*. The actual logarithmic gain when $\omega = 1/\tau$ is -3 dB for this factor. The phase angle is $\phi(\omega) = -\tan^{-1} \omega\tau$ for the denominator factor. The Bode diagram of a pole factor $(1 + j\omega\tau)^{-1}$ is shown in Fig. 8.9.

FIGURE 8.9
Bode diagram for
$(1 + j\omega\tau)^{-1}$.

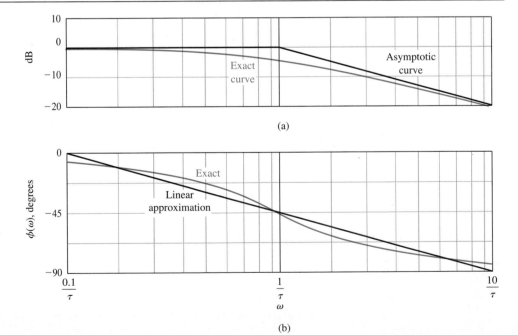

TABLE 8.2

$\omega\tau$	0.10	0.50	0.76	1	1.31	2	5	10
$20 \log |(1 + j\omega\tau)^{-1}|$, dB	−0.04	−1.0	−2.0	−3.0	−4.3	−7.0	−14.2	−20.04
Asymptotic approximation, dB	0	0	0	0	−2.3	−6.0	−14.0	−20.0
$\phi(\omega)$, degrees	−5.7	−26.6	−37.4	−45.0	−52.7	−63.4	−78.7	−84.3
Linear approximation, degrees	0	−31.5	−39.5	−45.0	−50.3	−58.5	−76.5	−90.0

The Bode diagram of a zero factor $(1 + j\omega\tau)$ is obtained in the same manner as that of the pole. However, the slope is positive at +20 dB/decade, and the phase angle is $\phi(\omega) = +\tan^{-1}\omega\tau$.

A linear approximation to the phase angle curve can be obtained as shown in Fig. 8.9. This linear approximation, which passes through the correct phase at the break frequency, is within 6° of the actual phase curve for all frequencies. This approximation will provide a useful means for readily determining the form of the phase angle curves of a transfer function $G(s)$. However, often the accurate phase angle curves are required and the actual phase curve for the first-order factor must be drawn. Therefore it is often worthwhile to prepare a cardboard (or plastic) template for drawing the multiple phase curves for the individual factors. The exact values of the frequency response for the pole $(1 + j\omega\tau)^{-1}$ as well as the values obtained by using the approximation for comparison are given in Table 8.2.

Complex Conjugate Poles or Zeros $[1 + (2\zeta/\omega_n)j\omega + (j\omega/\omega_n)^2]$.   The quadratic factor for a pair of complex conjugate poles can be written in normalized form as

$$[1 + j2\zeta u - u^2]^{-1}, \tag{8.33}$$

where $u = \omega/\omega_n$. Then, the logarithmic magnitude is

$$20 \log |G(\omega)| = -10 \log ((1 - u^2)^2 + 4\zeta^2 u^2), \tag{8.34}$$

and the phase angle is

$$\phi(\omega) = -\tan^{-1}\left(\frac{2\zeta u}{1 - u^2}\right). \tag{8.35}$$

When $u \ll 1$, the magnitude is

$$20 \log|G| = -10 \log 1 = 0 \text{ dB}$$

and the phase angle approaches 0°. When $u \gg 1$, the logarithmic magnitude approaches

$$20 \log|G| = -10 \log u^4 = -40 \log u,$$

which results in a curve with a slope of −40 dB/decade. The phase angle, when $u \gg 1$, approaches −180°. The magnitude asymptotes meet at the 0-dB line when $u = \omega/\omega_n = 1$.

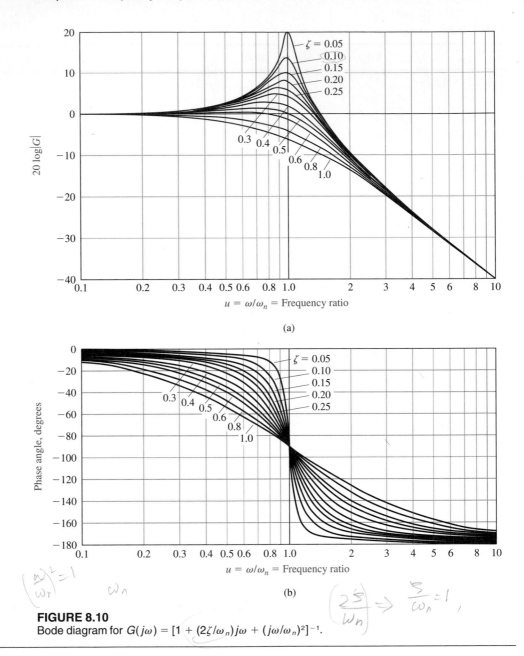

FIGURE 8.10
Bode diagram for $G(j\omega) = [1 + (2\zeta/\omega_n)j\omega + (j\omega/\omega_n)^2]^{-1}$.

However, the difference between the actual magnitude curve and the asymptotic approximation is a function of the damping ratio and must be accounted for when $\zeta < 0.707$. The Bode diagram of a quadratic factor due to a pair of complex conjugate poles is shown in Fig. 8.10. The maximum value of the frequency response, $M_{p\omega}$, occurs at the *resonant frequency* ω_r. When the damping ratio approaches zero, then ω_r approaches ω_n, the natural frequency. The resonant frequency is determined by taking the derivative of the magnitude

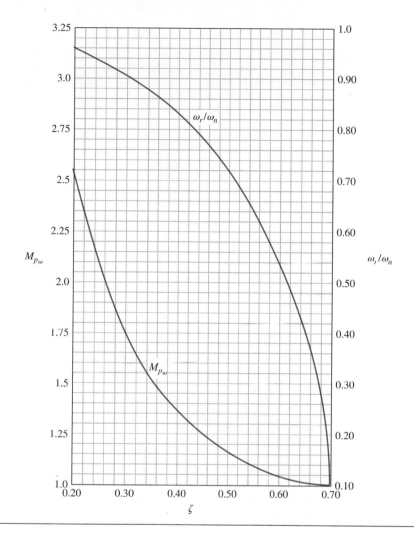

FIGURE 8.11
The maximum of the frequency response, M_{p_ω}, and the resonant frequency, ω_r, versus ζ for a pair of complex conjugate poles.

of Eq. (8.33) with respect to the normalized frequency, u, and setting it equal to zero. The resonant frequency is represented by the relation

$$\omega_r = \omega_n \sqrt{1 - 2\zeta^2}, \qquad \zeta < 0.707, \tag{8.36}$$

and the maximum value of the magnitude $|G(\omega)|$ is

$$M_{p_\omega} = |G(\omega_r)| = (2\zeta\sqrt{1 - \zeta^2})^{-1}, \qquad \zeta < 0.707, \tag{8.37}$$

for a pair of complex poles. The maximum value of the frequency response M_{p_ω} and the resonant frequency ω_r are shown as a function of the damping ratio ζ for a pair of complex poles in Fig. 8.11. Assuming the dominance of a pair of complex conjugate closed-loop poles, we find that these curves are useful for estimating the damping ratio of a system from an experimentally determined frequency response.

The frequency response curves can be evaluated on the s-plane by determining the vector lengths and angles at various frequencies ω along the ($s = +j\omega$)-axis. For example,

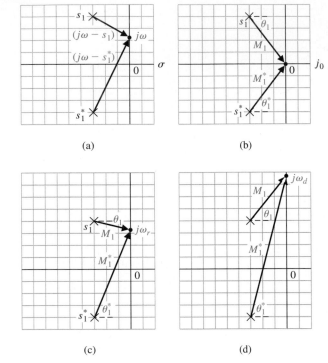

(c)   (d)

FIGURE 8.12
Vector evaluation
of the frequency
response for
selected values
of ω.

considering the second-order factor with complex conjugate poles, we have

$$G(s) = \frac{1}{(s/\omega_n)^2 + 2\zeta s/\omega_n + 1} = \frac{\omega_n^2}{s^2 + 2\zeta\omega_n s + \omega_n^2}. \qquad (8.38)$$

The poles lie on a circle of radius ω_n and are shown for a particular ζ in Fig. 8.12(a). The transfer function evaluated for real frequency $s = j\omega$ is written as

$$G(j\omega) = \frac{\omega_n^2}{(s - s_1)(s - s_1^*)}\bigg|_{s=j\omega} = \frac{\omega_n^2}{(j\omega - s_1)(j\omega - s_1^*)}. \qquad (8.39)$$

where s_1 and s_1^* are the complex conjugate poles. The vectors $(j\omega - s_1)$ and $(j\omega - s_1^*)$ are the vectors from the poles to the frequency $j\omega$, as shown in Fig. 8.12(a). Then the magnitude and phase may be evaluated for various specific frequencies. The magnitude is

$$|G(\omega)| = \frac{\omega_n^2}{|j\omega - s_1|\,|j\omega - s_1^*|}, \qquad (8.40)$$

and the phase is

$$\phi(\omega) = -\underline{/(j\omega - s_1)} - \underline{/(j\omega - s_1^*)}.$$

The magnitude and phase may be evaluated for three specific frequencies:

$$\omega = 0, \qquad \omega = \omega_r, \qquad \omega = \omega_d,$$

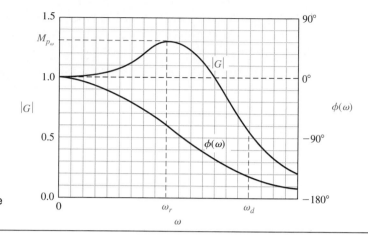

FIGURE 8.13
Bode diagram for complex conjugate poles.

as shown in Fig. 8.12 in parts (b), (c), and (d), respectively. The magnitude and phase corresponding to these frequencies are shown in Fig. 8.13.

EXAMPLE 8.4   **Bode diagram of twin-T network**

As an example of the determination of the frequency response using the pole–zero diagram and the vectors to $j\omega$, consider the twin-T network shown in Fig. 8.14 [6]. The transfer function of this network is

$$G(s) = \frac{V_o(s)}{V_{in}(s)} = \frac{(s\tau)^2 + 1}{(s\tau)^2 + 4s\tau + 1}, \tag{8.41}$$

where $\tau = RC$. The zeros are at $\pm j1$ and the poles are at $-2 \pm \sqrt{3}$ in the $s\tau$-plane, as shown in Fig. 8.15(a). Clearly, at $\omega = 0$, we have $|G| = 1$ and $\phi(\omega) = 0°$. At $\omega = 1/\tau$, $|G| = 0$ and the phase angle of the vector from the zero at $s\tau = j1$ passes through a transition of 180°. When ω approaches ∞, $|G| = 1$ and $\phi(\omega) = 0$ again. Evaluating several intermediate frequencies, we can readily obtain the frequency response, as shown in Fig. 8.15(b). ■

A summary of the asymptotic curves for basic terms of a transfer function is provided in Table 8.3.

In the previous examples the poles and zeros of $G(s)$ have been restricted to the left-hand plane. However, a system may have zeros located in the right-hand s-plane and may

FIGURE 8.14
Twin-T network.

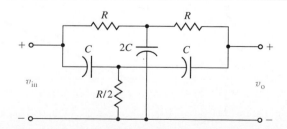

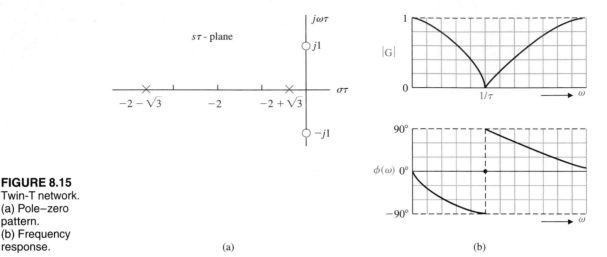

FIGURE 8.15
Twin-T network.
(a) Pole–zero
pattern.
(b) Frequency
response.

(a)   (b)

still be stable. Transfer functions with zeros in the right-hand s-plane are classified as *non-minimum phase-shift* transfer functions. If the zeros of a transfer function are all reflected about the $j\omega$-axis, there is no change in the magnitude of the transfer function, and the only difference is in the phase-shift characteristics. If the phase characteristics of the two system functions are compared, it can be readily shown that the net phase shift over the frequency range from zero to infinity is less for the system with all its zeros in the left-hand s-plane. Thus the transfer function $G_1(s)$, with all its zeros in the left-hand s-plane, is called a *minimum phase transfer function*. The transfer function $G_2(s)$, with $|G_2(j\omega)| = |G_1(j\omega)|$ and all the zeros of $G_1(s)$ reflected about the $j\omega$-axis into the right-hand s-plane, is called a *nonminimum phase transfer function*. Reflection of any zero or pair of zeros into the right half-plane results in a nonminimum phase transfer function.

> **A transfer function is called a minimum phase transfer function if all its zeros lie in the left-hand s-plane. It is called a nonminimum phase transfer function if it has zeros in the right-hand s-plane.**

The two pole–zero patterns shown in Figs. 8.16(a) and (b) have the same amplitude characteristics as can be deduced from the vector lengths. However, the phase characteristics are different for Figs. 8.16(a) and (b). The minimum phase characteristic of Fig. 8.16(a)

FIGURE 8.16
Pole–zero patterns
giving the same
amplitude
response and
different phase
characteristics.

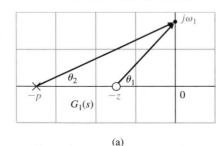

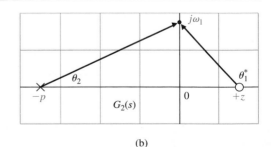

(a)   (b)

TABLE 8.3   Asymptotic Curves for Basic Terms of a Transfer Function

Term	Magnitude $20 \log |G|$	Phase, $\phi(\omega)$
1. Gain, $G(j\omega) = K$		
2. Zero, $G(j\omega) = (1 + j\omega/\omega_1)$		
3. Pole, $G(j\omega) = (1 + j\omega/\omega_1)^{-1}$		
4. Pole at the origin, $G(j\omega) = 1/j\omega$		
5. Two complex poles, $0.1 < \zeta < 1$, $G(j\omega) = (1 + j2\zeta u - u^2)^{-1}$, $u = \omega/\omega_n$		

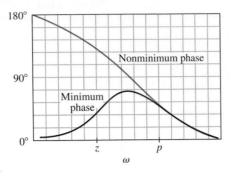

FIGURE 8.17
The phase characteristics for the minimum phase and nonminimum phase transfer function.

and the nonminimum phase characteristic of Fig. 8.16(b) are shown in Fig. 8.17. Clearly, the phase shift of

$$G_1(s) = \frac{s + z}{s + p}$$

ranges over less than 80°, whereas the phase shift of

$$G_2(s) = \frac{s - z}{s + p}$$

ranges over 180°. The meaning of the term *minimum phase* is illustrated by Fig. 8.17. The range of phase shift of a minimum phase transfer function is the least possible or minimum corresponding to a given amplitude curve, whereas the range of the nonminimum phase curve is greater than the minimum possible for the given amplitude curve.

A particularly interesting nonminimum phase network is the *all-pass* network, which can be realized with a symmetrical lattice network [8]. A symmetrical pattern of poles and

FIGURE 8.18
The all-pass network (a) pole–zero pattern and (b) frequency response and (c) a lattice network.

(a)

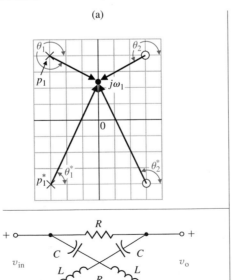

(c)

(b)

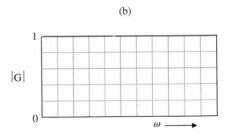

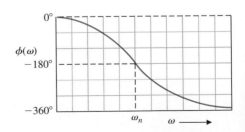

zeros is obtained as shown in Fig. 8.18(a). Again, the magnitude $|G|$ remains constant; in this case, it is equal to unity. However, the angle varies from $0°$ to $-360°$. Because $\theta_2 = 180° - \theta_1$ and $\theta_2^* = 180° - \theta_1^*$, the phase is given by $\phi(\omega) = -2(\theta_1 + \theta_1^*)$. The magnitude and phase characteristic of the all-pass network is shown in Fig. 8.18(b). A nonminimum phase lattice network is shown in Fig. 8.18(c).

8.3 AN EXAMPLE OF DRAWING THE BODE DIAGRAM

The Bode diagram of a transfer function $G(s)$, which contains several zeros and poles, is obtained by adding the plot due to each individual pole and zero. The simplicity of this method will be illustrated by considering a transfer function that possesses all the factors considered in the preceding section. The transfer function of interest is

$$G(j\omega) = \frac{5(1 + j0.1\omega)}{j\omega(1 + j0.5\omega)(1 + j0.6(\omega/50) + (j\omega/50)^2)}. \tag{8.42}$$

The factors, in order of their occurrence as frequency increases, are as follows:

1. A constant gain $K = 5$ 4. A zero at $\omega = 10$
2. A pole at the origin   5. A pair of complex poles
3. A pole at $\omega = 2$ at $\omega = \omega_n = 50$

First, we plot the magnitude characteristic for each individual pole and zero factor and the constant gain.

1. The constant gain is $20 \log 5 = 14$ dB, as shown in Fig. 8.19.
2. The magnitude of the pole at the origin extends from zero frequency to infinite frequencies and has a slope of -20 dB/decade intersecting the 0-dB line at $\omega = 1$, as shown in Fig. 8.19.
3. The asymptotic approximation of the magnitude of the pole at $\omega = 2$ has a slope of -20 dB/decade beyond the break frequency at $\omega = 2$. The asymptotic magnitude below the break frequency is 0 dB, as shown in Fig. 8.19.

FIGURE 8.19
Magnitude
asymptotes of
poles and zeros
used in the
example.

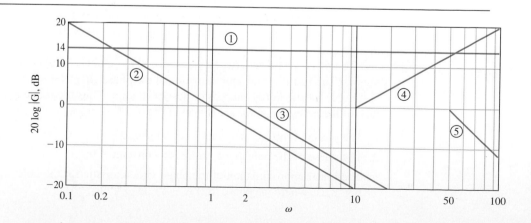

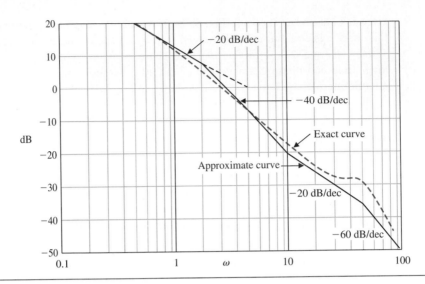

FIGURE 8.20
Magnitude
characteristic.

4. The asymptotic magnitude for the zero at $\omega = +10$ has a slope of $+20$ dB/decade beyond the break frequency at $\omega = 10$, as shown in Fig. 8.19.

5. The asymptotic approximation for the pair of complex poles at $\omega = \omega_n = 50$ has a slope of -40 dB/decade due to the quadratic forms. The break frequency is $\omega = \omega_n = 50$, as shown in Fig. 8.19. This approximation must be corrected to the actual magnitude because the damping ratio is $\zeta = 0.3$ and the magnitude differs appreciably from the approximation, as shown in Fig. 8.20.

Therefore the total asymptotic magnitude can be plotted by adding the asymptotes due to each factor, as shown by the solid line in Fig. 8.20. Examining the asymptotic curve of Fig. 8.20, one notes that the curve can be obtained directly by plotting each asymptote in order as frequency increases. Thus the slope is -20 dB/decade due to $(j\omega)^{-1}$ intersecting 14 dB at $\omega = 1$. Then at $\omega = 2$, the slope becomes -40 dB/decade due to the pole at $\omega = 2$. The slope changes to -20 dB/decade due to the zero at $\omega = 10$. Finally, the slope becomes -60 dB/decade at $\omega = 50$ due to the pair of complex poles at $\omega_n = 50$.

The exact magnitude curve is then obtained by utilizing Table 8.2, which provides the difference between the actual and asymptotic curves for a single pole or zero. The exact magnitude curve for the pair of complex poles is obtained by utilizing Fig. 8.10(a) for the quadratic factor. The exact magnitude curve for $G(j\omega)$ is shown by a dashed colored line in Fig. 8.20.

The phase characteristic can be obtained by adding the phase due to each individual factor. Usually, the linear approximation of the phase characteristic for a single pole or zero is suitable for the initial analysis or design attempt. Thus the individual phase characteristics for the poles and zeros are shown in Fig. 8.21.

1. The phase of the constant gain is, of course, $0°$.

2. The phase of the pole at the origin is a constant $-90°$.

3. The linear approximation of the phase characteristic for the pole at $\omega = 2$ is shown in Fig. 8.21, where the phase shift is $-45°$ at $\omega = 2$.

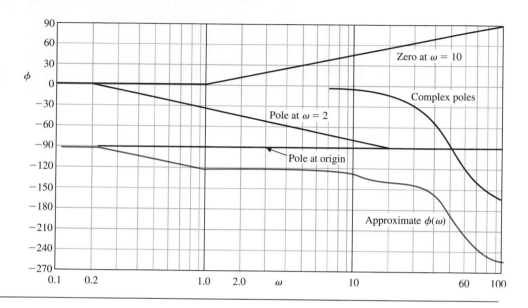

FIGURE 8.21
Phase
characteristic.

4. The linear approximation of the phase characteristic for the zero at $\omega = 10$ is also shown in Fig. 8.21, where the phase shift is $+45°$ at $\omega - 10$.

5. The actual phase characteristic for the pair of complex poles is obtained from Fig. 8.10 and is shown in Fig. 8.21.

Therefore the total phase characteristic, $\phi(\omega)$, is obtained by adding the phase due to each factor as shown in Fig. 8.21. While this curve is an approximation, its usefulness merits consideration as a first attempt to determine the phase characteristic. Thus, for example, a frequency of interest, as we shall note in the following section, is that frequency for which $\phi(\omega) = -180°$. The approximate curve indicates that a phase shift of $-180°$ occurs at $\omega = 46$. The actual phase shift at $\omega = 46$ can be readily calculated as

$$\phi(\omega) = -90° - \tan^{-1} \omega\tau_1 + \tan^{-1} \omega\tau_2 - \tan^{-1} \frac{2\zeta u}{1 - u^2}, \qquad (8.43)$$

where

$$\tau_1 = 0.5, \qquad \tau_2 = 0.1, \qquad u = \omega/\omega_n = \omega/50.$$

Then we find that

$$\phi(46) = -90° - \tan^{-1} 23 + \tan^{-1} 4.6 - \tan^{-1} 3.55 = -175°, \qquad (8.44)$$

and the approximate curve has an error of $5°$ at $\omega = 46$. However, once the approximate frequency of interest is ascertained from the approximate phase curve, the accurate phase shift for the neighboring frequencies is readily determined by using the exact phase shift relation (Eq. 8.43). This approach is usually preferable to the calculation of the exact phase shift for all frequencies over several decades. In summary, one may obtain approximate curves for the magnitude and phase shift of a transfer function $G(j\omega)$ in order to determine the important frequency ranges. Then, within the relatively small important frequency

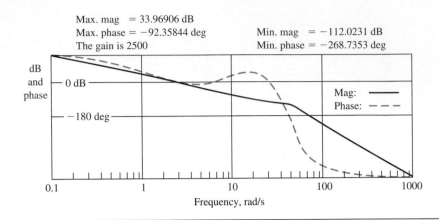

Max. mag = 33.96906 dB
Max. phase = −92.35844 deg   Min. mag = −112.0231 dB
The gain is 2500   Min. phase = −268.7353 deg

FIGURE 8.22
The Bode plot of
the $G(j\omega)$ of
Eq. (8.42).

ranges, the exact magnitude and phase shift can be readily evaluated by using the exact equations, such as Eq. (8.43).

The frequency response of $G(j\omega)$ can be calculated and plotted using a computer program such as *Matlab*. The Bode plot for the example in this section (Eq. 8.42) can be readily obtained, as shown in Fig. 8.22. The plot is generated for four decades and the 0-dB line is indicated as well as the −180° line. The data above the plot indicate that the magnitude is 34 dB and that the phase is −92.36° at $\omega = 0.1$. Similarly, the data indicate that the magnitude is −43 dB and that the phase is −243° at $\omega = 100$. Using the tabular data provided, one finds that the magnitude is 0 dB at $\omega = 3.0$ and the phase is −180° at $\omega = 50$.

8.4 FREQUENCY RESPONSE MEASUREMENTS

A sine wave can be used to measure the open-loop frequency response of a control system. In practice, a plot of amplitude versus frequency and phase versus frequency will be obtained [1, 3, 6]. From these two plots, the open-loop transfer function $GH(j\omega)$ can be deduced. Similarly, the closed-loop frequency response of a control system, $T(j\omega)$, may be obtained and the actual transfer function deduced.

A device called a wave analyzer can be used to measure the amplitude and phase variations as the frequency of the input sine wave is altered. Also, a device called a transfer function analyzer can be used to measure the open-loop and closed-loop transfer functions [6].

A signal analyzer instrument, the Hewlett-Packard 3562A Dynamic Signal Analyzer, performs frequency response measurements from dc to 100 kHz. Built-in analysis and modeling capabilities can derive poles and zeros from measured frequency responses or construct phase and magnitude responses from user-supplied models. This device can also synthesize the frequency response of a model of a system, allowing a comparison with an actual response (see Fig. E8.10).

As an example of determining the transfer function from the Bode plot, let us consider the plot shown in Fig. 8.23. The system is a stable circuit consisting of resistors and capacitors. Because the magnitude declines at about −20 dB/decade as ω increases between

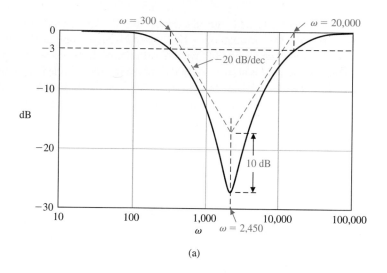

(a)

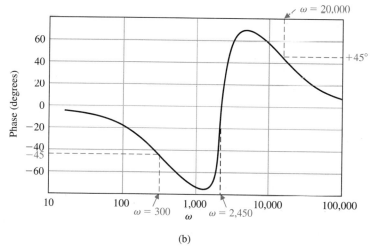

FIGURE 8.23
A Bode diagram for a system with an unidentified transfer function.

(b)

100 and 1000, and because the phase is $-45°$ and the magnitude is -3 dB at 300 rad/s, we can deduce that one factor is a pole at $p_1 = 300$. Next, we deduce that a pair of quadratic zeros with $\zeta = 0.16$ exist at $\omega_n = 2450$. This is inferred by noting that the phase changes abruptly by nearly $+180°$, passing through $0°$ at $\omega_n = 2450$. Also, the slope of the magnitude changes from -20 dB/decade to $+20$ dB/decade at $\omega_n = 2450$. We draw the asymptotes for the pole, p_1, and the numerator of the proposed transfer function $T(s)$ of Eq. (8.45), as shown in Fig. 8.23(a).

$$T(s) = \frac{(s/\omega_n)^2 + (2\zeta/\omega_n)s + 1}{(s/p_1 + 1)(s/p_2 + 1)} \tag{8.45}$$

The difference in magnitude from the corner frequency ($\omega_n = 2450$) of the asymptotes to the minimum response is 10 dB, which, from Eq. (8.37), indicates that $\zeta = 0.16$. (Compare the plot of the quadratic zeros to the plot of the quadratic poles in Fig. 8.9. Note that the plots need to be turned "upside down" for the quadratic zeros and that the phase goes from 0° to +180° instead of −180°.) Because the slope of the magnitude returns to 0 dB/decade as ω exceeds 50,000, we determine that there is a second pole as well as two zeros. This second pole is at $p_2 = 20{,}000$ because the magnitude is −3 dB from the asymptote and the phase is +45° at this point (−90° for the first pole, +180° for the pair of quadratic zeros, and −45° for the second pole). Therefore the transfer function is

$$T(s) = \frac{(s/2450)^2 + (0.32/2450)s + 1}{(s/300 + 1)(s/20{,}000 + 1)}.$$

This frequency response is actually obtained from a bridged-T network (see Problems 2.8 and 8.3 and Fig. 8.14).

8.5 PERFORMANCE SPECIFICATIONS IN THE FREQUENCY DOMAIN

We must continually ask the question: How does the frequency response of a system relate to the expected transient response of the system? In other words, given a set of time-domain (transient performance) specifications, how do we specify the frequency response? For a simple second-order system, we have already answered this question by considering the time-domain performance in terms of overshoot, settling time, and other performance criteria such as integral squared error. For the second-order system shown in Fig. 8.24, the closed-loop transfer function is

$$T(s) = \frac{\omega_n^2}{s^2 + 2\zeta\omega_n s + \omega_n^2}. \tag{8.46}$$

The frequency response of this feedback system will appear as shown in Fig. 8.25. Because this is a second-order system, the damping ratio of the system is related to the maximum magnitude $M_{p\omega}$, which occurs at the frequency ω_r as shown in Fig. 8.25.

At the resonant frequency, ω_r, a maximum value of the frequency response, $M_{p\omega}$, is attained.

The bandwidth, ω_B, is a measure of a system's ability to faithfully reproduce an input signal.

The bandwidth is the frequency, ω_B, at which the frequency response has declined 3 dB from its low frequency value.

The resonant frequency ω_r and the −3-dB *bandwidth* can be related to the speed of the transient response. Thus, as the bandwidth ω_B increases, the rise time of the step response of the system will decrease. Furthermore, the overshoot to a step input can be related

---

FIGURE 8.24
A second-order closed-loop system.

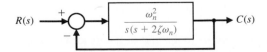

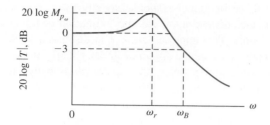

FIGURE 8.25
Magnitude characteristic of the second-order system.

to M_{p_ω} through the damping ratio ζ. The curves of Fig. 8.11 relate the resonance magnitude and frequency to the damping ratio of the second-order system. Then the step response overshoot may be estimated from Fig. 5.8 or may be calculated by utilizing Eq. (5.15). Thus we find as the resonant peak M_{p_ω} increases in magnitude, the overshoot to a step input increases. In general, the magnitude M_{p_ω} indicates the relative stability of a system.

The bandwidth of a system ω_B as indicated on the frequency response can be approximately related to the natural frequency of the system. Figure 8.26 shows the normalized bandwidth, ω_B/ω_n, versus ζ for the second-order system of Eq. (8.46). The response of the second-order system to a unit step input is

$$c(t) = 1 + e^{-\zeta\omega_n t} \cos(\omega_1 t + \theta). \tag{8.47}$$

The greater the magnitude of ω_n when ζ is constant, the more rapid the response approaches the desired steady-state value. Thus desirable frequency-domain specifications are as follows:

1. Relatively small resonant magnitudes: $M_{p_\omega} < 1.5$, for example.

2. Relatively large bandwidths so that the system time constant $\tau = 1/\zeta\omega_n$ is sufficiently small.

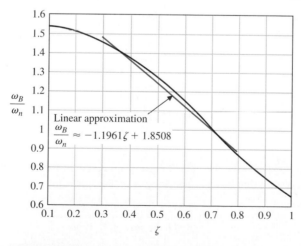

FIGURE 8.26
Normalized bandwidth, ω_B/ω_n, versus ζ for a second-order system (Eq. 8.46). The linear approximation $\omega_B/\omega_n = -1.1961\zeta + 1.8508$ is accurate for $0.3 \leq \zeta \leq 0.8$.

The usefulness of these frequency-response specifications and their relation to the actual transient performance depend upon the approximation of the system by a second-order pair of complex poles. This approximation was discussed in Section 7.3, and the second-order poles of $T(s)$ are called the *dominant roots*. Clearly, if the frequency response is dominated by a pair of complex poles, the relationships between the frequency response and the time response discussed in this section will be valid. Fortunately, a large proportion of control systems satisfy this dominant second-order approximation in practice.

The steady-state error specification can also be related to the frequency response of a closed-loop system. As we found in Section 5.4, the steady-state error for a specific test input signal can be related to the gain and number of integrations (poles at the origin) of the open-loop transfer function. Therefore, for the system shown in Fig. 8.24, the steady-state error for a ramp input is specified in terms of K_v, the velocity constant. The steady-state error for the system is

$$\lim_{t \to \infty} e(t) = \frac{A}{K_v},$$

where A = magnitude of the ramp input. The velocity constant for the closed-loop system of Fig. 8.24 is

$$K_v = \lim_{s \to 0} sG(s) = \lim_{s \to 0} s\left(\frac{\omega_n^2}{s(s + 2\zeta\omega_n)}\right) = \frac{\omega_n}{2\zeta}. \tag{8.48}$$

In Bode diagram form (in terms of time constants), the transfer function $G(s)$ is written as

$$G(s) = \frac{(\omega_n/2\zeta)}{s(s/(2\zeta\omega_n) + 1)} = \frac{K_v}{s(\tau s + 1)}, \tag{8.49}$$

and the gain constant is K_v for this type-one system. For example, reexamining the example of Section 8.3, we had a type-one system with an open-loop transfer function

$$G(j\omega) = \frac{5(1 + j\omega\tau_2)}{j\omega(1 + j\omega\tau_1)(1 + j0.6u - u^2)}, \tag{8.50}$$

where $u = \omega/\omega_n$. Therefore in this case we have $K_v = 5$. In general, if the open-loop transfer function of a feedback system is written as

$$G(j\omega) = \frac{K \prod_{i=1}^{M} (1 + j\omega\tau_i)}{(j\omega)^N \prod_{k=1}^{Q} (1 + j\omega\tau_k)}, \tag{8.51}$$

then the system is type N and the gain K is the gain constant for the steady-state error. Thus for a type-zero system that has an open-loop transfer function

$$G(j\omega) = \frac{K}{(1 + j\omega\tau_1)(1 + j\omega\tau_2)}. \tag{8.52}$$

$K = K_p$ (the position error constant) that appears as the low frequency gain on the Bode diagram.

Furthermore, the gain constant $K = K_v$ for the type-one system appears as the gain of the low frequency section of the magnitude characteristic. Considering only the pole and gain of the type-one system of Eq. (8.50), we have

$$G(j\omega) = \left(\frac{5}{j\omega}\right) = \left(\frac{K_v}{j\omega}\right), \quad \omega < 1/\tau_1, \tag{8.53}$$

and the K_v is equal to the frequency when this portion of the magnitude characteristic intersects the 0-dB line. For example, the low frequency intersection of $(K_v/j\omega)$ in Fig. 8.20 is equal to $\omega = 5$, as we expect.

Therefore the frequency response characteristics represent the performance of a system quite adequately, and with some experience they are quite useful for the analysis and design of feedback control systems.

8.6 LOG MAGNITUDE AND PHASE DIAGRAMS

There are several alternative methods of presenting the frequency response of a function $GH(j\omega)$. We have seen that suitable graphical presentations of the frequency response are (1) the polar plot and (2) the Bode diagram. An alternative approach to portraying the frequency response graphically is to plot the logarithmic magnitude in dB versus the phase angle for a range of frequencies. Because this information is equivalent to that portrayed by the Bode diagram, it is normally easier to obtain the Bode diagram and transfer the information to the coordinates of the log magnitude versus phase diagram. Alternatively, one can construct templates for first- and second-order factors and work directly on the log-magnitude–phase diagram. The gain and phase of cascaded transfer functions can then be added vectorially directly on the diagram.

An illustration will best portray the use of the log-magnitude–phase diagram. This diagram for a transfer function

$$GH_1(j\omega) = \frac{5}{j\omega(0.5j\omega + 1)((j\omega/6) + 1)} \tag{8.54}$$

is shown in Fig. 8.27. The numbers indicated along the curve are for values of frequency, ω.

The log-magnitude–phase curve for the transfer function

$$GH_2(j\omega) = \frac{5(0.1j\omega + 1)}{j\omega(0.5j\omega + 1)(1 + j0.6(\omega/50) + (j\omega/50)^2)} \tag{8.55}$$

considered in Section 8.3 is shown in Fig. 8.28. This curve is obtained most readily by utilizing the Bode diagrams of Figs. 8.20 and 8.21 to transfer the frequency response information to the log magnitude and phase coordinates. The shape of the locus of the frequency response on a log-magnitude–phase diagram is particularly important as the phase approaches $-180°$ and the magnitude approaches 0 dB. Clearly, the locus of Eq. (8.54) and Fig. 8.27 differs substantially from the locus of Eq. (8.55) and Fig. 8.28. Therefore, as the correlation between the shape of the locus and the transient response of a system is established, we will obtain another useful portrayal of the frequency response of a system. In Chapter 9, we will establish a stability criterion in the frequency domain for which it will

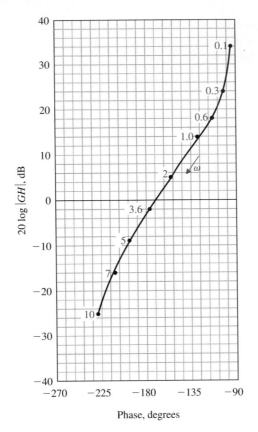

FIGURE 8.27
Log-magnitude–phase curve for $GH_1(j\omega)$.

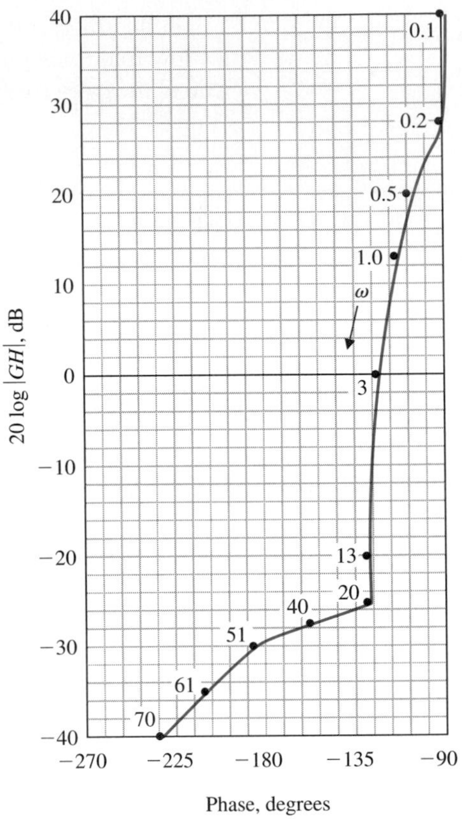

FIGURE 8.28   Log-magnitude–phase curve for $GH_2(j\omega)$.

be useful to utilize the logarithmic-magnitude–phase diagram to investigate the relative stability of closed-looped feedback control systems.

8.7 DESIGN EXAMPLE: ENGRAVING MACHINE CONTROL SYSTEM

The engraving machine shown in Fig. 8.29(a) uses two drive motors and associated lead screws to position the engraving scribe in the x direction [7]. A separate motor is used for both the y- and z-axes as shown. The block diagram model for the x-axis position control system is shown in Fig. 8.29(b). The goal is to select an appropriate gain K, utilizing frequency response methods, so that the time response to step commands is acceptable.

To represent the frequency response of the system, we will first obtain the open-loop and closed-loop Bode diagrams. Then we will use the closed-loop Bode diagram to predict the time response of the system and check the predicted results with the actual results.

To plot the frequency response, we arbitrarily select $K = 2$ and proceed with obtaining the Bode diagram. If the resulting system is not acceptable, we will later adjust the gain.

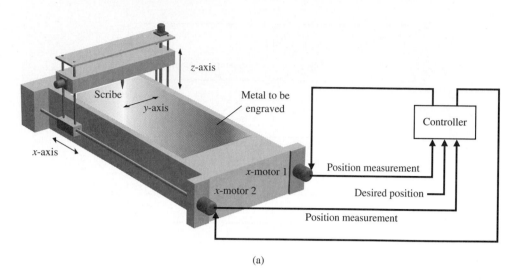

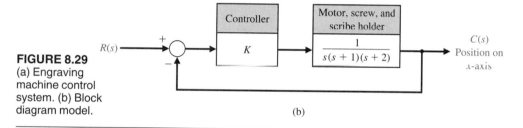

FIGURE 8.29
(a) Engraving
machine control
system. (b) Block
diagram model.

The frequency response of $G(j\omega)$ is partially listed in Table 8.4 and is plotted in Fig. 8.30. We need the frequency response of the closed-loop transfer function

$$T(s) = \frac{2}{s^3 + 3s^2 + 2s + 2}.$$ (8.56)

Therefore we let $s = j\omega$, obtaining

$$T(j\omega) = \frac{2}{(2 - 3\omega^2) + j\omega(2 - \omega^2)}.$$ (8.57)

The Bode diagram of the closed-loop system is shown in Fig. 8.31, where $20 \log |T| = 5$ dB at $\omega_r = 0.8$. Therefore

$$20 \log M_{p\omega} = 5 \quad \text{or} \quad M_{p\omega} = 1.78.$$

TABLE 8.4 Frequency Response for $G(j\omega)$

ω	0.2	0.4	0.8	1.0	1.4	1.8
$20 \log |G|$	14	7	-1	-4	-9	-13
ϕ	$-107°$	$-123°$	$-150.5°$	$-162°$	$-179.5°$	$-193°$

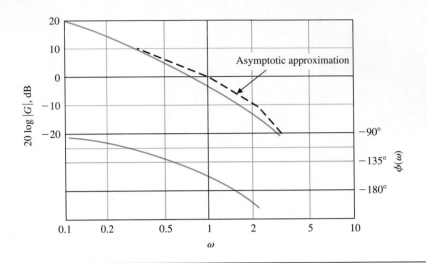

FIGURE 8.30
Bode diagram for
$G(j\omega)$.

If we assume that the system has dominant second-order roots, we can approximate the system with a second-order frequency response of the form shown in Fig. 8.10. Since $M_{p\omega} = 1.78$, we use Fig. 8.11 to estimate ζ to be 0.29. Using this ζ and $\omega_r = 0.8$, we can use Fig. 8.11 to estimate $\omega_r/\omega_n = 0.91$. Therefore

$$\omega_n = \frac{0.8}{0.91} = 0.88.$$

FIGURE 8.31
Bode diagram for
closed-loop
system.

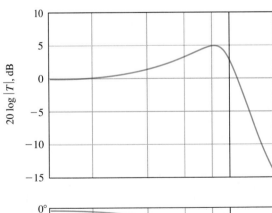

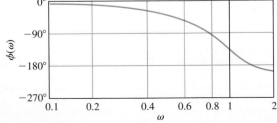

Since we are now approximating $T(s)$ as a second-order system, we have

$$T(s) \cong \frac{\omega_n^2}{s^2 + 2\zeta\omega_n s + \omega_n^2} = \frac{0.774}{s^2 + 0.51s + 0.774}. \tag{8.58}$$

We use Fig. 5.8 to predict the overshoot to a step input as 37% for $\zeta = 0.29$. The settling time is estimated as

$$T_s = \frac{4}{\zeta\omega_n} = \frac{4}{(0.29)0.88} = 15.7 \text{ seconds.}$$

The actual overshoot for a step input is 34%, and the actual settling time is 17 seconds. Clearly, the second-order approximation is reasonable in this case and can be used to determine suitable parameters on a system. If we require a system with lower overshoot, we would reduce K to 1 and repeat the procedure.

8.8 FREQUENCY RESPONSE METHODS USING *MATLAB*

This section begins with an introduction to the Bode diagram and then discusses the connection between the frequency response and performance specifications in the time domain. The section concludes with an illustrative example of designing a control system in the frequency domain.

The functions covered are bode and logspace. The bode function is used to generate a Bode diagram, and the logspace function generates a logarithmically spaced vector of frequencies utilized by the bode function.

Bode Diagram.   Consider the transfer function

$$G(s) = \frac{5(1 + 0.1s)}{s(1 + 0.5s)(1 + (0.6/50)s + (1/50^2)s^2)}. \tag{8.59}$$

The Bode diagram corresponding to Eq. (8.59) is shown in Fig. 8.32. The diagram consists of the logarithmic gain in dB versus ω in one plot and the phase $\phi(\omega)$ versus ω in a second plot. As with the root locus plots, it will be tempting to rely exclusively on *Matlab* to obtain your Bode diagrams. Treat *Matlab* as one tool in a tool kit that can be used to design and analyze control systems. It is essential to develop the capability to obtain approximate Bode diagrams manually. There is no substitute for a clear understanding of the underlying theory.

A Bode diagram is obtained with the bode function shown in Fig. 8.33. The Bode diagram is automatically generated if the bode function is invoked without left-hand arguments. Otherwise, the magnitude and phase characteristics are placed in the workspace through the variables *mag* and *phase*. A Bode diagram is obtained with the plot function using mag, phase, and ω. The vector ω contains the values of the frequency in rad/s at which the Bode diagram will be calculated. If ω is not specified, *Matlab* will automatically choose the frequency values by placing more points in regions where the frequency response is changing quickly. Since the Bode diagram is a log scale, if you choose to specify the frequencies explicitly, it is desirable to generate the vector ω using the logspace function. The logspace function is shown in Fig. 8.34.

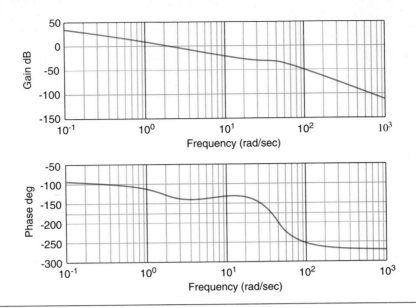

FIGURE 8.32
The Bode plot associated with Eq. (8.59).

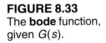

FIGURE 8.33
The **bode** function, given $G(s)$.

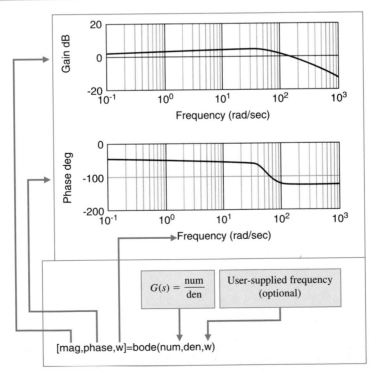

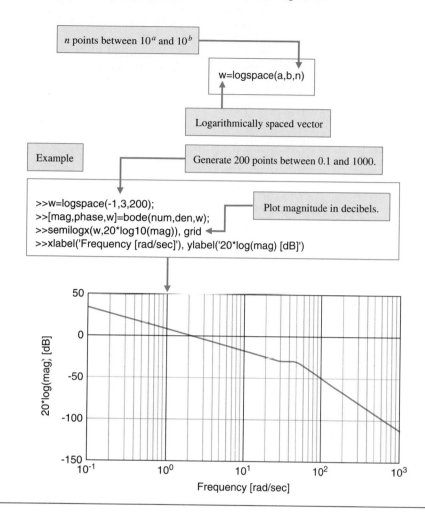

FIGURE 8.34
The **logspace** function.

The Bode diagram in Fig. 8.32 is generated using the script shown in Fig. 8.35. The **bode** function automatically selected the frequency range as $\omega = 0.1$ to 1000 rad/s. This range is user-selectable utilizing the **logspace** function. The **bode** function can be utilized with a state variable model, as shown in Fig. 8.36.

FIGURE 8.35
The script for the Bode diagram in Fig. 8.32.

bodescript.m

```
% Bode plot script for Figure 8.22
%
num=5*[0.1 1];
f1=[1 0]; f2=[0.5 1]; f3=[1/2500 .6/50 1];
den=conv(f1,conv(f2,f3));
%
bode(num,den)
```

Compute

$$s(1 + 0.5s)(1 + \frac{0.6}{50}s + \frac{1}{50^2}s^2)$$

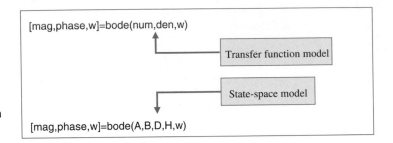

FIGURE 8.36
The **bode** function with a state variable model.

Keeping in mind our goal of designing control systems that satisfy certain performance specifications given in the time domain, we must establish a connection between the frequency response and the transient time response of a system. The relationship between specifications given in the time domain to those given in the frequency domain depends upon approximation of the system by a second-order system with the poles being the system *dominant roots*.

Consider the second-order system shown in Fig. 8.24 on page 412. The closed-loop transfer function is

$$T(s) = \frac{\omega_n^2}{s^2 + 2\zeta\omega_n s + \omega_n^2}. \tag{8.60}$$

The Bode diagram magnitude characteristic associated with the closed-loop transfer function in Eq. (8.60) is shown in Fig. 8.25 on page 413. The relationship between the resonant frequency, ω_r, the maximum of the frequency response, M_{p_ω}, and the damping ratio, ζ, and the natural frequency, ω_n, is shown in Fig. 8.37 (and in Fig. 8.11). The information in

FIGURE 8.37
(a) The relationship between (M_{p_ω}, ω_r) and (ζ, ω_n) for a second-order system. (b) *Matlab* script.

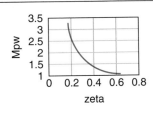 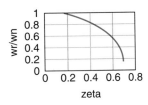

(a)

relation.m

```
zeta=[0.15:0.01:0.7];                         ← zeta ranges from 0.15 to 0.70
wr_over_wn=sqrt(ones(1,length(zeta))-2*zeta.^2);
Mp=(2*zeta .* sqrt(ones(1,length(zeta))-zeta.^2)).^(-1);
%
subplot(211),plot(zeta,Mp),grid
xlabel('zeta'), ylabel('Mpw')                 ← Generate plots
subplot(212),plot(zeta,wr_over_wn),grid
xlabel('zeta'), ylabel('wr/wn')
```

(b)

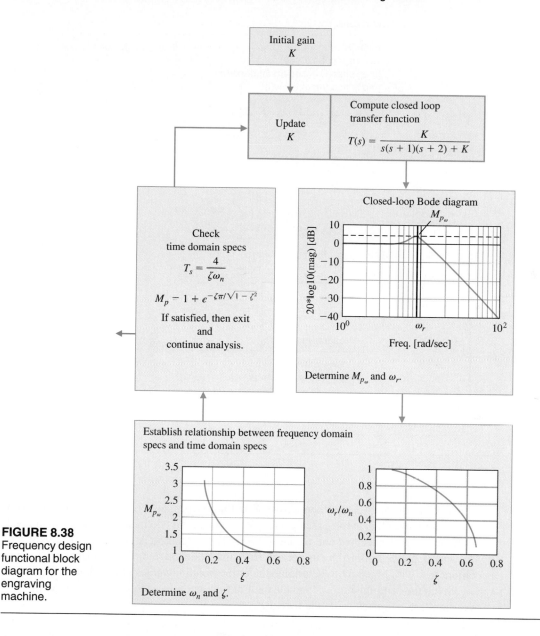

FIGURE 8.38
Frequency design functional block diagram for the engraving machine.

Fig. 8.37 will be quite helpful in designing control systems in the frequency domain while satisfying time-domain specifications.

EXAMPLE 8.5   **Engraving machine system**

Consider the block diagram model in Fig. 8.29 on page 417. Our objective is to select K so that the closed-loop system has an acceptable time response to a step command. A functional block diagram describing the frequency-domain design process is shown in Fig. 8.38.

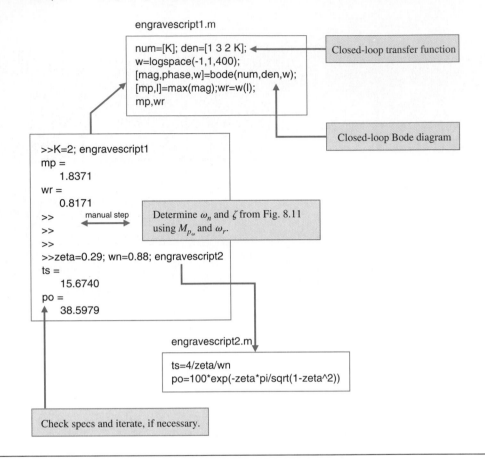

FIGURE 8.39
Script for the design of an engraving machine.

First we choose $K = 2$ and then iterate K if the performance is unacceptable. A script, shown in Fig. 8.39, is used in the design. The value of K is defined at the command level. Then the script is executed and the closed-loop Bode diagram is generated. The values of M_{p_ω} and ω_r are determined by inspection from the Bode diagram. Those values are used in conjunction with Fig. 8.37 to determine the corresponding values of ζ and ω_n.

Given the damping ratio, ζ, and the natural frequency, ω_n, the settling time and percent overshoot are estimated using the formulas

$$T_s \approx \frac{4}{\zeta\omega_n}, \qquad \text{P.O.} \approx 100 \exp\frac{-\zeta\pi}{\sqrt{1 - \zeta^2}}.$$

If the time-domain specifications are not satisfied, then we adjust K and iterate.

The values for ζ and ω_n corresponding to $K = 2$ are $\zeta = 0.29$ and $\omega_n = 0.88$. This leads to a prediction of P.O. = 37% and $T_s = 15.7$ seconds. The step response, shown in Fig. 8.40, is a verification that the performance predictions are quite accurate and that the closed-loop system performs adequately.

In this example, the second-order system approximation is reasonable and leads to an acceptable design. However, the second-order approximation may not always lead directly

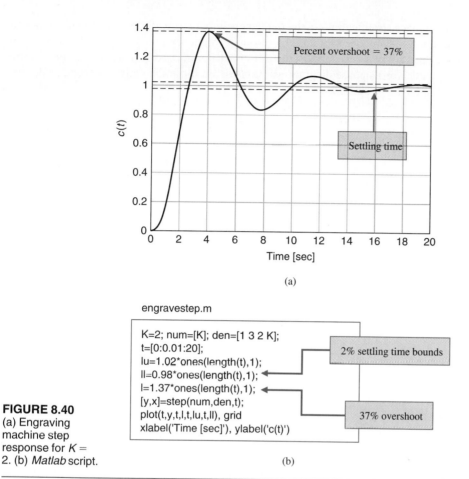

FIGURE 8.40
(a) Engraving machine step response for $K = 2$. (b) *Matlab* script.

engravestep.m

```
K=2; num=[K]; den=[1 3 2 K];
t=[0:0.01:20];
lu=1.02*ones(length(t),1);
ll=0.98*ones(length(t),1);
l=1.37*ones(length(t),1);
[y,x]=step(num,den,t);
plot(t,y,t,l,t,lu,t,ll), grid
xlabel('Time [sec]'), ylabel('c(t)')
```

2% settling time bounds

37% overshoot

(b)

to a good design. Fortunately, with *Matlab* we can construct an interactive design facility to assist in the design process by reducing the manual computational loads while providing easy access to a host of classical and modern control tools. ∎

8.9 SUMMARY

In this chapter, we have considered the representation of a feedback control system by its frequency response characteristics. The frequency response of a system was defined as the steady-state response of the system to a sinusoidal input signal. Several alternative forms of frequency response plots were considered, including the polar plot of the frequency response of a system $G(j\omega)$ and logarithmic plots, often called Bode plots, and the value of the logarithmic measure was illustrated. The ease of obtaining a Bode plot for the various factors of $G(j\omega)$ was noted, and an example was considered in detail. The asymptotic approximation for drawing the Bode diagram simplifies the computation considerably. Several performance specifications in the frequency domain were discussed; among them were the maximum magnitude, $M_{p\omega}$ and the resonant frequency ω_r. The relationship between the Bode diagram plot and the system error constants (K_p and K_v) was noted. Finally, the log magnitude versus phase diagram was considered for graphically representing the frequency response of a system.

EXERCISES

E8.1   Increased track densities for computer disk drives necessitate careful design of the head positioning control [1]. The transfer function is

$$G(s) = \frac{K}{(s + 1)^2}.$$

Plot the polar plot for this system when $K = 4$. Calculate the phase and magnitude at $\omega = 0.5, 1, 2$, and so on.

E8.2   The tendon-operated robotic hand shown in Fig. 1.13 uses a pneumatic actuator [8]. The actuator can be represented by

$$G(s) = \frac{2572}{s^2 + 386s + 15{,}434}$$

$$= \frac{2572}{(s + 45.3)(s + 341)}.$$

Plot the frequency response of $G(j\omega)$. Show that the magnitude of $G(j\omega)$ is -15.6 dB at $\omega = 10$ and -30 dB at $\omega = 200$. Also show that the phase is $-150°$ at $\omega = 700$.

E8.3   A robot arm has a joint control open-loop transfer function

$$G(s) = \frac{300(s + 100)}{s(s + 10)(s + 40)}.$$

Prove that the frequency equals 28.3 rad/s when the phase angle of $(j\omega)$ is $-180°$. Find the magnitude of $G(j\omega)$ at that frequency.

E8.4   The frequency response for a process of the form

$$G(s) = \frac{Ks}{(s + a)(s^2 + 20s + 100)}$$

is shown in Fig. E8.4. Determine K and a by examining the frequency response curves.

E8.5   The magnitude plot of a transfer function

$$G(s) = \frac{K(1 + 0.5s)(1 + as)}{s(1 + s/8)(1 + bs)(1 + s/36)}$$

is shown in Fig. E8.5. Determine K, a, and b from the plot.

Answers: $K = 8$, $a = \frac{1}{4}$, $b = \frac{1}{24}$

FIGURE E8.4
Bode diagram.

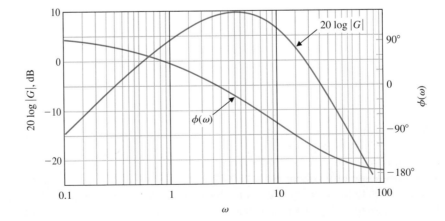

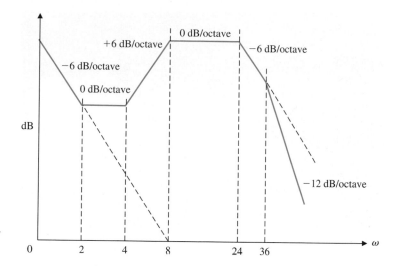

FIGURE E8.5
Bode diagram.
Note that −6 dB/
octave = −20 dB/
decade.

E8.6 Several studies have proposed an extravehicular robot that could move about a NASA space station and perform physical tasks at various worksites [9]. The arm is controlled by a unity feedback control with

$$G(s) = \frac{K}{s(s/10 + 1)(s/100 + 1)}.$$

Draw the Bode diagram for $K = 100$, and determine the frequency when $20 \log |G|$ is zero dB.

E8.7 Consider a system with a closed-loop transfer function

$$T(s) = \frac{C(s)}{R(s)} = \frac{4}{(s^2 + s + 1)(s^2 + 0.4s + 4)}.$$

This system will have no steady-state error for a step input. (a) Plot the frequency response, noting the two peaks in the magnitude response. (b) Predict the time response to a step input, noting that the system has four poles and cannot be represented as a dominant second-order system. (c) Plot the step response.

E8.8 A feedback system has a loop transfer function

$$G(s)H(s) = \frac{50}{s^2 + 11s + 10}.$$

(a) Determine the corner frequencies (break points) for the Bode plot. (b) Determine the slope of the

asymptotic plot at very low frequencies and at high frequencies. (c) Sketch the Bode magnitude plot.

E8.9 The Bode diagram of a system is shown in Fig. E8.9. Determine the transfer function $G(s)$.

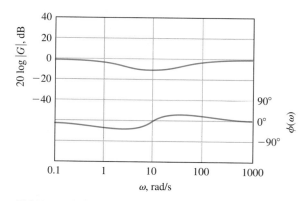

FIGURE E8.9

E8.10 The dynamic analyzer shown in Fig. E8.10(a) can be used to display the frequency response of a selected $G(j\omega)$ model. Also shown is a head positioning mechanism for a disk drive, which uses a linear motor to position the head. Figure E8.10(b) shows the actual frequency response of the head positioning mechanism. Estimate the poles and zeros of the device. Note $X = 1.37$ kHz at the first cursor and $\Delta X = 1.257$ kHz to the second cursor.

(a)

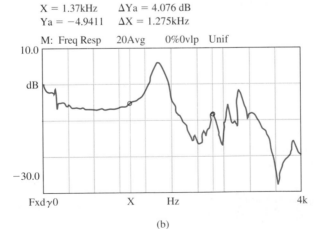

$X = 1.37\text{kHz}$ $\Delta Ya = 4.076\ \text{dB}$
$Ya = -4.9411$ $\Delta X = 1.275\text{kHz}$

M: Freq Resp   20Avg   0%0vlp   Unif

FIGURE E8.10
(a) Dual-exposure
photo showing the
head positioner
and the Signal
Analyzer 3562A.
(b) Frequency
response.
(Courtesy of
Hewlett-
Packard Co.)

(b)

PROBLEMS

P8.1 Sketch the polar plot of the frequency response for the following transfer functions:

(a) $GH(s) = \dfrac{1}{(1 + 0.5s)(1 + 2s)}$

(b) $GH(s) = \dfrac{(1 + 0.5s)}{s^2}$

(c) $GH(s) = \dfrac{(s + 4)}{(s^2 + 5s + 25)}$

(d) $GH(s) = \dfrac{30(s + 8)}{s(s + 2)(s + 4)}$

P8.2 Draw the Bode diagram representation of the frequency response for the transfer functions given in Problem 8.1.

P8.3 A rejection network that can be utilized instead of the twin-T network of Example 8.4 is the bridged-T network shown in Fig. P8.3. The transfer function of this network is

$$G(s) = \dfrac{s^2 + \omega_n^2}{s^2 + 2(\omega_n s/Q) + \omega_n^2}$$

(can you show this?), where $\omega_n^2 = 2/LC$ and $Q = \omega_n L/R_1$ and R_2 is adjusted so that $R_2 = (\omega_n L)^2/4R_1$

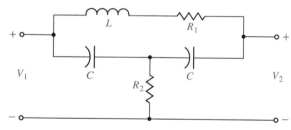

FIGURE P8.3 Bridged-T network.

FIGURE P8.4
(a) Pressure controller. (b) Flow graph model.

[3]. (a) Determine the pole–zero pattern and, utilizing the vector approach, evaluate the approximate frequency response. (b) Compare the frequency response of the twin-T and bridged-T networks when $Q = 10$.

P8.4 A control system for controlling the pressure in a closed chamber is shown in Fig. P8.4. The transfer function for the measuring element is

$$H(s) = \dfrac{450}{s^2 + 90s + 900},$$

and the transfer function for the valve is

$$G_1(s) = \dfrac{1}{(0.1s + 1)(s/15 + 1)}.$$

The controller transfer function is

$$G_c(s) = (10 + 2s).$$

Obtain the frequency response characteristics for the loop transfer function

$$G_c(s)G_1(s)H(s) \cdot [1/s].$$

P8.5 The robot industry in the United States is growing at a rate of 30% a year [8]. A typical industrial robot has six axes or degrees of freedom. A position control system for a force-sensing joint has a transfer function

$$G(s) = \dfrac{K}{(1 + s/2)(1 + s)(1 + s/9)(1 + s/40)},$$

where $H(s) = 1$ and $K = 10$. Plot the Bode diagram of this system.

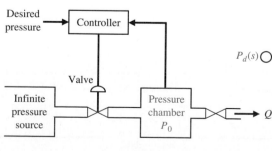

(a)   (b)

P8.6 The asymptotic logarithmic magnitude curves for two transfer functions are given in Fig. P8.6. Sketch the corresponding asymptotic phase shift curves for each system. Determine the transfer function for each system. Assume that the systems have minimum phase transfer functions.

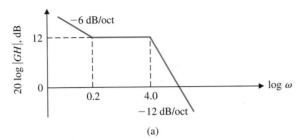

(a)

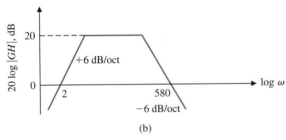

(b)

FIGURE P8.6 *Note* that -6 dB/octave $= -20$ dB/decade.

P8.7 Driverless vehicles can be used in warehouses, airports, and many other applications. These vehicles follow a wire imbedded in the floor and ad-

FIGURE P8.7
Steerable wheel
control.

just the steerable front wheels in order to maintain proper direction, as shown in Fig. P8.7(a) [10]. The sensing coils, mounted on the front wheel assembly, detect an error in the direction of travel and adjust the steering. The overall control system is shown in Fig. P8.7(b). The open-loop transfer function is

$$GH(s) = \frac{K}{s(s + \pi)^2} = \frac{K_v}{s(s/\pi + 1)^2}.$$

It is desired to have the bandwidth of the closed-loop system exceed 2π rad/s. (a) Set $K_v = 2\pi$ and plot the Bode diagram. (b) Using the Bode diagram, obtain the logarithmic magnitude versus phase angle curve.

P8.8 A feedback control system is shown in Fig. P8.8. The specification for the closed-loop system requires that the overshoot to a step input be less than 16%. (a) Determine the corresponding specification in the frequency domain $M_{p\omega}$ for the closed-loop transfer function

$$\frac{C(j\omega)}{R(j\omega)} = T(j\omega).$$

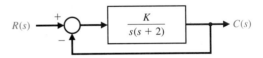

FIGURE P8.8

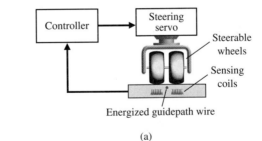

(a)

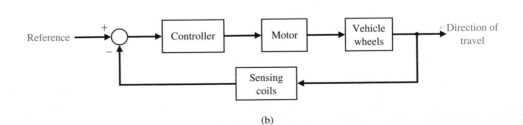

(b)

(b) Determine the resonant frequency, ω_r. (c) Determine the bandwidth of the closed-loop system.

P8.9 Draw the logarithmic magnitude versus phase angle curves for the transfer functions a and b of Problem 8.1.

P8.10 A linear actuator is utilized in the system shown in Fig. P8.10 to position a mass M. The actual position of the mass is measured by a slide wire resistor and thus $H(s) = 1.0$. The amplifier gain is to be selected so that the steady-state error of the system is less than 1% of the magnitude of the position reference $R(s)$. The actuator has a field coil with a resistance $R_f = 0.1$ ohm and $L_f = 0.2$ henries. The mass of the load is 0.1 kg, and the friction is 0.2 ns/m. The spring constant is equal to 0.4 n/m. (a) Determine the gain K necessary to maintain a steady-state error for a step input less than 1%. That is, K_p must be greater than 99. (b) Draw the Bode diagram of the loop transfer function $GH(s)$.

(c) Draw the logarithmic magnitude versus phase angle curve for $GH(j\omega)$. (d) Draw the Bode diagram for the closed-loop transfer function $Y(j\omega)/R(j\omega)$. Determine $M_{p\omega}$, ω_r, and the bandwidth.

P8.11 Automatic steering of a ship would be a particularly useful application of feedback control theory [20]. In the case of heavily traveled seas, it is important to maintain the motion of the ship along an accurate track. An automatic system is able to maintain a much smaller error from the desired heading than is a helmsman who recorrects at infrequent intervals. A mathematical model of the steering system has been developed for a ship moving at a constant velocity and for small deviations from the desired track. For a large tanker, the transfer function of the ship is

$$G(s) = \frac{E(s)}{\delta(s)} = \frac{0.164(s + 0.2)(-s + 0.32)}{s^2(s + 0.25)(s - 0.009)},$$

where $E(s)$ is the Laplace transform of the devia-

FIGURE P8.10
Linear actuator control.

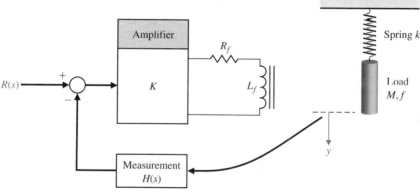

FIGURE P8.11
Frequency response of ship control system.

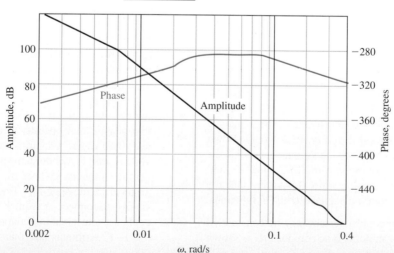

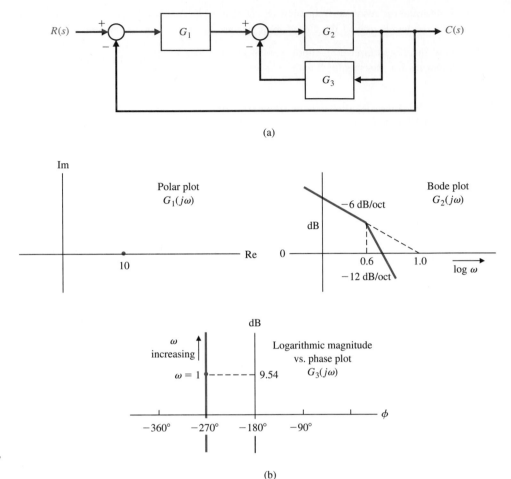

FIGURE P8.12
Feedback system.
Note that −6 dB/
octave = −20 dB/
decade.

tion of the ship from the desired heading and $\delta(s)$ is the Laplace transform of the angle of deflection of the steering rudder.

Verify that the frequency response of the ship, $E(j\omega)/\delta(j\omega)$, is that shown in Fig. P8.11.

P8.12 The block diagram of a feedback control system is shown in Fig. P8.12(a). The transfer functions of the blocks are represented by the frequency response curves shown in Fig. P8.12(b). (a) When G_3 is disconnected from the system, determine the damping ratio ζ of the system. (b) Connect G_3 and determine the damping ratio ζ. Assume that the systems have minimum phase transfer functions.

P8.13 A position control system may be constructed by using an ac motor and ac components as shown

in Fig. P8.13. The syncro and control transformer may be considered to be a transformer with a rotating winding. The syncro position detector rotor turns with the load through an angle θ_0. The syncro motor is energized with an ac reference voltage, for example, 115 volts, 60 cps. The input signal or command is $R(s) = \theta_{in}(s)$ and is applied by turning the rotor of the control transformer. The ac two-phase motor operates as a result of the amplified error signal. The advantages of an ac control system are (1) freedom from dc drift effects and (2) the simplicity and accuracy of ac components. To measure the open-loop frequency response, one simply disconnects X from Y and X' from Y', then applies a sinusoidal modulation signal generator to the $Y–Y'$ terminals and measures the response at

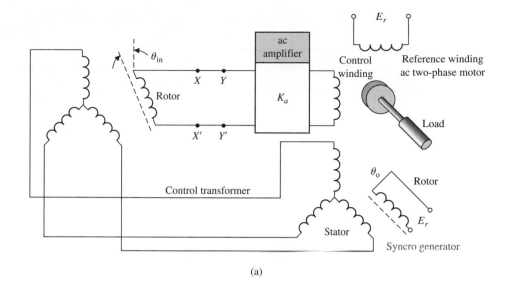

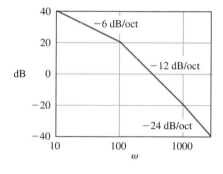

FIGURE P8.13
(a) ac motor control.
(b) Frequency response.

(a)

(b)

X–X'. [The error $(\theta_0 - \theta_i)$ will be adjusted to zero before applying the ac generator.] The resulting frequency response of the loop, $GH(j\omega)$, is shown in Fig. P8.13(b). Determine the transfer function $GH(j\omega)$. Assume that the system has a minimum phase transfer function.

P8.14 A bandpass amplifier may be represented by the circuit model shown in Fig. P8.14 [3]. When $R_1 = R_2 = 1 \text{ k}\Omega$, $C_1 = 100 \text{ pf}$, $C_2 = 1 \ \mu\text{f}$, and $K = 100$, show that

$$G(s) = \frac{10^9 s}{(s + 1000)(s + 10^7)}.$$

(a) Sketch the Bode diagram of $G(j\omega)$. (b) Find the

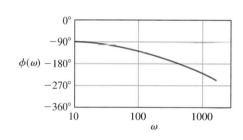

FIGURE P8.14 Bandpass amplifier.

midband gain (in dB). (c) Find the high and low frequency -3 dB points.

P8.15 To determine the transfer function of a plant $G(s)$, the frequency response may be measured using a sinusoidal input. One system yields the following data. Determine the transfer function $G(s)$.

ω, rad/s	$|G(j\omega)|$	Phase, degrees
0.1	50	−90
1	5.02	−92.4
2	2.57	−96.2
4	1.36	−100
5	1.17	−104
6.3	1.03	−110
8	0.97	−120
10	0.97	−143
12.5	0.74	−169
20	0.13	−245
31	0.026	−258

P8.16 The space shuttle has been used to repair satellites and the Hubble telescope. Fig. P8.16 illustrates how a crew member, with his feet strapped to the platform on the end of the shuttle's robot arm, used his arms to stop the satellite's spin and ignite the engine switch. The control system has the form shown in Fig. 4.14, where $G_1 = K = 8$ and $H(s) = 1$. The control system of the robot arm has a closed-loop transfer function

$$\frac{C(s)}{R(s)} = \frac{8}{s^2 + 4s + 8}.$$

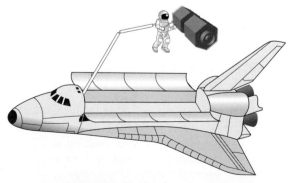

FIGURE P8.16 Satellite repair.

(a) Determine the response $c(t)$ to a unit step disturbance. (b) Determine the bandwidth of the system.

P8.17 The experimental Oblique Wing Aircraft (OWA) has a wing that pivots as shown in Fig. P8.17. The wing is in the normal unskewed position for low speeds and can move to a skewed position for improved supersonic flight [11]. The aircraft control system has $H(s) = 1$ and

$$G(s) = \frac{4(0.5s + 1)}{s(2s + 1)\left[\left(\dfrac{s}{8}\right)^2 + \left(\dfrac{s}{20}\right) + 1\right]}.$$

(a) Determine the Bode diagram. (b) Find the frequency, ω_1, when the magnitude is 0 dB and the frequency, ω_2, when the phase is −180°.

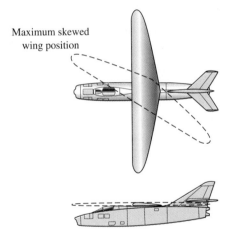

Maximum skewed wing position

FIGURE P8.17 The Oblique Wing Aircraft, top and side view.

P8.18 Remote operation plays an important role in hostile environments, such as those in nuclear or high-temperature environments and in deep space. In spite of the efforts of many researchers, a teleoperation system that is comparable to the human's direct operation has not been developed. Research engineers have been trying to improve teleoperations by feeding back rich sensory information acquired by the robot to the operator with a sensation of presence. This concept is called tele-existence or telepresence [9].

The tele-existence master-slave system consists of a master system with a visual and auditory sensation of presence, a computer control system, and an anthropomorphic slave robot mechanism with an arm having seven degrees of freedom and a locomotion mechanism. The operator's head movement, right arm movement, right hand movement, and other auxiliary motion are measured by the master system. A specially designed stereo visual and auditory input system mounted on the neck mechanism of the slave robot gathers visual and auditory information of the remote environment. These pieces of information are sent back to the master system and are applied to the specially designed stereo display system to evoke the sensation of presence of the operator. The locomotion control system has the loop transfer

$$GH(s) = \frac{12(s + 0.5)}{s^2 + 13s + 30}.$$

Obtain the Bode diagram for $GH(j\omega)$ and determine the frequency when $20 \log |GH|$ is very close to 0 dB.

P8.19 A dc motor controller used extensively in automobiles is shown in Fig. P8.19(a). The measured plot of $\Theta(s)/I(s)$ is shown in Fig. P8.19(b). Determine the transfer function of $\Theta(s)/I(s)$.

P8.20 Space robotics is an emerging field. For the successful development of space projects, robotics and automation will be a key technology. Autonomous and dexterous space robots can reduce the workload of astronauts and increase operational efficiency in many missions. Figure P8.20 shows a concept called a free-flying robot [9, 13]. A major characteristic of space robots, which clearly distinguishes them from robots operated on earth, is the lack of a fixed base. Any motion of the manipulator arm will induce reaction forces and moments in the base, which disturb its position and attitude.

FIGURE P8.20 A space robot with three arms, shown capturing a satellite.

FIGURE P8.19
(a) Motor controller.
(b) Measured plot.

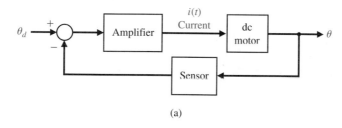

(a)

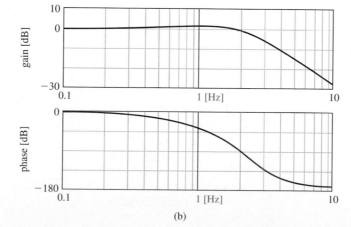

(b)

The control of one of the joints of the robot can be represented by the loop transfer function

$$GH(s) = \frac{781(s + 8)}{s^2 + 25s + 625}.$$

(a) Plot the Bode diagram of $GH(j\omega)$. (b) Determine the maximum value of $20 \log |GH|$, the frequency at which it occurs, and the phase at that frequency.

P8.21 Low-altitude wind shear is a major cause of air carrier accidents in the United States. Most of these accidents have been caused by either microbursts (small-scale, low-altitude, intense thunderstorm downdrafts that impact the surface and cause strong divergent outflows of wind) or by the gust front at the leading edge of expanding thunderstorm outflows. A microburst encounter is a serious problem for either *landing* or *departing* aircraft, since the aircraft is at low altitudes and is traveling at just over 25% above its stall speed [12].

The design of the control of an aircraft encountering wind shear after takeoff may be treated as a problem of stabilizing the climb rate about a desired value of the climb rate. The resulting controller utilizes only climb rate information.

The standard negative unity feedback system of Fig. 8.24 has a loop transfer function

$$G(s) = \frac{-200s^2}{s^3 + 14s^2 + 44s + 40}.$$

Note the negative gain in $G(s)$. This system represents the control system for climb rate. Draw the Bode diagram and determine gain (in dB) when the phase is $-180°$.

P8.22 The frequency response of a process $G(j\omega)$ is shown in Fig. P8.22. Determine $G(s)$.

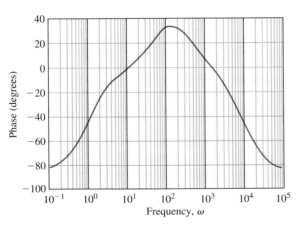

FIGURE P8.22

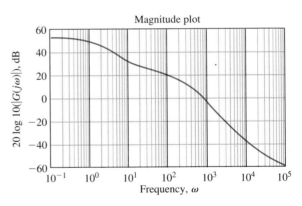

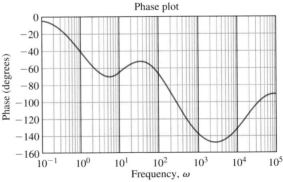

FIGURE P8.23

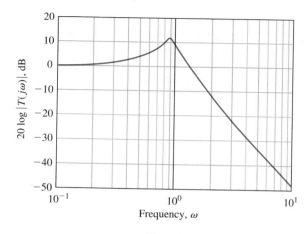

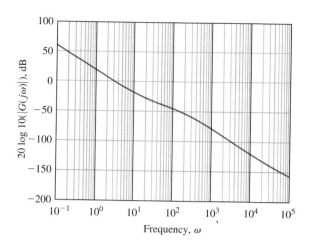

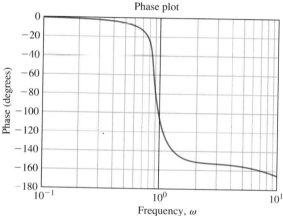

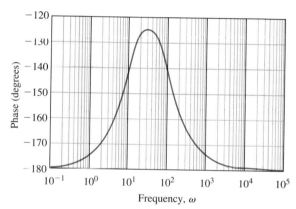

FIGURE P8.24

FIGURE P8.25

P8.23 The frequency response of a process $G(j\omega)$ is shown in Fig. P8.23. Deduce the type number (number of integrations) for the system. Determine the transfer function of the system, $G(s)$. Calculate the error to a unit step input.

P8.24 The Bode diagram of a closed-loop film transport system, $T(s)$, is shown in Fig. P8.24 [17]. Assume that the system transfer function, $T(s)$, has two dominant complex conjugate poles. (a) Deter-

mine the best second-order model for the system. (b) Determine the system bandwidth. (c) Predict the percent overshoot and settling time for a step input.

P8.25 A unity feedback closed-loop system has a steady-state error equal to $A/10$ where the input is $r(t) = At^2/2$. The Bode plot of the magnitude and phase angle versus ω is shown in Fig. P8.25 for $G(j\omega)$. Determine the transfer function $G(s)$.

ADVANCED PROBLEMS

AP8.1 A spring-mass-damper system is shown in Fig. AP8.1(a). The Bode diagram obtained by experimental means using a sinusoidal forcing func-

tion is shown in Fig. AP8.1(b). Determine the numerical values of m, f, and k.

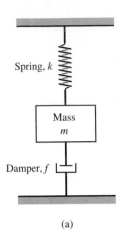

(a)

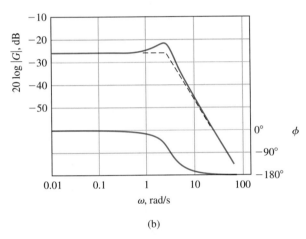

ω, rad/s

(b)

FIGURE AP8.1

AP8.2 A system is shown in Fig. AP8.2. The nominal value of the parameter b is 4.0. Determine the sensitivity S_b^T and plot $20 \log |S_b^T|$, the Bode magnitude diagram for $K = 2$.

AP8.3 As an automobile moves along the road, the vertical displacements at the tires act as the motion

excitation to the automobile suspension system [16]. Figure AP8.3 is a schematic diagram of a simplified automobile suspension system, for which we assume the input is sinusoidal. Determine the transfer function $X(s)/R(s)$, and plot the Bode diagram when $M = 1$ kg, $f = 4$ N-s/m, and $k = 18$ N/m.

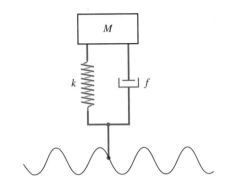

FIGURE AP8.3   Auto suspension system model.

AP8.4 A helicopter with a load on the end of a cable is shown in Fig. AP8.4(a). The position control system is shown in Fig. AP8.4(b), where the visual feedback is represented by $H(s)$. Draw the Bode diagram of $GH(j\omega)$.

AP8.5 A closed-loop system with unity feedback has a transfer function

$$T(s) = \frac{10(s + 1)}{s^2 + 9s + 10}.$$

(a) Determine the open-loop transfer function $G(s)$. (b) Plot the log-magnitude-phase (similar to Fig. 8.27), and identify the frequency points for ω equal to 1, 10, 50, 110, and 500. (c) Is the open-loop system stable? Is the closed-loop system stable?

FIGURE AP8.2

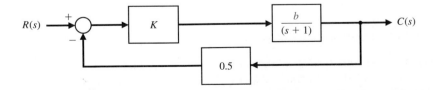

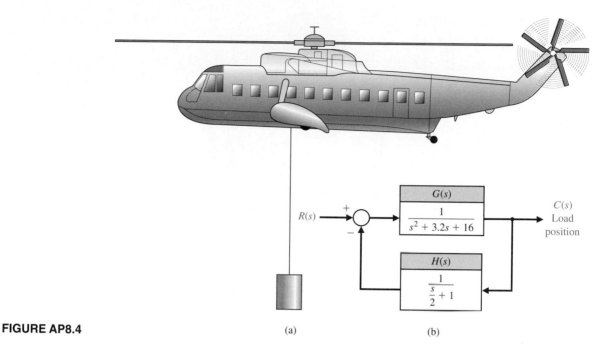

FIGURE AP8.4   (a)   (b)

DESIGN PROBLEMS

DP8.1 The behavior of a human steering an automobile remains interesting [14, 15, 16, 21]. The design and development of systems for four-wheel steering, active suspensions, active, independent braking, and "drive-by-wire" steering provide the engineer with considerably more freedom in altering vehicle-handling qualities than existed in the past.

    The vehicle and the driver are represented by the model in Fig. DP8.1, where the driver develops anticipation of the vehicle deviation from the center line. For $K = 1$, plot the Bode diagram of (a) the open-loop transfer function $G_c(s)G(s)$ and (b) the closed-loop transfer function $T(s)$. (c) Repeat parts

(a) and (b) when $K = 10$. (d) A driver can select the gain K. Determine the appropriate gain so that $M_{p\omega} \le 2$ and the bandwidth is the maximum attainable for the closed-loop system. (e) Determine the steady-state error of the system for a ramp input, $r(t) = t$.

DP8.2 The unmanned exploration of planets such as Mars requires a high level of autonomy because of the communication delays between robots in space and their Earth-based stations. This impacts all the components of the system: planning, sensing, and mechanism. In particular, such a level of autonomy can be achieved only if each robot has a percep-

FIGURE DP8.1

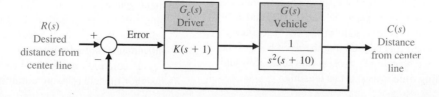

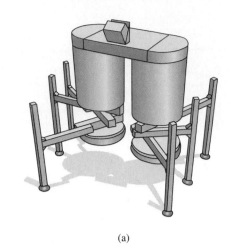

(a)

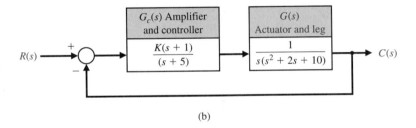

FIGURE DP8.2
(a) The six-legged
Ambler. (b) Block
diagram of the
control system for
one leg.

$R(s)$

$+$

$-$

$G_c(s)$ Amplifier and controller
$\dfrac{K(s + 1)}{(s + 5)}$

$G(s)$ Actuator and leg
$\dfrac{1}{s(s^2 + 2s + 10)}$

$C(s)$

(b)

tion system that can reliably build and maintain models of the environment. The Robotics Institute of Carnegie-Mellon University has proposed a perception system that is designed for application to autonomous planetary exploration. The perception system is a major part of the development of a complete system that includes planning and mechanism design. The target vehicle is the Ambler, a six-legged walking machine being developed at CMU, shown in Fig. DP8.2(a) [18]. The control system of one leg is shown in Fig. DP8.2(b).

(a) Draw the Bode diagram for $G_c(s)G(s)$ for $0.1 \leq \omega \leq 100$ when $K = 20$. Determine (1) the frequency when the phase is $-180°$ and (2) the frequency when $20 \log |GG_c| = 0$ dB. (b) Plot the Bode diagram for the closed-loop transfer function $T(s)$ when $K = 20$. (c) Determine M_{p_ω}, ω_r, and ω_B for the closed-loop system when $K = 20$ and $K = 40$. (d) Select the best gain of the two specified in part (c) when it is desired that the overshoot of the system to a step input, $r(t)$, is less than 35% and the settling time is as short as feasible.

DP8.3 A table is used to position vials under a dispenser head as shown in Fig. DP8.3(a). The objective is speed, accuracy, and smooth motion in order to eliminate spilling. The position control system is shown in Fig. DP8.3(b). Since we want small overshoot for a step input and yet desire a short settling time, we will limit $20 \log M_{p_\omega}$ to 3 dB for $T(j\omega)$. Plot the Bode diagram for a gain K that will result in a stable system. Then, adjust K until $20 \log M_{p\omega} = 3$ dB, and determine the closed-loop system bandwidth. Determine the steady-state error for the system for the gain K selected to meet the requirement for $M_{p\omega}$.

DP8.4 Anesthesia can be administered automatically by a control system. For certain operations, such as brain and eye surgery, involuntary muscle movements can be disastrous. Therefore, to ensure adequate operating conditions for the surgeon, muscle relaxant drugs, which block involuntary muscle movements, are administered.

A conventional method used by anesthesiologists for muscle relaxant administration is to inject

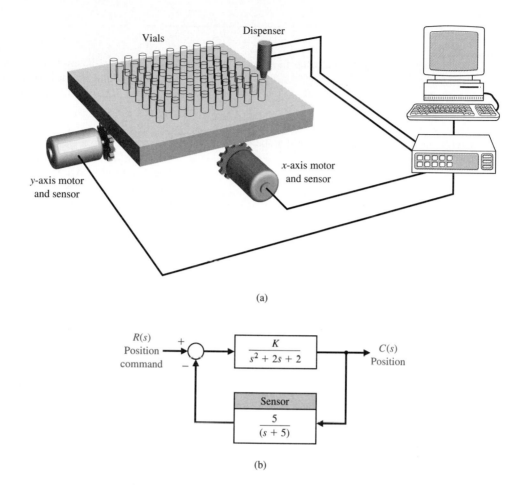

(a)

FIGURE DP8.3
Automatic
table and
dispenser.

(b)

a bolus dose whose size is determined by experience and to inject supplements as required. However, an anesthesiologist may sometimes fail to maintain a steady level of relaxation, resulting in a large drug consumption by the patient. Significant improvements may be achieved by introducing the concept of automatic control, which results in

considerable reduction in the total relaxant drug consumed [19].

A model of the anesthesia process is shown in Fig. DP8.4. Select a gain K and a controller constant τ so that the bandwidth of the closed-loop system is maximized while $M_{p\omega} \leq 1.5$. Determine the bandwidth attained for your design.

FIGURE DP8.4
Model of an
anesthesia control
system.

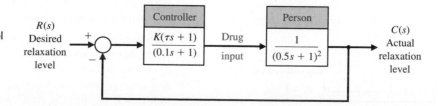

MATLAB PROBLEMS

MP8.1   Consider the closed-loop transfer function

$$T(s) = \frac{25}{s^2 + s + 25}.$$

Using *Matlab*, obtain the Bode plot and verify that the resonant frequency is 5 rad/s and that the peak magnitude $M_{p\omega}$ is 14 dB.

MP8.2   For the following transfer functions, sketch the Bode plots, then verify with *Matlab:*

(a) $G(s) = \dfrac{1}{(s + 1)(s + 10)}$

(b) $G(s) = \dfrac{s + 10}{(s + 1)(s + 20)}$

(c) $G(s) = \dfrac{1}{s^2 + 2s + 50}$

(d) $G(s) = \dfrac{s + 5}{(s + 1)(s^2 + 12s + 50)}$

MP8.3   A unity negative feedback system has the open-loop transfer function

$$G(s) = \frac{25}{s(s + 2)}.$$

Determine the closed-loop system bandwidth by using *Matlab* to obtain the Bode plot, and estimate the bandwidth from the plot. Label the plot with the (approximate) bandwidth.

MP8.4   A block diagram of a second-order system is shown in Fig. MP8.4.

(a) Determine the resonant peak, $M_{p\omega}$, the resonant frequency, ω_r, and the bandwidth, ω_B, of the system from the closed-loop Bode plot. Generate the Bode plot with *Matlab* for $\omega = 0.1$ to $\omega = 1000$ rad/s using the logspace function. (b) Estimate the system damping ratio, ζ, and natural frequency, ω_n, utilizing Fig. 8.11 in Section 8.2. (c) From the closed-loop transfer function, compute the actual ζ and ω_n and compare with your results in part (b).

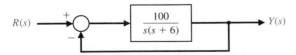

FIGURE MP8.4   A second-order feedback control system.

MP8.5   Consider the problem of controlling an inverted pendulum on a moving base, as shown in Fig. MP8.5(a). The transfer function of the open-loop system is

$$G(s) = \frac{-1/M_b L}{s^2 - (M_b + M_s)g/(M_b L)}.$$

The design objective is to balance the pendulum (i.e., $\theta(t) \approx 0$) in the presence of disturbance inputs. A block diagram representation of the system is depicted in Fig. MP8.5(b). Let $M_s = 10$ kg, $M_b = 100$ kg, $L = 1$ m, $g = 9.81$ m/s², $a = 5$, and $b = 10$. The design specifications, based on a unit step disturbance, are

1. settling time less than 10 seconds,
2. percent overshoot less than 40%, and
3. steady-state tracking error less than 0.1° in the presence of the disturbance.

Develop a set of interactive *Matlab* scripts to aid in the control system design. The first script should accomplish at least the following:

1. Compute the closed-loop transfer function from the disturbance to the output with K as an adjustable parameter,
2. Draw the Bode plot of the closed-loop system,
3. Automatically compute and output $M_{p\omega}$ and ω_r.

As an intermediate manual step, use $M_{p\omega}$ and ω_r and Fig. 8.11 in Section 8.2 to estimate ζ and ω_n. The second script should perform at least the following function: Estimate the settling time and percent overshoot using ζ and ω_n as input variables.

If the performance specifications are not satisfied, change K and iterate on the design using the first two scripts. After completion of the first two steps, the final step is to test the design by simulation. The function of the third script is to

1. plot the response, $\theta(t)$, to a unit step disturbance with K as an adjustable parameter, and
2. label the plot appropriately.

Utilizing the interactive scripts, design the controller to meet the specifications using frequency response Bode methods. To start the design process, use analytic methods to compute the minimum value of K to meet the steady-state tracking error specification. Use the minimum K as the first guess in the design iteration.

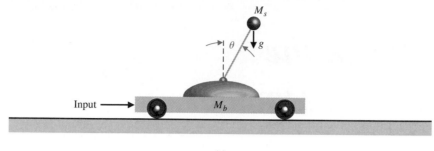

(a)

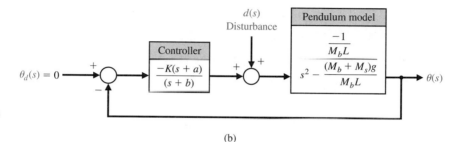

FIGURE MP8.5
(a) An inverted
pendulum on a
moving base. (b) A
block diagram
representation.

(b)

TERMS AND CONCEPTS

Bandwidth   The frequency at which the frequency response has declined 3 dB from its low frequency value.

Bode plot   The logarithm of the magnitude of the transfer function is plotted versus the logarithm of ω, the frequency. The phase, ϕ, of the transfer function is separately plotted versus the logarithm of the frequency.

Break frequency   The frequency at which the asymptotic approximation of the frequency response for a pole (or zero) changes slope.

Corner frequency   *See* Break frequency.

Decibel (dB)   The units of the logarithmic gain.

Fourier transform   The transformation of a function of time, $f(t)$, into the frequency domain.

Frequency response   The steady-state response of a system to a sinusoidal input signal.

Logarithmic magnitude   The logarithm of the magnitude of the transfer function, $20 \log_{10} |G|$.

Logarithmic plot   *See* Bode plot.

Maximum value of the frequency response   A pair of complex poles will result in a maximum value for the frequency response occurring at the resonant frequency.

Minimum phase   All the zeros of a transfer function lie in the left-hand side of the s-plane.

Natural frequency   The frequency of natural oscillation that would occur for two complex poles if the damping were equal to zero.

Nonminimum phase   Transfer functions with zeros in the right-hand s-plane.

Polar plot   A plot of the real part of $G(j\omega)$ versus the imaginary part of $G(j\omega)$.

Resonant frequency   The frequency, ω_r, at which the maximum value of the frequency response of a complex pair of poles is attained.

Transfer function in the frequency domain   The ratio of the output to the input signal where the input is a sinusoid. It is expressed as $G(j\omega)$.

Stability in the Frequency Domain

PREVIEW

As noted in earlier chapters, it is important to determine whether a system is stable. If it is stable, then the degree of stability is important to determine. We can use the frequency response of a transfer function around a feedback loop $GH(j\omega)$ to provide answers to our inquiry about the system's relative stability.

We will use some concepts developed in the theory of complex variables to obtain a stability criterion in the frequency domain. Then this criterion can be extended to indicate relative stability by indicating how close we come to operating at the edge of instability.

We will then demonstrate how to examine the frequency response of the closed-loop transfer function, $T(j\omega)$, as well as the loop transfer function $GH(j\omega)$, to predict the performance of the system.

Finally, we will use these methods to analyze the response and performance of a system with a pure time delay, without attenuation, located within the feedback loop of a closed-loop control system.

9.1 INTRODUCTION

For a control system, it is necessary to determine whether the system is stable. Furthermore, if the system is stable, it is often necessary to investigate the relative stability. In Chapter 6, we discussed the concept of stability and several methods of determining the absolute and relative stability of a system. The Routh–Hurwitz method, discussed in Chapter 6, is useful for investigating the characteristic equation expressed in terms of the complex variable $s = \sigma + j\omega$. Then in Chapter 7, we investigated the relative stability of a system utilizing the root locus method, which is also expressed in terms of the complex variable s. In this chapter, we are concerned with investigating the stability of a system in the real frequency domain, that is, in terms of the frequency response discussed in Chapter 8.

The frequency response of a system represents the sinusoidal steady-state response of a system and provides sufficient information for the determination of the relative stability of the system. The frequency response of a system can readily be obtained experimentally by exciting the system with sinusoidal input signals; therefore it can be utilized to investigate the relative stability of a system when the system parameter values have not been determined. Furthermore, a frequency-domain stability criterion would be useful for determining suitable approaches to adjusting the parameters of a system in order to increase its relative stability.

A frequency domain stability criterion was developed by H. Nyquist in 1932 and remains a fundamental approach to the investigation of the stability of linear control systems [1, 2]. The *Nyquist stability criterion* is based on a theorem in the theory of the function of a complex variable due to Cauchy. Cauchy's theorem is concerned with *mapping contours* in the complex s-plane, and fortunately the theorem can be understood without a formal proof, which uses complex variable theory.

To determine the relative stability of a closed-loop system, we must investigate the characteristic equation of the system:

$$F(s) = 1 + L(s) = 0. \tag{9.1}$$

For the single-loop control system of Fig. 9.1, $L(s) = G(s)H(s)$. For a multiloop system, we found in Section 2.7 that, in terms of signal-flow graphs, the characteristic equation is

$$F(s) = \Delta(s) = 1 - \Sigma L_n + \Sigma L_m L_q \cdots,$$

where $\Delta(s)$ is the graph determinant. Therefore we can represent the characteristic equation of single-loop or multiple-loop systems by Eq. (9.1), where $L(s)$ is a rational function of s. To ensure stability, we must ascertain that all the zeros of $F(s)$ lie in the left-hand s-plane. Nyquist thus proposed a mapping of the right-hand s-plane into the $F(s)$-plane. Therefore, to utilize and understand Nyquist's criterion, we shall first consider briefly the mapping of contours in the complex plane.

FIGURE 9.1
Single-loop
feedback control
system.

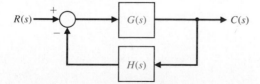

9.2 MAPPING CONTOURS IN THE *s*-PLANE

We are concerned with the mapping of contours in the *s*-plane by a function $F(s)$. A *contour map* is a contour or trajectory in one plane mapped or translated into another plane by a relation $F(s)$. Since *s* is a complex variable, $s = \sigma + j\omega$, the function $F(s)$ is itself complex and can be defined as $F(s) = u + jv$ and can be represented on a complex $F(s)$-plane with coordinates *u* and *v*. As an example, let us consider a function $F(s) = 2s + 1$ and a contour in the *s*-plane as shown in Fig. 9.2(a). The mapping of the *s*-plane unit square contour to the $F(s)$-plane is accomplished through the relation $F(s)$, and so

$$u + jv = F(s) = 2s + 1 = 2(\sigma + j\omega) + 1. \tag{9.2}$$

Therefore, in this case, we have

$$u = 2\sigma + 1 \tag{9.3}$$

and

$$v = 2\omega. \tag{9.4}$$

Thus the contour has been mapped by $F(s)$ into a contour of an identical form, a square, with the center shifted by one unit and the magnitude of a side multiplied by 2. This type of mapping, which retains the angles of the *s*-plane contour on the $F(s)$-plane, is called a *conformal mapping*. We also note that a closed contour in the *s*-plane results in a closed contour in the $F(s)$-plane.

The points *A*, *B*, *C*, and *D*, as shown in the *s*-plane contour, map into the points *A*, *B*, *C*, and *D* shown in the $F(s)$-plane. Furthermore, a direction of traversal of the *s*-plane contour can be indicated by the direction *ABCD* and the arrows shown on the contour. Then a similar traversal occurs on the $F(s)$-plane contour as we pass *ABCD* in order, as shown by the arrows. By convention, the area within a contour to the right of the traversal of the contour is considered to be the *area enclosed* by the contour. Therefore we will assume *clockwise traversal* of a contour to be positive and the area enclosed within the

FIGURE 9.2
Mapping a square contour by
$F(s) = 2s + 1 = 2(s + 1/2)$.

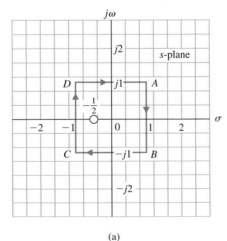

(a)

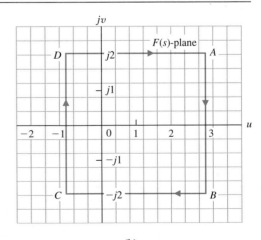

(b)

TABLE 9.1   Values of F(s)

$s = \sigma + j\omega$	Point A $1 + j1$	1	Point B $1 - j1$	$-j1$	Point C $-1 - j1$	-1	Point D $-1 + j1$	$j1$
$F(s) = u + jv$	$\dfrac{4 + 2j}{10}$	$\dfrac{1}{3}$	$\dfrac{4 - 2j}{10}$	$\dfrac{1 - 2j}{5}$	$-j$	-1	$+j$	$\dfrac{1 + 2j}{5}$

contour to be on the right. This convention is opposite to that usually employed in complex variable theory but is equally applicable and is generally used in control system theory. Readers might consider the area on the right as they walk along the contour in a clockwise direction and call this rule "clockwise and eyes right."

Typically, we are concerned with an $F(s)$ that is a rational function of s. Therefore it will be worthwhile to consider another example of a mapping of a contour. Let us again consider the unit square contour for the function

$$F(s) = \frac{s}{s + 2}. \tag{9.5}$$

Several values of $F(s)$ as s traverses the square contour are given in Table 9.1, and the resulting contour in the $F(s)$-plane is shown in Fig. 9.3(b). The contour in the $F(s)$-plane encloses the origin of the $F(s)$-plane because the origin lies within the enclosed area of the contour in the $F(s)$-plane.

Cauchy's theorem is concerned with mapping a function $F(s)$ that has a finite number of poles and zeros within the contour so that we may express $F(s)$ as

$$F(s) = \frac{K \prod_{i=1}^{n} (s + s_i)}{\prod_{k=1}^{M} (s + s_k)}, \tag{9.6}$$

FIGURE 9.3
Mapping for
$F(s) = s/(s + 2)$.

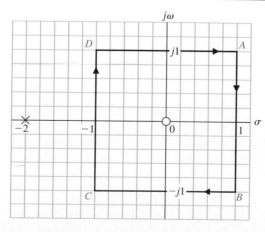

(a)

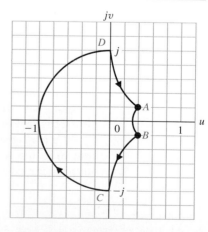

(b)

where s_i are the zeros of the function $F(s)$ and s_k are the poles of $F(s)$. The function $F(s)$ is the characteristic equation, and so

$$F(s) = 1 + L(s), \tag{9.7}$$

where

$$L(s) = \frac{N(s)}{D(s)}.$$

Therefore we have

$$F(s) = 1 + L(s) = 1 + \frac{N(s)}{D(s)} = \frac{D(s) + N(s)}{D(s)} = \frac{K \prod_{i=1}^{n} (s + s_i)}{\prod_{k=1}^{M} (s + s_k)}, \tag{9.8}$$

and the poles of $L(s)$ are the poles of $F(s)$. However, it is the zeros of $F(s)$ that are the characteristic roots of the system and that indicate its response. This is clear if we recall that the output of the system is

$$C(s) = T(s)R(s) = \frac{\Sigma P_k \Delta_k}{\Delta(s)} R(s) = \frac{\Sigma P_k \Delta_k}{F(s)} R(s), \tag{9.9}$$

where P_k and Δ_k are the path factors and cofactors as defined in Section 2.7.

Now, reexamining the example when $F(s) = 2(s + 1/2)$, we have one zero of $F(s)$ at $s = -1/2$, as shown in Fig. 9.2. The contour that we chose (that is, the unit square) enclosed and encircled the zero once within the area of the contour. Similarly, for the function $F(s) = s/(s + 2)$, the unit square encircled the zero at the origin but did not encircle the pole at $s = -2$. The encirclement of the poles and zeros of $F(s)$ can be related to the encirclement of the origin in the $F(s)$-plane by a *theorem of Cauchy,* commonly known as the *principle of the argument,* which states [3, 4]:

> If a contour Γ_s in the s-plane encircles Z zeros and P poles of $F(s)$ and does not pass through any poles or zeros of $F(s)$ as the traversal is in the clockwise direction along the contour, the corresponding contour Γ_F in the $F(s)$-plane encircles the origin of the $F(s)$-plane $N = Z - P$ times in the clockwise direction.

Thus for the examples shown in Figs. 9.2 and 9.3, the contour in the $F(s)$-plane encircles the origin once, because $N = Z - P = 1$, as we expect. As another example, consider the function $F(s) = s/(s + 1/2)$. For the unit square contour shown in Fig. 9.4(a), the resulting contour in the $F(s)$ plane is shown in Fig. 9.4(b). In this case, $N = Z - P = 0$ as is the case in Fig. 9.4(b), since the contour Γ_F does not encircle the origin.

Cauchy's theorem can be best comprehended by considering $F(s)$ in terms of the angle due to each pole and zero as the contour Γ_s is traversed in a clockwise direction. Thus let us consider the function

$$F(s) = \frac{(s + z_1)(s + z_2)}{(s + p_1)(s + p_2)}, \tag{9.10}$$

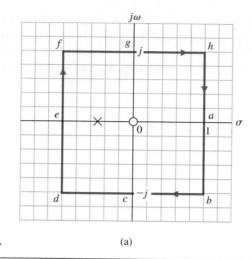

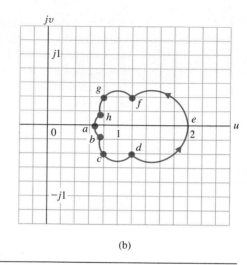

FIGURE 9.4
Mapping for
$F(s) = s/(s + 1/2)$.

(a)   (b)

where z_i is a zero of $F(s)$ and p_k is a pole of $F(s)$. Equation (9.10) can be written as

$$F(s) = |F(s)|\underline{/F(s)}$$

$$= \frac{|s + z_1| \, |s + z_2|}{|s + p_1| \, |s + p_2|}(\underline{/s + z_1} + \underline{/s + z_2} - \underline{/s + p_1} - \underline{/s + p_2}) \quad (9.11)$$

$$= |F(s)|(\phi_{z_1} + \phi_{z_2} - \phi_{p_1} - \phi_{p_2}).$$

Now, considering the vectors as shown for a specific contour Γ_s (Fig. 9.5a), we can determine the angles as s traverses the contour. Clearly, the net angle change as s traverses along Γ_s a full rotation of 360° for ϕ_{p_1}, ϕ_{p_2} and ϕ_{z_2} is zero degrees. However, for ϕ_{z_1} as s traverses

FIGURE 9.5
Evaluation of the
net angle of Γ_F.

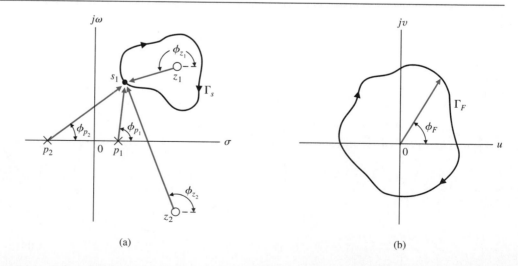

(a)   (b)

360° around Γ_s, the angle ϕ_{z_1} traverses a full 360° clockwise. Thus, as Γ_s is completely traversed, the net angle of $F(s)$ is equal to 360°, since only one zero is enclosed. If Z zeros were enclosed within Γ_s, then the net angle would be equal to $\phi_z = 2\pi(Z)$ rad. Following this reasoning, if Z zeros and P poles are encircled as Γ_s is traversed, then $2\pi(Z) - 2\pi(P)$ is the net resultant angle of $F(s)$. Thus the net angle of Γ_F of the contour in the $F(s)$-plane, ϕ_F, is simply

$$\phi_F = \phi_Z - \phi_P$$

or

$$2\pi N = 2\pi Z - 2\pi P, \tag{9.12}$$

and the net number of encirclements of the origin of the $F(s)$-plane is $N = Z - P$. Thus for the contour shown in Fig. 9.5(a), which encircles one zero, the contour Γ_F shown in Fig. 9.5(b) encircles the origin once in the clockwise direction.

As an example of the use of Cauchy's theorem, consider the pole–zero pattern shown in Fig. 9.6(a) with the contour Γ_s to be considered. The contour encloses and encircles three zeros and one pole. Therefore we obtain

$$N = 3 - 1 = +2,$$

and Γ_F completes two clockwise encirclements of the origin in the $F(s)$-plane as shown in Fig. 9.6(b).

For the pole and zero pattern shown and the contour Γ_s as shown in Fig. 9.7(a), one pole is encircled and no zeros are encircled. Therefore we have

$$N = Z - P = -1,$$

and we expect one encirclement of the origin by the contour Γ_F in the $F(s)$-plane. However, since the sign of N is negative, we find that the encirclement moves in the counterclockwise direction, as shown in Fig. 9.7(b).

FIGURE 9.6
Example of
Cauchy's theorem
with three zeros
and one pole
within Γ_s.

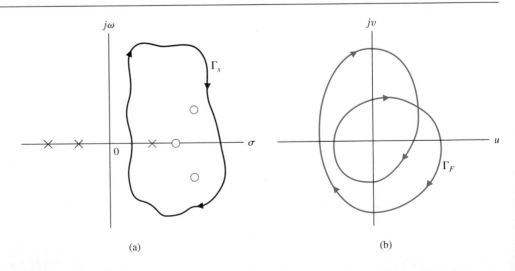

(a)   (b)

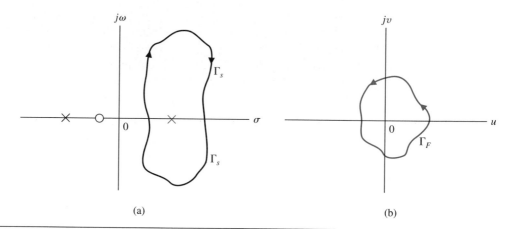

FIGURE 9.7
Example of
Cauchy's theorem
with one pole
within Γ_s.

Now that we have developed and illustrated the concept of mapping of contours through a function $F(s)$, we are ready to consider the stability criterion proposed by Nyquist.

9.3 THE NYQUIST CRITERION

To investigate the stability of a control system, we consider the characteristic equation, which is $F(s) = 0$, so that

$$F(s) = 1 + L(s) = \frac{K \prod_{i=1}^{n} (s + s_i)}{\prod_{k=1}^{M} (s + s_k)} = 0. \tag{9.13}$$

For a system to be stable, all the zeros of $F(s)$ must lie in the left-hand s-plane. Thus we find that the roots of a stable system [the zeros of $F(s)$] must lie to the left of the $j\omega$-axis in the s-plane. Therefore we chose a contour Γ_s in the s-plane that encloses the entire right-hand s-plane, and we determine whether any zeros of $F(s)$ lie within Γ_s by utilizing Cauchy's theorem. That is, we plot Γ_F in the $F(s)$-plane and determine the number of encirclements of the origin N. Then the number of zeros of $F(s)$ within the Γ_s contour [and therefore unstable zeros of $F(s)$] is

$$Z = N + P. \tag{9.14}$$

Thus if $P = 0$, as is usually the case, we find that the number of unstable roots of the system is equal to N, the number of encirclements of the origin of the $F(s)$-plane.

The Nyquist contour that encloses the entire right-hand s-plane is shown in Fig. 9.8. The contour Γ_s passes along the $j\omega$-axis from $-j\infty$ to $+j\infty$, and this part of the contour provides the familiar $F(j\omega)$. The contour is completed by a semicircular path of radius r where r approaches infinity.

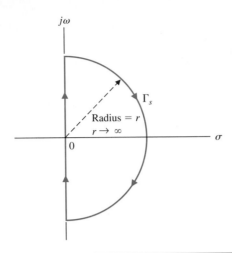

FIGURE 9.8
Nyquist contour.

Now, the Nyquist criterion is concerned with the mapping of the characteristic equation

$$F(s) = 1 + L(s) \tag{9.15}$$

and the number of encirclements of the origin of the $F(s)$-plane. Alternatively, we may define the function $F'(s)$ so that

$$F'(s) = F(s) - 1 = L(s). \tag{9.16}$$

The change of functions represented by Eq. (9.16) is very convenient because $L(s)$ is typically available in factored form, while $[1 + L(s)]$ is not. Then the mapping of Γ_s in the s-plane will be through the function $F'(s) = L(s)$ into the $L(s)$-plane. In this case, the number of clockwise encirclements of the origin of the $F(s)$-plane becomes the number of clockwise encirclements of the -1 point in the $F'(s) = L(s)$-plane because $F'(s) = F(s) - 1$. Therefore the *Nyquist stability criterion* can be stated as follows:

> **A feedback system is stable if and only if the contour Γ_L in the $L(s)$-plane does not encircle the $(-1, 0)$ point when the number of poles of $L(s)$ in the right-hand s-plane is zero ($P = 0$).**

When the number of poles of $L(s)$ in the right-hand s-plane is other than zero, the Nyquist criterion is stated:

> **A feedback control system is stable if and only if, for the contour Γ_L, the number of counterclockwise encirclements of the $(-1, 0)$ point is equal to the number of poles of $L(s)$ with positive real parts.**

The basis for the two statements is the fact that for the $F'(s) = L(s)$ mapping, the number of roots (or zeros) of $1 + L(s)$ in the right-hand s-plane is represented by the expression

$$Z = N + P.$$

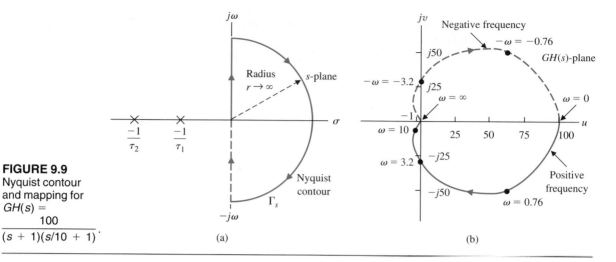

FIGURE 9.9
Nyquist contour and mapping for
$GH(s) =$
$$\frac{100}{(s + 1)(s/10 + 1)}.$$

Clearly, if the number of poles of $L(s)$ in the right-hand s-plane is zero ($P = 0$), we require for a stable system that $N = 0$ and the contour Γ_p must not encircle the -1 point. Also, if P is other than zero and we require for a stable system that $Z = 0$, then we must have $N = -P$, or P counterclockwise encirclements.

It is best to illustrate the use of the Nyquist criterion by completing several examples.

EXAMPLE 9.1   System with two real poles

A single-loop control system is shown in Fig. 9.1, where

$$GH(s) = \frac{K}{(\tau_1 s + 1)(\tau_2 s + 1)}. \tag{9.17}$$

In this case, $L(s) = GH(s)$, and we utilize a contour $\Gamma_L = \Gamma_{GH}$ in the $GH(s)$-plane. The contour Γ_s in the s-plane is shown in Fig. 9.9(a). and the contour Γ_{GH} is shown in Fig. 9.9(b) for $\tau_1 = 1$, $\tau_2 = 1/10$, and $K = 100$. The magnitude and phase of $GH(j\omega)$ for selected values of ω are given in Table 9.2. We use these values to obtain the polar plot of Fig. 9.9(b).

The $+j\omega$-axis is mapped into the solid colored line as shown in Fig. 9.9. The $-j\omega$-axis is mapped into the dashed colored line as shown in Fig. 9.9. The semicircle with $r \to \infty$ in the s-plane is mapped into the origin of the $GH(s)$-plane.

We note that the number of poles of $GH(s)$ in the right-hand s-plane is zero and thus $P = 0$. Therefore, for this system to be stable, we require $N = Z = 0$ and the contour must

TABLE 9.2   Magnitude and Phase of $GH(j\omega)$

ω	0	0.1	0.76	1	2	10	20	100	∞
$\|GH(j\omega)\|$	100	96	79.6	70.7	50.2	6.8	2.24	0.10	0
$\underline{/GH(j\omega)}$ (degrees)	0	-5.7	-41.5	-50.7	-74.7	-129.3	-150.5	-173.7	-180

not encircle the -1 point in the $GH(s)$-plane. Examining Fig. 9.9(b) and Eq. (9.17), we find that, irrespective of the value of K, the contour does not encircle the -1 point and the system is always stable for all K greater than zero. ∎

EXAMPLE 9.2   System with a pole at the origin

A single-loop control system is shown in Fig. 9.1, where

$$GH(s) = \frac{K}{s(\tau s + 1)}.$$

In this single-loop case, $L(s) = GH(s)$ and we determine the contour $\Gamma_L = \Gamma_{GH}$ in the $GH(s)$-plane. The contour Γ_s in the s-plane is shown in Fig. 9.10(a), where an infinitesimal detour around the pole at the origin is effected by a small semicircle of radius ϵ, where $\epsilon \to 0$. This detour is a consequence of the condition of Cauchy's theorem, which requires that the contour cannot pass through the pole at the origin. A sketch of the contour Γ_{GH} is shown in Fig. 9.10(b). Clearly, the portion of the contour Γ_{GH} from $\omega = 0^+$ to $\omega = +\infty$ is simply $GH(j\omega)$, the real frequency polar plot. Let us consider each portion of the Nyquist contour Γ_s in detail and determine the corresponding portions of the $GH(s)$-plane contour Γ_{GH}.

    (a) The Origin of the s-Plane.   The small semicircular detour around the pole at the origin can be represented by setting $s = \epsilon e^{j\phi}$ and allowing ϕ to vary from $-90°$ at $\omega = 0^-$ to $+90°$ at $\omega = 0^+$. Because ϵ approaches zero, the mapping for $GH(s)$ is

$$\lim_{\epsilon \to 0} GH(s) = \lim_{\epsilon \to 0} \left(\frac{K}{\epsilon e^{j\phi}}\right) = \lim_{\epsilon \to 0} \left(\frac{K}{\epsilon}\right) e^{-j\phi}. \qquad (9.18)$$

Therefore the angle of the contour in the $GH(s)$-plane changes from $90°$ at $\omega = 0^-$ to $-90°$ at $\omega = 0^+$, passing through $0°$ at $\omega = 0$. The radius of the contour in the $GH(s)$-plane for this portion of the contour is infinite, and this portion of the contour is shown in Fig. 9.10(b).

FIGURE 9.10
Nyquist contour and mapping for $GH(s) = K/s(\tau s + 1)$.

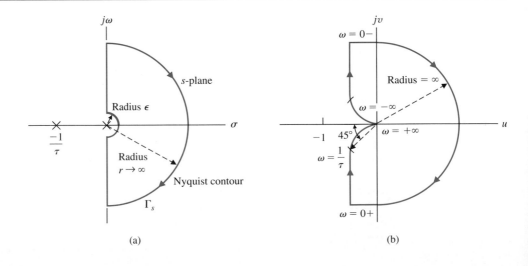

(a)

(b)

(b) The Portion from $\omega = 0^+$ to $\omega = +\infty$. The portion of the contour Γ_s from $\omega = 0^+$ to $\omega = +\infty$ is mapped by the function $GH(s)$ as the real frequency polar plot because $s = j\omega$ and

$$GH(s)|_{s=j\omega} = GH(j\omega) \tag{9.19}$$

for this part of the contour. This results in the real frequency polar plot shown in Fig. 9.10(b). When ω approaches $+\infty$, we have

$$\lim_{\omega \to +\infty} GH(j\omega) = \lim_{\omega \to +\infty} \frac{K}{+j\omega(j\omega\tau + 1)} \tag{9.20}$$

$$= \lim_{\omega \to \infty} \left| \frac{K}{\tau\omega^2} \right| \underline{/-(\pi/2) - \tan^{-1} \omega\tau}.$$

Therefore the magnitude approaches zero at an angle of $-180°$.

(c) The Portion from $\omega = +\infty$ to $\omega = -\infty$. The portion of Γ_s from $\omega = +\infty$ to $\omega = -\infty$ is mapped into the point zero at the origin of the $GH(s)$-plane by the function $GH(s)$. The mapping is represented by

$$\lim_{r \to \infty} GH(s)|_{s=re^{j\phi}} = \lim_{r \to \infty} \left| \frac{K}{r^2} \right| e^{-2j\phi} \tag{9.21}$$

as ϕ changes from $\phi = +90°$ at $\omega = +\infty$ to $\phi = -90°$ at $\omega = -\infty$. Thus the contour moves from an angle of $-180°$ at $\omega = +\infty$ to an angle of $+180°$ at $\omega = -\infty$. The magnitude of the $GH(s)$ contour when r is infinite is always zero or a constant.

(d) The Portion from $\omega = -\infty$ to $\omega = 0^-$. The portion of the contour Γ_s from $\omega = -\infty$ to $\omega = 0^-$ is mapped by the function $GH(s)$ as

$$GH(s)|_{s=-j\omega} = GH(-j\omega). \tag{9.22}$$

Thus we obtain the complex conjugate of $GH(j\omega)$, and the plot for the portion of the polar plot from $\omega = -\infty$ to $\omega = 0^-$ is symmetrical to the polar plot from $\omega = +\infty$ to $\omega = 0^+$. This symmetrical polar plot is shown on the $GH(s)$-plane in Fig. 9.10(b).

Now, to investigate the stability of this second-order system, we first note that the number of poles P within the right-hand s-plane is zero. Therefore, for this system to be stable, we require $N = Z = 0$, and the contour Γ_{GH} must not encircle the -1 point in the GH-plane. Examining Fig. 9.10(b), we find that irrespective of the value of the gain K and the time constant τ, the contour does not encircle the -1 point, and the system is always stable. As in Chapter 7, we are considering positive values of gain K. If negative values of gain are to be considered, one should use $-K$, where $K \geq 0$.

We may draw two general conclusions from this example as follows:

1. The plot of the contour Γ_{GH} for the range $-\infty < \omega < 0^-$ will be the complex conjugate of the plot for the range $0^+ < \omega < +\infty$ and the polar plot of $GH(s)$ will be symmetrical in the $GH(s)$-plane about the u-axis. Therefore *it is sufficient to construct the contour Γ_{GH} for the frequency range $0^+ < \omega < +\infty$ in order to investigate the stability.*

2. The magnitude of $GH(s)$ as $s = re^{j\phi}$ and $r \to \infty$ will normally approach zero or a constant. ∎

EXAMPLE 9.3   **System with three poles**

Let us again consider the single-loop system shown in Fig. 9.1 when

$$GH(s) = \frac{K}{s(\tau_1 s + 1)(\tau_2 s + 1)}. \tag{9.23}$$

The Nyquist contour Γ_s is shown in Fig. 9.10(a). Again this mapping is symmetrical for $GH(j\omega)$ and $GH(-j\omega)$ so that it is sufficient to investigate the $GH(j\omega)$-locus. The origin of the s-plane maps into a semicircle of infinite radius as in Example 9.2. Also, the semicircle $re^{j\phi}$ in the s-plane maps into the point $GH(s) = 0$, as we expect. Therefore, to investigate the stability of the system, it is sufficient to plot the portion of the contour Γ_{GH} that is the real frequency polar plot $GH(j\omega)$ for $0^+ < \omega < +\infty$. Therefore, when $s = +j\omega$, we have

$$GH(j\omega) = \frac{K}{j\omega(j\omega\tau_1 + 1)(j\omega\tau_2 + 1)}$$

$$= \frac{-K(\tau_1 + \tau_2) - jK(1/\omega)(1 - \omega^2\tau_1\tau_2)}{1 + \omega^2(\tau_1^2 + \tau_2^2) + \omega^4\tau_1^2\tau_2^2} \tag{9.24}$$

$$= \frac{K}{[\omega^4(\tau_1 + \tau_2)^2 + \omega^2(1 - \omega^2\tau_1\tau_2)^2]^{1/2}}$$

$$\times \; \underline{/-\tan^{-1} \omega\tau_1 - \tan^{-1} \omega\tau_2 - (\pi/2)}.$$

When $\omega = 0^+$, the magnitude of the locus is infinite at an angle of $-90°$ in the $GH(s)$-plane. When ω approaches $+\infty$, we have

$$\lim_{\omega \to \infty} GH(j\omega) = \lim_{\omega \to \infty} \left| \frac{1}{\omega^3\tau_1\tau_2} \right| \; \underline{/-(\pi/2) - \tan^{-1} \omega\tau_1 - \tan^{-1} \omega\tau_2} \tag{9.25}$$

$$= \left(\lim_{\omega \to \infty} \left| \frac{1}{\omega^3\tau_1\tau_2} \right| \right) \underline{/-(3\pi/2)}.$$

Therefore $GH(j\omega)$ approaches a magnitude of zero at an angle of $-270°$. To approach at an angle of $-270°$, the locus must cross the u-axis in the $GH(s)$-plane, as shown in Fig. 9.11. Thus it is possible to encircle the -1 point as is shown in Fig. 9.11. The number of encirclements, when the -1 point lies within the locus as shown in Fig. 9.11, is equal to two, and the system is unstable with two roots in the right-hand s-plane. The point where the $GH(s)$-locus intersects the real axis can be found by setting the imaginary part of $GH(j\omega) = u + jv$ equal to zero. We then have from Eq. (9.24)

$$v = \frac{-K(1/\omega)(1 - \omega^2\tau_1\tau_2)}{1 + \omega^2(\tau_1^2 + \tau_2^2) + \omega^4\tau_1^2\tau_2^2} = 0. \tag{9.26}$$

Thus $v = 0$ when $1 - \omega^2\tau_1\tau_2 = 0$ or $\omega = 1/\sqrt{\tau_1\tau_2}$. The magnitude of the real part, u, of $GH(j\omega)$ at this frequency is

$$u = \frac{-K(\tau_1 + \tau_2)}{1 + \omega^2(\tau_1^2 + \tau_2^2) + \omega^4\tau_1^2\tau_2^2} \Big|_{\omega^2 = 1/\tau_1\tau_2} \tag{9.27}$$

$$= \frac{-K(\tau_1 + \tau_2)\tau_1\tau_2}{\tau_1\tau_2 + (\tau_1^2 + \tau_2^2) + \tau_1\tau_2} = \frac{-K\tau_1\tau_2}{\tau_1 + \tau_2}.$$

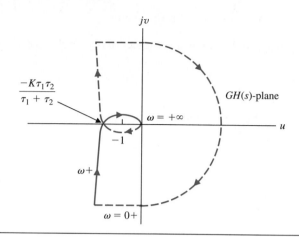

FIGURE 9.11
Nyquist diagram
for $GH(s) = K/s(\tau_1 s + 1)(\tau_2 s + 1)$.
The tic mark shown
to the left of the origin
is the -1 point.

Therefore the system is stable when

$$\frac{-K\tau_1\tau_2}{\tau_1 + \tau_2} \geq -1$$

or

$$K \leq \frac{\tau_1 + \tau_2}{\tau_1\tau_2}. \tag{9.28}$$

Consider the case where $\tau_1 = \tau_2 = 1$ so that

$$G(s)H(s) = \frac{K}{s(s + 1)^2}.$$

Using Eq. (9.28), we expect stability when

$$K \leq 2.$$

The Nyquist diagram for three values of K is shown in Fig. 9.12. ∎

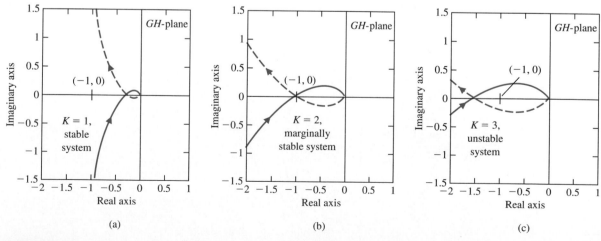

FIGURE 9.12
Nyquist plot for $G(s)H(s) = \dfrac{K}{s(s + 1)^2}$ when (a) $K = 1$, (b) $K = 2$, and (c) $K = 3$.

EXAMPLE 9.4   **System with two poles at the origin**

Again let us determine the stability of the single-loop system shown in Fig. 9.1 when

$$GH(s) = \frac{K}{s^2(\tau s + 1)}. \tag{9.29}$$

The real frequency polar plot is obtained when $s = j\omega$, and we have

$$GH(j\omega) = \frac{K}{-\omega^2(j\omega\tau + 1)} = \frac{K}{[\omega^4 + \tau^2\omega^6]^{1/2}}\underline{/-\pi - \tan^{-1}\omega\tau}. \tag{9.30}$$

We note that the angle of $GH(j\omega)$ is always $-180°$ or greater, and the locus of $GH(j\omega)$ is above the u-axis for all values of ω. As ω approaches 0^+, we have

$$\lim_{\omega \to 0+} GH(j\omega) = \left(\lim_{\omega \to 0+}\left|\frac{K}{\omega^2}\right|\right)\underline{/-\pi}. \tag{9.31}$$

As ω approaches $+\infty$, we have

$$\lim_{\omega \to +\infty} GH(j\omega) = \left(\lim_{\omega \to +\infty}\frac{K}{\omega^3}\right)\underline{/-3\pi/2}. \tag{9.32}$$

At the small semicircular detour at the origin of the s-plane where $s = \epsilon e^{j\phi}$, we have

$$\lim_{\epsilon \to 0} GH(s) = \lim_{\epsilon \to 0}\frac{K}{\epsilon^2}e^{-2j\phi}, \tag{9.33}$$

where $-\pi/2 \le \phi \le \pi/2$. Thus the contour Γ_{GH} ranges from an angle of $+\pi$ at $\omega = 0^+$ to $-\pi$ at $\omega = 0^+$ and passes through a full circle of 2π rad as ω changes from $\omega = 0^-$ to $\omega = 0^+$. The complete contour plot of Γ_{GH} is shown in Fig. 9.13. Because the contour encircles the -1 point twice, there are two roots of the closed-loop system in the right-hand plane and the system, irrespective of the gain K, is unstable. ∎

FIGURE 9.13
Nyquist contour
plot for $GH(s) = K/s^2(\tau s + 1)$.

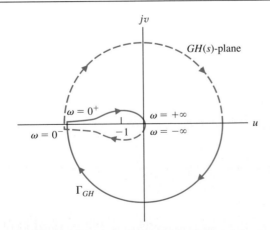

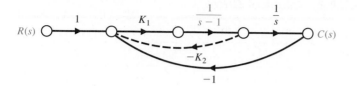

FIGURE 9.14
Second-order
feedback control
system.

EXAMPLE 9.5   **System with a pole in the right-hand s-plane**

Let us consider the control system shown in Fig. 9.14 and determine the stability of the system. First, let us consider the system without derivative feedback so that $K_2 = 0$. We then have the open-loop transfer function (without feedback)

$$GH(s) = \frac{K_1}{s(s - 1)}. \tag{9.34}$$

Thus the open-loop transfer function has one pole in the right-hand s-plane, and therefore $P = 1$. For this system to be stable, we require $N = -P = -1$, one counterclockwise encirclement of the -1 point. At the semicircular detour at the origin of the s-plane, we let $s = \epsilon e^{j\phi}$ when $-\pi/2 \leq \phi \leq \pi/2$. Then we have, when $s = \epsilon e^{j\phi}$,

$$\lim_{\epsilon \to 0} GH(s) = \lim_{\epsilon \to 0} \frac{K_1}{-\epsilon e^{j\phi}} = \left(\lim_{\epsilon \to 0} \left|\frac{K_1}{\epsilon}\right|\right) \underline{/-180° - \phi}. \tag{9.35}$$

Therefore this portion of the contour Γ_{GH} is a semicircle of infinite magnitude in the left-hand GH-plane, as shown in Fig. 9.15. When $s = j\omega$, we have

$$GH(j\omega) = \frac{K_1}{j\omega(j\omega - 1)} = \frac{K_1}{(\omega^2 + \omega^4)^{1/2}} \underline{/(-\pi/2) - \tan^{-1}(-\omega)}$$

$$= \frac{K_1}{(\omega^2 + \omega^4)^{1/2}} \underline{/(+\pi/2) + \tan^{-1}\omega}. \tag{9.36}$$

FIGURE 9.15
Nyquist diagram
for $GH(s) =$
$K_1/s(s - 1)$.

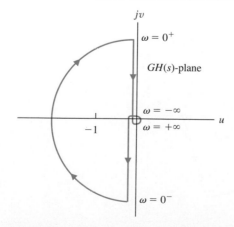

TABLE 9.3   Values of GH(s)

s	$j0^-$	$j0^+$	$j1$	$+j\infty$	$-j\infty$
$\lvert GH \rvert / K_1$	∞	∞	$1/\sqrt{2}$	0	0
$\underline{/GH}$	$-90°$	$+90°$	$+135°$	$+180°$	$-180°$

Finally, for the semicircle of radius r as r approaches infinity, we have

$$\lim_{r\to\infty} GH(s)\big|_{s=re^{j\phi}} = \left(\lim_{r\to\infty} \left\lvert \frac{K_1}{r^2} \right\rvert\right) e^{-2j\phi}, \tag{9.37}$$

where ϕ varies from $\pi/2$ to $-\pi/2$ in a clockwise direction. Therefore the contour Γ_{GH}, at the origin of the GH-plane, varies 2π rad in a counterclockwise direction, as shown in Fig. 9.15. Several important values of the $GH(s)$-locus are given in Table 9.3. The contour Γ_{GH} in the $GH(s)$-plane encircles the -1 point once in the clockwise direction and $N = +1$. Therefore

$$Z = N + P = 2. \tag{9.38}$$

and the system is unstable because two zeros of the characteristic equation, irrespective of the value of the gain K_1, lie in the right half of the s-plane.

Let us now reconsider the system when the derivative feedback is included in the system shown in Fig. 9.14 ($K_2 > 0$). Then the open-loop transfer function is

$$GH(s) = \frac{K_1(1 + K_2 s)}{s(s - 1)}. \tag{9.39}$$

The portion of the contour Γ_{GH} when $s = \epsilon e^{j\phi}$ is the same as the system without derivative feedback, as is shown in Fig. 9.16. However, when $s = re^{j\phi}$ as r approaches infinity, we have

$$\lim_{r\to\infty} GH(s)\big|_{s=re^{j\phi}} = \lim_{r\to\infty} \left\lvert \frac{K_1 K_2}{r} \right\rvert e^{-j\phi}, \tag{9.40}$$

FIGURE 9.16
Nyquist diagram for
$GH(s) = K_1(1 + K_2 s)/s(s - 1)$.
The tic mark shown
to the left of the origin
is the -1 point.

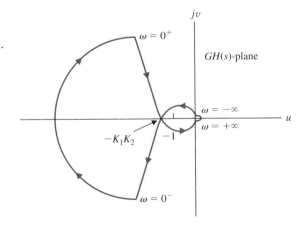

and the Γ_{GH}-contour at the origin of the *GH*-plane varies π rad in a counterclockwise direction, as shown in Fig. 9.16. The frequency locus $GH(j\omega)$ crosses the *u*-axis and is determined by considering the real frequency transfer function

$$GH(j\omega) = \frac{K_1(1 + K_2 j\omega)}{-\omega^2 - j\omega}$$

$$= \frac{-K_1(\omega^2 + \omega^2 K_2) + j(\omega - K_2 \omega^3)K_1}{\omega^2 + \omega^4}.$$

(9.41)

The $GH(j\omega)$-locus intersects the *u*-axis at a point where the imaginary part of $GH(j\omega)$ is zero. Therefore

$$\omega - K_2 \omega^3 = 0$$

at this point, or $\omega^2 = 1/K_2$. The value of the real part of $GH(j\omega)$ at the intersection is then

$$u\big|_{\omega^2 = 1/K_2} = \frac{-\omega^2 K_1(1 + K_2)}{\omega^2 + \omega^4}\bigg|_{\omega^2 = 1/K_2} = -K_1 K_2.$$

(9.42)

Therefore, when $-K_1 K_2 < -1$ or $K_1 K_2 > 1$, the contour Γ_{GH} encircles the -1 point once in a counterclockwise direction, and therefore $N = -1$. Then *Z*, the number of zeros of the system in the right-hand plane, is

$$Z - N + P = -1 + 1 = 0.$$

Thus the system is stable when $K_1 K_2 > 1$. Often, it may be useful to utilize a computer or calculator program to calculate the Nyquist diagram [5]. ∎

EXAMPLE 9.6   **System with a zero in the right-hand *s*-plane**

Let us consider the feedback control system shown in Fig. 9.1 when

$$GH(s) = \frac{K(s - 2)}{(s + 1)^2}.$$

Note this system is stable in the open-loop (no feedback) configuration. We have

$$GH(j\omega) = \frac{K(j\omega - 2)}{(j\omega + 1)^2} = \frac{K(j\omega - 2)}{(1 - \omega^2) + j2\omega}.$$

(9.43)

As ω approaches $+\infty$ on the $+j\omega$ axis, we have

$$\lim_{\omega \to +\infty} GH(j\omega) = \left(\lim_{\omega \to +\infty} \frac{K}{\omega}\right)\underline{/-\pi/2}.$$

When $\omega = \sqrt{3}$, we have $GH(j\omega) = K/2$. At $\omega = 0^+$, we have $GH(j\omega) = -2K$. The Nyquist diagram for $G(j\omega)/K$ is shown in Fig. 9.17. $GH(j\omega)$ intersects the $(-1 + j0)$ point when $K = 1/2$. Thus the system is stable for the limited range of gain $0 < K \le 1/2$. When $K > 1/2$, the number of encirclements of the -1 point is $N = 1$. The number of poles of $GH(s)$ in the right half *s*-plane is $P = 0$. Therefore we have

$$Z = N + P = 1$$

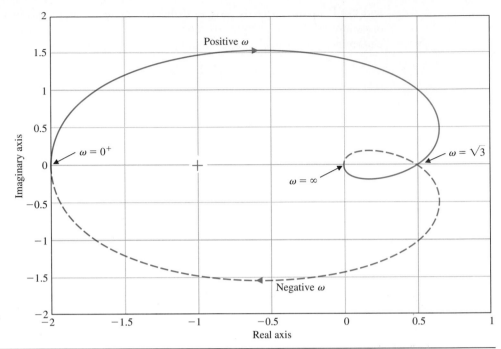

FIGURE 9.17
Nyquist diagram
for Example 9.6 for
$GH(j\omega)/K$.

and the system is unstable. Examining the Nyquist diagram of Fig. 9.17, which is plotted for $GH(j\omega)/K$, we conclude that the system is unstable for all $K > 1/2$. ∎

9.4 RELATIVE STABILITY AND THE NYQUIST CRITERION

We discussed the relative stability of a system in terms of the s-plane in Section 6.3. For the s-plane, we defined the relative stability of a system as the property measured by the relative settling time of each root or pair of roots. We would like to determine a similar measure of relative stability useful for the frequency response method. The Nyquist criterion provides us with suitable information concerning the absolute stability and, furthermore, can be utilized to define and ascertain the relative stability of a system.

The Nyquist stability criterion is defined in terms of the $(-1, 0)$ point on the polar plot or the 0-dB, $-180°$ point on the Bode diagram or log-magnitude–phase diagram. Clearly, the proximity of the $GH(j\omega)$-locus to this stability point is a measure of the relative stability of a system. The polar plot for $GH(j\omega)$ for several values of K and

$$GH(j\omega) = \frac{K}{j\omega(j\omega\tau_1 + 1)(j\omega\tau_2 + 1)} \tag{9.44}$$

is shown in Fig. 9.18. As K increases, the polar plot approaches the -1 point and eventually encircles the -1 point for a gain $K = K_3$. We determined in Section 9.3 that the locus intersects the u-axis at a point

$$u = \frac{-K\tau_1\tau_2}{\tau_1 + \tau_2}. \tag{9.45}$$

Therefore the system has roots on the $j\omega$-axis when

$$u = -1 \quad \text{or} \quad K = \left(\frac{\tau_1 + \tau_2}{\tau_1 \tau_2}\right).$$

As K is decreased below this marginal value, the stability is increased and the margin between the gain $K = (\tau_1 + \tau_2)/\tau_1\tau_2$ and a gain $K = K_2$ is a measure of the relative stability. This measure of relative stability is called the *gain margin* and is defined as *the reciprocal of the gain* $|GH(j\omega)|$ *at the frequency at which the phase angle reaches* $-180°$ (that is, $v = 0$). The gain margin is a measure of the factor by which the system gain would have to be increased for the $GH(j\omega)$ locus to pass through the $u = -1$ point. Thus, for a gain $K = K_2$ in Fig. 9.18, the gain margin is equal to the reciprocal of $GH(j\omega)$ when $v = 0$. Because $\omega = 1/\sqrt{\tau_1\tau_2}$ when the phase shift is $-180°$, we have a gain margin equal to

$$\frac{1}{|GH(j\omega)|} = \left[\frac{K_2\tau_1\tau_2}{\tau_1 + \tau_2}\right]^{-1} = \frac{1}{d}. \tag{9.46}$$

The gain margin can be defined in terms of a logarithmic (decibel) measure as

$$20 \log \left(\frac{1}{d}\right) = -20 \log d \text{ dB}. \tag{9.47}$$

For example, when $\tau_1 = \tau_2 = 1$, the system is stable when $K \le 2$. Thus when $K = K_2 = 0.5$, the gain margin is equal to

$$\frac{1}{d} = \left[\frac{K_2\tau_1\tau_2}{\tau_1 + \tau_2}\right]^{-1} = 4, \tag{9.48}$$

or, in logarithmic measure,

$$20 \log 4 = 12 \text{ dB}. \tag{9.49}$$

Therefore the gain margin indicates that the system gain can be increased by a factor of four (12 dB) before the stability boundary is reached.

FIGURE 9.18
Polar plot for
$GH(j\omega)$ for three
values of gain.

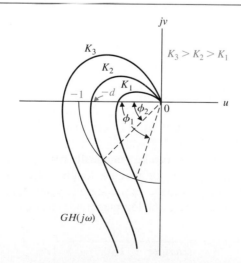

The gain margin is the increase in the system gain when phase $= -180°$ that will result in a marginally stable system with intersection of the $-1 + j0$ point on the Nyquist diagram.

An alternative measure of relative stability can be defined in terms of the phase angle margin between a specific system and a system that is marginally stable. Several roots of the characteristic equation lie on the $j\omega$-axis when the $GH(j\omega)$-locus intersects the $u = -1$, $v = 0$ point in the GH-plane. Therefore a measure of relative stability, the *phase margin,* is defined as *the phase angle through which the $GH(j\omega)$ locus must be rotated so that the unity magnitude $|GH(j\omega)| = 1$ point will pass through the $(-1, 0)$ point in the $GH(j\omega)$ plane.* This measure of relative stability is called the phase margin and is equal to the additional phase lag required before the system becomes unstable. This information can be determined from the Nyquist diagram shown in Fig. 9.18. For a gain $K = K_2$, an additional phase angle, ϕ_2, may be added to the system before the system becomes unstable. Furthermore, for the gain K_1, the phase margin is equal to ϕ_1, as shown in Fig. 9.18.

The phase margin is the amount of phase shift of the $GH(j\omega)$ at unity magnitude that will result in a marginally stable system with intersection of the $-1 + j0$ point on the Nyquist diagram.

The gain and phase margins are easily evaluated from the Bode diagram, and because it is preferable to draw the Bode diagram in contrast to the polar plot, it is worthwhile to illustrate the relative stability measures for the Bode diagram. The critical point for stability is $u = -1$, $v = 0$ in the $GH(j\omega)$-plane, which is equivalent to a logarithmic magnitude of 0 dB and a phase angle of $180°$ (or $-180°$) on the Bode diagram.

It is relatively straightforward to examine the Nyquist diagram of a minimum phase system. Special care is required with a nonminimum phase system, however, and the complete Nyquist diagram should be studied to determine stability.

The gain margin and phase margin can be readily calculated by utilizing a computer program, assuming the system is minimum phase. In contrast, for nonminimum phase systems, the complete Nyquist diagram must be constructed.

The Bode diagram of

$$GH(j\omega) = \frac{1}{j\omega(j\omega + 1)(0.2j\omega + 1)} \tag{9.50}$$

is shown in Fig. 9.19. The phase angle when the logarithmic magnitude is 0 dB is equal to $137°$. Thus the phase margin is $180° - 137° = 43°$, as shown in Fig. 9.19. The logarithmic magnitude when the phase angle is $-180°$ is -15 dB, and therefore the gain margin is equal to 15 dB, as shown in Fig. 9.19.

The frequency response of a system can be graphically portrayed on the logarithmic-magnitude–phase-angle diagram. For the log-magnitude–phase diagram, the critical stability point is the 0 dB, $-180°$ point, and the gain margin and phase margin can be easily determined and indicated on the diagram. The log-magnitude–phase locus of

$$GH_1(j\omega) = \frac{1}{j\omega(j\omega + 1)(0.2j\omega + 1)} \tag{9.51}$$

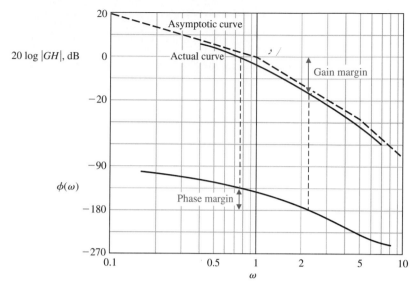

FIGURE 9.19
Bode diagram for
$GH_1(j\omega) = 1/j\omega(j\omega + 1)(0.2j\omega + 1)$.

is shown in Fig. 9.20. The indicated phase margin is 43°, and the gain margin is 15 dB. For comparison, the locus for

$$GH_2(j\omega) = \frac{1}{j\omega(j\omega + 1)^2} \tag{9.52}$$

is also shown in Fig. 9.20. The gain margin for GH_2 is equal to 5.7 dB, and the phase margin for GH_2 is equal to 20°. Clearly, the feedback system $GH_2(j\omega)$ is relatively less stable than the system $GH_1(j\omega)$. However, the question still remains: How much less stable is the system $GH_2(j\omega)$ in comparison to the system $GH_1(j\omega)$? In the following paragraph, we shall answer this question for a second-order system, and the general usefulness of the relation that we develop will depend on the presence of dominant roots.

Let us now determine the phase margin of a second-order system and relate the phase margin to the damping ratio ζ of an underdamped system. Consider the loop-transfer function of the system shown in Fig. 9.1 where

$$GH(s) = \frac{\omega_n^2}{s(s + 2\zeta\omega_n)}. \tag{9.53}$$

The characteristic equation for this second-order system is

$$s^2 + 2\zeta\omega_n s + \omega_n^2 = 0.$$

Therefore the closed-loop roots are

$$s = -\zeta\omega_n \pm j\omega_n\sqrt{1 - \zeta^2}.$$

The frequency domain form of Eq. (9.53) is

$$GH(j\omega) = \frac{\omega_n^2}{j\omega(j\omega + 2\zeta\omega_n)}. \tag{9.54}$$

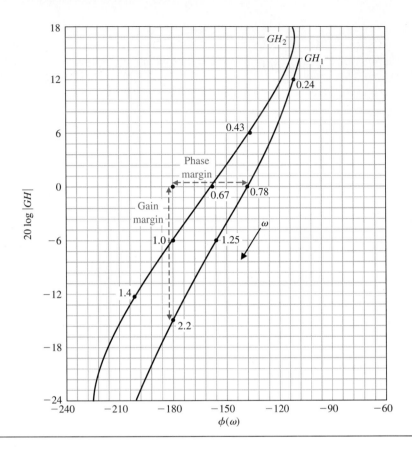

FIGURE 9.20
Log-magnitude–phase curve for GH_1 and GH_2.

The magnitude of the frequency response is equal to 1 at a frequency ω_c, and thus

$$\frac{\omega_n^2}{\omega_c(\omega_c^2 + 4\zeta^2\omega_n^2)^{1/2}} = 1. \tag{9.55}$$

Rearranging Eq. (9.55), we obtain

$$(\omega_c^2)^2 + 4\zeta^2\omega_n^2(\omega_c^2) - \omega_n^4 = 0. \tag{9.56}$$

Solving for ω_c, we find that

$$\frac{\omega_c^2}{\omega_n^2} = (4\zeta^4 + 1)^{1/2} - 2\zeta^2.$$

The phase margin for this system is

$$\phi_{pm} = 180° - 90° - \tan^{-1}\left(\frac{\omega_c}{2\zeta\omega_n}\right)$$

$$= 90° - \tan^{-1}\left(\frac{1}{2\zeta}[(4\zeta^4 + 1)^{1/2} - 2\zeta^2]^{1/2}\right) \tag{9.57}$$

$$= \tan^{-1}\left(2\zeta\left[\frac{1}{(4\zeta^4 + 1)^{1/2} - 2\zeta^2}\right]^{1/2}\right).$$

Equation (9.57) is the relationship between the damping ratio ζ and the phase margin ϕ_{pm} that provides a correlation between the frequency response and the time response. A plot of ζ versus ϕ_{pm} is shown in Fig. 9.21. The actual curve of ζ versus ϕ_{pm} can be approximated by the dashed colored line shown in Fig. 9.21. The slope of the linear approximation is equal to 0.01, and therefore an approximate linear relationship between the damping ratio and the phase margin is

$$\zeta = 0.01\phi_{pm}, \tag{9.58}$$

where the phase margin is measured in degrees. This approximation is reasonably accurate for $\zeta \leq 0.7$, and is a useful index for correlating the frequency response with the transient performance of a system. Equation (9.58) is a suitable approximation for a second-order system and may be used for higher-order systems if one can assume that the transient response of the system is primarily due to a pair of dominant underdamped roots. The approximation of a higher-order system by a dominant second-order system is a useful approximation indeed! Although it must be used with care, control engineers find this approach to be a simple, yet fairly accurate, technique of setting the specifications of a control system.

Therefore, for the system with a loop transfer function

$$GH(j\omega) = \frac{1}{j\omega(j\omega + 1)(0.2j\omega + 1)}, \tag{9.59}$$

we found that the phase margin was 43°, as shown in Fig. 9.19. Thus the damping ratio is approximately

$$\zeta \simeq 0.01\phi_{pm} = 0.43. \tag{9.60}$$

Then the peak response to a step input for this system is approximately

$$M_{p_t} = 1.22, \tag{9.61}$$

as obtained from Fig. 5.8 for $\zeta = 0.43$.

FIGURE 9.21
Damping ratio versus phase margin for a second-order system.

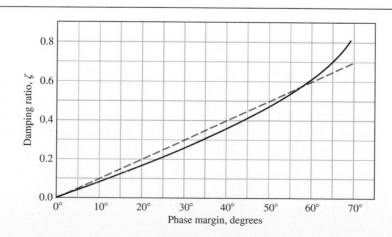

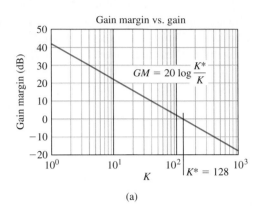

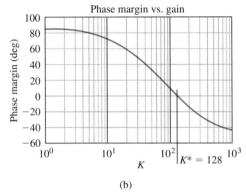

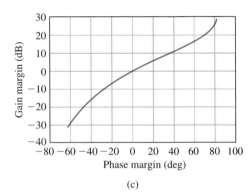

FIGURE 9.22
(a) Gain margin versus gain K, (b) phase margin versus gain K, (c) gain margin versus phase margin.

It is feasible to develop a computer program to calculate and plot phase margin and gain margin versus the gain K for a specified $GH(j\omega)$. One such program is available for use with gain plots[†] (see Section 7.7).

Consider the system of Fig. 9.1 with

$$GH(s) = \frac{K}{s(s + 4)^2}.$$

The gain for which the system is marginally stable is $K = K^* = 128$. The gain margin and the phase margin plotted versus K are shown in Figs. 9.22(a) and (b), respectively. The gain margin is plotted versus the phase margin as shown in Fig. 9.22(c). Note that either the phase margin or the gain margin is a suitable measure of the performance of the system. We will normally emphasize phase margin as a frequency-domain specification.

The phase margin of a system is a quite suitable frequency response measure for indicating the expected transient performance of a system. Another useful index of performance in the frequency domain is $M_{p\omega}$, the maximum magnitude of the closed-loop frequency response, and we shall now consider this practical index.

[†]Software is available from professors T. Kurfess and M. Nagurka, Carnegie-Mellon University, Pittsburgh, PA 15213.

9.5 TIME-DOMAIN PERFORMANCE CRITERIA SPECIFIED IN THE FREQUENCY DOMAIN

The transient performance of a feedback system can be estimated from the closed-loop frequency response. The *closed-loop frequency response* is the frequency response of the closed-loop transfer function $T(j\omega)$. The open- and closed-loop frequency responses for a single-loop system are related as follows:

$$\frac{C(j\omega)}{R(j\omega)} = T(j\omega) = \frac{G(j\omega)}{1 + GH(j\omega)}. \tag{9.62}$$

The Nyquist criterion and the phase margin index are defined for the open-loop transfer function $GH(j\omega)$. However, as we found in Section 8.2, the maximum magnitude of the closed-loop frequency response can be related to the damping ratio of a second-order system of

$$M_{p\omega} = |T(\omega_r)| - (2\zeta\sqrt{1 - \zeta^2})^{-1}, \qquad \zeta < 0.707. \tag{9.63}$$

This relation is graphically portrayed in Fig. 8.11. Because this relationship between the closed-loop frequency response and the transient response is a useful one, we would like to be able to determine $M_{p\omega}$ from the plots completed for the investigation of the Nyquist criterion. That is, it is desirable to be able to obtain the closed-loop frequency response (Eq. 9.62) from the open-loop frequency response. Of course, we could determine the closed-loop roots of $1 + GH(s)$ and plot the closed-loop frequency response. However, once we have invested all the effort necessary to find the closed-loop roots of a characteristic equation, then a closed-loop frequency response is not necessary.

The relation between the closed-loop and open-loop frequency response is easily obtained by considering Eq. (9.62) when $H(j\omega) = 1$. If the system is not in fact a unity feedback system where $H(j\omega) = 1$, we will simply redefine the system output to be equal to the output of $H(j\omega)$. Then Eq. (9.62) becomes

$$T(j\omega) = M(\omega)e^{j\phi(\omega)} = \frac{G(j\omega)}{1 + G(j\omega)}. \tag{9.64}$$

The relationship between $T(j\omega)$ and $G(j\omega)$ is readily obtained in terms of complex variables utilizing the $G(j\omega)$-plane. The coordinates of the $G(j\omega)$-plane are u and v, and we have

$$G(j\omega) = u + jv. \tag{9.65}$$

Therefore the magnitude of the closed-loop response $M(\omega)$ is

$$M = \left| \frac{G(j\omega)}{1 + G(j\omega)} \right| = \left| \frac{u + jv}{1 + u + jv} \right| = \frac{(u^2 + v^2)^{1/2}}{((1 + u)^2 + v^2)^{1/2}}. \tag{9.66}$$

Squaring Eq. (9.66) and rearranging, we obtain

$$(1 - M^2)u^2 + (1 - M^2)v^2 - 2M^2u = M^2. \tag{9.67}$$

Dividing Eq. (9.67) by $(1 - M^2)$ and adding the term $[M^2/(1 - M^2)]^2$ to both sides of Eq. (9.67), we have

$$u^2 + v^2 - \frac{2M^2u}{1 - M^2} + \left(\frac{M^2}{1 - M^2}\right)^2 = \left(\frac{M^2}{1 - M^2}\right) + \left(\frac{M^2}{1 - M^2}\right)^2. \tag{9.68}$$

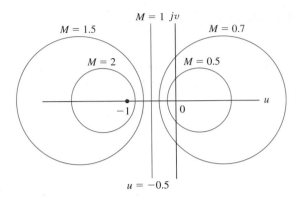

FIGURE 9.23
Constant *M* circles.

Rearranging, we obtain

$$\left(u - \frac{M^2}{1 - M^2} \right)^2 + v^2 = \left(\frac{M}{1 - M^2} \right)^2, \tag{9.69}$$

which is the equation of a circle on the (u, v)-plane with the center at

$$u = \frac{M^2}{1 - M^2}, \qquad v = 0.$$

The radius of the circle is equal to $|M/(1 - M^2)|$. Therefore we can plot several circles of constant magnitude M in the $[G(j\omega) = u + jv]$-plane. Several constant M circles are shown in Fig. 9.23. The circles to the left of $u = -1/2$ are for $M > 1$, and the circles to the right of $u = -1/2$ are for $M < 1$. When $M = 1$, the circle becomes the straight line $u = -1/2$, which is evident from inspection of Eq. (9.67).

The open-loop frequency response for a system is shown in Fig. 9.24 for two gain values where $K_2 > K_1$. The frequency response curve for the system with gain K_1 is tangent to magnitude circle M_1 at a frequency ω_{r_1}. Similarly, the frequency response curve for gain K_2 is tangent to magnitude circle M_2 at the frequency ω_{r_2}. Therefore the closed-loop frequency response magnitude curves are estimated as shown in Fig. 9.25. Hence we can

FIGURE 9.24
Polar plot of $G(j\omega)$
for two values of a
gain $(K_2 > K_1)$.

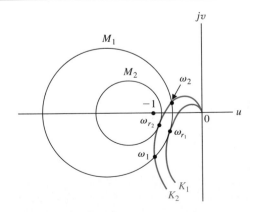

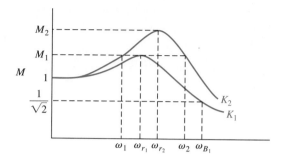

FIGURE 9.25
Closed-loop
frequency
response of
$T(j\omega) = G(j\omega)/$
$1 + G(j\omega)$. Note
that $K_2 > K_1$.

obtain the closed-loop frequency response of a system from the $(u + jv)$-plane. If the maximum magnitude, $M_{p\omega}$, is the only information desired, then it is sufficient to read this value directly from the polar plot. The maximum magnitude of the closed-loop frequency response, $M_{p\omega}$, is the value of the M circle that is tangent to the $G(j\omega)$-locus. The point of tangency occurs at the frequency ω_r, the resonant frequency. The complete closed-loop frequency response of a system can be obtained by reading the magnitude M of the circles that the $G(j\omega)$-locus intersects at several frequencies. Therefore the system with a gain $K = K_2$ has a closed-loop magnitude M_1 at the frequencies ω_1 and ω_2. This magnitude is read from Fig. 9.24 and is shown on the closed-loop frequency response in Fig. 9.25. The *bandwidth* for K_1 is shown as ω_{B_1}.

It may be empirically shown that the crossover frequency on the open-loop Bode diagram, ω_c, is related to the closed-loop system bandwidth, ω_B, by the approximation $\omega_B = 1.6\omega_c$ for ζ in the range 0.2 to 0.8.

In a similar manner, we can obtain circles of constant closed-loop phase angles. Thus, for Eq. (9.64), the angle relation is

$$\phi = \underline{/T(j\omega)} = \underline{/(u + jv)/(1 + u + jv)}$$

$$= \tan^{-1}\left(\frac{v}{u}\right) - \tan^{-1}\left(\frac{v}{1 + u}\right). \tag{9.70}$$

Taking the tangent of both sides and rearranging, we have

$$u^2 + v^2 + u - \frac{v}{N} = 0, \tag{9.71}$$

where $N = \tan\phi = $ constant. Adding the term $1/4[1 + (1/N^2)]$ to both sides of the equation and simplifying, we obtain

$$(u + 0.5)^2 + \left(v - \frac{1}{2N}\right)^2 = \frac{1}{4}\left(1 + \frac{1}{N^2}\right), \tag{9.72}$$

which is the equation of a circle with its center at $u = -0.5$ and $v = +(1/2N)$. The radius of the circle is equal to $1/2[1 + (1/N^2)]^{1/2}$. Therefore the constant phase angle curves can be obtained for various values of N in a manner similar to the M circles.

The constant M and N circles can be used for analysis and design in the polar plane. However, it is much easier to obtain the Bode diagram for a system, and it would be preferable if the constant M and N circles were translated to a logarithmic gain phase. N. B. Nichols transformed the constant M and N circles to the log-magnitude–phase diagram,

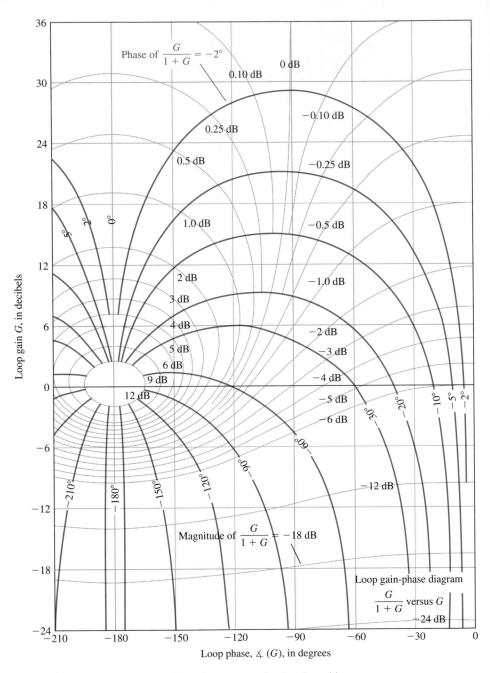

FIGURE 9.26 Nichols chart. The phase curves for the closed-loop system are shown in color.

and the resulting chart is called the *Nichols chart* [3, 7]. The M and N circles appear as contours on the Nichols chart shown in Fig. 9.26. The coordinates of the log-magnitude–phase diagram are the same as those used in Section 8.5. However, superimposed on the log-magnitude–phase plane we find constant M and N lines. The constant M lines are given in decibels and the N lines in degrees. An example will illustrate the use of the Nichols chart to determine the closed-loop frequency response.

EXAMPLE 9.7   Stability using the Nichols chart

Consider a feedback system with a loop transfer function

$$G(j\omega) = \frac{1}{j\omega(j\omega + 1)(0.2j\omega + 1)}. \tag{9.73}$$

The $G(j\omega)$-locus is plotted on the Nichols chart and is shown in Fig. 9.27. The maximum magnitude, $M_{p\omega}$, is equal to $+2.5$ dB and occurs at a frequency $\omega_r = 0.8$. The closed-loop phase angle at ω_r is equal to $-72°$. The 3-dB closed-loop bandwidth where the closed-loop magnitude is -3 dB is equal to $\omega_B = 1.33$, as shown in Fig. 9.27. The closed-loop phase angle at ω_B is equal to $-142°$. ∎

FIGURE 9.27
Nichols diagram for
$G(j\omega) = 1/j\omega(j\omega + 1)(0.2j\omega + 1)$.
Three points on curve are shown for $\omega = 0.5$, 0.8, and 1.35, respectively.

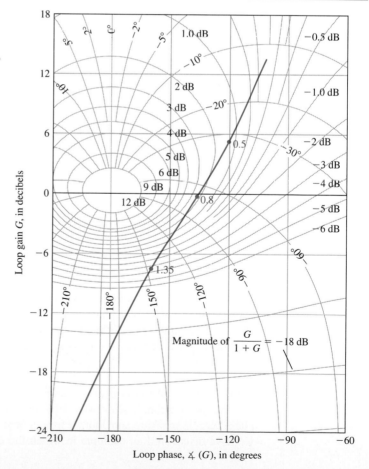

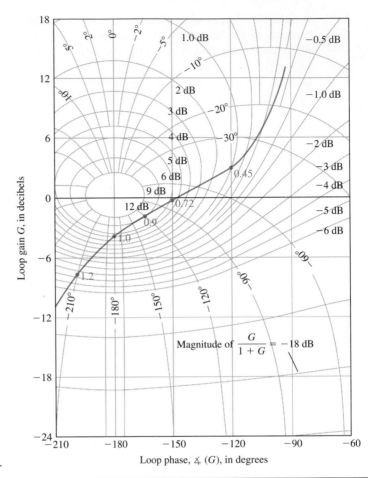

FIGURE 9.28
Nichols diagram for
$G(j\omega) = 0.64/j\omega[(j\omega)^2 + j\omega + 1]$.

EXAMPLE 9.8   **Third-order system**

Let us consider a system with an open-loop transfer function

$$G(j\omega) = \frac{0.64}{j\omega[(j\omega)^2 + j\omega + 1]}, \qquad (9.74)$$

where $\zeta = 0.5$ for the complex poles and $H(j\omega) = 1$. The Nichols diagram for this system is shown in Fig. 9.28. The phase margin for this system as it is determined from the Nichols chart is 30°. On the basis of the phase, we estimate the system damping ratio as $\zeta = 0.30$. The maximum magnitude is equal to $+9$ dB occurring at a frequency $\omega_r = 0.88$. Therefore

$$20 \log M_{p\omega} = 9 \text{ dB} \qquad \text{or} \qquad M_{p\omega} = 2.8.$$

Utilizing Fig. 8.11 to estimate the damping ratio, we find that $\zeta \approx 0.175$.

We are confronted with two conflicting damping ratios, where one is obtained from a phase margin measure and another from a peak frequency response measure. In this case, we have discovered an example in which the correlation between the frequency domain and the time domain is unclear and uncertain. This apparent conflict is caused by the nature of the $G(j\omega)$-locus, which slopes rapidly toward the 180° line from the 0-dB axis. If we determine the roots of the characteristic equation for $1 + GH(s)$, we obtain

$$q(s) = (s + 0.77)(s^2 + 0.225s + 0.826) = 0. \tag{9.75}$$

The damping ratio of the complex conjugate roots is equal to 0.124, where the complex roots do not dominate the response of the system. Therefore the real root will add some damping to the system and one might estimate the damping ratio as being approximately the value determined from the $M_{p\omega}$ index; that is, $\zeta = 0.175$. A designer must use the frequency-domain to time-domain correlations with caution. However, one is usually safe if the lower value of the damping ratio resulting from the phase margin and the $M_{p\omega}$ relation is utilized for analysis and design purposes. ∎

The Nichols chart can be used for design purposes by altering the $G(j\omega)$-locus in a suitable manner in order to obtain a desirable phase margin and $M_{p\omega}$. The system gain K is readily adjusted to provide a suitable phase margin and $M_{p\omega}$ by inspecting the Nichols chart. For example, let us reconsider Example 9.8, where

$$G(j\omega) - \frac{K}{j\omega[(j\omega)^2 + j\omega + 1]}. \tag{9.76}$$

The $G(j\omega)$-locus on the Nichols chart for $K = 0.64$ is shown in Fig. 9.28. Let us determine a suitable value for K so that the system damping ratio is greater than 0.30. Examining Fig. 8.11, we find that it is required that $M_{p\omega}$ be less than 1.75 (4.9 dB). From Fig. 9.28, we find that the $G(j\omega)$-locus will be tangent to the 4.9-dB curve if the $G(j\omega)$-locus is lowered by a factor of 2.2 dB. Therefore K should be reduced by 2.2 dB or the factor antilog $(2.2/20) = 1.28$. Thus the gain K must be less than $0.64/1.28 = 0.50$ if the system damping ratio is to be greater than 0.30.

9.6 SYSTEM BANDWIDTH

The bandwidth of the closed-loop control system is an excellent measurement of the range of fidelity of response of the system. In systems where the low frequency magnitude is 0 dB on the Bode diagram, the bandwidth is measured at the -3-dB frequency. The speed of response to a step input will be roughly proportional to ω_B and the settling time is inversely proportional to ω_B. Thus we seek a large bandwidth consistent with reasonable system components [12].

Consider the following two closed-loop system transfer functions:

$$T_1(s) = \frac{1}{s + 1}$$

and

$$T_2(s) = \frac{1}{5s + 1}. \tag{9.77}$$

The frequency response of the two systems is contrasted in part (a) of Fig. 9.29, and the step response of the systems is shown in part (b). Also, the response to a ramp is shown in

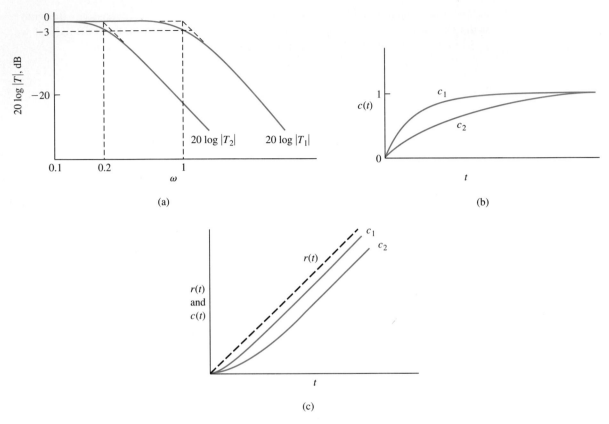

FIGURE 9.29   Response of two first-order systems.

part (c) of that figure. Clearly, the system with the larger bandwidth provides the faster step response and higher fidelity ramp response.

Consider the two second-order systems with closed-loop transfer functions:

$$T_3(s) = \frac{100}{s^2 + 10s + 100}$$

and

$$T_4(s) = \frac{900}{s^2 + 30s + 900}. \tag{9.78}$$

Both systems have a ζ of 0.5. The frequency response of both closed-loop systems is shown in Fig. 9.30(a). The natural frequency is 10 and 30 for systems T_3 and T_4, respectively. The bandwidth is 15 and 40 for systems T_3 and T_4, respectively. Both systems have a 15% overshoot, but T_4 has a peak time of 0.12 second compared to 0.36 for T_3, as shown in Fig. 9.30(b). Also, note that the settling time for T_4 is 0.37 second, while it is 0.9 second for T_3. Clearly, the system with a larger bandwidth provides a faster response. In general, we will pursue the design of systems with good stability and large bandwidth.

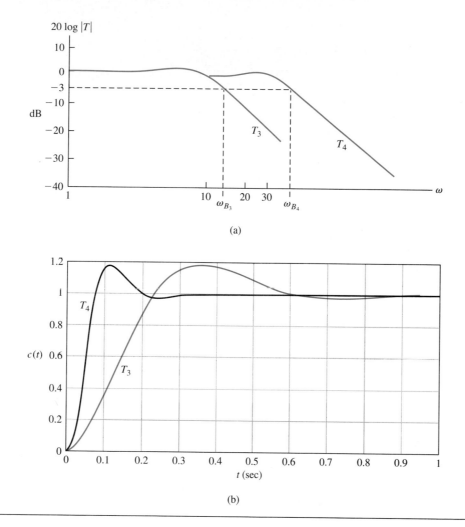

FIGURE 9.30
Response of two
second-order
systems.

9.7 THE STABILITY OF CONTROL SYSTEMS WITH TIME DELAYS

The Nyquist stability criterion has been discussed and illustrated in the previous sections for control systems whose transfer functions are rational polynomials of $j\omega$. There are many control systems that have a time delay within the closed loop of the system that affects the stability of the system. A *time delay* is the time interval between the start of an event at one point in a system and its resulting action at another point in the system. Fortunately, the Nyquist criterion can be utilized to determine the effect of the time delay on the relative stability of the feedback system. A pure time delay, without attenuation, is represented by the transfer function

$$G_d(s) = e^{-sT}, \tag{9.79}$$

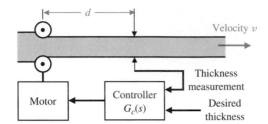

FIGURE 9.31
Steel rolling mill
control system.

where T is the delay time. The Nyquist criterion remains valid for a system with a time delay because the factor e^{-sT} does not introduce any additional poles or zeros within the contour. The factor adds a phase shift to the frequency response without altering the magnitude curve.

This type of time delay occurs in systems that have a movement of a material that requires a finite time to pass from an input or control point to an output or measured point [8, 9].

For example, a steel rolling mill control system is shown in Fig. 9.31. The motor adjusts the separation of the rolls so that the thickness error is minimized. If the steel is traveling at a velocity v, then the time delay between the roll adjustment and the measurement is

$$T = \frac{d}{v}.$$

Therefore, to have a negligible time delay, we must decrease the distance to the measurement and increase the velocity of the flow of steel. Usually, we cannot eliminate the effect of time delay and thus the loop transfer function is [10]

$$G(s)G_c(s)e^{-sT}. \tag{9.80}$$

However, one notes that the frequency response of this system is obtained from the loop transfer function

$$GH(j\omega) = GG_c(j\omega)e^{-j\omega T}. \tag{9.81}$$

The usual loop transfer function is plotted on the $GH(j\omega)$-plane and the stability ascertained relative to the -1 point. Alternatively, we can plot the Bode diagram including the delay factor and investigate the stability relative to the 0-dB, $-180°$ point. The delay factor $e^{-j\omega T}$ results in a phase shift

$$\phi(\omega) = -\omega T \tag{9.82}$$

and is readily added to the phase shift resulting from $GG_c(j\omega)$. Note that the angle is in radians in Eq. (9.82). An example will show the simplicity of this approach on the Bode diagram.

EXAMPLE 9.9   **Liquid level control system**

A level control system is shown in Fig. 9.32(a) and the block diagram in Fig. 9.32(b) [11]. The time delay between the valve adjustment and the fluid output is $T = d/v$. Therefore, if

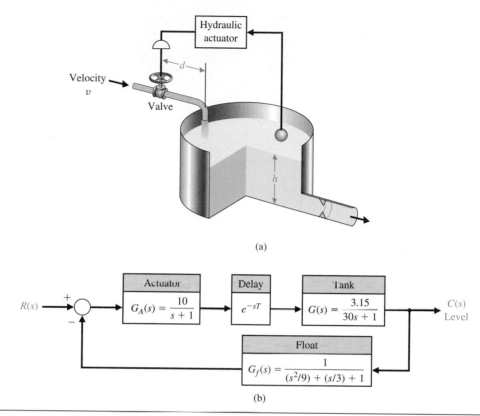

FIGURE 9.32
(a) Liquid level
control system.
(b) Block diagram.

the flow rate is 5 m³/s, the cross-sectional area of the pipe is 1 m², and the distance is equal to 5 m, then we have a time delay $T = 1$ second. The loop transfer function is then

$$GH(s) = G_A(s)G(s)G_f(s)e^{-sT} \tag{9.83}$$

$$= \frac{31.5}{(s + 1)(30s + 1)[(s^2/9) + (s/3) + 1]}e^{-sT}.$$

The Bode diagram for this system is shown in Fig. 9.33. The phase angle is shown both for the denominator factors alone and with the additional phase lag due to the time delay. The logarithmic gain curve crosses the 0-dB line at $\omega = 0.8$. Therefore the phase margin of the system without the pure time delay would be 40°. However, with the time delay added, we find that the phase margin is equal to $-3°$, and the system is unstable. Therefore the system gain must be reduced in order to provide a reasonable phase margin. To provide a phase margin of 30°, the gain would have to be decreased by a factor of 5 dB, to $K = 31.5/ 1.78 = 17.7$.

A time delay, e^{-sT}, in a feedback system introduces an additional phase lag and results in a less stable system. Therefore, as pure time delays are unavoidable in many systems, it is often necessary to reduce the loop gain in order to obtain a stable response. However, the

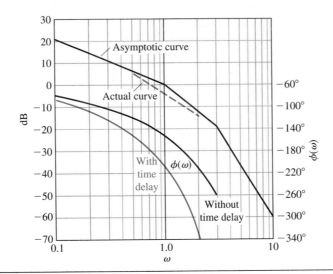

FIGURE 9.33
Bode diagram for
level control
system.

cost of stability is the resulting increase in the steady-state error of the system as the loop gain is reduced. ∎

9.8 DESIGN EXAMPLE: REMOTELY CONTROLLED RECONNAISSANCE VEHICLE

The use of remotely controlled vehicles for reconnaissance for U.N. peacekeeping missions may be an idea whose time has come. One concept of a roving vehicle is shown in Fig. 9.34(a), and a proposed speed control system is shown in Fig. 9.34(b). The desired speed $R(s)$ is transmitted by radio to the vehicle; the disturbance $D(s)$ represents hills and rocks. The goal is to achieve good overall control with low steady-state error and low-overshoot response to step commands, $R(s)$ [13].

First, to achieve low steady-state error for a unit step command, we calculate e_{ss} as

$$e_{ss} = \lim_{s \to 0} sE(s)$$

$$= \lim_{s \to 0} s \left[\frac{R(s)}{1 + G_c G(s)} \right] \qquad (9.84)$$

$$= \frac{1}{1 + G_c G(s)} = \frac{1}{1 + K/2}.$$

If we select $K = 20$, we will obtain a steady-state error of 9% of the magnitude of the input command. Using $K = 20$, we reformulate $G(s)$ for Bode diagram calculations, obtaining

$$G(s) = \frac{10(1 + s/2)}{(1 + s)(1 + s/2 + s^2/4)}. \qquad (9.85)$$

The calculations for $0 \leq \omega \leq 6$ provide the data summarized in Table 9.4. The Nichols diagram for $K = 20$ is shown in Fig. 9.35. Examining the Nichols chart, we find that $M_{p\omega}$ is 12 dB and the phase margin is 15 degrees. The step response of this system is under-damped and we predict an excessive overshoot of approximately 61%.

(a)

FIGURE 9.34
(a) Remotely controlled reconnaissance vehicle. (b) Speed control system. This vehicle could be used for United Nations peacekeeping missions such as in Somalia and Bosnia.

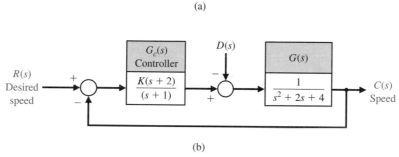

(b)

To reduce the overshoot to a step input, we can reduce the gain to achieve a predicted overshoot. To limit the overshoot to 25%, we select a desired ζ of the dominant roots as 0.4 (from Fig. 5.8) and thus require $M_{p\omega} = 1.35$ (from Fig. 8.11) or $20 \log M_{p\omega} = 2.6$ dB. To lower the gain, we will move the frequency response vertically down on the Nichols chart, as shown in Fig. 9.35. At $\omega_1 = 2.8$, we just intersect the 2.6-dB closed-loop curve. The reduction (vertical drop) in gain is equal to 13 dB or a factor of 4.5. Thus $K = 20/4.5 = 4.44$. For this reduced gain, the steady-state error is

$$ e_{ss} = \frac{1}{(1 + 4.4/2)} = 0.31, $$

or we have a 31% steady-state error.

The actual step response when $K = 4.44$, as shown in Fig. 9.36, has an overshoot of 32%. If we use a gain of 10, we have an overshoot of 48% with a steady-state error of 17%. The performance of the system is summarized in Table 9.5. As a suitable compromise, we

TABLE 9.4   Frequency Response Data for Design Example

ω	0	1.2	1.6	2.0	2.8	4	6
dB	20	18.4	17.8	16.0	10.5	2.7	−5.2
Degrees	0	−65	−86	−108	−142	−161	−170°

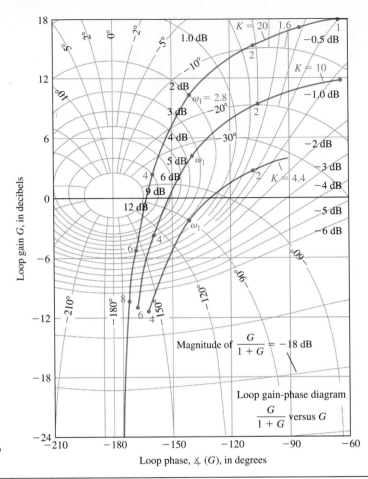

FIGURE 9.35
Nichol's diagram
for the design
example when
$K = 20$ and for two
reduced gains.

select $K = 10$ and draw the frequency response on the Nichols chart by moving the response for $K = 20$ down by $20 \log 2 = 6$ dB, as shown on Fig. 9.35.

Examining the Nichols chart for $K = 10$, we have $M_{p\omega} = 7$ dB, and a phase margin of 26 degrees. Thus we estimate a ζ for the dominant roots of 0.34, which should result in an overshoot to a step input of 30%. The actual response is recorded in Table 9.5. The band-

TABLE 9.5   Actual Response for Selected Gains

K	4.44	10	20
Percent overshoot	32.4	48.4	61.4
Settling time (seconds)	4.94	5.46	6.58
Peak time (seconds)	1.19	0.88	0.67
e_{ss}	31%	16.7%	9.1%

Note: Percent overshoot is defined by Eq. (5.12).

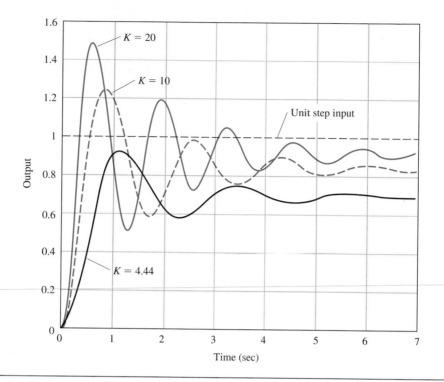

FIGURE 9.36
The response of
the system for
three values of K
for a unit step
input, $r(t)$.

width of the system is $\omega_B \cong 5$. Thus we predict a settling time

$$T_s = \frac{4}{\zeta \omega_n} = \frac{4}{(0.34)(\omega_B/1.4)} = 3.3 \text{ seconds,}$$

since $\omega_B \cong 1.4\,\omega_n$ for $\zeta = 0.34$. The actual settling time is approximately 5.4 seconds, as shown in Fig. 9.36.

The steady-state effect of a unit step disturbance can be determined by using the final value theorem with $R(s) = 0$, as follows:

$$c(\infty) = \lim_{s \to 0} s \left[\frac{G(s)}{1 + GG_c(s)} \right]\left(\frac{1}{s}\right) = \frac{1}{4 + 2K}. \tag{9.86}$$

Thus the unit disturbance is reduced by the factor $(4 + 2K)$. For $K = 10$, we have $c(\infty) = 1/24$, or the steady-state disturbance is reduced to 4% of the disturbance magnitude. Thus we have achieved a reasonable result with $K = 10$.

The best compromise design would be $K = 10$, since we achieve a compromise steady-state error of 16.7%. If the overshoot and settling time are excessive, then we need to reshape the $G(j\omega)$-locus on the Nichols chart by methods we will describe in Chapter 10.

9.9 STABILITY IN THE FREQUENCY DOMAIN USING *MATLAB*

We now approach the issue of stability using *Matlab* as a tool. This section revisits the Nyquist diagram, the Nichols chart, and the Bode diagram in our discussions on relative stability. Two examples will illustrate the frequency-domain design approach. We will

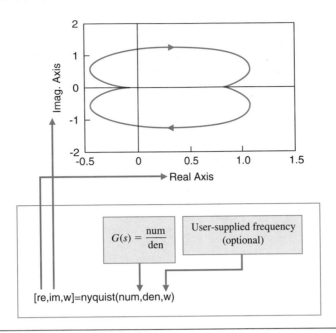

FIGURE 9.37
The **nyquist**
function.

make use of the frequency response of the closed-loop transfer function, $T(j\omega)$, as well as the loop transfer function $GH(j\omega)$. We also present an illustrative example that shows how to deal with a time delay in the system by utilizing a Padé approximation [6].

The *Matlab* functions covered in this section are nyquist, nichols, margin, pade, and ngrid.

It is generally more difficult to generate the Nyquist plot manually than the Bode diagram. However, we can use *Matlab* to generate the Nyquist plot rather easily. The Nyquist plot is generated with the nyquist function, as shown in Fig. 9.37. When nyquist is used without left-hand arguments, the Nyquist plot is automatically generated; otherwise, we must use the plot function to generate the plot using the vectors re and im.

FIGURE 9.38
The **nyquist**
function with
manual scaling.

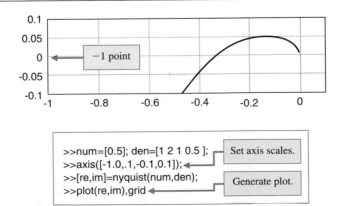

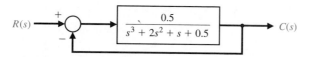

FIGURE 9.39
A closed-loop control system example for Nyquist and
Bode with relative stability.

One cautionary remark: Some time in the course of using the nyquist function, we may find that a Nyquist plot looks nontraditional or that some information appears to be missing. It may be necessary in these cases to use the axis function to override the automatic scaling and use the nyquist function with left-hand arguments in conjunction with the plot function. In this way, we can focus in on the −1 point region for our stability analysis, as illustrated in Fig. 9.38.

As discussed in Section 9.4, relative stability measures of *gain* and *phase* margins can be determined from both the Nyquist plot and the Bode diagram. The gain margin is a measure of how much the system gain would have to be increased for the $GH(j\omega)$ locus to pass through the $(-1, 0)$ point, thus resulting in an unstable system. The phase margin is a measure of the additional phase lag required before the system becomes unstable. Gain and phase margins can be determined from both the Nyquist plot and the Bode diagram.

Consider the system shown in Fig. 9.39. Relative stability can be determined from the Bode diagram using the margin function, which is shown in Fig. 9.40. The margin function

FIGURE 9.40
The **margin**
function.

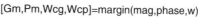

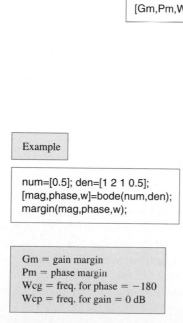

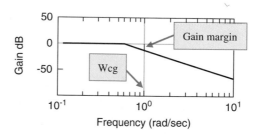

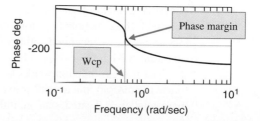

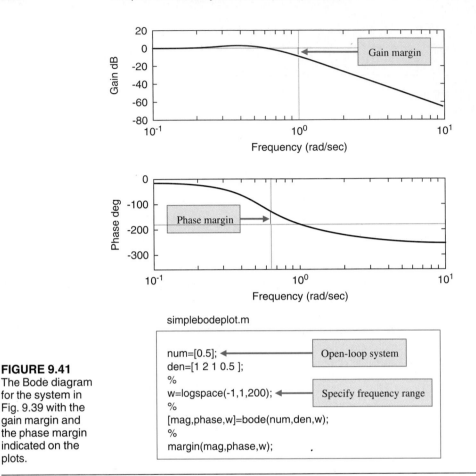

FIGURE 9.41
The Bode diagram for the system in Fig. 9.39 with the gain margin and the phase margin indicated on the plots.

is invoked in conjunction with the **bode** function to compute the gain and phase margins. If the **margin** function is invoked without left-hand arguments, the Bode diagram is automatically generated with the gain and phase margins labeled on the diagram. This is illustrated in Fig. 9.41 for the system shown in Fig. 9.39.

The script to generate the Nyquist plot for the system in Fig. 9.34 is shown in Fig. 9.42. In this case, the number of poles of $GH(s)$ with positive real parts is zero and the number of counterclockwise encirclements of -1 is zero; hence the closed-loop system is stable. We can also determine the gain margin and phase margin, as indicated in Fig. 9.42.

Nichols Chart.   Nichols charts can be generated using the **nichols** function, shown in Fig. 9.43. If the **nichols** function is invoked without left-hand arguments, the Nichols chart is automatically generated, otherwise one must use **nichols** in conjunction with the **plot** function. A Nichols chart grid is drawn on the existing plot with the **ngrid** function.

The **margin** function works best in conjunction with the **bode** function. It is possible to use the **margin** function after executing **nichols**. Unless a Bode plot with gain and phase margin labels is desired, one should invoke **margin** with left-hand arguments and place the

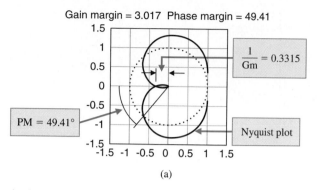

Gain margin = 3.017   Phase margin = 49.41

$$\frac{1}{Gm} = 0.3315$$

PM = 49.41°

Nyquist plot

(a)

nyquistplot.m

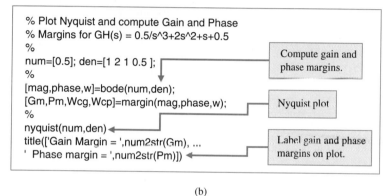

```
% Plot Nyquist and compute Gain and Phase
% Margins for GH(s) = 0.5/s^3+2s^2+s+0.5
%
num=[0.5]; den=[1 2 1 0.5 ];
%
[mag,phase,w]=bode(num,den);
[Gm,Pm,Wcg,Wcp]=margin(mag,phase,w);
%
nyquist(num,den)
title(['Gain Margin = ',num2str(Gm), ...
' Phase margin = ',num2str(Pm)])
```

Compute gain and phase margins.

Nyquist plot

Label gain and phase margins on plot.

FIGURE 9.42
(a) The Nyquist plot for the system in Fig. 9.39 with gain and phase margins.
(b) *Matlab* script.

(b)

FIGURE 9.43
The **nichols** function.

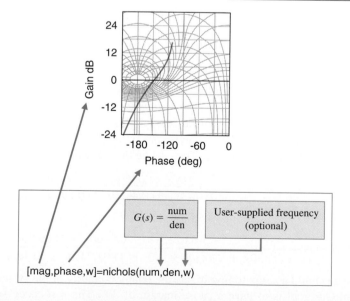

$$G(s) = \frac{num}{den}$$

User-supplied frequency (optional)

[mag,phase,w]=nichols(num,den,w)

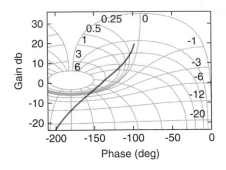

FIGURE 9.44
Nichols chart for
the system of
Eq. (9.87).

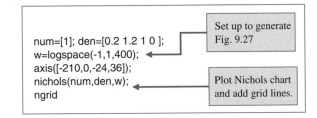

gain and phase margin values in the workspace. The Nichols chart, shown in Fig. 9.44, is
for the system

$$G(j\omega) = \frac{1}{j\omega(j\omega + 1)(0.2j\omega + 1)}. \qquad (9.87)$$

EXAMPLE 9.10   **Liquid level control system**

Consider a liquid level control system described by the block diagram shown in Fig. 9.32
on page 479 (see Example 9.9). Notice that this system has a time delay. The loop transfer
function is given by

$$GH(s) = \frac{31.5e^{-sT}}{(s + 1)(30s + 1)(s^2/9 + s/3 + 1)}. \qquad (9.88)$$

Since we want to use *Matlab* in our analysis, we need to change Eq. (9.88) in such a
way that $GH(s)$ has a transfer function form with polynomials in the numerator and the
denominator. To do this we, can make an approximation to e^{-sT} with the pade function,
shown in Fig. 9.45. For example, suppose our time delay is $T = 1$ second and we want a
second-order approximation $n = 2$. Then, using the pade function, we find that

$$e^{-s} \approx \frac{0.0743s^2 - 0.4460s + 0.8920}{0.0743s^2 + 0.4460s + 0.8920}. \qquad (9.89)$$

Substituting Eq. (9.89) into Eq. (9.88), we have

$$GH(s) \approx \frac{31.5(0.0743s^2 - 0.4460s + 0.8920)}{(s + 1)(30s + 1)(s^2/9 + s/3 + 1)(0.0743s^2 + 0.4460s + 0.8920)}.$$

Now we can build a script to investigate the relative stability of the system using the Bode
diagram. Our goal is to have a phase margin of 30°. The associated script is shown in
Fig. 9.46. To make the script interactive, we let the gain K (now set at $K = 31.5$) be

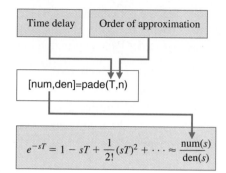

FIGURE 9.45
The **pade** function.

$$e^{-sT} = 1 - sT + \frac{1}{2!}(sT)^2 + \cdots \approx \frac{num(s)}{den(s)}$$

FIGURE 9.46
(a) Bode diagram
for the liquid level
control system.
(b) *Matlab* script.

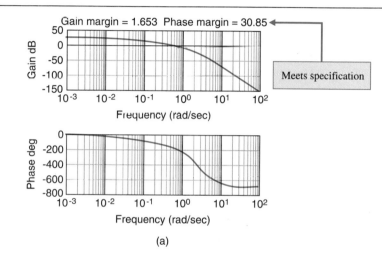

(a)

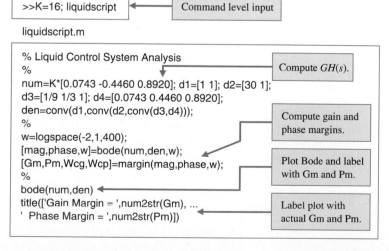

(b)

adjustable and defined outside the script at the command level. Then we set K and run the script to check the phase margin and iterate if necessary. The final selected gain is $K = 16$. Remember that we have utilized a second-order Padé approximation of the time delay in our analysis. ■

EXAMPLE 9.11   **Remotely controlled reconnaissance vehicle**

Consider the speed control system for a remotely controlled reconnaissance vehicle shown in Fig. 9.34 on page 481. The design objective is to achieve good control with low steady-state error and low overshoot to a step command. Building a script will allow us to perform many design iterations quickly and efficiently. First, we investigate the steady-state error specification. The steady-state error, e_{ss}, to a unit step command is

$$e_{ss} = \frac{1}{1 + K/2}. \tag{9.90}$$

The effect of the gain K on the steady-state error is clear from Eq. (9.90): If $K = 20$, the error is 9% of the input magnitude; if $K = 10$, the error is 17% of the input magnitude.

FIGURE 9.47
Remotely
controlled vehicle
(a) closed-loop
system Bode
diagram,
(b) *Matlab* script.

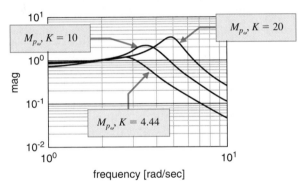

(a)

vehiclescript.m

```
w=logspace(0,1,200); K=20;        ◄——  Start with K = 20.
%
for i=1:3    ◄——
 numgc=K*[1 2]; dengc=[1 1];            Loop for three gains
 numg=[1]; deng=[1 2 4];               K = 20, 10, 4.44.
 [nums,dens]=series(numgc,dengc,numg,deng);
 [num,den]=cloop(nums,dens);
 [mag,phase,w]=bode(num,den,w);
 if i==1, mag1=mag; phase1=phase; K=10; end
 if i==2, mag2=mag; phase2=phase; K=4.44; end
 if i==3, mag3=mag; phase3=phase; end
end
%
loglog(w,mag1,'-',w,mag2,'-',w,mag3,'-'),grid
xlabel('frequency [rad/sec]'), ylabel('mag')
```

(b)

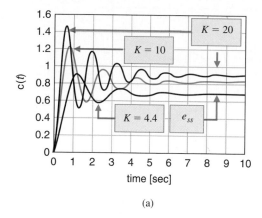

(a)

vehiclestep.m

```
t=[0:0.1:10]; K=20;
%
for i=1:3
  numgc=K*[1 2], dengc=[1 1];
  numg=[1]; deng=[1 2 4];
  [nums,dens]=series(numgc,dengc,numg,deng);
  [num,den]=cloop(nums,dens);
  [y,x]=step(num,den,t);
  if i==1, y1=y; K=10; end
  if i==2, y2=y; K=4.44; end
  if i==3, y3=y; end
end
%
plot(t,y1,'-',t,y2,'-',t,y3,'-'),grid
xlabel('time [sec]'), ylabel('c(t)')
```

Loop for three gains $K = 20, 10, 4.44$.

Compute step response.

FIGURE 9.48
Remotely
controlled vehicle
(a) step response,
(b) *Matlab* script.

(b)

Now we can investigate the overshoot specification in the frequency domain. Suppose we require that the percent overshoot be less than 50%. Solving

$$\text{P.O.} \approx 100 \, \exp^{-\zeta\pi/\sqrt{1-\zeta^2}} \leq 50$$

for ζ yields

$$\zeta \geq 0.215.$$

Referring to Fig. 8.11, we find that $M_{p\omega} \leq 2.45$. We must keep in mind that the information in Fig. 8.11 is for second-order systems only and can be used here only as a guideline. We now compute the closed-loop Bode diagram and check the values of $M_{p\omega}$. Any gain K for which $M_{p\omega} \leq 2.45$ may be a valid gain for our design, but we will have to investigate further to include step responses to check the actual overshoot. The script in Fig. 9.47 aids us in this task. In keeping with the spirit of the design steps in Section 9.8, we investigate further the gains $K = 20, 10$, and 4.44 (even though $M_{p\omega} > 2.45$ for $K = 20$). We can plot the step responses to quantify the overshoot, as shown in Fig. 9.48.

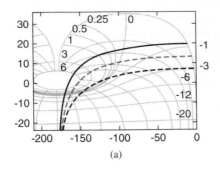

vehiclenichols.m

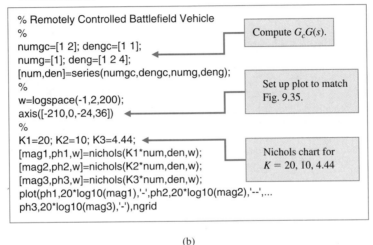

FIGURE 9.49
Remotely
controlled vehicle
(a) Nichols chart,
(b) *Matlab* script.

(b)

Additionally, we could have used a Nichols chart to aid the design process, as shown in Fig. 9.49.

The results of the analysis are summarized in Table 9.5 on page 482 for $K = 20$, 10, and 4.44. We choose $K = 10$ as our design gain. Then we obtain the Nyquist plot and check relative stability, as shown in Fig. 9.50. The gain margin is $GM = 49.56$ dB and the phase margin is $PM = 26.11°$. ■

9.10 SUMMARY

The stability of a feedback control system can be determined in the frequency domain by utilizing Nyquist's criterion. Furthermore, Nyquist's criterion provides us with two relative stability measures: (1) gain margin and (2) phase margin. These relative stability measures can be utilized as indices of the transient performance on the basis of correlations established between the frequency domain and the transient response. The magnitude and phase

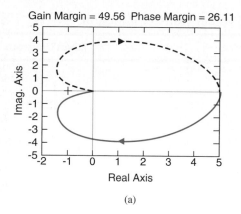

Gain Margin = 49.56 Phase Margin = 26.11

(a)

vehiclenyquist.m

```
% Remotely Controlled Vehicle
% Nyquist plot for K=10
%
numgc=10*[1 2]; dengc=[1 1];
numg=[1]; deng=[1 2 4];
[num,den]=series(numgc,dengc,numg,deng);
%
[mag,phase,w]=bode(num,den);
[Gm,Pm,Wcg,Wcp]=margin(mag,phase,w);
%
nyquist(num,den);
title(['Gain Margin = ',num2str(Gm), ...
'  Phase Margin = ',num2str(Pm)])
```

FIGURE 9.50
(a) Nyquist chart
for the remotely
controlled vehicle
with $K = 10$.
(b) *Matlab* script.

(b)

of the closed-loop system can be determined from the frequency response of the open-loop transfer function by utilizing constant magnitude and phase circles on the polar plot. Alternatively, we can utilize a log-magnitude–phase diagram with closed-loop magnitude and phase curves superimposed (called the Nichols chart) to obtain the closed-loop frequency response. A measure of relative stability, the maximum magnitude of the closed-loop frequency response, M_{p_ω}, is available from the Nichols chart. The frequency response, M_{p_ω}, can be correlated with the damping ratio of the time response and is a useful index of performance. Finally, a control system with a pure time delay can be investigated in a manner similar to that for systems without time delay. A summary of the Nyquist criterion, the relative stability measures, and the Nichols diagram are given in Table 9.6 for several transfer functions.

TABLE 9.6   Transfer Function Plots for Typical Transfer Function

$G(s)$	Polar Plot	Bode Diagram
1. $\dfrac{K}{s\tau_1 + 1}$		
2. $\dfrac{K}{(s\tau_1 + 1)(s\tau_2 + 1)}$		
3. $\dfrac{K}{(s\tau_1 + 1)(s\tau_2 + 1)(s\tau_3 + 1)}$		
4. $\dfrac{K}{s}$		

TABLE 9.6   *Continued*

Nichols Diagram	Root Locus	Comments

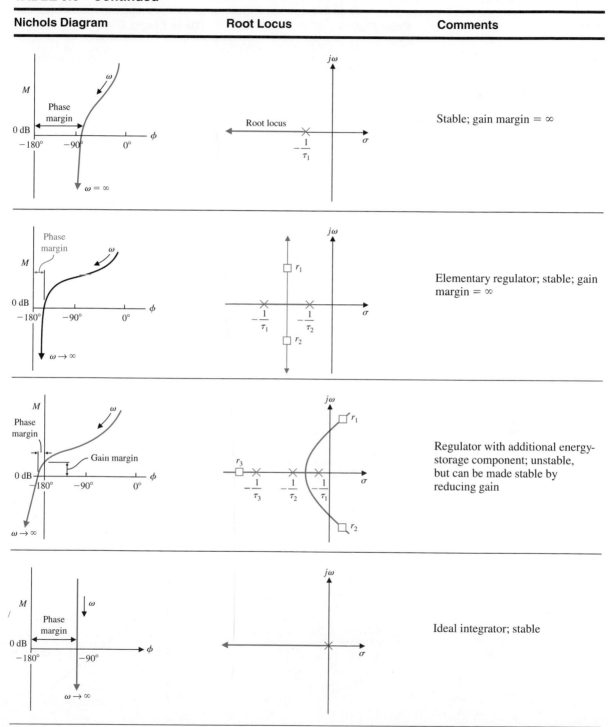

Stable; gain margin $= \infty$

Elementary regulator; stable; gain margin $= \infty$

Regulator with additional energy-storage component; unstable, but can be made stable by reducing gain

Ideal integrator; stable

(*continued*)

TABLE 9.6 *Continued*

G(s)	Polar Plot	Bode Diagram

5. $\dfrac{K}{s(s\tau_1 + 1)}$

6. $\dfrac{K}{s(s\tau_1 + 1)(s\tau_2 + 1)}$

7. $\dfrac{K(s\tau_a + 1)}{s(s\tau_1 + 1)(s\tau_2 + 1)}$

8. $\dfrac{K}{s^2}$

TABLE 9.6   *Continued*

Nichols Diagram	Root Locus	Comments

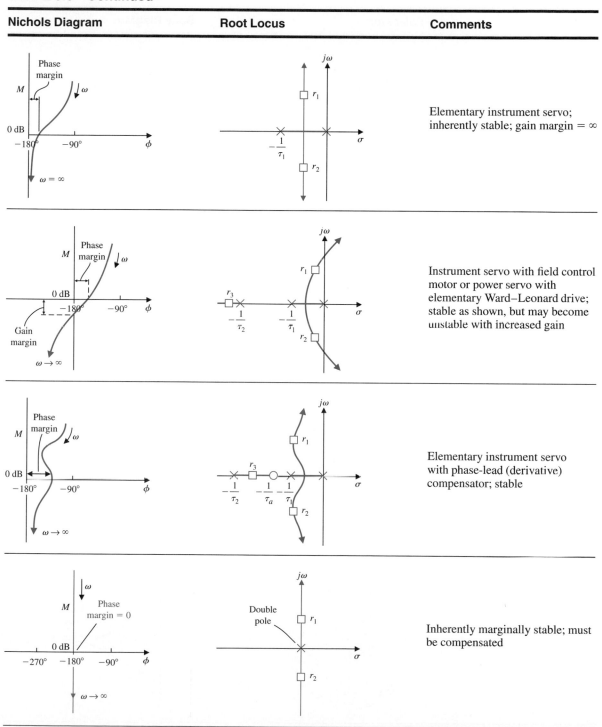

Elementary instrument servo; inherently stable; gain margin $= \infty$

Instrument servo with field control motor or power servo with elementary Ward–Leonard drive; stable as shown, but may become unstable with increased gain

Elementary instrument servo with phase-lead (derivative) compensator; stable

Inherently marginally stable; must be compensated

(continued)

TABLE 9.6   *Continued*

$G(s)$	Polar Plot	Bode Diagram

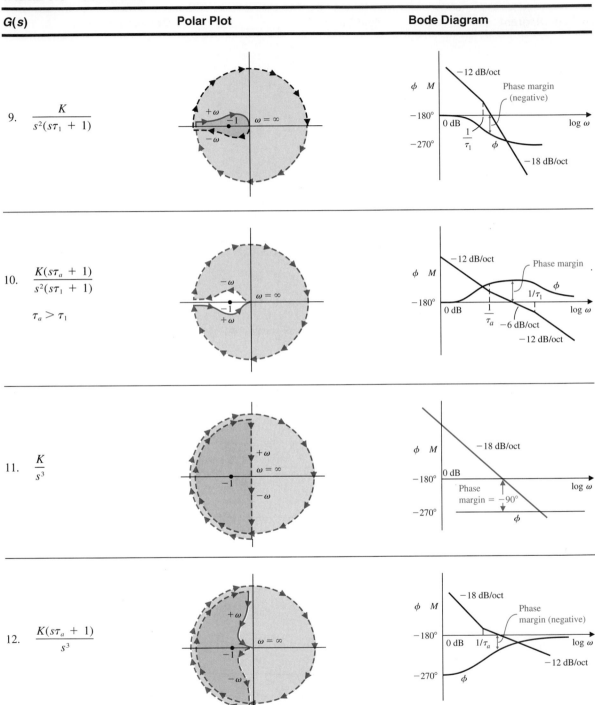

9. $\dfrac{K}{s^2(s\tau_1 + 1)}$

10. $\dfrac{K(s\tau_a + 1)}{s^2(s\tau_1 + 1)}$

    $\tau_a > \tau_1$

11. $\dfrac{K}{s^3}$

12. $\dfrac{K(s\tau_a + 1)}{s^3}$

TABLE 9.6   *Continued*

Nichols Diagram	Root Locus	Comments

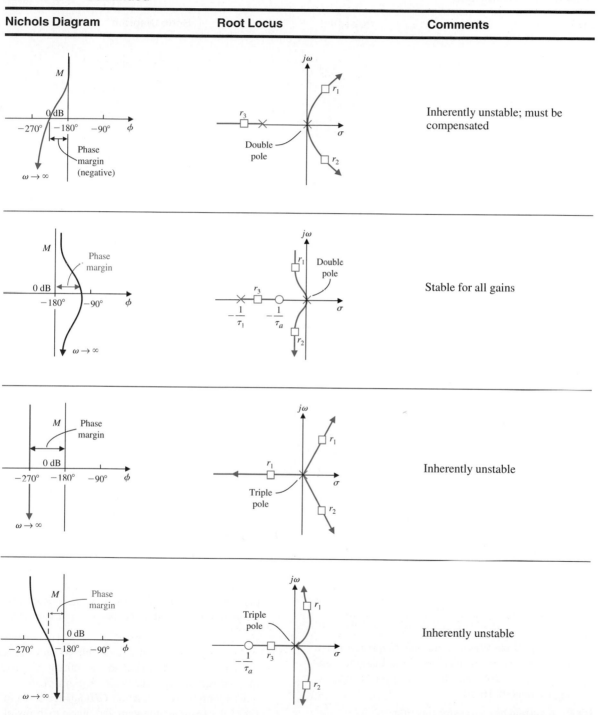

Inherently unstable; must be compensated

Stable for all gains

Inherently unstable

Inherently unstable

(*continued*)

TABLE 9.6 *Continued*

$G(s)$	Polar Plot	Bode Diagram
13. $\dfrac{K(s\tau_a + 1)(s\tau_b + 1)}{s^3}$		
14. $\dfrac{K(s\tau_a + 1)(s\tau_b + 1)}{s(s\tau_1 + 1)(s\tau_2 + 1)(s\tau_3 + 1)(s\tau_4 + 1)}$		
15. $\dfrac{K(s\tau_a + 1)}{s^2(s\tau_1 + 1)(s\tau_2 + 1)}$		

EXERCISES

E9.1   A system has a transfer function

$$G(s) = \frac{4(1 + s/3)}{s(1 + 2s)(1 + s/7 + s^2/49)}.$$

Plot the Bode diagram for the frequency range of 0.1 to 10. Show that the phase margin is approximately 30° and that the gain margin is approximately 16 dB.

E9.2   A system has a transfer function

$$G(s) = \frac{K(1 + s/5)}{s(1 + s/2)(1 + s/10)},$$

where $K = 6.14$. Using a computer or calculator program, show that the system crossover (0 dB) frequency is 3.5 rad/s and that the phase margin is 45°.

E9.3   An integrated circuit is available to serve as a feedback system to regulate the output voltage of a power supply. The Bode diagram of the required loop transfer function $GH(j\omega)$ is shown in Fig. E9.3 Estimate the gain and phase margins of the regulator.

Answer: $GM = 25$ dB, $PM = 75°$

TABLE 9.6 *Continued*

Nichols Diagram	Root Locus	Comments
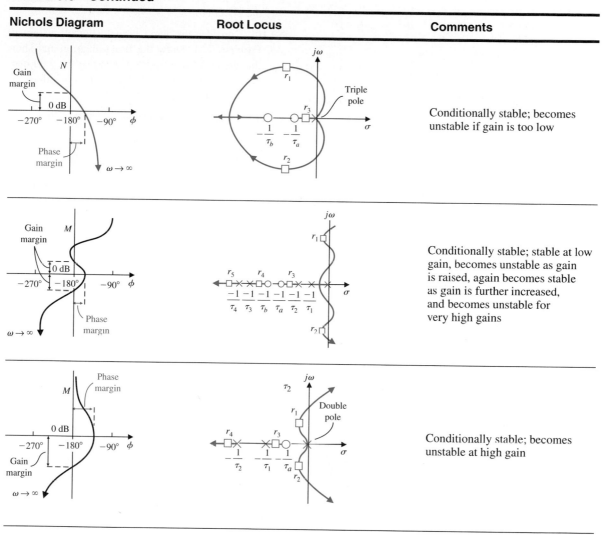		Conditionally stable; becomes unstable if gain is too low
		Conditionally stable; stable at low gain, becomes unstable as gain is raised, again becomes stable as gain is further increased, and becomes unstable for very high gains
		Conditionally stable; becomes unstable at high gain

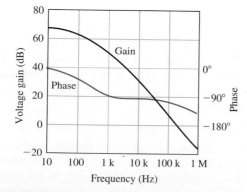

FIGURE E9.3 Power supply regulator.

E9.4 Consider a system with a loop transfer function

$$G(s) = \frac{100}{s(s + 10)},$$

where $H(s) = 1$. We wish to obtain a resonant peak $M_{p\omega} = 3.0$ dB for the closed-loop system. The peak occurs between 8 and 9 rad/s, and is only 1.25 dB at 8.66 rad/s. Plot the Nichols chart for the range of frequency from 8 to 15 rad/s. Show that the system gain needs to be raised by 4.6 dB, to 171. Determine the resonant frequency for the adjusted system.

E9.5 An integrated CMOS digital circuit can be represented by the Bode diagram shown in Fig. E9.5.

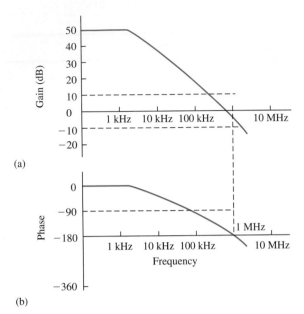

(a)

(b)

FIGURE E9.5   CMOS circuit.

(a) Find the gain and phase margins of the circuit.
(b) Estimate how much you would need to reduce
the system gain (dB) to obtain a phase margin
of 60°.

E9.6   A system has an open-loop transfer function

$$G(s) = \frac{K(s + 100)}{s(s + 10)(s + 40)}.$$

When $K = 500$, the system is unstable. Show that
if we reduce the gain to 50, the resonant peak is
3.5 dB. Find the phase margin of the system with
$K = 50$.

E9.7   A unity feedback system has a process

$$G(s) = \frac{K}{(s - 1)}.$$

Determine the range of K for which the system is
stable by drawing the polar plot.

E9.8   Consider a unity feedback system with

$$G(s) = \frac{K}{s(s + 1)(s + 2)}.$$

(a) For $K = 4$, show that the gain margin is 3.5 dB.
(b) If we wish to achieve a gain margin equal to
16 dB, determine the value of the gain K.

Answer: (b) $K = 0.98$

E9.9   For the system of E9.8, find the phase margin of
the system for $K = 3$.

E9.10   Consider the wind tunnel control system of
Problem 7.31. Draw the Bode diagram and show
that the phase margin is 25° and that the gain margin is 10 dB. Also show that the bandwidth of the
closed-loop system is 6 rad/s.

E9.11   Consider a unity feedback system with

$$G(s) = \frac{40(1 + 0.4s)}{s(1 + 2s)(1 + 0.24s + 0.04s^2)}.$$

(a) Using the *Matlab* program or equivalent, plot
the Bode diagram. (b) Find the gain margin and the
phase margin.

E9.12   A closed-loop system, as shown in Fig. 9.1, has
$H(s) = 1$ and

$$G(s) = \frac{K}{s(\tau_1 s + 1)(\tau_2 s + 1)},$$

where $\tau_1 = 0.02$ and $\tau_2 = 0.2$ second. (a) Select a
gain K so that the steady-state error for a ramp input is 10% of the magnitude of the ramp function
A, where $r(t) = At, t \geq 0$. (b) Plot the Bode plot of
$G(s)$, and determine the phase and gain margins.
(c) Using the Nichols chart, determine the band-
width ω_B, the resonant peak $M_{p\omega}$, and the resonant
frequency ω_r of the closed-loop system.

Answers: (a) $K = 10$
  (b) $PM = 32°$, $GM = 15$ dB
  (c) $\omega_B = 10.3$, $M_{p\omega} = 1.84$, $\omega_r = 6.5$

E9.13   A unit feedback system has a process

$$G(s) = \frac{150}{s(s + 5)}.$$

(a) Find the maximum magnitude of the closed-
loop frequency response using the Nichols chart.
(b) Find the bandwidth and the resonant frequency
of this system. (c) Use these frequency measures
to estimate the overshoot of the system to a step
response.

Answers: (a) 7.5 dB, (b) $\omega_B = 19$, $\omega_r = 12.6$

E9.14   A Nichols chart is given in Fig. E9.14 for a system where $G(j\omega)$ is plotted. Using the following
table, find (a) the peak resonance $M_{p\omega}$ in dB; (b) the
resonant frequency ω_r; (c) the 3-dB bandwidth;

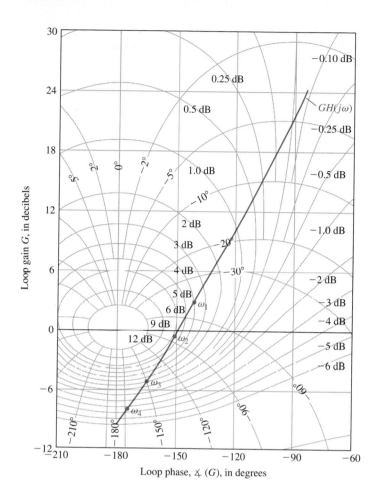

FIGURE E9.14
Nichols chart for
$G(j\omega)$.

Loop phase, $\angle (G)$, in degrees

(d) the phase margin of the system.

	ω_1	ω_2	ω_3	ω_4
rad/s	1	3	6	10

E9.15 Consider a unity feedback system with

$$G(s) = \frac{1000}{(s + 100)}.$$

Find the bandwidth of the open-loop system and the closed-loop system and compare the results.

Answers: open-loop $\omega_B = 100$, closed-loop $\omega_B = 1100$

E9.16 The pure time delay e^{-sT} may be approximated by a transfer function as

$$e^{-sT} \cong \frac{(1 - Ts/2)}{(1 + Ts/2)}$$

for $0 < \omega < 2/T$. Obtain the Bode diagram for the actual transfer function and the approximation for $T = 2$ for $0 < \omega < 1$.

E9.17 A unity feedback system has a plant

$$G(s) = \frac{K(s + 3)}{s(s + 1)(s + 5)}.$$

(a) Plot the Bode diagram and (b) determine the gain K required to obtain a phase margin of 40°. What is the steady-state error for a ramp input for the gain of part (b)?

E9.18 An actuator for a disk drive uses a shock mount to absorb vibrational energy at approximately 60 Hz [14]. The Bode diagram of $G(s)$ of the control system is shown in Fig. E9.18. (a) Find the expected percent overshoot for a step input for the closed-loop system, (b) estimate the bandwidth of

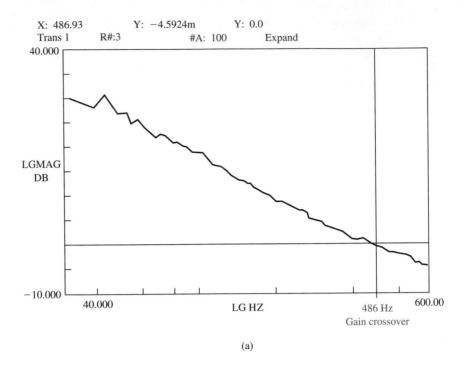

(a)

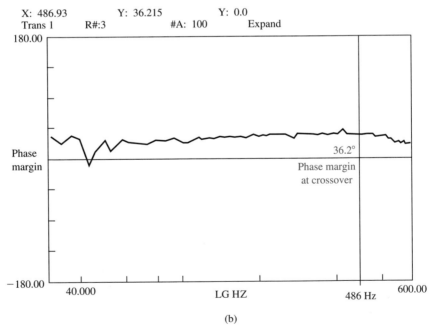

FIGURE E9.18
Bode diagram of
the disk drive $G(s)$.

(b)

the closed-loop system, and (c) estimate the settling time of the system.

E9.19 A unity feedback system with $G_c(s) = K$ has

$$G = (s)\frac{e^{-0.1s}}{(s + 4)}.$$

Select a gain K so that the phase margin of the system is 40°. Determine the gain margin for the selected gain, K.

E9.20 Consider a simple model of an automobile driver following another car on the highway at high speed. The model shown in Fig. E9.20 incorporates the driver's reaction time, T. One driver has $T = 1$ second and another has $T = 1.5$ seconds. Determine the time response, $c(t)$, of the system for both drivers for a step change in the command signal, $R(s) = -A/s$, due to the braking of the lead car.

E9.21 A unity feedback control system has a plant

$$G(s) = \frac{K}{s(s + 2)(s + 50)}.$$

Determine the phase margin, the crossover frequency, and the gain margin when $K = 1300$.

Answers: $PM = 16.6°$, $\omega_c = 4.9$, $GM = 4$ or 12 dB

E9.22 A unity feedback system has a plant

$$G(s) = \frac{K(s + 1)}{(s - 2)(s - 4)}.$$

(a) Using a Bode diagram for $K = 6$, determine the system phase margin. (b) Select a gain K so that the phase margin is at least 30°.

E9.23 Reconsider the system of E9.21 when $K = 438$. Determine the closed-loop system bandwidth, resonant frequency, and $M_{p\omega}$ using the Nichols chart.

Answers: $\omega_B = 4.25$ rad/s, $\omega_r = 2.7$, $M_{p\omega} = 1.7$

E9.24 A unity feedback system has a plant

$$G(s) = \frac{K}{(-1 + \tau s)},$$

where $K = \frac{1}{2}$ and $\tau = 1$. The polar plot for $G(j\omega)$ is shown in Fig. E9.24. Determine whether the system is stable by using the Nyquist criterion.

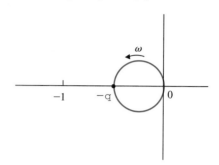

FIGURE E9.24

E9.25 A unity feedback system has a plant

$$G(s) = \frac{11.7}{s(1 + 0.05s)(1 + 0.1s)}.$$

Determine the phase margin and the crossover frequency.

Answers: $PM = 27.7°$, $\omega_c = 8.31$ rad/s

E9.26 For the system of E9.25, determine $M_{p\omega}$, ω_r, and ω_B for the closed-loop frequency response by using the Nichols chart.

E9.27 A unity feedback system has

$$G(s) = \frac{K}{s(s + 5)^2}.$$

Determine the maximum gain, K, for which the phase margin is at least 45° and the gain margin is at least 6 dB. What is the gain margin and phase margin for this value of K?

FIGURE E9.20

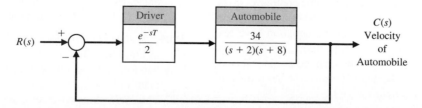

E9.28   A unity feedback system has

$$G(s) = \frac{K}{s(s + 0.2)}.$$

(a) Determine the phase margin of the system when $K = 0.16$. (b) Use the phase margin to estimate ζ and predict the overshoot. (c) Calculate the actual response for this second-order system, and compare the result with the part (b) estimate.

E9.29   A transfer function is

$$GH(s) = \frac{1}{s + 2}.$$

Using the contour in the s-plane shown in Fig. E9.29, determine the corresponding contour in the $F(s)$-plane ($B = -1 + j$).

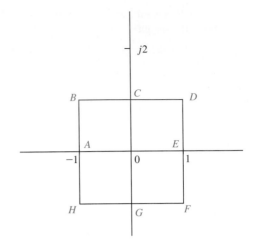

FIGURE E9.29

PROBLEMS

P9.1   For the polar plots of Problem P8.1, use the Nyquist criterion to ascertain the stability of the various systems. In each case, specify the values of N, P, and Z.

P9.2   Sketch the polar plots of the following loop transfer functions $GH(s)$, and determine whether the system is stable by utilizing the Nyquist criterion.

(a) $GH(s) = \dfrac{K}{s(s^2 + s + 4)}$

(b) $GH(s) = \dfrac{K(s + 2)}{s^2(s + 4)}$

If the system is stable, find the maximum value for K by determining the point where the polar plot crosses the u-axis.

P9.3   (a) Find a suitable contour Γ_s in the s-plane that can be used to determine whether all roots of the characteristic equation have damping ratios greater than ζ_1. (b) Find a suitable contour Γ_s in the s-plane that can be used to determine whether all the roots of the characteristic equation have real parts less than $s = -\sigma_1$. (c) Using the contour of part (b) and Cauchy's theorem, determine whether the following characteristic equation has roots with real parts less than $s = -1$:

$$q(s) = s^3 + 8s^2 + 30s + 36.$$

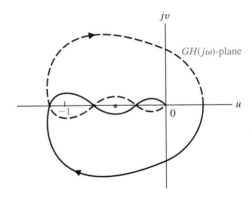

FIGURE P9.4   Polar plot of conditionally stable system.

P9.4   The polar plot of a conditionally stable system is shown in Fig. P9.4 for a specific gain K. (a) Determine whether the system is stable, and find the number of roots (if any) in the right-hand s-plane. The system has no poles of $GH(s)$ in the right half-plane. (b) Determine whether the system is stable if the -1 point lies at the colored dot on the axis.

P9.5   A speed control for a gasoline engine is shown in Fig. P9.5. Because of the restriction at the carburetor intake and the capacitance of the reduction manifold, the lag τ_t occurs and is equal to 1 second. The engine time constant τ_e is equal to $J/f = 3$ seconds. The speed measurement time constant is

FIGURE P9.5
Engine speed
control.

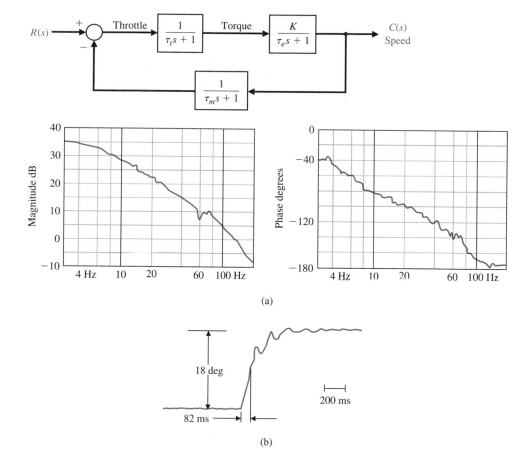

FIGURE P9.6
The MIT arm:
(a) frequency
response,
(b) position
response.

$\tau_m = 0.4$ second. (a) Determine the necessary gain K if the steady-state speed error is required to be less than 10% of the speed reference setting. (b) With the gain determined from part (a), utilize the Nyquist criterion to investigate the stability of the system. (c) Determine the phase and gain margins of the system.

P9.6 A direct-drive arm is an innovative mechanical arm in which no reducers are used between motors and their loads. Because the motor rotors are directly coupled to the loads, the drive systems have no backlash, small friction, and high mechanical stiffness, which are all important features for fast and accurate positioning and dexterous handling using sophisticated torque control.

The goal of the MIT direct-drive arm project is to achieve arm speeds of 10 m/s [15]. The arm has torques of up to 660 n-m (475 ft-lb). Feedback and a set of position and velocity sensors are used

with each motor. The frequency response of one joint of the arm is shown in Fig. P9.6(a). The two poles appear at 3.7 Hz and 68 Hz. Figure P9.6(b) shows the step response with position and velocity feedback used. The time constant of the closed-loop system is 82 ms. Develop the block diagram of the drive system and prove that 82 ms is a reasonable result.

P9.7 A vertical takeoff (VTOL) aircraft is an inherently unstable vehicle and requires an automatic stabilization system. An attitude stabilization system for the K-16B U.S. Army VTOL aircraft has been designed and is shown in block diagram form in Fig. P9.7 [16]. At 40 knots, the dynamics of the vehicle are approximately represented by the transfer function

$$G(s) = \frac{10}{(s^2 + 0.36)}.$$

FIGURE P9.7
VTOL aircraft
stabilization
system.

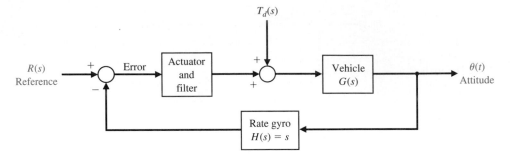

The actuator and filter is represented by the transfer function

$$G_1(s) = \frac{K_1(s + 7)}{(s + 3)}.$$

(a) Draw the Bode diagram of the loop transfer function $G_1(s)G(s)H(s)$ when the gain is $K_1 = 2$. (b) Determine the gain and phase margins of this system. (c) Determine the steady-state error for a

wind disturbance of $T_d(s) = 1/s$. (d) Determine the maximum amplitude of the resonant peak of the closed-loop frequency response and the frequency of the resonance. (e) Estimate the damping ratio of the system from M_{p_ω} and the phase margin.

P9.8 Electrohydraulic servomechanisms are utilized in control systems requiring a rapid response for a large mass. An electrohydraulic servomechanism can provide an output of 100 kW or greater [17]. A

FIGURE P9.8
(a) A servovalve
and actuator.
(Courtesy of Moog,
Inc., Industrial
Division). (b) Block
diagram.

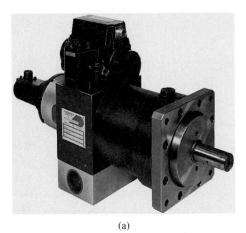

(a)

(b)

photo of a servovalve and actuator is shown in Fig. P9.8(a). The output sensor yields a measurement of actuator position, which is compared with V_{in}. The error is amplified and controls the hydraulic valve position, thus controlling the hydraulic fluid flow to the actuator. The block diagram of a closed-loop electrohydraulic servomechanism using pressure feedback to obtain damping is shown in Fig. P9.8(b) [17, 18]. Typical values for this system are $\tau = 0.02$ second and for the hydraulic system are $\omega_2 = 7(2\pi)$ and $\zeta_2 = 0.05$. The structural resonance ω_1 is equal to $10(2\pi)$ and the damping is $\zeta_1 = 0.05$. The loop gain is $K_A K_1 K_2 = 1.0$. (a) Sketch the Bode diagram and determine the phase margin of the system. (b) The damping of the system can be increased by drilling a small hole in the piston so that $\zeta_2 = 0.25$. Sketch the Bode diagram and determine the phase margin of this system.

P9.9 The key to future exploration and use of space is the reusable earth-to-orbit transport system, popularly known as the space shuttle. The shuttle, shown in Fig. P9.9(a), carries large payloads into space and returns them to earth for reuse [19]. The shuttle, roughly the size of a DC-9 with an empty weight of 75,000 kg, uses elevons at the trailing edge of the wing and a brake on the tail to control the flight. The block diagram of a pitch rate control system is shown in Fig. P9.9(b). The sensor is represented by a gain, $H(s) = 0.5$, and the vehicle by the transfer function

$$G(s) = \frac{0.30(s + 0.05)(s^2 + 1600)}{(s^2 + 0.05s + 16)(s + 70)}.$$

The controller $G_c(s)$ can be a gain or any suitable transfer function. (a) Draw the Bode diagram of the system when $G_c(s) = 2$ and determine the stability margin. (b) Draw the Bode diagram of the system when

$$G_c(s) = K_1 + K_2/s \quad \text{and} \quad K_2/K_1 = 0.5.$$

The gain K_1 should be selected so that the gain margin is 10 dB.

P9.10 Machine tools are often automatically controlled as shown in Fig. P9.10. These automatic systems are often called numerical machine con-

FIGURE P9.9
(a) The Earth orbiting space shuttle *Columbia* against the blackness of space on June 22, 1983. The remote manipulator robot is shown with the cargo bay doors open in this top view, taken by a satellite. (b) Pitch rate control system. (Courtesy of NASA.)

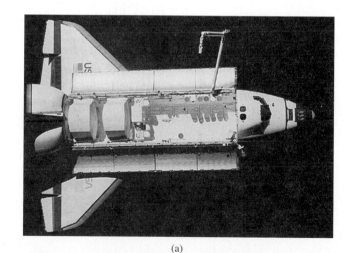

(a)

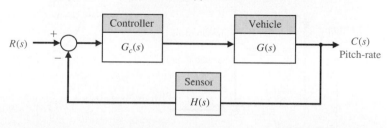

(b)

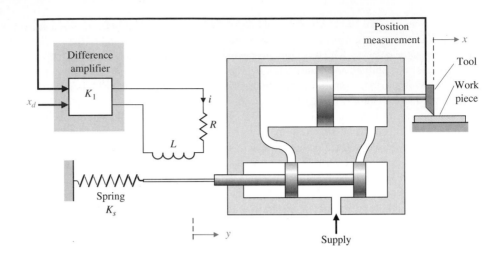

FIGURE P9.10
Machine tool
control.

trols [9, 20]. Considering one axis, the desired position of the machine tool is compared with the
actual position and is used to actuate a solenoid coil
and the shaft of a hydraulic actuator. The transfer
function of the actuator (see Table 2.7) is

$$G_a(s) = \frac{X(s)}{Y(s)} = \frac{K_a}{s(\tau_a s + 1)},$$

where $K_a = 1$ and $\tau_a = 0.4$ second. The output
voltage of the difference amplifier is

$$E_0(s) = K_1(X(s) - X_d(s)),$$

where $x_d(t)$ is the desired position input. The force
on the shaft is proportional to the current i so that

$F = K_2 i(t)$, where $K_2 = 3.0$. The spring constant
K_s is equal to 1.5 and $R = 0.1$ and $L = 0.2$.

(a) Determine the gain K_1 that results in a system with a phase margin of 30°. (b) For the gain K_1
of part (a), determine $M_{p\omega}$, ω_r, and the closed-loop
system bandwidth. (c) Estimate the percent over-
shoot of the transient response for a step input,
$X_d(s) = 1/s$, and the settling time.

P9.11   A control system for a chemical concentration
control system is shown in Fig. P9.11. The system
receives a granular feed of varying composition,
and it is desired to maintain a constant composition
of the output mixture by adjusting the feed-flow
valve. The transfer function of the tank and output

FIGURE P9.11
Chemical
concentration
control.

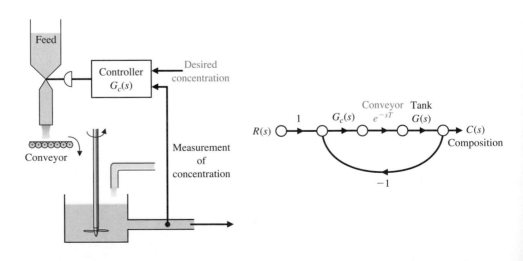

valve is

$$G(s) = \frac{5}{5s + 1},$$

and the controller is

$$G_c(s) = K_1 + \frac{K_2}{s}.$$

The transport of the feed along the conveyor requires a transport (or delay) time, $T = 1.5$ seconds. (a) Sketch the Bode diagram when $K_1 = K_2 = 1$, and investigate the stability of the system. (b) Sketch the Bode diagram when $K_1 = 0.1$ and $K_2 = 0.04$, and investigate the stability of the system. (c) When $K_1 = 0$, use the Nyquist criterion to calculate the maximum allowable gain K_2 for the system to remain stable.

P9.12   A simplified model of the control system for regulating the pupillary aperture in the human eye is shown in Fig. P9.12 [20]. The gain K represents

the pupillary gain and τ is the pupil time constant which is 0.5 second. The time delay T is equal to 0.1 second. The pupillary gain is equal to 4.0.
(a) Assuming the time delay is negligible, draw the Bode diagram for the system. Determine the phase margin of the system. (b) Include the effect of the time delay by adding the phase shift due to the delay. Determine the phase margin of the system with the time delay included.

P9.13   A controller is used to regulate the temperature of a mold for plastic part fabrication, as shown in Fig. P9.13. The value of the delay time is estimated as 1.2 seconds. (a) Utilizing the Nyquist criterion, determine the stability of the system for $K_a = K = 1$. (b) Determine a suitable value for K_a for a stable system when $K = 1$ that will yield a phase margin greater than 50°.

P9.14   Electronics and computers are being used to control automobiles. Figure P9.14 is an example of an automobile control system, the steering control

FIGURE P9.12
Human pupil
aperature control.

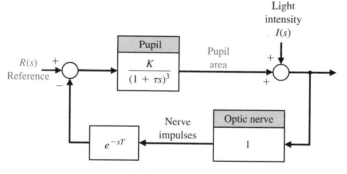

FIGURE P9.13
Temperature
controller.

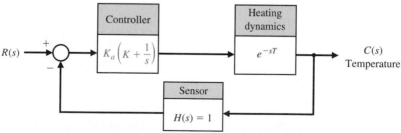

FIGURE P9.14
Automobile
steering control.

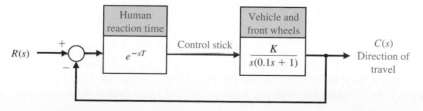

for a research automobile. The control stick is used for steering. A typical driver has a reaction time of $T = 0.2$ second.

(a) Using the Nichols chart, determine the magnitude of the gain K that will result in a system with a peak magnitude of the closed-loop frequency response $M_{p\omega}$ less than or equal to 2 dB.
(b) Estimate the damping ratio of the system based on (1) $M_{p\omega}$ and (2) the phase margin. Compare the results and explain the difference, if any.
(c) Determine the closed-loop 3-dB bandwidth of the system.

P9.15   Consider the automatic ship steering system discussed in Problem P8.11. The frequency response of the open-loop portion of the ship steering control system is shown in Fig. P8.11. The deviation of the tanker from the straight track is measured by radar and is used to generate the error signal as shown in Fig. P9.15. This error signal is used to control the rudder angle $\delta(s)$.

(a) Is this system stable? Discuss what an unstable ship steering system indicates in terms of the transient response of the system. Recall that the system under consideration is a ship attempting to follow a straight track.
(b) Is it possible to stabilize this system by lowering the gain of the transfer function $G(s)$?
(c) Is it possible to stabilize this system? Can you suggest a suitable feedback compensator?
(d) Repeat parts (a), (b), and (c) when switch S is closed.

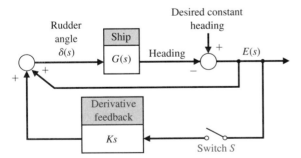

FIGURE P9.15   Automatic ship steering.

P9.16   An electric carrier that automatically follows a tape track laid out on a factory floor is shown in Fig. P9.16(a) [15]. Closed-loop feedback systems

are used to control the guidance and speed of the vehicle. The cart senses the tape path by means of an array of 16 phototransistors. The block diagram of the steering system is shown in Fig. P9.16(b). Select a gain K so that the phase margin is approximately 30°.

P9.17   The primary objective of many control systems is to maintain the output variable at the desired or reference condition when the system is subjected to a disturbance [23]. A typical chemical reactor control scheme is shown in Fig. P9.17. The disturbance is represented by $U(s)$ and the chemical process by G_3 and G_4. The controller is represented by G_1 and the valve by G_2. The feedback sensor is $H(s)$ and will be assumed to be equal to one. We will assume that G_2, G_3, and G_4 are all of the form

$$G_i(s) = \frac{K_i}{1 + \tau_i s},$$

where $\tau_3 = \tau_4 = 4$ seconds and $K_3 = K_4 = 0.1$. The valve constants are $K_2 = 20$ and $\tau_2 = 0.5$ second. It is desired to maintain a steady-state error less than 5% of the desired reference position.

(a) When $G_1(s) = K_1$, find the necessary gain to satisfy the error constant requirement. For this condition, determine the expected overshoot to a step change in the reference signal $r(t)$.
(b) If the controller has a proportional term plus an integral term so that $G_1(s) = K_1(1 + 1/s)$, determine a suitable gain to yield a system with an overshoot less than 30% but greater than 5%. For parts (a) and (b), use the approximation of damping ratio as a function of phase margin that yields $\zeta = 0.01 \ \phi_{pm}$. For these calculations, assume that $U(s) = 0$.
(c) Estimate the settling time of the step response of the system for the controller of parts (a) and (b).
(d) The system is expected to be subjected to a step disturbance $U(s) = A/s$. For ease, assume that the desired reference is $r(t) = 0$ when the system has settled. Determine the response of the system of part (b) to the disturbance.

P9.18   A model of a driver of an automobile attempting to steer a course is shown in Fig. P9.18, where $K = 5.3$. (a) Find the frequency response and the gain and phase margins when the reaction time T is zero. (b) Find the phase margin when the reaction time is 0.1 second. (c) Find the reaction time

(a)

FIGURE P9.16
(a) An electric
carrier vehicle
(photo courtesy
of Control
Engineering
Corporation).
(b) Block diagram.

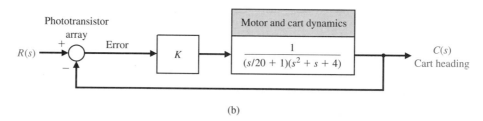

(b)

FIGURE P9.17
Chemical reactor
control.

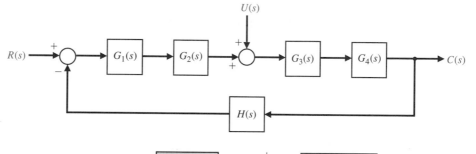

FIGURE P9.18
Automobile and
driver control.

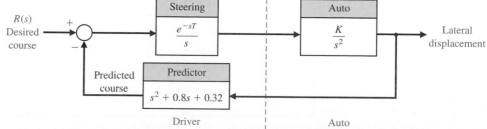

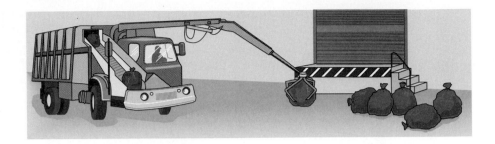

FIGURE P9.19
Waste collection
system.

that will cause the system to be borderline stable (phase margin = 0°).

P9.19 In the United States over $4.3 billion are spent annually for solid waste collection and disposal. One system, which uses a remote control pick-up arm for collecting waste bags, is shown in Fig. P9.19. The open-loop transfer of the remote pick-up arm is

$$GH(s) = \frac{0.3}{s(1 + 3s)(1 + s)}.$$

(a) Plot the Nichols chart and show that the gain margin is approximately 11.5 dB. (b) Determine the phase margin and the $M_{p\omega}$ for the closed loop. Also determine the closed-loop bandwidth.

P9.20 The Bell-Boeing V-22 Osprey Tiltrotor is both an airplane and a helicopter. Its advantage is the ability to rotate its engines to 90° from a vertical position, as shown in Fig. P7.33(a), for takeoffs and landings and then switch the engines to horizontal for cruising as an airplane. The altitude control system in the helicopter mode is shown in Fig. P9.20. (a) Obtain the frequency response of the system for $K = 100$. (b) Find the gain margin and the phase margin for this system. (c) Select a suitable gain K so that the phase margin is 40°. (Increase the gain above $K = 100$.) (d) Find the response $c(t)$ of the system for the gain selected in part (c).

P9.21 Consider a unity feedback system with

$$G(s) = \frac{K}{s(s + 1)(s + 4)}.$$

(a) Draw the Bode diagram accurately for $K = 4$. Determine (b) the gain margin, (c) the value of K required to provide a gain margin equal to 12 dB, and (d) the value of K to yield a steady-state error of 25% of the magnitude A for the ramp input $r(t) = At, t > 0$. Can this gain be utilized and achieve acceptable performance?

P9.22 The Nichols diagram for a $GH(j\omega)$ of a closed-loop system is shown in Fig. P9.22. The frequency for each point on the graph is given in the following table:

Point	1	2	3	4	5	6	7	8	9
ω	1	2.0	2.6	3.4	4.2	5.2	6.0	7.0	8.0

Determine (a) the resonant frequency, (b) the bandwidth, (c) the phase margin, and (d) the gain margin. (e) Estimate the overshoot and settling time of the response to a step input.

P9.23 A closed-loop system has a loop transfer function

$$GH(s) = \frac{K}{s(s + 1)(s + 10)}.$$

(a) Determine the gain K so that the phase margin is 60°. (b) For the gain K selected in part (a), determine the gain margin of the system.

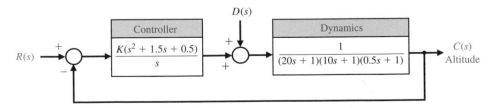

FIGURE P9.20
Tiltrotor aircraft
control.

$D(s)$

Controller
$$\frac{K(s^2 + 1.5s + 0.5)}{s}$$

Dynamics
$$\frac{1}{(20s + 1)(10s + 1)(0.5s + 1)}$$

$R(s)$

$C(s)$
Altitude

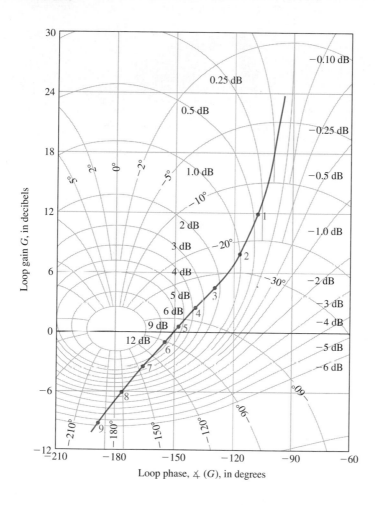

Loop gain G, in decibels

Loop phase, $\angle\,(G)$, in degrees

FIGURE P9.22
Nichols chart.

P9.24 A closed-loop with unity feedback system has a loop transfer function

$$G(s)H(s) = \frac{K(s + 4)}{s^2}.$$

(a) Determine the gain K so that the phase margin is 60°. (b) For the gain K selected in part (a), determine the gain margin. (c) Predict the bandwidth of the closed-loop system.

P9.25 A closed-loop system has the loop transfer function

$$GH(s) = \frac{Ke^{-0.1s}}{s}.$$

(a) Determine the gain K so that the phase margin is 60°. (b) Determine the gain K required if the phase margin required is 40°.

P9.26 A specialty machine shop is improving the efficiency of its surface grinding process [22]. The existing machine is sound mechanically but manually operated. Automating the machine will free the operator for other tasks and thus increase overall throughput of the machine shop. The grinding machine is shown in Fig. P9.26(a) with all three axes automated with motors and feedback systems. The control system for the y-axis is shown in Fig. P9.26(b). To achieve a low steady-state error to a ramp command, we choose $K = 10$. Draw the Bode diagram of the open-loop system and obtain the Nichols chart plot. Determine the gain and phase margin of the system and the bandwidth of the closed-loop system. Estimate the ζ of the system and the predicted overshoot and settling time.

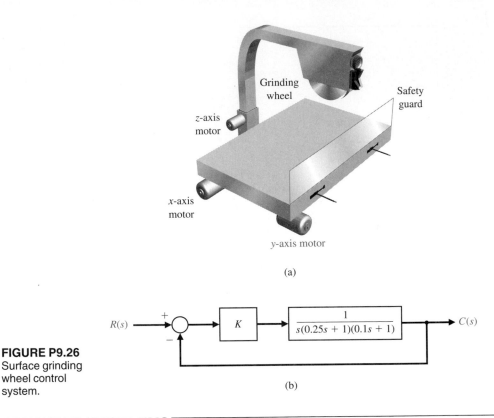

(a)

FIGURE P9.26
Surface grinding
wheel control
system.

(b)

ADVANCED PROBLEMS

AP9.1 Evolutionary spacecraft undergo substantial mass property and configuration changes during their operational lifetime [26]. For example, the inertias change considerably during operations. Consider the orientation control system shown in Fig. AP9.1.

(a) Plot the Bode diagram, and determine the gain and phase margins when $\omega_n^2 = 15{,}267$. (b) Repeat part (a) when $\omega_n^2 = 9500$. Note the effect of changing ω_n^2 by 38%.

AP9.2 Anesthesia is used in surgery to induce unconsciousness. One problem with drug-induced unconsciousness is large differences in patient responsiveness. Furthermore, the patient response changes during an operation. A model of drug-induced anesthesia control is shown in Fig. AP9.2. The proxy for unconsciousness is the arterial blood pressure.

(a) Plot the Bode diagram and determine the gain margin and the phase margin when $T = 0.05$ second. (b) Repeat part (a) when $T = 0.1$ second. Describe the effect of the 100% increase in the time delay T. (c) Using the phase margin, predict the overshoot for a step input for parts (a) and (b).

FIGURE AP9.1
Spacecraft-oriented control.

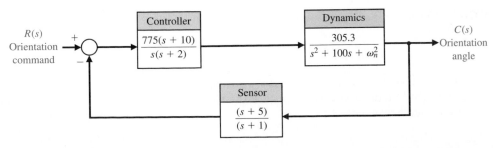

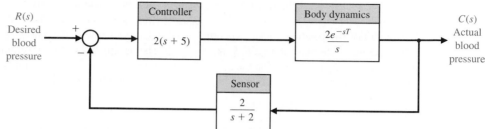

FIGURE AP9.2
Control of blood
pressure with
anesthesia.

AP9.3 Welding processes have been developed into an automated operation over the past decades. Weld quality features such as final metallurgy and joint mechanics typically are not measurable on-line for control. Therefore some indirect way of controlling the weld quality is necessary. A comprehensive approach to in-process control of welding includes both geometric features of the bead (such as the cross-sectional features of width, depth, and height) and thermal characteristics (such as the heat-affected zone width and cooling rate). The weld bead depth, which is the key geometric attribute of a major class of welds, is very difficult to measure directly, but a method to estimate the depth using temperature measurement has been developed [27]. A model of the weld control system is shown in Fig. AP9.3.

(a) Determine the phase margin and gain margin for the system when $K = 1$. (b) Repeat part (a) when $K = 1.5$. (c) Determine the bandwidth of the system for $K = 1$ and $K = 1.5$ by using the Nichols

chart. (d) Predict the settling time of a step response for $K = 1$ and $K = 1.5$.

AP9.4 The control of a paper-making machine is quite complex [28]. The goal is to deposit the proper amount of fiber suspension (pulp) at the right speed and in a uniform way. Dewatering, fiber deposition, rolling, and drying then take place in sequence. Control of the paper weight per unit area is very important. For the control system shown in Fig. AP9.4, select K so that the phase margin $\geq 40°$ and the gain margin ≥ 10 dB. Plot the step response for the selected gain. Determine the bandwidth of the closed-loop system.

AP9.5 NASA is planning a Mars mission with a rover vehicle in 1997. The solar-powered, six-wheeled, 22-pound vehicle will see where it is going with two tiny TV cameras and will measure distance to objects with five laser range finders. It will be able to climb a 30° slope in dry sand and will carry a spectrometer that can determine the chemical com-

FIGURE AP9.3
Weld bead depth
control.

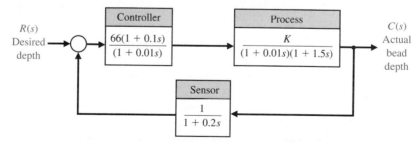

FIGURE AP9.4
Paper machine
control.

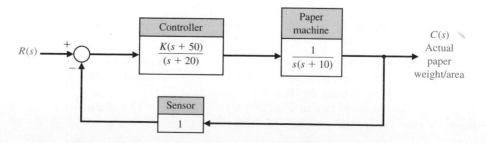

position of surface rocks. It will be controlled remotely from Earth.

For the model of the position control system shown in Fig. AP9.5, determine the gain K that maximizes the phase margin. Determine the overshoot for a step input with the selected gain.

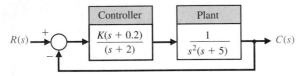

FIGURE AP9.5

AP9.6   The acidity of water draining from a coal mine is often controlled by adding lime to the water. A valve controls the lime addition and a sensor is downstream. For the model of the system shown in Fig. AP9.6, determine K and the distance D to maintain stability. We require $D > 2$ meters in order to allow full mixing before sensing.

AP9.7   Building elevators are limited to about 800 meters. Above that height, elevator cables become too thick and too heavy for practical use. One solution is to eliminate the cable. The key to the cordless elevator is the linear motor technology now being applied to the development of magnetically levitated rail transportation systems. Under consideration is a linear synchronous motor that propels a passenger car along the tracklike guideway running the length of the elevator shaft. The motor works by the interaction of an electromagnetic field from electric coils on the guideway with magnets on the car [29].

If we assume that the motor has negligible friction, the system may be represented by the model shown in Fig. AP9.7. Determine K so that the phase margin of the system is 45°. For the gain K selected, determine the system bandwidth. Also, calculate the maximum value of the output for a unit step disturbance for the selected gain.

AP9.8   A control system is shown in Fig. AP9.8. The gain K is greater than 500 and less than 3000. Select a gain that will cause the system step response to have an overshoot of less than 18%. Plot the Nichols diagram, and calculate the phase margin.

FIGURE AP9.6
Mine water
acidity
control

FIGURE AP9.7
Elevator
position
control

FIGURE AP9.8
Gain selection

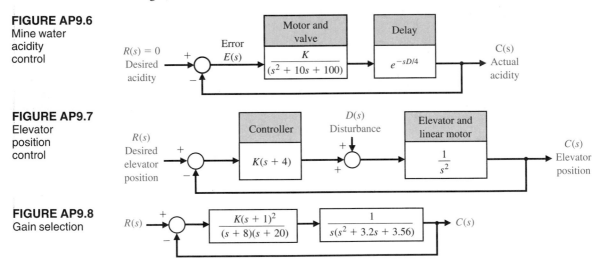

DESIGN PROBLEMS

DP9.1   A mobile robot for toxic waste cleanup is shown in Fig. DP9.1(a) [24]. The closed-loop speed control is represented by Fig. 9.1 with $H(s) = 1$. The Nichols chart in Fig. DP9.1(b) shows the plot of $G(j\omega)/K$ versus ω. The value of

the frequency at points indicated is recorded in the following table:

Point	1	2	3	4	5
ω	2	5	10	20	50

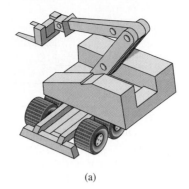

(a)

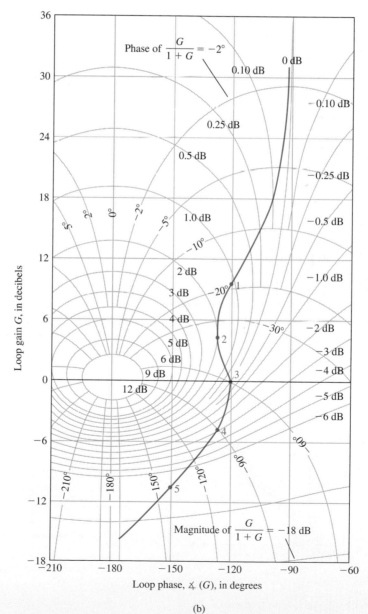

FIGURE DP9.1
(a) Mobile robot for toxic waste cleanup. (b) Nichols chart.

(b)

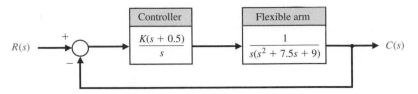

FIGURE DP9.2
Control of a flexible
robot arm.

(a) Determine the gain and phase margins of the closed-loop system when $K = 1$. (b) Determine the resonant peak in dB and the resonant frequency for $K = 1$. (c) Determine the system bandwidth and estimate the settling time and percent overshoot of this system for a step input. (d) Determine the appropriate gain K so that the overshoot to a step input is 30%, and estimate the settling time of the system.

DP9.2 Flexible-joint robot arms are constructed of lightweight materials and exhibit lightly damped open-loop dynamics [15]. A feedback control system for a flexible arm is shown in Fig. DP9.2. Select K so that the system has maximum phase margin. Predict the overshoot for a step input based on the phase margin attained, and compare it to the actual overshoot for a step input. Determine the bandwidth of the closed-loop system, and predict the settling time of the system to a step input and compare it to the actual settling time. Discuss the suitability of this control system.

DP9.3 An automatic drug delivery system is used in the regulation of critical care patients suffering from cardiac failure [25]. The goal is to maintain stable patient status within narrow bounds. Consider the use of a drug delivery system for the regulation of blood pressure by the infusion of a drug. The feedback control system is shown in Fig. DP9.3. Select an appropriate gain K that maintains narrow deviation for blood pressure while achieving a good dynamic response.

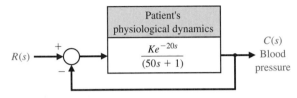

FIGURE DP9.3 Automatic drug delivery.

DP9.4 A robot tennis player is shown in Fig. DP9.4(a), and a simplified control system for $\theta_2(t)$ is shown in Fig. DP9.4(b). The goal of the control system is to attain the best step response while attaining a high K_v for the system. Select $K_{v_1} = 0.325$ and $K_{v_2} = 0.45$, and determine the phase margin, gain margin, and closed-loop bandwidth for each case. Estimate the step response for each case and select the best value for K.

DP9.5 An electrohydraulic actuator is used to actuate large loads for a robot manipulator as shown in Fig. DP9.5 [17]. The system is subjected in step input, and it is desired that the steady-state error be minimized. However, we wish to keep the overshoot less than 10%. Let $T = 0.8$ second.

FIGURE DP9.4
(a) An articulated
two-link tennis
player robot.
(b) Simplified
control system.

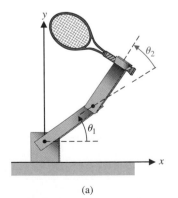

(a)

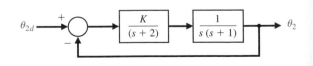

(b)

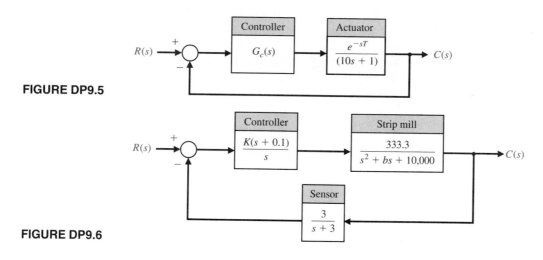

FIGURE DP9.5

FIGURE DP9.6

(a) Select the gain K when $G_c(s) = K$, and determine the resulting overshoot, settling time, and steady-state error. (b) Repeat part (a) when $G_c(s) = K_1 + K_2/s$ by selecting K_1 and K_2. Draw the Nichols chart for the selected gains K_1 and K_2.

DP9.6 The physical representation of a steel strip-rolling mill is a damped-spring system [8]. The output thickness sensor is located a negligible distance from the output of the mill, and the objective is to keep the thickness as close to a reference value as possible. Any change of the input strip thickness is regarded as a disturbance. The system is a non-unity feedback system as shown in Fig. DP9.6. Depending on the maintenance of the mill, the parameter varies as $80 \leq b < 300$.

Determine the phase margin and gain margin for the two extreme values of b when the normal value of the gain is $K = 170$. Recommend a reduced value for K so that the phase margin is

greater than 40° and the gain margin is greater than 8 dB for the range of b.

DP9.7 Vehicles for lunar construction and exploration work will face conditions unlike anything found on Earth. Furthermore, the vehicle will be controlled via remote control. A model of the vehicle and the control is shown in Fig. DP9.7. Select a suitable gain K when $T = 0.5$ second. The goal is to achieve a fast step response with an overshoot less than 20%.

DP9.8 The control of a high-speed steel-rolling mill is a challenging problem. The goal is to keep the strip thickness accurate and readily adjustable. The model of the control system is shown in Fig. DP9.8. Design a control system by selecting K so that the step response of the system is as fast as possible with an overshoot less than 0.5% and a settling time less than 4 seconds. Use the root locus

FIGURE DP9.7
Lunar vehicle
control.

FIGURE DP9.8
Steel-rolling
mill control.

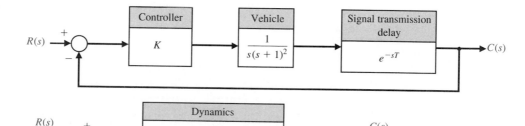

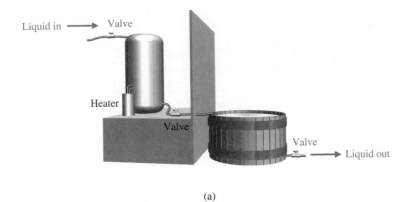

(a)

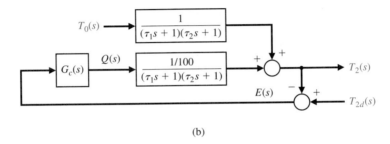

FIGURE DP9.9
Two-tank
temperature
control.

(b)

FIGURE DP9.10
Hot ingot robot
control.

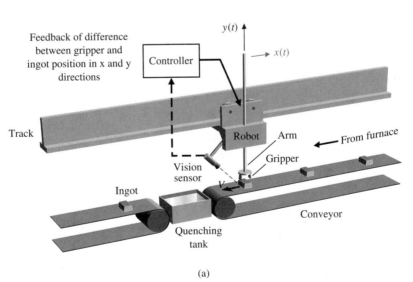

(a)

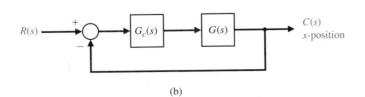

(b)

to select *K*, and calculate the roots for the selected *K*. Describe the dominant root(s) of the system.

DP9.9 A two-tank system containing a heated liquid has the model shown in Fig. DP9.9(a), where T_0 is the temperature of the fluid flowing into the first tank and T_2 is the temperature of the liquid flowing out of the second tank. The block diagram model is shown in Fig. DP9.9(b). The system of the two tanks has a heater in tank 1 with a controllable heat input, *Q*. The time constants are $\tau_1 = 10$ seconds and $\tau_2 = 50$ seconds.

(a) Determine $T_2(s)$ in terms of $T_0(s)$ and $T_{2d}(s)$.
(b) If $T_{2d}(s)$, the desired output temperature, is changed instantaneously from $T_{2d}(s) = A/s$ to $T_{2d}(s) = 2A/s$, determine the transient response of $T_2(t)$ when $G_c(s) = K = 500$. Assume that prior to the abrupt temperature change the system is at steady state.
(c) Find the steady-state error, e_{ss}, for the system of part (b), where $E(s) = T_{2d}(s) - T_2(s)$.
(d) Let $G_c(s) = K/s$ and repeat parts (b) and (c). Use a gain *K* such that the percent overshoot is less than 10%.
(e) Design a controller that will result in a system

with a settling time of $T_s < 150$ seconds and a percent overshoot of less than 10%, while maintaining a zero steady-state error when

$$G_c(s) = \left(K_1 + \frac{K_2}{s} \right).$$

(f) Prepare a table comparing the percent overshoot, settling time, and steady-state error for the designs of parts (b) through (e).

DP9.10 A computer controller for a robot that picks up hot ingots and places them in a quenching tank is shown in Fig. DP9.10(a). The robot places itself over the ingot and then moves down in the *y*-axis. The control system is shown in Fig. DP9.10(b), where

$$G(s) = \frac{e^{-sT}}{(s + 1)^2}$$

and $T = \pi/4$ seconds. Design a controller that reduces the steady-state error for a step input to 10% of the input magnitude while maintaining an overshoot of less than 10% when

$$G_c(s) = \left(K_1 + \frac{K_2}{s} \right).$$

MATLAB PROBLEMS

MP9.1 Consider a unity negative feedback control system with

$$G(s) = \frac{100}{s^2 + 4s + 10}.$$

Using *Matlab*, verify that the gain margin is ∞ and that the phase margin is 24°.

MP9.2 Using the nyquist function, obtain the polar plot for the following transfer functions:

(a) $G(s) = \dfrac{1}{s + 1}$

(b) $G(s) = \dfrac{15}{s^2 + 8s + 5}$

(c) $G(s) = \dfrac{10}{s^3 + 3s^2 + 3s + 1}$

MP9.3 Using the nichols, ngrid('new'), and logspace functions, obtain the Nichols chart with a

grid for the following transfer functions where $0.1 \le \omega \le 10$:

(a) $G(s) = \dfrac{1}{s + 0.1}$

(b) $G(s) = \dfrac{1}{s^2 + 2s + 1}$

(c) $G(s) = \dfrac{24}{s^3 + 9s^2 + 26s + 24}$

Determine the approximate phase and gain margins from the Nichols charts and label the charts accordingly.

MP9.4 A block diagram of the yaw acceleration control system for a bank-to-turn missile is shown in Fig. MP9.4. The input is yaw acceleration command (in *g*'s) and the output is missile yaw acceleration (in *g*'s). The controller is specified to be a

FIGURE MP9.4
A feedback control system for the yaw acceleration control of a bank-to-turn missile.

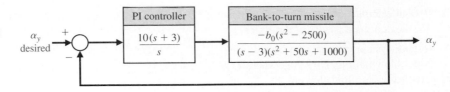

proportional, integral (PI) controller. The nominal value of b_0 is 0.5.

(a) Using the margin function, compute the phase margin, gain margin, and system crossover frequency (0 dB), assuming the nominal value of b_0. (b) Using the gain margin from part (a), determine the maximum value of b_0 for a stable system. Verify your answer with a Routh–Hurwitz analysis of the characteristic equation.

MP9.5   An engineering laboratory has presented a plan to operate an Earth orbiting satellite that is to be controlled from a ground station. A block diagram of the proposed system is shown in Fig. MP9.5. It takes T seconds for a signal to reach the spacecraft from the ground station and the identical delay for a return signal. The proposed ground based controller is a proportional-derivative (PD) controller, where

$$C(s) = K_1 + K_2 s.$$

(a) Assume no transmission time delay (i.e., $T = 0$), and design the controller to the following specifications: (1) percent overshoot less than 20% to a unit step input and (2) time to peak less than 30 seconds.
(b) Compute the phase margin with the controller in the loop but assuming a zero transmission time delay. Estimate the amount of allowable time delay for a stable system from the phase margin calculation.
(c) Using a second-order Padé approximation to the time delay, determine the maximum allowable delay, T_{max} for system stability by developing a *Matlab* script that employs the pade function and computes the closed-loop system poles as a function of the time delay T. Compare your answer with the one obtained in part (b).

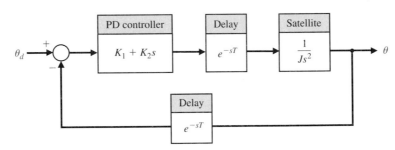

FIGURE MP9.5
A block diagram of a ground-controlled satellite.

TERMS AND CONCEPTS

Cauchy's theorem   If a contour encircles Z zero and P poles of $F(s)$ traversing clockwise, the corresponding contour in the $F(s)$-plane encircles the origin of the $F(s)$-plane $N = Z - P$ times clockwise.

Closed-loop frequency response   The frequency response of the closed-loop transfer function $T(j\omega)$.

Conformal mapping   A contour mapping that retains the angles on the s-plane on the $F(s)$-plane.

Contour map   A contour or trajectory in one plane is mapped into another plane by a relation $F(s)$.

Gain margin   The reciprocal of the gain $|GH|$ at the frequency at which the phase angle reaches 180°.

Nichols chart   A chart displaying the curves for the relationship between the open-loop and closed-loop frequency response.

Nyquist stability criterion   A feedback system is stable if, and only if, the contour in the $G(s)$-plane does not encircle the $(-1, 0)$ point when the number of poles of $G(s)$ in the right-hand s-plane is zero. If $G(s)$ has P poles in the right-hand plane, then the number of counterclockwise encirclements of the $(-1, 0)$ point must be equal to P for a stable system.

Phase margin   The phase angle through which the $GH(j\omega)$-locus must be rotated in order that the unity magnitude point passes through the $(-1, 0)$ point in the $GH(j\omega)$-plane.

Principle of the argument   *See* Cauchy's theorem.

Time delay   A pure time delay, T, so that events occurring at time t at one point in the system occur at another point in the system at a later time, $t + T$.

The Design of
Feedback Control
Systems

PREVIEW

Thus far we have striven to achieve the desired performance of a system by adjusting one or two parameters. However, parameter adjustment may not result in the desired performance. Thus it may be necessary to introduce a new block within the feedback loop to compensate for the original system's limitation. This block, with a transfer function $G_c(s)$, is called a compensator. It is the purpose of this chapter to develop several design techniques in the frequency and time domain that enable us to achieve the desired system performance.

We will discuss various candidates for service as compensators and show how they help to achieve improved performance. The design of control systems using state variable feedback will be considered in Chapter 11.

10.1 INTRODUCTION

The performance of a feedback control system is of primary importance. This subject was discussed at length in Chapter 5 and quantitative measures of performance were developed. We have found that a suitable control system is stable and that it results in an acceptable response to input commands, is less sensitive to system parameter changes, results in a minimum steady-state error for input commands, and, finally, is able to reduce the effect of undesirable disturbances. A feedback control system that provides an optimum performance without any necessary adjustments is rare indeed. Usually we find it necessary to compromise among the many conflicting and demanding specifications and to adjust the system parameters to provide a suitable and acceptable performance when it is not possible to obtain all the desired optimum specifications.

At several points in the preceding chapters, we have considered the question of design and adjustment of the system parameters in order to provide a desirable response and performance. In Chapter 5, we defined and established several suitable measures of performance. Then, in Chapter 6, we determined a method of investigating the stability of a control system, since we recognized that a system is unacceptable unless it is stable. In Chapter 7, we utilized the root locus method to effect a design of a self-balancing scale (Section 7.4) and then illustrated a method of parameter design by using the root locus method (Section 7.5). Furthermore, in Chapters 8 and 9, we developed suitable measures of performance in terms of the frequency variable ω and utilized them to design several suitable control systems. Thus we have been considering the problems of the design of feedback control systems as an integral part of the subjects of the preceding chapters. It is now our purpose to study the question somewhat further and to point out several significant design and compensation methods.

We have found in the preceding chapters that it is often possible to adjust the system parameters in order to provide the desired system response. However, we often find that it is not sufficient to adjust a system parameter and thus obtain the desired performance. Rather we are required to reconsider the structure of the system and redesign the system in order to obtain a suitable one. That is, we must examine the scheme or plan of the system and obtain a new design or plan that results in a suitable system. Thus *the design of a control system is concerned with the arrangement, or the plan, of the system structure and the selection of suitable components and parameters.* For example, if we desire a set of performance measures to be less than some specified values, we often encounter a conflicting set of requirements. Thus if we wish a system to have a percent overshoot less than 20% and $\omega_n T_p = 3.3$, we obtain a conflicting requirement on the system damping ratio, ζ, as can be seen by examining Fig. 5.8 again. Now, if we are unable to relax these two performance requirements, we must alter the system in some way. The alteration or adjustment of a control system in order to provide a suitable performance is called *compensation;* that is, compensation is the adjustment of a system in order to make up for deficiencies or inadequacies.

In redesigning a control system to alter the system response, an additional component

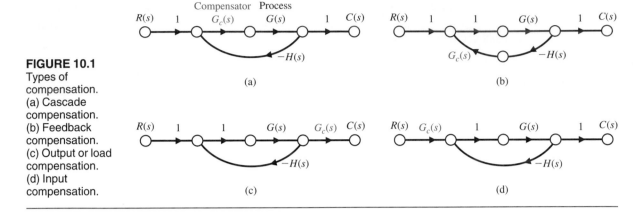

FIGURE 10.1
Types of
compensation.
(a) Cascade
compensation.
(b) Feedback
compensation.
(c) Output or load
compensation.
(d) Input
compensation.

is inserted within the structure of the feedback system. It is this additional component or device that equalizes or compensates for the performance deficiency. The compensating device may be an electric, a mechanical, a hydraulic, a pneumatic, or other type of device or network and is often called a *compensator*. Commonly, an electric circuit serves as a compensator in many control systems.

A compensator is an additional component or circuit that is inserted into a control system to equalize or compensate for a deficient performance.

The transfer function of a compensator is designated as $G_c(s) = E_o(s)/E_{in}(s)$, and the compensator can be placed in a suitable location within the structure of the system. Several types of compensation are shown in Fig. 10.1 for a simple single-loop feedback control system. The compensator placed in the feedforward path is called a *cascade* or series compensator (Fig. 10.1a). Similarly, the other compensation schemes are called feedback, output or load, and input compensation, as shown in Figs. 10.1(b), (c), and (d), respectively. The selection of the compensation scheme depends upon a consideration of the specifications, the power levels at various signal nodes in the system, and the networks available for use. Usually the output $C(s)$ is a direct output of the process $G(s)$ and the output compensation of Fig. 10.1(c) is not physically realizable. It will not be possible to consider all the possibilities in this chapter, and the reader is referred to Chapters 11 and 12 following the introductory material of this chapter.

10.2 APPROACHES TO SYSTEM DESIGN

The performance of a control system can be described in terms of the time-domain performance measures or the frequency-domain performance measures. The performance of a system can be specified by requiring a certain peak time, T_p, maximum overshoot, and settling-time for a step input. Furthermore, it is usually necessary to specify the maximum allowable steady-state error for several test signal inputs and disturbance inputs. These performance specifications can be defined in terms of the desirable location of the poles and zeros of the closed-loop system transfer function, $T(s)$. Thus the location of the s-plane poles and zeros of $T(s)$ can be specified. As we found in Chapter 7, the locus of the roots of the closed-loop system can be readily obtained for the variation of one system parameter. However, when the locus of roots does not result in a suitable root configuration, we must add a compensating network (Fig. 10.1) to alter the locus of the roots as the parameter is varied. Therefore we can use the root locus method and determine a suitable compensator network transfer function so that the resultant root locus results in the desired closed-loop root configuration.

Alternatively, we can describe the performance of a feedback control system in terms of frequency performance measures. Then a system can be described in terms of the peak of the closed-loop frequency response, M_{p_ω}, the resonant frequency ω_r, the bandwidth, and the phase margin of the system. We can add a suitable compensation network, if necessary, in order to satisfy the system specifications. The design of the network $G_c(s)$, is developed in terms of the frequency response as portrayed on the polar plane, the Bode diagram, or the Nichols chart. Because a cascade transfer function is readily accounted for on a Bode plot by adding the frequency response of the network, we usually prefer to approach the frequency response methods by utilizing the Bode diagram.

Thus the design of a system is concerned with the alteration of the frequency response or the root locus of the system in order to obtain a suitable system performance. For frequency response methods, we are concerned with altering the system so that the frequency response of the compensated system will satisfy the system specifications. Thus, in the case of the frequency response approach, we use compensation networks to alter and reshape the system characteristics represented on the Bode diagram and Nichols chart.

Alternatively, the design of a control system can be accomplished in the s-plane by root locus methods. For the case of the s-plane, the designer wishes to alter and reshape the root locus so that the roots of the system will lie in the desired position in the s-plane.

We have illustrated several of the aforementioned approaches in the preceding chapters. In Example 7.5, we utilized the root locus method in considering the design of a feedback network in order to obtain a satisfactory performance. In Chapter 9, we considered the selection of the gain in order to obtain a suitable phase margin and, therefore, a satisfactory relative stability.

Quite often, in practice, the best and simplest way to improve the performance of a control system is to alter, if possible, the process itself. That is, if the system designer is able to specify and alter the design of the process that is represented by the transfer function $G(s)$, then the performance of the system can be readily improved. For example, to improve the transient behavior of a servomechanism position controller, we often can choose a better motor for the system. In the case of an airplane control system, we might be able to alter the aerodynamic design of the airplane and thus improve the flight transient characteristics. Thus a control system designer should recognize that an alteration of the process may result in an improved system. However, often the process is fixed and unalterable or has been altered as much as possible and is still found to result in an unsatisfactory performance. Then the addition of compensation networks becomes useful for improving the performance of the system.

In the following sections, we will assume that the process has been improved as much as possible and that the $G(s)$ representing the process is unalterable. First, we shall consider the addition of a so-called phase-lead compensation network and describe the design of the network by root locus and frequency response techniques. Then, using both the root locus and frequency response techniques, we will describe the design of the integration compensation networks in order to obtain a suitable system performance.

10.3 CASCADE COMPENSATION NETWORKS

In this section, we will consider the design of a cascade or feedback network, as shown in Figs. 10.1(a) and (b), respectively. The compensation network, $G_c(s)$, is cascaded with the unalterable process $G(s)$ in order to provide a suitable loop transfer function $G_c(s)G(s)H(s)$. The compensator $G_c(s)$ can be chosen to alter either the shape of the root locus or the frequency response. In either case, the network may be chosen to have a transfer function

$$G_c(s) = \frac{K\prod_{i=1}^{M}(s + z_i)}{\prod_{j=1}^{N}(s + p_j)}. \tag{10.1}$$

Then the problem reduces to the judicious selection of the poles and zeros of the compensator. To illustrate the properties of the compensation network, we will consider a first-order compensator. The compensation approach developed on the basis of a first-order compensator can then be extended to higher-order compensators, for example, by cascading several first-order compensators.

A compensator, $G_c(s)$, is used with a plant $G(s)$ so that the overall loop gain can be set to satisfy the steady-state error requirement, and then $G_c(s)$ is used to adjust the system dynamics favorably without affecting the steady-state error.

Consider the first-order compensator with the transfer function

$$G_c(s) = \frac{K(s + z)}{(s + p)}. \tag{10.2}$$

The design problem becomes, then, the selection of z, p, and K in order to provide a suitable performance. When $|z| < |p|$, the network is called a *phase-lead network* and has a pole–zero s-plane configuration, as shown in Fig. 10.2. If the pole was negligible, that is, $|p| \gg |z|$, and the zero occurred at the origin of the s-plane, we would have a differentiator so that

$$G_c(s) \simeq \left(\frac{K}{p}\right)s. \tag{10.3}$$

Thus a compensation network of the form of Eq. (10.2) is a differentiator-type network. The differentiator network of Eq. (10.3) has a frequency characteristic as

$$G_c(j\omega) = j\left(\frac{K}{p}\right)\omega = \left(\frac{K}{p}\omega\right)e^{+j90°} \tag{10.4}$$

FIGURE 10.2
Pole–zero diagram
of the phase-lead
network.

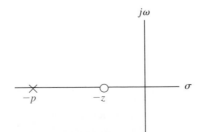

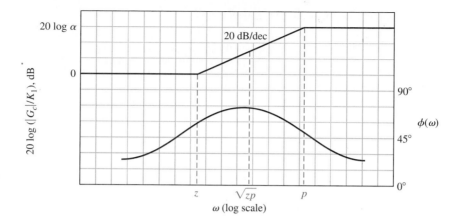

FIGURE 10.3
Bode diagram of
the phase-lead
network.

and phase angle of $+90°$, often called a phase-lead angle. Similarly, the frequency response of the differentiating network of Eq. (10.2) is

$$G_c(j\omega) = \frac{K(j\omega + z)}{(j\omega + p)} = \frac{(Kz/p)(j(\omega/z) + 1)}{(j(\omega/p) + 1)} = \frac{K_1(1 + j\omega\alpha\tau)}{(1 + j\omega\tau)}, \quad (10.5)$$

where $\tau = 1/p$, $p = \alpha z$, and $K_1 = K/\alpha$. The frequency response of this phase-lead network is shown in Fig. 10.3. The angle of the frequency characteristic is

$$\phi(\omega) = \tan^{-1}\alpha\omega\tau - \tan^{-1}\omega\tau. \quad (10.6)$$

Since the zero occurs first on the frequency axis, we obtain a phase-lead characteristic as shown in Fig. 10.3. The slope of the asymptotic magnitude curve is $+20$ dB/decade.

The phase-lead compensation transfer function can be obtained with the network shown in Fig. 10.4. The transfer function of this network is

$$G_c(s) = \frac{V_2(s)}{V_1(s)} = \frac{R_2}{R_2 + \{R_1(1/Cs)/[R_1 + (1/Cs)]\}}$$

$$= \left(\frac{R_2}{R_1 + R_2}\right) \frac{(R_1 Cs + 1)}{\{[R_1 R_2/(R_1 + R_2)]Cs + 1\}}. \quad (10.7)$$

Therefore we let

$$\tau = \frac{R_1 R_2}{R_1 + R_2}C \quad \text{and} \quad \alpha = \frac{R_1 + R_2}{R_2}$$

FIGURE 10.4
Phase-lead
network.

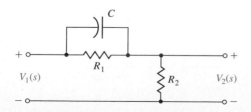

and obtain the transfer function

$$G_c(s) = \frac{(1 + \alpha\tau s)}{\alpha(1 + \tau s)}, \tag{10.8}$$

which is equal to Eq. (10.5) when an additional cascade gain K is inserted.

The maximum value of the phase lead occurs at a frequency ω_m, where ω_m is the geometric mean of $p = 1/\tau$ and $z = 1/\alpha\tau$; that is, the maximum phase lead occurs halfway between the pole and zero frequencies on the logarithmic frequency scale. Therefore

$$\omega_m = \sqrt{zp} = \frac{1}{\tau\sqrt{\alpha}}.$$

To obtain an equation for the maximum phase-lead angle, we rewrite the phase angle of Eq. (10.5) as

$$\phi = \tan^{-1}\frac{\alpha\omega\tau - \omega\tau}{1 + (\omega\tau)^2\alpha}. \tag{10.9}$$

Then, substituting the frequency for the maximum phase angle, $\omega_m = 1/\tau\sqrt{\alpha}$, we have

$$\tan\phi_m = \frac{(\alpha/\sqrt{\alpha}) - (1/\sqrt{\alpha})}{1 + 1} = \frac{\alpha - 1}{2\sqrt{\alpha}}. \tag{10.10}$$

Because the $\tan\phi_m$ equals $(\alpha - 1)/2\sqrt{\alpha}$, we utilize the triangular relationship and note that

$$\sin\phi_m = \frac{\alpha - 1}{\alpha + 1}. \tag{10.11}$$

Equation (10.11) is very useful for calculating a necessary α ratio between the pole and zero of a compensator in order to provide a required maximum phase lead. A plot of ϕ_m versus α is shown in Fig. 10.5. Clearly, the phase angle readily obtainable from this network is not much greater than 70°. Also, since $\alpha = (R_1 + R_2)/R_2$, there are practical limitations on the maximum value of α that one should attempt to obtain. Therefore, if one required a maximum angle greater than 70°, two cascade compensation networks would be utilized. Then the equivalent compensation transfer function would be $G_{c_1}(s)G_{c_2}(s)$ when the loading effect of $G_{c_2}(s)$ on $G_{c_1}(s)$ is negligible.

FIGURE 10.5
Maximum phase
angle ϕ_m versus α
for a lead network.

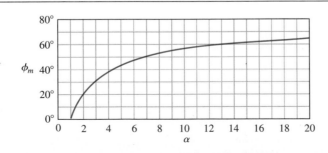

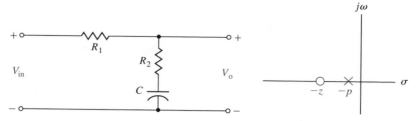

FIGURE 10.6
Phase-lag network.

FIGURE 10.7
Pole–zero diagram of the
phase-lag network.

It is often useful to add a cascade compensation network that provides a phase-lag characteristic. The *phase-lag network* is shown in Fig. 10.6. The transfer function of the phase-lag network is

$$G_c(s) = \frac{V_o(s)}{V_{in}(s)} = \frac{R_2 + (1/Cs)}{R_1 + R_2 + (1/Cs)} = \frac{R_2Cs + 1}{(R_1 + R_2)Cs + 1}. \qquad (10.12)$$

When $\tau = R_2C$ and $\alpha = (R_1 + R_2)/R_2$, we have

$$G_c(s) = \frac{1 + \tau s}{1 + \alpha\tau s} = \frac{1}{\alpha}\frac{(s + z)}{(s + p)}, \qquad (10.13)$$

where $z = 1/\tau$ and $p = 1/\alpha\tau$. In this case, because $\alpha > 1$, the pole lies closest to the origin of the s-plane, as shown in Fig. 10.7. This type of compensation network is often called an integrating network because it has a frequency response like an integrator over a finite range of frequencies. The Bode diagram of the phase-lag network is obtained from the transfer function

$$G_c(j\omega) = \frac{1 + j\omega\tau}{1 + j\omega\alpha\tau} \qquad (10.14)$$

and is shown in Fig. 10.8. The form of the Bode diagram of the lag network is similar to that of the phase-lead network; the difference is the resulting attenuation and phase-lag angle instead of amplification and phase-lead angle. However, note that the shape of the diagrams of Figs. 10.3 and 10.8 are similar. Therefore it can be shown that the maximum phase lag occurs at $\omega_m = \sqrt{zp}$.

In the succeeding sections, we wish to utilize these compensation networks in order to obtain a desired system frequency response or s-plane root location. The lead network is utilized to provide a phase-lead angle and thus a satisfactory phase margin for a system. Alternatively, the use of the phase-lead network can be visualized on the s-plane as enabling us to reshape the root locus and thus provide the desired root locations. The phase-lag network is utilized not to provide a phase-lag angle, which is normally a destabilizing influence, but rather to provide an attenuation and increase the steady-state error constant [3]. These approaches to design utilizing the phase-lead and phase-lag networks will be the subject of the following six sections.

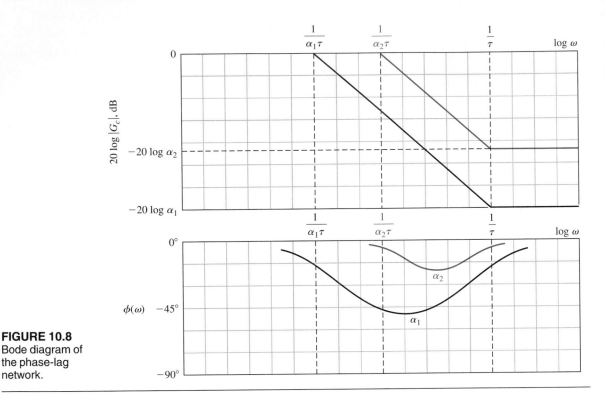

FIGURE 10.8
Bode diagram of
the phase-lag
network.

10.4 PHASE-LEAD DESIGN USING THE BODE DIAGRAM

The Bode diagram is used to design a suitable phase-lead network in preference to other frequency response plots. The frequency response of the cascade compensation network is added to the frequency response of the uncompensated system. That is, because the total loop transfer function of Fig. 10.1(a) is $G_c(j\omega)G(j\omega)H(j\omega)$, we will first plot the Bode diagram for $G(j\omega)H(j\omega)$. Then we can examine the plot for $G(j\omega)H(j\omega)$ and determine a suitable location for p and z of $G_c(j\omega)$ in order to satisfactorily reshape the frequency response. The uncompensated $G(j\omega)$ is plotted with the desired gain to allow an acceptable steady-state error. Then the phase margin and the expected $M_{p\omega}$ are examined to find whether they satisfy the specifications. If the phase margin is not sufficient, phase lead can be added to the phase angle curve of the system by placing the $G_c(j\omega)$ in a suitable location. To obtain maximum additional phase lead, we desire to place the network such that the frequency ω_m is located at the frequency where the magnitude of the compensated magnitude curves crosses the 0-dB axis. (Recall the definition of phase margin.) The value of the added phase lead required allows us to determine the necessary value for α from Eq. (10.11) or Fig. 10.5. The zero $\omega = 1/\alpha\tau$ is located by noting that the maximum phase lead should occur at $\omega_m = \sqrt{zp}$, halfway between the pole and the zero. Because the total magnitude gain for the network is 20 log α, we expect a gain of 10 log α at ω_m. Thus we determine the compensation network by completing the following steps:

1. Evaluate the uncompensated system phase margin when the error constants are satisfied.
2. Allowing for a small amount of safety, determine the necessary additional phase lead, ϕ_m.
3. Evaluate α from Eq. (10.11).
4. Evaluate $10 \log \alpha$ and determine the frequency where the uncompensated magnitude curve is equal to $-10 \log \alpha$ dB. Because the compensation network provides a gain of $10 \log \alpha$ at ω_m; this frequency is the new 0-dB crossover frequency and ω_m simultaneously.
5. Calculate the pole $p = \omega_m \sqrt{\alpha}$ and $z = p/\alpha$.
6. Draw the compensated frequency response, check the resulting phase margin, and repeat the steps if necessary. Finally, for an acceptable design, raise the gain of the amplifier in order to account for the attenuation $(1/\alpha)$.

EXAMPLE 10.1   **A lead compensator for a type 2 system**

Let us consider a single-loop feedback control system as shown in Fig. 10.1(a), where

$$G(s) = \frac{K_1}{s^2} \tag{10.15}$$

and $H(s) = 1$. The uncompensated system is a type 2 system and at first appears to possess a satisfactory steady-state error for both step and ramp input signals. However, uncompensated, the response of the system is an undamped oscillation because

$$T(s) = \frac{C(s)}{R(s)} = \frac{K_1}{s^2 + K_1}. \tag{10.16}$$

Therefore the compensation network is added so that the loop transfer function is $G_c(s)G(s)H(s)$. The specifications for the system are

$$\text{Settling time, } T_s \leq 4 \text{ seconds,}$$

$$\text{System damping constant } \zeta \geq 0.45.$$

The settling time requirement is

$$T_s = \frac{4}{\zeta \omega_n} = 4,$$

and therefore

$$\omega_n = \frac{1}{\zeta} = \frac{1}{0.45} = 2.22.$$

Perhaps the simplest way to check the value of ω_n for the frequency response is to relate ω_n to the bandwidth, ω_B, and evaluate the -3-dB bandwidth of the closed-loop system. For a closed-loop system with $\zeta = 0.45$, we estimate from Fig. 8.26 that $\omega_B = 1.33\omega_n$. Therefore we require a closed-loop bandwidth $\omega_B = 1.33(2.22) = 3.00$. The bandwidth can be checked following compensation by utilizing the Nichols chart. For the uncompensated system, the bandwidth of the system is $\omega_B = 1.33\omega_n$ and $\omega_n = \sqrt{K}$. Therefore a loop gain equal to $K = \omega_n^2 \approx 5$ would be sufficient. To provide a suitable margin for the settling time, we will select $K = 10$ in order to draw the Bode diagram of

$$GH(j\omega) = \frac{K}{(j\omega)^2}.$$

The Bode diagram of the uncompensated system is shown as solid lines in Fig. 10.9.

By using Eq. (9.58), the phase margin of the system is required to be approximately

$$\phi_{pm} = \frac{\zeta}{0.01} = \frac{0.45}{0.01} = 45°. \tag{10.17}$$

The phase margin of the uncompensated system is 0° because the double integration results in a constant 180° phase lag. Therefore we must add a 45° phase-lead angle at the crossover (0-dB) frequency of the compensated magnitude curve. Evaluating the value of α, we have

$$\frac{\alpha - 1}{\alpha + 1} = \sin \phi_m = \sin 45°, \tag{10.18}$$

and therefore $\alpha = 5.8$. To provide a margin of safety, we will use $\alpha = 6$. The value of $10 \log \alpha$ is then equal to 7.78 dB. Then the lead network will add an additional gain of 7.78 dB at the frequency ω_m, and it is desired to have ω_m equal to compensated slope near the 0-dB axis (the dashed line) so that the new crossover is ω_m and the dashed magnitude curve is 7.78 dB above the uncompensated curve at the crossover frequency. Thus the compensated crossover frequency is located by evaluating the frequency where the uncompensated magnitude curve is equal to -7.78 dB, which in this case is $\omega = 4.95$. Then the maximum phase-lead angle is added to $\omega = \omega_m = 4.95$, as shown in Fig. 10.9. Using step 5, we determine the pole $p = \omega_m \sqrt{\alpha} = 12.0$ and the zero $z = p/\alpha = 2.0$.

The bandwidth of the compensated system can be obtained from the Nichols chart. For estimating the bandwidth, we can simply examine Fig. 9.26 and note that the -3-dB line for the closed-loop system occurs when the magnitude of $GH(j\omega)$ is -6 dB and the phase shift of $GH(j\omega)$ is approximately $-140°$. Therefore, to estimate the bandwidth from the open-loop diagram, we will approximate the bandwidth as the frequency for which $20 \times$

FIGURE 10.9
Bode diagram for Example 10.1.

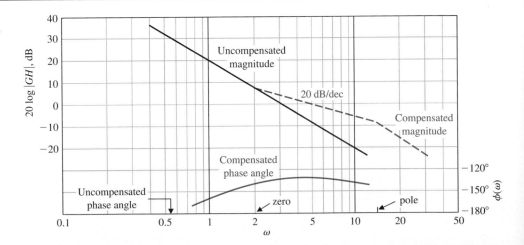

$\log|GH|$ is equal to -6 dB. Thus the bandwidth of the uncompensated system is approximately equal to $\omega_B = 4.4$, while the bandwidth of the compensated system is equal to $\omega_B = 8.4$. The lead compensation doubles the bandwidth in this case and the specification that $\omega_B > 3.00$ is satisfied. Therefore the compensation of the system is completed and the system specifications are satisfied. The total compensated loop transfer function is

$$G_c(j\omega)G(j\omega)H(j\omega) = \frac{10[(j\omega/2.0) + 1]}{(j\omega)^2[(j\omega/12.0) + 1]}. \tag{10.19}$$

The transfer function of the compensator is

$$G_c(s) = \frac{(1 + \alpha\tau s)}{\alpha(1 + \tau s)} = \frac{1}{6}\frac{[1 + (s/2.0)]}{[1 + (s/12.0)]} \tag{10.20}$$

in the form of Eq. (10.8). Because an attenuation of ⅙ results from the passive RC network, the gain of the amplifier in the loop must be raised by a factor of six so that the total dc loop gain is still equal to 10 as required in Eq. (10.19). When we add the compensation network Bode diagram to the uncompensated Bode diagram as in Fig. 10.9, we are assuming that we can raise the amplifier gain to account for this $1/\alpha$ attenuation. The pole and zero values can be read from Fig. 10.9, noting that $p = \alpha z$.

The total loop transfer function is (recall $H(s) = 1$)

$$GG_c(s) = \frac{10(1 + s/2)}{s^2(1 + s/12)} = \frac{60(s + 2)}{s^2(s + 12)}.$$

The closed-loop transfer function is

$$T(s) = \frac{60(s + 2)}{s^3 + 12s^2 + 60s + 120} = \frac{60(s + 2)}{(s^2 + 6s + 20)(s + 6)},$$

and the effects of the zero at $s = -2$ and the third pole at $s = -6$ will affect the transient response. Plotting the step response, we find an overshoot of 34% and a settling time of 1.4 seconds. ∎

EXAMPLE 10.2 A lead compensator for a second-order system

A feedback control system has a loop transfer function

$$GH(s) = \frac{K}{s(s + 2)}. \tag{10.21}$$

It is desired to have a steady-state error for a ramp input equal to 5% of the magnitude of the ramp. Therefore we require that

$$K_v = \frac{A}{e_{ss}} = \frac{A}{0.05A} = 20. \tag{10.22}$$

Furthermore, we desire that the phase margin of the system be at least 45°. The first step is to plot the Bode diagram of the uncompensated transfer function

$$GH(j\omega) = \frac{K_v}{j\omega(0.5j\omega + 1)} = \frac{20}{j\omega(0.5j\omega + 1)}, \tag{10.23}$$

as shown in Fig. 10.10(a). The frequency at which the magnitude curve crosses the 0-dB line is 6.2 rad/s, and the phase margin at this frequency is determined readily from the equation of the phase of $GH(j\omega)$, which is

$$\underline{/GH(j\omega)} = \phi(\omega) = -90° - \tan^{-1}(0.5\omega). \tag{10.24}$$

At the crossover frequency, $\omega = \omega_c = 6.2$ rad/s, we have

$$\phi(\omega) = -162°, \tag{10.25}$$

and therefore the phase margin is 18°. Using Eq. (10.24) to evaluate the phase margin is often easier than drawing the complete phase angle curve, which is shown in Fig. 10.10(a). Thus we need to add a phase-lead network so that the phase margin is raised to 45° at the new crossover (0-dB) frequency. Because the compensation crossover frequency is greater than the uncompensated crossover frequency, the phase lag of the uncompensated system is greater also. We shall account for this additional phase lag by attempting to obtain a maximum phase lead of 45° − 18° = 27° plus a small increment (10%) of phase lead to account for the added lag. Thus we will design a compensation network with a maximum phase lead equal to 27° + 3° = 30°. Then, calculating α, we obtain

$$\frac{\alpha - 1}{\alpha + 1} = \sin 30° = 0.5, \tag{10.26}$$

and therefore $\alpha = 3$.

The maximum phase lead occurs at ω_m, and this frequency will be selected so that the new crossover frequency and ω_m coincide. The magnitude of the lead network at ω_m is $10 \log \alpha = 10 \log 3 = 4.8$ dB. The compensated crossover frequency is then evaluated where the magnitude of $GH(j\omega)$ is -4.8 dB and thus $\omega_m = \omega_c = 8.4$. Drawing the compensated magnitude line so that it intersects the 0-dB axis at $\omega = \omega_c = 8.4$, we find that $z = 4.8$ and $p = \alpha z = 14.4$. Therefore the compensation network is

$$G_c(s) = \frac{1}{3} \frac{(1 + s/4.8)}{(1 + s/14.4)}. \tag{10.27}$$

The total dc loop gain must be raised by a factor of 3 in order to account for the factor $1/\alpha = \frac{1}{3}$. Then the compensated loop transfer function is

$$G_c(s)GH(s) = \frac{20[(s/4.8) + 1]}{s(0.5s + 1)[(s/14.4) + 1]}. \tag{10.28}$$

To verify the final phase margin, we can evaluate the phase of $G_c(j\omega)GH(j\omega)$ at $\omega = \omega_c = 8.4$ and therefore obtain the phase margin. The phase angle is then

$$\phi(\omega_c) = -90° - \tan^{-1} 0.5\omega_c - \tan^{-1} \frac{\omega_c}{14.4} + \tan^{-1} \frac{\omega_c}{4.8}$$

$$= -90° - 76.5° - 30.0° + 60.2° \tag{10.29}$$

$$= -136.3°.$$

Therefore the phase margin for the compensated system is 43.7°. If we desire to have exactly a 45° phase margin, we would repeat the steps with an increased value of α—for example, with $\alpha = 3.5$. In this case, the phase lag increased by 7° between $\omega = 6.2$ and

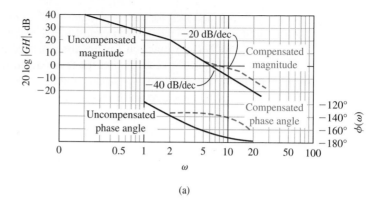

(a)

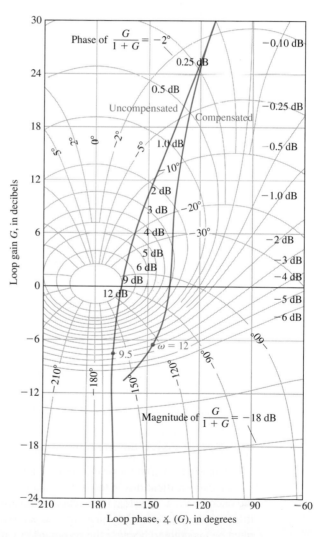

(b)

FIGURE 10.10
(a) Bode diagram for Example 10.2.
(b) Nichols diagram for Example 10.2.

$\omega = 8.4$, and therefore the allowance of 3° in the calculation of α was not sufficient. The step response of this system yields a 28% overshoot with a settling time of 0.75 second.

The Nichols diagram for the compensated and uncompensated system is shown on Fig. 10.10(b). The reshaping of the frequency response locus is clear on this diagram. One notes the increased phase margin for the compensated system as well as the reduced magnitude of $M_{p\omega}$, the maximum magnitude of the closed-loop frequency response. In this case, $M_{p\omega}$ has been reduced from an uncompensated value of $+12$ dB to a compensated value of approximately $+3.2$ dB. Also, we note that the closed-loop 3-dB bandwidth of the compensated system is equal to 12 rad/s compared with 9.5 rad/s for the uncompensated system. ∎

Looking again at Examples 10.1 and 10.2, we note that the system design is satisfactory when the asymptotic curve for the magnitude $20 \log|GG_c|$ crosses the 0-dB line with a slope of -6 dB/octave.

10.5 PHASE-LEAD DESIGN USING THE ROOT LOCUS

The design of the phase-lead compensation network can also be readily accomplished using the root locus. The phase-lead network has a transfer function

$$G_c(s) = \frac{[s + (1/\alpha\tau)]}{[s + (1/\tau)]} = \frac{(s + z)}{(s + p)}, \tag{10.30}$$

where α and τ are defined for the RC network in Eq. (10.7). The locations of the zero and pole are selected so as to result in a satisfactory root locus for the compensated system. The specifications of the system are used to specify the desired location of the dominant roots of the system. The s-plane root locus method is as follows:

1. List the system specifications and translate them into a desired root location for the dominant roots.
2. Sketch the uncompensated root locus, and determine whether the desired root locations can be realized with an uncompensated system.
3. If a compensator is necessary, place the zero of the phase-lead network directly below the desired root location (or to the left of the first two real poles).
4. Determine the pole location so that the total angle at the desired root location is 180° and therefore is on the compensated root locus.
5. Evaluate the total system gain at the desired root location and then calculate the error constant.
6. Repeat the steps if the error constant is not satisfactory.

Therefore we first locate our desired dominant root locations so that the dominant roots satisfy the specifications in terms of ζ and ω_n, as shown in Fig. 10.11(a). The root locus of the uncompensated system is sketched as illustrated in Fig. 10.11(b). Then the zero is added to provide a phase lead by placing it to the left of the first two real poles. Some caution must be maintained because the zero must not alter the dominance of the desired roots; that is, the zero should not be placed nearer the origin than the second pole on the real axis, or

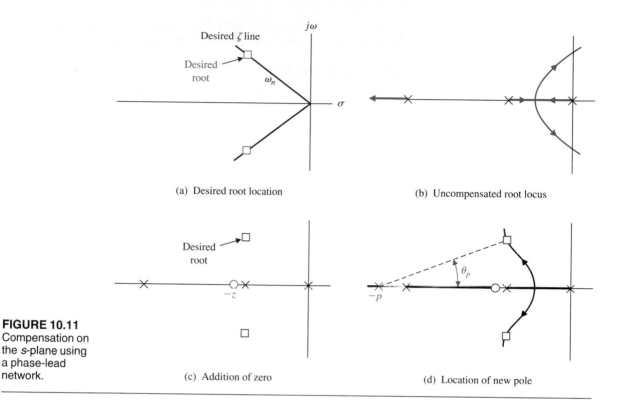

FIGURE 10.11
Compensation on the *s*-plane using a phase-lead network.

(a) Desired root location

(b) Uncompensated root locus

(c) Addition of zero

(d) Location of new pole

a real root near the origin will result and will dominate the system response. Thus, in Fig. 10.11(c), we note that the desired root is directly above the second pole, and we place the zero *z* somewhat to the left of the second real pole.

Consequently, the real root may be near the real zero and the coefficient of this term of the partial fraction expansion may be relatively small. Thus the response due to this real root may have very little effect on the overall system response. Nevertheless, the designer must be continually aware that the compensated system response will be influenced by the roots and zeros of the system and that the dominant roots will not by themselves dictate the response. It is usually wise to allow for some margin of error in the design and to test the compensated system using a computer simulation.

Because the desired root is a point on the root locus when the final compensation is accomplished, we expect the algebraic sum of the vector angles to be 180° at that point. Thus we calculate the angle from the pole of compensator, θ_p, in order to result in a total angle of 180°. Then, locating a line at an angle θ_p intersecting the desired root, we are able to evaluate the compensator pole, *p*, as shown in Fig. 10.11(d).

The advantage of the root locus method is the ability of the designer to specify the location of the dominant roots and therefore the dominant transient response. The disadvantage of the method is that one cannot directly specify an error constant (for example, K_v) as in the Bode diagram approach. After the design is completed, one evaluates the gain of the system at the root location, which depends on *p* and *z*, and then calculates the error constant for the compensated system. If the error constant is not satisfactory, one must repeat the design steps and alter the location of the desired root as well as the location of the compensator pole and zero. We shall reconsider Examples 10.1 and 10.2 and design a compensation network using the root locus (s-plane) approach.

EXAMPLE 10.3   Lead compensator using the root locus

Let us reconsider the system of Example 10.1 where the open-loop uncompensated transfer function is

$$GH(s) = \frac{K_1}{s^2}. \tag{10.31}$$

The characteristic equation of the uncompensated system is

$$1 + GH(s) = 1 + \frac{K_1}{s^2} = 0, \tag{10.32}$$

and the root locus is the $j\omega$-axis. Therefore we desire to compensate this system with a network, $G_c(s)$, where

$$G_c(s) = \frac{s + z}{s + p} \tag{10.33}$$

and $|z| < |p|$. The specifications for the system are

Settling time, $T_s \leq 4$ seconds

Percent overshoot for a step input $\leq 35\%$.

Therefore the damping ratio should be $\zeta \geq 0.32$. The settling time requirement is

$$T_s = \frac{4}{\zeta\omega_n} = 4,$$

and therefore $\zeta\omega_n = 1$. Thus we will choose a desired dominant root location as

$$r_1, \hat{r}_1 = -1 \pm j2 \tag{10.34}$$

as shown in Fig. 10.12 (thus $\zeta = 0.45$).

   Now we place the zero of the compensator directly below the desired location at $s = -z = -1$, as shown in Fig. 10.12. Then, measuring the angle at the desired root, we have

$$\phi = -2(116°) + 90° = -142°.$$

Therefore, to have a total of 180° at the desired root, we evaluate the angle from the undetermined pole, θ_p, as

$$-180° = -142° - \theta_p \tag{10.35}$$

or $\theta_p = 38°$. Then a line is drawn at an angle $\theta_p = 38°$ intersecting the desired root location and the real axis, as shown in Fig. 10.12. The point of intersection with the real axis is then $s = -p = -3.6$. Therefore the compensator is

$$G_c(s) = \frac{s + 1}{s + 3.6}, \tag{10.36}$$

and the compensated transfer function for the system is

$$GH(s)G_c(s) = \frac{K_1(s + 1)}{s^2(s + 3.6)}. \tag{10.37}$$

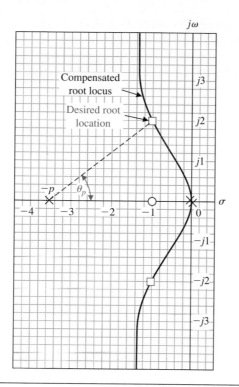

FIGURE 10.12
Phase-lead design
for Example 10.3.

The gain K_1 is evaluated by measuring the vector lengths from the poles and zeros to the root location. Hence

$$K_1 = \frac{(2.23)^2(3.25)}{2} = 8.1. \tag{10.38}$$

Finally, the error constants of this system are evaluated. We find that this system with two open-loop integrations will result in a zero steady-state error for a step and ramp input signal. The acceleration constant is

$$K_a = \frac{8.1}{3.6} = 2.25. \tag{10.39}$$

The steady-state performance of this system is quite satisfactory, and therefore the compensation is complete. When we compare the compensation network evaluated by the s-plane method with the network obtained by using the Bode diagram approach, we find that the magnitudes of the poles and zeros are different. However, the resulting system will have the same performance and we need not be concerned with the difference. In fact, the difference arises from the arbitrary design step (number 3), which places the zero directly below the desired root location. If we placed the zero at $s = -2.0$, we would find that the pole evaluated by the s-plane method is approximately equal to the pole evaluated by the Bode diagram approach.

The specifications for the transient response of this system were originally expressed in terms of the overshoot and the settling time of the system. These specifications were translated, on the basis of an approximation of the system by a second-order system, to an equivalent ζ and ω_n and therefore a desired root location. However, the original specifications will be satisfied only if the roots selected are dominant. The zero of the compensator and the root resulting from the addition of the compensator pole result in a third-order system with a zero. The validity of approximating this system with a second-order system without a zero is dependent upon the validity of the dominance assumption. Often the designer will simulate the final design by using a digital computer and obtain the actual transient response of the system. In this case, a computer simulation of the system resulted in an overshoot of 46% and a settling time of 3.8 seconds for a step input. These values compare moderately well with the specified values of 35% and 4 seconds and justify the utilization of the dominant root specifications. The difference in the overshoot from the specified value is due to the zero, which is not negligible. Thus again we find that the specification of dominant roots is a useful approach but must be utilized with caution and understanding. A second attempt to obtain a compensated system with an overshoot of 30% would utilize a prefilter to eliminate the effect of the zero in the closed-loop transfer function. The use of a prefilter will be described in Section 10.10. ∎

EXAMPLE 10.4   **Lead compensator for a type 1 system**

Now let us reconsider the system of Example 10.2 and design a compensator based on the root locus approach. The open-loop system transfer function is

$$GH(s) = \frac{K}{s(s + 2)}. \tag{10.40}$$

It is desired that the damping ratio of the dominant roots of the system be $\zeta = 0.45$ and that the velocity error constant be equal to 20. To satisfy the error constant requirement, the gain of the uncompensated system must be $K = 40$. When $K = 40$, the roots of the uncompensated system are

$$s^2 + 2s + 40 = (s + 1 + j6.25)(s + 1 - j6.25). \tag{10.41}$$

The damping ratio of the uncompensated roots is approximately 0.16, and therefore a compensation network must be added. To achieve a rapid settling time, we will select the real part of the desired roots as $\zeta\omega_n = 4$ and therefore $T_s = 1$ second. Also, the natural frequency of these roots is fairly large, $\omega_n = 9$; hence the velocity constant should be reasonably large. The location of the desired roots is shown on Fig. 10.13(a) for $\zeta\omega_n = 4$, $\zeta = 0.45$, and $\omega_n = 9$.

The zero of the compensator is placed at $s = -z = -4$, directly below the desired root location. Then the angle at the desired root location is

$$\phi = -116° - 104° + 90° = -130°. \tag{10.42}$$

Therefore the angle from the undetermined pole is determined from

$$-180° = -130° - \theta_p,$$

and thus $\theta_p = 50°$. This angle is drawn to intersect the desired root location, and p is evaluated as $s = -p = -10.6$, as shown in Fig. 10.13(a). The gain of the compensated system is then

$$K = \frac{9(8.25)(10.4)}{8} = 96.5. \tag{10.43}$$

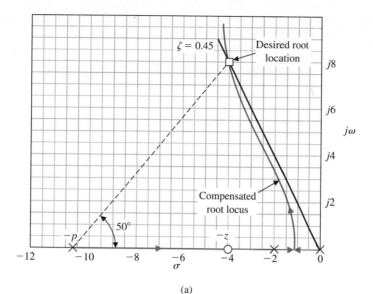

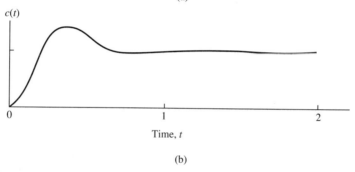

FIGURE 10.13
(a) Design of a phase-lead network on the s-plane for Example 10.4. (b) Step response of the compensated system of Example 10.4.

The compensated system is then

$$G_c(s)GH(s) = \frac{96.5(s + 4)}{s(s + 2)(s + 10.6)}. \qquad (10.44)$$

Therefore the velocity constant of the compensated system is

$$K_v = \lim_{s \to 0} s\{G(s)H(s)G_c(s)\} = \frac{96.5(4)}{2(10.6)} = 18.2. \qquad (10.45)$$

The velocity constant of the compensated system is less than the desired value of 20. Therefore we must repeat the design procedure for a second choice of a desired root. If we choose $\omega_n = 10$, the process can be repeated and the resulting gain K will be increased. The compensator pole and zero location will also be altered. Then the velocity constant can be again evaluated. We will leave it as an exercise for the reader to show that for $\omega_n = 10$, the velocity constant is $K_v = 22.7$ when $z = 4.5$ and $p = 11.6$.

Finally, for the compensation network of Eq. (10.44), we have

$$G_c(s) = \frac{s + 4}{s + 10.6} = \frac{(s + 1/\alpha\tau)}{(s + 1/\tau)}. \qquad (10.46)$$

The design of an RC-lead network to implement $G_c(s)$ as shown in Fig. 10.4 follows directly from Eqs. (10.46) and (10.7):

$$G_c(s) = \left(\frac{R_2}{R_1 + R_2}\right) \frac{(R_1 Cs + 1)}{(R_1 R_2/(R_1 + R_2)Cs + 1)}. \qquad (10.47)$$

Thus in this case we have

$$\frac{1}{R_1 C} = 4 \qquad \text{and} \qquad \alpha = \frac{R_1 + R_2}{R_2} = \frac{10.6}{4}.$$

Then, choosing $C = 1$ μf, we obtain $R_1 = 250{,}000$ ohms and $R_2 = 152{,}000$ ohms. The step response of the compensated system yields a 32% overshoot with a settling time of 0.8 second, as shown in Fig. 10.13(b). As shown here, we may use a computer to verify the actual transient response. ∎

The phase-lead compensation network is a useful compensator for altering the performance of a control system. The phase-lead network adds a phase-lead angle to provide adequate phase margin for feedback systems. Using an s-plane design approach, we can choose the phase-lead network in order to alter the system root locus and place the roots of the system in a desired position in the s-plane. When the design specifications include an error constant requirement, the Bode diagram method is more suitable, because the error constant of a system designed on the s-plane must be ascertained following the choice of a compensator pole and zero. Therefore the root locus method often results in an iterative design procedure when the error constant is specified. On the other hand, the root locus is a very satisfactory approach when the specifications are given in terms of overshoot and settling time, thus specifying the ζ and ω_n of the desired dominant roots in the s-plane. The use of a lead network compensator always extends the bandwidth of a feedback system, which may be objectionable for systems subjected to large amounts of noise. Also, lead networks are not suitable for providing high steady-state accuracy systems requiring very high error constants. To provide large error constants, typically K_p and K_v, we must consider the use of integration-type compensation networks, and therefore this will be the subject of concern in the following section.

10.6 SYSTEM DESIGN USING INTEGRATION NETWORKS

For a large percentage of control systems, the primary objective is to obtain a high steady-state accuracy. Another goal is to maintain the transient performance of these systems within reasonable limits. As we found in Chapters 4 and 5, the steady-state accuracy of many feedback systems can be increased by increasing the amplifier gain in the forward channel. However, the resulting transient response may be totally unacceptable, if not even unstable. Therefore it is often necessary to introduce a compensation network in the forward path of a feedback control system in order to provide a sufficient steady-state accuracy.

Consider the single-loop control system shown in Fig. 10.14. The compensation network is to be chosen so as to provide a large error constant. The steady-state error of this system is

$$\lim_{t \to \infty} e(t) = \lim_{s \to 0} s \left[\frac{R(s)}{1 + G_c(s)G(s)H(s)} \right]. \qquad (10.48)$$

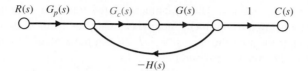

FIGURE 10.14
Single-loop
feedback control
system.

We found in Section 4.5 that the steady-state error of a system depends on the number of poles at the origin for $G_c(s)G(s)H(s)$. A pole at the origin can be considered an integration, and therefore the steady-state accuracy of a system ultimately depends on the number of integrations in the transfer function $G_c(s)G(s)H(s)$. If the steady-state accuracy is not sufficient, we will introduce an *integration-type network* $G_c(s)$ in order to compensate for the lack of integration in the original transfer function $G(s)H(s)$.

One form of controller widely used is the proportional plus integral (PI) controller, which has a transfer function

$$G_c(s) = K_P + \frac{K_I}{s}.\tag{10.49}$$

For an example, let us consider a temperature control system where the transfer function $H(s) = 1$, and the transfer function of the heat process is

$$G(s) = \frac{K_1}{(\tau_1 s + 1)(\tau_2 s + 1)}.$$

The steady-state error of the uncompensated system is then

$$\lim_{t \to \infty} e(t) = \lim_{s \to 0} s \left\{ \frac{A/s}{1 + G(s)H(s)} \right\} = \frac{A}{1 + K_1},\tag{10.50}$$

where $R(s) = A/s$, a step input signal. Clearly, to obtain a small steady-state error (less than 0.05 A, for example), the magnitude of the gain K_1 must be quite large. However, when K_1 is quite large, the transient performance of the system will very likely be unacceptable. Therefore we must consider the addition of a compensation transfer function $G_c(s)$, as shown in Fig. 10.14. To eliminate the steady-state error of this system, we might choose the compensation as

$$G_c(s) = K_2 + \frac{K_3}{s} = \frac{K_2 s + K_3}{s}.\tag{10.51}$$

This PI compensation can be readily constructed by using an integrator and an amplifier and adding their output signals. Now, the steady-state error for a step input of the system is always zero, because

$$\lim_{t \to \infty} e(t) = \lim_{s \to 0} s \frac{A/s}{1 + G_c(s)GH(s)}$$

$$= \lim_{s \to 0} \frac{A}{1 + [(K_2 s + K_3)/s]\{K_1/[(\tau_1 s + 1)(\tau_2 s + 1)]\}}\tag{10.52}$$

$$= 0.$$

The transient performance can be adjusted to satisfy the system specifications by adjusting the constants K_1, K_2, and K_3. The adjustment of the transient response is perhaps best accomplished by using the root locus methods of Chapter 7 and drawing a root locus for the gain K_2K_1 after locating the zero $s = -K_3/K_2$ on the s-plane by the method outlined for the s-plane in the preceding section.

The addition of an integration as $G_c(s) = K_2 + (K_3/s)$ can also be used to reduce the steady-state error for a ramp input, $r(t) = t$, $t \geq 0$. For example, if the uncompensated system $GH(s)$ possessed one integration, the additional integration due to $G_c(s)$ would result in a zero steady-state error for a ramp input. To illustrate the design of this type of integration compensator, we will consider a temperature control system in some detail.

EXAMPLE 10.5   Temperature control system

The uncompensated loop transfer function of a temperature control system is

$$GH(s) = \frac{K_1}{(2s + 1)(0.5s + 1)}, \tag{10.53}$$

where K_1 can be adjusted. To maintain zero steady-state error for a step input, we will add the PI compensation network

$$G_c(s) = K_2 + \frac{K_3}{s} = K_2\left(\frac{s + K_3/K_2}{s}\right). \tag{10.54}$$

Furthermore, the transient response of the system is required to have an overshoot less than or equal to 10%. Therefore the dominant complex roots must be on (or below) the $\zeta = 0.6$ line, as shown in Fig. 10.15. We will adjust the compensator zero so that the real part of the complex roots is $\zeta\omega_n = 0.75$ and thus the settling time is $T_s = 4/\zeta\omega_n = 1\frac{1}{3}$ seconds. Now, as in the preceding section, we will determine the location of the zero, $z = -K_3/K_2$, by ensuring that the angle at the desired root is $-180°$. Therefore the sum of the angles at the desired root is

$$-180° = -127° - 104° - 38° + \theta_z,$$

where θ_z is the angle from the undetermined zero. Therefore we find that $\theta_z = +89°$ and the location of the zero is $z = -0.75$. Finally, to determine the gain at the desired root, we

FIGURE 10.15
The s-plane design of an integration compensator.

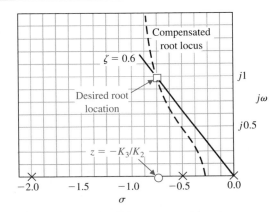

evaluate the vector lengths from the poles and zeros and obtain

$$K = K_1 K_2 = \frac{1.25(1.03)1.6}{1.0} = 2.08.$$

The compensated root locus and the location of the zero are shown in Fig. 10.15. It should be noted that the zero, $z = -K_3/K_2$, should be placed to the left of the pole at $s = -0.5$ to ensure that the complex roots dominate the transient response. In fact, the third root of the compensated system of Fig. 10.15 can be determined as $s = -1.0$, and therefore this real root is only $\frac{4}{3}$ times the real part of the complex roots. Thus, although complex roots dominate the response of the system, the equivalent damping of the system is somewhat less than $\zeta = 0.60$ due to the real root and zero.

The closed-loop transfer function of the system of Fig. 10.14 is

$$T(s) = \frac{G_p G_c G(s)}{1 + G_c G(s)} = \frac{2.08(s + 0.75)G_p(s)}{(s + 1)(s + r_1)(s + \hat{r}_1)} \tag{10.55}$$

where $r_1 = -0.75 + j1$. The effect of the zero is to increase the overshoot to a step input (see Fig. 5.13). Thus, if we wish to attain an overshoot of 5%, we may use a prefilter $G_p(s)$ so that the zero is eliminated in $T(s)$ by setting

$$G_p(s) = \frac{0.75}{(s + 0.75)}. \tag{10.56}$$

Note that the overall dc gain (set $s = 0$) is $T(0) = 1.0$ when $G_p(s) = 1$ or if we use the prefilter of Eq. (10.56). The overshoot without the prefilter is 17.6% and with the prefilter it is 2%. Further discussion of the use of a prefilter is provided in Section 10.10. ∎

10.7 PHASE-LAG DESIGN USING THE ROOT LOCUS

The phase-lag RC network of Fig. 10.6 is an integration-type network and can be used to increase the error constant of a feedback control system. We found in Section 10.3 that the transfer function of the RC phase-lag network is of the form

$$G_c(s) = \frac{1}{\alpha} \frac{(s + z)}{(s + p)}, \tag{10.57}$$

as given in Eq. (10.13), where

$$z = \frac{1}{\tau} = \frac{1}{R_2 C}, \qquad \alpha = \frac{R_1 + R_2}{R_2}, \qquad p = \frac{1}{\alpha \tau}.$$

The steady-state error of an uncompensated system is

$$\lim_{t \to \infty} e(t) = \lim_{s \to 0} s \left\{ \frac{R(s)}{1 + GH(s)} \right\}. \tag{10.58}$$

Then, for example, the velocity constant of a type-one system is

$$K_v = \lim_{s \to 0} s\{GH(s)\}, \tag{10.59}$$

as shown in Section 5.7. Therefore, if $GH(s)$ is written as

$$GH(s) = \frac{K\prod\limits_{i=1}^{M}(s + z_i)}{s\prod\limits_{j=1}^{Q}(s + p_j)}, \tag{10.60}$$

we obtain the velocity constant

$$K_v = \frac{K\prod\limits_{i=1}^{M} z_i}{\prod\limits_{j=1}^{Q} p_j}. \tag{10.61}$$

We will now add the integration-type phase-lag network as a compensator and determine the compensated velocity constant. If the velocity constant of the uncompensated system (Eq. 10.61) is designated as $K_{v_{\text{uncomp}}}$, we have

$$K_{v_{\text{comp}}} = \lim_{s \to 0} s\{G_c(s)GH(s)\} = \lim_{s \to 0} (G_c(s))K_{v_{\text{uncomp}}} \tag{10.62}$$

$$= \left(\frac{z}{p}\right)\left(\frac{1}{\alpha}\right)K_{v_{\text{uncomp}}} = \left(\frac{z}{p}\right)\left(\frac{K}{\alpha}\right)\left(\frac{\prod z_i}{\prod p_j}\right).$$

The gain on the compensated root locus at the desired root location will be (K/α). Now, if the pole and zero of the compensator are chosen so that $|z| = \alpha|p| < 1$, the resultant K_v will be increased at the desired root location by the ratio $z/p = \alpha$. Then, for example, if $z = 0.1$ and $p = 0.01$, the velocity constant of the desired root location will be increased by a factor of 10. However, if the compensator pole and zero appear relatively close together on the s-plane, their effect on the location of the desired root will be negligible. Therefore the compensator pole–zero combination near the origin of the s-plane compared to ω_n can be used to increase the error constant of a feedback system by the factor α while altering the root location very slightly. The factor α does have an upper limit, typically about 100, because the required resistors and capacitors of the network become excessively large for a higher α. For example, when $z = 0.1$ and $\alpha = 100$, we find from Eq. (10.57) that

$$z = 0.1 = \frac{1}{R_2 C} \quad \text{and} \quad \alpha = 100 = \frac{R_1 + R_2}{R_2}.$$

If we let $C = 10$ μf, then $R_1 = 1$ megohm and $R_1 = 99$ megohms. As we increase α, we increase the required magnitude of R_1. However, we should note that an attenuation, α, of 1000 or more may be obtained by utilizing pneumatic process controllers, which approximate a phase-lag characteristic (Fig. 10.8).

The steps necessary for the design of a phase-lag network on the s-plane are as follows:

1. Obtain the root locus of the uncompensated system.
2. Determine the transient performance specifications for the system, and locate suitable dominant root locations on the uncompensated root locus that will satisfy the specifications.

3. Calculate the loop gain at the desired root location and thus the system error constant.

4. Compare the uncompensated error constant with the desired error constant, and calculate the necessary increase that must result from the pole–zero ratio of the compensator, α.

5. With the known ratio of the pole–zero combination of the compensator, determine a suitable location of the pole and zero of the compensator so that the compensated root locus will still pass through the desired root location. Locate the pole and zero near the origin of the s-plane in comparison to ω_n.

The fifth requirement can be satisfied if the magnitude of the pole and zero is significantly less than ω_n of the dominant roots and they appear to merge as measured from the desired root location. The pole and zero will appear to merge at the root location if the angles from the compensator pole and zero are essentially equal as measured to the root location. One method of locating the zero and pole of the compensator is based on the requirement that the difference between the angle of the pole and the angle of the zero as measured at the desired root is less than 2°. An example will illustrate this approach to the design of a phase-lag compensator.

EXAMPLE 10.6   **Design of a phase-lag compensator**

Consider the uncompensated system of Example 10.2, where the uncompensated open-loop transfer function is

$$GH(s) = \frac{K}{s(s+2)}. \tag{10.63}$$

It is required that the damping ratio of the dominant complex roots be 0.45, while a system velocity constant equal to 20 is attained. The uncompensated root locus is a vertical line at $s = -1$ and results in a root on the $\zeta = 0.45$ line at $s = -1 \pm j2$, as shown in Fig. 10.16. Measuring the gain at this root, we have $K = (2.24)^2 = 5$. Therefore the velocity constant of the uncompensated system is

$$K_v = \frac{K}{2} = \frac{5}{2} = 2.5.$$

Thus the required ratio of the zero to the pole of the compensator is

$$\left| \frac{z}{p} \right| = \alpha = \frac{K_{v_{comp}}}{K_{v_{uncomp}}} = \frac{20}{2.5} = 8. \tag{10.64}$$

Examining Fig. 10.17, we find that we might set $z = -0.1$ and then $p = -0.1/8$. The difference of the angles from p and z at the desired root is approximately 1°, and therefore $s = -1 \pm j2$ is still the location of the dominant roots. A sketch of the compensated root locus is shown as a heavy colored line in Fig. 10.17. Therefore the compensated system transfer function is

$$G_c(s)GH(s) = \frac{5(s+0.1)}{s(s+2)(s+0.0125)}, \tag{10.65}$$

where $(K/\alpha) = 5$ or $K = 40$ in order to account for the attenuation of the lag network. ∎

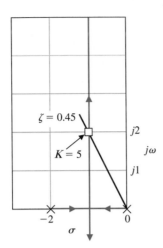

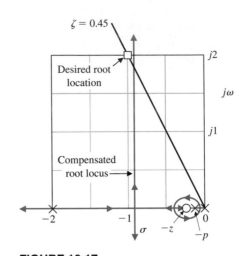

FIGURE 10.16
Root locus of the uncompensated system of Example 10.6.

FIGURE 10.17
Root locus of the compensated system of Example 10.6. Note that the actual root will differ from the desired root by a slight amount. The vertical portion of the locus leaves the σ axis at $\sigma = -0.95$.

EXAMPLE 10.7 **Design of a phase-lag compensator**

Let us now consider a system that is difficult to design using a phase-lead network. The open-loop transfer function of the uncompensated system is

$$GH(s) = \frac{K}{s(s + 10)^2}.$$ (10.66)

It is specified that the velocity constant of this system be equal to 20, while the damping ratio of the dominant roots is equal to 0.707. The gain necessary for a K_v of 20 is

$$K_v = 20 = \frac{K}{(10)^2}$$

or $K = 2000$. However, using Routh's criterion, we find that the roots of the characteristic equation lie on the $j\omega$-axis at $\pm j10$ when $K = 2000$. Clearly, the roots of the system when the K_v requirement is satisfied are a long way from satisfying the damping ratio specification, and it would be difficult to bring the dominant roots from the $j\omega$-axis to the $\zeta = 0.707$ line by using a phase-lead compensator. Therefore we will attempt to satisfy the K_v and ζ requirements by using a phase-lag network. The uncompensated root locus of this system is shown in Fig. 10.18, and the roots are shown when $\zeta = 0.707$ and $s = -2.9 \pm j2.9$. Measuring the gain at these roots, we find that $K = 236$. Therefore the necessary ratio of zero to pole of the compensator (use Eq. 10.64) is

$$\alpha = \left| \frac{z}{p} \right| = \frac{2000}{236} = 8.5.$$

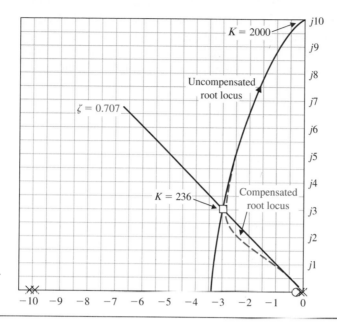

FIGURE 10.18
Design of a phase-lag compensator on the *s*-plane.

Thus we will choose $z = 0.1$ and $p = 0.1/9$ in order to allow a small margin of safety. Examining Fig. 10.18, we find that the difference between the angle from the pole and zero of $G_c(s)$ is negligible. Therefore the compensated system is

$$G_c(s)GH(s) = \frac{236(s + 0.1)}{s(s + 10)^2(s + 0.0111)}. \qquad (10.67)$$

where $(K/\alpha) = 236$ and $\alpha = 9$. ∎

The design of an integration compensator to increase the error constant of an uncompensated control system is particularly illustrative using *s*-plane and root locus methods. We shall now turn to similarly useful methods of designing integration compensation using Bode diagrams.

10.8 PHASE-LAG DESIGN USING THE BODE DIAGRAM

The design of a phase-lag *RC* network suitable for compensating a feedback control system can be readily accomplished on the Bode diagram. The advantage of the Bode diagram is again apparent, for we will simply add the frequency response of the compensator to the Bode diagram of the uncompensated system in order to obtain a satisfactory system frequency response. The transfer function of the phase-lag network written in Bode diagram form is

$$G_c(j\omega) = \frac{1 + j\omega\tau}{1 + j\omega\alpha\tau}, \qquad (10.68)$$

as we found in Eq. (10.14). The Bode diagram of the phase-lag network is shown in Fig. 10.8 for two values of α. On the Bode diagram, the pole and zero of the compensator have a magnitude much smaller than the smallest pole of the uncompensated system. Thus the phase lag is not the useful effect of the compensator, but rather it is the attenuation $-20 \log \alpha$ that is the useful effect for compensation. The phase-lag network is used to provide an attenuation and therefore to lower the 0-dB (crossover) frequency of the system. However, at lower crossover frequencies, we usually find that the phase margin of the system is increased and our specifications can be satisfied. The design procedure for a phase-lag network on the Bode diagram is as follows:

1. Draw the Bode diagram of the uncompensated system with the gain adjusted for the desired error constant.

2. Determine the phase margin of the uncompensated system and, if it is insufficient, proceed with the following steps.

3. Determine the frequency where the phase margin requirement would be satisfied if the magnitude curve crossed the 0-dB line at this frequency, ω'_c. (Allow for 5° phase lag from the phase-lag network when determining the new crossover frequency.)

4. Place the zero of the compensator one decade below the new crossover frequency ω'_c and thus ensure only 5° of additional phase lag at ω'_c (see Fig. 10.8) due to the lag network.

5. Measure the necessary attenuation at ω'_c to ensure that the magnitude curve crosses at this frequency.

6. Calculate α by noting that the attenuation introduced by the phase-lag network is $-20 \log \alpha$ at ω'_c.

7. Calculate the pole as $\omega_p = 1/\alpha\tau = \omega_z/\alpha$ and the design is completed.

An example of this design procedure will illustrate that the method is simple to carry out in practice.

EXAMPLE 10.8   Design of a phase-lag network

Let us reconsider the system of Example 10.6 and design a phase-lag network so that the desired phase margin is obtained. The uncompensated transfer function is

$$GH(j\omega) = \frac{K}{j\omega(j\omega + 2)} = \frac{K_v}{j\omega(0.5j\omega + 1)}, \tag{10.69}$$

where $K_v = K/2$. It is desired that $K_v = 20$ while a phase margin of 45° is attained. The uncompensated Bode diagram is shown as a solid line in Fig. 10.19. The uncompensated system has a phase margin of 20°, and the phase margin must be increased. Allowing 5° for the phase-lag compensator, we locate the frequency ω where $\phi(\omega) = -130°$, which is to be our new crossover frequency ω'_c. In this case, we find that $\omega'_c = 1.5$, which allows for a small margin of safety. The attenuation necessary to cause ω'_c to be the new crossover frequency is equal to 20 dB. Both the compensated and uncompensated magnitude curves are an asymptotic approximation. Both the actual curves are 2 dB lower than shown. Thus $\omega'_c = 1.5$ and the required attenuation is 20 dB.

Then we find that 20 dB = 20 log α, or $\alpha = 10$. Therefore the zero is one decade

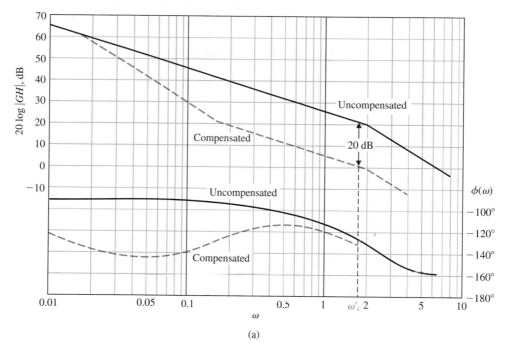

(a)

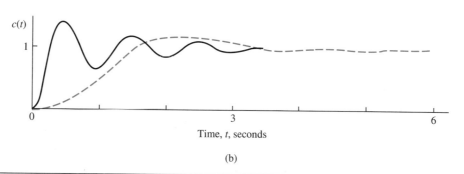

(b)

FIGURE 10.19
(a) Design of a phase-lag network on the Bode diagram for Example 10.8. (b) Time response to a step input for the uncompensated system (solid line) and the compensated system (dashed line) of Example 10.8.

below the crossover, or $\omega_z = \omega_c'/10 = 0.15$, and the pole is at $\omega_p = \omega_z/10 = 0.015$. The compensated system is then

$$G_c(j\omega)GH(j\omega) = \frac{20(6.66j\omega + 1)}{j\omega(0.5j\omega + 1)(66.6j\omega + 1)}. \tag{10.70}$$

The frequency response of the compensated system is shown in Fig. 10.19(a) with dashed lines. It is evident that the phase lag introduces an attenuation that lowers the crossover frequency and therefore increases the phase margin. Note that the phase angle of the lag network has almost totally disappeared at the crossover frequency ω_c'. As a final check, we numerically evaluate the phase margin at $\omega_c' = 1.5$ and find that $\phi_{pm} = 45°$, which is the desired result. Using the Nichols chart, we find that the closed-loop bandwidth of the system has been reduced from $\omega = 10$ rad/s for the uncompensated system to $\omega = 2.5$ rad/s for the compensated system. Due to the reduced bandwidth, we expect a slower time response to a step command.

The time response of the system is shown in Fig. 10.19(b). Note that the overshoot is 25% and the peak time is 2 seconds. Thus the response is within the specifications. ∎

EXAMPLE 10.9   **Design of a phase-lag compensator**

Let us reconsider the system of Example 10.7, which is

$$GH(j\omega) = \frac{K}{j\omega(j\omega + 10)^2} = \frac{K_v}{j\omega(0.1j\omega + 1)^2} \tag{10.71}$$

where $K_v = K/100$. A velocity constant of K_v equal to 20 is specified. Furthermore, a damping ratio of 0.707 for the dominant roots is required. From Fig. 9.21 we estimate that a phase margin of 65° is required. The frequency response of the uncompensated system is shown in Fig. 10.20. The phase margin of the uncompensated system is zero degrees. Allowing 5° for the lag network, we locate the frequency where the phase is −110°. This frequency is equal to 1.74, and therefore we will attempt to locate the new crossover frequency at $\omega_c' = 1.5$. Measuring the necessary attenuation at $\omega = \omega_{c_1}'$, we find that 23 dB is required; $23 = 20 \log \alpha$, or $\alpha = 14.2$. The zero of the compensator is located one decade below the crossover frequency, and thus

$$\omega_z = \frac{\omega_c'}{10} = 0.15.$$

The pole is then

$$\omega_p = \frac{\omega_z}{\alpha} = \frac{0.15}{14.2}.$$

FIGURE 10.20
Design of a phase-lag network on the Bode diagram for Example 10.9.

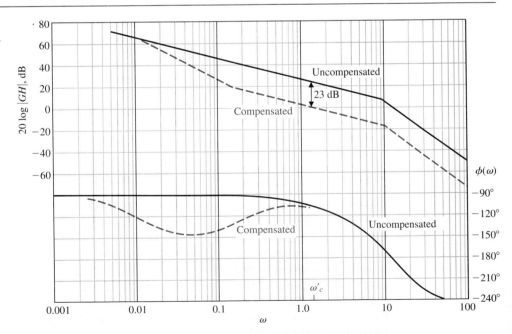

Therefore the compensated system is

$$G_c(j\omega)GH(j\omega) = \frac{20(6.66j\omega + 1)}{j\omega(0.1j\omega + 1)^2(94.6j\omega + 1)}. \tag{10.72}$$

The compensated frequency response is shown in Fig. 10.20. As a final check, we numerically evaluate the phase margin at $\omega_c' = 1.5$ and find that $\phi_{pm} = 67°$, which is within the specifications. ∎

Therefore a phase-lag compensation network can be used to alter the frequency response of a feedback control system in order to attain satisfactory system performance. Examining both Examples 10.8 and 10.9, we note again that the system design is satisfactory when the asymptotic curve for the magnitude of the compensated system crosses the 0-dB line with a slope of -6 dB/octave. The attenuation of the phase-lag network reduces the magnitude of the crossover (0-dB) frequency to a point where the phase margin of the system is satisfactory. Thus, in contrast to the phase-lead network, the phase-lag network reduces the closed-loop bandwidth of the system as it maintains a suitable error constant.

One might ask, Why do we not place the compensator zero more than one decade below the new crossover ω_c' (see item 4 of the design procedure) and thus ensure less than 5° of lag at ω_c' due to the compensator? This question can be answered by considering the requirements placed on the resistors and capacitors of the lag network by the values of the poles and zeros (see Eq. 10.12). As the magnitudes of the pole and zero of the lag network are decreased, the magnitudes of the resistors and the capacitor required increase proportionately. The zero of the lag compensator in terms of the circuit components is $z = 1/R_2C$, and the α of the network is $\alpha = (R_1 + R_2)/R_2$. Thus, considering Example 10.9, we require a zero at $z = 0.15$, which can be obtained with $C = 1$ μf and $R_2 = 6.66$ megohms. However, for $\alpha = 14$ we require a resistance R_1 of $R_1 = R_2(\alpha - 1) = 88$ megohms. Clearly, a designer does not wish to place the zero z further than one decade below ω_c' and thus require larger values of R_1, R_2, and C.

The phase-lead compensation network alters the frequency response of a network by adding a positive (leading) phase angle and therefore increases the phase margin at the crossover (0-dB) frequency. It becomes evident that a designer might wish to consider using a compensation network that provided the attenuation of a phase-lag network and the lead-phase angle of a phase-lead network. Such a network does exist. It is called a *lead-lag network* and is shown in Fig. 10.21. The transfer function of this network is

$$\frac{V_2(s)}{V_1(s)} = \frac{(R_1C_1s + 1)(R_2C_2s + 1)}{R_1R_2C_1C_2s^2 + (R_1C_1 + R_1C_2 + R_2C_2)s + 1}. \tag{10.73}$$

FIGURE 10.21
An *RC* lead-lag
network.

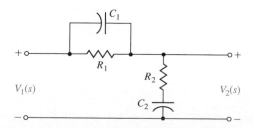

When $\alpha\tau_1 = R_1C_1$, $\beta\tau_2 = R_2C_2$, $\tau_1\tau_2 = R_1R_2C_1C_2$, we note that $\alpha\beta = 1$ and then Eq. (10.73) is

$$\frac{V_2(s)}{V_1(s)} = \frac{(1 + \alpha\tau_1 s)(1 + \beta\tau_2 s)}{(1 + \tau_1 s)(1 + \tau_2 s)}, \tag{10.74}$$

where $\alpha > 1$ and $\beta < 1$. The first terms in the numerator and denominator, which are a function of τ_1, provide the phase-lead portion of the network. The second terms, which are a function of τ_2, provide the phase-lag portion of the compensation network. The parameter β is adjusted to provide suitable attenuation of the low frequency portion of the frequency response, and the parameter α is adjusted to provide an additional phase lead at the new crossover (0-dB) frequency. Alternatively, the compensation can be designed on the s-plane by placing the lead pole and zero compensation in order to locate the dominant roots in a desired location. Then the phase-lag compensation is used to raise the error constant at the dominant root location by a suitable ratio, $1/\beta$. The design of a phase lead-lag compensator follows the procedures already discussed, and the reader is referred to further literature illustrating the utility of lead-lag compensation [2, 3, 29].

10.9 SYSTEM DESIGN ON THE BODE DIAGRAM USING ANALYTICAL AND COMPUTER METHODS

It is desirable to use computers, when appropriate, to assist the designer in the selection of the parameters of a compensator. The development of algorithms for computer-aided design is an important alternative approach to the trial-and-error methods considered in earlier sections. Computer programs have been developed for the selection of suitable parameter values for compensators based on satisfaction of frequency response criteria such as phase margin [3, 4].

An analytical technique of selecting the parameters of a lead or lag network has been developed for Bode diagrams [4, 5]. For a single-stage compensator

$$G_c(s) = \frac{1 + \alpha\tau s}{1 + \tau s}, \tag{10.75}$$

where $\alpha < 1$ yields a lag compensator and $\alpha > 1$ yields a lead compensator. The phase contribution of the compensator at the desired crossover frequency ω_c (see Eq. 10.9) is

$$p = \tan\phi = \frac{\alpha\omega_c\tau - \omega_c\tau}{1 + (\omega_c\tau)^2\alpha}. \tag{10.76}$$

The magnitude M (in dB) of the compensator at ω_c is

$$c = 10^{M/10} = \frac{1 + (\omega_c\alpha\tau)^2}{1 + (\omega_c\tau)^2}. \tag{10.77}$$

Eliminating $\omega_c\tau$ from Eqs. (10.76) and (10.77), we obtain the nontrivial solution equation for α as

$$(p^2 - c + 1)\alpha^2 + 2p^2c\alpha + p^2c^2 + c^2 - c = 0. \tag{10.78}$$

For a single-stage compensator, it is necessary that $c > p^2 + 1$. If we solve for α from

Eq. (10.78), we can obtain τ from

$$\tau = \frac{1}{\omega_c} \sqrt{\frac{1 - c}{c - \alpha^2}}. \qquad (10.79)$$

The design steps for a lead compensator are:

1. Select the desired ω_c.
2. Determine the phase margin desired and therefore the required phase ϕ for Eq. (10.76).
3. Verify that the phase lead is applicable, $\phi > 0$ and $M > 0$.
4. Determine whether a single stage will be sufficient when $c > p^2 + 1$.
5. Determine α from Eq. (10.78).
6. Determine τ from Eq. (10.79).

If we need to design a single-lag compensator, then $\phi < 0$ and $M < 0$ (step 3). Also, step 4 will require $c < [1/(1 + p^2)]$. Otherwise the method is the same.

EXAMPLE 10.10   Design using an analytical technique

Let us reconsider the system of Example 10.1 and design a lead network by the analytical technique. Examine the uncompensated curves in Fig. 10.9. We select $\omega_c = 5$. Then, as before, we desire a phase margin of 45°. The compensator must yield this phase, so

$$p = \tan 45° = 1. \qquad (10.80)$$

The required magnitude contribution is 8 dB or $M = 8$, so that

$$c = 10^{8/10} = 6.31. \qquad (10.81)$$

Using c and p, we obtain

$$-4.31\alpha^2 + 12.62\alpha + 73.32 = 0. \qquad (10.82)$$

Solving for α, we obtain $\alpha = 5.84$. Solving Eq. (10.79), we obtain $\tau = 0.087$. Therefore the compensator is

$$G_c(s) = \frac{1 + 0.515s}{1 + 0.087s}. \qquad (10.83)$$

The pole is equal to 11.5 and the zero is 1.94. This design is similar to that obtained by the iteration technique of Section 10.4. ∎

10.10 SYSTEMS WITH A PREFILTER

In the earlier sections of this chapter, we utilized compensators of the form

$$G_c(s) = \frac{(s + z)}{(s + p)}$$

that alter the roots of the characteristic equation of the closed-loop system. However, the closed-loop transfer function, $T(s)$, will contain the zero of $G_c(s)$ as a zero of $T(s)$. This zero will significantly affect the response of the system $T(s)$.

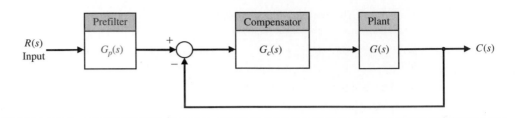

FIGURE 10.22
Control system
with a prefilter,
$G_p(s)$.

Let us consider the system shown in Fig. 10.22, where

$$G(s) = \frac{1}{s}.$$

We will introduce a PI compensator so that

$$G_c(s) = K_1 + \frac{K_2}{s} = \frac{K_1 s + K_2}{s}.$$

The closed-loop transfer function of the system with a prefilter (Fig. 10.22) is

$$T(s) = \frac{(K_1 s + K_2)G_p(s)}{s^2 + K_1 s + K_2}. \tag{10.84}$$

For illustrative purposes, the specifications require a settling time of 0.5 second and an overshoot of approximately 4%. Then, we use $\zeta = 1/\sqrt{2}$ and note that

$$T_s = \frac{4}{\zeta \omega_n}.$$

Thus we require $\zeta \omega_n = 8$ or $\omega_n = 8\sqrt{2}$. We now obtain

$$K_1 = 2\zeta \omega_n = 16 \qquad \text{and} \qquad K_2 = \omega_n^2 = 128.$$

The closed-loop transfer function when $G_p(s) = 1$ is then

$$T(s) = \frac{16(s + 8)}{s^2 + 16s + 128}.$$

The effect of the zero on the step response is significant. Using Fig. 5.13(a), we have $a/\zeta \omega_n = 1$ and $\zeta = 1/\sqrt{2}$ and the overshoot to a step as predicted from Fig. 5.13(a) is 21%.

We use a prefilter $G_p(s)$ to eliminate the zero from $T(s)$ while maintaining the dc gain of 1, thus requiring

$$G_p(s) = \frac{8}{s + 8}.$$

Then we have

$$T(s) = \frac{128}{s^2 + 16s + 128},$$

and the overshoot of this system is 4.5%, as expected.

Reviewing Fig. 5.13(a), we note that the zero at $s = -a$ has a significant effect when $a/\zeta\omega_n < 5$ where $-a$ is the zero and $-\zeta\omega_n$ is the real part of the dominant roots of the characteristic equation of $T(s)$.

Let us now reconsider Example 10.3, which includes the design of a lead compensator. The resulting closed-loop transfer (using Fig. 10.22) was determined to be

$$T(s) = \frac{8.1(s + 1)G_p(s)}{(s + 1 + j2)(s + 1 - j2)(s + 1.62)}.$$

If $G_p(s) = 1$ (no prefilter), then we obtain a response with an overshoot of 46.6% and a settling time of 3.8 seconds. If we use a prefilter,

$$G_p(s) = \frac{1}{s + 1},$$

we obtain an overshoot of 6.7% and a settling time of 3.8 seconds. The real root at $s = -1.62$ helps to damp the step response. The prefilter is very useful in permitting the designer to introduce a compensator with a zero to adjust the root locations (poles) of the closed-loop transfer function while eliminating the effect of the zero incorporated in $T(s)$.

In general, we will add a prefilter for systems with lead networks or PI compensators. We will not use a prefilter for a system with a lag network, since we expect the effect of the zero to be insignificant. To check this assertion, let us reconsider the design obtained in Example 10.6. The system with a phase-lag controller is

$$G(s)G_c(s) = \frac{5(s + 0.1)}{s(s + 2)(s + 0.0125)}.$$

The closed-loop transfer is then

$$T(s) \cong \frac{5(s + 0.1)}{(s^2 + 1.98s + 5.1)(s + 0.095)} \approx \frac{5}{(s^2 + 1.98s + 5.1)},$$

since the zero at $s = -0.1$ and the pole at $s = -0.095$ approximately cancel. We expect an overshoot of 20% and a settling time of 4.0 seconds for the design parameters ($\zeta = 0.45$ and $\zeta\omega_n = 1$). However, the actual response has an overshoot of 26% and a longer settling time of 5.8 seconds due to the effect of the real pole of $T(s)$ at $s = -0.095$. Thus we usually do not use a prefilter with systems that utilize lag compensators.

EXAMPLE 10.11   **Design of a third-order system**

Consider a system of the form shown in Fig. 10.22 with

$$G(s) = \frac{1}{s(s + 1)(s + 5)}.$$

Let us design a system that will yield a step response with an overshoot less than 2% and a settling time less than 3 seconds by using both $G_c(s)$ and $G_p(s)$ to achieve the desired response.

We use a lead compensation network

$$G_c(s) = \frac{K(s + 1.2)}{(s + 10)}.$$

TABLE 10.1   Effect of a Prefilter on the Step Response

	$G_p(s) = 1$	$p = 1.20$	$p = 2.4$
Percent overshoot	9.9%	0%	4.8%
90% rise time (seconds)	1.05	2.30	1.60
Settling time (seconds)	2.9	3.0	3.2

and select K to find the complex roots with $\zeta = 1/\sqrt{2}$. Then, with $K = 78.7$, the closed-loop transfer function is

$$T(s) = \frac{78.7(s + 1.2)G_p(s)}{(s + 1.71 + j1.71)(s + 1.71 - j1.71)(s + 1.45)(s + 11.1)}$$

$$\cong \frac{7.1(s + 1.2)G_p(s)}{(s^2 + 3.4s + 5.85)(s + 1.45)},$$

where

$$G_p(s) = \frac{p}{(s + p)}. \qquad (10.85)$$

Then the closed-loop transfer function is

$$T(s) \cong \frac{7.1p(s + 1.2)}{(s^2 + 3.45s + 5.85)(s + 1.45)(s + p)}.$$

If $p = 1.2$, we cancel the effect of the zero. The response of the system with a prefilter is summarized in Table 10.1. We choose the appropriate value for p to achieve the response desired. Note that $p = 2.40$ will provide a response that may be desirable, since it effects a faster rise time than $p = 1.20$. The prefilter provides an additional parameter to select for design purposes. ■

10.11 DESIGN FOR DEADBEAT RESPONSE

Often the goal for a control system is to achieve a fast response to a step command with minimal overshoot. We define a *deadbeat response* as a response that proceeds rapidly to the desired level and holds at that level with minimal overshoot. We use the $\pm 2\%$ band at the desired level as the acceptable range of variation from the desired response. Then, if the response enters the band at time, T_s, it has satisfied the settling time T_s upon entry to the band, as illustrated in Fig. 10.23. A deadbeat response has the following characteristics:

1. Steady-state error $= 0$

2. Fast response → minimum rise time and settling time

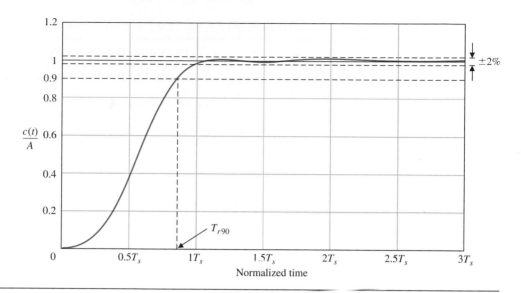

FIGURE 10.23
The deadbeat response. A is the magnitude of the step input.

3. $0.1\% \leq$ percent overshoot $< 2\%$

4. Percent undershoot $< 2\%$

Characteristics (3) and (4) require that the response remain within the $\pm 2\%$ band so that the entry to the band occurs at the settling time.

We consider the transfer function of a closed-loop system, $T(s)$. To determine the coefficients that yield the optimal deadbeat response, the standard transfer function is first normalized. An example of this for a third-order system is

$$T(s) = \frac{\omega_n^3}{s^3 + \alpha\omega_n s^2 + \beta\omega_n^2 s + \omega_n^3}. \tag{10.86}$$

Dividing the numerator and denominator by ω_n^3 yields

$$T(s) = \frac{1}{\dfrac{s^3}{\omega_n^3} + \alpha\dfrac{s^2}{\omega_n^2} + \beta\dfrac{s}{\omega_n} + 1}. \tag{10.87}$$

Let $\bar{s} = s/\omega_n$ to obtain

$$T(s) = \frac{1}{\bar{s}^3 + \alpha\bar{s}^2 + \beta\bar{s} + 1}. \tag{10.88}$$

Equation (10.88) is the normalized, third-order, closed-loop transfer function. For a higher-order system, the same method is used to derive the normalized equation. The coefficients of the equation—α, β, γ, and so on—are then assigned the values necessary to meet the requirement of deadbeat response. The coefficients recorded in Table 10.2 were selected to achieve deadbeat response and minimize settling time and rise time to 100% of the desired command, T_r. The form of Eq. (10.88) is normalized since $\bar{s} = s/\omega_n$. Thus we choose ω_n based on the desired settling time or rise time. Therefore, if we have a third-order system with a required settling time of 1.2 seconds, we note from Table 10.2 that the normalized settling time is

$$\omega_n T_s = 4.04.$$

TABLE 10.2   Coefficients and Response Measures of a Deadbeat System

System Order	Coefficients					Percent Over-shoot, P.O.	Percent Under-shoot, P.U.	90% Rise Time, T_{r90}	100% Rise Time, T_r	Settling Time, T_s
	α	β	γ	δ	ϵ					
2nd	1.82					0.10%	0.00%	3.47	6.58	4.82
3rd	1.90	2.20				1.65%	1.36%	3.48	4.32	4.04
4th	2.20	3.50	2.80			0.89%	0.95%	4.16	5.29	4.81
5th	2.70	4.90	5.40	3.40		1.29%	0.37%	4.84	5.73	5.43
6th	3.15	6.50	8.70	7.55	4.05	1.63%	0.94%	5.49	6.31	6.04

Note: All time is normalized.

Therefore we require

$$\omega_n = \frac{4.04}{T_s} = \frac{4.04}{1.2} = 3.37.$$

Once ω_n is chosen, the complete closed-loop transfer function is known, having the form of Eq. (10.86). When designing a system to obtain a deadbeat response, the compensator is chosen and the closed-loop transfer function is found. This compensated transfer function is then set equal to Eq. (10.86) and the required compensator can be determined.

EXAMPLE 10.12   Design of a system with a deadbeat response

Let us consider a unity feedback system with a compensator $G_c(s)$ and a prefilter $G_p(s)$ as shown in Fig. 10.22. The plant is

$$G(s) = \frac{K}{s(s + 1)}$$

and the compensator is

$$G_c(s) = \frac{(s + z)}{(s + p)}.$$

Using the necessary prefilter,

$$G_p(s) = \frac{z}{(s + z)}.$$

The closed-loop transfer function is

$$T(s) = \frac{Kz}{s^3 + (1 + p)s^2 + (K + p)s + Kz}.$$

We use Table 10.2 to determine the required coefficients, $\alpha = 1.90$ and $\beta = 2.20$. If we select a settling time of 2 seconds, then $\omega_n T_s = 4.04$ and thus $\omega_n = 2.02$. The required closed-loop system has the characteristic equation

$$q(s) = s^3 + \alpha\omega_n s^2 + \beta\omega_n^2 s + \omega_n^3 = s^3 + 3.84s^2 + 4.44s + 8.24.$$

Then we determine that $p = 2.84$, $z = 1.34$, and $K = 6.14$. The response of this system will have $T_s = 2$ seconds, $T_r = 2.14$ seconds, and $T_{r90} = 1.72$ seconds. ∎

10.12 DESIGN EXAMPLE: ROTOR WINDER CONTROL SYSTEM

Our goal is to replace a manual operation using a machine to wind copper wire onto the rotors of small motors. Each motor has three separate windings of several hundred turns of wire. It is important that the windings be consistent and that the throughput of the process be high. The operator simply inserts an unwound rotor, pushes a start button, and then removes the completely wound rotor. The dc motor is used to achieve accurate, rapid windings. Thus the goal is to achieve high steady-state accuracy for both position and velocity. The control system is shown in Fig. 10.24(a) and the block diagram in Fig. 10.24(b). This system has zero steady-state error for a step input, and the steady-state error for a ramp input is

$$e_{ss} = A/K_v,$$

where

$$K_v = \lim_{s \to 0} \frac{G_c(s)}{50}.$$

FIGURE 10.24
(a) Rotor winder
control system.
(b) Block diagram.

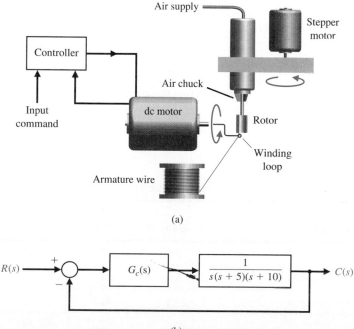

(a)

(b)

When $G_c(s) = K$, we have $K_v = K/50$. If we select $K = 500$, we will have $K_v = 10$, but the overshoot to a step is 70% and the settling time is 8 seconds.

First, let us try a lead compensator so that

$$G_c(s) = \frac{K(s + z_1)}{(s + p_1)}. \tag{10.89}$$

Selecting $z_1 = 4$ and the pole p_1 so that the complex roots have a ζ of 0.6, we have

$$G_c(s) = \frac{191.2(s + 4)}{(s + 7.3)}. \tag{10.90}$$

Therefore the response to a step input has a 3% overshoot and a settling time of 1.5 seconds. However,

$$K_v = \frac{191.2(4)}{7.3(50)} = 2.1,$$

which is inadequate.

If we use a phase-lag design, we select

$$G_c(s) = \frac{K(s + z_2)}{(s + p_2)}$$

in order to achieve $K_v = 38$. Thus the velocity constant of the lag-compensated system is

$$K_v = \frac{Kz_2}{50p_2}.$$

Using a root locus, we select $K = 105$ in order to achieve a reasonable uncompensated step response with an overshoot of less than or equal to 10%. We select $\alpha = z/p$ to achieve the desired K_v. We then have

$$\alpha = \frac{50K_v}{K} = \frac{50(38)}{105} = 18.1.$$

Selecting $z_2 = 0.1$ to avoid impacting the uncompensated root locus, we have $p_2 = 0.0055$. We then obtain a step response with a 12% overshoot and a settling time of 2.5 seconds.

The results for the simple gain, the lead network, and the lag network are summarized in Table 10.3.

TABLE 10.3   Design Example Results

Controller	Gain, K	Lead Network	Lag Network	Lead-Lag Network
Step overshoot	70%	3%	12%	5%
Settling time (seconds)	8	1.5	2.5	2.0
Steady-state error for ramp	10%	48%	2.6%	4.8%
K_v	10	2.1	38	21

Let us return to the lead-network system and add a cascade lag network so that the compensator is

$$G_c(s) = \frac{K(s + z_1)(s + z_2)}{(s + p_1)(s + p_2)}.$$

(10.91)

The lead compensator of Eq. (10.90) requires $K = 191.2$, $z_1 = 4$, and $p_1 = 7.3$. The root locus for the system is shown in Fig. 10.25. We recall that this lead network resulted in $K_v = 2.1$ (see Table 10.3). To obtain $K_v = 21$, we use $\alpha = 10$ and select $z_2 = 0.1$ and $p_2 = 0.01$. Then the total system is

$$G(s)G_c(s) = \frac{191.2(s + 4)(s + 0.1)}{s(s + 5)(s + 10)(s + 7.28)(s + 0.01)}.$$

(10.92)

The step response and ramp response of this system are shown in Fig. 10.26 in parts (a) and (b), respectively, and are summarized in Table 10.3. Clearly, the lead-lag design is suitable for satisfaction of the design goals.

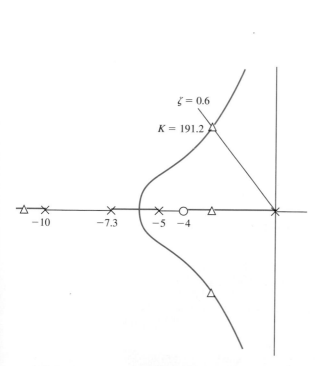

FIGURE 10.25
Root locus for lead compensator.

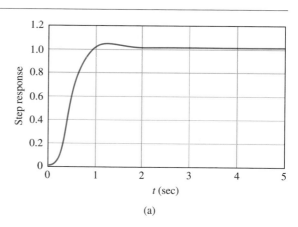

(a)

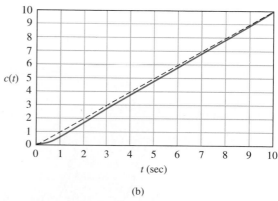

(b)

FIGURE 10.26
(a) Step response and (b) ramp response for rotor winder system.

10.13 DESIGN EXAMPLE: THE *X-Y* PLOTTER

Many physical phenomena are characterized by parameters that are transient or slowly varying. If recorded, these changes can be examined at leisure and stored for future reference or comparison. To accomplish such a recording, a number of electromechanical instruments have been developed, among them the *X-Y* recorder. In this instrument, the displacement along the *X*-axis represents a variable of interest or time and the displacement along the *Y*-axis varies as a function of yet another variable [6].

Such recorders can be found in many laboratories recording experimental data such as changes in temperature, variations in transducer output levels, and stress versus applied strain, to name just a few. The HP 7090A plotting system is shown in Fig. 10.27.

The purpose of a plotter is to follow accurately the input signal as it varies. We will consider the design of the movement of one axis, since the movement dynamics of both axes are identical. Thus we will strive to control very accurately the position and the movement of the pen as it follows the input signal.

To achieve accurate results, our goal is to achieve (1) a step response with an overshoot of less than 5% and a settling time less than 0.5 second, and (2) a percentage steady-state error for a step equal to zero. If we achieve these specifications, we will have a fast and accurate response.

Since we wish to move the pen, we select a dc motor as the actuator. The feedback sensor will be a 500-line optical encoder. By detecting all state changes of the two-channel quadrature output of the encoder, 2000 encoder counts per revolution of the motor shaft can be detected. This yields an encoder resolution of 0.001 inch at the pen tip. The encoder is mounted on the shaft of the motor. Since the encoder provides digital data, it is compared with the input signal by using a microprocessor. Next, we propose to use the difference signal calculated by the microprocessor as the error signal and then use the microprocessor to calculate the necessary algorithm to obtain the designed compensator. The output of the compensator is then converted to an analog signal that will drive the motor.

The model of the feedback position control system is shown in Fig. 10.28. Since the

FIGURE 10.27
The HP 7090A Measurement Plotting System®. Copyright 1986, Hewlett-Packard Co. Reproduced with permission.

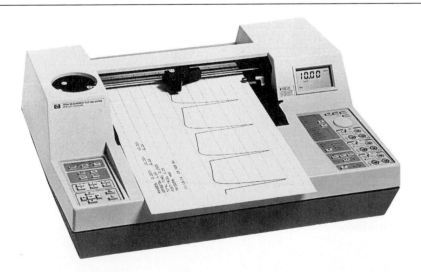

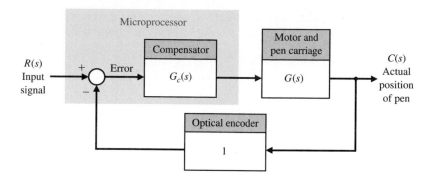

FIGURE 10.28
Model of the pen-
plotter control
system.

microprocessor calculation speed is very fast compared to the rate of change of the encoder and input signals, we assume that the continuous signal model is very accurate.

The model for the motor and pen carriage is

$$G(s) = \frac{1}{s(s + 10)(s + 1000)},$$ (10.93)

and our initial attempt at a specification of a compensator is to use a simple gain so that

$$G_c(s) - K.$$

In this case, we have only one parameter to adjust—that is, K. To achieve a fast response, we have to adjust K so that it will provide two dominant s-plane roots with a damping ratio of 0.707, which will result in a step response overshoot of about 4.5%. A sketch of the root locus (note the break in the real axis) is shown in Fig. 10.29.

Adjusting the gain to $K = 47,200$, we obtain a system with an overshoot of 3.6% to a step input and a settling time of 0.8 second. Since the transfer function has a pole at the origin, we have a steady-state error of zero for a step input.

FIGURE 10.29
Root locus for
the pen plotter,
showing the roots
with a damping
ratio of $1\sqrt{2}$. The
dominant roots are
$s = -4.9 \pm j4.9$.

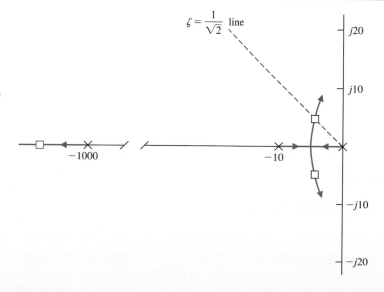

This system does not meet our specifications, so we select a compensator that will reduce the settling time. Let us select a lead compensator so that

$$G_c(s) = \frac{K\alpha(s + z)}{(s + p)},$$ (10.94)

where $p = \alpha z$. Let us use the method of Section 10.5, which selects the phase-lead compensator on the s-plane. Hence we place the zero at $s = -20$ and determine the location of the pole, p, that will place the dominant roots on the line that has the damping ratio of $1\sqrt{2}$. Thus we find that $p = 60$ and that $\alpha = 3$, so that

$$G(s)G_c(s) = \frac{142{,}600(s + 20)}{s(s + 10)(s + 60)(s + 1000)}.$$ (10.95)

Obtaining the actual step response, we determine that the percent overshoot is 2% and that the settling time is 0.35 second, which meet the specifications. The third design approach is to recognize that the encoder can be used to generate a velocity signal by counting the rate at which encoder lines pass by a fixed point. Thus we can use the microprocessor. Since the position signal and the velocity signal are available, we can describe the compensator as

$$G_c(s) = K_1 + K_2 s,$$ (10.96)

where K_1 is the gain for the error signal and K_2 is the gain for the velocity signal. Then we can write

$$G(s)G_c(s) = \frac{K_2(s + K_1/K_2)}{s(s + 10)(s + 1000)}.$$

If we set $K_1/K_2 = 10$, we cancel the pole at $s = -10$ and obtain

$$G(s)G_c(s) = \frac{K_2}{s(s + 1000)}.$$

The characteristic equation for this system is

$$s^2 + 1000s + K_2 = 0,$$ (10.97)

and we desire that $\zeta = 1/\sqrt{2}$. Noting that $2\zeta\omega_n = 1000$, we have $\omega_n = 707$ and $K_2 = \omega_n^2$. Therefore we obtain $K_2 = 5 \times 10^5$ and the compensated system is

$$G_c G(s) = \frac{5 \times 10^5}{s(s + 1000)}.$$ (10.98)

The response of this system will provide an overshoot of 4.3% and a settling time of 8 milliseconds.

The results for the three approaches to system design are compared in Table 10.4. Clearly, the best design uses the velocity feedback, and this is the actual design adopted for the HP 7090A.

10.14 SYSTEM DESIGN USING *MATLAB*

In this section, the compensation of control systems is illustrated using frequency response and s-plane methods. We reconsider the rotor winder design example of Section 10.12 to illustrate the use of *Matlab* scripts in designing and developing control systems with good

TABLE 10.4   Results for Three Designs

System	Step Response	
	Percent Overshoot	Settling Time (milliseconds)
Gain adjustment	3.6	800
Gain and lead compensator	2.0	350
Gain adjustment plus velocity signal multiplied by gain K_2	4.3	8

performance characteristics. We examine both the lead and lag compensators for this design example and display the system response using *Matlab*.

EXAMPLE 10.13   Rotor winder control system

Let us reconsider the rotor winder control system shown in Fig. 10.24. The design objective is to achieve high steady-state accuracy to a ramp input. The steady-state error to a unit ramp input, $R(s) = 1/s^2$, is

$$e_{ss} = \frac{1}{K_v},$$

where

$$K_v = \lim_{s \to 0} \frac{G_c(s)}{50}.$$

Of course, the performance specification of overshoot and settling time must be considered as well as steady-state tracking error. In all likelihood, a simple gain will not be satisfactory. So we will also consider compensation utilizing lead and lag compensators using both Bode diagram and root locus plot design methods. Our approach is to develop a series of *Matlab* scripts to aid in the compensator designs.

Consider first a simple gain controller, $G_c(s)$, where

$$G_c(s) = K.$$

Then,

$$e_{ss} = \frac{50}{K}.$$

Clearly, the larger we make K, the smaller is the steady-state error e_{ss}. However, we must consider the effect that increasing K has on the transient response, as shown in Fig. 10.30. When $K = 500$, our steady-state error for a ramp is 10% but the overshoot is 70% and the settling time is approximately 8 seconds for a step input. We consider this to be unacceptable performance and thus turn to compensation. The two important compensator types that we consider are lead and lag compensators.

First we try a lead compensator

$$G_c(s) = \frac{K(s + z)}{(s + p)},$$

where $|z| < |p|$. The lead compensator will give us the capability to improve the transient response. First, we will use a frequency domain approach to design the lead compensator.

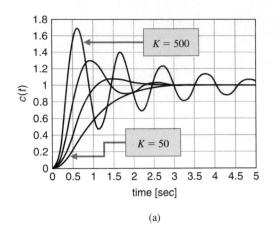

(a)

```
rotorgain.m

K=[50 100 200 500];          Compute response
%                            for four gains.
numg=[1]; deng=[1 15 50 0];
t=[0:0.1:5];
%
for i=1:4                    Closed-loop
[nums,dens]=series(K(i),1,numg,deng);   transfer function
[num,den]=cloop(nums,dens);
[y,x]=step(num,den,t);
Ys(:,i)=y;                   Store response for
end                          ith gain in Ys.
%
plot(t,Ys(:,1),'-',t,Ys(:,2),'-',t,Ys(:,3),'-',t,Ys(:,4),'-')
xlabel('time [sec]'), ylabel('c(t)')
```

FIGURE 10.30
(a) Transient
response for
simple gain
controller.
(b) *Matlab* script.

(b)

We desire a steady-state error of less than 10% to a ramp input and $K_v = 10$. In addition to the steady-state specifications, we desire to meet certain performance specifications: (1) settling time $T_s \leq 3$ seconds, and (2) percent overshoot for a step input $\leq 10\%$. Solving the approximate formulas

$$\text{P.O.} \approx 100 \ \exp^{-\zeta\pi/\sqrt{1-\zeta^2}} = 10 \quad \text{and} \quad T_s \approx \frac{4}{\zeta\omega_n} = 3$$

for ζ and ω_n, respectively, yields $\zeta = 0.59$ and $\omega_n = 2.26$, and we obtain the phase margin requirement:

$$\phi_{pm} \approx \frac{\zeta}{0.01} = 60°.$$

The steps leading to the final design are as follows:

1. Obtain the uncompensated system Bode diagram with $K = 500$, and compute the phase margin.

2. Determine the amount of necessary phase lead.

3. Evaluate α from $\sin \phi_m = (\alpha - 1)/(\alpha + 1)$.

4. Compute $10 \log \alpha$ and find the frequency ω_m on the uncompensated Bode diagram where the magnitude curve is equal to $-10 \log \alpha$.

5. In the neighborhood of ω_m on the uncompensated Bode, draw a line through the 0-dB point at ω_m with slope equal to the current slope plus 20 dB/decade. Locate the intersection of the line with the uncompensated Bode to determine the lead compensation zero location. Then calculate the lead compensator pole location as $p = \alpha z$.

6. Draw the compensated Bode, and check the phase margin. Repeat any steps if necessary.

7. Raise the gain to account for attenuation $(1/\alpha)$.

8. Verify the final design with simulation using step functions, and repeat any steps if necessary.

We utilize three scripts in the design. The design scripts are shown in Figs. 10.31–10.33. The first script is for the Bode diagram of the uncompensated system, as shown in

FIGURE 10.31
(a) Bode diagram.
(b) *Matlab* script.

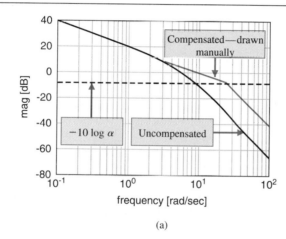

(a)

rotorlead.m

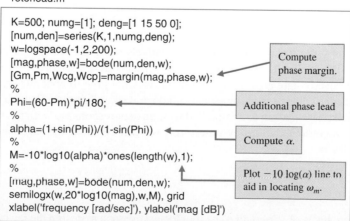

(b)

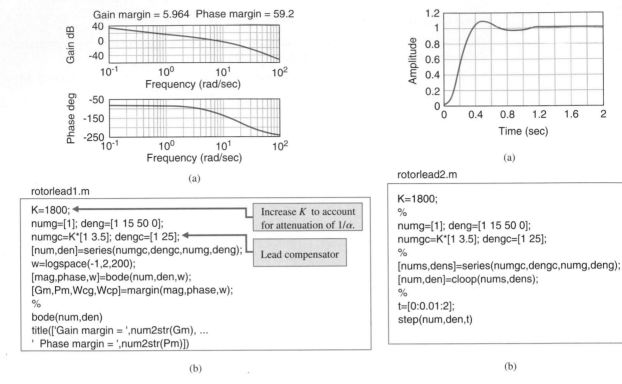

FIGURE 10.32
Lead compensator: (a) compensated Bode diagram, (b) *Matlab* script.

FIGURE 10.33
Lead compensator: (a) step response,
(b) *Matlab* script.

Fig. 10.31. The detailed Bode diagram of the compensated system is shown in Fig. 10.32. Finally, Fig. 10.33 displays the *Matlab* script for the step response analysis. The final lead compensator design is

$$G_c(s) = \frac{1800(s + 3.5)}{(s + 25)},$$

where $K = 1800$ was selected after iteratively using the *Matlab* script.

The settling time and overshoot specifications are satisfied, but $K_v = 5$, resulting in a 20% steady-state error to a ramp input. It is possible to continue the design iteration and refine the compensator somewhat, although it should be clear that the lead compensator has added phase margin and improved the transient response as anticipated.

To reduce the steady-state error, we can consider the lag compensator, which has the form

$$G_c(s) = \frac{K(s + z)}{(s + p)},$$

where $|p| < |z|$. We will use a root locus approach to design the lag compensator, although it can be done using Bode as well. The desired root location region of the dominant roots is specified by

$$\zeta = 0.59 \qquad \text{and} \qquad \omega_n = 2.26.$$

The steps in the design are as follows:

1. Obtain the root locus of the uncompensated system.
2. Locate suitable root locations on the uncompensated system that lie in the region defined by $\zeta = 0.59$ and $\omega_n = 2.26$.
3. Calculate the loop gain at the desired root location and the system error constant, $K_{v_{uncomp}}$.
4. Compute $\alpha = K_{v_{comp}}/K_{v_{uncomp}}$ where $K_{v_{comp}} = 10$.
5. With α known, determine suitable locations of the compensator pole and zero so that the compensated root locus still passes through desired location.
6. Verify with simulation and repeat any steps if necessary.

The design methodology is illustrated in Figs. 10.34–10.36. Using the rlocfind function, we can compute the gain K associated with the roots of our choice on the uncompensated

FIGURE 10.34
Lag compensator;
(a) uncompen-
sated root locus,
(b) *Matlab* script.

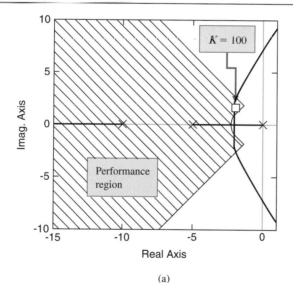

(a)

rotorlag.m

```
numg=[1]; deng=[1 15 50 0];
axis([-15,1,-10,10]);
clg; rlocus(numg,deng); hold on
%
zeta=0.5912; wn=2.2555;
%
x=[-10:0.1:-zeta*wn]; y=-(sqrt(1-zeta^2)/zeta)*x;
xc=[-10:0.1:-zeta*wn]; c=sqrt(wn^2-xc.^2);
%
plot(x,y,':',x,-y,':',xc,c,':',xc,-c,':')
```

Plot performance
regions on locus.

(b)

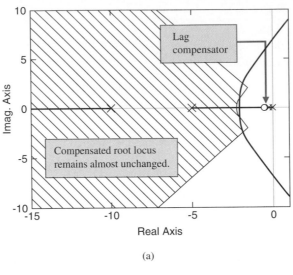

(a)

rotorlag1.m

```
numg=[1]; deng=[1 15 50 0];
numgc=[1 0.1]; dengc=[1 0.01];
[num,den]=series(numgc,dengc,numg,deng);
axis([-15,1,-10,10]);
clg; rlocus(num,den); hold on
%
zeta=0.5912; wn=2.2555;
x=[-10:0.1:-zeta*wn]; y=-(sqrt(1-zeta^2)/zeta)*x;
xc=[-10:0.1:-zeta*wn];c=sqrt(wn^2-xc.^2);
plot(x,y,':',x,-y,':',xc,c,':',xc,-c,':')
```

FIGURE 10.35
Lag compensator:
compensated root
locus.

(b)

root locus that lie in the performance region. We then compute α to ensure that we achieve the desired K_v. We place the lag compensator pole and zero to avoid impacting the uncompensated root locus. In Fig. 10.35, the lag compensator pole and zero are very near the origin, at $z = -0.1$ and $p = -0.01$.

TABLE 10.5   Compensator Design Results Using *Matlab*

Controller	Gain, *K*	Lead	Lag
Step overshoot	70%	8%	13%
Settling time (seconds)	8	1	4
Steady-state error for ramp	10%	20%	10%
K_v	10	5	10

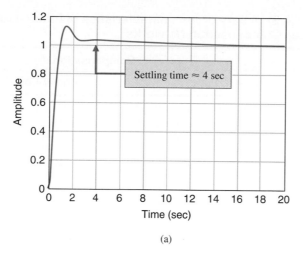

FIGURE 10.36
Lag compensator:
(a) step response,
(b) *Matlab*
response.

rotorlag2.m

```
K=100;
%
numg_[1]; deng=[1 15 50 0];
numgc=K*[1 0.1]; dengc=[1 0.01];
%
[nums,dens]=series(numgc,dengc,numg,deng);
[num,den]=cloop(nums,dens);
%
step(num,den)
```

(b)

The settling time and overshoot specifications are nearly satisfied and $K_v = 10$, as desired. It is possible to continue the design iteration and refine the compensator somewhat, although it should be clear that the lag compensator has improved the steady-state errors to a ramp input relative to the lead compensator design. The final lag compensator design is

$$G_c(s) = \frac{100(s + 0.1)}{(s + 0.01)}.$$

The resulting performance is summarized in Table 10.5. ∎

10.15 SUMMARY

In this chapter we have considered several alternative approaches to the design of feedback control systems. In the first two sections, we discussed the concepts of design and compensation and noted the several design cases that we completed in the preceding chapters. Then we examined the possibility of introducing cascade compensation networks within the feedback loops of control systems. The cascade compensation networks are useful for altering the shape of the root locus or frequency response of a system. The phase-lead network and the phase-lag network were considered in detail as candidates for system

TABLE 10.6   A Summary of the Characteristics of Phase-Lead and Phase-Lag Compensation Networks

	Compensation	
	Phase-Lead	**Phase-Lag**
Approach	Addition of phase-lead angle near crossover frequency on Bode diagram. Add lead network to yield desired dominant roots in s-plane	Addition of phase-lag to yield an increased error constant while maintaining desired dominant roots in s-plane or phase margin on Bode diagram
Results	1. Increases system bandwidth 2. Increases gain at higher frequencies	1. Decreases system bandwidth
Advantages	1. Yields desired response 2. Improves dynamic response	1. Suppresses high frequency noise 2. Reduces steady-state error
Disadvantages	1. Requires additional amplifier gain 2. Increases bandwidth and thus susceptibility to noise 3. May require large values of components for RC network	1. Slows down transient response 2. May require large values of components for RC network
Applications	1. When fast transient response is desired	1. When error constants are specified
Not applicable	1. When phase decreases rapidly near crossover frequency	1. When no low frequency range exists where phase is equal to desired phase margin

compensators. Then system compensation was studied by using a phase-lead s-plane network on the Bode diagram and the root locus s-plane. We noted that the phase-lead compensator increases the phase margin of the system and thus provides additional stability. When the design specifications include an error constant, the design of a phase-lead network is more readily accomplished on the Bode diagram. Alternatively, when an error constant is not specified but the settling time and overshoot for a step input are specified, the design of a phase-lead network is more readily carried out on the s-plane. When large error constants are specified for a feedback system, it is usually easier to compensate the system by using integration (phase-lag) networks. We also noted that the phase-lead compensation increases the system bandwidth, whereas the phase-lag compensation decreases the system bandwidth. The bandwidth often may be an important factor when noise is present at the input and generated within the system. Also we noted that a satisfactory system is obtained when the asymptotic course for magnitude of the compensated system crosses the 0-dB line with a slope of -6 dB/octave. The characteristics of the phase-lead and phase-lag compensation networks are summarized in Table 10.6. Operational amplifier circuits for phase-lead and phase-lag and for PI and PD compensators are summarized in Table 10.7 [1].

TABLE 10.7 Operational Amplifier Circuits for Compensators

Type of Controller	$G_c(s) = \dfrac{V_0(s)}{V_1(s)}$	
PD	$G_c = \dfrac{R_4 R_2}{R_3 R_1}(R_1 C_1 s + 1)$	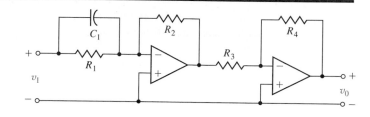
PI	$G_c = \dfrac{R_4 R_2 (R_2 C_2 s + 1)}{R_3 R_1 (R_2 C_2 s)}$	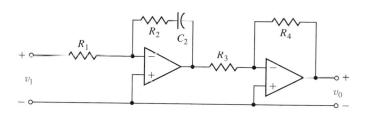
Lead or lag $\quad$ Lead if $R_1 C_1 > R_2 C_2$ $\quad$ Lag if $R_1 C_1 < R_2 C_2$	$G_c = \dfrac{R_4 R_2 (R_1 C_1 s + 1)}{R_3 R_1 (R_2 C_2 s + 1)}$	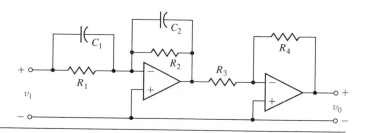

EXERCISES

E10.1 A negative feedback control system has a transfer function

$$G(s) = \frac{K}{(s + 3)}.$$

We select a compensator,

$$G_c(s) = \frac{s + a}{s},$$

in order to achieve zero steady-state error for a step input. Select a and K so that the overshoot to a step is approximately 5% and the settling time is approximately 1 second.

Answer: $K = 5$, $a = 6.4$

E10.2 A control system with negative unity feedback has a process

$$G(s) = \frac{400}{s(s + 40)},$$

and we wish to use a proportional plus integral compensation where

$$G_c(s) = K_1 + \frac{K_2}{s}.$$

Note that the steady-state error of this system for a ramp input is zero. (a) Set $K_2 = 1$ and find a suitable value of K_1 so the step response will have an overshoot of approximately 20%. (b) What is the expected settling time of the compensated system?

E10.3   A unity negative feedback control system in a manufacturing system has a process transfer function

$$G(s) = \frac{e^{-s}}{s + 1},$$

and it is proposed to use a compensator to achieve a 5% overshoot to a step input. The compensator is [4]

$$G_c(s) = K\left(1 + \frac{1}{\tau s}\right),$$

which provides proportional plus integral control. Show that one solution is $K = 0.5$ and $\tau = 1$.

E10.4   Consider a unity negative feedback system with

$$G(s) = \frac{K}{s(s + 2)(s + 3)},$$

where K is set equal to 20 in order to achieve a specified $K_v = 3.33$. We wish to add a lead-lag compensator

$$G_c(s) = \frac{(s + 0.15)(s + 0.7)}{(s + 0.015)(s + 7)}.$$

Show that the gain margin of the compensated system is 24 dB and that the phase margin is 75°.

E10.5   Consider a unity feedback system with the transfer function

$$G(s) = \frac{K}{s(s + 2)(s + 4)}.$$

It is desired to obtain the dominant roots with $\omega_n = 3$ and $\zeta = 0.5$. We wish to obtain a $K_v = 2.7$. Show that we require a compensator

$$G_c(s) = \frac{7.53(s + 2.2)}{(s + 16.4)}.$$

Determine the value of K that should be selected.

E10.6   Reconsider the wind tunnel control system of Problem 7.31. When $K = 326$, find $T(s)$ and estimate the expected overshoot and settling time. Compare your estimates with the actual overshoot of 60% and a settling time of 4 seconds. Explain the discrepancy in your estimates.

E10.7   NASA astronauts retrieved a satellite and brought it into the cargo bay of the space shuttle, as shown in Fig. E10.7(a). An astronaut within the shuttle controlled the robot arm upon which Joseph

FIGURE E10.7
Retrieval of a
satellite.

(a)

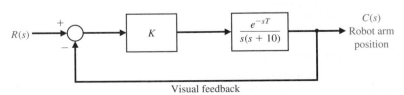

Visual feedback

(b)

Allen is seen standing. A model of the feedback control system is shown in Fig. E10.7(b). Determine the value of K that will result in a phase margin of 50° when $T = 0.1$ second.

Answer: $K = 37$

E10.8 A negative unity feedback system has a plant

$$G(s) = \frac{2257}{s(\tau s + 1)},$$

where $\tau = 2.8$ ms. Select a compensator

$$G_c(s) = K_1 + K_2/s$$

so that the dominant roots of the characteristic equation have ζ equal to $1/\sqrt{2}$. Plot $c(t)$ for a step input.

E10.9 A control system with a controller is shown in Fig. E10.9. Select K_1 and K_2 so that the overshoot to a step input is equal to 5% and the velocity constant, K_v, is equal to 5. Use a simulation program to verify the results of your design.

E10.10 A control system with a controller is shown in Fig. E10.10. We will select $K_2 = 4$ in order to provide a reasonable steady-state error to a step [8]. Using a simulation program, find K_1 to obtain a phase margin of 60°. Find the peak time and percent overshoot of this system.

E10.11 A unity feedback system has

$$G(s) = \frac{1350}{s(s + 2)(s + 30)}.$$

A lead network is selected so that

$$G_c(s) = \frac{(1 + 0.25s)}{(1 + 0.025s)}.$$

Determine the peak magnitude and the bandwidth of the closed-loop frequency response using the

(a) Nichols chart, and (b) a plot of the closed-loop frequency response.

Answer: $M_{p\omega} = 2.3$ dB, $\omega_B = 22$

E10.12 The control of an automobile ignition system has unity negative feedback and a loop transfer function $G_c(s)G(s)$, where

$$G(s) = \frac{K}{s(s + 10)} \quad \text{and}$$

$$G_c(s) = K_1 + K_2/s.$$

A designer selects $K_2/K_1 = 0.5$ and asks you to determine KK_1 so that the dominant roots have a ζ of $1/\sqrt{2}$ and the settling time is less than 2 seconds.

E10.13 The design of Example 10.3 determined a lead network in order to obtain desirable dominant root locations using a cascade compensator $G_c(s)$ in the system configuration shown in Fig. 10.1(a). The same lead network would be obtained if we used the feedback compensation configuration of Fig. 10.1(b). Determine the closed-loop transfer function, $T(s) = C(s)/R(s)$, of both the cascade and feedback configurations, and show how the transfer function of each configuration differs. Explain how the response to a step, $R(s)$, will be different for each system.

E10.14 A robot will be operated by NASA to build a permanent lunar station. The position control system for the gripper tool is shown in Fig. 10.1(a), where $H(s) = 1$ and

$$G(s) = \frac{3}{s(s + 1)(0.5s + 1)}.$$

Determine a compensator lag network, $G_c(s)$, that will provide a phase margin of 45°.

Answer: $G_c(s) = \dfrac{1 + 20s}{1 + 106s}$

FIGURE E10.9
Design of a controller.

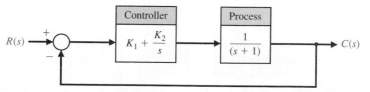

FIGURE E10.10
Design of a PI controller.

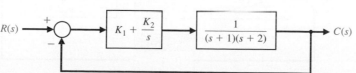

E10.15   A unity feedback control system has a plant transfer function

$$G(s) = \frac{40}{s(s + 2)}.$$

It is desired to attain a steady-state error to a ramp, $r(t) = At$, of less than $0.05A$, and a phase margin of 30°. It is desired to have a crossover frequency, ω_c, of 10 rad/s. Use the methods of Section 10.9 to determine whether a lead or a lag compensator is required.

E10.16   Reconsider the system and specifications of Exercise 10.15 when the required crossover frequency is 2 rad/s.

E10.17   Reconsider the system of Exercise 10.9. Select K_1 and K_2 so that the step response is deadbeat and the settling time is less than 2 seconds.

PROBLEMS

P10.1   The design of a lunar excursion module (LEM) is an interesting control problem. The attitude control system for the lunar vehicle is shown in Fig. P10.1. The vehicle damping is negligible and the attitude is controlled by gas jets. The torque, as a first approximation, will be considered to be proportional to the signal $V(s)$ so that $T(s) = K_2 V(s)$. The loop gain may be selected by the designer in order to provide a suitable damping. A damping ratio of $\zeta = 0.6$ with a settling time of less than 2.5 seconds is required. Using a lead-network compensation, select the necessary compensator $G_c(s)$ by using (a) frequency response techniques and (b) root locus methods.

P10.2   A magnetic tape recorder transport for modern computers requires a high-accuracy, rapid-response control system. The requirements for a specific transport are as follows: (1) The tape must stop or start in 10 ms; (2) it must be possible to read 45,000 characters per second. This system was dis-

cussed in Problem P7.11. It is desired to set $J = 5 \times 10^{-3}$, and K_1 is set on the basis of the maximum error allowable for a velocity input. In this case, it is desired to maintain a steady-state speed error of less than 5%. We will use a tachometer in this case and set $K_a = 50,000$ and $K_2 = 1$. To provide a suitable performance, a compensator $G_c(s)$ is inserted immediately following the photocell transducer. Select a compensator $G_c(s)$ so that the overshoot of the system for a step input is less than 25%. We will assume that $\tau_1 = \tau_a = 0$.

P10.3   A simplified version of the attitude rate control for the F-94 or X-15 type aircraft is shown in Fig. P10.3. When the vehicle is flying at four times the speed of sound (Mach 4) at an altitude of 100,000 ft, the parameters are [30]

$$\frac{1}{\tau_a} = 1.0, \qquad K_1 = 0.5,$$

$$\zeta\omega_a = 1.0, \qquad \omega_a = 2.$$

FIGURE P10.1
Attitude control system for a lunar excursion module.

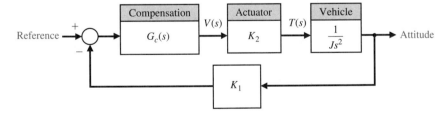

FIGURE P10.3
Aircraft attitude control.

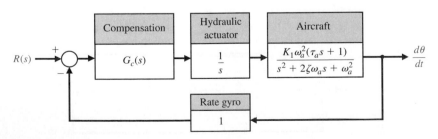

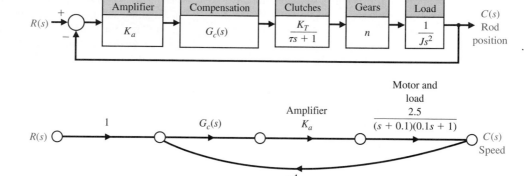

FIGURE P10.4
Nuclear reactor
rod control.

$R(s)$ — Amplifier K_a — Compensation $G_c(s)$ — Clutches $\dfrac{K_T}{\tau s + 1}$ — Gears n — Load $\dfrac{1}{Js^2}$ — $C(s)$ Rod position

FIGURE P10.5
Stabilized rate
table.

$R(s)$ — 1 — $G_c(s)$ — Amplifier K_a — Motor and load $\dfrac{2.5}{(s + 0.1)(0.1s + 1)}$ — $C(s)$ Speed — -1

Design a compensator, $G_c(s)$, so that the response to a step input has an overshoot of less than 5% and a settling time of less than 5 seconds.

P10.4 Magnetic particle clutches are useful actuator devices for high power requirements because they can typically provide a 200-w mechanical power output. The particle clutches provide a high torque-to-inertia ratio and fast time constant response. A particle clutch positioning system for nuclear reactor rods is shown in Fig. P10.4. The motor drives two counter-rotating clutch housings. The clutch housings are geared through parallel gear trains, and the direction of the servo output is dependent on the clutch that is energized. The time constant of a 200-W clutch is $\tau = \frac{1}{40}$ second. The constants are such that $K_T n/J = 1$. It is desired that the maximum overshoot for a step input be in the range of 10% to 20%. Design a compensating network so that the system is adequately stabilized. The settling time of the system should be less than or equal to 2 seconds.

P10.5 A stabilized precision rate table uses a precision tachometer and a dc direct-drive torque motor, as shown in Fig. P10.5. It is desired to maintain a high steady-state accuracy for the speed control. To obtain a zero steady-state error for a step command design, select a proportional plus integral compensator as discussed in Section 10.6. Select the appropriate gain constants so that the system has an overshoot of approximately 10% and a settling time in the range of 0.4 to 0.6 second.

P10.6 Repeat Problem 10.5 by using a lead network compensator, and compare the results.

P10.7 The primary control loop of a nuclear power plant includes a time delay due to the need to transport the fluid from the reactor to the measurement point (see Fig. P10.7). The transfer function of the controller is

$$G_1(s) = \left(K_1 + \frac{K_2}{s} \right).$$

FIGURE P10.7
Nuclear reactor
control.

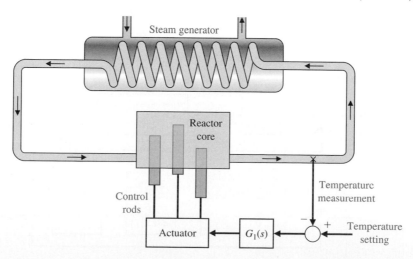

The transfer function of the reactor and time delay is

$$G(s) = \frac{e^{-sT}}{\tau s + 1},$$

where $T = 0.4$ second and $\tau = 0.2$ second. Using frequency response methods, design the controller so that the overshoot of the system is less than 10%. Estimate the settling time of the system designed. Determine the actual overshoot and settling time.

P10.8   A chemical reactor process whose production rate is a function of catalyst addition is shown in block diagram form in Fig. P10.8 [11]. The time delay is $T = 50$ seconds and the time constant, τ, is approximately 40 seconds. The gain of the process is $K = 1$. Design a compensation by using Bode diagram methods in order to provide a suitable system response. It is desired to have a steady-state error for a step input, $R(s) = A/s$, less than $0.10A$. For the system with the compensation added, estimate the settling time of the system.

P10.9   A numerical path-controlled machine turret lathe is an interesting problem in attaining sufficient accuracy [2, 26]. A block diagram of a turret lathe control system is shown in Fig. P10.9. The gear ratio is $n = 0.1$, $J = 10^{-3}$, and $f = 10^{-2}$. It is necessary to attain an accuracy of 5×10^{-4} in., and therefore a steady-state position accuracy of 2.5% is specified for a ramp input. Design a cascade compensator to be inserted before the silicon-controlled rectifiers in order to provide a response to a step command with an overshoot of less than 5%. A suitable damping ratio for this system is 0.7. The gain of the silicon-controlled rectifiers is $K_R = 5$. Design a suitable lag compensator by using the (a) Bode diagram method and (b) s-plane method.

P10.10   The HS *Denison*, shown in Fig. P10.10(a), is a large, 80-ton hydrofoil seacraft built by Grumman Corp. for the U.S. Maritime Administration. The *Denison* is capable of operating in seas ranging to 9 ft in amplitude at a speed of 60 knots as a result of the utilization of an automatic stabilization control system. Stabilization is achieved by means of flaps on the main foils and the adjustment of the aft foil. The stabilization control system maintains a level flight through rough seas. Thus a system that minimizes deviations from a constant lift force or, equivalently, that minimizes the pitch angle θ has been designed. A block diagram of the lift control system is shown in Fig. P10.10(b). The desired response of the system to wave disturbance is a constant-level travel of the craft. Establish a set of reasonable specifications and design a compensator $G_c(s)$ so that the performance of the system is suitable. Assume that the disturbance is due to waves with a frequency $\omega = 6$ rad/s.

P10.11   A unity feedback system of the form shown in Fig. 10.1(a) has a plant

$$G(s) = \frac{4.19}{s(s + 1)(s + 5)}.$$

(a) Determine the step response when $G_c(s) = 1$, and calculate the settling time and steady state for a ramp input $r(t) = t, t > 0$. (b) Design a lag network using the root locus method so that the velocity constant is increased to 8.4. Determine the settling time of the compensated system.

P10.12   A unity feedback control system of the form shown in Fig. 10.1(a) has a plant

$$G(s) = \frac{160}{s^2}.$$

Select a lead-lag compensator so that the percent overshoot for a step input is less than 5% and the

FIGURE P10.8
Chemical reactor
control.

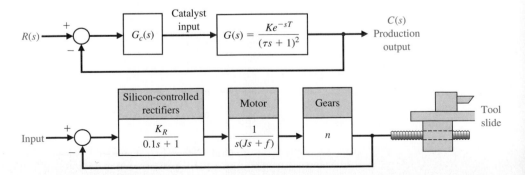

FIGURE P10.9
Path-controlled
turret lathe.

(a)

FIGURE P10.10
(a) The HS *Denison* hydrofoil. (Photo courtesy of Grumman Aircraft Engineering Corp.)
(b) A block diagram of the lift control system.

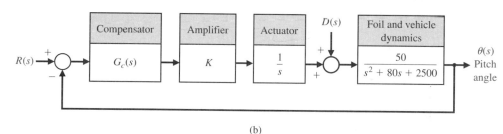

(b)

settling time is less than 1 second. It also is desired that the acceleration constant, K_a, be greater than 7500 (see Table 5.4).

P10.13 A unity feedback system has a plant

$$G(s) = \frac{20}{s(1 + 0.1s)(1 + 0.05s)}.$$

Select a compensator $G_c(s)$ so that the phase margin is at least 75°. Use a two-stage lead compensator

$$G_c(s) = \frac{K(1 + s/\omega_1)(1 + s/\omega_3)}{(1 + s/\omega_2)(1 + s/\omega_4)}.$$

FIGURE P10.14

It is required that the error for a ramp input be ½% of the magnitude of the ramp input ($K_v = 200$).

P10.14 Materials testing requires the design of control systems that can faithfully reproduce normal specimen operating environments over a range of specimen parameters [26]. From the control system design viewpoint, a materials-testing machine system can be considered a servomechanism in which we want to have the load waveform track the reference signal. The system is shown in Fig. P10.14.

(a) Determine the phase margin of the system with $G_c(s) = K$, choosing K so that a phase margin of

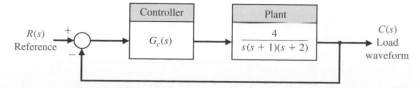

50° is achieved. Determine the system bandwidth for this design.

(b) The additional requirement introduced is that the velocity constant, K_v, be equal to 2.0. Design a lag network so that the phase margin is 50° and $K_v = 2$.

P10.15 For the system described in Problem 10.14, the goal is to achieve a phase margin of 50° with the additional requirement that the settling time be less than 4 seconds. Design a lead network to meet the specifications. As before, we require $K_v = 2$.

P10.16 A robot with an extended arm has a heavy load, whose effect is a disturbance, as shown in Fig. P10.16 [25]. For ease, use $R(s) = 0$ and design $G_c(s)$ so that the effect of the disturbance is less than 10% of the open-loop system effect.

P10.17 A driver and car may be represented by the simplified model shown in Fig. P10.17 [18]. The goal is to have the speed adjust to a step input with less than 10% overshoot and a settling time of 1 second. Select a proportional plus integral (PI) controller to yield these specifications. For the selected controller, determine the actual response (a) for $G_p(s) = 1$ and (b) with a prefilter $G_p(s)$ that removes the zero from the closed-loop transfer function $T(s)$.

P10.18 A unity feedback control system for a robot submarine has a plant with a third-order transfer function [22]:

$$G(s) = \frac{K}{s(s + 10)(s + 50)}.$$

It is desired that the overshoot be approximately 7.5% for a step input and the settling time of the

system be 400 ms. Find a suitable phase-lead compensator by using root locus methods. Let the zero of the compensator be located at $s = -15$, and determine the compensator pole. Determine the resulting system K_v.

P10.19 Now that expenditures for manned space exploration have leveled off, NASA is developing remote manipulators. These manipulators can be used to extend the hand and the power of humankind through space by means of radio. A concept of a remote manipulator is shown in Fig. P10.19(a) [12, 25]. The closed loop control is shown schematically in Fig. P10.19(b). Assuming an average distance of 238,855 miles from earth to the moon, the time delay in transmission of a communication signal T is 1.28 seconds. The operator uses a control stick to control remotely the manipulator placed on the moon to assist in geological experiments, and the TV display to access the response of the manipulator. The time constant of the manipulator is ¼ second.

(a) Set the gain K_1 so that the system has a phase margin of approximately 30°. Evaluate the percentage steady-state error for this system for a step input. (b) To reduce the steady-state error for a position command input to 5%, add a lag compensation network in cascade with K_1. Plot the step response.

P10.20 There have been significant developments in the application of robotics technology to nuclear power plant maintenance problems. Thus far, robotics technology in the nuclear industry has been used primarily on spent-fuel reprocessing and waste management. Today, the industry is beginning to apply the technology to such areas as pri-

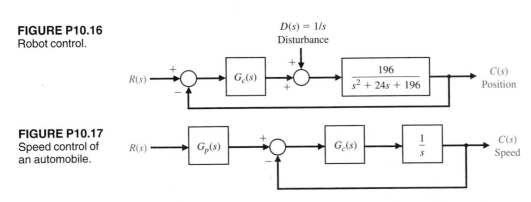

FIGURE P10.16
Robot control.

FIGURE P10.17
Speed control of an automobile.

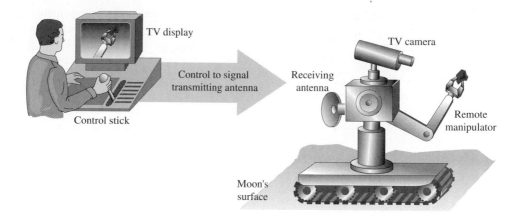

FIGURE P10.19
(a) Conceptual diagram of a remote manipulator on the moon controlled by a person on the earth.
(b) Feedback diagram of the remote manipulator control system with $\tau =$ transmission time delay of the video signal.

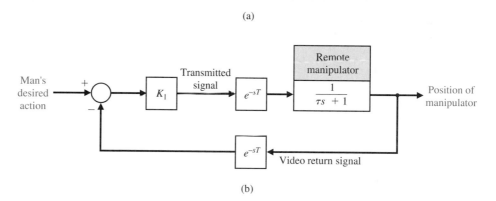

(b)

mary containment inspection, reactor maintenance, facility decontamination, and accident recovery activities. These developments suggest that the application of remotely operated devices can significantly reduce radiation exposure to personnel and improve maintenance-program performance.

Currently, an operational robotic system is under development to address particular operational problems within a nuclear power plant. This device, IRIS (Industrial Remote Inspection System), is a general-purpose surveillance system that conducts particular inspection and handling tasks with the goal of significantly reducing personnel exposure to high radiation fields [13]. The device is shown in Fig. P10.20.

FIGURE P10.20
Remotely controlled robot for nuclear plants.

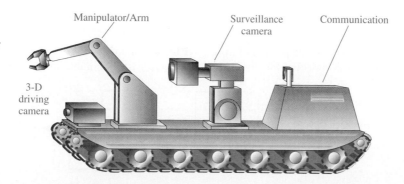

The open-loop transfer function is

$$G(s) = \frac{Ke^{-sT}}{(s + 1)(s + 3)}.$$

(a) Determine a suitable gain K for the system when $T = 0.5$ second so that the overshoot to a step input is less than 30%. Determine the steady-state error. (b) Design a compensator

$$G_c(s) = \frac{s + 2}{s + b}$$

to improve the step response for the system in part (a) so that the steady-state error is less than 12%. Assume the closed-loop system of Fig. 10.1(a).

P10.21   An uncompensated control system with unity feedback has a plant transfer function

$$G(s) = \frac{K}{s(s/2 + 1)(s/6 + 1)}.$$

It is desired to have a velocity error constant of $K_v = 20$. It is also desired to have a phase margin approximately to 45° and a closed-loop bandwidth greater than $\omega = 4$ rad/s. Use two identical cascaded phase-lead networks to compensate the system.

P10.22   For the system of Problem 10.21, design a phase-lag network to yield the desired specifications, with the exception that a bandwidth equal to or greater than 2 rad/s will be acceptable.

P10.23   For the system of Problem 10.21, we wish to achieve the same phase margin and K_v, but in addition we wish to limit the bandwidth to less than 10 rad/s but greater than 2 rad/s. Utilize a lead-lag compensation network to compensate the system. The lead-lag network could be of the form

$$G_c(s) = \frac{(1 + s/10a)(1 + s/b)}{(1 + s/a)(1 + s/10b)}.$$

where a is to be selected for the lag portion of the compensator and b is to be selected for the lead

portion of the compensator. The ratio α is chosen to be 10 for both the lead and lag portions.

P10.24   A system of the form of Fig. 10.1(a) with unity feedback has

$$G(s) = \frac{K}{(s + 2)^2}.$$

It is desired that the steady-state error to a step input be approximately 4% and that the phase margin of the system be approximately 50°. Design a lag network to meet these specifications.

P10.25   The stability and performance of the rotation of a robot (similar to waist rotation) is a challenging control problem. The system requires high gains in order to achieve high resolution; yet a large overshoot of the transient response cannot be tolerated. The block diagram of an electro-hydraulic system for rotation control is shown in Fig. P10.25 [16]. The arm-rotating dynamics are represented by

$$G(s) = \frac{360}{s(s^2/6400 + s/50 + 1)}.$$

It is desired to have $K_v = 10$ for the compensated system. Design a compensator that results in an overshoot to a step input of less than 15%.

P10.26   The possibility of overcoming wheel friction, wear, and vibration by contactless suspension for passenger-carrying mass-transit vehicles is being investigated throughout the world. One design uses a magnetic suspension with an attraction force between the vehicle and the guideway with an accurately controlled airgap. A system is shown in Fig. P10.26, which incorporates feedback compensation. Using root locus methods, select a suitable value for K_1 and b so the system has a damping ratio for the underdamped roots of $\zeta = 0.50$. Assume, if appropriate, that the pole of the airgap feedback loop ($s = -200$) can be neglected.

P10.27   A computer uses a printer as a fast output device. It is desirable to maintain accurate position

FIGURE P10.25
Robot position
control.

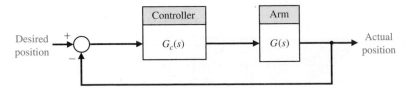

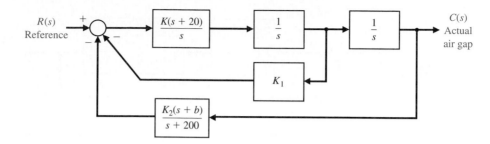

FIGURE P10.26
Airgap control
of train.

control while moving the paper rapidly through the printer. Consider a system with unity feedback and a transfer function for the motor and amplifier as

$$G(s) = \frac{0.15}{s(s + 1)(5s + 1)}.$$

Design a lead network compensator so that the system bandwidth is 0.75 rad/s and the phase margin is 30°. Use a lead network with $\alpha = 10$.

P10.28 An engineering design team is attempting to control a process shown in Fig. P10.28. The system has a controller $G_c(s)$, but the design team is unable to select $G_c(s)$ appropriately. It is agreed that a system with a phase margin of 50° is acceptable, but $G_c(s)$ is unknown. You are asked to determine $G_c(s)$.

First, let $G_c(s) = K$ and, using a computer program, find (a) a value of K that yields a phase margin of 50° and the system's step response for this value of K. (b) Determine the settling time, percent overshoot, and the peak time. (c) Obtain the system's closed-loop frequency response, and determine $M_{p\omega}$ and the bandwidth.

The team has decided to let

$$G_c(s) = \frac{K(s + 12)}{(s + 20)}$$

and to repeat parts (a), (b), and (c). Determine the gain K that results in a phase margin of 50° and then proceed to evaluate the time response and the closed-loop frequency response. Prepare a table contrasting the results of the two selected control-

lers for $G_c(s)$ by comparing settling time, percent overshoot, peak time, $M_{p\omega}$, and bandwidth.

P10.29 An adaptive suspension vehicle uses a legged locomotion principle. The control of the leg can be represented by a unity feedback system with [10]

$$G(s) = \frac{K}{s(s + 10)(s + 14)}.$$

It is desired to achieve a steady-state error for a ramp input of 10% and a damping ratio of the dominant roots of 0.707. Determine a suitable lag compensator, and determine the actual overshoot and settling time of this system.

P10.30 A liquid level control system (see Fig. 9.32) has a loop transfer function

$$G_c(s)G(s)H(s)$$

where $H(s) = 1$, $G_c(s)$ is a compensator, and the plant is

$$G(s) = \frac{10e^{-sT}}{s^2(s + 10)}$$

where $T = 50$ ms. Design a compensator so that $M_{p\omega}$ does not exceed 3.5 dB and ω_r is approximately 1.4 rad/s. Predict the overshoot and settling time of the compensated system when the input is a step. Plot the actual response.

P10.31 An automated guided vehicle (AGV) can be considered as an automated mobile conveyor designed to transport materials. Most AGVs require some type of guide path. The steering stability of

FIGURE P10.28
Controller design.

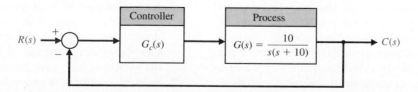

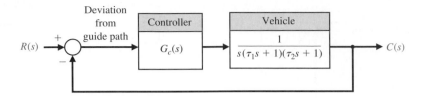

FIGURE P10.31
Steering control
for vehicle.

the guidance control system has not been fully solved. The slight "snaking" of the AGV about the track generally has been acceptable, although it indicates instability of the steering guidance control system [9].

Most AGVs have a specification of maximum speed of about 1 m/s, although in practice they are usually operated at half that speed. In a fully automated manufacturing environment, there should be few personnel in the production area; therefore the AGV should be able to be run at full speed. As the speed of the AGV increases, so does the difficulty in designing stable and smooth tracking controls.

A steering system for an AGV is shown in Fig. P10.31, where $\tau_1 = 40$ ms and $\tau_2 = 1$ ms. It is required that the velocity constant, K_v, is 100 so that the steady-state error for a ramp input is 1% of the slope of the ramp. Neglect τ_2 and design a lead compensator so that the phase margin is

$$45° \leq \text{PM} \leq 65°.$$

Attempt to obtain the two limiting cases for phase margin, and compare your results for the two designs by determining the actual percent overshoot and settling time for a step input.

P10.32   For the systems of Problem 10.31, use a phase-lag compensator and attempt to achieve a phase margin of approximately 50°. Determine the actual overshoot and peak time for the compensated system.

P10.33   When a motor drives a flexible structure, the structure's natural frequencies, as compared to the bandwidth of the servodrive, determine the contribution of the structural flexbility to the errors of the resulting motion. In current industrial robots, the drives are often relatively slow and the structures are relatively rigid, so that overshoots and other errors are caused mainly by the servodrive. However, depending on the accuracy required, the structural deflections of the driven members can become significant. Structural flexibility must be considered the major source of motion errors in space

structures and manipulators. Because of weight restrictions in space, large arm lengths result in flexible structures. Furthermore, future industrial robots should require lighter and more flexible manipulators.

To investigate the effects of structural flexibility and how different control schemes can reduce unwanted oscillations, an experimental apparatus consisting of a dc motor driving a slender aluminum beam was constructed. The purpose of the experiments was to identify simple and effective control strategies to deal with the motion errors that occur when a servomotor is driving a very flexible structure [14].

The experimental apparatus is shown in Fig. P10.33(a) and the control system is shown in Fig. P10.33(b). The goal is that the system will have a K_v of 100. (a) When $G_c(s) = K$, determine K and plot the Bode diagram. Find the phase margin and gain margin. (b) Using the Nichols chart, find ω_r, $M_{p\omega}$, and ω_B. (c) Select a compensator so that the phase margin is greater than 35° and find ω_r, $M_{p\omega}$, and ω_B for the compensated system.

P10.34   A human's ability to perform physical tasks is limited not by intellect but by physical strength. If in an appropriate environment a machine's mechanical power is closely integrated with a human arm's mechanical strength under the control of the human intellect, the resulting system will be superior to a loosely integrated combination of a human and a fully automated robot.

Extenders are defined as a class of robot manipulators that extend the strength of the human arm while maintaining human control of the task [15]. The defining characteristic of an extender is the transmission of both power and information signals. The extender is worn by the human; the physical contact between the extender and the human allows direct transfer of mechanical power and information signals. Because of this unique interface, control of the extender trajectory can be accomplished without any type of joystick, key-

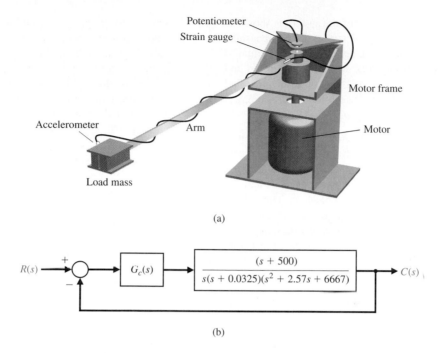

(a)

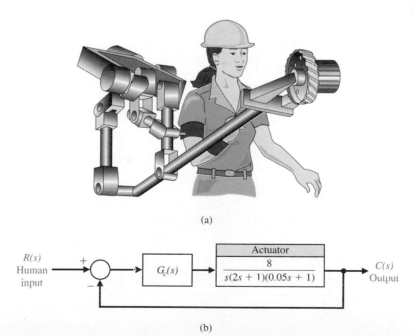

(b)

FIGURE P10.33
Flexible arm
control.

board, or master-slave system. The human provides a control system for the extender, while the
extender actuators provide most of the strength
necessary for the task. The human becomes a part
of the extender and "feels" a scaled-down version
of the load that the extender is carrying. The ex-

tender is distinguished from a conventional master-
slave system; in that type of system, the human
operator is either at a remote location or close to
the slave manipulator but is not in direct physical
contact with the slave in the sense of transfer of
power. An extender is shown in Fig. P10.34(a)

FIGURE P10.34
Extender robot
control.

(a)

(b)

[15]. The block diagram of the system is shown in Fig. P10.34(b). The goal is that the compensated system will have a velocity constant, K_v, equal to 80, that the settling time will be 1.6 seconds, and that the overshoot will be 16% so that the dominant roots have a ζ of 0.5. Determine a lead-lag compensator using root locus methods.

P10.35   A magnetically levitated train is operating in Berlin, Germany. The M-Bahn 1600-m line represents the current state of worldwide systems. Fully automated trains can run at short intervals and operate with excellent energy efficiency. The control system for the levitation of the car is shown in Fig. P10.35. Select a compensator so that the phase margin of the system is $55° \leq PM \leq 65°$. Predict the response of the system to a step command, and determine the actual step response for comparison.

P10.36   Engineers are developing new sensors for machining and other manufacturing processes. A new sensing technique gleans information about the cutting process from acoustic emission (AE) signals. AE is a low-amplitude, high-frequency stress wave from a rapid release of strain energy in a continuous medium. AE is sensitive to material, tool geometry, tool wear, and cutting parameters such as feed and speed. AE sensors are commonly piezoelectric crystals sensitive in the 100 kHz to 1 MHz range. They are cost effective and can be mounted on most machine tools.

Current research has been extracting depth-of-cut information from AE signals. One study links sensitivity of the AE power signal to small depth-of-cut changes in diamond turning [16, 20, 21]. The power of the AE signals is sensitive to small, instantaneous changes in depth of cut during diamond turning and conventional milling. A milling machine using an AE sensor is shown in Fig. P10.36(a). The system for controlling the depth of cut of a milling machine is shown in

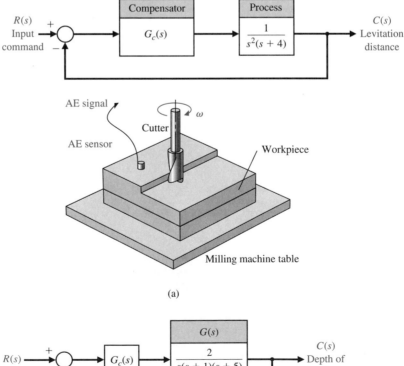

FIGURE P10.35
Magnetically levitated train control.

FIGURE P10.36
Milling machine control.

(a)

(b)

Fig. P10.36(b). Design a compensator so that the percent overshoot to a step input is less than or equal to 20% while the velocity constant is greater than 8.

P10.37 A system's open-loop transfer function is a pure time delay of 0.5 second so that $G(s) = e^{-s/2}$. Select a compensator $G_c(s)$ so that the steady-state error for a step input is less than 2% of the magnitude of the step and the phase margin is greater than 30°. Determine the bandwidth of the compensated system, and plot the step response.

P10.38 A unity-negative feedback system has

$$G(s) = \frac{1}{s(s + 5)(s + 10)}.$$

The objective is that the dominant roots have a ζ equal to 0.707 while achieving zero steady state error for a ramp input. Select a proportional plus integral (PI) controller so that the requirements are met. Determine the resulting peak time and settling time of the system.

P10.39 A unity feedback system of the form shown in Fig. 10.1(a) has

$$G(s) = \frac{1}{(s + 1)(s + 10)}.$$

Design a compensator $G_c(s)$ so that the overshoot for a step input $R(s)$ is less than 10% and the steady-state error is less than 5%. Determine the bandwidth of the system.

P10.40 A unity feedback system has a plant

$$G(s) = \frac{40}{s(s + 2)}.$$

It is desired to have a phase margin of 30° and a relatively large bandwidth. Select the crossover frequency $\omega_c = 10$ rad/s, and design a lead compensator using the analytical method of Section 10.9. Verify the results by plotting the compensated Bode diagram.

P10.41 A unity feedback system has a plant

$$G(s) = \frac{40}{s(s + 2)}.$$

It is desired that the phase margin be equal to 30°. For a ramp input, $r(t) = t$, it is desired that the steady-state error be equal to 0.05. Design a lag compensator to satisfy the requirements using the methods of Section 10.9. Verify the results by plotting the Bode diagram.

P10.42 For the system and requirements of Problem 10.41, determine the required compensator when the steady-state error for the ramp input must be equal to 0.02.

P10.43 Repeat Example 10.12 when it is desired that the 100% rise time, T_r, is 1 second.

P10.44 Reconsider the design for Example 10.4. Using a system as shown in Fig. 10.22 and the compensator determined in Eq. (10.44), select an appropriate prefilter. Compare the response of the system with and without the prefilter.

ADVANCED PROBLEMS

AP10.1 A three-axis pick-and-place application requires the precise movement of a robotic arm in three-dimensional space, as shown in Fig. AP10.1 (on page 584) for joint 2. The arm has specific linear paths it must follow to avoid other pieces of machinery. The overshoot for a step input should be less than 13%.

(a) Let $G_c(s) = K$, and determine the gain K that satisfies the requirement. Determine the resulting settling time. (b) Use a lead network and reduce the settling time to less than 3 seconds.

AP10.2 The system of Advanced Problem 10.1 is to have a percent overshoot less than 13%. In addition, it is desired that the steady-state error for a unit ramp input will be less than 0.125 ($K_v = 8$) [28]. Design a lag network to meet the specifications. Check the resulting percent overshoot and settling time for the design.

AP10.3 The system of Advanced Problem 10.1 is required to have a percent overshoot less than 13% with a steady-state error for a unit ramp input less

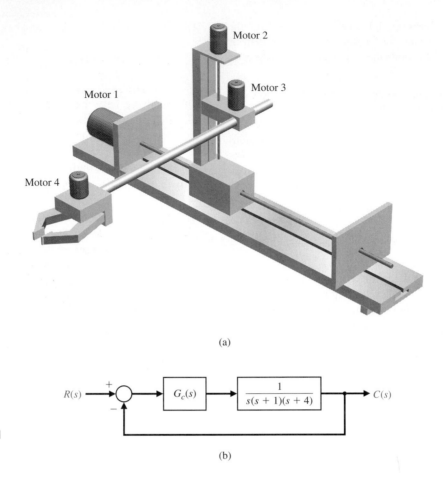

(a)

(b)

FIGURE AP10.1
Pick-and-place
robot.

than 0.125 ($K_v = 8$). Design a proportional plus
integral (PI) controller to meet the specifications.

AP10.4 A dc motor control system with unity feedback has the form shown in Fig. AP10.4. Select K_1
and K_2 so that the system response has a settling
time less than 0.6 second and an overshoot less
than 5% for a step input.

AP10.5 A unity feedback system is shown in
Fig. AP10.5. It is desired that the step response of

the system have an overshoot of about 16%, a fast
response, and a settling time of about 1.8 seconds.

(a) Design a lead compensator $G_c(s)$ to
achieve the dominant roots desired. (b) Determine
the step response of the system when $G_p(s) = 1$.
(c) Select a prefilter $G_p(s)$ and determine the step
response of the system with the prefilter.

AP10.6 Reconsider Example 10.12 when we wish to
minimize the settling time of the system while re-

FIGURE AP10.4
Motor control
system.

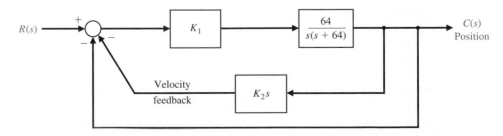

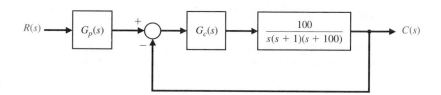

FIGURE AP10.5

quiring that $K < 52$. Determine the appropriate compensator that will minimize the settling time. Plot the system response.

AP10.7 A system has the form shown in Fig. 10.22 with

$$G(s) = \frac{1}{s(s + 2)(s + 8)}.$$

A lead compensator is used with

$$G_c(s) = \frac{K(s + 3)}{(s + 28)}.$$

Determine K so that the complex roots have $\zeta = 1/\sqrt{2}$. The prefilter is

$$G_p(s) = \frac{p}{(s + p)}.$$

(a) Determine the overshoot and rise time for $G_p(s) = 1$ and for $p = 3$. (b) Select an appropriate value for p that will give an overshoot of 1%, and compare the results.

AP10.8 The Manutec robot has large inertia and arm length resulting in a challenging control problem, as shown in Fig. AP10.8(a). The block diagram model of the system is shown in AP10.8(b). The plant dynamics are represented by

$$G(s) = \frac{250}{s(s + 2)(s + 40)(s + 45)}.$$

The percentage overshoot for a step input should be less than 20% with a rise time less than ½ second and a settling time less than 1.2 seconds. Also, it is desired that for a ramp input $K_v \geq 10$. Determine a suitable lead compensator.

FIGURE AP10.8
(a) Manutec robot.
(b) Block diagram.

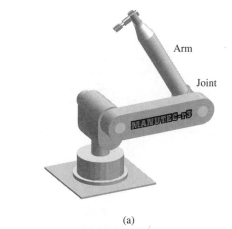

(a)

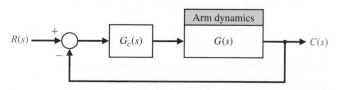

(b)

DESIGN PROBLEMS

DP10.1   Two robots are shown in Fig. DP10.1 cooperating with each other to manipulate a long shaft

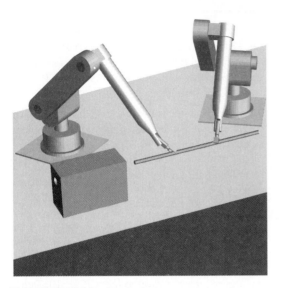

FIGURE DP10.1   Two robots cooperate to insert a shaft.

prior to inserting it into the hole in the block resting on the table. Long part insertion is a good example of a task that can benefit from cooperative control. The feedback control system of one robot joint of the form shown in Fig. 10.1 has $H(s) = 1$ and

$$G(s) = \frac{4}{s(s + 0.5)}.$$

The specifications require a steady-state error for a unit ramp input of 0.0125, and the step response has an overshoot of less than 25% with a settling time of less than 2 seconds. Determine a lead-lag compensator that will meet the specifications, and plot the compensated and uncompensated response for the ramp and step inputs.

DP10.2   The heading control of the traditional bi-wing aircraft shown in Fig. DP10.2(a) is represented by the block diagram of Fig. DP10.2(b).

(a) Determine the minimum value of the gain K when $G_c(s) = K$ so that the steady-state effect of a unit step disturbance, $D(s) = 1/s$, is less than or equal to 5% of the unit step $[c(\infty) = 0.05]$.

FIGURE DP10.2
(a) Bi-wing aircraft.
(*Source: The Illustrated London News,* October 9, 1920.) (b) Control system.

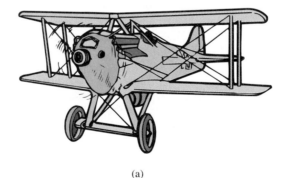

(a)

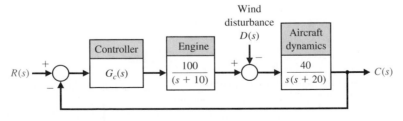

(b)

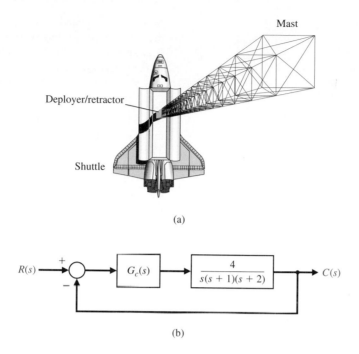

FIGURE DP10.3
Mast flight system.

(a)

(b)

(b) Determine whether the system using the gain of part (a) is stable.

(c) Design a compensator using one stage of lead compensation so that the phase margin is 30°.

(d) Design a two-stage lead compensator so that the phase margin is 55°.

(e) Compare the bandwidth of the systems of parts (c) and (d).

(f) Plot the step response $c(t)$ for the systems of parts (c) and (d) and compare percent overshoot, settling time, and peak time.

DP10.3 NASA has identified the need for large deployable space structures, which will be constructed of lightweight materials and will contain large numbers of joints or structural connections. This need is evident for programs such as the space station. These deployable space structures may have precision shape requirements and a need for vibrations suppression during on-orbit operations [17].

One such structure is the mast flight system, which is shown in Fig. DP10.3(a). The intent of the

system is to provide an experimental test bed for controls and dynamics. The basic element in the mast flight system is a 60.7-m-long truss beam structure, which is attached to the shuttle orbiter. Included at the tip of the truss structure are the primary actuators and collocated sensors. A deployment/retraction subsystem, which also secures the stowed beam package during launch and landing, is provided.

The system uses a large motor to move the structure and has the block diagram shown in Fig. DP10.3(b). The goal is an overshoot to a step response of less than or equal to 16%; thus we estimate the system ζ as 0.5 and the required phase margin as 50°. Design for $0.1 < K < 1$ and record overshoot, rise time, and phase margin for selected gains.

DP10.4 A mobile robot using a vision system as the measurement device is shown in Fig. DP10.4 [22]. The control system is of the form shown in Fig. 10.14, where

$$G(s) = \frac{1}{(s + 1)(0.5s + 1)}$$

Motion subsystem

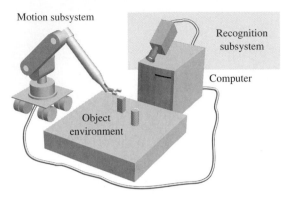

Recognition
subsystem

Computer

Object
environment

FIGURE DP10.4   A robot and vision system.

and $G_c(s)$ is selected as a PI controller so that the steady-state error for a step input is equal to zero. We then have

$$G_c(s) = K_1 + \frac{K_2}{s} = \frac{K_1 s + K_2}{s}.$$

Determine a suitable G_c so that (a) the percent overshoot for a step input is 5% or less; (b) the settling time is less than 6 seconds; (c) the system K_v is greater than 0.9; and (d) the peak time for a step input is minimized.

DP10.5   A high-speed train is under development in Texas [23] with a design based on the French *Train à Grande Vitesse* (TGV). Train speeds of 186 miles per hour are foreseen. To achieve these speeds on tight curves, the train may use independent axles combined with the ability to tilt the train. Hydraulic cylinders connecting the passenger compartments to their wheeled bogies allow the train to lean into curves like a motorcycle. A pendulumlike device on the leading bogie of each car senses when it is entering a curve and feeds this information to the hydraulic system. Tilting does not make the train safer, but it does make passengers more comfortable.

Consider the tilt control as shown in Fig. DP10.5. Design a compensator $G_c(s)$ for a

step-input command so that the overshoot is less than 5% and the settling time less than 0.6 second. It is also desired that the steady-state error for a velocity (ramp) input be less than 0.15A where $r(t) = At$, $t > 0$. Verify the results for the design.

DP10.6   A large antenna, as shown in Fig. DP10.6(a), is used to receive satellite signals and must accurately track the satellite as it moves across the sky. The control system uses an armature-controlled motor and a controller to be selected, as shown in Fig. DP10.6(b). The system specifications require a steady-state error for a ramp input, $r(t) = Bt$, less than or equal to 0.01B. We also seek an overshoot to a step input less than 5% with a settling time less than 2 seconds.

(a) Design a controller $G_c(s)$ and plot the resulting time response. (b) Determine the effect of the disturbance $D(s) = Q/s$ on the output $C(s)$. (For ease, let $R(s) = 0$.)

DP10.7   High-performance tape transport systems are designed with a small capstan to pull the tape past the read/write heads and with take-up reels turned by dc motors. The tape is to be controlled at speeds up to 200 inches per second, with start-up as fast as possible, while preventing permanent distortion of the tape. Since we wish to control the speed and the tension of the tape, we will use a dc tachometer for the speed sensor and a potentiometer for the position sensor. We will use a dc motor for the actuator. Then the linear model for the system is a feedback system with $H(s) = 1$ and

$$\frac{C(s)}{E(s)} = G(s) =$$

$$\frac{K(s + 4000)}{s(s + 1000)(s + 3000)(s + p_1)(s + p_1^*)},$$

where $p_1 = +2000 + j2000$ and $C(s)$ is position.

The specifications for the system are (1) settling time of less than 12 ms, (2) an overshoot to a step position command of less than 10%, and (3) a

FIGURE DP10.5

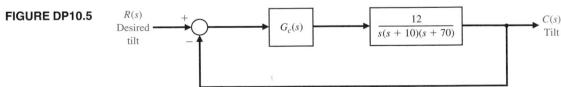

$R(s)$
Desired
tilt

$+$

$-$

$G_c(s)$

$\dfrac{12}{s(s + 10)(s + 70)}$

$C(s)$
Tilt

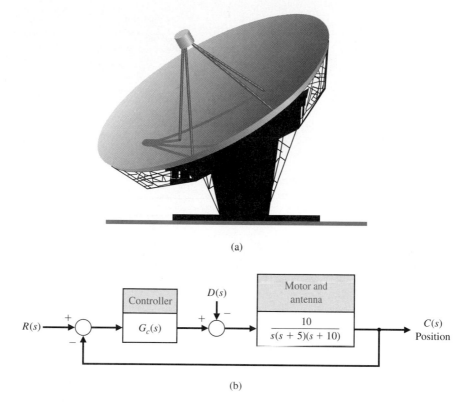

(a)

FIGURE DP10.6
Antenna position
control.

(b)

steady-state velocity error of less than .5%. Determine a compensator scheme to achieve these stringent specifications.

DP10.8 The past several years have witnessed a significant engine model building activity in the automotive industry in a category referred to as "control-oriented" or "control design" models. These models contain representations of the throttle body, engine pumping phenomena, induction process dynamics, fuel system, engine torque generation, and rotating inertia.

The control of the fuel-to-air ratio in an automobile carburetor became of prime importance in

the 1980s as automakers worked to reduce exhaust-pollution emissions. Thus auto engine designers turned to the feedback control of the fuel-to-air ratio. Operation of an engine at or near a particular air-to-fuel ratio requires management of both air and fuel flow into the manifold system. The fuel command is considered the input and the engine speed as the output [9, 11].

The block diagram of the system is shown in Fig. DP10.8, where $T = 0.066$ second. A compensator is required to yield zero steady-state error for a step input and an overshoot of less than 10%. It is also desired that the settling time not exceed 10 seconds.

FIGURE DP10.8
Engine control
system.

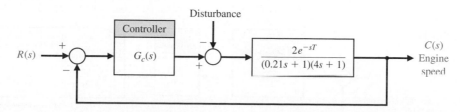

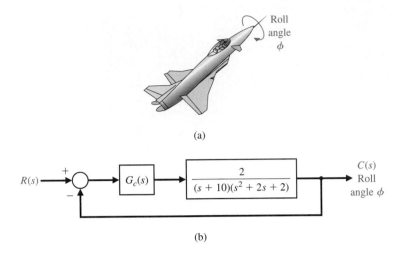

(a)

$$R(s) \xrightarrow{\;+\;} \bigotimes \xrightarrow{-} \boxed{G_c(s)} \longrightarrow \boxed{\dfrac{2}{(s+10)(s^2+2s+2)}} \longrightarrow \begin{array}{l} C(s) \\ \text{Roll} \\ \text{angle } \phi \end{array}$$

FIGURE DP10.9
Roll angle control
of a jet airplane.

(b)

DP10.9   A high-performance jet airplane is shown in
Fig. DP10.9(a), and the roll-angle control system is
shown in Fig. DP10.9(b). Design a controller $G_c(s)$

so that the step response is well-behaved and the
steady-state error is zero.

MATLAB PROBLEMS

MP10.1   Consider the control system in Fig. MP10.1,
where

$$G(s) = \frac{10}{s+1} \quad \text{and} \quad G_c(s) = \frac{9}{s+1}.$$

Using *Matlab,* show that the phase margin is approximately 12° and that the percent overshoot to a
unit step input is 70%.

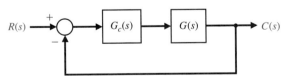

FIGURE MP10.1   A feedback control system with
compensation.

MP10.2   A fighter aircraft has the transfer function

$$\frac{\dot{\theta}}{\delta} = \frac{-10(s+1)(s+0.01)}{(s^2+2s+2)(s^2+0.02s+0.0101)},$$

where $\dot{\theta}$ is the pitch rate (rad/s) and δ is the elevator deflection (rad). The four poles represent the
phugoid and short-period modes. The phugoid
mode has a natural frequency of 0.1 rad/s, and the
short period mode is 1.4 rad/s. The block diagram
is shown in Fig. MP10.2.

(a) Let G_c be the lead compensator

$$G_c = K \frac{s+z}{s+p},$$

where $|z| < |p|$. Using Bode plot methods, design
the lead compensator to meet the following specifications: (1) settling time to a unit step less than 2

FIGURE MP10.2
An aircraft pitch
rate feedback
control system.

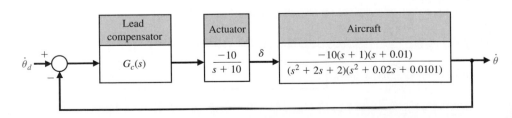

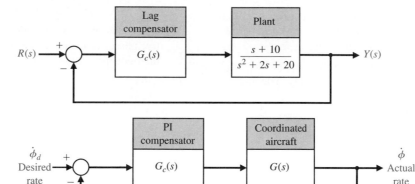

FIGURE MP10.4
A unity feedback
control system.

FIGURE MP10.5
A lateral beam
guidance system
inner loop.

seconds, and (2) percent overshoot less than 10%.
(b) Simulate the closed-loop system with a step input of $10°$/second, and show the time history of $\dot{\theta}$.

MP10.3 The pitch attitude motion of a rigid spacecraft is described by

$$J\ddot{\theta} = u,$$

where J is the principal moment of inertia and u is the input torque on the vehicle [7]. Consider the PD controller

$$C(s) = K_1 + K_2 s.$$

(a) Draw a block diagram of the control system. Design a control system to meet the following specifications: (1) closed-loop system bandwidth about 10 rad/s, and (2) percent overshoot less than 20% to a $10°$ step input. Complete the design by developing and using an interactive *Matlab* script. (b) Verify the design by simulating the response to a $10°$ step input. (c) Include a closed-loop transfer function Bode plot to verify the bandwidth requirement is satisfied.

MP10.4 Consider the control system shown in Fig. MP10.4. Your objective is to design a lag compensator using root locus methods to meet the following specifications: (1) steady-state error less

than 10% for a step input, (2) phase margin greater than 45°, and (3) settling time less than 5 seconds for a unit step input.

(a) Design a lag compensator utilizing root locus methods to meet the design specifications. Develop a set of *Matlab* scripts to assist in the design process. (b) Test the controller developed in part (a) by simulating the closed-loop system response to unit step input. Provide the time histories of the output $y(t)$. (c) Compute the phase margin using the margin function.

MP10.5 A lateral beam guidance system has an inner loop as shown in Fig. MP10.5, where the transfer function for the coordinated aircraft is [30]

$$G(s) = \frac{23}{s + 23}.$$

Consider the PI controller

$$G_c(s) = K_1 + \frac{K_2}{s}.$$

(a) Design a control system to meet the following specifications: (1) settling time to a unit step input of less than 1 second, and (2) steady-state tracking error for a unit ramp input of less than 0.1. (b) Verify the design by simulation.

TERMS AND CONCEPTS

Cascade compensation network A compensator network placed in cascade or series with the system process.

Compensation The alteration or adjustment of a control system in order to provide a suitable performance.

Compensator   An additional component or circuit that is inserted into the system to equalize or compensate for the performance deficiency.

Deadbeat response   A system with a rapid response, minimal overshoot, and zero steady-state error for a step input.

Design of a control system   The arrangement or the plan of the system structure and the selection of suitable components and parameters.

Integration network   A network that acts, in part, like an integrator.

Lag network   *See* Phase-lag network.

Lead-lag network   A network with the characteristics of both a lead network and a lag network.

Lead network   *See* Phase-lead network.

Phase-lag network   A network that provides a negative phase angle and a significant attenuation over the frequency range of interest.

Phase-lead network   A network that provides a positive phase angle over the frequency range of interest. Thus phase lead can be used to cause a system to have an adequate phase margin.

PI controller   Controller with a proportional term and an integral term (Proportional-Integral).

Prefilter   A transfer function, $G_p(s)$, that filters the input signal $R(s)$ prior to calculating the error signal.

The Design of State Variable Feedback Systems

PREVIEW

In this chapter, we will consider the design of control systems using the state equation model. We first describe the system test for controllability and observability and proceed to describe one procedure for determining an optimal control system. We then consider the use of state variable feedback to design systems with prescribed performance. Finally, we describe the use of internal model design to achieve prescribed steady-state response to selected input commands.

11.1 INTRODUCTION

The time-domain method, expressed in terms of state variables, can also be utilized to design a suitable compensation scheme for a control system. Typically, we are interested in controlling the system with a control signal, $\mathbf{u}(t)$, which is a function of several measurable state variables. Then we develop a state variable controller that operates on the information available in measured form. This type of system compensation is quite useful for system optimization and will be considered in this chapter.

We first determine an optimal controller for a system described in terms of state variables. We then describe the design of state variable feedback systems with specified characteristic root locations. We also briefly consider the internal model design method and describe the limitations of the state variable feedback methods.

11.2   CONTROLLABILITY

A system described by the matrices $(\mathbf{A}, \mathbf{B})$ can be said to be controllable if there exists an unconstrained control $\mathbf{u}$ that can transfer any initial state $\mathbf{x}(0)$ to any other desired location $\mathbf{x}(t)$. For the system

$$\dot{\mathbf{x}} = \mathbf{A}\mathbf{x} + \mathbf{B}\mathbf{u},$$

we can determine whether the system is controllable by examining the algebraic condition

$$\text{rank } [\mathbf{B} \ \ \mathbf{A}\mathbf{B} \ \ \mathbf{A}^2\mathbf{B} \ \cdots \ \mathbf{A}^{n-1} \ \mathbf{B}] = n. \qquad (11.1)$$

For a single-input, single-output system, the controllability matrix $\mathbf{P}_c$ is described in terms of $\mathbf{A}$ and $\mathbf{b}$,

$$\mathbf{P}_c = [\mathbf{b} \ \ \mathbf{A}\mathbf{b} \ \ \mathbf{A}^2\mathbf{b} \ \cdots \ \mathbf{A}^{n-1}\mathbf{b}], \qquad (11.2)$$

which is an $n \times n$ matrix. Therefore, if the determinant of $\mathbf{P}_c$ is nonzero, the system is controllable [12].

Another method of determining whether a system is controllable is to draw the state variable flow graph model and determine whether the control signal, u, has a path to each state variable. If a path to each state exists, then each state variable may be controlled and the system is controllable.

A system that may be described in the phase variable format is always controllable.

EXAMPLE 11.1   **Controllability of a system**

Let us consider a system described as the transfer function

$$\frac{C(s)}{U(s)} = T(s) = \frac{1}{s^3 + a_2 s^2 + a_1 s + a_0}.$$

The flow graph model for this system is shown in Fig. 11.1. We note that there exists a path from $u(t)$ to all state variables and that the system is thus controllable.

The matrix differential equation is

$$\dot{\mathbf{x}} = \begin{bmatrix} 0 & 1 & 0 \\ 0 & 0 & 1 \\ -a_0 & -a_1 & -a_2 \end{bmatrix} \mathbf{x} + \begin{bmatrix} 0 \\ 0 \\ 1 \end{bmatrix} u.$$

FIGURE 11.1
Third-order system
flow graph model.

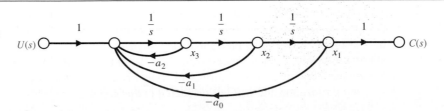

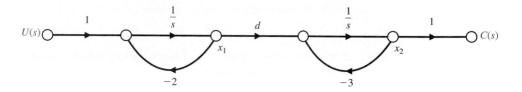

FIGURE 11.2
Flow graph model
for Example 11.2.

Then, we have

$$\mathbf{b} = \begin{bmatrix} 0 \\ 0 \\ 1 \end{bmatrix}, \qquad \mathbf{Ab} = \begin{bmatrix} 0 \\ 1 \\ -a_2 \end{bmatrix}, \qquad \mathbf{A}^2\mathbf{b} = \begin{bmatrix} 1 \\ -a_2 \\ (a_2^2 - a_1) \end{bmatrix}.$$

Therefore we obtain

$$\mathbf{P}_c = \begin{bmatrix} 0 & 0 & 1 \\ 0 & 1 & -a_2 \\ 1 & -a_2 & (a_2^2 - a_1) \end{bmatrix}.$$

The determinant of $\mathbf{P}_c$ is nonzero, and again we show that this system is controllable. ∎

EXAMPLE 11.2   Controllability of a two-state system

Let us consider a system represented by the two state equations

$$\dot{x}_1 = -2x_1 + u, \qquad \dot{x}_2 = -3x_2 + dx_1$$

and determine the condition for controllability. Also, we have $c = x_2$ as shown in the flow graph model of Fig. 11.2. Examination of the flow model shows that the system is controllable for $d \neq 0$. When $d = 0$, the input u does not have a path to x_2.

We can confirm the requirement on the parameter d by generating the matrix $\mathbf{P}_c$:

$$\mathbf{b} = \begin{bmatrix} 1 \\ 0 \end{bmatrix} \qquad \text{and} \qquad \mathbf{Ab} = \begin{bmatrix} -2 & 0 \\ d & -3 \end{bmatrix}\begin{bmatrix} 1 \\ 0 \end{bmatrix} = \begin{bmatrix} -2 \\ d \end{bmatrix}.$$

Therefore we have

$$\mathbf{P}_c = \begin{bmatrix} 1 & -2 \\ 0 & d \end{bmatrix},$$

and the determinant of $\mathbf{P}_c$ is equal to d, which is nonzero only when d is nonzero. ∎

11.3 OBSERVABILITY

All the roots of the characteristic equation can be placed where desired in the s-plane if, and only if, a system is *observable* and controllable. Observability refers to the ability to estimate a state variable. Thus we say a system is observable if the output has a component due to each state variable. Another way of stating the requirement is to prescribe that a path exist between each state variable and the output variable on the system flow graph model [13].

A system is observable if, and only if, there exists a finite time T such that the initial state $\mathbf{x}(0)$ can be determined from the observation history $c(t)$ given the control $u(t)$.

Consider the single-input, single-output system

$$\dot{\mathbf{x}} = \mathbf{Ax} + \mathbf{b}u \qquad \text{and} \qquad c = \mathbf{dx},$$

where $\mathbf{d}$ is a row vector and $\mathbf{x}$ is a column vector. This system is observable when the determinant of $\mathbf{Q}$ is nonzero where

$$\mathbf{Q} = \begin{bmatrix} \mathbf{d} \\ \mathbf{dA} \\ \vdots \\ \mathbf{dA}^{n-1} \end{bmatrix}, \tag{11.3}$$

which is an $n \times n$ matrix.

Alternatively, we can draw the flow graph model of a system and determine whether every state variable has a path to the output $c(t)$. A system that can be described in the phase variable format is always observable.

EXAMPLE 11.3   **Observability of a system**

Reconsider the system of Example 11.1. The flow graph model is shown in Fig. 11.1. Examining the flow graph, we note that a path to $c(t)$ exists for each state variable and thus the system is observable.

Alternatively, to construct $\mathbf{Q}$, we use

$$\mathbf{A} = \begin{bmatrix} 0 & 1 & 0 \\ 0 & 0 & 1 \\ -a_0 & -a_1 & -a_2 \end{bmatrix} \qquad \text{and} \qquad \mathbf{d} = [1, 0, 0].$$

Therefore

$$\mathbf{dA} = [0, 1, 0] \qquad \text{and} \qquad \mathbf{dA}^2 = [0, 0, 1].$$

Thus we obtain

$$\mathbf{Q} = \begin{bmatrix} 1 & 0 & 0 \\ 0 & 1 & 0 \\ 0 & 0 & 1 \end{bmatrix}.$$

Clearly, det $\mathbf{Q} = 1$ and the system is observable. ∎

EXAMPLE 11.4   **Observability of a two-state system**

Consider the system given by

$$\dot{\mathbf{x}} = \begin{bmatrix} 2 & 0 \\ -1 & 1 \end{bmatrix} \mathbf{x} + \begin{bmatrix} 1 \\ -1 \end{bmatrix} u \qquad \text{and} \qquad c = [1, 1]\mathbf{x}.$$

We first draw the state variable flow graph model, as shown in Fig. 11.3. The system appears to be controllable and observable from the flow graph model. So let us check the system by using the $\mathbf{P}_c$ and $\mathbf{Q}$ matrices.

The matrices for $\mathbf{P}_c$ are

$$\mathbf{b} = \begin{bmatrix} 1 \\ -1 \end{bmatrix} \qquad \text{and} \qquad \mathbf{Ab} = \begin{bmatrix} 2 \\ -2 \end{bmatrix}.$$

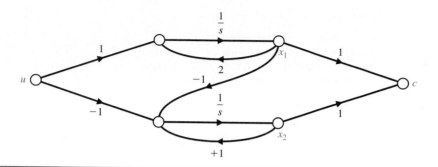

FIGURE 11.3
Signal-flow model
for Example 11.4.

Therefore we have

$$\mathbf{P}_c = \begin{bmatrix} 1 & 2 \\ -1 & -2 \end{bmatrix}$$

and det $\mathbf{P}_c = 0$. Thus the system is not controllable.

To determine $\mathbf{Q}$, we obtain

$$\mathbf{d} = [1, 1] \qquad \text{and} \qquad \mathbf{dA} = [1, 1].$$

Therefore we obtain

$$\mathbf{Q} = \begin{bmatrix} 1 & 1 \\ 1 & 1 \end{bmatrix}$$

and det $\mathbf{Q} = 0$. Therefore the system is not observable.

If we look again at the state model, we note that

$$c = x_1 + x_2.$$

However,

$$\dot{x}_1 + \dot{x}_2 = 2x_1 + (x_2 - x_1) + u - u = x_1 + x_2.$$

Thus the system state variables do not depend on u and the system is not controllable. Similarly, the output $(x_1 + x_2)$ depends on $x_1(0)$ plus $x_2(0)$ and does not allow us to determine $x_1(0)$ and $x_2(0)$ independently. Thus the system is not observable. ∎

11.4 OPTIMAL CONTROL SYSTEMS

The design of automatic control systems is an important function of control engineering. The purpose of design is to realize a system with practical components that will provide the desired operating performance. The desired performance can be readily stated in terms of time-domain performance indices. For example, the maximum overshoot and rise time for a step input are valuable time-domain indices. In the case of steady-state and transient performance, the performance indices are normally specified in the time domain and therefore it is natural that we wish to develop design procedures in the time domain.

The performance of a control system can be represented by integral performance measures, as we found in Section 5.5. Therefore the design of a system must be based on

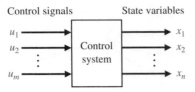

FIGURE 11.4
A control system in terms of *x* and *u*.

minimizing a performance index such as the integral of the squared error (ISE), as in Section 5.5. Systems that are adjusted to provide a minimum performance index are often called *optimal control systems*. In this section, we will consider the design of an optimal control system where the system is described by a state variable formulation. We will consider the measurement of the state variables and their use in developing a control signal $u(t)$ so that the performance of the system is optimized.

The performance of a control system, written in terms of the state variables of a system, can be expressed in general as

$$J = \int_0^{t_f} g(\mathbf{x},\ \mathbf{u},\ t)dt, \tag{11.4}$$

where $\mathbf{x}$ equals the state vector, $\mathbf{u}$ equals the control vector, and t_f equals the final time.[†]

We are interested in minimizing the error of the system; therefore when the desired state vector is represented as $\mathbf{x}_d = \mathbf{0}$, we are able to consider the error as identically equal to the value of the state vector. That is, we desire the system to be at equilibrium, $\mathbf{x} = \mathbf{x}_d = \mathbf{0}$, and any deviation from equilibrium is considered an error. Therefore in this section we will consider the design of optimal control systems using *state variable feedback* and error-squared performance indices [1–3].

The control system we will consider is shown in Fig. 11.4 and can be represented by the vector differential equation

$$\dot{\mathbf{x}} = \mathbf{A}\mathbf{x} + \mathbf{B}\mathbf{u}. \tag{11.5}$$

We will select a feedback controller so that $\mathbf{u}$ is some function of the measured state variables $\mathbf{x}$ and therefore

$$\mathbf{u} = \mathbf{h}(\mathbf{x}).$$

For example, we might use

$$u_1 = k_1 x_1, \qquad u_2 = k_2 x_2, \qquad \ldots, \qquad u_m = k_m x_m. \tag{11.6}$$

Alternatively, we might choose the control vector as

$$u_1 = k_1(x_1 + x_2), \qquad u_2 = k_2(x_2 + x_3), \qquad \ldots \tag{11.7}$$

The choice of the control signals is somewhat arbitrary and depends partially on the actual desired performance and the complexity of the feedback structure allowable. Often we are limited in the number of state variables available for feedback, since we are only able to utilize measurable state variables.

[†]Note that to denote the performance index, J is used instead of I, as in Chapter 5. This will enable the reader to distinguish readily the performance index from the identity matrix, which is represented by the boldfaced capital $\mathbf{I}$.

Now, in our case we limit the feedback function to a linear function so that $\mathbf{u} = \mathbf{Hx}$, where $\mathbf{H}$ is an $m \times n$ matrix. Therefore, in expanded form, we have

$$
\begin{bmatrix} u_1 \\ u_2 \\ \vdots \\ u_m \end{bmatrix} = \begin{bmatrix} h_{11} \cdots h_{1n} \\ \vdots \quad \vdots \\ h_{m1} \cdots h_{mn} \end{bmatrix} \begin{bmatrix} x_1 \\ x_2 \\ \vdots \\ x_n \end{bmatrix}.
\tag{11.8}
$$

Then, substituting Eq. (11.8) into Eq. (11.5), we obtain

$$
\dot{\mathbf{x}} = \mathbf{Ax} + \mathbf{BHx} = \mathbf{Dx},
\tag{11.9}
$$

where $\mathbf{D}$ is the $n \times n$ matrix resulting from the addition of the elements of $\mathbf{A}$ and $\mathbf{BH}$.

Now, returning to the error-squared performance index, we recall from Section 5.5 that the index for a single state variable, x_1, is written as

$$
J = \int_0^{t_f} (x_1(t))^2 \, dt.
\tag{11.10}
$$

A performance index written in terms of two state variables would then be

$$
J = \int_0^{t_f} (x_1^2 + x_2^2) \, dt.
\tag{11.11}
$$

Therefore, because we wish to define the performance index in terms of an integral of the sum of the state variables squared, we will utilize the matrix operation

$$
\mathbf{x}^T \mathbf{x} = [x_1, x_2, x_3, \ldots, x_n] \begin{bmatrix} x_1 \\ x_2 \\ \vdots \\ x_n \end{bmatrix} = (x_1^2 + x_2^2 + x_3^2 + \cdots + x_n^2), \tag{11.12}
$$

where $\mathbf{x}^T$ indicates the transpose of the $\mathbf{x}$ matrix.[†] Then the specific form of the performance index, in terms of the state vector, is

$$
J = \int_0^{t_f} (\mathbf{x}^T \mathbf{x}) \, dt.
\tag{11.13}
$$

The general form of the performance index (Eq. 11.4) incorporates a term with $\mathbf{u}$ that we have not included at this point, but we will do so later in this section.

Again considering Eq. (11.13), we will let the final time of interest be $t_f = \infty$. To obtain the minimum value of J, we postulate the existence of an exact differential so that

$$
\frac{d}{dt} (\mathbf{x}^T \mathbf{P} \mathbf{x}) = -\mathbf{x}^T \mathbf{x},
\tag{11.14}
$$

where $\mathbf{P}$ is to be determined. A symmetric $\mathbf{P}$ matrix will be used to simplify the algebra without any loss of generality. Then, for a symmetric $\mathbf{P}$ matrix, $p_{ij} = p_{ji}$. Completing the differentiation indicated on the left-hand side of Eq. (11.14), we have

$$
\frac{d}{dt} (\mathbf{x}^T \mathbf{P} \mathbf{x}) = \dot{\mathbf{x}}^T \mathbf{P} \mathbf{x} + \mathbf{x}^T \mathbf{P} \dot{\mathbf{x}}.
$$

[†]The matrix operation $\mathbf{x}^T\mathbf{x}$ is discussed in Appendix C, Section C.4.

Then, substituting Eq. (11.9), we obtain

$$\frac{d}{dt}(\mathbf{x}^T\mathbf{Px}) = (\mathbf{Dx})^T\mathbf{Px} + \mathbf{x}^T\mathbf{P}(\mathbf{Dx})$$

$$= \mathbf{x}^T\mathbf{D}^T\mathbf{Px} + \mathbf{x}^T\mathbf{PDx} \tag{11.15}$$

$$= \mathbf{x}^T(\mathbf{D}^T\mathbf{P} + \mathbf{PD})\mathbf{x},$$

where $(\mathbf{Dx})^T = \mathbf{x}^T\mathbf{D}^T$ by the definition of the transpose of a product. If we let $(\mathbf{D}^T\mathbf{P} + \mathbf{PD})$ $= -\mathbf{I}$, then Eq. (11.15) becomes

$$\frac{d}{dt}(\mathbf{x}^T\mathbf{Px}) = -\mathbf{x}^T\mathbf{x}, \tag{11.16}$$

which is the exact differential we are seeking. Substituting Eq. (11.16) into Eq. (11.13), we obtain

$$J = \int_0^\infty -\frac{d}{dt}(\mathbf{x}^T\mathbf{Px})\,dt = -\mathbf{x}^T\mathbf{Px}\Big|_0^\infty = \mathbf{x}^T(0)\mathbf{Px}(0). \tag{11.17}$$

In the evaluation of the limit at $t = \infty$, we have assumed that the system is stable and hence $\mathbf{x}(\infty) = 0$, as desired. Therefore, to minimize the performance index J, we consider the two equations

$$J = \int_0^\infty \mathbf{x}^T\mathbf{x}\,dt = \mathbf{x}^T(0)\mathbf{Px}(0) \tag{11.18}$$

and

$$\mathbf{D}^T\mathbf{P} + \mathbf{PD} = -\mathbf{I}. \tag{11.19}$$

The design steps are then as follows:

1. Determine the matrix $\mathbf{P}$ that satisfies Eq. (11.19), where $\mathbf{D}$ is known.
2. Minimize J by determining the minimum of Eq. (11.18) by adjusting one or more unspecified system parameters.

EXAMPLE 11.5   State variable feedback

Consider the control system shown in Fig. 11.5 in signal-flow graph form. The state variables are identified as x_1 and x_2. The performance of this system is quite unsatisfactory because an undamped response results for a step input or disturbance signal. The vector differential equation of this system is

$$\frac{d}{dt}\begin{bmatrix} x_1 \\ x_2 \end{bmatrix} = \begin{bmatrix} 0 & 1 \\ 0 & 0 \end{bmatrix}\begin{bmatrix} x_1 \\ x_2 \end{bmatrix} + \begin{bmatrix} 0 \\ 1 \end{bmatrix}u(t), \tag{11.20}$$

FIGURE 11.5
Signal-flow graph
of the control
system of
Example 11.5.

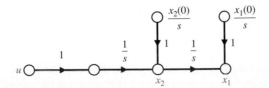

where

$$\mathbf{A} = \begin{bmatrix} 0 & 1 \\ 0 & 0 \end{bmatrix} \text{ and } \mathbf{b} = \begin{bmatrix} 0 \\ 1 \end{bmatrix}.$$

We will choose a feedback control system so that

$$u(t) = -k_1 x_1 - k_2 x_2, \qquad (11.21)$$

and therefore the control signal is a linear function of the two state variables. The sign of the feedback is negative in order to provide negative feedback. Then Eq. (11.20) becomes

$$\dot{x}_1 = x_2,$$
$$\dot{x}_2 = -k_1 x_1 - k_2 x_2, \qquad (11.22)$$

or, in matrix form, we have

$$\dot{\mathbf{x}} = \mathbf{Dx}$$
$$= \begin{bmatrix} 0 & 1 \\ -k_1 & -k_2 \end{bmatrix} \mathbf{x}. \qquad (11.23)$$

We note that x_1 would represent the position of a position control system and the transfer function of the system would be $G(s) = 1/Ms^2$, where $M = 1$ and the friction is negligible. In any case, to avoid needless algebraic manipulation, we will let $k_1 = 1$ and determine a suitable value for k_2 so that the performance index is minimized. Then, writing Eq. (11.19), we have

$$\mathbf{D}^T \mathbf{P} + \mathbf{P} \mathbf{D} = -\mathbf{I},$$

$$\begin{bmatrix} 0 & -1 \\ 1 & -k_2 \end{bmatrix} \begin{bmatrix} p_{11} & p_{12} \\ p_{12} & p_{22} \end{bmatrix} + \begin{bmatrix} p_{11} & p_{12} \\ p_{12} & p_{22} \end{bmatrix} \begin{bmatrix} 0 & 1 \\ -1 & -k_2 \end{bmatrix} = \begin{bmatrix} -1 & 0 \\ 0 & -1 \end{bmatrix}. \qquad (11.24)$$

Completing the matrix multiplication and addition, we have

$$-p_{12} - p_{12} = -1,$$
$$p_{11} - k_2 p_{12} - p_{22} = 0, \qquad (11.25)$$
$$p_{12} - k_2 p_{22} + p_{12} - k_2 p_{22} = -1.$$

Then, solving these simultaneous equations, we obtain

$$p_{12} = \frac{1}{2}, \qquad p_{22} = \frac{1}{k_2}, \qquad p_{11} = \frac{k_2^2 + 2}{2k_2}.$$

The integral performance index is then

$$J = \mathbf{x}^T(0)\mathbf{P}\mathbf{x}(0), \qquad (11.26)$$

and we will consider the case where each state is initially displaced one unit from equilibrium so that $\mathbf{x}^T(0) = [1, 1]$. Therefore Eq. (11.26) becomes

$$J = [1, 1] \begin{bmatrix} p_{11} & p_{12} \\ p_{12} & p_{22} \end{bmatrix} \begin{bmatrix} 1 \\ 1 \end{bmatrix}$$

$$= [1, 1] \begin{bmatrix} (p_{11} + p_{12}) \\ (p_{12} + p_{22}) \end{bmatrix} \qquad (11.27)$$

$$= (p_{11} + p_{12}) + (p_{12} + p_{22}) = p_{11} + 2p_{12} + p_{22}.$$

Substituting the values of the elements of **P**, we have

$$J = \frac{k_2^2 + 2}{2k_2} + 1 + \frac{1}{k_2} = \frac{k_2^2 + 2k_2 + 4}{2k_2}. \tag{11.28}$$

To minimize as a function of k_2, we take the derivative with respect to k_2 and set it equal to zero:

$$\frac{\partial J}{\partial k_2} = \frac{2k_2(2k_2 + 2) - 2(k_2^2 + 2k_2 + 4)}{(2k_2)^2} = 0. \tag{11.29}$$

Therefore $k_2^2 = 4$ and $k_2 = 2$ when J is a minimum. The minimum value of J is obtained by substituting $k_2 = 2$ into Eq. (11.28). Thus we obtain

$$J_{min} = 3.$$

The system matrix **D**, obtained for the compensated system, is then

$$\mathbf{D} = \begin{bmatrix} 0 & 1 \\ -1 & -2 \end{bmatrix}. \tag{11.30}$$

The characteristic equation of the compensated system is therefore

$$\det [\lambda \mathbf{I} - \mathbf{D}] = \det \begin{bmatrix} \lambda & -1 \\ 1 & \lambda + 2 \end{bmatrix} = \lambda^2 + 2\lambda + 1. \tag{11.31}$$

Because this is a second-order system, we note that the characteristic equation is of the form $(s^2 + 2\zeta\omega_n s + \omega_n^2)$, and therefore the damping ratio of the compensated system is $\zeta = 1.0$. This compensated system is considered to be an optimal system in that the compensated system results in a minimum value for the performance index. Of course, we recognize that this system is optimal only for the specific set of initial conditions that were assumed. The compensated system is shown in Fig. 11.6. A curve of the performance index as a function of k_2 is shown in Fig. 11.7. It is clear that this system is not very sensitive to changes in k_2 and will maintain a near minimum performance index if the k_2 is altered some percentage. We define the sensitivity of an optimal system as

$$S_k^{opt} = \frac{\Delta J/J}{\Delta k/k}, \tag{11.32}$$

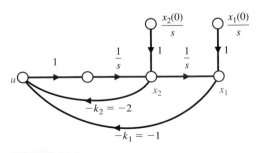

FIGURE 11.6
Compensated control system of Example 11.5.

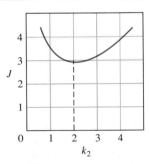

FIGURE 11.7
Performance index versus the parameter k_2.

where k is the design parameter. Then for this example we have $k = k_2$, and therefore

$$S_{k_2}^{\text{opt}} \simeq \frac{0.08/3}{0.5/2} = 0.107. \ \blacksquare \tag{11.33}$$

EXAMPLE 11.6   Determination of an optimal system

Now let us reconsider the system of Example 11.5, where both the feedback gains, k_1 and k_2, are unspecified. To simplify the algebra without any loss in insight into the problem, let us set $k_1 = k_2 = k$. The reader can prove that if k_1 and k_2 are unspecified, then $k_1 = k_2$ when the minimum of the performance index (Eq. 11.18) is obtained. Then for the system of Example 11.5, Eq. (11.23) becomes

$$\dot{\mathbf{x}} = \mathbf{D}\mathbf{x} = \begin{bmatrix} 0 & 1 \\ -k & -k \end{bmatrix} \mathbf{x}. \tag{11.34}$$

To determine the **P** matrix, we utilize Eq. (11.19), which is

$$\mathbf{D}^T\mathbf{P} + \mathbf{PD} = -\mathbf{I}. \tag{11.35}$$

Solving the set of simultaneous equations resulting from Eq. (11.35), we find that

$$p_{12} = \frac{1}{2k}, \qquad p_{22} = \frac{(k+1)}{2k^2}, \qquad p_{11} = \frac{(1+2k)}{2k}.$$

Let us consider the case where the system is initially displaced one unit from equilibrium so that $\mathbf{x}^T(0) = [1, 0]$. Then the performance index (Eq. 11.18) becomes

$$J = \int_0^\infty \mathbf{x}^T\mathbf{x} \ dt = \mathbf{x}^T(0)\mathbf{P}\mathbf{x}(0) = p_{11}. \tag{11.36}$$

Thus the performance index to be minimized is

$$J = p_{11} = \frac{(1+2k)}{2k} = 1 + \frac{1}{2k}. \tag{11.37}$$

Clearly, the minimum value of J is obtained when k approaches infinity; the result is $J_{\min} = 1$. A plot of J versus k, shown in Fig. 11.8, illustrates that the performance index approaches a minimum asymptotically as k approaches an infinite value. Now we recognize that in providing a very large gain k, we can cause the feedback signal

$$u(t) = -k(x_1(t) + x_2(t))$$

FIGURE 11.8
Performance
index versus the
feedback gain k for
Example 11.6.

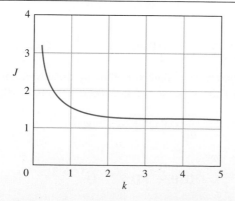

to be very large. However, we are restricted to realizable magnitudes of the control signal $u(t)$. Therefore we must introduce a *constraint* on $u(t)$ so that the gain k is not made too large. Then, for example, if we establish a constraint on $u(t)$ so that

$$|u(t)| \leq 50, \tag{11.38}$$

we require that the maximum acceptable value of k in this case is

$$k_{max} = \frac{|u|_{max}}{x_1(0)} = 50. \tag{11.39}$$

Then the minimum value of J is

$$J_{min} = 1 + \frac{1}{2k_{max}} = 1.01, \tag{11.40}$$

which is sufficiently close to the absolute minimum of J to satisfy our requirements.

Upon the examination of the performance index (Eq. 11.13), we recognize that the reason the magnitude of the control signal is not accounted for in the original calculations is that $u(t)$ is not included within the expression for the performance index. However, in many cases we are concerned with the expenditure of the control signal energy. For example, in an electric vehicle control system, $[u(t)]^2$ represents the expenditure of battery energy and must be restricted to conserve the energy for long periods of travel. To account for the expenditure of the energy of the control signal, we will utilize the performance index

$$J = \int_0^\infty (\mathbf{x}^T \mathbf{I} \mathbf{x} + \lambda \mathbf{u}^T \mathbf{u}) \, dt, \tag{11.41}$$

where λ is a scalar weighting factor and $\mathbf{I} =$ identity matrix. The weighting factor λ will be chosen so that the relative importance of the state variable performance is contrasted with the importance of the expenditure of the system energy resource that is represented by $\mathbf{u}^T \mathbf{u}$. As in the previous paragraphs, we will represent the state variable feedback by the matrix equation

$$\mathbf{u} = \mathbf{H} \mathbf{x} \tag{11.42}$$

and the system with this state variable feedback as

$$\dot{\mathbf{x}} = \mathbf{A} \mathbf{x} + \mathbf{B} \mathbf{u} = \mathbf{D} \mathbf{x}. \tag{11.43}$$

Now, substituting Eq. (11.42) into Eq. (11.41), we have

$$J = \int_0^\infty (\mathbf{x}^T \mathbf{I} \mathbf{x} + \lambda (\mathbf{H} \mathbf{x})^T (\mathbf{H} \mathbf{x})) \, dt$$

$$= \int_0^\infty [\mathbf{x}^T (\mathbf{I} + \lambda \mathbf{H}^T \mathbf{H}) \mathbf{x}] \, dt = \int_0^\infty \mathbf{x}^T \mathbf{Q} \mathbf{x} \, dt, \tag{11.44}$$

where $\mathbf{Q} = (\mathbf{I} + \lambda \mathbf{H}^T \mathbf{H})$ is an $n \times n$ matrix. Following the development of Eqs. (11.13) through (11.17), we postulate the existence of an exact differential so that

$$\frac{d}{dt}(\mathbf{x}^T \mathbf{P} \mathbf{x}) = -\mathbf{x}^T \mathbf{Q} \mathbf{x}. \tag{11.45}$$

Then in this case we require that

$$\mathbf{D}^T \mathbf{P} + \mathbf{P} \mathbf{D} = -\mathbf{Q}. \tag{11.46}$$

and thus we have as before (Eq. 11.17)

$$J = \mathbf{x}^T(0)\mathbf{P}\mathbf{x}(0). \tag{11.47}$$

Now the design steps are exactly as for Eqs. (11.18) and (11.19) with the exception that the left side of Eq. (11.46) equals $-\mathbf{Q}$ instead of $-\mathbf{I}$. Of course, if $\lambda = 0$, Eq. (11.46) reduces to Eq. (11.19). Now let us reconsider Example 11.5 when λ is other than zero and account for the expenditure of control signal energy. ■

EXAMPLE 11.7   **Optimal system with energy constraint**

Let us reconsider the system of Example 11.5, which is shown in Fig. 11.5. For this system we use a state variable feedback so that

$$\mathbf{u} = \mathbf{Hx} = \begin{bmatrix} k & 0 \\ 0 & k \end{bmatrix}\begin{bmatrix} x_1 \\ x_2 \end{bmatrix} = k\mathbf{Ix}. \tag{11.48}$$

Therefore the matrix $\mathbf{Q}$ is

$$\mathbf{Q} = (\mathbf{I} + \lambda\mathbf{II}^T\mathbf{H}) = (\mathbf{I} + \lambda k^2\mathbf{I}) = (1 + \lambda k^2)\mathbf{I}. \tag{11.49}$$

As in Example 11.6 we will let $\mathbf{x}^T(0) = [1, 0]$ so that $J = p_{11}$. We evaluate p_{11} from Eq. (11.46) as

$$\mathbf{D}^T\mathbf{P} + \mathbf{PD} = -\mathbf{Q} = -(1 + \lambda k^2)\mathbf{I}. \tag{11.50}$$

Thus we find that

$$J = p_{11} = (1 + \lambda k^2)\left(1 + \frac{1}{2k}\right), \tag{11.51}$$

and we note that the right-hand side of Eq. (11.51) reduces to Eq. (11.37) when $\lambda = 0$. Now the minimum of J is found by taking the derivative of J, which is

$$\frac{dJ}{dk} = 2\lambda k + \frac{\lambda}{2} - \frac{1}{2k^2} = \frac{4\lambda k^3 + \lambda k^2 - 1}{2k^2} = 0. \tag{11.52}$$

Therefore the minimum of the performance index occurs when $k = k_{\min}$, where $k_{\min}$ is the solution of Eq. (11.52).

A simple method of solution for Eq. (11.52) is the Newton–Raphson method, illustrated in Section 6.5. Let us complete this example for the case where the control energy and the state variables squared are equally important so that $\lambda = 1$. Then Eq. (11.52) becomes $4k^3 + k^2 - 1 = 0$, and using the Newton–Raphson method, we find that $k_{\min} = 0.555$. The value of the performance index J obtained with $k_{\min}$ is considerably greater than that of the previous example, because the expenditure of energy is equally weighted as a cost. The plot of J versus k for this case is shown in Fig. 11.9. The plot of J versus k for Example 11.6 is also shown for comparison in Fig. 11.9. ■

It has become clear from the examples in this chapter that the actual minimum obtained depends on the initial conditions, the definition of the performance index, and the value of the scalar factor λ.

The design of several parameters can be accomplished in a manner similar to that illustrated in the examples. Also, the design procedure can be carried out for higher-order

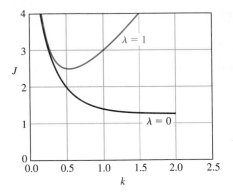

FIGURE 11.9
Performance
index versus the
feedback gain k for
Example 11.7.

systems. However, we must then consider the use of a digital computer to determine the solution of Eq. (11.19) in order to obtain the **P** matrix. Also, the computer may provide a suitable approach for evaluating the minimum value of J for one or more parameters. However, the solution of Eq. (11.46) may be difficult, especially when the system order is quite high ($n > 3$). An alternative method suitable for computer calculation can be stated without proof in the following paragraph.

Consider the uncompensated single-input, single-output system with

$$\dot{\mathbf{x}} = A\mathbf{x} + \mathbf{b}u$$

and feedback

$$u = -\mathbf{K}^T\mathbf{x} = -[k_1, k_2, \ldots, k_n]\mathbf{x}.$$

The performance index is

$$J = \int_0^\infty (\mathbf{x}^T\mathbf{Q}\mathbf{x} + ru^2)\, dt,$$

where r is the scalar weighting factor. This index is minimized when

$$\mathbf{K} = \mathbf{P}\mathbf{b}r^{-1}.$$

The $n \times n$ matrix **P** is determined from the solution of the equation

$$A^T\mathbf{P} + \mathbf{P}A - \mathbf{P}\mathbf{b}\mathbf{b}^T\mathbf{P}r^{-1} + \mathbf{Q} = \mathbf{0}, \tag{11.53}$$

where $\mathbf{Q} = \mathbf{I}$, the identity matrix in many cases. Equation (11.53) can be easily programmed for a computer, or solved using *Matlab*. Equation (11.53) is often called the Riccati equation. This optimal control is called the linear quadratic regulator (LQR) [14, 22].

11.5 POLE PLACEMENT USING STATE FEEDBACK

In Section 11.4 we considered the use of state variable feedback in achieving optimization of a performance index. In this section we will use state variable feedback to achieve the desired pole locations of the closed-loop transfer function $T(s)$. The approach is based on the feedback of all the state variables, and therefore

$$\mathbf{u} = \mathbf{Hx}. \tag{11.54}$$

When using this state variable feedback, the roots of the characteristic equation are placed where the transient performance meets the desired response.

EXAMPLE 11.8   **Design of a third-order system**

Let us consider the third-order system with the differential equation

$$\frac{d^3y}{dt^3} + 5\frac{d^2y}{dt^2} + 3\frac{dy}{dt} + 2y = u.$$

We can select the state variables as the phase variables (see Section 3.4) so that $x_1 = y$, $x_2 = dy/dt$, $x_3 = d^2y/dt^2$ and then

$$\dot{\mathbf{x}} = \begin{bmatrix} 0 & 1 & 0 \\ 0 & 0 & 1 \\ -2 & -3 & -5 \end{bmatrix} x + \begin{bmatrix} 0 \\ 0 \\ 1 \end{bmatrix} u = \mathbf{Ax} + \mathbf{b}u,$$

where $c = x_1$. If the state variable feedback matrix $\mathbf{H}$ is

$$\mathbf{H} = [k_1 \quad k_2 \quad k_3]$$

and

$$u = -\mathbf{Hx},$$

then

$$\dot{\mathbf{x}} = \mathbf{Ax} - \mathbf{bHx} = (\mathbf{A} - \mathbf{bH})\mathbf{x}.$$

The state feedback matrix is

$$[\mathbf{A} - \mathbf{bH}] = \begin{bmatrix} 0 & 1 & 0 \\ 0 & 0 & 1 \\ (-2 - k_1) & (-3 - k_2) & (-5 - k_3) \end{bmatrix},$$

and the characteristic equation is

$$\det[\mathbf{A} - \mathbf{bH}] = s^3 + (5 + k_3)s^2 + (3 + k_2)s + (2 + k_1).$$

If we seek a rapid response with a low overshoot, we choose a desired characteristic equation as (see Eq. 5.17 and Table 5.3)

$$(s^2 + 2\zeta\omega_n s + \omega_n^2)(s + \zeta\omega_n).$$

We choose $\zeta = 0.8$ for minimal overshoot and ω_n to meet the settling time requirement. If we want a settling time equal to 1 second, then

$$T_s = \frac{4}{\zeta\omega_n} = \frac{4}{(0.8)\omega_n} \cong 1.$$

If we choose $\omega_n = 6$, the desired characteristic equation is

$$(s^2 + 9.6s + 36)(s + 4.8) = s^3 + 14.4s^2 + 82.1s + 172.8.$$

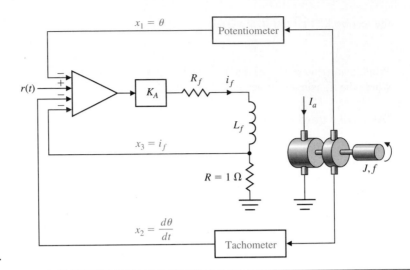

FIGURE 11.10
Position control
system with state
variable feedback.

Therefore we require $k_3 = 9.4$, $k_2 = 79.1$, and $k_3 = 170.8$. Using a simulation, we find no overshoot and a settling time of 1 second, as desired for a step input. ∎

As a practical example of state variable feedback, consider the feedback system shown in Fig. 11.10. This position control uses a field-controlled motor, and the transfer function was obtained in Section 2.5 as

$$G(s) = \frac{K}{s(s + f/J)(s + R_f/L_f)},$$

where $K = K_a K_m / J L_f$. For our purposes, we will assume that $f/J = 1$ and $R_f/L_f = 5$. As shown in Fig. 11.10, the system has feedback of the three state variables: position, velocity, and field current. We will assume that the feedback constant for the position is equal to 1, as shown in Fig. 11.11, which provides a signal-flow graph representation of the system. Without state variable feedback of x_2 and x_3, we set $K_3 = K_2 = 0$ and we have

$$G(s) = \frac{K}{s(s + 1)(s + 5)}. \tag{11.55}$$

This system will become unstable when $K \geq 30$. However, with state feedback of all the state variables, we can ensure that the system is stable and set the transient performance of the system to a desired performance.

FIGURE 11.11
Signal-flow graph
of the state
variable feedback
system.

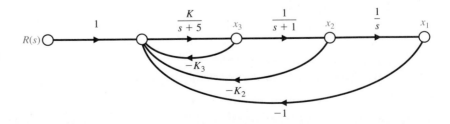

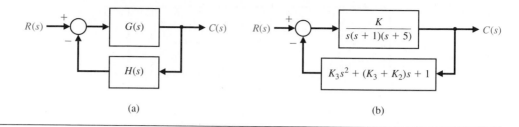

FIGURE 11.12
Equivalent
block diagram
representation of
the state variable
feedback system.

(a)   (b)

The signal-flow graph of Fig. 11.11 can be converted to the block diagram form shown in Fig. 11.12. The transfer function $G(s)$ remains unaffected (as in Eq. 11.55) and the $H(s)$ accounts for the state variable feedback. Therefore

$$H(s) = K_3\left[s^2 + \left(\frac{K_3 + K_2}{K_3}\right)s + \frac{1}{K_3} \right] \tag{11.56}$$

and

$$G(s)H(s) = \frac{M[s^2 + Qs + (1/K_3)]}{s(s + 1)(s + 5)}, \tag{11.57}$$

where $M = KK_3$ and $Q = (K_3 + K_2)/K_3$. Since K_3 and K_2 can be set independently, the designer can select the location of the zeros of $G(s)H(s)$.

As an illustration, let us choose the zeros of $GH(s)$ so that they cancel the real poles of $G(s)$. We set the numerator polynomial

$$H(s) = K_3\left(s^2 + Qs + \frac{1}{K_3}\right) = K_3(s + 1)(s + 5). \tag{11.58}$$

This requires $K_3 = \frac{1}{5}$ and $Q = 6$, which sets $K_2 = 1$. Then

$$GH(s) = \frac{M(s + 1)(s + 5)}{s(s + 1)(s + 5)}, \tag{11.59}$$

where $M = KK_3$. The closed-loop transfer function is then

$$\frac{C(s)}{R(s)} = T(s) = \frac{G(s)}{1 + G(s)H(s)} = \frac{K}{(s + 1)(s + 5)(s + M)}. \tag{11.60}$$

Therefore, although we could choose $M = 10$, which would ensure the stability of the system, the closed-loop response of the system will be dictated by the poles at $s = -1$ and $s = -5$. Hence we will usually choose the zeros of $GH(s)$ in order to achieve closed-loop roots in a desirable location in the left-hand plane and to ensure system stability.

EXAMPLE 11.9   State variable feedback design

Let us again consider the system of Fig. 11.12(b) and set the zeros of $GH(s)$ at $s = -4 + j2$ and $s = -4 - j2$. Then the numerator of $GH(s)$ will be

$$H(s) = K_3\left(s^2 + Qs + \frac{1}{K_3}\right)$$

$$= K_3(s + 4 + j2)(s + 4 - j2) = K_3(s^2 + 8s + 20). \tag{11.61}$$

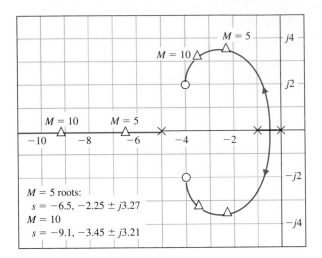

FIGURE 11.13
Compensated system root locus.

Therefore $K_3 = \frac{1}{20}$ and $Q = 8$, resulting in $K_2 = \frac{7}{20}$. The resulting root locus for

$$G(s)H(s) = \frac{M(s^2 + 8s + 20)}{s(s + 1)(s + 5)}$$

is shown in Fig. 11.13. The system is stable for all values of gain $M = KK_3$. For $M = 10$ the complex roots have $\zeta = 0.73$, so that we expect an overshoot for a step input of approximately 5%. The settling time will be approximately 1 second. The closed-loop transfer function is

$$\frac{C(s)}{R(s)} = T(s) = \frac{G(s)}{1 + G(s)(H(s)}$$

$$= \frac{200}{(s + 3.45 + j3.2)(s + 3.45 - j3.2)(s + 9.1)}.$$

An alternative approach is to set the closed-loop roots of $1 + G(s)H(s) = 0$ at desired locations and then solve for the gain values of K, K_3, and K_2 that are required. For example, if we desire closed-loop roots at $s = -10$, $s = -5 + j$ and $s = -5 - j$, we have the characteristic equation

$$q(s) = (s + 10)(s^2 + 10s + 26) = s^3 + 20s^2 + 126s + 260 = 0. \quad (11.62)$$

Because

$$1 + G(s)H(s) = s(s + 1)(s + 5) + M\left(s^2 + Qs + \frac{1}{K_3}\right) = 0, \quad (11.63)$$

we equate Eq. (11.62) and Eq. (11.63), obtaining $M = 14$, $Q = 121/14$, $K_3 = 14/260$, and $K_2 = 0.41$. ■

In many cases the state variables are available and we can use state variable feedback to obtain a stable, well-compensated system.

EXAMPLE 11.10   Inverted pendulum control

A control system can be designed so that if $u(t)$ is a function of the state variables, a stable system will result. The design of a stable feedback control system is based on a suitable selection of a feedback system structure. Therefore, considering the control of the cart and the unstable inverted pendulum shown in Fig. 3.19, we must measure and utilize the state variables of the system in order to control the cart (see Example 3.5). Thus if we desire to measure the state variable $x_3 = \theta$, we could use a potentiometer connected to the shaft of the pendulum hinge. Similarly, we could measure the rate of change of the angle, $x_4 = \dot{\theta}$, by using a tachometer generator. The state variables, x_1 and x_2, which are the position and velocity of the cart, can also be measured by suitable sensors. If the state variables are all measured, then they can be utilized in a feedback controller so that $u = \mathbf{hx}$, where $\mathbf{h}$ is the feedback matrix. The state vector $\mathbf{x}$ represents the state of the system; therefore knowledge of $\mathbf{x}(t)$ and the equations describing the system dynamics provide sufficient information for control and stabilization of a system. This design approach is called state feedback control [4, 5, 7].

To illustrate the utilization of state variable feedback, let us reconsider the unstable portion of the inverted pendulum system and design a suitable state variable feedback control system. We begin by considering a reduced system. If we assume that the control signal is an acceleration signal and that the mass of the cart is negligible, we can focus on the unstable dynamics of the pendulum. When $u(t)$ is an acceleration signal, Eq. (3.61) becomes

$$gx_3 - l\dot{x}_4 = \dot{x}_2 = \ddot{y} = u(t).$$

For the reduced system, where the control signal is an acceleration signal, the position and velocity of the cart are integral functions of $u(t)$. The portion of the state vector under consideration is $[x_3, x_4] = [\theta, \dot{\theta}]$. Thus the state vector differential equation reduces to

$$\frac{d}{dt}\begin{bmatrix} x_3 \\ x_4 \end{bmatrix} = \begin{bmatrix} 0 & 1 \\ g/l & 0 \end{bmatrix}\begin{bmatrix} x_3 \\ x_4 \end{bmatrix} + \begin{bmatrix} 0 \\ -(1/l) \end{bmatrix} u(t). \tag{11.64}$$

Clearly, the $\mathbf{A}$ matrix of Eq. (11.64) is simply the lower right-hand portion of the $\mathbf{A}$ matrix of Eq. (3.65) and the system has the characteristic equation $[\lambda^2 - (g/l)]$ with one root in the right-hand s-plane. To stabilize the system, we generate a control signal that is a function of the two state variables x_3 and x_4. Then we have

$$u(t) = \mathbf{hx} = [h_1, \ h_2]\begin{bmatrix} x_3 \\ x_4 \end{bmatrix} = h_1 x_3 + h_2 x_4.$$

Substituting this control signal relationship into Eq. (11.64), we have

$$\begin{bmatrix} \dot{x}_3 \\ \dot{x}_4 \end{bmatrix} = \begin{bmatrix} 0 & 1 \\ g/l & 0 \end{bmatrix}\begin{bmatrix} x_3 \\ x_4 \end{bmatrix} + \begin{bmatrix} 0 \\ -(1/l)(h_1 x_3 + h_2 x_4) \end{bmatrix}.$$

Combining the two additive terms on the right side of the equation, we find

$$\begin{bmatrix} \dot{x}_3 \\ \dot{x}_4 \end{bmatrix} = \begin{bmatrix} 0 & 1 \\ (1/l)(g - h_1) & -(h_2/l) \end{bmatrix}\begin{bmatrix} x_3 \\ x_4 \end{bmatrix}.$$

Therefore, obtaining the characteristic equation, we have

$$\det \begin{bmatrix} +\lambda & -1 \\ -(1/l)(g - h_1) & (\lambda + h_2/l) \end{bmatrix} = \lambda\left(\lambda + \frac{h_2}{l}\right) - \frac{1}{l}(g - h_1)$$

$$= \lambda^2 + \left(\frac{h_2}{l}\right)\lambda + \frac{1}{l}(h_1 - g). \tag{11.65}$$

Thus for the system to be stable, we require that $(h_2/l) > 0$ and $h_1 > g$. Hence we have stabilized an unstable system by measuring the state variables x_3 and x_4 and using the control function $u = h_1 x_3 + h_2 x_4$ to obtain a stable system. If we wish to achieve a rapid response with modest overshoot, we select $\omega_n = 10$ and $\zeta = 0.8$. Then we require

$$\frac{h_2}{l} = 16 \quad \text{and} \quad \frac{(h_1 - g)}{l} = 100.$$

The step response would have an overshoot of 1.5% and a settling time of 0.5 second. ■

In this section we have established an approach for the design of a feedback control system by using the state variables as the feedback variables in order to increase the stability of the system and obtain the desired system response.

11.6 ACKERMANN'S FORMULA

For a single-input, single-output system, Ackermann's formula is useful for determining the state variable feedback matrix

$$\mathbf{H} = [k_1, k_2, \ldots, k_n],$$

where

$$\mathbf{u} = -\mathbf{Hx}.$$

Given the desired characteristic equation,

$$q(s) = s^n + \alpha_1 s^{n-1} + \cdots + \alpha_n.$$

The state feedback gain matrix is given by

$$\mathbf{H} = [0, 0, \ldots, 1]\mathbf{P}_c^{-1}\, q(\mathbf{A}), \tag{11.66}$$

where

$$q(\mathbf{A}) = \mathbf{A}^n + \alpha_1 \mathbf{A}^{n-1} + \cdots \alpha_{n-1}\mathbf{A} + \alpha_n\mathbf{I}$$

and $\mathbf{P}_c$ is the controllability matrix of Eq. (11.2).

EXAMPLE 11.11   **Second-order system**

Consider the system

$$G(s) = \frac{1}{s^2}$$

and determine the feedback gain to place the closed-loop poles at $s = -1 \pm j$. Therefore we require

$$q(s) = s^2 + 2s + 2,$$

and $\alpha_1 = \alpha_2 = 2$. The matrix equation for the system is

$$\dot{\mathbf{x}} = \begin{bmatrix} 0 & 1 \\ 0 & 0 \end{bmatrix} \mathbf{x} + \begin{bmatrix} 0 \\ 1 \end{bmatrix} u.$$

The controllability matrix is

$$\mathbf{P}_c = [\mathbf{b} \quad \mathbf{Ab}] = \begin{bmatrix} 0 & 1 \\ 1 & 0 \end{bmatrix}.$$

We therefore obtain

$$\mathbf{H} = [0 \quad 1] \begin{bmatrix} 0 & 1 \\ 1 & 0 \end{bmatrix}^{-1} q(\mathbf{A}),$$

where

$$\mathbf{P}_c^{-1} = \frac{1}{-1} \begin{bmatrix} 0 & -1 \\ -1 & 0 \end{bmatrix} = \begin{bmatrix} 0 & 1 \\ 1 & 0 \end{bmatrix}$$

and

$$q(\mathbf{A}) = \begin{bmatrix} 0 & 1 \\ 0 & 0 \end{bmatrix}^2 + 2 \begin{bmatrix} 0 & 1 \\ 0 & 0 \end{bmatrix} + 2 \begin{bmatrix} 1 & 0 \\ 0 & 1 \end{bmatrix} = \begin{bmatrix} 2 & 2 \\ 0 & 2 \end{bmatrix}.$$

Then, we have

$$\mathbf{H} = [0 \quad 1] \begin{bmatrix} 0 & 1 \\ 1 & 0 \end{bmatrix} \begin{bmatrix} 2 & 2 \\ 0 & 2 \end{bmatrix} = [0 \quad 1] \begin{bmatrix} 0 & 2 \\ 2 & 2 \end{bmatrix} = [2, 2]. \quad \blacksquare$$

11.7 LIMITATIONS OF STATE VARIABLE FEEDBACK

State feedback is not usually practical, for two reasons. The first is that state feedback leads to PD-type or PID compensators, which have infinite bandwidth, whereas real components and compensators always have finite bandwidth. The second reason is that it is simply not possible or practical to sense all the states and feed them back. In reality, only certain states or combinations of them are measurable as outputs. Consequently, any practical compensator must rely only on system outputs, inputs, and a few state variables for compensation.

11.8 INTERNAL MODEL DESIGN

In this section we consider the problem of designing a compensator that provides asymptotic tracking of a reference input with zero steady-state error. The reference inputs considered can include steps, ramps, and other persistent signals, such as sinusoids. For a step input, we know that zero steady-state tracking errors can be achieved with a type-one system. This idea is formalized here by introducing an *internal model* of the reference input in the compensator [5, 20].

Let us consider a state variable model of the plant given by

$$\dot{\mathbf{x}} = \mathbf{A}\mathbf{x} + \mathbf{b}u, \qquad c = \mathbf{d}\mathbf{x}, \tag{11.67}$$

where $\mathbf{x}$ is the state vector, u is the input, and c is the output. We will consider reference inputs to be generated by linear systems of the form

$$\dot{\mathbf{x}}_r = \mathbf{A}_r\mathbf{x}_r, \qquad r = \mathbf{d}_r\mathbf{x}_r, \tag{11.68}$$

with unknown initial conditions. An equivalent model of the reference input, $r(t)$, is

$$r^{(n)} = \alpha_{n-1}r^{(n-1)} + \alpha_{n-2}r^{(n-2)} + \cdots + \alpha_1\dot{r} + \alpha_0 r, \tag{11.69}$$

where $r^{(n)}$ is the nth derivative of $r(t)$.

We begin by considering a familiar design problem, namely, the design of a controller to enable the tracking of a step reference input with zero steady-state error. In this case, the reference input is generated by

$$\dot{x}_r = 0, \qquad r = x_r, \tag{11.70}$$

or, equivalently,

$$\dot{r} = 0, \tag{11.71}$$

and the tracking error, e, is defined as

$$e = c - r.$$

Taking the time derivative yields

$$\dot{e} = \dot{c} = \mathbf{d}\dot{\mathbf{x}},$$

where we have utilized the reference input model of Eq. (11.71) and the plant model of Eq. (11.67). If we define two intermediate variables, $\mathbf{z}$ and w, as

$$\mathbf{z} = \dot{\mathbf{x}}, \qquad w = \dot{u},$$

we have

$$\begin{pmatrix} \dot{e} \\ \dot{\mathbf{z}} \end{pmatrix} = \begin{bmatrix} 0 & \mathbf{d} \\ 0 & \mathbf{A} \end{bmatrix} \begin{pmatrix} e \\ \mathbf{z} \end{pmatrix} + \begin{bmatrix} 0 \\ \mathbf{b} \end{bmatrix} w. \tag{11.72}$$

If the system in Eq. (11.72) is controllable, we can find a feedback of the form

$$w = -K_1 e - \mathbf{K}_2\mathbf{z} \tag{11.73}$$

such that Eq. (11.72) is stable. This implies that the tracking error e is stable; thus we will have achieved the objective of asymptotic tracking with zero steady-state error. The control input, found by integrating Eq. (11.73), is

$$u(t) = -K_1 \int_0^t e(\tau)d\tau - \mathbf{K}_2\mathbf{x}(t).$$

The corresponding block diagram is shown in Fig. 11.14. We see that the compensator includes an *internal model* (i.e., an integrator) of the reference step input.

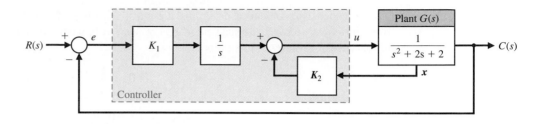

FIGURE 11.14
Internal model
design for a step
input.

EXAMPLE 11.12   **Internal model design for a unit step input**

Let us consider a plant given by

$$\dot{\mathbf{x}} = \begin{bmatrix} 0 & 1 \\ -2 & -2 \end{bmatrix} \mathbf{x} + \begin{bmatrix} 0 \\ 1 \end{bmatrix} u, \qquad c = [1 \quad 0]\mathbf{x}. \qquad (11.74)$$

We want to design a controller for this system to track a reference step input with zero steady-state error. From Eq. (11.72), we have

$$\begin{pmatrix} \dot{e} \\ \dot{\mathbf{z}} \end{pmatrix} = \begin{bmatrix} 0 & 1 & 0 \\ 0 & 0 & 1 \\ 0 & -2 & -2 \end{bmatrix} \begin{pmatrix} e \\ \mathbf{z} \end{pmatrix} + \begin{bmatrix} 0 \\ 0 \\ 1 \end{bmatrix} w. \qquad (11.75)$$

A check of controllability shows that Eq. (11.75) is completely controllable. We choose

$$K_1 = 20, \qquad \mathbf{K}_2 = [20 \quad 10],$$

in order to attain the roots of the characteristic equation of Eq. (11.75) at $s = -1 \pm j$, -10. With w given in Eq. (11.73), we have Eq. (11.75) as asymptotically stable. So, for any initial tracking error, $e(0)$, we are guaranteed that $e(t) \rightarrow 0$ as $t \rightarrow \infty$. The asymptotic stability of the tracking error is illustrated in Fig. 11.15 for a step input. ∎

Consider the block diagram model of Fig. 11.14 where the plant is represented by $G(s)$ and the cascade controller is $G_c(s) = K_1/s$. The *internal model principle* states that if $G(s)G_c(s)$ contains $R(s)$, then $c(t)$ will track $r(t)$ asymptotically. In this case, $R(s) = 1/s$, which is contained in $GG_c(s)$, as we expect.

FIGURE 11.15
Internal model
design response to
an initial tracking
error for a unit step
input.

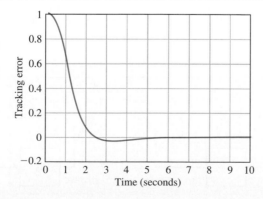

Consider the problem of designing a controller to provide asymptotic tracking of a *ramp* input with zero steady-state error, $r(t) = Mt$, $t \geq 0$, where M is the ramp magnitude. In this case, the reference input model is

$$\dot{\mathbf{x}}_r = \mathbf{A}_r \mathbf{x}_r = \begin{bmatrix} 0 & 1 \\ 0 & 0 \end{bmatrix} \mathbf{x}_r$$

$$r = \mathbf{d}_r \mathbf{x}_r = [1 \quad 0] \mathbf{x}_r.$$

(11.76)

In input–output form, the reference model in Eq. (11.76) is given by

$$\ddot{r} = 0.$$

Proceeding as before, we define the tracking error as

$$e = c - r,$$

and taking the time-derivative *twice* yields

$$\ddot{e} = \ddot{c} = \mathbf{d}\ddot{\mathbf{x}}.$$

With the definitions

$$\mathbf{z} = \ddot{\mathbf{x}}, \qquad w = \ddot{u},$$

we have

$$\begin{pmatrix} \dot{e} \\ \ddot{e} \\ \dot{\mathbf{z}} \end{pmatrix} = \begin{bmatrix} 0 & 1 & 0 \\ 0 & 0 & \mathbf{d} \\ 0 & 0 & \mathbf{A} \end{bmatrix} \begin{pmatrix} e \\ \dot{e} \\ \mathbf{z} \end{pmatrix} + \begin{bmatrix} 0 \\ 0 \\ \mathbf{b} \end{bmatrix} w.$$

(11.77)

So, if Eq. (11.77) is controllable, then we can compute K_i, $i = 1, 2, 3$, such that with

$$w = -[K_1 \quad K_2 \quad \mathbf{K}_3] \begin{bmatrix} e \\ \dot{e} \\ \mathbf{z} \end{bmatrix},$$

(11.78)

the system represented by Eq. (11.77) is asymptotically stable, hence the tracking error $e(t) \rightarrow 0$ as $t \rightarrow \infty$, as desired. The control, u, is found by integrating Eq. (11.78) twice. In Fig. 11.16, we see that the resulting controller has a double integrator, which is the *internal model* of the reference ramp input.

The internal model approach can be extended to other reference inputs by following the same general procedure outlined for the step and ramp inputs. In addition, the internal model design can be used to reject persistent disturbances by including models of the disturbances in the compensator.

FIGURE 11.16
Internal model design for a ramp input. Note that $GG_c(s)$ contains $(1/s^2)$, the reference input $R(s)$.

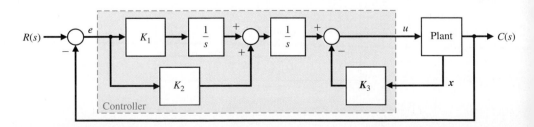

11.9 DESIGN EXAMPLE: AUTOMATIC TEST SYSTEM

An automatic test and inspection system uses a dc motor to move a set of test probes, as shown in Fig. 11.17. Low throughput and a high degree of error are possible from manually testing various panels of switches, relay, and indicator lights. Automating the test from a controller requires placing a plug across the leads of a part and testing for continuity, resistance, or functionality [19]. The system uses a dc motor with an encoded disk to measure position and velocity, as shown in Fig. 11.18. The parameters of the system are shown in Fig. 11.19 with K representing the required power amplifier.

We select the state variables as $x_1 = \theta$, $x_2 = d\theta/dt$, and $x_3 = i_f$, as shown in Fig. 11.20. State variable feedback is available, and we let

$$u = [-K_1, \ -K_2, \ -K_3]\mathbf{x}$$

FIGURE 11.17
Automatic test
system.

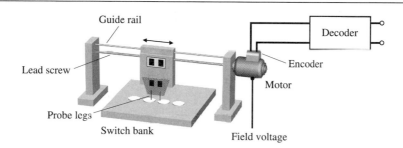

FIGURE 11.18
A dc motor with
mounted encoder
wheel.

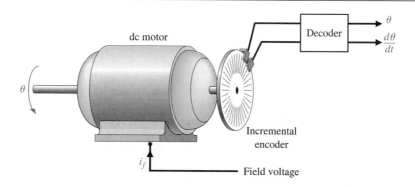

FIGURE 11.19
Block diagram.

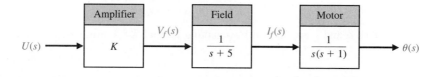

FIGURE 11.20
Signal-flow graph.

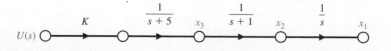

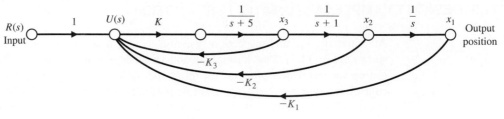

FIGURE 11.21
Feedback system.

or

$$u = -K_1 x_1 - K_2 x_2 - K_3 x_3,$$ (11.79)

as shown in Fig. 11.21. The goal is to select the gains so that the response to a step command has a settling time of less than 2 seconds and an overshoot of less than 4.0%.

To achieve accurate output position, we let $K_1 = 1$ and determine K, K_2, and K_3. The characteristic equation of the system may be obtained several ways.

Since $u = \mathbf{hx}$ is already given in Eq. (11.79), we use Fig. 11.20 to obtain

$$\dot{\mathbf{x}} = \mathbf{Ax} + \mathbf{b}u = \begin{bmatrix} 0 & 1 & 0 \\ 0 & -1 & 1 \\ 0 & 0 & -5 \end{bmatrix} \mathbf{x} + \begin{bmatrix} 0 \\ 0 \\ K \end{bmatrix} u.$$ (11.80)

Adding u as defined by Eq. (11.79), we have

$$\dot{\mathbf{x}} = \begin{bmatrix} 0 & 1 & 0 \\ 0 & -K & 1 \\ -K & -KK_2 & -(5 + K_3 K) \end{bmatrix} \mathbf{x}$$ (11.81)

when $K_1 = 1$. The characteristic equation can also be readily obtained from Fig. 11.21 by letting $G(s)$ be the forward path transfer function and letting $H(s)$ be the equivalent feedback transfer function of Fig. 11.22. Then

$$G(s) = \frac{K}{s(s + 1)(s + 5)}$$

and

$$H(s) = K_1 + sK_2 + K_3(s + 1)s = K_3 \left[s^2 + \frac{K_2 + K_3}{K_3} s + \frac{K_1}{K_2} \right].$$ (11.82)

Thus we can plot a root locus for $K_3 K$ as

$$1 + \frac{KK_3(s^2 + as + b)}{s(s + 1)(s + 5)} = 0,$$ (11.83)

where a and b are chosen by selecting K_2 and K_3. Letting $K_1 = 1$ and setting $a = 8$ and

FIGURE 11.22
Equivalent model
of feedback
system of
Fig. 11.21.

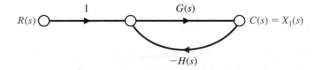

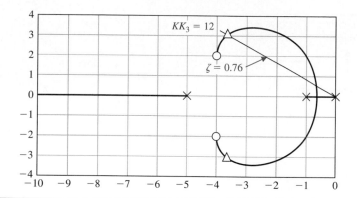

FIGURE 11.23
Root locus for
automatic test
system.

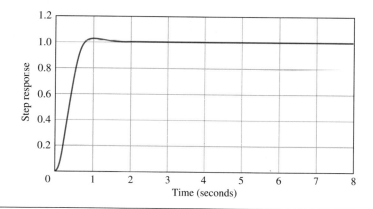

FIGURE 11.24
Step response.

$b = 30$, we place the zeros at $s = -4 \pm j2$ in order to pull the locus to the left in the s-plane. Then,

$$\frac{K_2 + K_3}{K_3} = 8 \qquad \text{and} \qquad \frac{1}{K_3} = 20.$$

Therefore $K_1 = 1$, $K_2 = 0.35$, and $K_3 = 0.05$. A plot of the root locus is shown in Fig. 11.23. When $KK_3 = 12$, the roots lie on the $\zeta = 0.76$ line, as shown in Fig. 11.23. Then, since $K_3 = 0.05$, we have $K = 240$. The roots at $K = 240$ are

$$s = -10.62, \qquad s = -3.69 \pm j3.00.$$

The step response of this system is shown in Fig. 11.24. The overshoot is 3% and the settling time is 1.8 seconds. Thus the design is quite acceptable.

11.10 STATE VARIABLE DESIGN USING *MATLAB*

Controllability and observability of a system in state variable feedback form can be checked using the *Matlab* functions ctrb and obsv, respectively. The inputs to the ctrb function, shown in Fig. 11.25, are the system matrix $\mathbf{A}$ and the input matrix $\mathbf{b}$; the output of ctrb is the controllability matrix $\mathbf{P}_c$. Similarly, the input to the obsv function, shown in

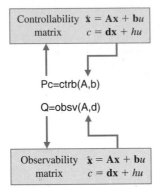

FIGURE 11.25
The **ctrb** and **obsv**
functions.

Fig. 11.25, is the system matrix **A** and the output matrix **d;** the output of obsv is the observability matrix **Q.**

Notice that the controllability matrix, $\mathbf{P}_c$, is a function only of **A** and **b,** while the observability matrix, **Q,** is a function only of **A** and **d.**

EXAMPLE 11.13   Satellite trajectory control

Let us consider a satellite in a circular, equatorial orbit at an altitude of 250 nautical miles above the earth, as illustrated in Fig. 11.26 [16, 27]. The satellite motion (in the orbit plane) is described by the normalized state variable model

$$\dot{\mathbf{x}} = \begin{bmatrix} 0 & 1 & 0 & 0 \\ 3\omega^2 & 0 & 0 & 0 \\ 0 & 0 & 0 & 1 \\ 0 & -2\omega & 0 & 0 \end{bmatrix} \mathbf{x} + \begin{bmatrix} 0 \\ 1 \\ 0 \\ 0 \end{bmatrix} u_r + \begin{bmatrix} 0 \\ 0 \\ 0 \\ 1 \end{bmatrix} u_t, \qquad (11.84)$$

FIGURE 11.26
The satellite in an
equatorial, circular
orbit.

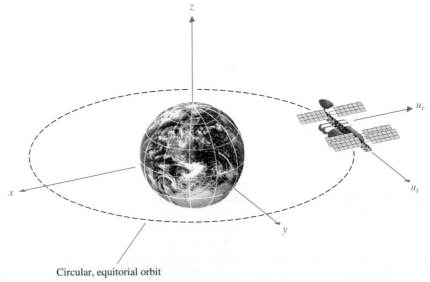

Circular, equitorial orbit

radial.m

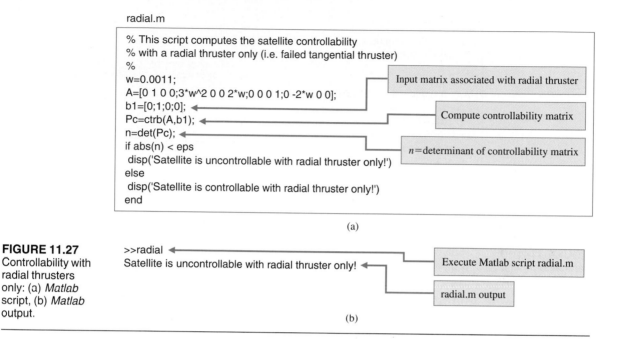

```
% This script computes the satellite controllability
% with a radial thruster only (i.e. failed tangential thruster)
%
w=0.0011;
A=[0 1 0 0;3*w^2 0 0 2*w;0 0 0 1;0 -2*w 0 0];
b1=[0;1;0;0];
Pc=ctrb(A,b1);
n=det(Pc);
if abs(n) < eps
  disp('Satellite is uncontrollable with radial thruster only!')
else
  disp('Satellite is controllable with radial thruster only!')
end
```

Input matrix associated with radial thruster

Compute controllability matrix

n=determinant of controllability matrix

(a)

FIGURE 11.27
Controllability with radial thrusters only: (a) *Matlab* script, (b) *Matlab* output.

```
>>radial
Satellite is uncontrollable with radial thruster only!
```

Execute Matlab script radial.m

radial.m output

(b)

where the state vector **x** represents normalized perturbations from the circular, equatorial orbit, u_r is the input from a radial thruster, u_t is the input from a tangential thruster, and $\omega = 0.0011$ rad/s (approximately one orbit of 90 minutes) is the orbital rate for the satellite at the specific altitude. In the absence of perturbations, the satellite will remain in the nominal circular, equatorial orbit. However, disturbances such as aerodynamic drag can cause the satellite to deviate from its nominal path. The problem is to design a controller that commands the satellite thrusters in such a manner that the actual orbit remains near the desired circular orbit. Before commencing with the design, we always check controllability. In this case, we investigate controllability using the radial and tangential thrusters independently.

Suppose the tangential thruster fails (i.e., $u_t = 0$) and only the radial thruster is operational. Is the satellite controllable from u_r only? We answer this question by using *Matlab* to determine the controllability. Using the script shown in Fig. 11.27, we find that the determinant $\mathbf{P}_c$ is zero; thus the satellite is not completely controllable when the tangential thruster fails.

Suppose now that the radial thruster fails (i.e., $u_r = 0$) and that the tangential thruster is functioning properly. Is the satellite controllable from u_t only? Using the script in Fig. 11.28, we find that the satellite is completely controllable using the tangential thruster only. ■

We conclude this section with a controller design for an automatic test system using state variable models. The design approach utilizes root locus methods and incorporates *Matlab* scripts to assist in the procedure.

EXAMPLE 11.14   Automatic test system

The state-space representation for an automatic test system is

$$\dot{\mathbf{x}} = \mathbf{A}\mathbf{x} + \mathbf{b}u, \qquad (11.85)$$

tangential.m

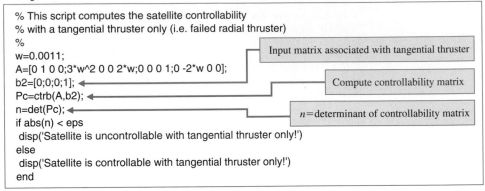

```
% This script computes the satellite controllability
% with a tangential thruster only (i.e. failed radial thruster)
%
w=0.0011;
A=[0 1 0 0;3*w^2 0 0 2*w;0 0 0 1;0 -2*w 0 0];
b2=[0;0;0;1];
Pc=ctrb(A,b2);
n=det(Pc);
if abs(n) < eps
 disp('Satellite is uncontrollable with tangential thruster only!')
else
 disp('Satellite is controllable with tangential thruster only!')
end
```

(a)

FIGURE 11.28
Controllability with
tangential thrusters
only: (a) *Matlab*
script, (b) *Matlab*
output.

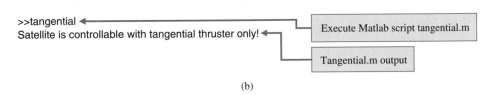

>>tangential
Satellite is controllable with tangential thruster only!

(b)

where

$$\mathbf{A} = \begin{bmatrix} 0 & 1 & 0 \\ 0 & -1 & 1 \\ 0 & 0 & -5 \end{bmatrix}, \qquad \mathbf{b} = \begin{bmatrix} 0 \\ 0 \\ K \end{bmatrix}.$$

Our design specifications are a step response with (1) a settling time less than 2 seconds and (2) an overshoot less than 4%. We assume that the state variables are available for feedback so that the control is given by

$$u = (-K_1, -K_2, -K_3)\mathbf{x} = \mathbf{Hx}. \tag{11.86}$$

We must select the gains K, K_1, K_2, and K_3 to meet the performance specifications. Using the design approximations

$$T_s \approx \frac{4}{\zeta \omega_n} < 2 \qquad \text{and} \qquad \text{P.O.} \approx 100 \ \exp^{-\zeta\pi/\sqrt{1-\zeta^2}} < 4,$$

we find that

$$\zeta > 0.72 \qquad \text{and} \qquad \omega_n > 2.8.$$

This defines a region in the complex plane in which our dominant roots must lie so that we expect to meet the design specifications, as shown in Fig. 11.29. Substituting Eq. (11.86) into Eq. (11.85) yields

$$\dot{\mathbf{x}} = \begin{bmatrix} 0 & 1 & 0 \\ 0 & -1 & 1 \\ -KK_1 & -KK_2 & -(5 + KK_3) \end{bmatrix} \mathbf{x} = \mathbf{Dx}, \tag{11.87}$$

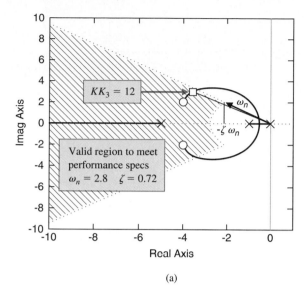

autolocus.m

```
% Root locus script for the Automatic Test System
% including performance specs regions
num=[1 8 20]; den=[1 6 5 0];
clg; rlocus(num,den); hold on       ← Hold plot to add
%                                      stability regions
zeta=0.72; wn=2.8;
x=[-10:0.1:-zeta*wn]; y=-(sqrt(1-zeta^2)/zeta)*x;
xc=[-10:0.1:-zeta*wn];c=sqrt(wn^2-xc.^2);
plot(x,y,':',x,-y,':',xc,c,':',xc,-c,':')
```

FIGURE 11.29
(a) Root locus for
the automatic test
system. (b) *Matlab*
script.

(b)

where $\mathbf{D} = \mathbf{A} - \mathbf{bH}$. The characteristic equation associated with Eq. (11.87) can be obtained by evaluating $\det(s\mathbf{I} - \mathbf{D}) = 0$, resulting in

$$s(s + 1)(s + 5) + KK_3\left(s^2 + \frac{K_3 + K_2}{K_3}s + \frac{K_1}{K_3}\right) = 0. \qquad (11.88)$$

If we view KK_3 as a parameter and let $K_1 = 1$, then we can write Eq. (11.88) as

$$1 + KK_3\frac{\left(s^2 + \dfrac{K_3 + K_2}{K_3}s + \dfrac{1}{K_3}\right)}{s(s + 1)(s + 5)} = 0.$$

We place the zeros at $s = -4 \pm 2j$ in order to pull the locus to the left in the s-plane. Thus our desired numerator polynomial is $s^2 + 8s + 20$. Comparing corresponding coefficients leads to

$$\frac{K_3 + K_2}{K_3} = 8 \qquad \text{and} \qquad \frac{1}{K_3} = 20.$$

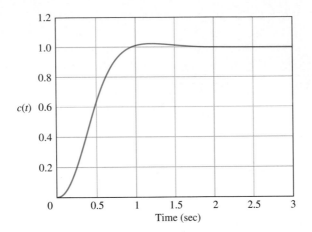

FIGURE 11.30
Step response for
the automatic test
system.

Therefore $K_2 = 0.35$ and $K_3 = 0.05$. We can now plot a root locus with KK_3 as the parameter, as shown in Fig. 11.29.

The characteristic equation, Eq. (11.88), is

$$1 + KK_3 \frac{s^2 + 8s + 20}{s(s + 1)(s + 5)} = 0.$$

The roots for the selected gain, $KK_3 = 12$, lie in the performance region, as shown in Fig. 11.29. The rlocfind function is used to determine the value of KK_3 at the selected point. The final gains are

$$K = 240.00,$$

$$K_1 = 1.00,$$

$$K_2 = 0.35,$$

$$K_3 = 0.05.$$

The controller design results in a settling time of about 1.8 seconds and an overshoot of 3%, as shown in Fig. 11.30. ∎

11.11 SUMMARY

The design of control systems in the time domain was briefly examined. Specifically, the optimal design of a system using state variable feedback and an integral performance index was considered. Also, the s-plane design of systems utilizing state variable feedback was examined. Finally, internal model design was discussed.

EXERCISES

E11.1 The ability to balance actively is a key ingredient in the mobility of a device that hops and runs on one springy leg, as shown in Fig. 11.1 [9]. The control of the attitude of the device uses a gyro-

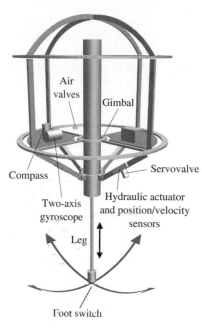

Air
valves
Gimbal

Compass   Servovalve

Hydraulic actuator
Two-axis   and position/velocity
gyroscope   sensors

Leg

Foot switch

FIGURE E11.1   Single-leg control.

scope and a feedback such that $\mathbf{u} = \mathbf{Hx}$ where

$$\mathbf{H} = \begin{bmatrix} -1 & 0 \\ 0 & -k \end{bmatrix} \quad \text{and}$$

$$\mathbf{A} = \begin{bmatrix} 0 & 1 \\ 1 & 0 \end{bmatrix} \text{ and } \mathbf{B} = \mathbf{I}.$$

Determine a value for k so that the response of each hop is critically damped.

E11.2   A magnetically suspended steel ball can be described by the linear equation

$$\dot{\mathbf{x}} = \begin{bmatrix} 0 & 1 \\ 3 & 0 \end{bmatrix} \mathbf{x} + \begin{bmatrix} 0 \\ 1 \end{bmatrix} u.$$

The state variables are $x_1 =$ position and $x_2 =$ velocity and are both measurable. Select a feedback format so that the system is critically damped and the settling time is 2 seconds.

E11.3   A system is described by the matrix equations

$$\dot{\mathbf{x}} = \begin{bmatrix} 0 & 1 \\ 0 & -3 \end{bmatrix} \mathbf{x} + \begin{bmatrix} 0 \\ 1 \end{bmatrix} u$$

and $c = 2x_2$. Determine whether the system is controllable and observable.

E11.4   A system is described by the matrix equations

$$\dot{\mathbf{x}} = \begin{bmatrix} -4 & 0 \\ 0 & -1 \end{bmatrix} \mathbf{x} + \begin{bmatrix} 0 \\ 1 \end{bmatrix} u$$

and $c = x_1$. Determine whether the system is controllable and observable.

E11.5   A system is described by the matrix equations

$$\dot{\mathbf{x}} = \begin{bmatrix} 0 & 1 \\ -1 & -2 \end{bmatrix} \mathbf{x} + \begin{bmatrix} 1 \\ -1 \end{bmatrix} u$$

and $c = x_1(t)$. Determine whether the system is controllable and observable.

E11.6   A system is described by the matrix equations

$$\dot{\mathbf{x}} = \begin{bmatrix} 0 & 1 \\ -1 & -2 \end{bmatrix} \mathbf{x} + \begin{bmatrix} 0 \\ 1 \end{bmatrix} u$$

and $c = x_1(t)$. Determine whether the system is controllable and observable.

PROBLEMS

P11.1   A first-order system is represented by the time-domain differential equation

$$\dot{x} = 3x + 2u.$$

A feedback controller is to be designed where

$$u(t) = -kx$$

and the desired equilibrium condition is $x(t) = 0$ as $t \to \infty$. The performance integral is defined as

$$J = \int_0^\infty x^2 \, dt,$$

and the initial value of the state variable is $x(0) = 1$. Obtain the value of k in order to make J a minimum. Is this k physically realizable? Select a practical value for the gain k and evaluate the performance index with that gain. Is the system stable without the feedback due to $u(t)$?

P11.2   To account for the expenditure of energy and resources, the control signal is often included in the performance integral. Then, the system may not utilize an unlimited control signal $u(t)$. One suitable performance index, which includes the effect of the magnitude of the control signal, is

$$J = \int_0^\infty (x^2(t) + \lambda u^2(t))\, dt.$$

(a) Repeat Problem 11.1 for the performance index. (b) If $\lambda = 1$, obtain the value of k that minimizes the performance index. Calculate the resulting minimum value of J.

P11.3   An unstable robot system is described by the vector differential equation [10]

$$\frac{d}{dt}\begin{bmatrix} x_1 \\ x_2 \end{bmatrix} = \begin{bmatrix} 1 & 0 \\ -1 & 2 \end{bmatrix}\begin{bmatrix} x_1 \\ x_2 \end{bmatrix} + \begin{bmatrix} 1 \\ 1 \end{bmatrix}u(t).$$

Both state variables are measurable, and so the control signal is set as $u(t) = -k(x_1 + x_2)$. Following the method of Section 11.4, design gain k so that the performance index is minimized. Evaluate the minimum value of the performance index. Determine the sensitivity of the performance to a change in k. Assume that the initial conditions are

$$\mathbf{x}(0) = \begin{bmatrix} 1 \\ 1 \end{bmatrix}.$$

Is the system stable without the feedback signals due to $u(t)$?

P11.4   Determine the feedback gain k of Example 11.6 that minimizes the performance index

$$J = \int_0^\infty \mathbf{x}^T\mathbf{x}\, dt$$

when $\mathbf{x}^T(0) = [1, 1]$. Plot the performance index J versus the gain k.

P11.5   Determine the feedback gain k of Example 11.7 that minimizes the performance index

$$J = \int_0^\infty (\mathbf{x}^T\mathbf{x} + \mathbf{u}^T\mathbf{u})\, dt$$

when $\mathbf{x}^T(0) = [1, 1]$. Plot the performance index J versus the gain k.

P11.6   For the solutions of Problems 11.3, 11.4, and 11.5, determine the roots of the closed-loop optimal control system. Note that the resulting closed-loop roots depend on the performance index selected.

P11.7   A system has the vector differential equation as given in Eq. (11.20). It is desired that both state

variables be used in the feedback so that $u(t) = -k_1x_1 - k_2x_2$. Also, it is desired to have a natural frequency, ω_n, for this system equal to 2. Find a set of gains k_1 and k_2 in order to achieve an optimal system when J is given by Eq. (11.13). Assume $\mathbf{x}^T(0) = [1, 0]$.

P11.8   For the system of Example 11.5, determine the optimum value for k_2 when $k_1 = 1$ and $\mathbf{x}^T(0) = [1, 0]$.

P11.9   An interesting mechanical system with a challenging control problem is the ball and beam shown in Fig. P11.9(a) [11]. It consists of a rigid beam that is free to rotate in the plane of the paper around a center pivot, with a solid ball rolling along a groove in the top of the beam. The control problem is to position the ball at a desired point on the beam using a torque applied to the beam as a control input at the pivot.

A linear model of the system with a measured value of the angle ϕ and its angular velocity $d\phi/dt = \omega$ is available. Select a feedback scheme so that the response of the closed-loop system has an overshoot of 4% and a settling time of 1 second for a step input.

P11.10   The dynamics of a rocket are represented by

$$\frac{C(s)}{U(s)} = G(s) = \frac{1}{s^2},$$

and state variable feedback is used where $x_1 = c(t)$ and $u = -x_2 - 0.5x_1$. Determine the roots of the characteristic equation of this system and the response of the system when the initial conditions are $x_1(0) = 0$ and $x_2(0) = 1$.

P11.11   The transfer function, $G(s)$, of a plant to be controlled is

$$\frac{C(s)}{U(s)} = G(s) = \frac{1}{s^2 + 5s + 4}.$$

Use state variable feedback and incorporate a command input $R(s)$ so that the steady-state error for a step input is zero. Also, select the gains so that the system has a rapid response with an overshoot of approximately 1% and a settling time less than 1 second.

P11.12   A dc motor has a transfer function

$$G(s) = \frac{4}{s(s^2 + 5s + 4)}.$$

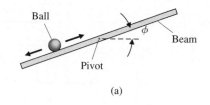

(a)

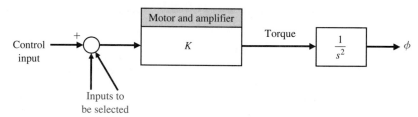

(b)

FIGURE P11.9
(a) Ball and beam.
(b) Model of the
ball and beam.

Determine whether this system is controllable and observable.

P11.13 A feedback system has a plant transfer function

$$\frac{C(s)}{R(s)} = G(s) = \frac{K}{s(s + 70)}.$$

It is desired that the velocity error constant, K_v, be 35 and the overshoot to a step be approximately 4% so that ζ is $1/\sqrt{2}$. The settling time desired is 0.11 second. Design an appropriate state variable feedback system.

P11.14 A plant has the transfer function

$$\frac{C(s)}{U(s)} = \frac{1}{s(s + 10)}.$$

Determine the state variable feedback gains to achieve a settling time of 1 second and an overshoot of about 10%. Also draw the flow graph of the resulting system and select an appropriate gain coefficient between the input $R(s)$ and the variable $U(s)$ in order to achieve a steady-state error equal to zero for a step input.

P11.15 A telerobot system has the matrix equations [18]

$$\dot{\mathbf{x}} = \begin{bmatrix} -1 & 0 & 0 \\ 0 & -2 & 0 \\ 0 & 0 & -3 \end{bmatrix} \mathbf{x} + \begin{bmatrix} 1 \\ 1 \\ 0 \end{bmatrix} u$$

and

$$c = [1 \quad 0 \quad 2]\mathbf{x}.$$

(a) Determine the transfer function $G(s) = C(s)/U(s)$. (b) Draw the signal-flow diagram indicating the state variables. (c) Determine whether the system is controllable. (d) Determine whether the system is observable.

P11.16 Hydraulic power actuators were used to drive the dinosaurs of the movie *Jurassic Park* [23]. The motions of the large monsters required high-power actuators requiring 1200 watts.

One specific limb motion has dynamics represented by

$$G(s) = \frac{1}{s(s + 1)}.$$

It is desired to place the closed-loop poles at $s = -2 + j2$. Determine the required state variable feedback using Ackermann's formula. Assume that the position and the velocity of the output motion are available for measurement.

P11.17 A system has a transfer function

$$\frac{C(s)}{R(s)} = \frac{(s + a)}{s^4 + 53s^3 + 10s^2 + 10s + 4}.$$

Determine a real value of a so that the system is either uncontrollable or unobservable.

P11.18 A system has a plant

$$\frac{C(s)}{U(s)} = G(s) = \frac{1}{(s + 1)^2}.$$

(a) Find the matrix differential equation to represent this system. Identify the state variables on

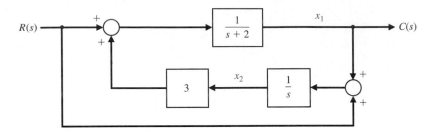

FIGURE P11.19

a signal-flow graph. (b) Select a state variable feedback structure using $u(t)$ and select the feedback
gains so that the response $c(t)$ of the unforced system is critically damped when the initial condition
$x_1(0) = 1$ and $x_2(0) = 0$, where $x_1 = c(t)$. The
repeated roots are at $s = -\sqrt{2}$.

P11.19   The block diagram of a system is shown in
Fig. P11.19. Determine whether the system is controllable and observable.

P11.20   Consider the automatic ship steering system
discussed in Problems 8.11 and 9.15. The state
variable form of the system differential equation is

$$\dot{\mathbf{x}}(t) = \begin{bmatrix} -0.05 & -6 & 0 & 0 \\ -10^{-3} & -0.15 & 0 & 0 \\ 1 & 0 & 0 & 13 \\ 0 & 1 & 0 & 0 \end{bmatrix} \mathbf{x}(t)$$

$$+ \begin{bmatrix} -0.2 \\ 0.03 \\ 0 \\ 0 \end{bmatrix} \delta(t),$$

where $\mathbf{x}^T(t) = [\dot{v}, \omega_s, y, \theta]$. The state variables are
$x_1 = \dot{v} = $ the transverse velocity; $x_2 = \omega_s = $ angular
rate of ship's coordinate frame relative to response
frame; $x_3 = y = $ deviation distance on an axis perpendicular to the track; $x_4 = \theta = $ deviation angle.
(a) Determine whether the system is stable.
(b) Feedback can be added so that

$$\delta(t) = -k_1 x_1 - k_3 x_3.$$

Determine whether this system is stable for suitable values of k_1 and k_3.

P11.21   An RL circuit is shown in Fig. P11.21. (a) Select the two stable variables and obtain the vector
differential equation where the output is $v_0(t)$.
(b) Determine whether the state variables are ob-

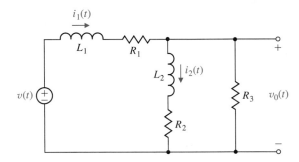

FIGURE P11.21 RLC circuit.

servable when $R_1/L_1 = R_2/L_2$. (c) Find the conditions when the system has two equal roots.

P11.22   A manipulator control system has a plant
transfer function of

$$G(s) = \frac{1}{s(s + 0.4)}$$

and negative unity feedback [17, 22]. Represent
this system by a state variable signal-flow graph
and a vector differential equation. (a) Plot the response of the closed-loop system to a step input.
(b) Use state variable feedback so that the overshoot is 5% and the settling time is 1.35 seconds.
(c) Plot the response of the state variable feedback
system to a step input.

P11.23   Reconsider the system of Example 11.11
when it is desired that the steady-state error for a
step input be zero and the desired roots of the characteristic equation be $s = -2 \pm j1$ and $s = -10$.

P11.24   Reconsider the system of Example 11.11
when it is desired that the steady-state error for a
ramp input be zero and the desired roots of the
characteristic equation be $s = -2 \pm j2$ and
$s = -20$.

ADVANCED PROBLEMS

AP11.1 A dc motor control system has the form shown in Fig. AP11.1 [6]. The three state variables are available for measurement; the output position is $x_1(t)$. Select the feedback gains so that the system has a steady-state error equal to zero for a step input and a response with a percent overshoot less than 3%.

AP11.2 A system has a plant

$$G(s) = \frac{3s^2 + 4s - 2}{s^3 + 3s^2 + 7s + 5}.$$

Add state variable feedback so that the closed-loop poles are $s = -4, -4$, and -5.

AP11.3 A system has a matrix differential equation

$$\dot{\mathbf{x}} = \begin{bmatrix} 1 & 0 \\ 0 & 2 \end{bmatrix} \mathbf{x} + \begin{bmatrix} b_1 \\ b_2 \end{bmatrix} u.$$

What values for b_1 and b_2 are required so that the system is controllable?

AP11.4 The vector differential equation describing the inverted pendulum of Example 3.3 (see page 130)

$$\frac{d\mathbf{x}}{dt} = \begin{bmatrix} 0 & 1 & 0 & 0 \\ 0 & 0 & -1 & 0 \\ 0 & 0 & 0 & 1 \\ 0 & 0 & 9.8 & 0 \end{bmatrix} \mathbf{x} + \begin{bmatrix} 0 \\ 1 \\ 0 \\ -1 \end{bmatrix} u.$$

Assume that all state variables are available for measurement, and use state variable feedback. Place the system characteristic roots at $s = (-2 \pm j)$, -5, and -5.

AP11.5 An automobile suspension system has three physical state variables, as shown in Fig. AP11.5 [15]. The state variable feedback structure is shown in the figure, with $K_1 = 1$. Select K_2 and K_3 so that the roots of the characteristic equation are three real roots lying between $s = -3$ and $s = -6$. Also select K_p so that the steady-state error for a step input is equal to zero.

AP11.6 A system is represented by the differential equation

$$\frac{d^2y}{dt^2} + 2\frac{dy}{dt} + y = \frac{du}{dt} + u,$$

where $y = $ output and $u = $ input.

(a) Develop the phase variable representation and show that it is a controllable system. (b) Define the state variables as $x_1 = y$ and $x_2 = dy/dt - u$, and determine whether the system is controllable. Note that the controllability of a system depends on the definition of the state variables.

AP11.7 The new Radisson Diamond uses pontoons and stabilizers to damp out the effect of waves hitting the ship, as shown in Fig. AP11.7(a). The block diagram of the ship roll control system is

FIGURE AP11.1
Field-controlled dc motor.

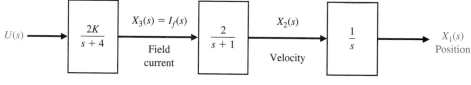

FIGURE AP11.5
Automobile suspension system.

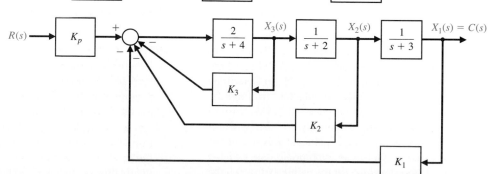

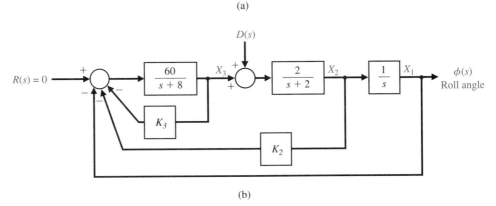

FIGURE AP11.7
(a) Radisson
Diamond (courtesy
of Conde-Nast
Traveler, July
1993, p. 23).
(b) Control system
to reduce the effect
of the disturbance.

shown in Fig. AP11.7(b). Determine the feedback
gains K_2 and K_3 so that the characteristic roots are
$s = -15$ and $s = -2 \pm j2$. Plot the roll output
$\phi(t)$ for a unit step disturbance.

AP11.8   Reconsider the liquid-level control system
described in Problem 3.36.
    (a) Design a state variable controller using
only $h(t)$ as the feedback variable so that the step
response has an overshoot less than 10% and a
settling time less than or equal to 5 seconds.
(b) Design a state variable controller feedback
using two state variables, level $h(t)$ and shaft position $\theta(t)$, to satisfy the specifications of part (a).
(c) Compare the results of parts (a) and (b).

AP11.9   The motion control of a lightweight hospital
transport vehicle can be represented by a system
of two masses as shown in Fig. AP11.9, where
$m_1 = m_2 = 1$ and $k_1 = k_2 = 1$ [24]. (a) Determine
the state vector differential equation. (b) Find the
roots of the characteristic equation. (c) We wish to
stabilize the system by letting $u = -kx_i$, where u
is the force on the lower mass and x_i is one of the
state variables. Select an appropriate state variable
x_i. (d) Choose a value for the gain k and sketch the
root locus as k varies.

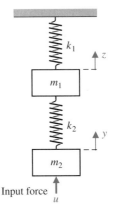

FIGURE AP11.9   Model of hospital vehicle.

AP11.10   Consider the inverted pendulum mounted to
a motor, as shown in Fig. AP11.10. The motor and
load are assumed to have no friction damping. The
pendulum to be balanced is attached to the horizontal shaft of a servomotor. The servomotor carries a tachogenerator, so that a velocity signal is
available, but there is no position signal. When the
motor is unpowered, the pendulum will hang vertically downward and if slightly disturbed will perform oscillations. If lifted to the top of its arc, the

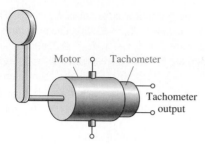

FIGURE AP11.10 Motor and inverted pendulum.

pendulum is unstable in that position. Devise a feedback compensator, $G_c(s)$, using only the velocity signal from the tachometer.

AP11.11 Determine an internal model controller $G_c(s)$ for the system shown in Fig. AP11.11. It is desired that the steady-state error to a step input be zero. It is also desired that the settling time be less than 5 seconds.

AP11.12 Repeat Advanced Problem 11.11 when it is desired that the steady-state error to a ramp input be zero and that the settling time of the ramp response be less than 6 seconds.

FIGURE AP11.11
Internal
model
control.

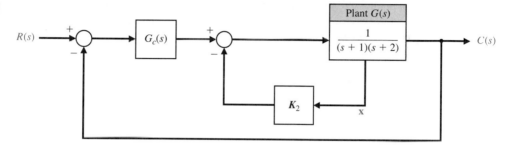

DESIGN PROBLEMS

DP11.1 Consider the device for the magnetic levitation of a steel ball, as shown in Figs. DP11.1(a) and (b). Obtain a design that will provide a stable response where the ball will remain within 10% of its desired position. Assume that y and dy/dt are measurable.

DP11.2 The control of the fuel-to-air ratio in an automobile carburetor became of prime importance in the 1980s as automakers worked to reduce exhaust-pollution emissions. Thus auto engine designers turned to the feedback control of the fuel-to-air ratio. A sensor was placed in the exhaust stream and used as an input to a controller. The controller actually adjusts the orifice that controls the flow of fuel into the engine [3].

Select the devices and develop a linear model for the entire system. Assume that the sensor measures the actual fuel-to-air ratio with a negligible delay. With this model, determine the optimum controller when a system is desired with a zero steady-state error to a step input and the overshoot for a step command is less than 10%.

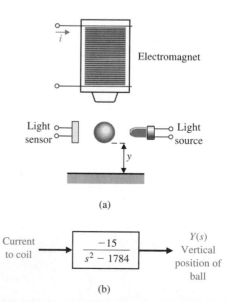

FIGURE DP11.1 (a) The levitation of a ball using an electromagnet. (b) The model of the electromagnet and the ball.

DP11.3 The efficiency of a diesel engine is very sensitive to the speed of rotation. Thus we wish to control the speed of the diesel engine that drives the electric motors of a diesel-electric locomotive for large railroad trains. The locomotive is powered by dc motors located on each of the axles. Consider the model of the electric drive shown in Fig. DP11.3 where the throttle position is set by moving the input potentiometer.

The controlled speed ω_0 is sensed by a tachometer, which supplies a feedback voltage v_0. The tachometer may be belt driven from the motor shaft. An electronic amplifier amplifies the error $(v_r - v_0)$ between the reference and feedback voltage signals and provides a voltage v_f that is supplied to the field winding of a dc generator.

The generator is run at a constant speed ω_d by the diesel engine and generates a voltage v_g that is supplied to the armature of a dc motor. The motor is armature controlled with a fixed current i supplied to its field. As a result, the motor produces a torque T and drives the load connected to its shaft so that the controlled speed ω_0 tends to equal the command speed ω_r.

It is known that the generator constant K_g is

100 and the motor constant is $K_m = 10$. The back emf constant is $K_b = 31/50$. The constants for the motor are $J = 1$, $f = 1$, $L_a = 0.2$, and $R_a = 1$. The generator has a field resistance $R_f = 1$ and a field inductance $L_f = 0.1$. Also the tachometer has a gain of $K_t = 1$ and the generator has $L_g = 0.1$ and $R_g = 1$. All three constants are in the appropriate SI units.

(a) Develop a linear model for the system and analyze the performance of the system as K is adjusted. What is the steady-state error for a fixed throttle position so that $v_r = A$? (b) Include the effect of a load torque disturbance, T_d, in the model of the system and determine the effect of a load disturbance torque on the speed, $\omega(s)$. For the design of part (a), what is the steady-state error introduced by the disturbance? (c) Select the three state variables that are readily measured and obtain the state vector differential equation. Then design a state variable feedback system to achieve favorable performance.

DP11.4 A high-performance helicopter has a model shown in Fig. DP11.4. The goal is to control the pitch angle, θ, of the helicopter by adjusting the rotor angle, δ.

FIGURE DP11.3
Diesel-electric
locomotive.

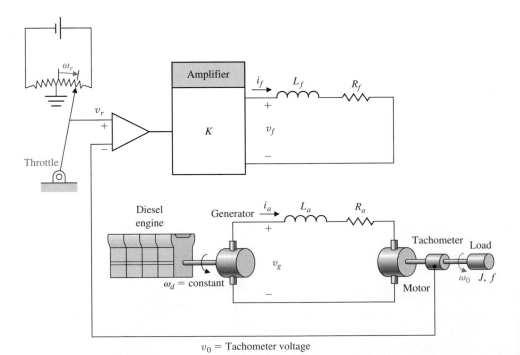

v_0 = Tachometer voltage

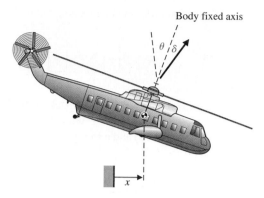

FIGURE DP11.4 Helicopter pitch angle, θ, control.

The equations of motion of the helicopter are

$$\frac{d^2\theta}{dt^2} = -\upsilon_1\frac{d\theta}{dt} - \alpha_1\frac{dx}{dt} + n\delta$$

$$\frac{d^2x}{dt^2} = g\theta - \alpha_2\frac{d\theta}{dt} - \sigma_2\frac{dx}{dt} + g\delta$$

where x is the translation in the horizontal direction. For a military, high-performance helicopter we find that

$$\begin{array}{ll} \sigma_1 = 0.415 & \alpha_2 = 1.43 \\ \sigma_2 = 0.0198 & n\ = 6.27 \\ \alpha_1 = 0.0111 & g\ = 9.8 \end{array}$$

all in appropriate SI units.

Find (a) the state variable representation of this system and (b) the transfer function representation for $\theta(s)/\delta(s)$. (c) Use state variable feedback to achieve adequate performances for the controlled system.

Desired specifications include (1) a steady-state for an input step command for $\theta_d(s)$, the desired pitch angle, less than 20% of the input step magnitude; (2) an overshoot for a step input command less than 20%; and (3) a settling time for a step command of less than 1.5 seconds.

DP11.5 The headbox process is used in the manufacture of paper to transform the pulp slurry flow into a jet of 2 cm and then spread it onto a mesh belt [25]. To achieve desirable paper quality, the pulp slurry must be distributed as evenly as possible on the belt and the relationship between the velocity of the jet and that of the belt, called the *jet/belt ratio*, must be maintained. One of the main control

variables is the pressure in the headbox, which in turn controls the velocity of the slurry at the jet. The total pressure in the headbox is the sum of the liquid level pressure and the air pressure that is pumped into the headbox. Since the pressurized headbox is a highly dynamic and coupled system, manual control would be difficult to maintain and could result in degradation in the sheet properties.

The state-space model of a typical headbox, linearized about a particular stationary point, is given by

$$\dot{\mathbf{x}} = \begin{bmatrix} -0.4 & 0.01 \\ -0.01 & 0 \end{bmatrix}\mathbf{x}$$

$$+ \begin{bmatrix} 3.4 \times 10^{-2} & 1 \\ 1 \times 10^{-3} & 0 \end{bmatrix}\mathbf{u}$$

and $\mathbf{c} = [x_1, x_2]$.

The state variables are $x_1 = $ liquid level and $x_2 = $ pressure. The control variables are $u_1 = $ pump current and $u_2 = $ valve opening. Design a state variable feedback system that has a characteristic equation with real roots with a magnitude greater than four.

DP11.6 A coupled-drive apparatus is shown in Fig. DP11.6. The coupled drives consist of two pulleys connected via an elastic belt, which is tensioned by a third pulley mounted on springs providing an underdamped dynamic mode. One of the main pulleys, pulley A, is driven by an electric dc motor. Both pulleys A and B are fitted with tachometers that generate measurable voltages pro-

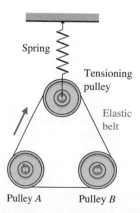

FIGURE DP11.6

portional to the rate of rotation of the pulley. When a voltage is applied to the dc motor, pulley A will accelerate at a rate governed by the total inertia experienced by the system. Pulley B, at the other end of the elastic belt, will also accelerate owing to the applied voltage or torque, but with a lagging effect caused by the elasticity of the belt. Integration of the velocity signals measured at each pulley will provide an angular position estimate for the pulley [26].

The second-order model of a coupled-drive is

$$\dot{\mathbf{x}} = \begin{bmatrix} 0 & 1 \\ -36 & -12 \end{bmatrix} \mathbf{x} + \begin{bmatrix} 0 \\ 1 \end{bmatrix} u$$

and $\mathbf{c} = x_1$.

Design a state variable feedback controller that will yield a step response with deadbeat response (Section 10.11) and a settling time less than 0.5 second.

MATLAB PROBLEMS

MP11.1   Consider the system

$$\dot{\mathbf{x}} = \begin{bmatrix} -1 & 1 & 0 \\ 4 & 0 & -3 \\ -6 & 8 & 10 \end{bmatrix} \mathbf{x} + \begin{bmatrix} 1 \\ 0 \\ -1 \end{bmatrix} u,$$

$$y = \begin{bmatrix} 1 & 2 & 1 \end{bmatrix} \mathbf{x}.$$

Using the ctrb and obsv functions, show that the system is controllable and observable.

MP11.2   The following model has been proposed to describe the motion of a constant velocity guided missile:

$$\dot{\mathbf{x}} = \begin{bmatrix} 0 & 1 & 0 & 0 & 0 \\ -0.1 & -0.5 & 0 & 0 & 0 \\ 0.5 & 0 & 0 & 0 & 0 \\ 0 & 0 & 10 & 0 & 0 \\ 0.5 & 1 & 0 & 0 & 0 \end{bmatrix} \mathbf{x} + \begin{bmatrix} 0 \\ 1 \\ 0 \\ 0 \\ 0 \end{bmatrix} u,$$

$$y = \begin{bmatrix} 0 & 0 & 0 & 1 & 0 \end{bmatrix} \mathbf{x}.$$

(a) Verify that the system is not controllable by analyzing the controllability matrix using the ctrb function.

(b) Develop a controllable state variable model by first computing the transfer function from u to y, then cancel any common factors in the numerator and denominator polynomials of the transfer function. With the modified transfer function just obtained, use the tf2ss function to determine a modified state variable model for the system.

(c) Verify that the modified state variable model in part (b) is controllable.

(d) Is the constant velocity guided missile stable?

(e) Comment on the relationship between controllability and complexity of the state variable model (where complexity is measured by the number of state variables).

MP11.3   A linearized model of a vertical takeoff and landing (VTOL) aircraft is [27]:

$$\dot{\mathbf{x}} = \mathbf{Ax} + \mathbf{b}_1 u_1 + \mathbf{b}_2 u_2,$$

where

$$\mathbf{A} = \begin{bmatrix} -0.0366 & 0.0271 & 0.0188 & -0.4555 \\ 0.0482 & -1.0100 & 0.0024 & -4.0208 \\ 0.1002 & 0.3681 & -0.7070 & 1.4200 \\ 0 & 0 & 1 & 0 \end{bmatrix},$$

and

$$\mathbf{b}_1 = \begin{bmatrix} 0.4422 \\ 3.5446 \\ -5.5200 \\ 0 \end{bmatrix}, \quad \mathbf{b}_2 = \begin{bmatrix} 0.1761 \\ -7.5922 \\ 4.4900 \\ 0 \end{bmatrix}.$$

The state vector components are (i) x_1 is the horizontal velocity (knots), (ii) x_2 is the vertical velocity (knots), (iii) x_3 is the pitch rate (degrees/second), and (iv) x_4 is the pitch angle (degrees). The input u_1 is used mainly to control the vertical motion, and u_2 is for the horizontal motion.

(a) Compute the eigenvalues of the system matrix $\mathbf{A}$. Is the system stable? (b) Determine the characteristic polynomial associated with $\mathbf{A}$ using the poly function. Compute the roots of the characteristic equation, and compare with the eigenvalues in part (a). (c) Is the system controllable from u_1 alone? What about from u_2 alone? Comment on the results.

MP11.4 In an effort to open up the far side of the moon to exploration, studies have been conducted to determine the feasibility of operating a communication satellite around the translunar equilibrium point in the earth-sun-moon system. The desired satellite orbit, known as a halo orbit, is shown in Fig. MP11.4. The objective of the controller is to keep the satellite on a halo orbit trajectory that can be seen from the earth so that the lines of communication are accessible at all times. The communication link is from the earth to the satellite and then to the far side of the moon.

The linearized (and normalized) equations of motion of the satellite around the translunar equilibrium point are [28]

$$\dot{\mathbf{x}} = \begin{bmatrix} 0 & 0 & 0 & 1 & 0 & 0 \\ 0 & 0 & 0 & 0 & 1 & 0 \\ 0 & 0 & 0 & 0 & 0 & 1 \\ 7.3809 & 0 & 0 & 0 & 2 & 0 \\ 0 & -2.1904 & 0 & -2 & 0 & 0 \\ 0 & 0 & -3.1904 & 0 & 0 & 0 \end{bmatrix} \mathbf{x}$$

$$+ \begin{bmatrix} 0 \\ 0 \\ 0 \\ 1 \\ 0 \\ 0 \end{bmatrix} u_1 + \begin{bmatrix} 0 \\ 0 \\ 0 \\ 0 \\ 1 \\ 0 \end{bmatrix} u_2 + \begin{bmatrix} 0 \\ 0 \\ 0 \\ 0 \\ 0 \\ 1 \end{bmatrix} u_3.$$

The state vector, $\mathbf{x}$, is the satellite position and velocity, and the inputs, u_i, $i = 1, 2, 3$, are the engine thrust accelerations in the ξ, η, and ζ directions, respectively.

(a) Is the translunar equilibrium point a stable location? (b) Is the system controllable from u_1 alone? (c) Repeat part (b) for u_2. (d) Repeat part (b) for u_3. (e) Suppose that we can observe the position in the η direction. Determine the transfer function from u_2 to η. (*Hint:* Let $y = [0\ 1\ 0\ 0\ 0\ 0]\mathbf{x}$.) (f) Compute a state space representation of the transfer function in part (e) using the tf2ss function. Verify that the system is controllable. (g) Using state feedback

$$u_2 = -\mathbf{Kx},$$

design a controller (i.e., find $\mathbf{K}$) for the system in part (f) such that the closed-loop system poles are at $s_{1,2} = -1 \pm j$ and $s_{3,4} = -10$.

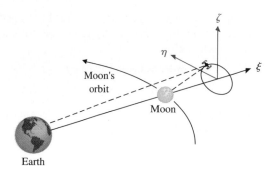

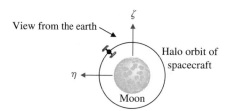

FIGURE MP11.4 The translunar satellite halo orbit.

MP11.5 Consider the system

$$\dot{\mathbf{x}}(t) = \begin{bmatrix} 0 & 1 & 0 \\ 0 & 0 & 1 \\ -2 & -4 & -6 \end{bmatrix} \mathbf{x}(t), \qquad (1)$$

$$y(t) = [1\ 0\ 0]\mathbf{x}(t).$$

Suppose that we are given three observations, $y(t_i)$, $i = 1, 2, 3$, as follows:

$$\begin{aligned} y(t_1) &= 1 & \text{at} \quad t_1 &= 0 \\ y(t_2) &= -0.0256 & \text{at} \quad t_2 &= 2 \\ y(t_3) &= -0.2522 & \text{at} \quad t_3 &= 4. \end{aligned}$$

(a) Using the three observations, develop a method to determine the initial value of the state vector, $\mathbf{x}(t_0)$, for the system in Eq. (1) that will reproduce the three observations when simulated using the lsim function. (b) With the observations given, compute $\mathbf{x}(t_0)$ and discuss the condition under which this problem can be solved in general. (c) Verify the result by simulating the system response to the computed initial condition. (*Hint:* Recall that $\mathbf{x}(t) = e^{\mathbf{A}(t-t_0)}\mathbf{x}(t_0)$ for the system in Eq. 1.)

MP11.6 A system is described by a single-input state equation with

$$\mathbf{A} = \begin{bmatrix} 0 & 1 \\ -1 & 0 \end{bmatrix} \qquad \mathbf{B} = \begin{bmatrix} 0 \\ 1 \end{bmatrix}.$$

Using the method of Section 11.4 (Eq. 11.19) and a negative unity feedback, determine the optimal system when $\mathbf{x}^T(0) = [1, 0]$.

MP11.7   A first-order system is given by

$$\dot{x} = -x + u$$

with the initial condition $x(0) = x_0$. We want to design a feedback controller

$$u = -kx$$

such that the performance index

$$J = \int_0^\infty x^2(t) + \lambda u^2(t)dt$$

is minimized.

(a) Let $\lambda = 1$. Develop a formula for J in terms of k, valid for any x_0, and use *Matlab* to plot J/x_0^2 versus k. From the plot, determine the approximate value of $k = k_{min}$ that minimizes J/x_0^2. (b) Verify the result in part (a) analytically. (c) Using the procedure developed in part (a), obtain a plot of k_{min} versus λ, where k_{min} is the gain that minimizes the performance index.

TERMS AND CONCEPTS

Controllable system   A system with unconstrained control u that transfers any initial state $\mathbf{x}(0)$ to any other state $\mathbf{x}(t)$.

Observable system   A system with an output that possesses a component due to each state variable.

Optimal control system   A system whose parameters are adjusted so that the performance index reaches an extremum value.

State variable feedback   Occurs when the control signal, u, for the process is a direct function of all the state variables.

Robust Control Systems

PREVIEW

We have described the benefits of using a compensator, $G_c(s)$, to achieve the desired performance of a feedback system. In this chapter we extend the concept of compensation and introduce a powerful controller, the proportional-integral-derivative (PID) device. In addition, we consider the use of the internal model design method and state variable feedback for the design of robust systems. Finally, we consider an additional method we call pseudo-quantitative feedback.

The design of highly accurate control systems in the presence of significant uncertainty requires the designer to seek a robust system. Thus in this chapter we will utilize five methods for robust design. The design methods include the root locus, frequency response, and ITAE methods for a robust PID system. In addition, we use the internal model and pseudo-quantitative feedback methods to achieve robust control.

12.1 INTRODUCTION

A control system designed using the methods and concepts of the preceding chapters assumes knowledge of the model of the plant and controller and constant parameters. The plant model will always be an inaccurate representation of the actual physical system because of

- Parameter changes
- Unmodeled dynamics
- Unmodeled time delays
- Changes in equilibrium point (operating point)
- Sensor noise
- Unpredicted disturbance inputs.

The goal of robust systems design is to retain assurance of system performance in spite of model inaccuracies and changes. A system is *robust* when the system has acceptable changes in performance due to model changes or inaccuracies.

A robust control system exhibits the desired performance despite the presence of significant plant (process) uncertainty.

The system structure that incorporates potential system uncertainties is shown in Fig. 12.1. This model includes the sensor noise $N(s)$, the unpredicted disturbance input $D(s)$, and a plant $G(s)$ with potentially unmodeled dynamics or parameter changes. The unmodeled dynamics and parameter changes may be significant or very large, and for these systems, the challenge is to design a system that retains the desired performance.

12.2 ROBUST CONTROL SYSTEMS AND SYSTEM SENSITIVITY

Designing highly accurate systems in the presence of significant plant uncertainty is a classical feedback design problem. The theoretical bases for the solution of this problem date back to the works of H. S. Black and H. W. Bode in the early 1930s, when this problem was referred to as the sensitivities design problem. However, a significant amount of literature has been published since then regarding the design of systems subject to large plant

---

FIGURE 12.1
Closed-loop
system structural
diagram.

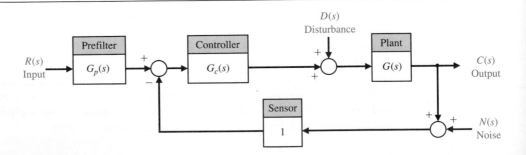

uncertainty. The designer seeks to obtain a system that performs adequately over a large range of uncertain parameters. A system is said to be *robust* when it is durable, hardy, and resilient.

A control system is robust when (1) it has low sensitivities, (2) it is stable over the range of parameter variations, and (3) the performance continues to meet the specifications in the presence of a set of changes in the system parameters [3, 4]. Robustness is the sensitivity to effects that are not considered in the analysis and design phase—for example, disturbances, measurement noise, and unmodeled dynamics. The system should be able to withstand these neglected effects when performing the tasks for which it was designed.

For small parameter perturbations, we may use, as a measure of robustness, the differential sensitivities discussed in Sections 4.2 (system sensitivity) and 7.6 (root sensitivity) [6].

System sensitivity is defined as

$$S_\alpha^T = \frac{\partial T/T}{\partial \alpha/\alpha},$$ (12.1)

where α is the parameter and T the transfer function of the system.

Root sensitivity is defined as

$$S_\alpha^{r_i} = \frac{\partial r_i}{\partial \alpha/\alpha}.$$ (12.2)

When the zeros of $T(s)$ are independent of the parameter α, we showed that

$$S_\alpha^T = -\sum_{i=1}^{n} S_\alpha^{r_i} \cdot \frac{1}{(s + r_i)}$$ (12.3)

for an nth order system. For example, if we have a closed-loop system, as shown in Fig. 12.2, where the variable parameter is α, then $T(s) = 1/[s + (\alpha + 1)]$ and

$$S_\alpha^T = \frac{-\alpha}{s + \alpha + 1}.$$ (12.4)

Furthermore, the root is $r_1 = +(\alpha + 1)$ and

$$-S_\alpha^{r_1} = -\alpha.$$ (12.5)

Therefore

$$S_\alpha^T = -S_\alpha^{r_1} \frac{1}{(s + \alpha + 1)}.$$ (12.6)

Let us examine the sensitivity of the second-order system shown in Fig. 12.3. The transfer function of the closed-loop system is

$$T(s) = \frac{K}{s^2 + s + K}.$$ (12.7)

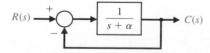

FIGURE 12.2
A first-order system.

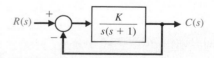

FIGURE 12.3
A second-order system.

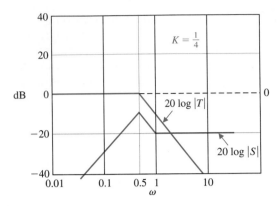

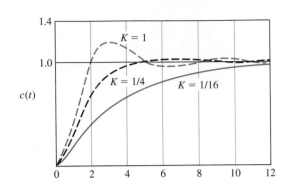

FIGURE 12.4   Sensitivity and $20 \log|T(j\omega)|$ for the second-order system in Fig. 12.3. The asymptotic approximations are shown for $K = \frac{1}{4}$.

FIGURE 12.5   Step response for selected gain K.

As we saw in Eq. (4.12), the system sensitivity for K is

$$S_K^T = \frac{1}{1 + GH(s)} = \frac{s(s + 1)}{s^2 + s + K}. \tag{12.8}$$

A Bode plot of the asymptotes of $20 \log|T(j\omega)|$ and $20 \log|S(j\omega)|$ is shown in Fig. 12.4 for $K = \frac{1}{4}$ (critical damping). Note that the sensitivity is small for lower frequencies, while the transfer function primarily passes low frequencies. Also, note for this case, $T(s) = 1 - S(s)$.

Of course, the sensitivity S only represents robustness for small changes in gain. If K changes from $\frac{1}{4}$ within the range $K = \frac{1}{16}$ to $K = 1$, the resulting range of step response is shown in Fig. 12.5. This system, with an expected wide range of K, may not be considered adequately robust. A robust system would be expected to yield essentially the same (within an agreed-upon variation) response to a selected input.

EXAMPLE 12.1   **Sensitivity of a controlled system**

Consider the system shown in Fig. 12.6, where $G(s) = 1/s^2$ and a PD controller $G_c(s) = b_1 + b_2 s$. Then, the sensitivity with respect to changes in $G(s)$ is

$$S_G^T = \frac{1}{1 + GG_c} = \frac{s^2}{s^2 + b_2 s + b_1} \tag{12.9}$$

FIGURE 12.6
A system with a
PD controller.

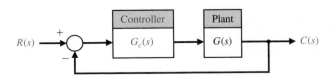

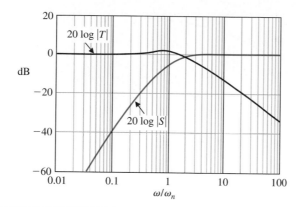

FIGURE 12.7
Sensitivity and
$T(s)$ for the
second-order
system in
Fig. 12.6.

and

$$T(s) = \frac{b_2 s + b_1}{s^2 + b_2 s + b_1}.$$ (12.10)

Consider the normal condition $\zeta = 1$ and $\omega_n = \sqrt{b_1}$. Then $b_2 = 2\omega_n$ to achieve $\zeta = 1$. Therefore we may plot $20 \log|S|$ and $20 \log|T|$ on a Bode diagram, as shown in Fig. 12.7. Note that the frequency ω_n is an indicator on the boundary between the frequency region in which the sensitivity is the important design criterion and the region in which the stability margin is important. Thus, if we specify ω_n properly to take into consideration the extent of modeling error and the frequency of external disturbance, we can expect that the system has an acceptable amount of robustness. Note that $G_c(s)$ is a proportional-derivative (PD) controller. ∎

EXAMPLE 12.2   System with a right-hand plane zero

Consider the system shown in Fig. 12.8, where the plant has a zero in the right-hand plane. The closed-loop transfer function is

$$T(s) = \frac{K(s - 1)}{s^2 + (2 + K)s + (1 - K)}.$$ (12.11)

The system is stable for a gain $-2 < K < 1$. The steady-state error is

$$e_{ss} = \frac{1 - 2K}{1 - K}$$ (12.12)

and $e_{ss} = 0$ when $K = \frac{1}{2}$. The response to a negative unit step input $R(s) = -1/s$ is shown in Fig. 12.9. Note the initial undershoot at $t = 1$ second. This system is sensitive to

FIGURE 12.8
A second-order
system.

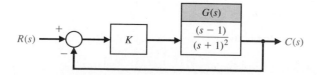

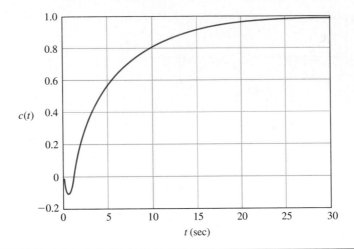

FIGURE 12.9
Step response of
the system in
Fig. 12.8 with
$K = \frac{1}{2}$.

TABLE 12.1   Results for Example 12.2

K	0.25	0.45	0.50	0.55	0.75
$\|e_{ss}\|$	0.67	0.18	0	0.22	1.0
Undershoot	5%	9%	10%	11%	15%
Settling time (seconds)	15	24	27	30	45

changes in K as recorded in Table 12.1. The performance of this system might be considered barely acceptable for a change of gain of only $\pm 10\%$. Thus this system would not be considered robust. The steady-state error of this system changes greatly as K changes. ■

12.3 ANALYSIS OF ROBUSTNESS

Consider the closed-loop system shown in Fig. 12.1. System goals include maintaining small tracking error $[e(t) = r(t) - c(t)]$ for an input $r(t)$ and keeping the output $c(t)$ small for a disturbance $d(t)$. The sensor noise $n(t)$ must be small with respect to $r(t)$ so that $|r| \gg |n|$.

The sensitivity function is

$$S(s) = [1 + G_c(s)G(s)]^{-1},$$

and the closed-loop transfer function is

$$T(s) = \frac{G_c(s)G(s)}{1 + G_c(s)G(s)}$$

when $G_p(s) = 1$. We then have

$$S(s) + T(s) = 1. \qquad (12.13)$$

Thus it is important to make $S(s)$ small. For physically realizable systems, the loop gain $L(s) = G_c G(s)$ must be small for high frequencies. This means that $S(j\omega)$ approaches 1 at high frequencies.

An *additive* perturbation characterizes the set of possible plants as (here we assume $G_c(s) = 1$)

$$G_a(s) = G(s) + A(s),$$

where $G(s)$ is the nominal plant and $A(s)$ is the perturbation that is bounded in magnitude. It is assumed that $G_a(s)$ and $G(s)$ have the same number of poles in the right-hand s-plane (if any) [36]. Then the system stability will not change if

$$|A(j\omega)| < |1 + G(j\omega)| \quad \text{for all } \omega. \tag{12.14}$$

Of course, this assures stability but not dynamic performance.

A multiplicative perturbation gives the plant

$$G_m(s) = G(s)[1 + M(s)].$$

The perturbation is bounded in magnitude, and again it is assumed that $G_m(s)$ and $G(s)$ have the same number of poles in the right-hand s-plane. Then the system stability will not change if

$$|M(j\omega)| < \left|1 + \frac{1}{G(j\omega)}\right| \quad \text{for all } \omega. \tag{12.15}$$

Equation (12.15) is called the *robust stability criterion*. This is a test for robustness with respect to a multiplicative perturbation. This form of perturbation is often used because it satisfies the intuitive properties of (1) being small at low frequencies, where the nominal plant model is usually well known, and (2) being large at high frequencies, where the nominal model is always inexact.

EXAMPLE 12.3   **System with multiplicative perturbation**

Consider the system of Fig. 12.1 with $G_p(s) = 1$, $G_c = K$, and

$$G(s) = \frac{170,000(s + 0.1)}{s(s + 3)(s^2 + 10s + 10,000)}.$$

The system is unstable with $K = 1$, but a reduction in gain to $K = 0.5$ will stabilize the system. Now consider the effect of an unmodeled pole at 50 rad/s. In this case, the multiplicative perturbation is determined from

$$[1 + M(s)] = \frac{50}{s + 50}$$

or $M(s) = -s/(s + 50)$. The magnitude bound is then

$$|M(j\omega)| = \left|\frac{-j\omega}{j\omega + 50}\right|.$$

$|M(j\omega)|$ and $|1 + 1/KG(j\omega)|$ are plotted in Fig. 12.10(a), where it is seen that the criterion of Eq. (12.15) is not satisfied. Thus the system may not be stable.

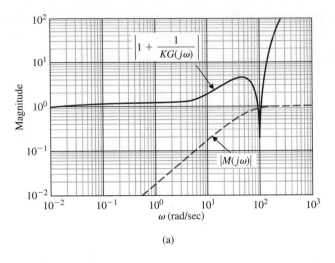

(a)

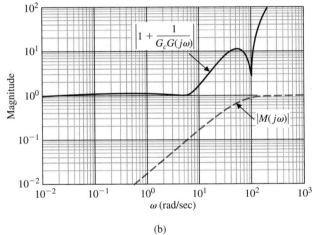

FIGURE 12.10
The robust stability
criterion for
Example 12.3.

(b)

If we use a lag compensator $G_c(s)$ where

$$G_c(s) = \frac{0.15(s + 25)}{(s + 2.5)},$$

the loop transfer function is $1 + G_c(s)G(s)$ and we reshape the function $G_cG(j\omega)$ in the frequency range $2 < \omega < 25$. Then we have the altered magnitude

$$\left| 1 + \frac{1}{G_cG(j\omega)} \right|,$$

as plotted in Fig. 12.10(b). Here the robustness inequality is satisfied and the system remains stable. ■

The control objective is to select a prefilter $G_p(s)$ and a compensator $G_c(s)$ in a two-degree-of-freedom feedback configuration (Fig. 12.1) so that the transient, steady-state, and frequency-domain specifications are achieved and the cost of feedback measured by the bandwidth of the compensator $G_c(j\omega)$ is sufficiently small. This bandwidth constraint is needed mainly because of noise that is inevitable in measuring the system output. A large noise amplification can saturate either the latter stages of $G_c(s)$ or the early plant stages.

12.4 SYSTEMS WITH UNCERTAIN PARAMETERS

Many systems have several parameters that are constants but uncertain within a range. For example, consider a system with a characteristic equation

$$s^n + a_{n-1}s^{n-1} + a_{n-2}s^{n-2} + \cdots + a_0 = 0 \tag{12.16}$$

with known coefficients within bounds

$$\alpha_i \le a_i \le \beta_i, \qquad i = 0, \ldots, n,$$

where $a_n = 1$.

To ascertain the stability of the system, one might have to investigate all possible combinations of parameters. Fortunately, it is possible to investigate a limited number of worst-case polynomials [22]. The analysis of only four polynomials is sufficient, and they are readily defined for a third-order system with a characteristic equation

$$s^3 + a_2s^2 + a_1s + a_0 = 0. \tag{12.17}$$

Then the four polynomials are

$$q_1(s) = s^3 + \alpha_2 s^2 + \beta_1 s + \beta_0,$$
$$q_2(s) = s^3 + \beta_2 s^2 + \alpha_1 s + \alpha_0,$$
$$q_3(s) = s^3 + \beta_2 s^2 + \beta_1 s + \alpha_0,$$
$$q_4(s) = s^3 + \alpha_2 s^2 + \alpha_1 s + \beta_0.$$

One of the four polynomials represents the *worst case* and may indicate unstable performance or at least the worst performance for the system in that case.

EXAMPLE 12.4   **Third-order system with uncertain coefficients**

Consider a third-order system with uncertain coefficients such that

$$8 \le a_0 \le 60 \Rightarrow \alpha_0 = 8, \qquad \beta_0 = 60,$$
$$12 \le a_1 \le 100 \Rightarrow \alpha_1 = 12, \qquad \beta_1 = 100,$$
$$7 \le a_2 \le 25 \Rightarrow \alpha_2 = 7, \qquad \beta_2 = 25.$$

The four polynomials are

$$q_1(s) = s^3 + 7s^2 + 100s + 60,$$
$$q_2(s) = s^3 + 25s^2 + 12s + 8,$$
$$q_3(s) = s^3 + 25s^2 + 100s + 8,$$
$$q_4(s) = s^3 + 7s^2 + 12s + 60.$$

We then proceed to check these four polynomials using the Routh criterion and hence determine that the system is stable for all the range of uncertain parameters. ∎

EXAMPLE 12.5   **Stability of uncertain system**

Consider a unity feedback system with a plant transfer function (nominal conditions)

$$G(s) = \frac{4.5}{s(s + 1)(s + 2)}.$$

The nominal characteristic equation is then

$$q(s) = s^3 + 3s^2 + 2s + 4.5 = 0.$$

Using Routh's criterion, we find that this system is nominally stable. However, if the system has uncertain coefficients such that

$$4 \le a_0 \le 5 \Rightarrow \alpha_0 = 4, \quad \beta_0 = 5,$$
$$1 \le a_1 \le 3 \Rightarrow \alpha_1 = 1, \quad \beta_1 = 3,$$
$$2 \le a_2 \le 4 \Rightarrow \alpha_2 = 2, \quad \beta_2 = 4,$$

then we must examine the four polynomials:

$$q_1(s) = s^3 + 2s^2 + 3s + 5,$$
$$q_2(s) = s^3 + 4s^2 + 1s + 4,$$
$$q_3(s) = s^3 + 4s^2 + 3s + 4,$$
$$q_4(s) = s^3 + 2s^2 + 1s + 5.$$

Using Routh's criteria, $q_1(s)$ and $q_3(s)$ are stable and $q_2(s)$ is marginally stable. For $q_4(s)$ we have

s^3	1	1
s^2	2	5
s^1	$-3/2$	
s^0	5	

Therefore, the system is unstable for the worst case where $\alpha_2 = $ minimum, $\alpha_1 = $ minimum, and $\beta_0 = $ maximum. This occurs when the plant has changed to

$$G(s) = \frac{5}{s(s + 1)(s + 1)}.$$

Note that the third pole has moved toward the $j\omega$-axis to its limit at $s = -1$ and that the gain has increased to its limit at $K = 5$. Often, we are able to examine the transfer function $G(s)$ and predict the worst-case conditions. ∎

12.5 THE DESIGN OF ROBUST CONTROL SYSTEMS

The design of robust control systems is based on two tasks: determining the structure of the controller and adjusting the controller's parameters to give an "optimal" system performance. This design process is normally done with "assumed complete knowledge" of the plant. Furthermore, the plant is normally described by a linear time-invariant continuous model. The structure of the controller is chosen such that the system's response can meet certain performance criteria.

One possible objective in the design of a control system is that the controlled system's output should exactly and instantaneously reproduce its input. That is, the system's transfer function should be unity:

$$T(s) = \frac{C(s)}{R(s)} = 1. \tag{12.18}$$

In other words, the system should be presentable on a Bode gain versus frequency diagram with a 0-dB gain of infinite bandwidth and zero phase shift. In practice, this is not possible, since every system will contain inductive- and capacitive-type components that store energy in some form. It is these elements and their interconnections with energy dissipative components that produce the system's dynamic response characteristics. Such systems reproduce some inputs almost exactly, while other inputs are not reproduced at all, signifying that the system's bandwidth is less than infinite.

Once it is recognized that the system's dynamics cannot be ignored, a new design objective is needed. One possible design objective is to maintain the magnitude response curve as flat and as close to unity for a large bandwidth as possible for a given plant and controller combination [22].

Another important goal of a control system design is that the effect on the output of the system due to disturbances is minimized. Thus we wish to minimize $C(s)/D(s)$ over a range of frequency.

Consider the control system shown in Fig. 12.11, where $G(s) = G_1(s)G_2(s)$ is the plant and $D(s)$ is the disturbance. We then have

$$T(s) = \frac{C(s)}{R(s)} = \frac{G_c G_1 G_2(s)}{1 + G_c G_1 G_2(s)} \tag{12.19}$$

and

$$\frac{C(s)}{D(s)} = \frac{G_2(s)}{1 + G_c G_1 G_2(s)}. \tag{12.20}$$

Note that both the reference and disturbance transfer functions have the same denominator, or, in other words, they have the same characteristic equation—that is,

$$1 + G_c(s)G_1(s)G_2(s) = 1 + L(s) = 0. \tag{12.21}$$

Furthermore, we recall that the sensitivity of $T(s)$ with respect to $G(s)$ is

$$S_G^T = \frac{1}{1 + G_c G_1 G_2(s)} \tag{12.22}$$

FIGURE 12.11
A system with a
disturbance.

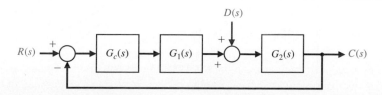

and the characteristic equation is the influencing factor on the sensitivity. Equation (12.22) shows that for low sensitivity S we require a high value of loop gain $L(j\omega)$, but it is known that high gain could cause instability or poor responsiveness of $T(s)$. Thus the designer seeks the following:

1. $T(s)$ with wide bandwidth and faithful reproduction of $R(s)$.
2. Large loop gain $L(s)$ in order to minimize sensitivity S.
3. Large loop gain $L(s)$ attained primarily by $G_c(s)G_1(s)$, since $C(s)/D(s) \approx 1/G_cG_1(s)$.

Setting the design of robust systems in frequency-domain terms, we are required to find a proper compensator, $G_c(s)$, such that the closed-loop sensitivity is less than some tolerance value, and sensitivity minimization involves finding a proper compensator such that the closed-loop sensitivity equals or is arbitrarily close to the minimal attainable sensitivity. Similarly, the gain margin problem is to find a proper compensator to achieve some prescribed gain margin, and gain margin maximization involves finding a proper compensator to achieve the maximal attainable gain margin. For the frequency domain specifications, we require the following for the Bode diagram of $G_cG(j\omega)$, shown in Fig. 12.12:

1. For relative stability, $G_cG(j\omega)$ must have, for an adequate range of ω, not more than -20 dB/decade slope at or near the crossover frequency ω_c.
2. Steady-state accuracy achieved by the low frequency gain.
3. Accuracy over a bandwidth ω_B by not allowing $|G_cG(\omega)|$ to fall below a prescribed level.
4. Disturbance rejection by high gain for $G_c(j\omega)$ over the system bandwidth.

Using the root sensitivity concept, we can state that we require that S_α^r be minimized while attaining $T(s)$ with dominant roots that will provide the appropriate response and minimize the effect of $D(s)$. Again we see that the goal is to have the gain of the loop primarily attained by $G_c(s)$. As an example, let $G_c(s) = K$, $G_1(s) = 1$, and $G_2(s) = 1/s(s + 1)$ for the system in Fig. 12.11. This system has two roots and we select a gain K so that $C(s)/D(s)$ is minimized, S_K^r is minimized, and $T(s)$ has desirable dominant roots. The sensitivity is

$$S_K^r = \frac{dr}{dK} \cdot \frac{K}{r} = \frac{ds}{dK}\bigg|_{s=r} \cdot \frac{K}{r}, \qquad (12.23)$$

FIGURE 12.12
Bode diagram for $20 \log |G_cG(j\omega)|$.

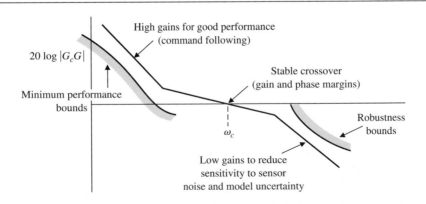

and the characteristic equation is

$$s(s + 1) + K = 0. \tag{12.24}$$

Therefore $dK/ds = -(2s + 1)$, since $K = -s(s + 1)$. We then obtain

$$S_K^r = \frac{-1}{(2s + 1)} \cdot \left. \frac{(-s(s + 1))}{s} \right|_{s=r}. \tag{12.25}$$

When $\zeta < 1$, the roots are complex and $r = -0.5 + j\omega$. Then,

$$|S_K^r| = \left(\frac{0.25 + \omega^2}{4\omega^2} \right)^{1/2}. \tag{12.26}$$

The magnitude of the sensitivity is plotted in Fig. 12.13 for $K = 0.2$ to $K = 5$. Also, the percent overshoot to a step is shown. Clearly, it is best to reduce the sensitivity but to limit K to 1.5 or less. We then attain the majority of the attainable reduction in sensitivity while maintaining good performance for the step response. In general, we can use the design procedure as follows:

1. Draw the root locus of the compensated system with $G_c(s)$ chosen to attain the desired location for the dominant roots.

2. Maximize the gain of $G_c(s)$ so that the effect of the disturbance is reduced.

3. Determine S_α^r and attain the minimum value of the sensitivity consistent with the transient response required as described in step 1.

EXAMPLE 12.6   Sensitivity and compensation

Let us reconsider the system in Example 10.1 when $G(s) = 1/s^2$, $G_1(s) = 1$, and $G_c(s)$ is to be selected by frequency response methods. Therefore the compensator is to be selected to achieve an appropriate gain and phase margin while minimizing sensitivity and the effect of the disturbance. Thus we choose

$$G_c(s) = \frac{K(s/z + 1)}{(s/p + 1)}. \tag{12.27}$$

FIGURE 12.13
Sensitivity and percent overshoot for a second-order system.

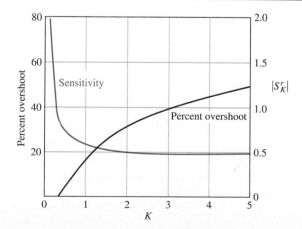

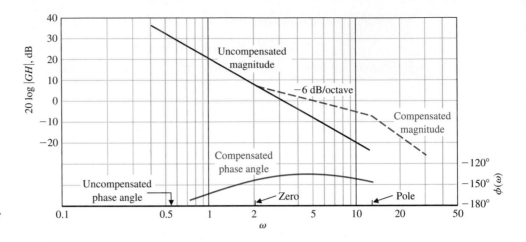

FIGURE 12.14
Bode diagram for
Example 12.6.

As in Example 10.1, we choose $K = 10$ to reduce the effect of the disturbance. To attain a phase margin of 45°, we select $z = 2.0$ and $p = 12.0$. We then attain the compensated diagram shown in Fig. 10.9 and repeated in Fig. 12.14. Recall that the closed-loop bandwidth is $\omega_B = 1.6\,\omega_c$. Thus we will increase the bandwidth by using the compensator and improve the fidelity of reproduction of the input signals.

The sensitivity may be ascertained at ω_c as

$$|S_G^T(\omega_c)| = \left|\frac{1}{1 + G_c G(j\omega)}\right|_{\omega_c}. \tag{12.28}$$

To estimate $|S_G^T|$, we recall that the Nichols chart enables us to obtain

$$|T(\omega)| = \left|\frac{G_c G(\omega)}{1 + G_c G(\omega)}\right|. \tag{12.29}$$

Thus we can plot a few points of $G_c G(j\omega)$ on the Nichols chart and then read $T(\omega)$ from the Nichols chart. Then,

$$|S_G^T(\omega_1)| = \frac{|T(\omega_1)|}{|G_c G(\omega_1)|}, \tag{12.30}$$

where ω_1 is chosen arbitrarily as $\omega_c/2.5$. In general, we choose a frequency below ω_c to determine the value of $|S(\omega_1)|$. Of course, we desire a low value of sensitivity. The Nichols chart for the compensated system is shown in Fig. 12.15. For $\omega_1 = \omega_c/2.5 = 2$, we have $20 \log T = 2.5$ dB and $20 \log G_c G = 9$ dB. Therefore

$$|S(\omega_1)| = \frac{|T|}{|G_c G|} = \frac{1.33}{2.8} = 0.47. \blacksquare$$

EXAMPLE 12.7   **Sensitivity with a lead compensator**

Let us again consider the system in Example 12.6, using the root locus design obtained in Example 10.3. The compensator was chosen as

$$G_c(s) = \frac{8.1(s + 1)}{(s + 3.6)}, \tag{12.31}$$

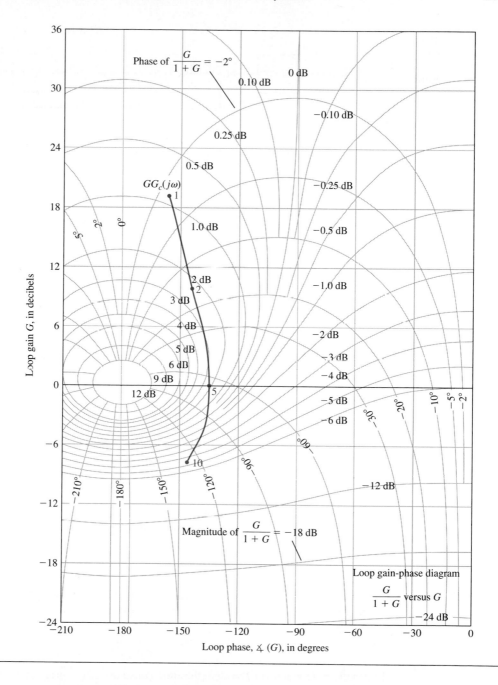

FIGURE 12.15
Nichols chart for
Example 12.7.

for the system of Fig. 12.16. The dominant roots are thus $s = -1 \pm j2$. Since the gain is 8.1, the effect of the disturbance is reduced and the time response meets the specifications. The sensitivity at a root r may be obtained by assuming that the system, with dominant roots, may be approximated by the second-order system

$$T(s) = \frac{K}{s^2 + 2\zeta\omega_n s + K} = \frac{K}{s^2 + 2s + K},$$

FIGURE 12.16
Feedback control
system with a
desired input $R(s)$
and an undesired
input $D(s)$.

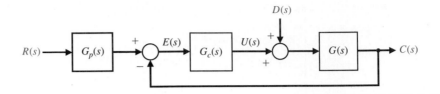

since $\zeta \omega_n = 1$. The characteristic equation is thus

$$s^2 + 2s + K = 0.$$

Then $dK/ds = -(2s + 2)$ since $K = -(s^2 + 2s)$. Therefore

$$S_K^r = \frac{-1}{(2s + 2)} \cdot \left. \frac{-(s^2 + 2s)}{s} \right|_{s=r} = \left. \frac{(s + 2)}{(2s + 2)} \right|_{s=r} \tag{12.32}$$

where $r = -1 + j2$. Then, substituting $s = r$, we obtain

$$|S_K^r| = 0.56.$$

If we raise the gain to $K = 10$, we expect $r \cong -1.1 \pm j2.2$. Then the sensitivity is

$$|S_K^r| = 0.53.$$

Thus as K increases the sensitivity decreases but the transient performance deteriorates. ■

12.6   THREE-TERM (PID) CONTROLLERS

One form of controller widely used in industrial process control is called a *three-term,* or *process controller.* This controller has a transfer function

$$\frac{U(s)}{E(s)} = G_c(s) = K_p + \frac{K_I}{s} + K_D s. \tag{12.33}$$

The controller provides a proportional term, an integration term, and a derivative term [2, 20, 35]. The equation for the output in the time domain is

$$u(t) = K_p e(t) + K_I \int e(t) \, dt + K_D \frac{de(t)}{dt}. \tag{12.34}$$

The three-mode controller is also called a PID controller because it contains a proportional, an integral, and a derivative term. The transfer function of the derivative term is actually

$$G_d(s) = \frac{K_D s}{\tau_d s + 1}, \tag{12.35}$$

but usually τ_d is much smaller than the time constants of the process itself and so may be neglected.

If we set $K_D = 0$, then we have the familiar PI controller discussed in Section 10.6. When $K_I = 0$, we have

$$G_c(s) = K_p + K_D s, \tag{12.36}$$

which is called a proportional plus derivative (PD) controller (Example 12.1).

Many industrial processes are controlled using proportional-integral-derivative (PID) controllers. The popularity of PID controllers can be attributed partly to their robust performance in a wide range of operating conditions and partly to their functional simplicity, which allows engineers to operate them in a simple, straightforward manner. To implement such a controller, three parameters must be determined for the given process: proportional gain, integral gain, and derivative gain [35].

Consider the PID controller

$$G_c(s) = K_1 + \frac{K_2}{s} + K_3 s = \frac{K_3 s^2 + K_1 s + K_2}{s}$$

$$= \frac{K_3(s^2 + as + b)}{s} = \frac{K_3(s + z_1)(s + z_2)}{s}$$

(12.37)

where $a = K_1/K_3$ and $b = K_2/K_3$. Therefore a PID controller introduces a transfer function with one pole at the origin and two zeros that can be located anywhere in the left-hand s-plane.

Recall that a root locus begins at the poles and ends at the zeros. If we have a system as shown in Fig. 12.16 with

$$G(s) = \frac{1}{(s + 2)(s + 3)}$$

and we use a PID controller with complex zeros, we can plot the root locus as shown in Fig. 12.17. As the gain, K_3, of the controller is increased, the complex roots approach the zeros. The closed-loop transfer function is

$$T(s) = \frac{G(s)G_c(s)G_p(s)}{1 + G(s)G_c(s)}$$

$$= \frac{K_3(s + z_1)(s + \hat{z}_1)}{(s + r_2)(s + r_1)(s + \hat{r}_1)} G_p(s)$$

(12.38)

$$\cong \frac{K_3 G_p(s)}{(s + r_2)},$$

because the zeros and the complex roots are approximately equal ($r_1 \approx z_1$). Setting $G_p(s) = 1$, we have

$$T(s) = \frac{K_3}{s + r_2} \approx \frac{K_3}{s + K_3}$$

(12.39)

when $K_3 \gg 1$. The only limiting factor is the allowable magnitude of $U(s)$ (Fig. 12.16) when K_3 is large. If K_3 is 100, the system has a fast response and zero steady-state error. Furthermore, the effect of the disturbance is reduced significantly.

In general, we note that PID controllers are particularly useful for reducing steady-state error and improving the transient response when $G(s)$ has one or two poles (or may be approximated by a second-order plant).

We may use frequency response methods to represent the addition of a PID controller. The PID controller, Eq. (12.37), may be rewritten as

$$G_c(s) = \frac{K_2 \left(\frac{K_3}{K_2} s^2 + \frac{K_1}{K_2} s + 1 \right)}{s} = \frac{K_2(\tau s + 1)\left(\frac{\tau}{\alpha} s + 1 \right)}{s}.$$

(12.40)

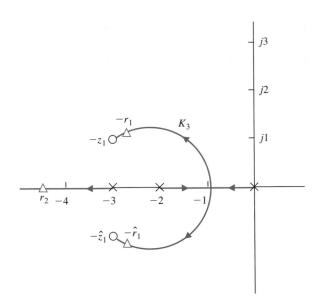

FIGURE 12.17
Root locus for
plant with a PID
controller with
complex zeros.

The Bode diagram of Eq. (12.40) is shown in Fig. 12.18 for $\omega\tau$, $K_2 = 2$, and $\alpha = 10$. The PID controller is a form of a lag-lead compensator with a variable gain, K_2. Of course, it is possible that the controller will have complex zeros and a Bode diagram that will be dependent on the ζ of the complex zeros. The contribution by the zeros to the Bode chart may be visualized by reviewing Fig. 8.10 for complex poles and noting that the phase and magnitude change as ζ changes. The PID controller with complex zeros is

$$G_c(\omega) = \frac{K_2[1 + (2\zeta/\omega_n)j\omega - (\omega/\omega_n)^2]}{j\omega}. \tag{12.41}$$

Normally, we choose $0.9 > \zeta > 0.7$.

FIGURE 12.18
Bode diagram for a
PID controller.

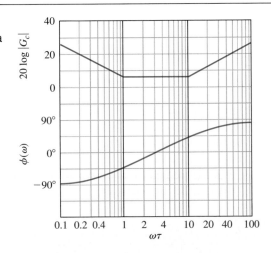

12.7 THE DESIGN OF ROBUST PID CONTROLLED SYSTEMS

The selection of the three coefficients of PID controllers is basically a search problem in a three-dimensional space. Points in the search space correspond to different selections of a PID controller's three parameters. By choosing different points of the parameter space, we can produce, for example, different step responses for a step input. A PID controller can be determined by moving in this search space on a trial and error basis.

The main problem in the selection of the three coefficients is that these coefficients do not readily translate into the desired performance and robustness characteristics that the control system designer has in mind. Several rules and methods have been proposed to solve this problem. In this section we consider several design methods using root locus and performance indices.

The first design method uses the ITAE performance index of Section 5.9 and the optimum coefficients of Table 5.6 for a step input or Table 5.7 for a ramp input. Hence we select the three PID coefficients to minimize the ITAE performance index, which produces an excellent transient response to a step (Fig. 5.30c) or a ramp. The design procedure consists of three steps:

1. Select the ω_n of the closed-loop system by specifying the settling time.

2. Determine the three coefficients using the appropriate optimum equation (Table 5.6) and the ω_n of step 1 to obtain $G_c(s)$.

3. Determine a prefilter $G_p(s)$ so that the closed-loop system transfer function, $T(s)$, does not have any zeros as required by Eq. (5.47).

EXAMPLE 12.8   **Robust control of temperature**

Consider a temperature controller with a control system as shown in Fig. 12.16 and a plant

$$G(s) = \frac{1}{(s + 1)^2}.$$ (12.42)

If $G_c(s) = 1$, the steady-state error is 50% and the settling time is 3.2 seconds for a step input. We desire to obtain an optimum ITAE performance for a step input and a settling time of less than 0.5 second. Using a PID controller, we have

$$G_c(s) = \frac{K_3 s^2 + K_1 s + K_2}{s}.$$ (12.43)

Therefore the closed-loop transfer function without prefiltering $[G_p(s) = 1]$ is

$$T_1(s) = \frac{C(s)}{R(s)} = \frac{G_c G(s)}{1 + G_c G(s)}$$

$$= \frac{K_3 s^2 + K_1 s + K_2}{s^3 + (2 + K_3)s^2 + (1 + K_1)s + K_2}.$$ (12.44)

The optimum coefficients of the characteristic equation for ITAE are obtained from Table 5.6 as

$$(s^3 + 1.75\omega_n s^2 + 2.15\omega_n^2 s + \omega_n^3).$$ (12.45)

TABLE 12.2   Results for Example 12.8

Controller	$G_c = 1$	PID and $G_p = 1$	PID with $G_p(s)$ Prefilter
Percent overshoot	0	31.7%	1.9%
Settling time (seconds)	3.2	0.20	0.45
Steady-state error	50.1%	0.0%	0.00%
$|c(t)/d(t)|_{maximum}$	52%	0.4%	0.4%

We need to select ω_n in order to meet the settling time requirement. Since $T_s = 4/\zeta\omega_n$ and ζ is unknown but near 0.8, we set $\omega_n = 10$. Then, equating the denominator of Eq. (12.44) to Eq. (12.45), we obtain the three coefficients as $K_1 = 214$, $K_3 = 15.5$, and $K_2 = 1000$. Then Eq. (12.44) becomes

$$T_1(s) = \frac{15.5s^2 + 214s + 1000}{s^3 + 17.5s^2 + 215s + 1000} \tag{12.46}$$

$$= \frac{15.5(s + 6.9 + j4.1)(s + 6.9 - j4.1)}{s^3 + 17.5s^2 + 215s + 1000}.$$

The response of this system to a step input has an overshoot of 32%, as recorded in Table 12.2.

We select a prefilter $G_p(s)$ so that we achieve the desired ITAE response with

$$T(s) = \frac{G_c G G_p(s)}{1 + G_c G(s)} = \frac{1000}{s^3 + 17.5s^2 + 215s + 1000}. \tag{12.47}$$

Therefore we require

$$G_p(s) = \frac{64.5}{(s^2 + 13.8s + 64.5)} \tag{12.48}$$

in order to eliminate the zeros in Eq. (12.46) and bring the overall numerator to 1000. The response of the system, $T(s)$, to a step input is indicated in Table 12.2. The system has a small overshoot, a settling time of less than ½ second, and zero steady-state error. Furthermore, for a disturbance $D(s) = 1/s$, the maximum value of $c(t)$ due to the disturbance is 0.4% of the magnitude of the disturbance. This is a very favorable design. ■

EXAMPLE 12.9   Robust system design

Let us reconsider the system in Example 12.8 when the plant varies significantly so that

$$G(s) = \frac{K}{(\tau s + 1)^2}, \tag{12.49}$$

where $0.5 \leq \tau \leq 1$, and $1 \leq K \leq 2$. It is desired to achieve robust behavior using an ITAE optimum system with a prefilter while attaining an overshoot of less than 4% and a settling time of less than 2 seconds while $G(s)$ can attain any value in the range indicated. We select $\omega_n = 8$ in order to attain the settling time and determine the ITAE coefficients for

TABLE 12.3   Results for Example 12.9 with $\omega_n = 8$

Plant Conditions	$\tau = 1$ $K = 1$	$\tau = 0.5,$ $K = 1$	$\tau = 1,$ $K = 2$	$\tau = 0.5,$ $K = 2$
Percent overshoot	2%	0%	0%	1%
Settling time (seconds)	1.25	0.8	0.8	0.9

$K = 1$ and $\tau = 1$. Then, completing the calculation, we obtain the system without a prefilter $[G_p(s) = 1]$ as

$$T_1(s) = \frac{12(s^2 + 11.38s + 42.67)}{s^3 + 14s^2 + 137.6s + 512} \tag{12.50}$$

and

$$G_c(s) = \frac{12(s^2 + 11.38s + 42.67)}{s}. \tag{12.51}$$

We select a prefilter

$$G_p(s) = \frac{42.67}{(s^2 + 11.38s + 42.67)} \tag{12.52}$$

to obtain the optimum ITAE transfer function

$$T(s) = \frac{512}{s^3 + 14s^2 + 137.6s + 512}. \tag{12.53}$$

We then obtain the step response for the four conditions: $\tau = 1$, $K = 1$; $\tau = 0.5$, $K = 1$; $\tau = 1$, $K = 2$; and $\tau = 0.5$, $K = 2$. The results are summarized in Table 12.3. Clearly, this is a very robust system. ■

The value of ω_n that can be chosen will be limited by considering the maximum allowable $u(t)$ where $u(t)$ is the output of the controller, as shown in Fig. 12.16. If the maximum value of $e(t)$ is 1, then $u(t)$ would normally be limited to 100 or less. As an example, consider the system in Fig. 12.16 with a PID controller, $G(s) = 1/s(s + 1)$, and the necessary prefilter $G_p(s)$ to achieve ITAE performance. If we select $\omega_n = 10$, 20, and 40, the maximum value of $u(t)$ is as recorded in Table 12.4. If we wish to limit $u(t)$ to a maximum equal to 100, we need to limit ω_n to 16. Thus we are limited in the settling time we can achieve.

TABLE 12.4   Maximum Value of Plant Input

ω_n	10	20	40
$u(t)$ maximum for $R(s) = 1/s$	35	135	550
Settling time (seconds)	0.9	0.5	0.3

668   Chapter 12   Robust Control Systems

Let us consider the design of a PID compensator using frequency response techniques for a system with a time delay so that

$$G(s) = \frac{Ke^{-Ts}}{(\tau s + 1)}. \qquad (12.54)$$

This type of plant represents many industrial processes that incorporate a time delay. We use a PID compensator to introduce two equal zeros so that

$$G_c(s) = \frac{K_2(\tau_1 s + 1)^2}{s}. \qquad (12.55)$$

The design method is as follows:

1. Plot the uncompensated Bode diagram for $K_2 G(s)/s$ with a gain K_2 that satisfies the steady-state error requirement.
2. Place the two equal zeros at or near the crossover frequency, ω_c.
3. Test the results and adjust K_2 or the zero locations, if necessary.

EXAMPLE 12.10 PID control of a system with a delay

Consider the system of Fig. 12.15 when

$$G(s) = \frac{20e^{-0.1s}}{(0.1s + 1)}, \qquad (12.56)$$

where $K = 20$ is selected to achieve a small steady-state error for a step input. We want an overshoot to a step input of less than 5%.

Plotting the Bode diagram for $G(j\omega)$, we find that the uncompensated system has a negative phase margin and that the system is unstable.

We will use a PID controller of the form of Eq. (12.55) to attain a desirable phase margin of 70°. Then the loop transfer function is

$$GG_c(s) = \frac{20e^{-0.1s}(\tau_1 s + 1)^2}{s(0.1s + 1)}, \qquad (12.57)$$

where $K_2 K = 20$. We plot the Bode diagram without the two zeros, as shown in Fig. 12.19. The phase margin is $-32°$, and the system is unstable prior to the introduction of the zeros.

Since we have introduced a pole at the origin because of the internal term in the PID compensator, we may reduce the gain $K_2 K$ because e_{ss} is now zero. We place the two zeros at or near the crossover, $\omega_c = 11$. We choose to set $\tau_1 = 0.06$ so that the two zeros are set at $\omega = 16.7$. Also, we reduce the gain to $K_2 K = 4.5$. Then we obtain the frequency response shown in Fig. 12.20, where

$$G_c G(s) = \frac{4.5(0.06s + 1)^2 e^{-0.1s}}{s(0.1s + 1)}. \qquad (12.58)$$

The new crossover frequency is $\omega_c' = 4.5$ and the phase margin is 70°. The step response of this system has no overshoot and has a settling time of 0.80 second. This response satisfies the requirements. However, if we wanted to adjust the system further, we could raise $K_2 K$ to 10 and achieve a somewhat faster response with an overshoot of less than 5%. ∎

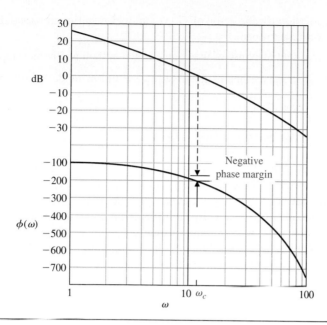

FIGURE 12.19
Bode diagram
for $G(s)/s$ for
Example 12.10.

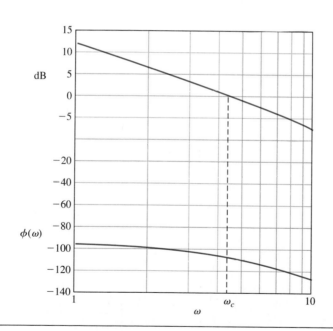

FIGURE 12.20
Bode diagram
for $G_c G(s)$ for
Example 12.10.

As a final consideration of the design of robust control systems using a PID controller, we turn to an s-plane root locus method. This design approach may be simply stated as follows:

1. Place the poles and zeros of $G(s)/s$ on the s-plane.

2. Select a location for the zeros of $G_c(s)$ that will result in an acceptable root locus and suitable dominant roots.

3. Test the transient response of the compensated system and iterate step 2, if necessary.

The root locus design method is illustrated by the following design example for an aircraft autopilot.

12.8 DESIGN EXAMPLE: AIRCRAFT AUTOPILOT

A typical aircraft autopilot control system consists of electrical, mechanical, and hydraulic devices that move the flaps, elevators, fuel-flow controllers, and other components that cause the aircraft to vary its flight. Sensors provide information on velocity, heading, rate of rotation, and other flight data. This information is combined with the desired flight characteristics (commands) electronically available to the autopilot. The autopilot should be able to fly the aircraft on a heading and under conditions set by the pilot. The command often consists of a predetermined heading. Design often focuses on a forward-moving aircraft that moves somewhat up or down without moving right or left and without rolling (rotating the wingtips). Such a study is called pitch axis design. The aircraft is represented by a plant [26]

$$G(s) = \frac{K}{(s + 1/\tau)(s^2 + 2\zeta_1\omega_1 s + \omega_1^2)},$$

(12.59)

where τ is the time constant of the actuator. Let $\tau = \frac{1}{4}$, $\omega_1 = 2$, and $\zeta = \frac{1}{2}$. Then, the s-plane plot has two complex poles—a pole at the origin and a pole at $s = -4$, as shown in Fig. 12.21. The complex poles, representing the aircraft dynamics, can vary within the

FIGURE 12.21
Root locus for aircraft autopilot. The complex poles can vary within the dashed-line box (in color).

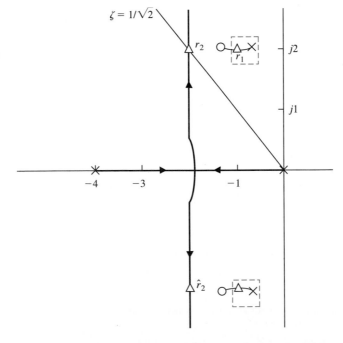

dashed-line box shown in the figure. We then choose the zeros of the controller as $s = -1.3 \pm j2$, as shown. We select the gain K so that the roots r_2 and $\hat{r}_2$ are complex with a ζ of $1/\sqrt{2}$. The other roots r_1 and $\hat{r}_1$ lie very near the zeros. Therefore the closed-loop transfer function is approximately

$$T(s) \cong \frac{\omega_n^2}{s^2 + 2\zeta\omega_n s + \omega_n^2} = \frac{5}{s^2 + 3.16s + 5}, \qquad (12.60)$$

since $\omega_n = \sqrt{5}$ and $\zeta = 1/\sqrt{2}$. The resulting response to a step input has an overshoot of 4.5% and a settling time of 2.5 seconds, as expected.

12.9 THE DESIGN OF A SPACE TELESCOPE CONTROL SYSTEM

Scientists have proposed the operation of a space vehicle as a space-based research laboratory and test bed for equipment to be used on a manned space station. The industrial space facility (ISF) would remain in space and the astronauts would be able to use it only when the shuttle is attached [16, 21]. The ISF will be the first permanent, human-operated commercial space facility designed for R&D, testing, and, eventually, processing in the space environment.

We will consider an experiment operated in space but controlled from earth. The goal is to manipulate and position a small telescope to accurately point at a planet. The goal is to have a steady-state error equal to zero, while maintaining a fast response to a step with an overshoot of less than 5%. The actuator chosen is a low-power actuator, and the model of the combined actuator and telescope is shown in Fig. 12.22. The command signal is received from an earth station with a delay of $\pi/16$ seconds. A sensor will measure the pointing direction of the telescope accurately. However, this measurement is relayed back to earth with a delay of $\pi/16$ seconds. Thus the total transfer function of the telescope, actuator, sensor, and round-trip delay (Fig. 12.23) is

$$G(s) = \frac{e^{-s\pi/8}}{(s + 1)^2}. \qquad (12.61)$$

We propose a controller that is the three-term controller described in Section 12.6, where

$$G_c(s) = K_1 + \frac{K_2}{s} + K_3 s = \frac{K_1 s + K_2 + K_3 s^2}{s}. \qquad (12.62)$$

Clearly, the use of only the proportional term will not be acceptable, since we require a steady-state error of zero for a step input. Thus we must use a finite value of K_2 and hence we may elect to use either proportional plus integral control (PI) or proportional plus integral plus derivative control (PID).

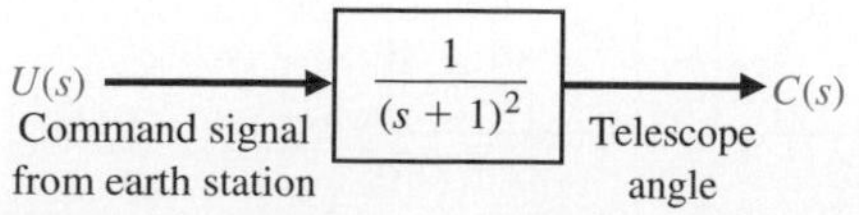

FIGURE 12.22
Model of a low-power actuator and telescope.

$U(s)$ Command signal from earth station → $\dfrac{1}{(s + 1)^2}$ Telescope angle → $C(s)$

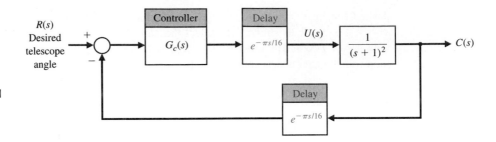

FIGURE 12.23
Feedback control system for the telescope experiment.

We will first try PI control, so that

$$G_c(s) = K_1 + \frac{K_2}{s} = \frac{K_1 s + K_2}{s}.$$ (12.63)

Since we have a pure delay, e^{-sT}, we use the frequency response methods for the design process. Thus we will translate the overshoot specification to the frequency domain. If we have two dominant characteristic roots, the overshoot to a step is 5% when $\zeta = 0.7$ or the phase margin requirement is about 70°.

If we choose $K_1 = 0.022$ and $K_2 = 0.22$, we have

$$G_c(s)G(s) = \frac{0.22(0.1s + 1)e^{-s\pi/8}}{s(s + 1)^2}$$ (12.64)

and the Bode diagram is shown in Fig. 12.24. The location of the zero at $s = 10$ was chosen to add a phase lead angle so that the desired phase margin is attained. An iterative procedure yields a series of trials for K_1 and K_2 until the desired phase margin is achieved. Note that we have achieved a phase margin of about 63°. The actual step response was plotted, and it is determined that the overshoot was 4.7% with a setting time of 16 seconds, as recorded in Table 12.5.

The proportional plus integral plus derivative controller is

$$G_c(s) = \frac{K_1 s + K_2 + K_3 s^2}{s}.$$ (12.65)

FIGURE 12.24
Bode diagram for the system with the PI controller.

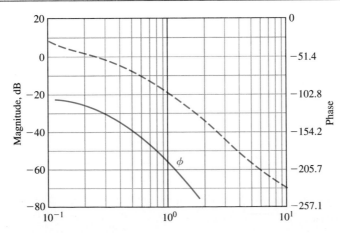

TABLE 12.5   Step Response of the Space Telescope for Two Controllers

	Steady-State Error	Percent Overshoot	Settling Time (seconds)
PI controller	0	4.7	16.0
PID controller	0	3.7	5.8

We now have three parameters to vary to achieve the desired phase margin. If we select, after some iteration, $K_1 = 0.8$, $K_2 = 0.5$, and $K_3 = 10^{-3}$, we obtain a phase margin of 64°. The percentage overshoot is 3.7% and the settling time is 5.8 seconds. Perhaps the easiest way to select the gain constants is to let K_3 be a small, but nonzero, number initially and $K_1 = K_2 = 0$, then plot the frequency response. In this case, we choose $K_3 = 10^{-3}$ and obtain a Bode plot. We then use $K_1 \approx K_2$ and iterate to obtain the appropriate values of these unspecified gains.

The performance of the PI and the PID compensated systems is recorded in Table 12.5. The PID controller is the most desirable, since it provides a shorter settling time.

12.10   THE DESIGN OF A ROBUST BOBBIN DRIVE

Monofilament nylon is produced by an extrusion process that outputs filament at a constant rate. The product is wound onto a bobbin that rotates at a maximum speed of 2000 rpm. The tension in the filament must be held between 0.2 and 0.6 pound to ensure that it is not stretched. The winding diameter varies between 2 to 4 inches.

The filament is laid onto the bobbin by a ballscrew-driven arm that oscillates back and forth at constant speed, as shown in Fig. 12.25(a). The arm must reverse rapidly at the end of the move. The required ballscrew speed is 60 rpm. The prime requirement of the bobbin drive is to provide a controlled tension. Since the winding diameter varies by 2 to 1, the tension will fall by 50% from start to finish.

The control system will have a system structure as shown in Fig. 12.25(b), for which we select a PID controller. The parameter variations are $1.5 \leq K_m \leq 2.5$ and $3 \leq p \leq 5$ with the nominal conditions $K_m = 2$ and $p = 4$. Furthermore, a third pole at $s = -50$ has been omitted from the model. The requirements are an overshoot less than 2.5% and a settling time less than 0.4 second. The magnitude of $u(t)$ must be less than 100.

Using a PID controller, the ITAE design, and the nominal parameters, we determine ω_n from the settling time requirement. Since we expect that $\zeta \cong 0.8$, we use

$$T_s = \frac{4}{0.8\omega_n} < 0.4.$$

We select $\omega_n = 23$ as the maximum allowable for $|u| < 100$. Then, for

$$G_c(s) = K_1 + \frac{K_2}{s} + K_3 s,$$

we obtain $K_1 = 568.68$, $K_2 = 6083.5$, and $K_3 = 18.13$. Using the appropriate prefilter, the response is as recorded in Table 12.6. Clearly, the system does not offer robust perfor-

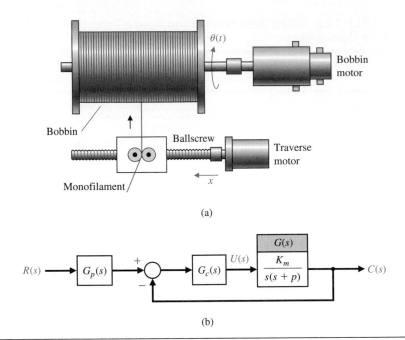

FIGURE 12.25
A monofilament
bobbin winder.

mance, since the overshoot requirement is not satisfied when the worst-case parameters are
considered.

We also examine the performance of the system with the nominal parameters but with
the unmodeled pole added so that the actual plant is

$$G(s) = \frac{2(50)}{s(s + 4)(s + 50)}. \qquad (12.66)$$

(Note that the dc gain, $\lim_{s \to 0} sG(s)$, remains 0.5.) The response of the PID controller with
the added pole is recorded in Table 12.6. Again, the system fails the requirement of robust
performance.

**TABLE 12.6   Response of the Bobbin Drive System for a Unit Step Input
(original design)**

	Parameters	Percent Overshoot	Settling Time	$\left\|\dfrac{u(t)}{r(t)}\right\|_{maximum}$
Nominal parameters	$K_m = 2,$ $p = 4$	1.96%	0.318	98
Worst-case parameters	$K_m = 1.5,$ $p = 3$	7.48%	0.375	95
Nominal parameters and added third pole at $s = -50$		9.82%	0.732	90

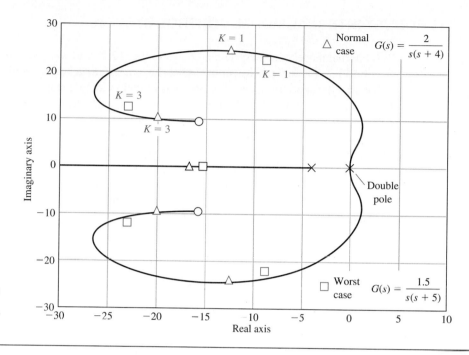

FIGURE 12.26
Root locus for the normal case and the worst case for $K = 1$ and $K = 3$.

We need to adjust the system so that the performance with the worst-case parameters is acceptable. Examine the root locus for the nominal parameters as shown in Fig. 12.26. Insert a cascade gain K prior to $G_c(s)$ so that we have $KG_cG(s)$. Then the roots for $K = 1$ and $K = 3$ are shown on the locus. Since the worst-case response occurs when the motor constant K_m drops to 1.5, we use the cascade gain $K = 3$ to move the roots to the left on the s-plane. Then, when the gain K_m drops to 1.5, the roots still are in the desired region. The response of the system with $K = 3$ is recorded in Table 12.7 for the nominal and worst-case conditions, and with the added pole. This system meets all the specifications. This approach uses a cascade gain that, when adjusted correctly, will drive the dominant roots near the complex zeros of the PID controller. Then, when the worst parameter change occurs, the system will still maintain the required performance.

TABLE 12.7   Response of the Bobbin Drive System for a Unit Step with Additional Cascade Gain $K = 3$

	Percent Overshoot	Settling Time (seconds)
Nominal parameters	0.12%	0.218
Worst-case parameters	0.47%	0.214
Nominal parameters and third pole	0.50%	0.242

12.11 THE ROBUST INTERNAL MODEL CONTROL SYSTEM

The internal model control system is shown in Fig. 12.27 and was previously considered in Section 11.7. We now reconsider the use of the internal model design with special attention to robust system performance. The **internal model principle** states that if $G_c(s)G(s)$ contains $R(s)$, then $c(t)$ will track $r(t)$ asymptotically (in the steady state) and the tracking is robust.

Examining the system of Fig. 12.27, we note that for lower-order plants, state variable feedback will not be required and a suitable $G_c(s)$ can be obtained. However, with higher-order systems, the feedback of all state variables may be required.

Consider a simple system with $G(s) = 1/s$, for which we seek a ramp response with a steady-state error of zero. A PI controller is sufficient and we let $\mathbf{K} = \mathbf{0}$ (no state variable feedback). Then we have

$$G_c G(s) = \left(K_1 + \frac{K_2}{s} \right)\frac{1}{s} = \frac{(K_1 s + K_2)}{s^2}. \tag{12.67}$$

Note that for a ramp, $R(s) = 1/s^2$, which is contained as a factor of Eq. (12.67), and the closed-loop transfer function is

$$T(s) = \frac{(K_1 s + K_2)}{s^2 + K_1 s + K_2}. \tag{12.68}$$

Using the ITAE specifications for a ramp response (Table 5.7), we require

$$T(s) = \frac{3.2\omega_n s + \omega_n^2}{s^2 + 3.2\omega_n s + \omega_n^2}. \tag{12.69}$$

We select ω_n to satisfy a specification for the settling time. For a settling time of 1 second, we select $\omega_n = 5$. Then, we require $K_1 = 16$ and $K_2 = 25$. The response of this system settles in 1 second and then tracks the ramp with zero steady-state error. If this system (designed for a ramp input) receives a step input, the response has an overshoot of 5% and a settling time of 1.5 seconds. This system is very robust to changes in the plant. For example, if $G(s) = K/s$ changes gain so that K shifts from $K = 1$ by $\pm 50\%$, the change in the ramp response is insignificant.

EXAMPLE 12.11 **Design of an internal model control system**

Consider the system of Fig. 12.28 with state variable feedback and a compensator $G_c(s)$. We wish to track a step input with zero steady-state error. Here we select a PID controller

---

FIGURE 12.27
The internal model
control system.

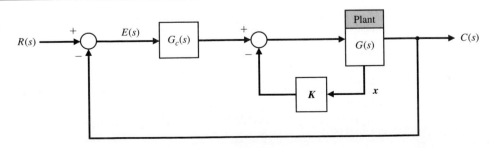

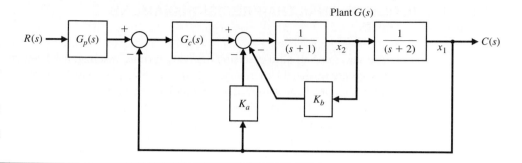

FIGURE 12.28
An internal model control with state variable feedback and $G_c(s)$.

for $G_c(s)$. We then have

$$G_c(s) = \frac{K_3 s^2 + K_1 s + K_2}{s},$$

and $G(s)G_c(s)$ will contain $R(s) = 1/s$, the input command. Note that we feed back both state variables and add these additional signals after $G_c(s)$ in order to retain the integrator in $G_c(s)$.

The goal is to achieve a settling time less than 1 second and a deadbeat response (see Section 10.11) while retaining a robust response. Here we assume that the two poles of $G(s)$ can change by $\pm 50\%$. Then the worst-case condition is

$$\hat{G}(s) = \frac{1}{(s + 0.5)(s + 1)}.$$

One design approach is to design the control for this worst-case condition. Another approach, which we use here, is to design for the nominal $G(s)$ and one-half the desired settling time. Then we expect to meet the settling time requirement and attain a very fast, highly robust system. Note that the prefilter $G_p(s)$ is used to attain the desired form for $T(s)$.

The response desired is deadbeat (see Table 10.2) so use a third-order transfer function as

$$T(s) = \frac{\omega_n^3}{s^3 + 1.9\omega_n s^2 + 2.20\omega_n^2 s + \omega_n^3}, \qquad (12.70)$$

and the settling time is $T_s = 4.04/\omega_n$. For a settling time of ½ second, we use $\omega_n = 8.08$.

The closed-loop transfer function of the system of Fig. 12.28 with the appropriate $G_p(s)$ is

$$T(s) = \frac{K_2}{s^3 + (3 + K_3 + K_b)s^2 + (2 + K_1 + K_a + 2K_b)s + K_2}. \qquad (12.71)$$

We let $K_a = 10$, $K_b = 2$, $K_1 = 127.6$, $K_2 = 527.5$, and $K_3 = 10.35$. Note that $T(s)$ could be achieved with other gains, including $K_b = 0$.

The step response of this system has a deadbeat response with an overshoot of 1.65% and a settling time of 0.5 second. When the plant poles of $G(s)$ change by $\pm 50\%$, the overshoot changes to 1.86% and the settling time becomes 0.95 second. This is an outstanding design of a very robust, deadbeat response system. ∎

12.12 THE DESIGN OF AN ULTRA-PRECISION DIAMOND TURNING MACHINE

The design of an ultra-precision diamond turning machine has been studied at Lawrence Livermore National Laboratory. This machine shapes optical devices such as mirrors with ultra-high precision using a diamond tool as the cutting device. In this discussion, we will consider only the z-axis control. Using frequency response identification with sinusoidal input to the actuator, it was determined that

$$G(s) = \frac{4500}{(s + 60)}. \qquad (12.72)$$

The system can accommodate high gains, such as 4500, since the input command, $r(t)$, is a series of step commands of very small magnitude (a fraction of a micron). The system has an outer loop for position feedback using a laser interferometer with an accuracy of 0.1 micron (10^{-7} m). An inner feedback loop is also used for velocity feedback, as shown in Fig. 12.29.

It is desired to select the controllers, $G_1(s)$ and $G_2(s)$, to obtain an overdamped, highly robust, high-bandwidth system. The robust system must accommodate changes in $G(s)$ due to varying loads, materials, and cutting requirements. Thus we seek large phase margin and gain margin for the inner and outer loops and low root sensitivity. The specifications are summarized in Table 12.8.

Since we desire zero steady-state error for the velocity loop, we use a velocity loop controller $G_2(s) = G_3(s)G_4(s)$, where $G_3(s)$ is a PI controller and $G_4(s)$ is a lead controller. We utilize

$$G_2(s) = G_3(s)G_4(s) = \frac{(1 + K_3 s)}{K_3 s} \cdot \frac{(1 + K_4 s)}{\alpha \left(1 + \dfrac{K_4}{\alpha} s\right)} \cdot K_2$$

and choose $K_3 = 0.00532$, $K_4 = 0.00272$, and $\alpha = 2.95$. We now have

$$G_2(s) = K_2 \frac{(s + 188)}{s} \cdot \frac{(s + 368)}{(s + 1085)}.$$

FIGURE 12.29
Turning machine control system.

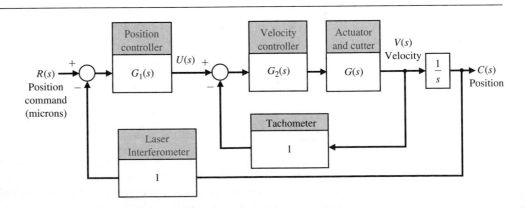

TABLE 12.8   Specifications for Turning Machine Control System

| Specification | Transfer Function | |
	Velocity, $V(s)/U(s)$	Position, $C(s)/R(s)$
Minimum bandwidth	950 rad/s	95 rad/s
Steady-state error to a step	0	0
Minimum damping ratio, ζ	0.8	0.9
Maximum root sensitivity, $|S_K^r|$	1.0	1.5
Minimum phase margin	90°	75°
Minimum gain margin	40 dB	60 dB

The root locus for $G_2(s)G(s)$ is shown in Fig. 12.30. When $K_2 = 2$, we have for the velocity closed-loop transfer function

$$T_2(s) = \frac{V(s)}{U(s)} = \frac{9000(s + 188)(s + 368)}{(s + 205)(s + 305)(s + 10^4)} \approx \frac{10^4}{(s + 10^4)}, \qquad (12.73)$$

which is a large bandwidth system. The actual bandwidth and root sensitivity are summa-

FIGURE 12.30
Root locus for velocity loop as K_2 varies.

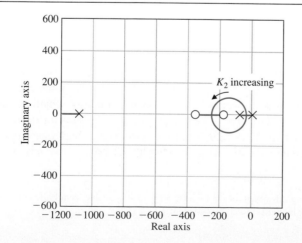

TABLE 12.9   Design Results for Turning Machine Control System

Achieved Result	Velocity Transfer Function	Position Transfer Function, $C(s)/R(s)$
Closed-loop bandwidth	4000 rad/s	1000 rad/s
Steady-state error	0	0
Damping ratio, ζ	1.0	1.0
Root sensitivity, $\lvert S_K^r \rvert$	0.92	1.2
Phase margin	93°	85°
Gain margin	infinite	76 dB

rized in Table 12.9. Note that we have exceeded the specifications for the velocity transfer function.

We will use a lead network for the position loop as

$$G_1(s) = K_1 \frac{(1 + K_5 s)}{\alpha \left(1 + \dfrac{K_5}{\alpha} s\right)},$$

and we choose $\alpha = 2.0$ and $K_5 = 0.0185$ so that

$$G_1(s) = \frac{K_1(s + 54)}{(s + 108)}.$$

We then plot the root locus for

$$G_1(s) \cdot T_2(s) \cdot \frac{1}{s}.$$

If we use the approximate $T_2(s)$ of Eq. (12.73), we have the root locus of Fig. 12.31(a). Using the actual $T_2(s)$, we get the close-up of the root locus shown in Fig. 12.31(b). Using he gain charts discussed in Section 7.7, we examine the root sensitivity and the root locations and select $K_2 = 1000$. Then, we achieve the actual results for the total system transfer function as recorded in Table 12.9. The total system has a high phase margin, a low sensitivity, and is overdamped with a large bandwidth. Clearly, this system is very robust.[†]

[†]Information regarding this design or the use of gain plots is available from Professor T. Kurfess at Carnegie-Mellon University (E-mail: Kurfess +@ CMU.EDU).

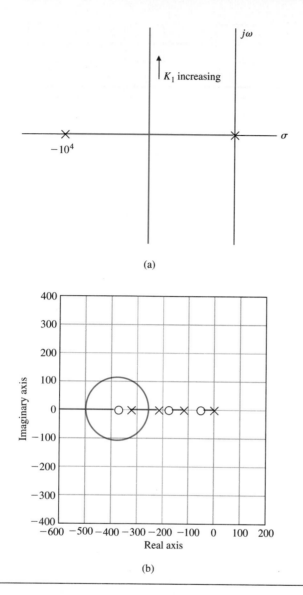

FIGURE 12.31
The root locus for $K_1 > 0$ for (a) overview and (b) close-up near origin of the s-plane.

12.13 THE PSEUDO-QUANTITATIVE FEEDBACK SYSTEM

Quantitative feedback theory (QFT) uses a controller, as shown in Fig. 12.32, to achieve robust performance. The goal is to achieve a wide bandwidth for the closed-loop transfer function with a high loop gain K. Typical QFT design methods use graphical and numerical methods in conjunction with the Nichols chart. Generally, QFT design seeks high loop gain and large phase margin so that robust performance is achieved [27–29, 32].

In this section we pursue a simple method of achieving the goals of QFT with an

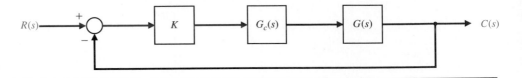

FIGURE 12.32
Feedback system.

s-plane, root locus approach to the selection of the gain K and the compensator $G_c(s)$. This approach, dubbed pseudo QFT, follows these steps:

1. Place the n poles and m zeros of $G(s)$ on the s-plane for the nth order $G(s)$. Also, add any poles of $G_c(s)$.

2. Starting near the origin, place the zeros of $G_c(s)$ immediately to the left of each of the $(n - 1)$ poles on the left-hand s-plane. This leaves one pole far to the left of the left-hand side of the s-plane.

3. Increase the gain K so that the roots of the characteristic equation (poles of the closed-loop transfer function) are close to the zeros of $G_c G(s)$.

This method introduces zeros so that all but one root loci end on finite zeros. If the gain K is sufficiently large, then the poles of $T(s)$ are almost equal to the zeros of $G_c G(s)$. This leaves one pole of $T(s)$ with a significant partial fraction residue and the system with a phase margin of approximately 90° (actually about 85°).

EXAMPLE 12.12　**Design using the pseudo-QFT method**

Consider the system of Fig. 12.32 with

$$G(s) = \frac{1}{(s + p_1)(s + p_2)},$$

where the nominal case is $p_1 = 1$ and $p_2 = 2$, with $\pm 50\%$ variation. The worst case is with $p_1 = 0.5$ and $p_2 = 1$. We wish to design the system for zero steady-state error for a step input, so we use the PID controller

$$G_c(s) = \frac{(s + z_1)(s + z_2)}{s}.$$

We then invoke the internal model principle, with $R(s) = 1/s$ incorporated within $G_c G(s)$. Using step 1, we place the poles of $G_c G(s)$ on the s-plane, as shown in Fig. 12.33. There are three poles, at $s = 0$, -1, and -2, as shown. Step 2 calls for placing a zero to the left of the pole at the origin and at the pole at $s = -1$, as shown in Fig. 12.33.

The compensator is thus

$$G_c(s) = \frac{(s + 0.8)(s + 1.8)}{s}. \tag{12.74}$$

We select $K = 100$, so that the roots of the characteristic equation are close to the zeros. The closed-loop transfer function is

$$T(s) = \frac{100(s + 0.80)(s + 1.80)}{(s + 0.798)(s + 1.797)(s + 100.4)} \approx \frac{100}{(s + 100)}. \tag{12.75}$$

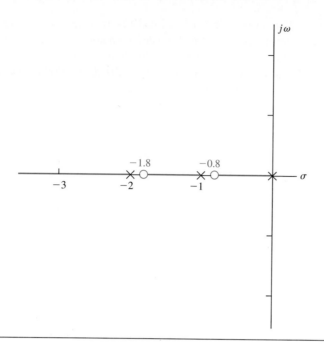

FIGURE 12.33
Root locus for
$KG_cG(s)$.

TABLE 12.10   Performance of Pseudo-QFT Design

	Percent Overshoot	Settling Time
Nominal $G(s)$	0.01%	40 ms
Worst-case $G(s)$	0.97%	40 ms

This closed-loop system provides a fast response and possesses a phase margin of approximately 85°. The performance is summarized in Table 12.10.

When the worst-case conditions are realized ($p_1 = 0.5$ and $p_2 = 1$), the performance remains essentially unchanged, as shown in Table 12.10. Pseudo-QFT design results in very robust systems. ■

12.14 ROBUST CONTROL SYSTEMS USING *MATLAB*

In this section, we will investigate robust control systems using *Matlab*. In particular, we will consider the commonly used PID controller in the feedback control system shown in Fig. 12.1. Notice that the system has a *prefilter* $G_p(s)$. The contribution of the prefilter to optimum performance is discussed in Section 10.10.

The PID controller has the form

$$G_c(s) = \frac{K_3s^2 + K_1s + K_2}{s}.$$

Notice that the PID controller is not a rational function (i.e., the degree of the numerator polynomial is greater than the degree of the denominator polynomial). One will experience difficulty if the PID controller is input into *Matlab* in the standard numerator and denominator fashion. The problem can be resolved by utilizing the conv function rather than the series function in the manipulations.

The objective is to choose the parameters K_1, K_2, and K_3 to meet the performance specifications and have desirable robustness properties. Unfortunately, it is not immediately clear how to choose the parameters in the PID controller to obtain certain robustness characteristics. An illustrative example will show that it is possible to choose the parameters iteratively and verify the robustness by simulation. Using *Matlab* helps in this process, since the entire design and simulation can be mechanized utilizing scripts and can easily be executed repeatedly.

EXAMPLE 12.13   Robust control of temperature

Consider the feedback control system in Fig. 12.1, where

$$G(s) = \frac{1}{(s + c_0)^2},$$

the nominal value is $c_0 = 1$, and $G_p(s) = 1$. We will design a compensator based on $c_0 = 1$ and check robustness by simulation. Our design specifications are as follows:

1. Settling time $T_s \leq 0.5$ second, and
2. Optimum ITAE performance for a step input.

For this design we will not utilize a prefilter to meet specification (2) but will instead show that acceptable performance (i.e., low overshoot) can be obtained by increasing a cascade gain.

The closed-loop transfer function is

$$T(s) = \frac{K_3 s^2 + K_1 s + K_2}{s^3 + (2 + K_3)s^2 + (1 + K_1)s + K_2}. \tag{12.76}$$

The associated root locus equation is

$$1 + \hat{K}\left(\frac{s^2 + as + b}{s^3}\right) = 0,$$

where

$$\hat{K} = K_3 + 2, \qquad a = \frac{1 + K_1}{2 + K_3}, \qquad b = \frac{K_2}{2 + K_3}.$$

The settling time requirement $T_s < 0.5$ second leads us to choose the roots of $(s^2 + as + b)$ to the left of the $s = -\zeta\omega_n = -8$ line in the s-plane, as shown in Fig. 12.34, to ensure that the locus travels into the required s-plane region. We have chosen $a = 16$ and $b = 70$ to ensure the locus travels past the $s = -8$ line. We select a point on the root locus in the performance region, and using the rlocfind function, we find the associated gain $\hat{K}$

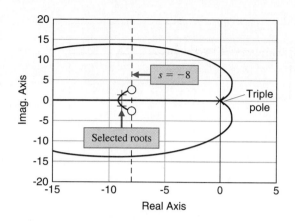

FIGURE 12.34
Root locus for the PID compensated temperature controller as $\hat{K}$ varies.

```
>>a=16; b=70;  num=[1 a b]; den=[1 0 0 0]; rlocus(num,den)
>>rlocfind(num,den)
```

and the associated value of ω_n. For our chosen point, we find that

$$\hat{K} = 118.$$

Then, with $\hat{K}$, a, and b, we can solve for the PID coefficients as follows:

$$K_3 = \hat{K} - 2 = 116,$$

$$K_1 = a(2 + K_3) - 1 = 1187,$$

$$K_2 = b(2 + K_3) = 8260.$$

To meet the overshoot performance requirements for a step input, we will utilize a cascade gain K that will be chosen by iterative methods using the **step** function, as illustrated in Fig. 12.35. The step response corresponding to $K = 5$ has an acceptable overshoot of 2%. With the addition of the gain $K = 5$, the final PID controller is

$$G_c(s) = K\frac{(K_3 s^2 + K_1 s + K_2)}{s} = 5\frac{(116 s^2 + 1187 s + 8260)}{s}. \qquad (12.77)$$

We do not use the prefilter as in Example 12.8. Instead we increase the cascade gain K to obtain satisfactory transient response. Now we can consider the question of robustness to changes in the plant parameter c_0.

The investigation into the robustness of the design consists of a step response analysis using the PID controller given in Eq. (12.77) for a range of plant parameter variations of $0.1 \le c_0 \le 10$. The results are displayed in Fig. 12.36. The script is written to compute the step response for a given c_0. It might be a good idea to place the input of c_0 at the command prompt level to make the script more interactive.

The simulation results indicate that the PID design is robust with respect to changes in c_0. The differences in the step responses for $0.1 \le c_0 \le 10$ are barely discernible on the plot. If the results showed otherwise, it would be possible to iterate on the design until an

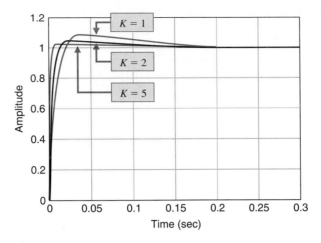

tempstep.m

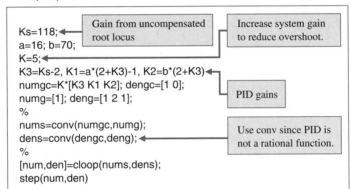

FIGURE 12.35
Step response
of the PID
temperature
controller.

acceptable performance was achieved. The interactive capability of *Matlab* allows us to check the robustness by simulation. ∎

12.15 SUMMARY

The design of highly accurate control systems in the presence of significant plant uncertainty requires the designer to seek a robust control system. A robust control system exhibits low sensitivities to parameter change and is stable over a wide range of parameter variations.

The three-mode, or PID, controller was considered as a compensator to aid in the design of robust control systems. The design issue for a PID controller is the selection of the gain and two zeros of the controller transfer function. We utilized three design methods for the selection of the controller: the root locus method, the frequency response method, and the ITAE performance index method. An operational amplifier circuit used for a PID controller is shown in Fig. 12.37. In general, the use of a PID controller will enable the designer to attain a robust control system.

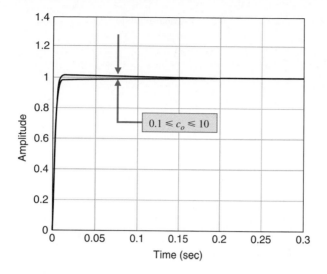

FIGURE 12.36
Robust PID
controller analysis
with variations
in c_0.

robust PID.m

```
c0=10   ◄——————  Specify plant parameter.
numg=[1]; deng=[1 2*c0 c0^2];
numgc=5*[116 1187 8260]; dengc=[1 0];
%
numa=conv(numg,numgc); dena=conv(deng,dengc);
%
[num,den]=cloop(numa,dena);
%
step(num,den)
```

$$G_c(s) = \frac{V_0(s)}{V_1(s)} = \frac{R_4 R_2 (R_1 C_1 s + 1)(R_2 C_2 s + 1)}{R_3 R_1 (R_2 C_2 s)}$$

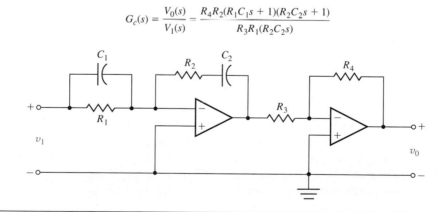

FIGURE 12.37
Operational
amplifier circuit
used for PID
controller.

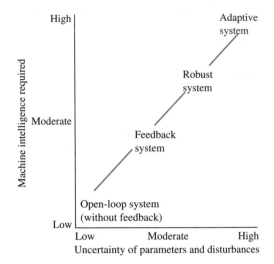

FIGURE 12.38
Intelligence required versus uncertainty for modern control systems.

The internal model control system with state variable feedback and a controller $G_c(s)$ was utilized to obtain a robust control system. Finally, the robust nature of a pseudo-QFT control system was demonstrated.

A robust control system provides stable, consistent performance as specified by the designer in spite of wide variation of plant parameters and disturbances. It also provides highly robust response to command inputs and a steady-state tracking error equal to zero.

For systems with uncertain parameters, the need for robust systems will require the incorporation of advanced machine intelligence, as shown in Fig. 12.38.

EXERCISES

E12.1   Consider a system of the form shown in Fig. 12.16 where

$$G(s) = \frac{1}{(s + 1)}.$$

Using the ITAE performance method for a step input, determine the required $G_c(s)$. Assume $\omega_n = 20$ for Table 5.7. Determine the step response with and without a prefilter $G_p(s)$.

E12.2   For the ITAE design obtained in Exercise 12.1, determine the response due to a disturbance $D(s) = 1/s$.

E12.3   A closed-loop unity feedback system has

$$G(s) = \frac{9}{s(s + p)},$$

where p is normally equal to 3. Determine S_p^T and plot $|T(j\omega)|$ and $|S(j\omega)|$ on a Bode plot.

E12.4   A PID controller is used in the system in Fig. 12.16 where

$$G(s) = \frac{1}{(s + 2)(s + 8)}.$$

The gain K_3 of the controller (Eq. 12.37) is limited

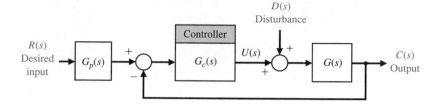

FIGURE E12.6
System with
controller.

to 180. Select a set of compensator zeros so that the pair of closed-loop roots is approximately equal to the zeros. Find the step response for the approximation in Eq. (12.39) and the actual response and compare them.

E12.5   A system has a plant

$$G(s) = \frac{15,900}{s\left(\dfrac{s}{100} + 1\right)\left(\dfrac{s}{200} + 1\right)}$$

and negative unity feedback with a PD compensator

$$G_c(s) = K_1 + K_2 s.$$

The objective is to design $G_c(s)$ so that the overshoot to a step is less than 20% and the settling time is less than 60 ms. Find a suitable $G_c(s)$.

E12.6   Consider the control system shown in Fig. E12.6 when $G(s) = 1/(s + 4)$, and select a PI controller so that the settling time is less than 1 second for an ITAE step response. Plot $c(t)$ for a step input $r(t)$ with and without a prefilter. Determine and plot $c(t)$ for a step disturbance. Discuss the effectiveness of the system.

E12.7   For the control system of Fig. E12.6 with $G(s) = 1/(s + 4)^2$, select a PID controller to achieve a settling time of less than 1.0 second for an ITAE

step response. Plot $c(t)$ for a step input $r(t)$ with and without a prefilter. Determine and plot $c(t)$ for a step disturbance. Discuss the effectiveness of the system.

E12.8   Repeat Exercise 12.6 while striving to achieve a minimum settling time while adding the constraint that $|u(t)| \le 80$ for $t > 0$ for a unit step input, $r(t) = 1, t \ge 0$.

E12.9   A system has the form shown in Fig. E12.6 with

$$G(s) = \frac{K}{s(s + 1)(s + 4)},$$

where $K = 1$. Design a PD controller to place the dominant closed-loop poles at $s = -1.5 \pm j2$. Determine the step response of the system. Predict the effect of a change in K of $\pm 50\%$. Estimate the step response of the worst-case system.

E12.10   A system has the form shown in Fig. E12.6 with

$$G(s) = \frac{K}{s(s + 1)(s + 4)},$$

where $K = 1$. Design a PI controller so that the dominant roots are at $s = -0.365 \pm j0.514$. Determine the step response of the system. Predict the effect of a change in K of $\pm 50\%$. Estimate the step response of the worst-case system.

PROBLEMS

P12.1   Interest in unmanned underwater vehicles (UUVs) has been increasing recently, with a large number of possible applications being considered. These include intelligence-gathering, mine detection, and surveillance. Regardless of the intended mission, a strong need exists for reliable and robust

control of the vehicle. The proposed vehicle is shown in Fig. P12.1(a) [13].

It is desired to control the vehicle through a range of operating conditions. The vehicle is 30 feet long with a vertical sail near the front. The control inputs are stern-plane, rudder, and shaft

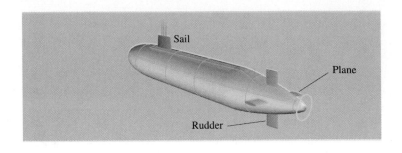

(a)

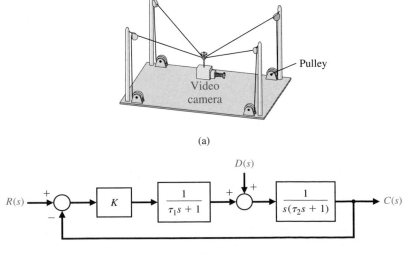

(b)

FIGURE P12.1
Control of an
underwater
vehicle.

speed commands. In this case, we wish to control the vehicle roll by using the stern planes. The control system is shown in Fig. P12.1(b), where $R(s) = 0$, the desired roll angle, and $D(s) = 1/s$. We select $G_c(s) = K(s + 1)$, where $K = 2$.

(a) Plot $20 \log|T|$ and $20 \log|S_K^T|$ on a Bode diagram. (b) Evaluate $|S_K^T|$ at ω_B, $\omega_{B/2}$, and $\omega_{B/4}$ ($T(s) = C(s)/R(s)$).

P12.2   A new suspended, mobile, remote-controlled video-camera system to bring three-dimensional mobility to professional NFL football is shown in Fig. P12.2(a) [24]. The camera can be moved over the field as well as up and down. The motor control on each pulley is represented by the system in Fig. P12.2(b), where $\tau_1 = 20$ ms and $\tau_2 = 2$ ms.

(a) Select K so that $M_{p\omega} = 1.84$. (b) Plot $20 \log|T|$ and $20 \log|S_K^T|$ on one Bode diagram. (c) Evaluate $|S_K^T|$ at ω_B, $\omega_{B/2}$, and $\omega_{B/4}$. (d) Let $R(s) = 0$ and determine the effect of $D(s) = 1/s$ for the gain K of part (a) by plotting $c(t)$.

FIGURE P12.2
Remote-controlled
TV camera.

(a)

(b)

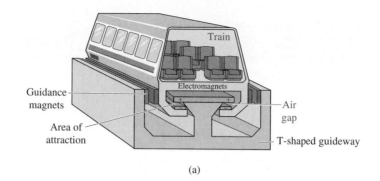

(a)

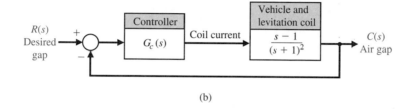

FIGURE P12.3
Maglev train
control.

(b)

P12.3 Magnetic levitation (maglev) trains may replace airplanes on routes shorter than 200 miles. The maglev train developed by a German firm uses electromagnetic attraction to propel and levitate heavy vehicles, carrying up to 400 passengers at 300-mph speeds. But the ¼-inch gap between car and track is hard to maintain [12, 17].

The air-gap control system is shown in Fig. P12.3(a). The block diagram of the air-gap control system is shown in Fig. P12.3(b). The compensator is

$$G_c(s) = \frac{K(s + 1)}{(s + 12)}.$$

(a) Find the range of K for a stable system. (b) Select a gain so that the steady-state error of the system is zero for a step input command. (c) Find $c(t)$ for the gain of part (b). (d) Find $c(t)$ when K varies ±10% from the gain of part (b).

P12.4 Computer control of a robot to spray-paint an automobile is shown by the system shown in Fig. P12.4(a) [1]. We wish to investigate the system when $K = 2$, 4, and 6. (a) For the three values of K, determine ζ, ω_n, percent overshoot, settling time, and steady-state error for a step input. Record your results in a table. (b) Determine the sensitivity $|S_K^r|$ for the three values of K. (c) Select the best of the three values of K. (d) For the value selected in

part (c), determine $c(t)$ for a disturbance $D(s) = 1/s$ when $R(s) = 0$.

P12.5 An automatically guided vehicle is shown in Fig. P12.5(a) and its control system is shown in Fig. P12.5(b). The goal is to track the guide wire accurately, to be insensitive to changes in the gain K_1, and to reduce the effect of the disturbance [15, 25]. The gain K_1 is normally equal to 1 and $\tau_1 = \frac{1}{25}$ second.

(a) Select a compensator, $G_c(s)$, so that the percent overshoot to a step input is less than or equal to 10%, the settling time is less than 100 ms, and the velocity constant, K_v, for a ramp input is 100.
(b) For the compensator selected in part (a), determine the sensitivity of the system to small changes in K_1 by determining $S_{K_1}^r$ or $S_{K_1}^T$.
(c) If K_1 changes to 2 while $G_c(s)$ of part (a) remains unchanged, find the step response of the system and compare selected performance figures with those obtained in part (a).
(d) Determine the effect of $D(s) = 1/s$ by plotting $c(t)$ when $R(s) = 0$.

P12.6 A roll-wrapping machine (RWM) receives, wraps, and labels large paper rolls produced in a paper mill [9, 16]. The RWM consists of several major stations: positioning station, waiting station, wrapping station, and so forth. We will focus on

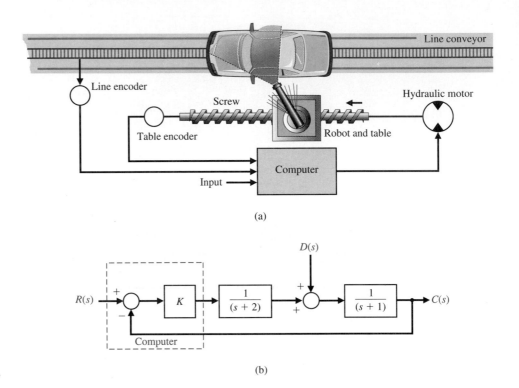

(a)

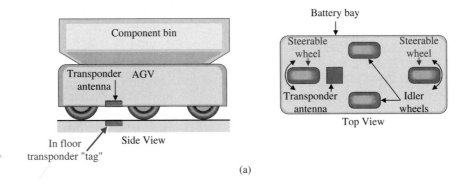

(b)

FIGURE P12.4
Spray paint robot.

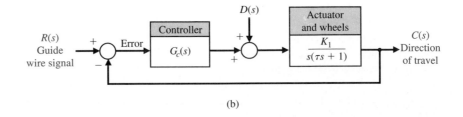

(a)

FIGURE P12.5
Automatically
guided vehicle.

(b)

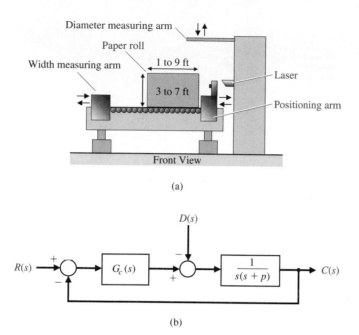

(a)

(b)

FIGURE P12.6
Roll-wrapping
machine control.

the positioning station shown in Fig. P12.6(a). The positioning station is the first station that sees a paper roll. This station is responsible for receiving and weighing the roll, measuring its diameter and width, determining the desired wrap for the roll, positioning it for down-stream processing, and finally ejecting it from the station.

Functionally, the RWM can be categorized as a complex operation because each functional step (e.g., measuring the width) involves a large number of field device actions and relies upon a number of accompanying sensors.

The control system for accurately positioning the width-measuring arm is shown in Fig. P12.6(b). The pole of the positioning arm, p, is normally equal to 2, but it is subject to change because of loading and misalignment of the machine. (a) For p = 2, design a compensator so that the complex roots are $s = -2 \pm j2\sqrt{3}$. (b) Plot $c(t)$ for a step input $R(s) = 1/s$. (c) Plot $c(t)$ for a disturbance $D(s) = 1/s$, with $R(s) = 0$. (d) Repeat parts (b) and (c) when p changes to 1 and $G_c(s)$ remains as designed in part (a). Compare the results for the two values of the pole p.

P12.7 The function of a steel plate mill is to roll reheated slabs into plates of scheduled thickness and dimension [5, 10]. The final products are of rectangular plane view shapes having a width of up to 3300 mm and a thickness of 180 mm.

A schematic layout of the mill is shown in Fig. P12.7(a). The mill has two major rolling stands, denoted No. 1 and No. 2. These are equipped with large rolls (up to 508 mm in diameter), which are driven by high-power electric motors (up to 4470 kW). Roll gaps and forces are maintained by large hydraulic cylinders.

Typical operation of the mill can be described as follows. Slabs coming from the reheating furnace initially go through the No. 1 stand, whose function is to reduce the slabs to the required width. The slabs proceed through the No. 2 stand, where finishing passes are carried out to produce required slab thickness, and finally through the hot plate leveller, which gives each plate a smooth finishing surface.

One of the key systems controls the thickness of the plates by adjusting the rolls. The block diagram of this control system is shown in Fig. P12.7(b). The plant is represented by

$$G(s) = \frac{1}{s(s^2 + 4s + 5)}.$$

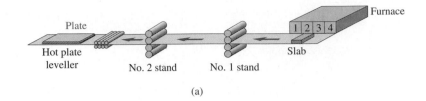

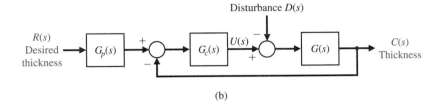

FIGURE P12.7
Steel rolling mill
control.

(a)

(b)

The controller is a PID with two equal real zeros.
(a) Select the PID zeros and the gains so that the
real parts of the four closed-loop roots are equal.
(b) For the design of part (a), obtain the step response without a prefilter [$G_p(s) = 1$]. (c) Repeat
part (b) for an appropriate prefilter. (d) For the system, determine the effect of a unit step disturbance
by evaluating $c(t)$ with $r(t) = 0$.

P12.8   A motor and load with negligible friction and a
voltage-to-current amplifier, K_a, is used in a feedback control system, as shown in Fig. P12.8. A designer selects a PID controller

$$G_c(s) = K_1 + \frac{K_2}{s} + K_3 s,$$

where $K_1 = 5$, $K_2 = 500$, and $K_3 = 0.0475$.
   (a) Determine the appropriate value of K_a so
that the phase margin of the system is 42°. (b) For
the gain K_a, plot the root locus of the system and
determine the roots of the system for the K_a of part
(a). (c) Determine the maximum value of $c(t)$
when $D(s) = 1/s$ and $R(s) = 0$ for the K_a of part
(a). (d) Determine the response to a step input, $r(t)$,
with and without a prefilter.

P12.9   A unity feedback system has a nominal characteristic equation

$$q(s) = s^3 + 2.5s^2 + 2s + 4 = 0.$$

The coefficients vary as follows:

$$2 \le a_2 \le 3 \qquad 1 \le a_1 \le 2.8$$
$$2.5 \le a_0 \le 4.5.$$

Determine whether the system is stable for these
uncertain coefficients.

P12.10   Future astronauts may drive on the moon in a
pressurized vehicle, shown in Fig. P12.10(a), that
would have a range of 620 miles and could be used
for missions of up to six months. Boeing Company
engineers first analyzed the Apollo-era Lunar Roving Vehicle, then designed the new vehicle, incorporating improvements in radiation and thermal
protection, shock and vibration control, and lubri-
cation and sealants. NASA plans to begin construction on a smaller, nonpressurized vehicle in 1999.
   The steering control of the moon buggy is
shown in Fig. P12.10(b). The objective of the
control design is to achieve a step response to a
steering command with zero steady-state error, an

FIGURE P12.8

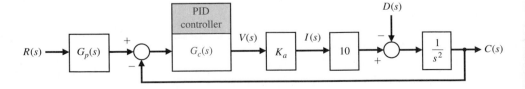

(a)

FIGURE P12.10
(a) A moon
vehicle.
(b) Steering control
for the moon
vehicle.

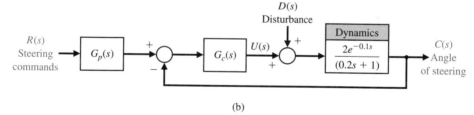

$R(s)$
Steering
commands

$G_p(s)$

$D(s)$
Disturbance

$G_c(s)$

$U(s)$

Dynamics
$$\frac{2e^{-0.1s}}{(0.2s + 1)}$$

$C(s)$
Angle
of steering

(b)

overshoot less than 20%, and a peak time less than 0.3 second with a $|u(t)| \leq 50$. It is also necessary to determine the effect of a step disturbance $D(s) = 1/s$ when $R(s) = 0$, to ensure the reduction of moon surface effects. Using (a) a PI and (b) a PID controller, design an acceptable controller and record in a table the results for each design. Compare the performance of each design. Use a prefilter, $G_p(s)$, if necessary.

P12.11 A plant has a transfer function

$$G(s) = \frac{1}{s^2}.$$

It is desired to use a negative unity feedback with a

PID controller and a prefilter. The goal is to achieve a peak time of 0.4 second with ITAE-type performance. Predict the system overshoot and settling time for a step input.

P12.12 A three-dimensional cam for generating a function of two variables is shown in Fig. P12.12(a). Both x and θ may be controlled using a position control system [19]. The control of x may be achieved with a dc motor and position feedback of the form shown in Fig. P12.12(b), with the dc motor and load represented by

$$G(s) = \frac{K}{s(s + p)(s + 4)},$$

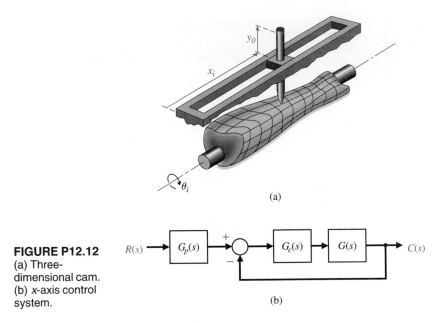

(a)

FIGURE P12.12
(a) Three-dimensional cam.
(b) *x*-axis control system.

(b)

where $1 \leq K \leq 3$ and $1 \leq p \leq 3$. Normally, $K = 2$ and $p = 2$. Design an ITAE system with a PID controller so that the peak time response to a step input is less than 2.5 seconds for the worst-case performance.

ADVANCED PROBLEMS

AP12.1 To minimize vibrational effects, a telescope is magnetically levitated. This method also eliminates friction in the azimuth magnetic drive system. The photodetectors for the sensing system require electrical connections, which can be represented by a spring with a constant of 1 kg/m. The mass of the telescope is 100 kg. The system block diagram is shown in Fig. AP12.1. Design a PID controller so that the velocity error constant is $K_v = 100$ and the maximum overshoot for a step input is less than 5%.

AP12.2 One promising solution to traffic gridlock is a magnetic levitation (maglev) system. Vehicles are suspended on a guideway above the highway and guided by magnetic forces instead of relying on wheels or aerodynamic forces. Magnets provide the propulsion for the vehicles [7, 12, 17]. Ideally, maglev can offer the environmental and safety advantages of a high-speed train, the speed and low friction of an airplane, and the convenience of an automobile. All these shared attributes notwithstanding, the maglev system is truly a new mode of travel and will enhance the other modes of travel by relieving congestion and providing connections among them. Maglev travel would be fast, operating at 150 to 300 miles per hour.

The tilt control of a maglev vehicle is illustrated in Figs. AP12.2(a) and (b). The dynamics of

FIGURE AP12.1
Magnetically levitated telescope position control system.

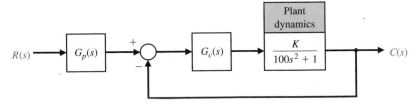

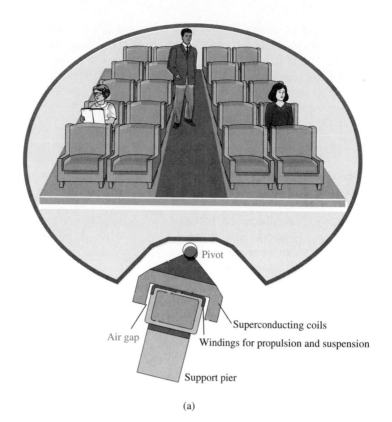

(a)

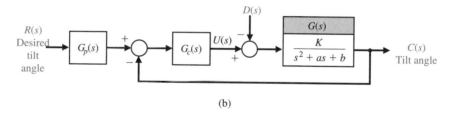

(b)

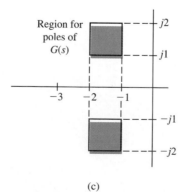

FIGURE AP12.2
(a) and (b) Tilt
control for a
maglev vehicle.
(c) Plant dynamics.

(c)

the plant $G(s)$ are subject to variation so that the poles will lie within the cross-hatched box shown in Fig. AP12.2(c) and $1 \le K \le 2$.

The objective is to achieve a robust system with a step response possessing an overshoot less than 10%, and a settling time less than 2 seconds when $|u(t)| \le 100$. Obtain a design with a PI, PD, and PID controller and compare the results. Use a prefilter, $G_p(s)$, if necessary.

AP12.3   Antiskid braking systems present a challenging control problem, since brake/automotive system parameter variations can vary significantly (e.g., due to brake-pad coefficient of friction changes or road slope variations) and environmental influences (e.g., due to adverse road conditions). The objective of the antiskid system is to regulate wheel slip to maximize the coefficient of friction between the tire and road for any given road surface [8]. As one would expect, the braking coefficient of friction is greatest for dry asphalt, slightly reduced for wet asphalt, and greatly reduced for ice.

One simplified model of the braking system is represented by a plant transfer function $G(s)$ with a system as shown in Fig. 12.16 with

$$G(s) = \frac{C(s)}{U(s)} = \frac{1}{(s + a)(s + b)},$$

where normally $a = 1$ and $b = 4$.

(a) Using a PID controller, design a very robust system where for a step input the overshoot is less

than 4% and the settling time is 1 second or less. The steady-state error must be less than 1% for a step. We expect a and b to vary by $\pm 50\%$.
(b) Design a system to yield the specifications of part (a) using an ITAE performance index. Predict the overshoot and settling time for your design.

AP12.4   A robot has been designed to aid in hip-replacement surgery. The device, called RoBoDoc, is used to orient and mill precisely the femoral cavity for acceptance of the prosthetic hip implant. Clearly, we want a very robust surgical tool control, since there is no second chance to redrill a bone [23, 30]. The control system will be as shown in Fig. 12.16 with

$$G(s) = \frac{b}{s^2 + as + b},$$

where $2 \le a \le 4$ and $5 \le b \le 9$.

Select a PID controller so that the system is very robust. Use the s-plane root locus method. Select the appropriate $G_p(s)$ and plot the response to a step input.

AP12.5   A spacecraft with a camera is shown in Fig. AP12.5. The camera slews about 16° in a canted plane relative to the base. Reaction jets stabilize the base against the reaction torques from the slewing motors. The rotational speed control for the camera slewing has a plant transfer function

$$G(s) = \frac{1}{(s + 1)(s + 2)(s + 4)}.$$

FIGURE AP12.5
Spacecraft with a camera.

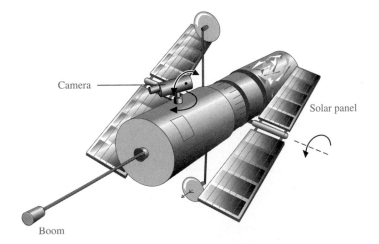

Camera

Solar panel

Boom

A PID controller is used in a system as shown in Fig. 12.16 where

$$G_c(s) = \frac{K(s + \sigma + j1)(s + \sigma - j1)}{s}.$$

Determine K and the resulting roots so that all the roots have ζ less than or equal to $1/\sqrt{2}$ when $1 < \sigma < 2$ (vary σ in increments of 0.25). Determine the root sensitivity to changes in K.

AP12.6   Consider the system of Fig. 12.16 with

$$G(s) = \frac{K_1}{s(s + 10)},$$

where $K_1 = 1$ under normal conditions. Design a PID controller to achieve a phase margin of 50°. The controller is

$$G_c = \frac{K(s^2 + 20s + b)}{s}$$

with complex zeros. Select an appropriate prefilter. Determine the effect of a change of $\pm 25\%$ in K_1 by developing a tabular record of the system performance.

AP12.7   Consider the system of Fig. 12.16 with

$$G(s) = \frac{K_1}{s(\tau s + 1)},$$

where $K_1 = 1$ and $\tau \approx 0.001$ second, which may be neglected. (Check this later in the design process.) Select a PID controller so that the settling time for a step input is less than 0.8 second and the

overshoot is less than 5%. Also, the effect of a disturbance at the output must be reduced to less than 5% of the magnitude of the disturbance. Select ω_n, and use the ITAE design method.

AP12.8   Consider the system of Fig. 12.16 with

$$G(s) = \frac{1}{s}.$$

The goal is to select a PI controller using the ITAE design criterion while constraining the control signal as $|u(t)| \le 1$ for a unit step input. Determine the appropriate PI controller and the settling time for a step input. Use a prefilter.

AP12.9   Consider the system of Fig. 12.16 with

$$G(s) = \frac{3}{s(s^2 + 4s + 5)}.$$

Design a PID controller to achieve (a) an acceleration constant $K_a = 2$, (b) a phase margin equal to 45°, and (c) a bandwidth greater than 2.8 rad/s. Select an appropriate prefilter and plot the response to a step input.

AP12.10   A machine tool control system is shown in Fig. AP12.10. The transfer function of the power amplifier, prime mover, moving carriage, and tool bit is

$$G(s) = \frac{50}{s(s + 1)(s + 4)(s + 5)}.$$

The goal is to have an overshoot for a step input less than 25% while achieving a peak time less than

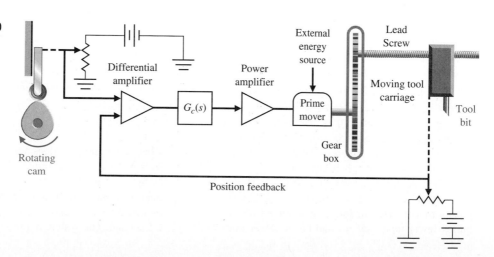

FIGURE AP12.10
A machine tool control system.

3 seconds. Determine a suitable controller using (a) PD control, (b) PI control, and (c) PID control. (d) Then select the best controller.

AP12.11   Consider a system with the structure shown in Fig. 12.16 with

$$G(s) = \frac{K}{s^2 + 2as + a^2},$$

where $1 \le a \le 3$ and $2 \le K \le 4$.

Use a PID controller and design the controller for the worst-case condition. It is desired that the settling time be less than 0.8 second with an ITAE performance.

AP12.12   A system of the form shown in Fig. 12.16 has

$$G(s) = \frac{s + r}{(s + p)(s + q)},$$

where $3 \le p \le 5$, $0 \le q \le 1$, and $1 \le r \le 2$. A compensator is used,

$$G_c(s) = \frac{K(s + z_1)(s + z_2)}{(s + p_1)(s + p_2)},$$

with all real poles and zeros. Select an appropriate compensator to achieve robust performance.

AP12.13   A system of the form shown in Fig. 12.32 has a plant

$$G(s) = \frac{1}{(s + 2)(s + 4)(s + 6)}.$$

It is desired to attain a steady-state error for a step input. Select a compensator $G_c(s)$ and gain K using the pseudo-QFT method, and determine the performance of the system when all the poles of $G(s)$ change by -50%. Describe the robust nature of the system.

DESIGN PROBLEMS

DP12.1   A position control system for a large turntable is shown in Fig. DP12.1(a) and the block diagram of the system is shown in Fig. DP12.1(b) [11, 14]. This system uses a large torque motor with $K_m = 15$. The objective is to reduce the steady-state effect of a step change in the load disturbance to 5% of the magnitude of the step disturbance while maintaining a fast response to a step input command, $R(s)$, with less than 5% overshoot. Select K_1 and the compensator when (a) $G_c(s) = K$ and (b) $G_c(s) = K_2 + K_3 s$ (a PD compensator). Plot the step response for the disturbance and the input for both compensators. Determine whether a prefilter is required to meet the overshoot requirement.

DP12.2   Digital audio tape (DAT) stores 1.3 gigabytes of data in a package the size of a credit card—roughly nine times more than half-inch-wide reel-to-reel tape or quarter-inch-wide cartridge tape. A DAT sells for a few dollars—about the same as a floppy disk even though it can store 1000 times more data. A DAT can record for two hours (longer than either reel-to-reel or cartridge tape), which means that it can run longer unattended and requires fewer tape changes and hence fewer inter-

ruptions of data transfer. DAT gives access to a given data file within 20 seconds, on the average, compared with a few to several minutes for either cartridge or reel-to-reel tape [2].

The tape drive electronically controls the relative speeds of the drum and tape so that the heads follow the tracks on the tape, as shown in Fig. DP12.2(a). The control system is much more complex than that for a CD ROM because more motors have to be accurately controlled: capstan, take-up and supply reels, drum, and tension control.

Let us consider the speed control system shown in Fig. DP12.2(b). The motor and load transfer function varies because the tape moves from one reel to the other. The transfer function is

$$G(s) = \frac{K_m}{(s + p_1)(s + p_2)},$$

where the nominal values are $K_m = 4$, $p_1 = 1$, and $p_2 = 4$. However, the range of variation is $3 \le K_m \le 5$; $0.5 \le p_1 \le 1.5$; $3.5 \le p_2 \le 4.5$. The system must respond rapidly with an overshoot to a step of less than 13% and a settling time of less than 0.5 second. The system requires a fast peak time so

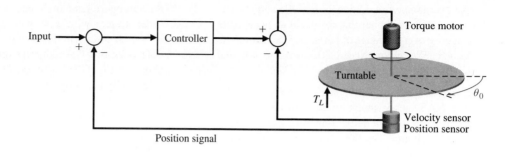

(a)

FIGURE DP12.1
Turntable control.

(b)

FIGURE DP12.2
Control of a DAT
player.

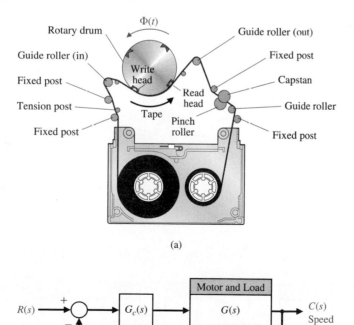

(a)

(b)

the overdamped condition is not allowed. Determine a PID controller that results in an acceptable response over the range of parameter variation. The gain $K_m K_3$ cannot exceed 20 when the nominal $K_m = 4$ and K_3 is defined in Eq. (12.37).

DP12.3 The Gamma-Ray Imaging Device (GRID) is a NASA experiment to be flown on a long-duration, high-altitude balloon during the coming solar maximum. The GRID on a balloon is an instrument that will qualitatively improve hard X-ray imaging and carry out the first gamma-ray imaging for the study of solar high-energy phenomena in the next phase of peak solar activity. From its long-duration balloon platform, GRID will observe numerous hard X-ray bursts, coronal hard X-ray sources, "superhot" thermal events, and microflares [2]. Figure DP12.3(a) depicts the GRID payload attached to the balloon. The major components of the GRID experiment consist of a 5.2-meter canister and mounting gondola, a high-altitude balloon, and a cable connecting the gondola and balloon. The instrument-sun pointing requirements of the experiment are 0.1 degree pointing accuracy and 0.2 arc second per 4 ms pointing stability.

An optical sun sensor provides a measure of

the sun-instrument angle and is modeled as a first-order system with a dc gain and a pole at $s = -500$. A torque motor actuates the canister/gondola assembly. The azimuth angle control system is shown in Fig. DP12.3(b). (a) The PID controller is selected by the design team so that

$$G_c(s) = \frac{K_3(s^2 + as + b)}{s},$$

where $a = 6$, and b is to be selected. A prefilter is used as shown in Fig. DP12.3(b). Determine the value of K_3 and b so that the dominant roots have a ζ of 0.8 and the overshoot to a step input is less than 3%. Determine the actual step response percent overshoot, t_p, and T_s. (b) Design a PID filter using the ITAE performance criteria when $\omega_n = 8$. Compare the actual results with those obtained in part (a).

DP12.4 Many university and government laboratories have constructed robot hands capable of grasping and manipulating objects. But teaching the artificial devices to perform even simple tasks required formidable computer programming. Now, however, the Dexterous Hand Master (DAM) can be worn over a human hand to record the side-to-side and bending motions of finger joints. Each joint is

FIGURE DP12.3
The GRID device.

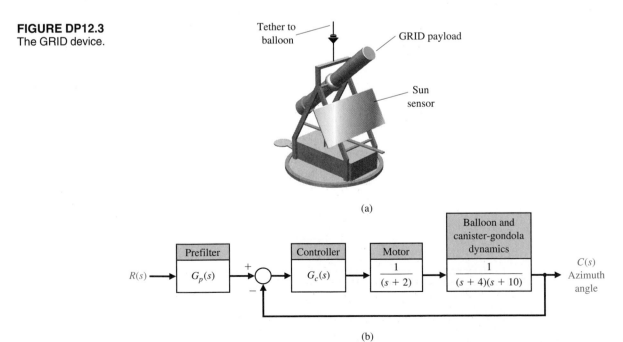

(a)

(b)

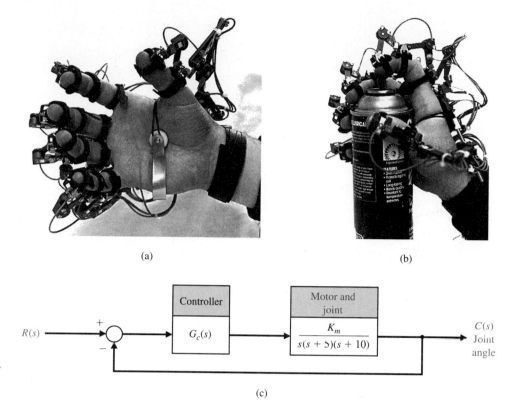

(a)

(b)

FIGURE DP12.4
Dexterous Hand
Master.

(c)

$R(s)$ → + − → Controller $G_c(s)$ → Motor and joint $\dfrac{K_m}{s(s+5)(s+10)}$ → $C(s)$ Joint angle

fitted with a sensor that changes its signal depending on position. The signals from all the sensors are translated into computer data and used to operate robot hands [1].

The DAM is shown in parts (a) and (b) of Fig. DP12.4. The joint angle control system is shown in part (c). The normal value of K_m is 1.0. The goal is to design a PID controller so that the steady-state error for a ramp input is zero. Also, the settling time must be less than 3 seconds for the ramp input. It is desired that the controller be

$$G_c(s) = \frac{K_3(s^2 + 6s + 18)}{s}.$$

(a) Select K_3 and obtain the ramp response. Plot the root locus as K_3 varies. (b) If K_m changes to one-half of its normal value, and $G_c(s)$ remains as designed in part (a), obtain the ramp response of the system. Compare the results of parts (a) and (b) and discuss the robustness of the system.

DP12.5 Objects smaller than the wavelengths of visible light are a staple of contemporary science

and technology. Biologists study single molecules of protein or DNA; materials scientists examine atomic-scale flaws in crystals; microelectronics engineers lay out circuit patterns only a few tenths of atoms thick. Until recently, this minute world could be seen only by cumbersome, often destructive methods such as electron microscopy and X-ray diffraction. It lay beyond the reach of any instrument as simple and direct as the familiar light microscope. New microscopes, typified by the scanning tunneling microscope (STM), are now available [3].

The precision of position control required is in the order of nanometers. The STM relies on piezoelectric sensors that change size when an electric voltage across the material is changed. The "aperture" in the STM is a tiny tungsten probe, its tip ground so fine that it may consist of only a single atom and measure just 0.2 nanometer in width. Piezoelectric controls maneuver the tip to within a nanometer or two of the surface of a conducting specimen—so close that the electron

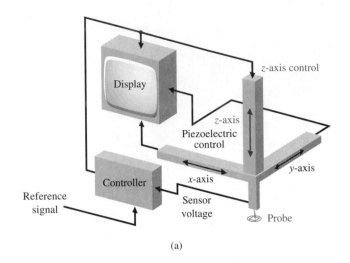

(a)

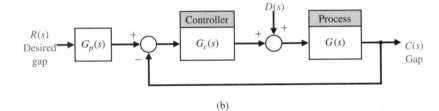

FIGURE DP12.5
Microscope
control.

(b)

clouds of the atom at the probe tip and of the nearest atom of the specimen overlap. A feedback mechanism senses the variations in tunneling current and varies the voltage applied to a third, z-axis, control. The z-axis piezoelectric moves the probe vertically to stabilize the current and to maintain a constant gap between the microscope's tip and the surface. The control system is shown in Fig. DP12.5(a), and the block diagram is shown in Fig. DP12.5(b). The process is

$$G(s) = \frac{17{,}640}{s(s^2 + 59.4s + 1764)},$$

and the controller is chosen to have two real, unequal zeros so that (Eq. 12.40) we have

$$G_c(s) = \frac{K_2(\tau_1 s + 1)(\tau_2 s + 1)}{s}.$$

(a) Use the ITAE design method to determine $G_c(s)$. (b) Determine the step response of the system with and without a prefilter $G_p(s)$. (c) Determine the response of the system to a disturbance when $D(s) = 1/s$. (d) Using the prefilter and $G_c(s)$

of parts (a) and (b), determine the actual response when the process changes to

$$G(s) = \frac{16{,}000}{s(s^2 + 40s + 1600)}.$$

DP12.6 The system described in DP12.5 is to be designed using the frequency response techniques described in Section 12.6 with

$$G_c(s) = \frac{K_2(\tau_1 s + 1)(\tau_2 s + 1)}{s}.$$

Select the coefficients of $G_c(s)$ so that the phase margin is approximately 70°. Obtain the step response of the system with and without a prefilter, $G_p(s)$.

DP12.7 The use of control theory to provide insight into neurophysiology has a long history. As early as the beginning of the century, many investigators described a muscle control phenomenon caused by the feedback action of muscle spindles and by sensors based on a combination of muscle length and rate of change of muscle length.

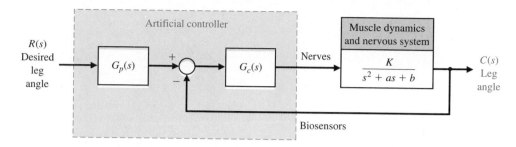

FIGURE DP12.7
Artificial control of standing and leg articulation.

This analysis of muscle regulation has been based on the theory of single-input, single-output control systems. An example is a proposal that the stretch reflex is an experimental observation of a motor control strategy, namely, control of individual muscle length by the spindles. Others later proposed the regulation of individual muscle stiffness (by sensors of both length and force) as the motor control strategy [34].

One model of the human standing-balance mechanism is shown in Fig. DP12.7. Let us consider the case of a paraplegic who has lost control of his standing mechanism. We propose to add an artificial controller to enable the person to stand and articulate his legs. (a) Design a controller when the normal values of the parameters are $K = 10$, $a = 12$, and $b = 100$, in order to achieve a step response with percent overshoot less than 10%, steady-state error less than 5%, and a settling time less than 2 seconds. Try a controller with proportional gain, PI, PD, and PID. (b) When the person is fatigued, the parameters may change to $K = 15$, $a = 8$, and $b = 144$. Examine the performance of this system with the controllers of part (a). Prepare a table contrasting the results of parts (a) and (b).

DP12.8 The goal is to design an elevator control system so that the elevator will move from floor to floor rapidly and stop accurately at the selected floor (Fig. DP12.8). The elevator will contain from one to three occupants. However the weight of the elevator should be greater than the weight of the occupants, and you may assume that the elevator weighs 1000 pounds and each occupant weighs 150 pounds. Design a system to accurately control the elevator to within one centimeter. Assume that the large dc motor is field controlled. Also assume that the time constant of the motor and load is one second, the time constant of the power amplifier driving the motor is one-half second, and the time

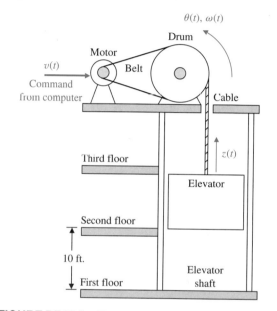

FIGURE DP12.8 Elevator position control.

constant of the field is negligible. We seek an overshoot less than 6% and a settling time less than 4 seconds.

DP12.9 Patients with a cardiological illness and less than normal heart muscle strength can benefit from an assistance device. An electric ventricular assist device (EVAD) converts electric power into blood flow by moving a pusher plate against a flexible blood sac. The pusher plate reciprocates to eject blood in systole and to allow the sac to fill in diastole. The EVAD will be implanted in tandem or in parallel with the intact natural heart as shown in Fig. DP12.9(a). The EVAD is driven by rechargeable batteries, and the electric power is transmitted inductively across the skin through a transmission system. The batteries and the transmission system limit the electric energy storage and the transmitted

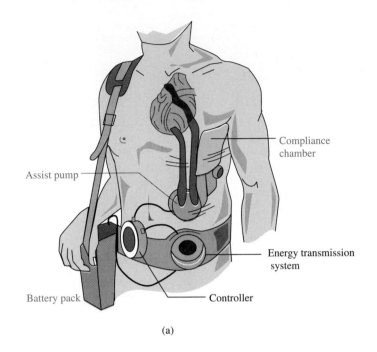

(a)

FIGURE DP12.9
(a) An electric ventricular assist device for cardiology patients.
(b) Feedback control system.

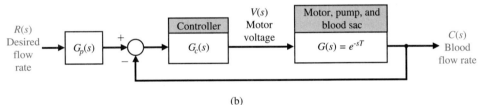

(b)

peak power. Consequently, it is desirable to drive the EVAD in a fashion that minimizes its electric power consumption [29].

The EVAD has a single input, the applied motor voltage, and a single output, the blood flow rate. The control system of the EVAD performs two main tasks: It adjusts the motor voltage to drive the pusher plate through its desired stroke, and it varies the EVAD's blood flow to meet the body's cardiac output demand. The blood flow controller adjusts the blood flow rate by varying the EVAD's beat rate.

A model of the feedback control system is shown in Fig. DP12.9(b). The motor, pump, and blood sac can be modeled by a time delay with $T = 1$ second. The goal is to achieve a step response with less than 5% steady-state error and less than 10% overshoot. Furthermore, to prolong the batteries, the voltage is limited to 30 V. Design a

controller using (a) $G_c(s) = K/s$, (b) a PI controller, and (c) a PID controller. Compare the results for the three controllers by recording in a table the percent overshoot, peak time, settling time and the maximum value of $v(t)$.

DP12.10 One arm of a space robot is shown in Fig. DP12.10(a). The block diagram for the control of the arm is shown in Fig. DP12.10(b). The transfer function of the motor and arm is

$$G(s) = \frac{1}{s(s + 10)}.$$

(a) If $G_c(s) = K$, determine the gain necessary for an overshoot of 4.5%, and plot the step response. (b) Design a proportional plus derivative (PD) controller using the ITAE method and $\omega_n = 10$. Determine the required prefilter $G_p(s)$. (c) Design a PI controller and a prefilter using the ITAE method.

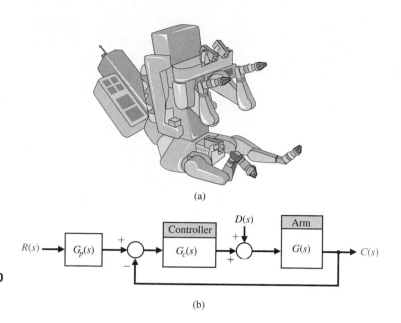

(a)

(b)

FIGURE DP12.10
Space robot
control.

(d) Design a PID controller and a prefilter using the ITAE method with $\omega_n = 10$. (e) Determine the effect of a unit step disturbance for each design. Record the maximum value of $c(t)$ and the final value of $c(t)$ for the disturbance input. (f) Determine the overshoot, peak time, and settling time for an input step $R(s)$ for each design above. (g) The plant is subject to variation due to load changes. Find the magnitude of the sensitivity at $\omega = 5$, $|S_G^T(j5)|$ where

$$T = \frac{GG_c}{1 + GG_c}.$$

(h) Based on the results of parts (e), (f), and (g), select the best controller.

DP12.11 A photovoltaic system is mounted on a space station in order to develop the power for the station. The photovoltaic panels should follow the sun with good accuracy in order to maximize the energy from the panels. The system uses a dc motor so that the transfer function of the panel mount and the motor is

$$G(s) = \frac{1}{s(s + 25)}.$$

We will select a controller $G_c(s)$ assuming that an optical sensor is available to accurately track the sun's position, and thus $H(s) = 1$.

The goal is to design $G_c(s)$ so that (1) the percent overshoot to a step is less than 7% and (2) the steady-state error to a ramp input should be less than or equal to 1%. Determine the best phase-lead controller. Examine the robustness of the system to a 10% variation in the motor time constant.

DP12.12 Electromagnetic suspension systems for air-cushioned trains are known as magnetic levitation (maglev) trains. One maglev train uses a super-conducting magnet system [17]. It uses superconducting coils and the levitation distance $x(t)$ is inherently unstable. The model of the levitation is

$$G(s) = \frac{X(s)}{V(s)} = \frac{K}{(s\tau_1 + 1)(s^2 - \omega_1^2)},$$

where $V(s)$ is the coil voltage; τ_1 is the magnet time constant; and ω_1 is the natural frequency. The system uses a position sensor with a negligible time constant. A train traveling at 300 km/h would have $\tau_1 = 1$ second and $\omega_1 = 100$ rad/s. Determine a controller that can maintain steady, accurate levitation when disturbances occur along the railway. Use the system model of Fig. 12.1.

DP12.13 Reconsider the Mars rover problem described in DP6.3. The system uses a PID controller, and a robust system is desired. The specifications are (1) maximum overshoot equal to 18%, (2) settling time less than 2 seconds, (3) rise time equal to

or greater than 0.20 to limit the power requirements, (4) phase margin greater than 65°, (5) gain
margin greater than 8 dB, (6) maximum root sensitivity (magnitude of real and imaginary parts)
less than one. Select the best value of the gain, K.

DP12.14 A benchmark problem consists of the mass-
spring system shown in Fig. DP12.14, which represents a flexible structure. Let $m_1 = m_2 = 1$ and
$0.5 \leq k \leq 2.0$ [33]. It is possible to measure x_1 and
x_2 and use a controller prior to $u(t)$. Obtain the system description, choose a control structure, and de-

sign a robust system. Determine the response of the
system to a unit step disturbance. Assume the output $x_2(t)$ is the variable to be controlled.

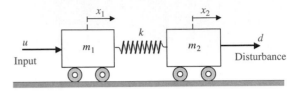

FIGURE DP12.14

MATLAB PROBLEMS

MP12.1 Consider the control system in Fig. MP12.1,
where

$$G(s) = \frac{1}{Js^2}.$$

The value of J is known to change slowly with
time, although, for design purposes, the nominal
value is chosen to be $J = 10$.

(a) Design a PID compensator (denoted by $G_c(s)$)
to achieve a phase margin greater than 45° and a
bandwidth less than 5 rad/s. (b) Using the PID controller designed in part (a), develop a *Matlab* script
to generate a plot of the phase margin as J varies
from $1 \leq J \leq 30$.

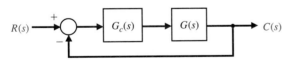

FIGURE MP12.1   A feedback control system with
compensation.

MP12.2 Consider the feedback control system in
Fig. MP12.2. The exact value of parameter b is unknown; however, for design purposes the nominal
value is taken to be $b = 4$. The value of $a = 8$ is
known very precisely.

(a) Design the proportional controller K such that
the closed-loop system response to a unit step input
has a settling time less than 5 seconds and an overshoot of less than 10%. Use the nominal value of b
in the design.
(b) Investigate the effects of variations in the parameter b on the closed-loop system unit step response. Let $b = 0, 1, 4,$ and 40, and co-plot the step
response associated with each value of b. In all
cases, use the proportional controller from part (a).
Discuss the results.

MP12.3 A model of a flexible structure is given by

$$G(s) = \frac{(1 + k\omega_n^2)s^2 + 2\zeta\omega_n s + \omega_n^2}{s^2(s^2 + 2\zeta\omega_n s + \omega_n^2)}$$

where ω_n is the natural frequency of the flexible
mode and ζ is the corresponding damping ratio. In
general, it is difficult to know the structural damping precisely, while the natural frequency can be
predicted more accurately using well-established
modeling techniques. Assume the nominal values
of $\omega_n = 2$ rad/s, $\zeta = 0.005$, and $k = 0.1$.

(a) Design a lead compensator to meet the following specifications: (1) a closed-loop system response to a unit step input with a settling time
less than 200 seconds and (2) an overshoot of less
than 50%.

FIGURE MP12.2
A feedback control
system with
uncertain
parameter b.

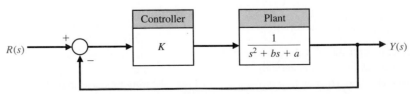

FIGURE MP12.4
An industrial
controlled process
with a time delay in
the loop.

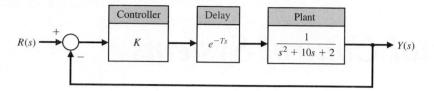

(b) With the controller from part (a), investigate the closed-loop system unit step response with $\zeta = 0, 0.005, 0.1,$ and 1. Co-plot the various unit step responses and discuss the results.

(c) From a control system point-of-view, is it preferable to have the actual flexible structure damping less than or greater than the design value? Explain.

MP12.4 The industrial process shown in Fig. MP12.4 is known to have a time delay in the loop. In practice, it is often the case that the magnitude of system time delays cannot be precisely determined. The magnitude of the time delay may change in an unpredictable manner depending on the process environment. A robust control system should be able to operate satisfactorily in the presence of the system time delays.

(a) Develop a *Matlab* script to compute and plot the phase margin for the industrial process in Fig. MP12.4 when the time delay, T, varies between 0 and 5 seconds. Use the **pade** function with a second-order approximation to approximate the time delay. Plot the phase margin as a function of the time delay.

(b) Determine the maximum time delay allowable for system stability. Use the plot generated in part (a) to compute the maximum time delay approximately.

MP12.5 A unity negative feedback loop has the loop transfer function

$$GH(s) = \frac{a(s-1)}{s^2 + 2s + 1}.$$

It is known from the underlying physics of the problem that the parameter a can vary only between $0 < a < 1$. Develop a *Matlab* script to generate the following plots:

(a) The steady-state tracking error due to a negative unit step input (i.e., $R(s) = -1/s$) versus the parameter a.

(b) The maximum % initial undershoot (or overshoot) versus parameter a.

(c) The gain margin versus the parameter a.

(d) Based on the results in parts (a)–(c), comment on the robustness of the system to changes in parameter a in terms of steady-state errors, stability, and transient time response.

TERMS AND CONCEPTS

PID controller A controller with three terms in which its output is the sum of a proportional term, an integrating term, and a differentiating term, with an adjustable gain for each term.

Prefilter A transfer function, $G_p(s)$, that filters the input signal $R(s)$ prior to the calculation of the error signal.

Process controller *See* PID controller.

Robust control A system that exhibits the desired performance in the presence of significant plant uncertainty.

Three-mode controller *See* PID controller.

Three-term controller *See* PID controller.

Digital Control Systems

PREVIEW

A digital computer may be used to serve as a compensator or controller in a feedback control system. Since the computer receives data only at specific intervals, it is necessary to develop a method for describing and analyzing the performance of computer control systems.

The computer system uses data sampled at prespecified intervals, resulting in a time series of signals. These time series, called sampled data, can be transformed to the *s*-domain and, ultimately, to the *z*-domain by the relation $z = e^{sT}$. The complex frequency domain in terms of *z* has properties similar to those of the Laplace *s*-domain.

We may use the *z*-transform of a transfer function to analyze the stability and transient response of a system. Thus we may readily determine the response of a closed-loop feedback system with a digital computer serving as the compensator (or controller) block. We also can utilize the root locus method to determine the location of the roots of the characteristic equation. These computer control systems are widely used in industry. They also play an important role in the control of industrial processes, which utilize a combination of a computer and an actuator working together to perform a series of tasks.

13.1 INTRODUCTION

The use of a digital computer as a compensator (controller) device has grown during the past two decades as the price and reliability of digital computers have improved dramatically [1, 2]. A block diagram of a single-loop digital control system is shown in Fig. 13.1. The digital computer in this system configuration receives the error in digital form and performs calculations in order to provide an output in digital form. The computer may be programmed to provide an output so that the performance of the process is near or equal to the desired performance. Many computers are able to receive and manipulate several inputs, so a digital computer control system can often be a multivariable system.

A digital computer receives and operates on signals in digital (numerical) form, as contrasted to continuous signals [3]. A *digital control system* uses digital signals and a digital computer to control a process. The measurement data are converted from analog form to digital form by means of the analog-to-digital converter shown in Fig. 13.1. After processing the inputs, the digital computer provides an output in digital form. This output is then converted to analog form by the digital-to-analog converter, shown in Fig. 13.1.

13.2 DIGITAL COMPUTER CONTROL SYSTEM APPLICATIONS

The total number of computer control systems installed in industry has grown over the past three decades [2]. Currently there are approximately 100 million control systems using a computer, although the computer size and power may vary significantly. If we consider only computer control systems of a relatively complex nature, such as chemical process control or aircraft control, the number of computer control systems is approximately 20 million.

A digital computer consists of a central processing unit (CPU), input–output units, and a memory unit. The size and power of a computer will vary according to the size, speed, and power of the CPU, as well as the size, speed, and organization of the memory unit. Small computers, called *minicomputers,* have become increasingly common since 1980. Powerful but inexpensive computers, called *microcomputers,* which use a 16-bit word or 32-bit word, have become readily available. These systems use a microprocessor as a CPU. Therefore the nature of the control task, the extent of the data required in memory, and the speed of calculation required will dictate the selection of the computer within the range of available computers.

The size of computers and the cost for the active logic devices used to construct them have both declined exponentially. The active components per cubic centimeter have increased so that the actual computer can be reduced in size. Also the speed of computers has

FIGURE 13.1
A block diagram of a computer control system including the signal converters. The signal is indicated as digital or analog.

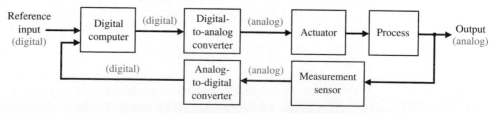

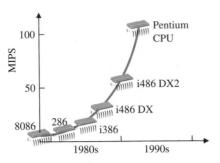

FIGURE 13.2
The development
of INTEL
microprocessors,
measured in
millions of
instructions
executed per
second.
(*Source:* INTEL.)

increased exponentially. It is estimated that the sale of minicomputers and microcomputers in 1994 accounted for over 50 billion dollars in the United States. With the availability of fast, low-priced, and small-sized microprocessors, much of the control of industrial and commercial processes is moving toward the use of computers within the control system. The performance of INTEL microprocessors, as shown in Fig. 13.2, is illustrative of the development of powerful computing devices.

Digital control systems are used in many applications: for machine tools, metalworking processes, chemical processes, aircraft control, and automobile traffic control, among others [4–8]. An example of a computer control system used in the aircraft industry is shown in Fig. 13.3. Automatic computer-controlled systems are used for purposes as diverse as measuring the objective refraction of the human eye and controlling the engine spark timing or air–fuel ratio of automobile engines. The latter innovations are necessary to reduce automobile emissions and increase fuel economy.

The advantages of using digital control include improved measurement sensitivity; the use of digitally coded signals, digital sensors and transducers, and microprocessors; reduced sensitivity to signal noise; and the capability to easily reconfigure the control algorithm in software. Improved sensitivity results from the low energy signals required by digital sensors and devices. The use of digitally coded signals permits the wide application of digital devices and communications. Digital sensors and transducers can effectively measure, transmit, and couple signals and devices. In addition, many systems are inherently digital because they send out pulse signals. Examples of such a digital system are a radar tracking system and a space satellite.

13.3 SAMPLED-DATA SYSTEMS

Computers used in control systems are interconnected to the actuator and the process by means of signal converters. The output of the computer is processed by a digital-to-analog converter. We will assume that all the numbers that enter or leave the computer do so at the same fixed period T, called the *sampling period*. Thus, for example, the reference input shown in Fig. 13.4 is a sequence of sample values $r(kT)$. The variables $r(kT)$, $m(kT)$, and $u(kT)$ are discrete signals in contrast to $m(t)$ and $c(t)$, which are continuous functions of time.

**Data obtained for the system variables only at discrete intervals, and
denoted as $x(kT)$, are called sampled data or a discrete signal.**

FIGURE 13.3
The flight deck of the Boeing 757 and 767 features digital control electronics, including an engine
indicating system and a crew alerting system. All systems controls are within reach of either pilot.
The system includes an inertial reference system making use of laser gyroscopes and an
electronic attitude director indicator. A flight-management computer system integrates
navigation, guidance, and performance data functions. When coupled with the automatic flight
control system (automatic pilot), the flight-management system provides accurate engine thrust
settings and flight-path guidance during all phases of flight from immediately after takeoff to final
approach and landing. The system can predict the speeds and altitudes that will result in the best
fuel economy and command the airplane to follow the most fuel-efficient, or the "least time," flight
paths. (Courtesy of Boeing Airplane Co.)

A sampler is basically a switch that closes every T seconds for one instant of time.
Consider an ideal sampler as shown in Fig. 13.5. The input is $r(t)$ and the output is $r*(t)$,
where nT is the current sample time and the current value of $r*(t)$ is $r(nT)$. We then have
$r*(t) = r(nT)\delta(t - nT)$ where δ is the impulse function.

Let us assume that we sample a signal $r(t)$, as shown in Fig. 13.5, and obtain $r*(t)$.

FIGURE 13.4
A digital control
system.

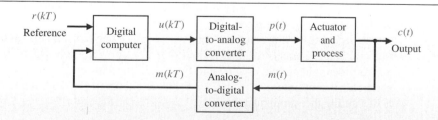

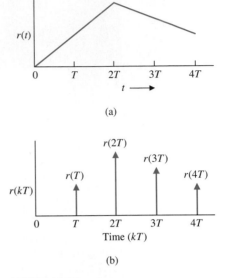

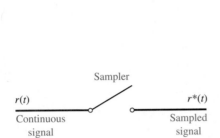

FIGURE 13.5
An ideal sampler with an input $r(t)$.

FIGURE 13.6
(a) An input signal $r(t)$ and (b) the sampled signal $r^*(t) = \sum_{k=0}^{\infty} r(kT)\delta(t - kT)$. The colored vertical arrow represents an impulse.

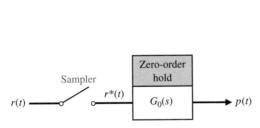

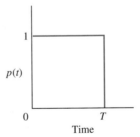

FIGURE 13.7
A sampler and zero-order hold circuit.

FIGURE 13.8
The response of a zero-order hold to an impulse input $r(kT)$, which equals 1 when $k = 0$ and equals zero when $k \neq 0$, so that $r^*(t) = r(0)\delta(t)$.

Then we portray the series for $r^*(t)$ as a string of impulses starting at $t = 0$, spaced at T seconds and of amplitude $r(kT)$. For example, consider the input signal $r(t)$ shown in Fig. 13.6(a). The sampled signal is shown in Fig. 13.6(b) with an impulse represented by a vertical arrow of magnitude $r(kT)$.

A digital-to-analog converter serves as a device that converts the sampled signal $r^*(t)$ to a continuous signal $p(t)$. The digital-to-analog converter can usually be represented by a zero-order hold circuit, as shown in Fig. 13.7. The zero-order hold takes the value $r(kT)$ and holds it constant for $kT \leq t < (k + 1)T$, as shown in Fig. 13.8 for $k = 0$. Thus we use $r(kT)$ during the sampling period.

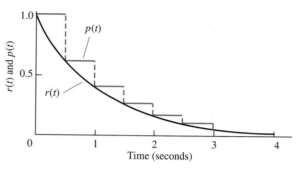

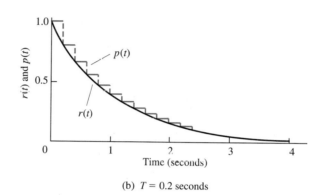

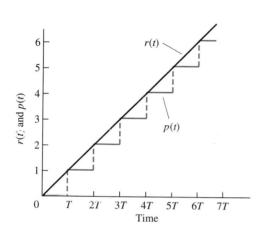

FIGURE 13.9
The response of a sampler and zero-order hold for a ramp input $r(t) = t$.

FIGURE 13.10
The response of a sampler and zero-order hold to an input $r(t) = e^{-t}$ for two values of sampling period T.

A sampler and zero-order hold can accurately follow the input signal if T is small compared to the transient changes in the signal. The response of a sampler and zero-order hold for a ramp input is shown in Fig. 13.9. Finally, the response of a sampler and zero-order hold for an exponentially decaying signal is shown in Fig. 13.10 for two values of sampling period. Clearly, the output $p(t)$ will approach the input $r(t)$ as T approaches zero (as we sample very often).

The impulse response of a zero-order hold is shown in Fig. 13.8. Then, the transfer function of the zero-order hold is

$$G_0(s) = \frac{1}{s} - \frac{1}{s}e^{-sT} = \frac{(1 - e^{-sT})}{s}. \tag{13.1}$$

The precision of the digital computer and the associated signal converters is limited (see Fig. 13.4). *Precision* is the degree of exactness or discrimination with which a quantity is stated. The precision of the computer is limited by a finite word length. The precision of the analog-to-digital converter is limited by an ability to store its output only in digital logic composed of a finite number of binary digits. The converted signal, $m(kT)$, is then said to include an *amplitude quantization error*. When the quantization error and the error due to a computer's finite word size are small relative to the amplitude of the signal [13, 19], the system is sufficiently precise and the precision limitations can be neglected.

13.4 THE z-TRANSFORM

Because the output of the ideal sampler, $r*(t)$, is a series of impulses with values $r(kT)$, we have

$$r*(t) = \sum_{k=0}^{\infty} r(kT)\delta(t - kT) \tag{13.2}$$

for a signal for $t > 0$. Using the Laplace transform, we have

$$\mathcal{L}\{r*(t)\} = \sum_{k=0}^{\infty} r(kT)e^{-ksT}. \tag{13.3}$$

We now have an infinite series that involves factors of e^{sT} and its powers. We define

$$z = e^{sT}, \tag{13.4}$$

where this relationship involves a conformal mapping from the s-plane to the z-plane. We then define a new transform, called the z-transform, so that

$$Z\{r(t)\} = Z\{r*(t)\} = \sum_{k=0}^{\infty} r(kT)z^{-k}. \tag{13.5}$$

As an example, let us determine the z-transform of the unit step function $u(t)$ [not to be confused with the control signal $u(t)$]. We obtain

$$Z\{u(t)\} = \sum_{k=0}^{\infty} u(kT)z^{-k} = \sum_{k=0}^{\infty} z^{-k}, \tag{13.6}$$

since $u(kT) = 1$ for $k \geq 0$. This series can be written in closed form as[†]

$$U(z) = \frac{1}{1 - z^{-1}} = \frac{z}{z - 1}. \tag{13.7}$$

In general we will define the z-transform of a function $f(t)$ as

$$Z\{f(t)\} = F(z) = \sum_{k=0}^{\infty} f(kT)z^{-k}. \tag{13.8}$$

EXAMPLE 13.1   **Transform of exponential**

Let us determine the z-transform of $f(t) = e^{-at}$ for $t \geq 0$. Then

$$Z\{e^{-at}\} = F(z) = \sum_{k=0}^{\infty} e^{-akT}z^{-k} = \sum_{k=0}^{\infty} (ze^{+aT})^{-k}. \tag{13.9}$$

Again this series can be written in closed form as

$$F(z) = \frac{1}{1 - (ze^{aT})^{-1}} = \frac{z}{z - e^{-aT}}. \tag{13.10}$$

[†]Recall that the infinite geometric series may be written $(1 - bx)^{-1} = 1 + bx + (bx)^2 + (bx)^3 + \cdots$, if $(bx)^2 < 1$.

In general we may show that

$$Z\{e^{-at}f(t)\} = F(e^{aT}z). \blacksquare$$

EXAMPLE 13.2   **Transform of sinusoid**

Let us determine the z-transform of $f(t) = \sin \omega t$ for $t \geq 0$. We can write $\sin \omega t$ as

$$\sin \omega t = \frac{e^{j\omega T} - e^{-j\omega T}}{2j}.$$

Therefore

$$\{\sin \omega t\} = \left\{\frac{e^{j\omega T}}{2j} - \frac{e^{-j\omega T}}{2j}\right\}. \tag{13.11}$$

Then

$$\begin{aligned}
F(z) &= \frac{1}{2j}\left(\frac{z}{z - e^{j\omega T}} - \frac{z}{z - e^{-j\omega T}}\right) \\
&= \frac{1}{2j}\left(\frac{z(e^{j\omega T} - e^{-j\omega T})}{z^2 - z(e^{j\omega T} + e^{j\omega T}) + 1}\right) \tag{13.12} \\
&= \frac{z \sin \omega T}{z^2 - 2z \cos \omega T + 1}. \blacksquare
\end{aligned}$$

A table of z-transforms is given in Table 13.1. A table of properties of the z-transform is given in Table 13.2. As in the case of Laplace transforms, we are ultimately interested in the output $c(t)$ of the system. Therefore we must use an inverse transform to obtain $c(t)$ from $C(z)$. We may obtain the output by (1) expanding $C(z)$ in a power series, (2) expanding $C(z)$ into partial fractions and using Table 13.1 to obtain the inverse of each term, or (3) obtaining the inverse z-transform by an inversion integral. We will limit our methods to (1) and (2) in this limited discussion.

EXAMPLE 13.3   **Transfer function of open-loop system**

Let us consider the system shown in Fig. 13.11 for $T = 1$. The transfer function of the zero-order hold (Eq. 13.1) is

$$G_0(s) = \frac{(1 - e^{-sT})}{s}.$$

Therefore the transfer function $C(s)/R^*(s)$ is

$$\frac{C(s)}{R^*(s)} = G_0(s)G_p(s) = G(s) = \frac{1 - e^{-sT}}{s^2(s + 1)}. \tag{13.13}$$

Expanding into partial fractions, we have

$$G(s) = (1 - e^{-sT})\left(\frac{1}{s^2} - \frac{1}{s} + \frac{1}{s + 1}\right), \tag{13.14}$$

$$G(z) = Z\{G(s)\} = (1 - z^{-1})Z\left(\frac{1}{s^2} - \frac{1}{s} + \frac{1}{s + 1}\right). \tag{13.15}$$

TABLE 13.1   z-Transforms

$x(t)$	$X(s)$	$X(z)$
$\delta(t) = \begin{cases} 1 & t = 0, \\ 0 & t = kT,\ k \neq 0 \end{cases}$	1	1
$\delta(t - kT) = \begin{cases} 1 & t = kT, \\ 0 & t \neq kT \end{cases}$	e^{-kTs}	z^{-k}
$u(t)$, unit step	$1/s$	$\dfrac{z}{z-1}$
t	$1/s^2$	$\dfrac{Tz}{(z-1)^2}$
e^{-at}	$\dfrac{1}{s+a}$	$\dfrac{z}{z - e^{-aT}}$
$1 - e^{-at}$	$\dfrac{1}{s(s+a)}$	$\dfrac{(1 - e^{-aT})z}{(z-1)(z - e^{-aT})}$
$\sin \omega t$	$\dfrac{\omega}{s^2 + \omega^2}$	$\dfrac{z \sin \omega T}{z^2 - 2z \cos \omega T + 1}$
$\cos \omega t$	$\dfrac{s}{s^2 + \omega^2}$	$\dfrac{z(z - \cos \omega T)}{z^2 - 2z \cos \omega T + 1}$
$e^{-at} \sin \omega t$	$\dfrac{\omega}{(s+a)^2 + \omega^2}$	$\dfrac{(ze^{-aT} \sin \omega T)}{z^2 - 2ze^{-aT} \cos \omega T + e^{-2aT}}$
$e^{-at} \cos \omega t$	$\dfrac{s+a}{(s+a)^2 + \omega^2}$	$\dfrac{z^2 - ze^{-aT} \cos \omega T}{z^2 - 2ze^{-aT} \cos \omega T + e^{-2aT}}$

TABLE 13.2   Properties of the z-Transform

	$x(t)$	$X(z)$
1.	$kx(t)$	$kX(z)$
2.	$x_1(t) + x_2(t)$	$X_1(z) + X_2(z)$
3.	$x(t + T)$	$zX(z) - zx(0)$
4.	$tx(t)$	$-Tz\dfrac{dX(z)}{dz}$
5.	$e^{-at}x(t)$	$X(ze^{aT})$
6.	$x(0)$, initial value	$\lim_{z \to \infty} X(z)$ if the limit exists
7.	$x(\infty)$, final value	$\lim_{z \to 1} (z - 1)X(z)$ if the limit exists and the system is stable; that is, if all poles of $(z - 1)X(z)$ are inside unit circle $\|z\| = 1$ on z-plane.

FIGURE 13.11
An open-loop
sampled data
system (without
feedback).

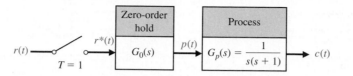

Using the entries of Table 13.1 to obtain the z-transform of each term, we have

$$G(z) = (1 - z^{-1}) \left(\frac{Tz}{(z - 1)^2} - \frac{z}{z - 1} + \frac{z}{z - e^{-T}} \right)$$

$$= \left(\frac{(ze^{-T} - z + Tz) + (1 - e^{-T} - Te^{-T})}{(z - 1)(z - e^{-T})} \right).$$

When $T = 1$, we obtain

$$G(z) = \frac{ze^{-1} + 1 - 2e^{-1}}{(z - 1)(z - e^{-1})}$$

$$= \frac{0.3678z + 0.2644}{(z - 1)(z - 0.3678)} = \frac{0.3678z + 0.2644}{z^2 - 1.3678z + 0.3678}.$$

(13.16)

The response of this system to a unit impulse is obtained for $R(z) = 1$ so that $C(z) = G(z) \cdot 1$. We may obtain $C(z)$ by dividing the denominator into the numerator as:

$$
\begin{array}{r}
0.3678z^{-1} + 0.7675z^{-2} + 0.9145z^{-3} + \cdots = C(z) \\
z^2 - 1.3678z + 0.3678 \overline{)0.3678z \quad + 0.2644} \\
\underline{0.3678z \quad - 0.5031 \quad + 0.1353z^{-1}} \\
+ 0.7675 \quad - 0.1353z^{-1} \\
\underline{+ 0.7675 \quad - 1.0497z^{-1} + 0.2823z^{-2}} \\
0.9145z^{-1} - 0.2823z^{-2}
\end{array}
$$

(13.17)

This calculation yields the response at the sampling instants and can be carried as far as is needed for $C(z)$. From Eq. (13.5), we have

$$C(z) = \sum_{k=0}^{\infty} c(kT)z^{-k}$$

In this case we have obtained $c(kT)$ as follows: $c(0) = 0$, $c(T) = 0.3678$, $c(2T) = 0.7675$, and $c(3T) = 0.9145$. Note that $c(kT)$ provides the values of $c(t)$ at $t = kT$. ∎

We have determined $C(z)$, the z-transform of the output sampled signal. The z-transform of the input sampled signal is $R(z)$. The transfer function in the z-domain is

$$\frac{C(z)}{R(z)} = G(z).$$

(13.18)

Since we determined the sampled output, we can use an output sampler to depict this condition, as shown in Fig. 13.12, which represents the system of Fig. 13.11 with the ad-

FIGURE 13.12
System with sampled output.

FIGURE 13.13
The z-transform transfer function in block diagram form.

ditional (fictitious) samples (shown in color). We assume that both samplers have the same sampling period and operate synchronously. Then,

$$C(z) = G(z)R(z), \qquad (13.19)$$

as required. We may represent Eq. (13.19), which is a z-transform equation, by the block diagram of Fig. 13.13.

13.5 CLOSED-LOOP FEEDBACK SAMPLED-DATA SYSTEMS

In this section, we consider closed-loop sampled-data control systems. Consider the system shown in Fig. 13.14(a). The sampled-data z-transform model of this figure with a sampled-output signal $C(z)$ is shown in Fig. 13.14(b). The closed-loop transfer function $T(z)$ (using block diagram reduction) is

$$\frac{C(z)}{R(z)} = T(z) = \frac{G(z)}{1 + G(z)}. \qquad (13.20)$$

Here we assume that the $G(z)$ is the z-transform of $G(s) = G_0(s)G_p(s)$, where $G_0(s)$ is the zero-order hold and $G_p(s)$ is the plant transfer function.

A feedback control system with a digital controller is shown in Fig. 13.15(a). The

FIGURE 13.14
Feedback control system with unity feedback. $G(z)$ is the z-transform of $G(s)$, which represents the plant and the zero-order hold.

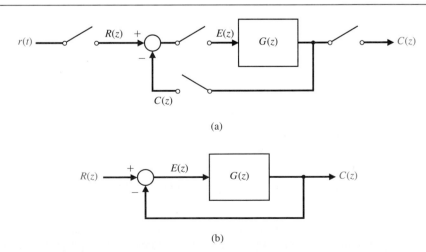

(a)

(b)

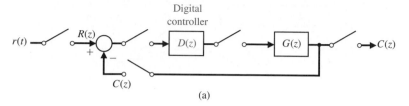

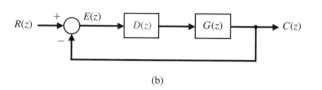

FIGURE 13.15
(a) Feedback
control system with
a digital controller.
(b) Block diagram
model. Note that
$G(z) =$
$Z\{G_o(s)G_p(s)\}$.

z-transform block diagram model is shown in Fig. 13.15(b). The closed-loop transfer function is

$$\frac{C(z)}{R(z)} = T(z) = \frac{G(z)D(z)}{1 + G(z)D(z)}. \tag{13.21}$$

EXAMPLE 13.4   Response of closed-loop system

Now let us consider the closed-loop system as shown in Fig. 13.16. We have obtained the z-transform model of this system as shown in Fig. 13.14. Therefore we have

$$\frac{C(z)}{R(z)} = \frac{G(z)}{1 + G(z)}. \tag{13.22}$$

In Example 13.3, we obtained $G(z)$ as Eq. (13.16) when $T = 1$ second. Substituting $G(z)$ into Eq. (13.22), we obtain

$$\frac{C(z)}{R(z)} = \frac{0.3678z + 0.2644}{z^2 - z + 0.6322}. \tag{13.23}$$

Since the input is a unit step

$$R(z) = \frac{z}{z - 1}, \tag{13.24}$$

then
$$C(z) = \frac{z(0.3678z + 0.2644)}{(z - 1)(z^2 - z + 0.6322)} = \frac{0.3678z^2 + 0.2644z}{z^3 - 2z^2 + 1.6322z - 0.6322}.$$

FIGURE 13.16
A closed-loop
sampled data
system.

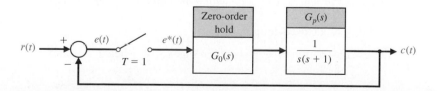

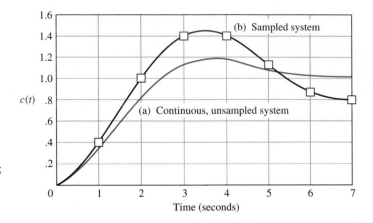

FIGURE 13.17
The response of
a second-order
system:
(a) continuous (T
= 0), not sampled;
(b) sampled
system, T =
1 second.

Completing the division, we have

$$C(z) = 0.3678z^{-1} + z^{-2} + 1.4z^{-3} + 1.4z^{-4} + 1.147z^{-5} \ldots \quad (13.25)$$

The values of $c(kT)$ are shown in Fig. 13.17, using the symbol □. The complete response of the sampled-data closed-loop system is shown (obtained by *Matlab;* see Section 13.12) and contrasted to the response of a continuous system (when $T = 0$). The overshoot of the sampled system is 45%, in contrast to 17% for the continuous system. Furthermore, the settling time of the sampled system is twice as long as that of the continuous system. ■

13.6 STABILITY ANALYSIS IN THE z-PLANE

A linear continuous feedback control system is stable if all poles of the closed-loop transfer function $T(s)$ lie in the left half of the s-plane. The z-plane is related to the s-plane by the transformation

$$z = e^{sT} = e^{(\sigma + j\omega)T}. \quad (13.26)$$

We may also write this relationship as

$$|z| = e^{\sigma T}$$

and

$$\underline{/z} = \omega T. \quad (13.27)$$

In the left-hand s-plane, $\sigma < 0$ and therefore the related magnitude of z varies between 0 and 1. Therefore the imaginary axis of the s-plane corresponds to the unit circle in the z-plane, and the inside of the unit circle corresponds to the left half of the s-plane [15].

Therefore we can state that a *sampled system is stable if all the poles of the closed-loop transfer function $T(z)$ lie within the unit circle of the z-plane.*

EXAMPLE 13.5   **Stability of a closed-loop system**

Let us consider the system shown in Fig. 13.18, when $T = 1$ and

$$G_p(s) = \frac{K}{s(s + 1)}. \quad (13.28)$$

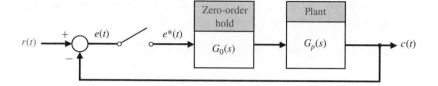

FIGURE 13.18
A closed-loop
sampled system.

Then, recalling Eq. (13.16), we note that

$$G(z) = \frac{K(0.3678z + 0.2644)}{z^2 - 1.3678z + 0.3678} = \frac{K(az + b)}{z^2 - (1 + a)z + a}, \qquad (13.29)$$

where $a = 0.3678$ and $b = 0.2644$.

The poles of the closed-loop transfer function $T(z)$ are the roots of the equation $[1 + G(z)] = 0$. We call $q(z) = 1 + G(z) = 0$ the characteristic equation. Therefore we obtain

$$q(z) = 1 + G(z) = z^2 - (1 + a)z + a + Kaz + Kb = 0. \qquad (13.30)$$

When $K = 1$, we have

$$q(z) = z^2 - z + 0.6322 \qquad (13.31)$$
$$= (z - 0.50 + j0.6182)(z - 0.50 - j0.6182) = 0.$$

Therefore the system is stable because the roots lie within the unit circle. When $K = 10$, we have

$$q(z) = z^2 + 2.310z + 3.012 \qquad (13.32)$$
$$= (z + 1.155 + j1.295)(z + 1.155 - j1.295)$$

and the system is unstable because both roots lie outside the unit circle. This system is stable for $0 < K < 2.39$. The locus of the roots as K varies is discussed in Section 13.10.

We notice that a second-order sampled system can be unstable with increasing gain where a second-order continuous system is stable for all values of gain (assuming both the poles of the open-loop system lie in the left half s-plane). ∎

13.7 PERFORMANCE OF A SAMPLED-DATA SECOND-ORDER SYSTEM

Let us consider the performance of a sampled second-order system with a zero-order hold, as shown in Fig. 13.18, when the plant is

$$G_p(s) = \frac{K}{s(\tau s + 1)}. \qquad (13.33)$$

We then obtain $G(z)$ for the unspecified sampling period T as

$$G(z) = \frac{K\{(z - E)[T - \tau(z - 1)] + \tau(z - 1)^2\}}{(z - 1)(z - E)}, \qquad (13.34)$$

where $E = e^{-T/\tau}$. The stability of the system is analyzed by considering the characteristic equation

$$q(z) = z^2 + z\{K[T - \tau(1 - E)] - (1 + E)\}$$
$$+ K[\tau(1 - E) - TE] + E = 0. \quad (13.35)$$

Because the polynomial $q(z)$ is a quadratic and has real coefficients, the necessary and sufficient conditions for $q(z)$ to have all its roots within the unit circle are

$$|q(0)| < 1, \qquad q(1) > 0, \qquad q(-1) > 0.$$

These stability conditions for a second-order system can be established by mapping the z-plane characteristic equation into the s-plane and checking for positive coefficients of $q(s)$. Using these conditions, we establish the necessary conditions from Eq. (13.35) as

$$K\tau < \frac{1 - E}{1 - E - (T/\tau)E}, \quad (13.36)$$

$$K\tau < \frac{2(1 + E)}{(T/\tau)(1 + E) - 2(1 - E)}, \quad (13.37)$$

and $K > 0$, $T > 0$. For this system, we can calculate the maximum gain permissible for a stable system. The maximum gain allowable is given in Table 13.3 for several values of T/τ. If the computer system has sufficient speed of computation and data handling, it is possible to set $T/\tau = 0.1$ and obtain system characteristics approaching those of a continuous (nonsampled) system.

The maximum overshoot of the second-order system for a unit step input is shown in Fig. 13.19.

The performance criterion, integral squared error, can be written as

$$I = \frac{1}{\tau} \int_0^\infty e^2(t)dt. \quad (13.38)$$

The loci of this criterion are given in Fig. 13.20 for constant values of I. For a given value of T/τ, one can determine the minimum value of I and the required value of $K\tau$. The optimal curve shown in Fig. 13.20 indicates the required $K\tau$ for a specified T/τ that minimizes I. For example, when $T/\tau = 0.75$, we require $K\tau = 1$ in order to minimize the performance criterion I.

The steady-state error for a unit ramp input $r(t) = t$ is shown in Fig. 13.21. For a given T/τ we can reduce the steady-state error, but then the system yields greater overshoot and settling time for a step input.

TABLE 13.3 Maximum Gain for a Second-Order Sampled System

	T/τ	0	0.1	0.5	1	2
Maximum	$K\tau$	∞	20.4	4.0	2.32	1.45

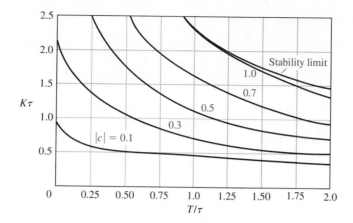

FIGURE 13.19
The maximum overshoot $|c|$ for a second-order sampled system for a unit step input.

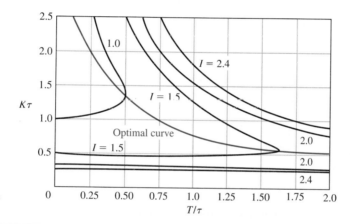

FIGURE 13.20
The loci of integral squared error for a second-order sampled system for constant values of I.

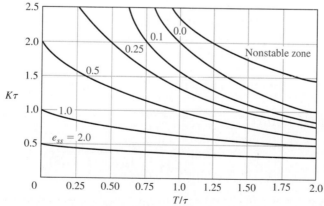

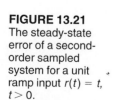

FIGURE 13.21
The steady-state error of a second-order sampled system for a unit ramp input $r(t) = t$, $t > 0$.

EXAMPLE 13.6   **Design of a sampled system**

Let us consider a closed-loop sampled system as shown in Fig. 13.18 when

$$G_p(s) = \frac{K}{s(0.1s + 1)(0.005s + 1)} \tag{13.39}$$

and we need to select T and K for suitable performance. As an approximation, we neglect the effects of the time constant $\tau_2 = 0.005$ second because it is only 5% of the primary time constant $\tau_1 = 0.1$. Then we can use Figs. 13.19, 13.20, and 13.21 to select K and T. Limiting the overshoot to 30% for the step input, we select $T/\tau = 0.25$, yielding $K\tau = 1.4$. For these values, the steady-state error for a unit ramp input is approximately 0.6 (see Fig. 13.21).

Because $\tau = 0.1$, we then set $T = 0.025$ second and $K = 14$. The sampling rate is then required to be 40 samples per second.

The overshoot to step input and the steady-state error for a ramp input may be reduced if we set T/τ to equal 0.1. The overshoot to a step input will be 25% for $K\tau = 1.6$. Using Fig. 13.21, we estimate that the steady-state error for a unit ramp input is 0.55 for $K\tau = 1.6$. ∎

13.8 CLOSED-LOOP SYSTEMS WITH DIGITAL COMPUTER COMPENSATION

A closed-loop sampled system with a digital computer used to improve the performance is shown in Fig. 13.15. The closed-loop transfer function is

$$\frac{C(z)}{R(z)} = T(z) = \frac{G(z)D(z)}{1 + G(z)D(z)}. \tag{13.40}$$

The transfer function of the computer is represented by

$$\frac{U(z)}{E(z)} = D(z). \tag{13.41}$$

In our prior calculations $D(z)$ was represented simply by a gain K. As an illustration of the power of the computer as a compensator, we will reconsider the second-order system with a zero-order hold and a plant

$$G_p(s) = \frac{1}{s(s + 1)} \quad \text{when } T = 1.$$

Then (see Eq. 13.16)

$$G(z) = \frac{0.3678(z + 0.7189)}{(z - 1)(z - 0.3678)}. \tag{13.42}$$

If we select

$$D(z) = \frac{K(z - 0.3678)}{(z + r)}, \tag{13.43}$$

we cancel the pole of $G(z)$ at $z = 0.3678$ and have two parameters, r and K, to set. If we select

$$D(z) = \frac{1.359(z - 0.3678)}{(z + 0.240)}, \tag{13.44}$$

we have

$$G(z)D(z) = \frac{0.50(z + 0.7189)}{(z - 1)(z + 0.240)}. \tag{13.45}$$

If we calculate the response of the system to a unit step, we find that the output is equal to the input at the fourth sampling instant and thereafter. The responses for both the uncompensated and the compensated system are shown in Fig. 13.22. The overshoot of the compensated system is 4%, whereas the overshoot of the uncompensated system is 45%. It is beyond the objective of this book to discuss all the extensive methods for the analytical selection of the parameters of $D(z)$, and the reader is referred to other texts [2–4]. However, we will consider two methods of compensator design: (1) the $G_c(s)$-to-$D(z)$ conversion method, in the following paragraphs, and (2) the root locus z-plane method, in Section 13.10.

One method for determining $D(z)$ first determines a controller $G_c(s)$ for a given plant $G_p(s)$ for the system shown in Fig. 13.23. Then the controller is converted to $D(z)$ for the given sampling period T. The methods described in Chapter 10 are used to determine $G_c(s)$. This design method is called the $G_c(s)$-to-$D(z)$ conversion method. It converts the $G_c(s)$ of Fig. 13.23 to $D(z)$ of Fig. 13.15.

We consider a first-order compensator

$$G_c(s) = K\frac{s + a}{s + b} \tag{13.46}$$

FIGURE 13.22
The response of a sampled-data second-order system to a unit step input.

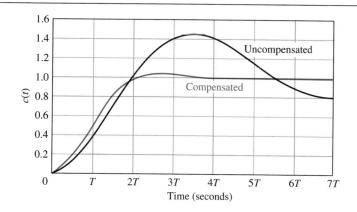

FIGURE 13.23
The continuous system model of a sampled system.

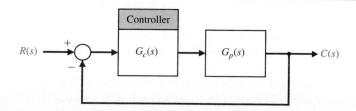

and a digital controller

$$D(z) = C\frac{z - A}{z - B}. \tag{13.47}$$

We determine the z-transform of $G_c(s)$ and set it equal to $D(z)$ as

$$Z\{G_c(s)\} = D(z) \tag{13.48}$$

Then the relationship between the two transfer functions is $A = e^{-aT}$, $B = e^{-bT}$, and when $s = 0$, we require

$$C\frac{(1 - A)}{(1 - B)} = K\frac{a}{b}. \tag{13.49}$$

EXAMPLE 13.7   Design to meet a phase margin specification

Consider a system with a plant

$$G_p(s) = \frac{1740}{s(0.25s + 1)}. \tag{13.50}$$

We will attempt to design $G_c(s)$ so that we achieve a phase margin of $45°$ with a crossover frequency $\omega_c = 125$ rad/s. Using the Bode diagram of $G_p(s)$, we find that the phase margin is $2°$. Using the method of Section 10.4, we find that the required pole–zero ratio is $\alpha = 6.25$. It is specified that $\omega_c = 125$, and so we note that $\omega_c = (ab)^{1/2}$. Therefore, $a = 50$ and $b = 312$. The lead compensator is then

$$G_c(s) = \frac{K(s + 50)}{(s + 312)}. \tag{13.51}$$

We select K in order to yield $|GG_c(j\omega)| = 1$ when $\omega = \omega_c = 125$ rad/s. Then we find $K = 5.6$. The compensator $G_c(s)$ is to be realized by $D(z)$, and so we solve the relationships with a selected sampling period. Setting $T = 0.001$ second, we have

$$A = e^{-0.05} = 0.95, \qquad B = e^{-0.312} = 0.73, \qquad \text{and} \qquad C = 4.85.$$

Then we have

$$D(z) = \frac{4.85(z - 0.95)}{(z - 0.73)}. \tag{13.52}$$

Of course, if we selected another value for the sampling period, then the coefficients of $D(z)$ would differ. ∎

In general, we select a small sampling period so that the design based on the continuous system will accurately carry over to the z-plane. However, we should not select too small a T, or the computation requirements may be more than necessary. In general, use a sampling period $T \cong 1/10f_B$ where $f_B = \omega_B/2\pi$ and ω_B is the bandwidth of the closed-loop continuous system.

The bandwidth of the system designed in Example 13.7 is $\omega_B = 180$ rad/s or $f_B = 28.6$ Hz. Then we select a period $T = 0.003$ second. Note that $T = 0.001$ second was used in Example 13.7.

13.9 THE DESIGN OF A WORKTABLE MOTION CONTROL SYSTEM

An important positioning system in manufacturing systems is a worktable motion control system. The system controls the motion of a worktable at a certain location [21]. We assume that the table is activated in each axis by a motor and lead screw, as shown in Fig. 13.24(a). We consider the x-axis and examine the motion control for a feedback system, as shown in Fig. 13.24(b). The goal is to obtain a fast response with a rapid rise time and settling time to a step command while not exceeding an overshoot of 5%.

The specifications are then (1) a percent overshoot equal to 5% and (2) a minimum settling time and rise time. Rise time is defined as the time to reach the magnitude of the command and is illustrated in Fig. 5.7 by T_R.

To configure the system, we choose a power amplifier and motor so that the system is described by Fig. 13.25. Then, obtaining the transfer function of the motor and power amplifier, we have

$$G_p(s) = \frac{1}{s(s + 10)(s + 20)}. \tag{13.53}$$

We will initially use a continuous system and design $G_c(s)$ as described in Section 13.8. We then obtain $D(z)$ from $G_c(s)$. First we select the controller as a simple gain, K, in order to determine the response that can be achieved without a compensator. Plotting the root

---

FIGURE 13.24
A table motion
control system:
(a) actuator and
table; (b) block
diagram.

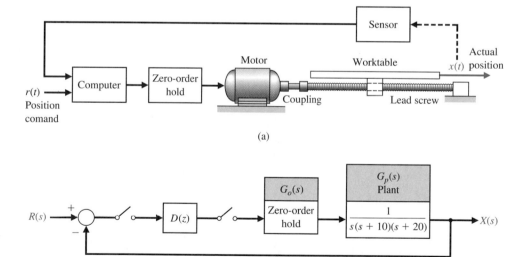

(a)

(b)

FIGURE 13.25
Model of the wheel
control for a mobile
robot.

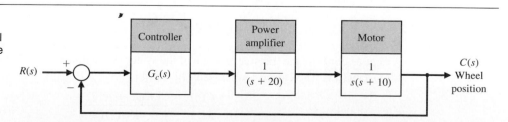

TABLE 13.4   Performance for Two Controllers

Compensator $G_c(s)$	K	Percent Overshoot	Settling Time (seconds)	Rise Time (seconds)
1. K	700	5.0	1.12	0.40
2. $K(s + 11)/(s + 62)$	8000	5.0	0.60	0.25

locus, we find that when $K = 700$, the dominant complex roots have a damping ratio of 0.707 and we expect a 5% overshoot. Then, using a simulation, we find that the overshoot is 5%, the rise time is 0.48 second, and the settling time is 1.12 seconds. These values are recorded as item 1 in Table 13.4.

The next step is to introduce a lead compensator, so that

$$G_c(s) = \frac{K(s + a)}{(s + b)}. \tag{13.54}$$

We will select the zero at $s = -11$ so that the complex roots near the origin dominate. Then, using the method of Section 10.5, we find that we require the pole at $s = -62$. Evaluating the gain at the roots, we find that $K = 8000$. Then the step response has a rise time of 0.25 second and a settling time of 0.60 second. This is an improved response, and we finalize this system as acceptable.

It now remains to select the sampling period and then use the method of Section 13.8 to obtain $D(z)$. The rise time of the compensated continuous system is 0.25 second. Then, we require $T \ll T_R$ in order to obtain a system response predicted by the design of the continuous system. Let us select $T = 0.01$ second. We have

$$G_c(s) = \frac{8000(s + 11)}{(s + 62)}.$$

Then,

$$D(z) = C\frac{z - A}{z - B},$$

where

$$A = e^{-11T} = 0.8958 \quad \text{and} \quad B = e^{-62T} = 0.5379.$$

We now have

$$C = K\frac{a(1 - B)}{b(1 - A)} = \frac{8000(11)(0.462)}{62(0.1042)} = 6293.$$

Then, using this $D(z)$, we expect a response very similar to that obtained for the continuous system model.

13.10   THE ROOT LOCUS OF DIGITAL CONTROL SYSTEMS

Let us consider the transfer function of the system shown in Fig. 13.26. Recall that $G(s) = G_0(s)G_p(s)$. The closed-loop transfer function is

$$\frac{C(z)}{R(z)} = \frac{KG(z)D(z)}{1 + KG(z)D(z)}. \tag{13.55}$$

FIGURE 13.26
Closed-loop
system with a
digital controller.

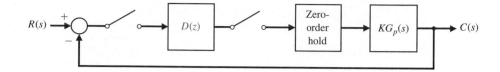

The characteristic equation is

$$1 + KG(z)D(z) = 0,$$

which is analogous to the characteristic equation for the s-plane analysis of $KG(s)$. Thus we can plot the root locus for the characteristic equation of the sampled system as K varies. The rules for obtaining the root locus are summarized in Table 13.5.

EXAMPLE 13.8   **Root-locus of second-order system**

Consider the system shown in Fig. 13.26 with $D(z) = 1$ and $G_p(s) = 1/s^2$. Then, we obtain

$$KG(z) = \frac{T^2}{2} \frac{K(z + 1)}{(z - 1)^2}.$$

Let $T = \sqrt{2}$ and plot the root locus. We now have

$$KG(z) = \frac{K(z + 1)}{(z - 1)^2},$$

TABLE 13.5   Root Locus in the z-Plane

1. The root locus starts at the poles and progresses to the zeros.

2. The root locus lies on a section of the real axis to the left of an odd number of poles and zeros.

3. The root locus is symmetrical with respect to the horizontal real axis.

4. The root locus may break away from the real axis and may reenter the real axis. The breakaway and entry points are determined from the equation

$$K = -\frac{N(z)}{D(z)} = F(z),$$

with $z = \sigma$. Then obtain the solution of $\dfrac{dF(\sigma)}{d\sigma} = 0.$

5. Plot the locus of roots that satisfy

$$1 + KG(z)D(z) = 0$$

or

$$|KG(z)D(z)| = 1$$

and

$$\underline{/G(z)D(z)} = 180° \pm k360°, \quad k = 0, 1, 2, \ldots.$$

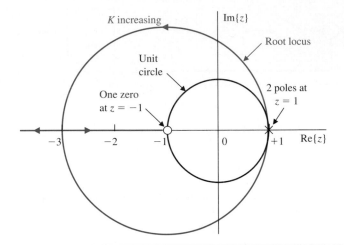

FIGURE 13.27
Root locus for
Example 13.8.

and the poles and zeros are shown on the z-plane in Fig. 13.27. The characteristic equation is

$$1 + KG(z) = 1 + \frac{K(z + 1)}{(z - 1)^2} = 0.$$

Let $z = \sigma$ and solve for K to obtain

$$K = -\frac{(\sigma - 1)^2}{(\sigma + 1)} = F(\sigma).$$

Then, obtain the derivative $dF(\sigma)/d\sigma = 0$ and calculate the roots as $\sigma_1 = -3$ and $\sigma_2 = 1$. The locus leaves the two poles at $\sigma_2 = 1$ and reenters at $\sigma_1 = -3$, as shown in Fig. 13.27. The unit circle is also shown in Fig. 13.27. Clearly, the system always has two roots outside the unit circle and is always unstable for all $K > 0$. ■

We now turn to the design of a digital controller $D(z)$ to achieve a specified response utilizing a root locus method. We will select a controller

$$D(z) = \frac{(z - a)}{(z - b)}.$$

Use $(z - a)$ to cancel one pole at $G(z)$ that lies on the positive real axis of the z-plane. Then select $(z - b)$ so that the locus of the compensated system will give a set of complex roots at a desired point within the unit circle on the z-plane.

EXAMPLE 13.9   **Design of a digital compensator**

Let us design a compensator $D(z)$ that will result in a stable system when $G_p(s)$ is as described in Example 13.8. With $D(z) = 1$, we have an unstable system. Select a $D(z)$ as

$$D(z) = \frac{z - a}{z - b}$$

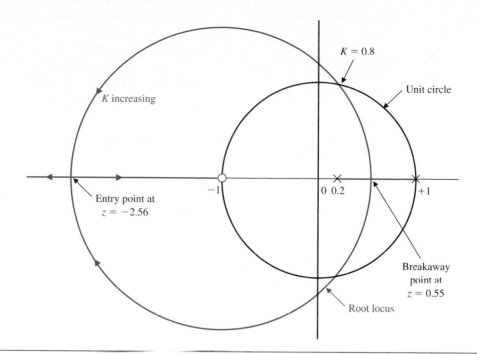

FIGURE 13.28
Root locus for
Example 13.9.

so that

$$KG(z)D(z) = \frac{K(z + 1)(z - a)}{(z - 1)^2 (z - b)}.$$

If we set $a = 1$ and $b = 0.2$, we have

$$KG(z)D(z) = \frac{K(z + 1)}{(z - 1)(z - 0.2)}.$$

Using the equation for $F(\sigma)$, we obtain the entry point as $z = -2.56$, as shown in Fig. 13.28. The root locus is on the unit circle at $K = 0.8$. Thus the system is stable for $K < 0.8$. If we select $K = 0.25$, we find that the step response has an overshoot of 20% and a settling time equal to 8.5 seconds.

If the system performance were inadequate, we would improve the root locus by selecting $a = 1$ and $b = -0.98$ so that

$$KG(z)D(z) = \frac{K(z + 1)}{(z - 1)(z + 0.98)} \cong \frac{K}{(z - 1)}.$$

Then, the root locus would lie on the real axis of the z-plane. When $K = 1$, the root of the characteristic equation is at the origin and $T(z) = 1/z = z^{-1}$. Then, the response of the sampled system (at the sampling instants) is the input step delayed by one sampling period. ■

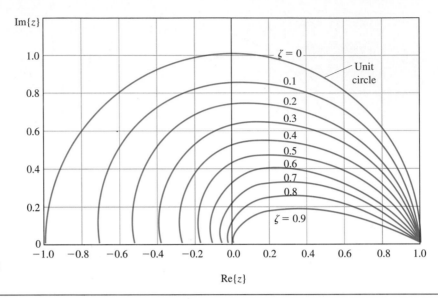

FIGURE 13.29
Curves of constant ζ on the z-plane.

We can draw lines of constant ζ on the z-plane. The mapping between the s-plane and the z-plane is obtained by the relation $z = e^{sT}$. The lines of constant ζ on the s-plane are radial lines with

$$\frac{\sigma}{\omega} = -\tan\theta = -\tan(\sin^{-1}\zeta) = -\frac{\zeta}{\sqrt{1-\zeta^2}}.$$

Then, since $s = \sigma + j\omega$, we have

$$z = e^{\sigma T}e^{j\omega T},$$

where

$$\sigma = -\frac{\zeta}{\sqrt{1-\zeta^2}}\,\omega.$$

The plot of these lines for constant ζ is shown on Fig. 13.29 for a range of T. A common value of ζ for many design specifications is $\zeta = 1/\sqrt{2}$. Then, we have $\sigma = -\omega$ and

$$z = e^{-\omega T}e^{j\omega T} = e^{-\omega T}\,\underline{/\theta},$$

where $\theta = \omega T$.

13.11   IMPLEMENTATION OF DIGITAL CONTROLLERS

We will consider the PID controller with an s-domain transfer function

$$\frac{U(s)}{X(s)} = G_c(s) = K_1 + \frac{K_2}{s} + K_3 s. \tag{13.56}$$

We can determine a digital implementation of this controller by using a discrete approximation for the derivative and integration. For the time derivative, we use the *backward difference rule*

$$u(kT) = \frac{dx}{dt}\bigg|_{t=kT} = \frac{1}{T}(x(kT) - x[(k-1)T]).$$ (13.57)

The *z*-transform of Eq. (13.57) is then

$$U(z) = \frac{(1 - z^{-1})}{T} X(z) = \frac{(z - 1)}{Tz} X(z).$$

The integration of $x(t)$ can be represented by the *forward-rectangular integration* at $t = kT$ as

$$u(kT) = u[(k-1)T] + Tx(kT)$$ (13.58)

where $u(kT)$ is the output of the integrator at $t = kT$. The *z*-transform of Eq. (13.58) is

$$U(z) = z^{-1}U(z) + TX(z),$$

and the transfer function is then

$$\frac{U(z)}{X(z)} = \frac{Tz}{(z - 1)}.$$

Hence, the *z*-domain transfer function of the PID controller is

$$G_c(z) = K_1 + \frac{K_2 Tz}{(z - 1)} + K_3 \frac{(z - 1)}{Tz}.$$ (13.59)

The complete difference equation algorithm that provides the PID controller is obtained by adding the three terms to obtain (we use $x(kT) = x(k)$)

$$u(k) = K_1 x(k) + K_2[u(k - 1) + Tx(k)] + (K_3/T)[x(k) - x(k - 1)]$$ (13.60)

$$= [K_1 + K_2 T + (K_3/T)]x(k) + K_3 Tx(k - 1) + K_2 u(k - 1).$$

Equation (13.60) can be implemented using a digital computer or microprocessor. Of course, we can obtain a PI or PD controller by setting an appropriate gain equal to zero.

13.12 DIGITAL CONTROL SYSTEMS USING *MATLAB*

The process of designing and analyzing sampled-data systems is enhanced with the use of interactive computer tools. Many of the *Matlab* functions for continuous-time control design have equivalent counterparts for sampled-data systems. Model conversion can be accomplished with the functions c2dm and d2cm, shown in Fig. 13.30. The function c2dm converts continuous-time systems to discrete-time systems; the function d2cm converts discrete-time systems to continuous-time systems. For example, consider the plant transfer function

$$G_p(s) = \frac{1}{s(s + 1)},$$

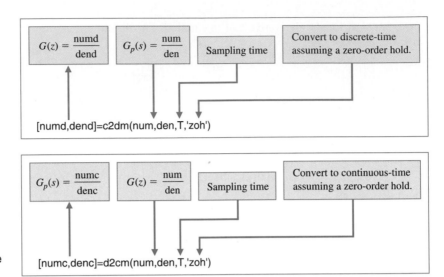

FIGURE 13.30
The **c2dm** and the **d2cm** functions.

as shown in Fig. 13.16. For a sampling period of $T = 1$ second, we know from Eq. (13.16) that

$$G(z) = \frac{0.3678(z + 0.7189)}{(z - 1)(z - 0.3680)} = \frac{0.3679z + 0.2644}{z^2 - 1.368z + 0.3680}. \qquad (13.61)$$

We can use *Matlab* to obtain the $G(z)$, as shown in Fig. 13.31.

The functions dstep, dimpulse, and dlsim are used for simulation of sampled-data systems. The unit step response is generated by dstep. The dstep function format is shown in Fig. 13.32. The unit impulse response is generated by the function dimpulse, and the response to an arbitrary input is obtained by the dlsim function. The dimpulse and dlsim functions are shown in Figs. 13.33 and 13.34, respectively. These sampled-data system

FIGURE 13.31
Using the **c2dm** function to convert $G(s) = G_0(s)G_p(s)$ to $G(z)$.

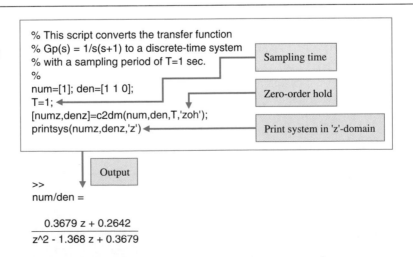

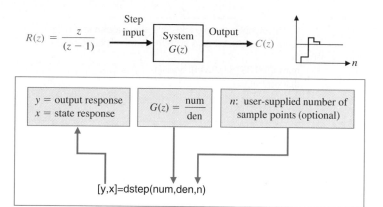

FIGURE 13.32
The **dstep** function generates the output $c(kT)$ for a step input.

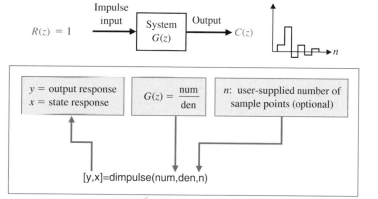

FIGURE 13.33
The **dimpulse** function generates the output $c(kT)$ for an impulse input.

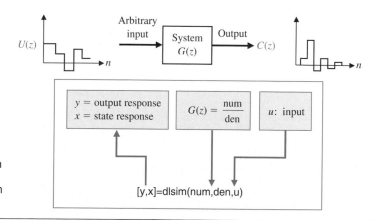

FIGURE 13.34
The **dlsim** function generates the output $c(kT)$ for an arbitrary input.

simulation functions operate in essentially the same manner as their counterparts for continuous-time (unsampled) systems. The output is $c(kT)$ and is shown as $c(kT)$ held constant for the period T.

We now reconsider Example 13.4 and approach the problem of obtaining a step response without utilizing long division.

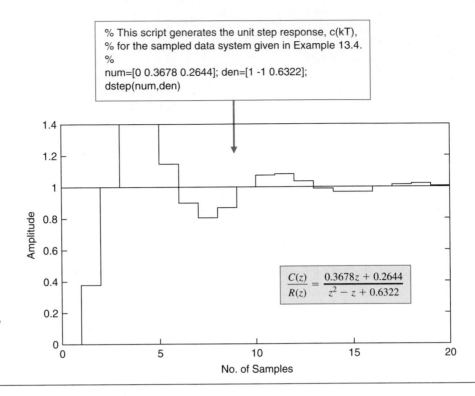

% This script generates the unit step response, c(kT),
% for the sampled data system given in Example 13.4.
%
num=[0 0.3678 0.2644]; den=[1 -1 0.6322];
dstep(num,den)

FIGURE 13.35
The discrete
response, $c(kT)$,
of a sampled
second-order
system to a
unit step.

$$\frac{C(z)}{R(z)} = \frac{0.3678z + 0.2644}{z^2 - z + 0.6322}$$

EXAMPLE 13.10 Unit step response

In Example 13.4, we considered the problem of computing the step response of a closed-loop sampled-data system. In that example, the response, $c(kT)$, was computed using long division. We can use *Matlab* to compute the response $c(kT)$ using the dstep function, shown in Fig. 13.32. With the closed-loop transfer function given by

$$\frac{C(z)}{R(z)} = \frac{0.3678z + 0.2644}{z^2 - z + 0.6322},$$

the associated closed-loop step response is shown in Fig. 13.35. The discrete step response shown in this figure is also shown in Fig. 13.17. To determine the actual continuous response, $c(t)$, we use the *Matlab* script as shown in Fig. 13.36. The zero-order hold is modeled by the transfer function $G_0(s)$, where

$$G_0(s) = \frac{1 - e^{-sT}}{s}.$$

In the *Matlab* script in Fig. 13.36, we approximate the e^{-sT} term using the pade function with a second-order approximation and a sampling time of 1 second. We then compute an approximation for $G_0(s)$ based on the Padé approximation of e^{-sT}. ∎

The subject of digital computer compensation was discussed in Section 13.8. In the next example, we reconsider the subject utilizing *Matlab*.

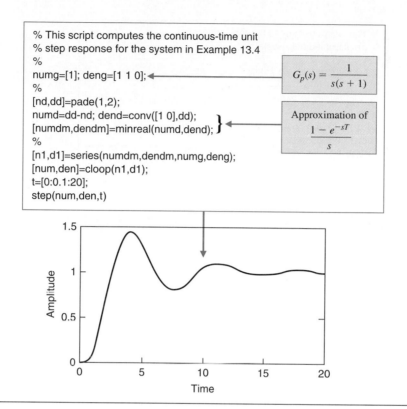

```
% This script computes the continuous-time unit
% step response for the system in Example 13.4
%
numg=[1]; deng=[1 1 0];          ◄──────  $G_p(s) = \dfrac{1}{s(s + 1)}$
%
[nd,dd]=pade(1,2);
numd=dd-nd; dend=conv([1 0],dd);     }    Approximation of
[numdm,dendm]=minreal(numd,dend);   }  ◄──  $\dfrac{1 - e^{-sT}}{s}$
%
[n1,d1]=series(numdm,dendm,numg,deng);
[num,den]=cloop(n1,d1);
t=[0:0.1:20];
step(num,den,t)
```

FIGURE 13.36
The continuous
response, $c(t)$,
to a unit step for
the system of
Fig. 13.16.

EXAMPLE 13.11   Root locus of a digital control system

Recall from Eq. (13.16) that a plant is given by

$$G(z) = \frac{0.3678(z + 0.7189)}{(z - 1)(z - 0.3680)}.$$

The compensator is selected to be

$$D(z) = \frac{K(z - 0.3678)}{z + 0.2400},$$

with the parameter K as a variable yet to be determined. When

$$G(z)D(z) = K\frac{0.3678(z + 0.7189)}{(z - 1)(z + 0.2400)} \tag{13.62}$$

we have the problem in a form for which the root locus method is directly applicable. The rlocus function works for discrete-time systems in the same way as for continuous-time systems. Using a *Matlab* script, the root locus associated with Eq. (13.62) is easily generated, as shown in Fig. 13.37. Remember that the stability region is defined by the unit circle in the complex plane. The *Matlab* function rlocfind can be used with the discrete-time system root locus in exactly the same way as for continuous-time systems to determine the

```
% This script generates the root locus for
% the sampled data system
%
%      K (0.3678) (z+0.7189)
%      -------------------------------
%        (z-1)(z+0.2400)
%
num=[0.3678 0.2644]; den=[1.0000   -0.7600   -0.2400];
rlocus(num,den);hold on
x=[-1:0.1:1];y=sqrt(ones(1,length(x))-x.^2);
plot(x,y,'--',x,-y,'--')
```

Plot unit circle.

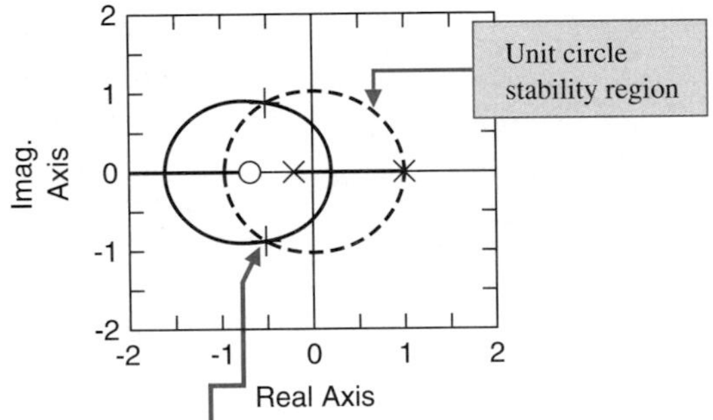

Unit circle stability region

Determine K at the unit circle boundary.

```
>>rlocfind(num,den)
Select a point in the graphics window

selected_point =
    -0.4787 + 0.8530i

ans =
    4.6390
```

$K = 4.639$

FIGURE 13.37
The **rlocus** function for sampled data systems.

value of the system gain associated with any point on the locus. Using rlocfind, we determine that $K = 4.639$ places the roots on the unit circle. ∎

13.13 SUMMARY

The use of a digital computer as the compensation device for a closed-loop control system has grown during the past two decades as the price and reliability of computers have improved dramatically. A computer can be used to complete many calculations during the sampling interval T and to provide an output signal that is used to drive an actuator of a process. Computer control is used today for chemical processes, aircraft control, machine tools, and many common processes.

The z-transform can be used to analyze the stability and response of a sampled system and to design appropriate systems incorporating a computer. Computer control systems have become increasingly common as low-cost computers have become readily available.

EXERCISES

E13.1 State whether the following signals are discrete or continuous.

(a) Elevation contours on a map.
(b) Temperature in a room.
(c) Digital clock display.
(d) The score of a basketball game.
(e) The output of a loudspeaker.

E13.2 (a) Find the values $c(kT)$ when

$$C(z) = \frac{z}{z^2 - 3z + 2}$$

for $k = 0$ to 4. (b) Obtain a closed form of solution for $c(kT)$ as a function of k.

E13.3 A system has a response $c(kT) = kT$ for $k \geq 0$. Find $C(z)$ for this response.

E13.4 We have a function

$$Y(s) = \frac{5}{s(s + 1)(s + 5)}.$$

Using a partial fraction expansion of $Y(s)$ and Table 13.1, find $Y(z)$, when $T = 0.2$ second.

E13.5 The space shuttle, with its robot arm, is shown in Fig. E13.5(a). An astronaut controls the robot arm and gripper by using a window and the TV cameras [9]. Discuss the use of digital control for this system and draw a block diagram for the system, including a computer for display generation and control.

E13.6 Computer control of a robot to spray paint an automobile is shown by the system in Fig. E13.6 [1]. The system is of the type shown in Fig. 13.25 where

$$KG_p(s) = \frac{20}{s(s/2 + 1)},$$

and we desire a phase margin of 45°. A compensator for this system was obtained in Section 10.8. Obtain the $D(z)$ required when $T = 0.001$ second.

FIGURE E13.5
(a) Space shuttle and robot arm and (b) astronaut control of the arm.

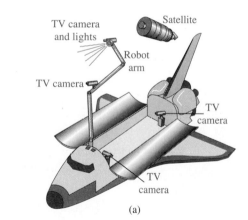

(a)

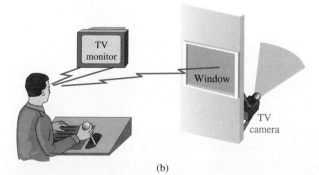

(b)

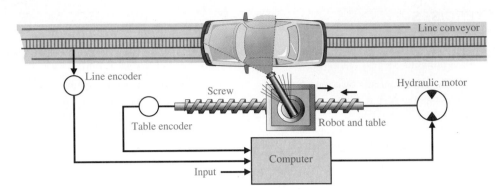

FIGURE E13.6
Automobile spray
paint system.

E13.7 Find the response for the first four sampling instants for

$$Y(z) = \frac{z^3 + 2z^2 + 1}{z^3 - 1.5z^2 + 0.5z}.$$

Then find $y(0)$, $y(1)$, $y(2)$ and $y(3)$.

E13.8 Determine whether the closed-loop system with $T(z)$ is stable when

$$T(z) = \frac{z^2 + z}{z^2 + 0.1z - 0.2}.$$

E13.9 (a) Determine $c(kT)$ for $k = 0$ to 3 when

$$C(z) = \frac{z + 1}{z^2 - 1}.$$

(b) Determine the closed form solution for $c(kT)$ as a function of k.

E13.10 A system has $G(z)$ as described by Eq. (13.34) with $T = 0.01$ second and $\tau = 0.008$ second. (a) Find K so that the overshoot is less than 40%. (b) Determine the steady-state error in response to a unit ramp input. (c) Determine K to minimize the integral squared error.

E13.11 A system has a plant transfer function

$$G_p(s) = \frac{100}{s^2 + 100}.$$

(a) Determine $G(z)$ for $G_p(s)$ preceded by a zero-order hold with $T = 0.05$ second. (b) Determine whether the digital system is stable. (c) Plot the impulse response of $G(z)$ for the first 15 samples. (d) Plot the response for a sine wave input with the same frequency as the natural frequency of the system.

E13.12 Find the z-transform of

$$X(s) = \frac{(s + 3)}{s^2 + 3s + 2}$$

when the sampling period is 2 seconds.

E13.13 The characteristic equation of a sampled system is

$$z^2 + (K - 1.5)z + 0.5 = 0.$$

Find the range of K so that the system is stable.

E13.14 A unity feedback system as shown in Fig. 13.18 has a plant

$$G_p(s) = \frac{K}{s(s + 3)},$$

with $T = 0.5$. Determine whether the system is stable when $K = 5$. Determine the maximum value of K for stability.

PROBLEMS

P13.1 The input to a sampler is $r(t) = \sin \omega t$, where $\omega = 1/\pi$. Plot the input to the sampler and the output $r^*(t)$ for the first 2 seconds when $T = 0.25$ second.

P13.2 The input to a sampler is $r(t) = \sin \omega t$, where $\omega = 1/\pi$. The output of the sampler enters a zero-order hold, as shown in Fig. 13.7. Plot the output of the hold circuit $p(t)$ for the first 2 seconds when $T = 0.25$ second.

P13.3 A unit ramp $r(t) = t$, $t > 0$, is used as an input to a process where $G(s) = 1/(s + 1)$, as shown in Fig. P13.3. Determine the output $c(kT)$ for the first four sampling instants.

FIGURE P13.3 Sampling system.

P13.4 A closed-loop system has a hold circuit and process as shown in Fig. 13.18. Determine $G(z)$ when $T = 1$ and

$$G_p(s) = \frac{1}{(s + 1)}.$$

P13.5 For the system in Problem 13.4, let $r(t)$ be a unit step input and calculate the response of the system by synthetic division.

P13.6 For the output of the system in Problem 13.4, find the initial and final values of the output directly from $C(z)$.

P13.7 A closed-loop system is shown in Fig. 13.18. This system represents the pitch control of an aircraft. The plant transfer function is $G_p(s) = K/[s(0.5s + 1)]$. Select a gain K and sampling period T so that the overshoot is limited to 0.3 for a unit step input and the steady-state error for a unit ramp input is less than 1.0.

P13.8 Consider the computer compensated system shown in Fig. 13.26 when $T = 1$ and

$$KG_p(s) = \frac{K}{s(s + 1)}.$$

Select the parameters K and r of $D(z)$ when

$$D(z) = \frac{(z - 0.3678)}{(z + r)}.$$

Select within the range: $1 < K < 2$ and $0 < r < 1$.
Determine the response of the compensated system and compare it with the uncompensated system.

P13.9 A new suspended, mobile, remote-controlled system to bring three-dimensional mobility to professional NFL football is shown in Fig. P13.9. The camera can be moved over the field as well as up and down. The motor control on each pulley is rep-

resented by Fig. 13.18 with

$$G_p(s) = \frac{10}{s(s + 1)(s/10 + 1)}.$$

We wish to achieve a phase margin of 45° using $G_c(s)$. Select a suitable crossover frequency and sampling period to obtain $D(z)$. Use the $G_c(s)$-to-$D(z)$ conversion method.

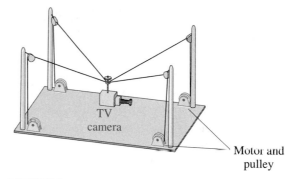

FIGURE P13.9 Mobile camera for football field.

P13.10 Consider a system as shown in Fig. 13.15 with a zero-order hold and a plant

$$G_p(s) = \frac{1}{s(s + 10)}$$

and $T = 0.1$ second.
(a) Let $D(z) = K$ and determine the transfer function $G(z)D(z)$. (b) Determine the characteristic equation of the closed-loop system. (c) Calculate the maximum value of K for a stable system. (d) Determine K such that the overshoot is less than 30%. (e) Calculate the closed-loop transfer function $T(z)$ for K of part (d) and plot the step response. (f) Determine the location of the closed-loop roots and the overshoot if K is one-half of the value determined in part (c). (g) Plot the step response for the K of part (f).

P13.11 (a) For the system described in Problem 13.10, design a lag compensator $G_c(s)$ using the methods of Chapter 10 to achieve an overshoot less than 30% and a steady-state error less than 0.01 for a ramp input. Assume a continuous nonsampled system with $G_p(s)$.
(b) Determine a suitable $D(z)$ to satisfy the requirements of part (a) with a sampling period $T = 0.1$ second. Assume a zero-order hold and sampler and use the $G_c(s)$-to-$D(z)$ conversion method.
(c) Plot the step response of the system with the

continuous-time compensator $G_c(s)$ of part (a) and of the digital system with the $D(z)$ of part (b). Compare the results.

(d) Repeat part (b) for $T = 0.01$ second and then repeat part (c).

(e) Plot the ramp response for $D(z)$ with $T = 0.1$ second and compare it with the continuous-system response.

P13.12   The transfer function of a plant and a zero-order hold (Fig. 13.18) is

$$G(z) = \frac{K(z + 0.1)}{z(z - 1)}.$$

(a) Plot the root locus. (b) Determine the range of gain K for a stable system.

P13.13   The space station orientation controller described in Exercise 7.6 is implemented with a sampler and hold and has the transfer function (Fig. 13.18)

$$G(z) = \frac{K(z^2 + 1.1206z - 0.0364)}{z^3 - 1.7358z^2 + 0.8711z - 0.1353}.$$

(a) Plot the root locus. (b) Determine the value of K so that two of the roots of the characteristic equation are equal. (c) Determine all the roots of the characteristic equation for the gain of part (b).

P13.14   A sampled-data system with a sampling period $T = 0.05s$ (Fig. 13.18) is

$$G(z) =$$

$$\frac{K(z^3 + 10.3614z^2 + 9.758z + 0.8353)}{z^4 - 3.7123z^3 + 5.1644z^2 - 3.195z + 0.7408}.$$

(a) Plot the root locus. (b) Determine K when the two real poles break away from the real axis. (c) Calculate the maximum K for stability.

P13.15   A closed-loop system with a sampler and hold, as shown in Fig. 13.18, has a plant transfer function

$$G_p(s) = \frac{4}{(s - 4)}.$$

Calculate and plot $c(kT)$ for $0 \le T \le 0.6$ when $T = 0.1$ second. The input signal is a unit step.

P13.16   A closed-loop system as shown in Fig. 13.18 has

$$G_p(s) = \frac{1}{s(s + 1)}.$$

Calculate and plot $c(kT)$ for $0 \le k \le 8$ when $T = 1$ second and the input is a unit step.

P13.17   A closed-loop system as shown in Fig. 13.18 has

$$G_p(s) = \frac{K}{s(s + 1)}$$

and $T = 1$ second. Plot the root locus for $K \ge 0$, and determine the gain K that results in the two roots of the characteristic equation on the z-circle (at the stability limit).

P13.18   A unity feedback system as shown in Fig. 13.18 has

$$G_p(s) = \frac{K}{s(s + 2)}.$$

If the system is continuous ($T = 0$), then $K = 2$ yields a step response with an overshoot of 4.5% and a settling time of 4 seconds. Plot the response for $0 \le T \le 1.2$, varying T by increments of 0.2 when $K = 2$. Complete a table recording overshoot and settling time versus T.

ADVANCED PROBLEMS

AP13.1   A closed-loop system as shown in Fig. 13.18 has a plant

$$G_p(s) = \frac{K(1 + as)}{s^2},$$

where a is adjustable to achieve a suitable response. Plot the root locus when $a = 10$. De-

termine the range of K for stability when $T = 1$ second.

AP13.2   Increasing constraints on weight, performance, fuel consumption, and reliability created a need for a new type of flight control system known as fly-by-wire. This approach implies that particular system components are interconnected elec-

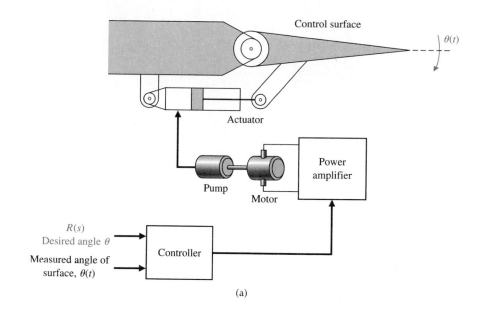

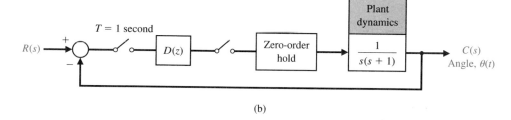

FIGURE AP13.2
(a) Fly-by-wire aircraft control surface system and (b) block diagram. The sampling period is 1 second.

trically rather than mechanically and that they operate under the supervision of a computer responsible for monitoring, control, and coordination tasks. The fly-by-wire principle allows for the implementation of totally digital and highly redundant control systems reaching a remarkable level of reliability and performance [24].

Operational characteristics of a flight control system greatly depend on the dynamic stiffness of an actuator, representing its ability to maintain the position of the control surface in spite of the disturbing effects of random external forces.

One flight actuator system consists of a special type of dc motor, driven by a power amplifier, which drives a hydraulic pump, connected to either side of a hydraulic cylinder. The ram of the hydraulic cylinder is directly connected to a control surface of an aircraft through some appropriate mechanical linkage, as shown in Fig. AP13.2(a).

The block diagram of the closed-loop system is shown in Fig. AP13.2(b). A step response with

an overshoot less than 5% and a settling time less than 4 seconds is desired. (a) Design $D(z)$ and plot the system root locus. (b) Plot the system response of the system design of part (a).

AP13.3 A manufacturer uses an adhesive to form a seam along the edge of the material, as shown in Fig. AP13.3. It is critical that the glue be applied evenly to avoid flaws; however, the speed at which the material passes beneath the dispensing head is not constant. The glue needs to be dispensed at a rate proportional to the varying speed of the material. The controller adjusts the valve that dispenses the glue [12].

The system can be represented by the block diagram shown in Fig. 13.15, where $G_p(s) = 2/(0.03s + 1)$ and a zero-order hold $G_0(s)$. Use a controller

$$D(z) = \frac{KT}{1 - z^{-1}} = \frac{KTz}{z - 1}$$

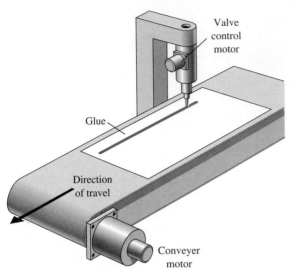

FIGURE AP13.3   A glue control system.

that represents an integral controller. Determine $G(z)D(z)$ for $T = 30$ ms, and plot the root locus. Select an appropriate gain K and plot the step response.

AP13.4   A system of the form shown in Fig. 13.15 has $D(z) = K$ and

$$G_p(s) = \frac{27}{s(s + 27)}.$$

When $T = 0.1$, find a suitable K for a rapid step response with an overshoot less than 10%.

AP13.5   A system of the form shown in Fig. 13.18 has

$$G_p(s) = \frac{10}{s + 1}.$$

Determine the range of sampling period T for which the system is stable. Select a sampling period T so that the system is stable and provides a rapid response.

DESIGN PROBLEMS

DP13.1   A temperature system as shown in Fig. 13.15 has a plant transfer function

$$G_p(s) = \frac{0.8}{3s + 1}$$

and a sampling period T of 0.5 second.
  (a) Using $D(z) = K,$ select a gain K so that the system is stable. (b) The system may be slow and overdamped, and thus we seek to design a lead network using the method of Section 10.8. Determine a suitable controller $G_c(s)$ and then calculate $D(z)$. (c) Verify the design obtained in part (b) by plotting the step response of the system for the selected $D(z)$.

DP13.2   A disk drive read-write head-positioning system has a system as shown in Fig. 13.15 [11]. The plant transfer function is

$$G_p(s) = \frac{5}{s^2 + 0.4s + 1000}.$$

Accurate control using a digital compensator is required. Let $T = 5$ ms and design a compensator, $D(z)$, using (a) the $G_c(s)$-to-$D(z)$ conversion method and (b) the root locus method.

DP13.3   Vehicle traction control, which includes antiskid braking and antispin acceleration, can enhance vehicle performance and handling. The

objective of this control is to maximize tire traction by preventing the wheels from locking during braking and from spinning during acceleration.

  Wheel slip, the difference between the vehicle speed and the wheel speed (normalized by the vehicle speed for braking and the wheel speed for acceleration), is chosen as the controlled variable for most of the traction-control algorithm because of its strong influence on the tractive force between the tire and the road [20].

  A model for one wheel is shown in Fig. DP13.3 when λ is the wheel slip. The goal is to minimize the slip when a disturbance occurs due to road conditions. Design a controller $D(z)$ so that the ζ of the system is $1/\sqrt{2}$, and determine the resulting K. Assume $T = 0.1$ second. Plot the resulting step response, and find the overshoot and settling time.

DP13.4   A machine-tool system has the form shown in Fig. 13.25 with [10]

$$G_p(s) = \frac{0.1}{s(s + 0.1)}.$$

The sampling rate is chosen as $T = 1$ second. It is desired that the step response have an overshoot of 16% or less and a settling time of 12 seconds or less. Also, the error to a unit ramp input, $r(t) = t,$

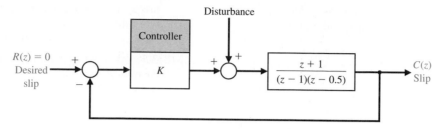

FIGURE DP13.3

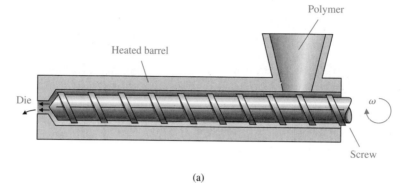

(a)

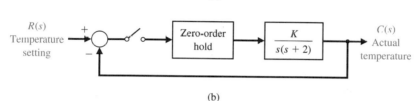

FIGURE DP13.5
Control system for
an extruder.

(b)

must be less than or equal to 1. Design a $D(z)$ to achieve these specifications.

DP13.5 Plastic extrusion is a well-established method widely used in the polymer processing industry [12]. Such extruders typically consist of a large barrel divided into several temperature zones, with a hopper at one end and a die at the other. Polymer is fed into the barrel in raw and solid form from the hopper and is pushed forward by a powerful screw. Simultaneously, it is gradually heated while passing through the various temperature zones set in gradually increasing temperatures. The heat produced by the heaters in the barrel, together with the heat released from the friction between the raw polymer and the surfaces of the barrel and the screw, eventually causes the melting of the polymer, which is then pushed by the screw out from the die, to be processed further for various purposes.

The output variables are the outflow from the die and the polymer temperature. The main controlling variable is the screw speed, since the response of the process to it is rapid.

The control system for the output polymer temperature is shown in Fig. DP13.5. Select a gain K and a sampling period T to obtain a step overshoot of 10% while reducing the steady-state error for a ramp input.

MATLAB PROBLEMS

MP13.1 Using *Matlab,* plot the unit step response of the system

$$G(z) = \frac{0.2145z + 0.1609}{z^2 - 0.75z + 0.125}.$$

Verify graphically that the steady-state value of the output is 1.

MP13.2 Convert the following continuous-time transfer functions to sampled-data systems using the c2dm function. Assume a sample period of 1 second and a zero-order hold, $G_0(s)$.

(a) $G_p(s) = \dfrac{1}{s}$

(b) $G_p(s) = \dfrac{s}{s^2 + 4}$

(c) $G_p(s) = \dfrac{s + 5}{s + 1}$

(d) $G_p(s) = \dfrac{1}{s(s + 1)}$

MP13.3 The closed-loop transfer function of a sampled-data system is given by

$$T(z) = \frac{C(z)}{R(z)} = \frac{1.7(z + 0.46)}{z^2 + z + 0.5}.$$

(a) Compute the unit step response of the system using the dstep function. (b) Determine the continuous-time transfer function equivalent of $T(z)$ using the d2cm function and assume a sampling period of $T = 0.1$ second. (c) Compute the unit step response of the continuous (nonsampled) system using the step function, and compare the plot with part (a).

MP13.4 Consider the sampled data system with the loop transfer function

$$G(z)D(z) = K\frac{z^2 + 4z + 4.25}{z^2 - 0.1z - 1.5}.$$

(a) Plot the root locus using the rlocus function. (b) From the root locus, determine the range of K for stability. Use the rlocfind function.

MP13.5 An industrial grinding process is given by the transfer function [17]

$$G_p(s) = \frac{10}{s(s + 5)}.$$

The objective is to use a digital computer to improve the performance, where the transfer function of the computer is represented by $D(z)$. The design specifications are (1) phase margin greater than 45°, and (2) settling time less than 1 second.
        (a) Design a controller

$$G_c(s) = K\frac{s + a}{s + b}$$

to meet the design specifications. (b) Assuming a sampling time of $T = 0.02$ second, convert $G_c(s)$ to $D(z)$. (c) Simulate the continuous-time closed-loop system with a unit step input. (d) Simulate the sampled data closed-loop system with a unit step input. (e) Compare the results in parts (c) and (d) and comment.

TERMS AND CONCEPTS

Amplitude quantization error The sampled signal available only with a limited precision. The error between the actual signal and the sampled signal.

Digital computer compensator A system that uses a digital computer as the compensator element.

Digital control system A control system using digital signals and a digital computer to control a process.

Sampled data Data obtained for the system variables only at discrete intervals. Data obtained once every sampling period.

Sampled-data system A system where part of the system acts on sampled data (sampled variables).

Sampling period The period when all the numbers leave or enter the computer. The period for which the sampled variable is held constant.

Stability of a sampled data system The stable condition exists when all the poles of the closed-loop transfer function $T(z)$ are within the unit circle on the z-plane.

z-plane The plane with the vertical axis equal to the imaginary part of z and the horizontal axis equal to the real part of z.

z-transform A conformal mapping from the s-plane to the z-plane by the relation $z = e^{sT}$. A transform from the s-domain to the z-domain.

Laplace Transform Pairs

TABLE A.1

$F(s)$	$f(t),\ t \geq 0$
1. 1	$\delta(t_0)$, unit impulse at $t = t_0$
2. $1/s$	1, unit step
3. $\dfrac{n!}{s^{n+1}}$	t^n
4. $\dfrac{1}{(s + a)}$	e^{-at}
5. $\dfrac{1}{(s + a)^n}$	$\dfrac{1}{(n - 1)!}\, t^{n-1} e^{-at}$
6. $\dfrac{a}{s(s + a)}$	$1 - e^{-at}$
7. $\dfrac{1}{(s + a)(s + b)}$	$\dfrac{1}{(b - a)}\,(e^{-at} - e^{-bt})$
8. $\dfrac{s + \alpha}{(s + a)(s + b)}$	$\dfrac{1}{(b - a)}\,[(\alpha - a)e^{-at} - (\alpha - b)e^{-bt}]$
9. $\dfrac{ab}{s(s + a)(s + b)}$	$1 - \dfrac{b}{(b - a)}\,e^{-at} + \dfrac{a}{(b - a)}\,e^{-bt}$
10. $\dfrac{1}{(s + a)(s + b)(s + c)}$	$\dfrac{e^{-at}}{(b - a)(c - a)} + \dfrac{e^{-bt}}{(c - a)(a - b)} + \dfrac{e^{-ct}}{(a - c)(b - c)}$
11. $\dfrac{s + \alpha}{(s + a)(s + b)(s + c)}$	$\dfrac{(\alpha - a)e^{-at}}{(b - a)(c - a)} + \dfrac{(\alpha - b)e^{-bt}}{(c - b)(a - b)} + \dfrac{(\alpha - c)e^{-ct}}{(a - c)(b - c)}$
12. $\dfrac{ab(s + \alpha)}{s(s + a)(s + b)}$	$\alpha - \dfrac{b(\alpha - a)}{(b - a)}\,e^{-at} + \dfrac{a(\alpha - b)}{(b - a)}\,e^{-bt}$
13. $\dfrac{\omega}{s^2 + \omega^2}$	$\sin \omega t$
14. $\dfrac{s}{s^2 + \omega^2}$	$\cos \omega t$
15. $\dfrac{s + \alpha}{s^2 + \omega^2}$	$\dfrac{\sqrt{\alpha^2 + \omega^2}}{\omega}\,\sin(\omega t + \phi),\ \phi = \tan^{-1} \omega/\alpha$

(continued)

TABLE A.1 *Continued*

$F(s)$	$f(t), t \geq 0$
16. $\dfrac{\omega}{(s + a)^2 + \omega^2}$	$e^{-at} \sin \omega t$
17. $\dfrac{(s + a)}{(s + a)^2 + \omega^2}$	$e^{-at} \cos \omega t$
18. $\dfrac{s + \alpha}{(s + a)^2 + \omega^2}$	$\dfrac{1}{\omega}[(\alpha - a)^2 + \omega^2]^{1/2} e^{-at} \sin(\omega t + \phi),$ $\phi = \tan^{-1} \dfrac{\omega}{\alpha - a}$
19. $\dfrac{\omega_n^2}{s^2 + 2\zeta\omega_n s + \omega_n^2}$	$\dfrac{\omega_n}{\sqrt{1 - \zeta^2}} e^{-\zeta\omega_n t} \sin \omega_n \sqrt{1 - \zeta^2}\, t, \quad \zeta < 1$
20. $\dfrac{1}{s[(s + a)^2 + \omega^2]}$	$\dfrac{1}{a^2 + \omega^2} + \dfrac{1}{\omega\sqrt{a^2 + \omega^2}} e^{-at} \sin(\omega t - \phi),$ $\phi = \tan^{-1} \omega/-a$
21. $\dfrac{\omega_n^2}{s(s^2 + 2\zeta\omega_n s + \omega_n^2)}$	$1 - \dfrac{1}{\sqrt{1 - \zeta^2}} e^{-\zeta\omega_n t} \sin(\omega_n\sqrt{1 - \zeta^2}\, t + \phi),$ $\phi = \cos^{-1} \zeta, \zeta < 1$
22. $\dfrac{(s + \alpha)}{s[(s + a)^2 + \omega^2]}$	$\dfrac{\alpha}{a^2 + \omega^2} + \dfrac{1}{\omega}\left[\dfrac{(\alpha - a)^2 + \omega^2}{a^2 + \omega^2}\right]^{1/2} e^{-at} \sin(\omega t + \phi),$ $\phi = \tan^{-1} \dfrac{\omega}{\alpha - a} - \tan^{-1} \dfrac{\omega}{-a}$
23. $\dfrac{1}{(s + c)[(s + a)^2 + \omega^2]}$	$\dfrac{e^{-ct}}{(c - a)^2 + \omega^2} + \dfrac{e^{-at} \sin(\omega t + \phi)}{\omega[(c - a)^2 + \omega^2]^{1/2}}, \phi = \tan^{-1} \dfrac{\omega}{c - a}$

Symbols, Units, and Conversion Factors

TABLE B.1   Symbols and Units

Parameter or Variable Name	Symbol	SI	English
Acceleration, angular	$\alpha(t)$	rad/s^2	rad/s^2
Acceleration, translational	$a(t)$	m/s^2	ft/s^2
Friction, rotational	f	$\dfrac{\text{n-m}}{\text{rad/s}}$	$\dfrac{\text{ft-lb}}{\text{rad/s}}$
Friction, translational	f	$\dfrac{\text{n}}{\text{m/s}}$	$\dfrac{\text{lb}}{\text{ft/s}}$
Inertia, rotational	J	$\dfrac{\text{n-m}}{\text{rad/s}^2}$	$\dfrac{\text{ft-lb}}{\text{rad/s}^2}$
Mass	M	kg	slugs
Position, rotational	$\Theta(t)$	rad	rad
Position, translational	$x(t)$	m	ft
Speed, rotational	$\omega(t)$	rad/s	rad/s
Speed, translational	$v(t)$	m/s	ft/s
Torque	$T(t)$	n-m	ft-lb

TABLE B.2 Conversion Factors

To Convert	Into	Multiply by	To Convert	Into	Multiply by
Btu	ft-lb	778.3	kw	Btu/min	56.92
Btu	joules	1054.8	kw	ft-lb/min	4.462×10^4
Btu/hr	ft-lb/s	0.2162	kw	hp	1.341
Btu/hr	watts	0.2931			
Btu/min	hp	0.02356	miles (statute)	ft	5280
Btu/min	kw	0.01757	mph	ft/min	88
Btu/min	watts	17.57	mph	ft/s	1.467
			mph	m/s	0.44704
cal	joules	4.182	mils	cms	2.540×10^{-3}
cm	ft	3.281×10^{-2}	mils	in.	0.001
cm	in.	0.3937	min (angles)	deg	0.01667
cm^3	ft^3	3.531×10^{-5}	min (angles)	rad	2.909×10^{-4}
deg (angle)	rad	0.01745	n-m	ft-lb	0.73756
deg/s	rpm	0.1667	n-m	dyne-cm	10^7
dynes	gm	1.020×10^{-3}	n-m-s	watt	1.0
dynes	lb	2.248×10^{-6}			
dynes	newtons	10^{-5}	oz	gm	28.349527
			oz-in.	dyne-cm	70,615.7
ft/s	miles/hr	0.6818	oz-in^2	gm-cm^2	1.829×10^2
ft/s	miles/min	0.01136	oz-in.	ft-lb	5.208×10^{-3}
ft-lb	gm-cm	1.383×10^4	oz-in.	gm-cm	72.01
ft-lb	oz-in.	192			
ft-lb/min	Btu/min	1.286×10^{-3}	lb(force)	newtons	4.4482
ft-lb/s	hp	1.818×10^{-3}	lb/ft^3	gm/cm^3	0.01602
ft-lb/s	kw	1.356×10^{-3}	lb-ft-s^2	oz-in^2	7.419×10^4
$\dfrac{\text{ft-lb}}{\text{rad/s}}$	$\dfrac{\text{oz-in.}}{\text{rpm}}$	20.11	rad	deg	57.30
			rad	min	3438
			rad	s	2.063×10^5
gm	dynes	980.7	rad/s	deg/s	57.30
gm	lb	2.205×10^{-3}	rad/s	rpm	9.549
gm-cm^2	oz-in^2	5.468×10^{-3}	rad/s	rps	0.1592
gm-cm	oz-in.	1.389×10^{-2}	rpm	deg/s	6.0
gm-cm	ft-lb	1.235×10^{-5}	rpm	rad/s	0.1047
hp	Btu/min	42.44	s (angle)	deg	2.778×10^{-4}
hp	ft-lb/min	33,000	s (angle)	rad	4.848×10^{-6}
hp	ft-lb/s	550.0	slugs (mass)	kg	14.594
hp	watts	745.7	slug-ft^2	km^2	1.3558
in.	meters	2.540×10^{-2}	watts	Btu/hr	3.413
in.	cm	2.540	watts	Btu/min	0.05688
			watts	ft-lb/min	44.27
joules	Btu	9.480×10^{-4}	watts	hp	1.341×10^{-3}
joules	ergs	10^7	watts	n-m/s	1.0
joules	ft-lb	0.7376	watts-hr	Btu	3.413
joules	watt-hr	2.778×10^{-4}			
kg	lb	2.205			
kg	slugs	6.852×10^{-2}			

An Introduction to Matrix Algebra

C.1 DEFINITIONS

In many situations, we must deal with rectangular arrays of numbers or functions. The rectangular array of numbers (or functions)

$$
\mathbf{A} = \begin{bmatrix}
a_{11} & a_{12} & \cdots & a_{1n} \\
a_{21} & a_{22} & \cdots & a_{2n} \\
\vdots & \vdots & & \vdots \\
a_{m1} & a_{m2} & \cdots & a_{mn}
\end{bmatrix} \tag{C.1}
$$

is known as a *matrix*. The numbers a_{ij} are called *elements* of the matrix, with the subscript i denoting the row and the subscript j denoting the column.

A matrix with m rows and n columns is said to be a matrix of *order* (m, n) or alternatively called an $m \times n$ (m by n) matrix. When the number of the columns equals the number of rows ($m = n$), the matrix is called a *square matrix* of order n. It is common to use boldfaced capital letters to denote an $m \times n$ matrix.

A matrix comprised of only one column, that is, an $m \times 1$ matrix, is known as a column matrix or, more commonly, a *column vector*. We will represent a column vector with boldfaced lowercase letters as

$$
\mathbf{y} = \begin{bmatrix}
y_1 \\
y_2 \\
\vdots \\
y_m
\end{bmatrix}. \tag{C.2}
$$

Analogously, a *row vector* is an ordered collection of numbers written in a row—that is, a $1 \times n$ matrix. We will use boldfaced lowercase letters to represent vectors. Therefore a row vector will be written as

$$
\mathbf{z} = [z_1, z_2, \ldots, z_n] \tag{C.3}
$$

with n elements.

A few matrices with distinctive characteristics are given special names. A square matrix in which all the elements are zero except those on the principal diagonal, $a_{11}, a_{22}, \ldots,$

a_{nn}, is called a *diagonal matrix*. Then, for example, a 3×3 diagonal matrix would be

$$\mathbf{B} = \begin{bmatrix} b_{11} & 0 & 0 \\ 0 & b_{22} & 0 \\ 0 & 0 & b_{33} \end{bmatrix}. \tag{C.4}$$

If all the elements of a diagonal matrix have the value 1, then the matrix is known as the *identity matrix* **I**, which is written as

$$\mathbf{I} = \begin{bmatrix} 1 & 0 & \ldots & 0 \\ 0 & 1 & \ldots & 0 \\ \vdots & \vdots & \ldots & \vdots \\ 0 & 0 & \ldots & 1 \end{bmatrix}. \tag{C.5}$$

When all the elements of a matrix are equal to zero, the matrix is called the *zero*, or *null matrix*. When the elements of a matrix have a special relationship so that $a_{ij} = a_{ji}$, it is called a *symmetrical* matrix. Thus, for example, the matrix

$$\mathbf{H} = \begin{bmatrix} 3 & -2 & 1 \\ -2 & 6 & 4 \\ 1 & 4 & 8 \end{bmatrix} \tag{C.6}$$

is a symmetrical matrix of order (3, 3).

C.2 ADDITION AND SUBTRACTION OF MATRICES

The addition of two matrices is possible only for matrices of the same order. The sum of two matrices is obtained by adding the corresponding elements. Thus if the elements of **A** are a_{ij} and the elements of **B** are b_{ij}, and if

$$\mathbf{C} = \mathbf{A} + \mathbf{B}, \tag{C.7}$$

then the elements of **C** that are c_{ij} are obtained as

$$c_{ij} = a_{ij} + b_{ij}. \tag{C.8}$$

Then, for example, the matrix addition for two 3×3 matrices is as follows:

$$\mathbf{C} = \begin{bmatrix} 2 & 1 & 0 \\ 1 & -1 & 3 \\ 0 & 6 & 2 \end{bmatrix} + \begin{bmatrix} 8 & 2 & 1 \\ 1 & 3 & 0 \\ 4 & 2 & 1 \end{bmatrix} = \begin{bmatrix} 10 & 3 & 1 \\ 2 & 2 & 3 \\ 4 & 8 & 3 \end{bmatrix}. \tag{C.9}$$

From the operation used for performing the operation of addition, we note that the process is commutative; that is

$$\mathbf{A} + \mathbf{B} = \mathbf{B} + \mathbf{A}. \tag{C.10}$$

Also, we note that the addition operation is associative, so that

$$(\mathbf{A} + \mathbf{B}) + \mathbf{C} = \mathbf{A} + (\mathbf{B} + \mathbf{C}). \tag{C.11}$$

To perform the operation of subtraction, we note that if a matrix **A** is multiplied by a constant α, then every element of the matrix is multiplied by this constant. Therefore we can write

$$\alpha \mathbf{A} = \begin{bmatrix} \alpha a_{11} & \alpha a_{12} & \cdots & \alpha a_{1n} \\ \alpha a_{12} & \alpha a_{22} & \cdots & \alpha a_{2n} \\ \cdot & \cdot & & \cdot \\ \cdot & \cdot & & \cdot \\ \cdot & \cdot & & \cdot \\ \alpha a_{m1} & \alpha a_{m2} & \cdots & \alpha a_{mn} \end{bmatrix}. \tag{C.12}$$

Then, to carry out a subtraction operation, we use $\alpha = -1$, and $-\mathbf{A}$ is obtained by multiplying each element of $\mathbf{A}$ by -1. Then, for example,

$$\mathbf{C} = \mathbf{B} - \mathbf{A} = \begin{bmatrix} 2 & 1 \\ 4 & 2 \end{bmatrix} - \begin{bmatrix} 6 & 1 \\ 3 & 1 \end{bmatrix} = \begin{bmatrix} -4 & 0 \\ 1 & 1 \end{bmatrix}. \tag{C.13}$$

C.3 MULTIPLICATION OF MATRICES

Matrix multiplication is defined in such a way as to assist in the solution of simultaneous linear equations. The multiplication of two matrices $\mathbf{AB}$ requires that the number of columns of $\mathbf{A}$ be equal to the number of rows of $\mathbf{B}$. Thus if $\mathbf{A}$ is of order $m \times n$ and $\mathbf{B}$ is of order $n \times q$, then the product is of order $m \times q$. The elements of a product

$$\mathbf{C} = \mathbf{AB} \tag{C.14}$$

are found by multiplying the ith row of $\mathbf{A}$ and the jth column of $\mathbf{B}$ and summing these products to give the element c_{ij}. That is,

$$c_{ij} = a_{i1}b_{1j} + a_{i2}b_{2j} + \cdots + a_{iq}b_{qj} = \sum_{k=1}^{q} a_{ik}b_{kj}. \tag{C.15}$$

Thus we obtain c_{11}, the first element of $\mathbf{C}$, by multiplying the first row of $\mathbf{A}$ by the first column of $\mathbf{B}$ and summing the products of the elements. We should note that, in general, matrix multiplication is not commutative, that is

$$\mathbf{AB} \neq \mathbf{BA}. \tag{C.16}$$

Also, we will note that the multiplication of a matrix of $m \times n$ by a column vector (order $n \times 1$) results in a column vector of order $m \times 1$.

A specific example of multiplication of a column vector by a matrix is

$$\mathbf{x} = \mathbf{Ay} = \begin{bmatrix} a_{11} & a_{12} & a_{13} \\ a_{21} & a_{22} & a_{23} \end{bmatrix} \begin{bmatrix} y_1 \\ y_2 \\ y_3 \end{bmatrix} = \begin{bmatrix} (a_{11}y_1 + a_{12}y_2 + a_{13}y_3) \\ (a_{21}y_1 + a_{22}y_2 + a_{23}y_3) \end{bmatrix}. \tag{C.17}$$

Note that $\mathbf{A}$ is of order 2×3 and $\mathbf{y}$ is of order 3×1. Therefore the resulting matrix $\mathbf{x}$ is of order 2×1, which is a column vector with two rows. There are two elements of $\mathbf{x}$, and

$$x_1 = (a_{11}y_1 + a_{12}y_2 + a_{13}y_3) \tag{C.18}$$

is the first element obtained by multiplying the first row of $\mathbf{A}$ by the first (and only) column of $\mathbf{y}$.

Another example, which the reader should verify, is

$$\mathbf{C} = \mathbf{AB} = \begin{bmatrix} 2 & -1 \\ -1 & 2 \end{bmatrix} \begin{bmatrix} 3 & 2 \\ -1 & -2 \end{bmatrix} = \begin{bmatrix} 7 & 6 \\ -5 & -6 \end{bmatrix}. \tag{C.19}$$

For example, the element c_{22} is obtained as $c_{22} = -1(2) + 2(-2) = -6$.

Now we are able to use this definition of multiplication in representing a set of simultaneous linear algebraic equations by a matrix equation. Consider the following set of algebraic equations:

$$3x_1 + 2x_2 + x_3 = u_1,$$

$$2x_1 + x_2 + 6x_3 = u_2, \tag{C.20}$$

$$4x_1 - x_2 + 2x_3 = u_3.$$

We can identify two column vectors as

$$\mathbf{x} = \begin{bmatrix} x_1 \\ x_2 \\ x_3 \end{bmatrix} \quad \text{and} \quad \mathbf{u} = \begin{bmatrix} u_1 \\ u_2 \\ u_3 \end{bmatrix}. \tag{C.21}$$

Then we can write the matrix equation

$$\mathbf{Ax} = \mathbf{u}, \tag{C.22}$$

where

$$\mathbf{A} = \begin{bmatrix} 3 & 2 & 1 \\ 2 & 1 & 6 \\ 4 & -1 & 2 \end{bmatrix}.$$

We immediately note the utility of the matrix equation as a compact form of a set of simultaneous equations.

The multiplication of a row vector and a column vector can be written as

$$\mathbf{xy} = [x_1, x_2, \ldots, x_n] \begin{bmatrix} y_1 \\ y_2 \\ \vdots \\ y_n \end{bmatrix} = x_1 y_1 + x_2 y_2 + \cdots + x_n y_n. \tag{C.23}$$

Thus we note that the multiplication of a row vector and a column vector results in a number that is a sum of a product of specific elements of each vector.

As a final item in this section, we note that the multiplication of any matrix by the identity matrix results in the original matrix, that is $\mathbf{AI} = \mathbf{A}$.

C.4 OTHER USEFUL MATRIX OPERATIONS AND DEFINITIONS

The *transpose* of a matrix $\mathbf{A}$ is denoted in this text as $\mathbf{A}^T$. One will often find the notation $\mathbf{A}'$ for $\mathbf{A}^T$ in the literature. The transpose of a matrix $\mathbf{A}$ is obtained by interchanging the rows and columns of **A**. Then, for example, if

$$\mathbf{A} = \begin{bmatrix} 6 & 0 & 2 \\ 1 & 4 & 1 \\ -2 & 3 & -1 \end{bmatrix},$$

then

$$\mathbf{A}^T = \begin{bmatrix} 6 & 1 & -2 \\ 0 & 4 & 3 \\ 2 & 1 & -1 \end{bmatrix}. \tag{C.24}$$

Therefore, we are able to denote a row vector as the transpose of a column vector and write

$$\mathbf{x}^T = [x_1, x_2, \ldots, x_n]. \tag{C.25}$$

Because $\mathbf{x}^T$ is a row vector, we obtain a matrix multiplication of $\mathbf{x}^T$ by $\mathbf{x}$ as follows:

$$\mathbf{x}^T\mathbf{x} - [x_1, x_2, \ldots, x_n] \begin{bmatrix} x_1 \\ x_2 \\ \vdots \\ x_n \end{bmatrix} = x_1^2 + x_2^2 + \cdots + x_n^2. \tag{C.26}$$

Thus the multiplication $\mathbf{x}^T\mathbf{x}$ results in the sum of the squares of each element of $\mathbf{x}$.

The transpose of the product of two matrices is the product in reverse order of their transposes, so that

$$(\mathbf{AB})^T = \mathbf{B}^T\mathbf{A}^T. \tag{C.27}$$

The sum of the main diagonal elements of a square matrix **A** is called the *trace* of **A**, written as

$$\text{tr } \mathbf{A} = a_{11} + a_{22} + \cdots + a_{nn}. \tag{C.28}$$

The *determinant* of a square matrix is obtained by enclosing the elements of the matrix **A** within vertical bars as, for example,

$$\det \mathbf{A} = \begin{vmatrix} a_{11} & a_{12} \\ a_{21} & a_{22} \end{vmatrix}. \tag{C.29}$$

If the determinant of **A** is equal to zero, then the determinant is said to be singular. The value of a determinant is determined by obtaining the minors and cofactors of the determinants. The *minor* of an element a_{ij} of a determinant of order n is a determinant of order $(n - 1)$ obtained by removing the row i and the column j of the original determinant. The cofactor of a given element of a determinant is the minor of the element with either a plus or minus sign attached; hence

$$\text{cofactor of } a_{ij} = \alpha_{ij} = (-1)^{i+j}M_{ij},$$

where M_{ij} is the minor of a_{ij}. For example, the cofactor of the element a_{23} of

$$\det \mathbf{A} = \begin{vmatrix} a_{11} & a_{12} & a_{13} \\ a_{21} & a_{22} & a_{23} \\ a_{31} & a_{32} & a_{33} \end{vmatrix} \tag{C.30}$$

is

$$\alpha_{23} = (-1)^5 M_{23} = - \begin{vmatrix} a_{11} & a_{12} \\ a_{31} & a_{32} \end{vmatrix}. \tag{C.31}$$

The value of a determinant of second order (2×2) is

$$\begin{vmatrix} a_{11} & a_{12} \\ a_{21} & a_{22} \end{vmatrix} = (a_{11}a_{22} - a_{21}a_{12}). \tag{C.32}$$

The general nth-order determinant has a value given by

$$\det \mathbf{A} = \sum_{j=1}^{n} a_{ij}\alpha_{ij} \quad \text{with } i \text{ chosen for one row,} \quad \text{or}$$
$$\tag{C.33}$$
$$\det \mathbf{A} = \sum_{i=1}^{n} a_{ij}\alpha_{ij} \quad \text{with } j \text{ chosen for one column.}$$

That is, the elements a_{ij} are chosen for a specific row (or column) and that entire row (or column) is expanded according to Eq. (C.33). For example, the value of a specific 3×3 determinant is

$$\det \mathbf{A} = \det \begin{bmatrix} 2 & 3 & 5 \\ 1 & 0 & 1 \\ 2 & 1 & 0 \end{bmatrix}$$

$$= 2 \begin{vmatrix} 0 & 1 \\ 1 & 0 \end{vmatrix} - 1 \begin{vmatrix} 3 & 5 \\ 1 & 0 \end{vmatrix} + 2 \begin{vmatrix} 3 & 5 \\ 0 & 1 \end{vmatrix} \tag{C.34}$$

$$= 2(-1) - (-5) + 2(3) = 9,$$

where we have expanded in the first column.

The *adjoint matrix* of a square matrix $\mathbf{A}$ is formed by replacing each element a_{ij} by the cofactor α_{ij} and transposing. Therefore

$$\text{adjoint } \mathbf{A} = \begin{bmatrix} \alpha_{11} & \alpha_{12} & \cdots & \alpha_{1n} \\ \alpha_{21} & \alpha_{22} & \cdots & \alpha_{2n} \\ \vdots & \vdots & & \vdots \\ \alpha_{n1} & \alpha_{n2} & \cdots & \alpha_{nn} \end{bmatrix}^T = \begin{bmatrix} \alpha_{11} & \alpha_{21} & \cdots & \alpha_{n1} \\ \alpha_{12} & \alpha_{22} & \cdots & \alpha_{n2} \\ \vdots & \vdots & & \vdots \\ \alpha_{1n} & \alpha_{2n} & \cdots & \alpha_{nn} \end{bmatrix}. \tag{C.35}$$

C.5 MATRIX INVERSION

The inverse of a square matrix $\mathbf{A}$ is written as $\mathbf{A}^{-1}$ and is defined as satisfying the relationship

$$\mathbf{A}^{-1}\mathbf{A} = \mathbf{A}\mathbf{A}^{-1} = \mathbf{I}. \tag{C.36}$$

The inverse of a matrix $\mathbf{A}$ is

$$\mathbf{A}^{-1} = \frac{\text{adjoint of } \mathbf{A}}{\det \mathbf{A}} \tag{C.37}$$

when the det $\mathbf{A}$ is not equal to zero. For a 2×2 matrix we have the adjoint matrix

$$\text{adjoint } \mathbf{A} = \begin{bmatrix} a_{22} & -a_{12} \\ -a_{21} & a_{11} \end{bmatrix} \tag{C.38}$$

and the det $\mathbf{A} = a_{11}a_{22} - a_{12}a_{21}$. Consider the matrix

$$\mathbf{A} = \begin{bmatrix} 1 & 2 & 3 \\ 2 & -1 & 4 \\ 0 & -1 & 1 \end{bmatrix}. \tag{C.39}$$

The determinant has a value det $\mathbf{A} = -7$. The cofactor α_{11} is

$$\alpha_{11} = (-1)^2 \begin{vmatrix} -1 & 4 \\ -1 & 1 \end{vmatrix} = 3. \tag{C.40}$$

In a similar manner we obtain

$$\mathbf{A}^{-1} = \frac{\text{adjoint } \mathbf{A}}{\text{det } \mathbf{A}} = \left(-\frac{1}{7}\right) \begin{bmatrix} 3 & -5 & 11 \\ -2 & 1 & 2 \\ -2 & 1 & -5 \end{bmatrix}. \tag{C.41}$$

C.6 MATRICES AND CHARACTERISTIC ROOTS

A set of simultaneous linear algebraic equations can be represented by the matrix equation

$$\mathbf{y} = \mathbf{A}\mathbf{x}, \tag{C.42}$$

where the $\mathbf{y}$ vector can be considered as a transformation of the vector $\mathbf{x}$. The question may be asked whether it may happen that a vector $\mathbf{y}$ may be a scalar multiple of $\mathbf{x}$. Trying $\mathbf{y} = \lambda\mathbf{x}$, where λ is a scalar, we have

$$\lambda\mathbf{x} = \mathbf{A}\mathbf{x}. \tag{C.43}$$

Alternatively, Eq. (C.43) can be written as

$$\lambda\mathbf{x} - \mathbf{A}\mathbf{x} = (\lambda\mathbf{I} - \mathbf{A})\mathbf{x} = \mathbf{0}, \tag{C.44}$$

where $\mathbf{I}$ = identity matrix. Thus the solution for $\mathbf{x}$ exists if and only if

$$\det (\lambda\mathbf{I} - \mathbf{A}) = 0. \tag{C.45}$$

This determinant is called the characteristic determinant of $\mathbf{A}$. Expansion of the determinant of Eq. (C.45) results in the *characteristic equation*. The characteristic equation is an nth-order polynomial in λ. The n roots of this characteristic equation are called the *characteristic roots*. For every possible value λ_i ($i = 1, 2, \ldots, n$) of the nth-order characteristic equation, we can write

$$(\lambda_i\mathbf{I} - \mathbf{A})\mathbf{x}_i = 0. \tag{C.46}$$

The vector $\mathbf{x}_i$ is the *characteristic vector* for the ith root. Let us consider the matrix

$$\mathbf{A} = \begin{bmatrix} 2 & 1 & 1 \\ 2 & 3 & 4 \\ -1 & -1 & -2 \end{bmatrix}. \tag{C.47}$$

The characteristic equation is found as follows:

$$\det \begin{bmatrix} (\lambda - 2) & -1 & -1 \\ -2 & (\lambda - 3) & -4 \\ 1 & 1 & (\lambda + 2) \end{bmatrix} = (-\lambda^3 + 3\lambda^2 + \lambda - 3) = 0. \quad \text{(C.48)}$$

The roots of the characteristic equation are $\lambda_1 = 1$, $\lambda_2 = -1$, $\lambda_3 = 3$. When $\lambda = \lambda_1 = 1$, we find the first characteristic vector from the equation

$$\mathbf{A}\mathbf{x}_1 = \lambda_1 \mathbf{x}_1, \quad \text{(C.49)}$$

and we have $\mathbf{x}_1^T = k[1, -1, 0]$, where k is an arbitrary constant usually chosen equal to 1. Similarly, we find

$$\mathbf{x}_2^T = [0, 1, -1]$$

and

$$\mathbf{x}_3^T = [2, 3, -1]. \quad \text{(C.50)}$$

C.7 THE CALCULUS OF MATRICES

The derivative of a matrix $\mathbf{A} = \mathbf{A}(t)$ is defined as

$$\frac{d}{dt}[\mathbf{A}(t)] = \begin{bmatrix} da_{11}(t)/dt & da_{12}(t)/dt & \cdots & da_{1n}(t)/dt \\ \vdots & \vdots & & \vdots \\ da_{n1}(t)/dt & da_{n2}(t)/dt & \cdots & da_{nn}(t)/dt \end{bmatrix}. \quad \text{(C.51)}$$

That is, the derivative of a matrix is simply the derivative of each element $a_{ij}(t)$ of the matrix.

The *matrix exponential function* is defined as the power series

$$\exp[\mathbf{A}] = e^{\mathbf{A}} = \mathbf{I} + \frac{\mathbf{A}}{1!} + \frac{\mathbf{A}^2}{2!} + \cdots + \frac{\mathbf{A}^k}{k!} + \cdots = \sum_{k=0}^{\infty} \frac{\mathbf{A}^k}{k!}, \quad \text{(C.52)}$$

where $\mathbf{A}^2 = \mathbf{A}\mathbf{A}$ and, similarly, $\mathbf{A}^k$ implies $\mathbf{A}$ multiplied k times. This series can be shown to be convergent for all square matrices. Also, a matrix exponential that is a function of time is defined as

$$e^{\mathbf{A}t} = \sum_{k=0}^{\infty} \frac{\mathbf{A}^k t^k}{k!}. \quad \text{(C.53)}$$

If we differentiate with respect to time, then we have

$$\frac{d}{dt}(e^{\mathbf{A}t}) = \mathbf{A}e^{\mathbf{A}t}. \quad \text{(C.54)}$$

Therefore, for a differential equation

$$\frac{d\mathbf{x}}{dt} = \mathbf{A}\mathbf{x}, \quad \text{(C.55)}$$

we might postulate a solution $\mathbf{x} = e^{\mathbf{A}t}\mathbf{c} = \phi\mathbf{c}$, where the matrix ϕ is $\phi = e^{\mathbf{A}t}$ and $\mathbf{c}$ is an unknown column vector. Then we have

$$\frac{d\mathbf{x}}{dt} = \mathbf{A}\mathbf{x} \tag{C.56}$$

or

$$\mathbf{A}e^{\mathbf{A}t} = \mathbf{A}e^{\mathbf{A}t}, \tag{C.57}$$

and we have in fact satisfied the relationship, Eq. (C.55). Then, the value of $\mathbf{c}$ is simply $\mathbf{x}(0)$, the initial value of $\mathbf{x}$, because, when $t = 0$, we have $\mathbf{x}(0) = \mathbf{c}$. Therefore the solution to Eq. (C.55) is

$$\mathbf{x}(t) = e^{\mathbf{A}t}\mathbf{x}(0). \tag{C.58}$$

Decibel Conversion

M	0	1	2	3	4	5	6	7	8	9
0.0	m =	−40.00	−33.98	−30.46	−27.96	−26.02	−24.44	−23.10	−21.94	−20.92
0.1	−20.00	−19.17	−18.42	−17.72	−17.08	−16.48	−15.92	−15.39	−14.89	−14.42
0.2	−13.98	−13.56	−13.15	−12.77	−12.40	−12.04	−11.70	−11.37	−11.06	−10.75
0.3	−10.46	−10.17	−9.90	−9.63	−9.37	−9.12	−8.87	−8.64	−8.40	−8.18
0.4	−7.96	−7.74	−7.54	−7.33	−7.13	−6.94	−6.74	−6.56	−6.38	−6.20
0.5	−6.02	−5.85	−5.68	−5.51	−5.35	−5.19	−5.04	−4.88	−4.73	−4.58
0.6	−4.44	−4.29	−4.15	−4.01	−3.88	−3.74	−3.61	−3.48	−3.35	−3.22
0.7	−3.10	−2.97	−2.85	−2.73	−2.62	−2.50	−2.38	−2.27	−2.16	−2.05
0.8	−1.94	−1.83	−1.72	−1.62	−1.51	−1.41	−1.31	−1.21	−1.11	−1.01
0.9	−0.92	−0.82	−0.72	−0.63	−0.54	−0.45	−0.35	−0.26	−0.18	−0.09
1.0	0.00	0.09	0.17	0.26	0.34	0.42	0.51	0.59	0.67	0.75
1.1	0.83	0.91	0.98	1.06	1.14	1.21	1.29	1.36	1.44	1.51
1.2	1.58	1.66	1.73	1.80	1.87	1.94	2.01	2.08	2.14	2.21
1.3	2.28	2.35	2.41	2.48	2.54	2.61	2.67	2.73	2.80	2.86
1.4	2.92	2.98	3.05	3.11	3.17	3.23	3.29	3.35	3.41	3.46
1.5	3.52	3.58	3.64	3.69	3.75	3.81	3.86	3.92	3.97	4.03
1.6	4.08	4.14	4.19	4.24	4.30	4.35	4.40	4.45	4.51	4.56
1.7	4.61	4.66	4.71	4.76	4.81	4.86	4.91	4.96	5.01	5.06
1.8	5.11	5.15	5.20	5.25	5.30	5.34	5.39	5.44	5.48	5.53
1.9	5.58	5.62	5.67	5.71	5.76	5.80	5.85	5.89	5.93	5.98
2.	6.02	6.44	6.85	7.23	7.60	7.96	8.30	8.63	8.94	9.25
3.	9.54	9.83	10.10	10.37	10.63	10.88	11.13	11.36	11.60	11.82
4.	12.04	12.26	12.46	12.67	12.87	13.06	13.26	13.44	13.62	13.80
5.	13.98	14.15	14.32	14.49	14.65	14.81	14.96	15.12	15.27	15.42
	0.	1.	2.	3.	4.	5.	6.	7.	8.	9.

M	0	1	2	3	4	5	6	7	8	9
6.	15.56	15.71	15.85	15.99	16.12	16.26	16.39	16.52	16.65	16.78
7.	16.90	17.03	17.15	17.27	17.38	17.50	17.62	17.73	17.84	17.95
8.	18.06	18.17	18.28	18.38	18.49	18.59	18.69	18.79	18.89	18.99
9.	19.08	19.18	19.28	19.37	19.46	19.55	19.65	19.74	19.82	19.91
	0.	1.	2.	3.	4.	5.	6.	7.	8.	9.

Decibels $= 20 \log_{10} M$.

Complex Numbers

E.1 A COMPLEX NUMBER

We all are familiar with the solution of the algebraic equation

$$x^2 - 1 = 0, \tag{E.1}$$

which is $x = 1$. However, we often encounter the equation

$$x^2 + 1 = 0. \tag{E.2}$$

A number that satisfies Eq. (E.2) is not a real number. We note that Eq. (E.2) may be written as

$$x^2 = -1, \tag{E.3}$$

and we denote the solution of Eq. (E.3) by the use of an imaginary number $j1$, so that

$$j^2 = -1 \tag{E.4}$$

and

$$j = \sqrt{-1}. \tag{E.5}$$

An *imaginary number* is defined as the product of the imaginary unit j with a real number. Thus we may, for example, write an imaginary number as jb. A complex number is the sum of a real number and an imaginary number, so that

$$c = a + jb, \tag{E.6}$$

where a and b are real numbers. We designate a as the real part of the complex number and b as the imaginary part and use the notation

$$\text{Re}\{c\} = a \tag{E.7}$$

and

$$\text{Im}\{c\} = b. \tag{E.8}$$

E.2 RECTANGULAR, EXPONENTIAL, AND POLAR FORMS

The complex number $a + jb$ may be represented on a rectangular coordinate place called a *complex plane*. The complex plane has a real axis and an imaginary axis, as shown in

Fig. E.1. The complex number c is the directed line identified as c with coordinates a, b. The *rectangular form* is expressed in Eq. (E.6) and pictured in Fig. E.1.

An alternative way to express the complex number c is to use the distance from the origin and the angle θ, as shown in Fig. E.2. The *exponential form* is written as

$$c = re^{j\theta}, \tag{E.9}$$

where

$$r = (a^2 + b^2)^{1/2} \tag{E.10}$$

and

$$\theta = \tan^{-1}(b/a). \tag{E.11}$$

Note that $a = r\cos\theta$ and $b = r\sin\theta$.

The number r is also called the *magnitude* of c, denoted as $|c|$. The angle θ can also be denoted by the form $\underline{/\theta}$. Thus we may represent the complex number in *polar form* as

$$c = |c|\underline{/\theta} = r\underline{/\theta}. \tag{E.12}$$

EXAMPLE E.1   **Exponential and polar forms**

Express $c = 4 + j3$ in exponential and polar form.

Solution   First, draw the complex plane diagram as shown in Fig. E.3. Then find r as

$$r = (4^2 + 3^2)^{1/2} = 5$$

and θ as

$$\theta = \tan^{-1}(3/4) = 36.9°.$$

The exponential form is then

$$c = 5e^{j36.9°}.$$

The polar form is

$$c = 5\underline{/36.9°}. \blacksquare$$

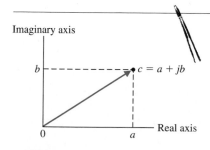

FIGURE E.1
Rectangular form of a complex number.

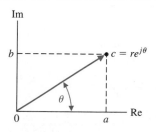

FIGURE E.2
Exponential form of a complex number.

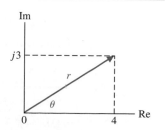

FIGURE E.3
Complex plane for Example E.1.

E.3 MATHEMATICAL OPERATIONS

The *conjugate* of the complex number $c = a + jb$ is called c^* and is defined as

$$c^* = a - jb. \tag{E.13}$$

In polar form, we have

$$c^* = r\underline{/-\theta}. \tag{E.14}$$

To add or subtract two complex numbers, we add (or subtract) their real parts and their imaginary parts. Therefore, if $c = a + jb$ and $d = f + jg$, then

$$c + d = (a + jb) + (f + jg) = (a + f) + j(b + g). \tag{E.15}$$

The multiplication of two complex numbers is obtained as follows (note $j^2 = -1$):

$$\begin{aligned} cd &= (a + jb)(f + jg) \\ &= af + jag + jbf + j^2bg \\ &= (af - bg) + j(ag + bf). \end{aligned} \tag{E.16}$$

Alternatively, we use the polar form to obtain

$$cd = (r_1\underline{/\theta_1})(r_2\underline{/\theta_2}) = r_1r_2\underline{/\theta_1 + \theta_2}, \tag{E.17}$$

where

$$c = r_1\underline{/\theta_1} \quad \text{and} \quad d = r_2\underline{/\theta_2}.$$

Division of one complex number by another complex number is easily obtained using the polar form as follows:

$$\frac{c}{d} = \frac{r_1\underline{/\theta_1}}{r_2\underline{/\theta_2}} = \frac{r_1}{r_2}\underline{/\theta_1 - \theta_2}. \tag{E.18}$$

It is easiest to add and subtract complex numbers in rectangular form and to multiply and divide them in polar form.

A few useful relations for complex numbers are summarized in Table E.1.

TABLE E.1

Useful Relationships for Complex Numbers

(1)	$\dfrac{1}{j} = -j$
(2)	$(-j)(j) = 1$
(3)	$j^2 = -1$
(4)	$1\underline{/\pi/2} = j$
(5)	$c^k = r^k\underline{/k\theta}$

EXAMPLE E.2   **Complex number operations**

Find $c + d$, $c - d$, cd, and c/d when $c = 4 + j3$ and $d = 1 - j$.

Solution   First, we will express c and d in polar form as

$$c = 5\underline{/36.9°} \qquad \text{and} \qquad d = \sqrt{2}\underline{/-45°}.$$

Then, for addition, we have

$$c + d = (4 + j3) + (1 - j) = 5 + j2.$$

For subtraction, we have

$$c - d = (4 + j3) - (1 - j) = 3 + j4.$$

For multiplication, we use the polar form to obtain

$$cd = (5\underline{/36.9°})(\sqrt{2}\underline{/-45°}) = 5\sqrt{2}\underline{/-8.1°}.$$

For division, we have

$$\frac{c}{d} = \frac{5\underline{/36.9°}}{\sqrt{2}\underline{/-45°}} = \frac{5}{\sqrt{2}}\underline{/81.9°}. \quad \blacksquare$$

APPENDIX F

Matlab Basics

F.1 INTRODUCTION

Matlab is an interactive program for scientific and engineering calculations. The *Matlab* family of programs includes the base program plus a variety of *toolboxes*, a collection of special files called *M-files* that extend the functionality of the base program [1–4]. Together, the base program plus the *Control System Toolbox* provide the capability to use *Matlab* for control system design and analysis. Whenever *Matlab* is referred to in this book, you can interpret that to mean the base program plus the *Control System Toolbox*.

Most of the statements, functions, and commands are computer platform independent. Regardless of what particular computer system you use, your interaction with *Matlab* is basically the same. This appendix concentrates on this computer platform–independent interaction. A typical session will utilize a variety of objects that allow you to interact with the program: (1) statements and variables, (2) matrices, (3) graphics, and (4) scripts. *Matlab* interprets and acts on input in the form of one or more of these objects. The goal in this appendix is to introduce each of the four objects in preparation for our ultimate goal of using *Matlab* for control system design and analysis.

The manner in which *Matlab* interacts with a specific computer system is computer platform dependent. Examples of computer-dependent functions include installation, the file structure, hard-copy generation of the graphics, the invoking and exiting of a session, and memory allocation. Questions related to platform-dependent issues are not addressed here. This is not to imply that they are not important, but rather that there are better sources of information such as the *Matlab User's Guide* or the local resident expert.

Before proceeding, make sure that you can start a *Matlab* session and then exit the session. To begin a session on a Macintosh, you will most likely double-click on the *Matlab* program icon. On an IBM PC compatible, you will most likely type matlab at the DOS prompt.

The remainder of this appendix consists of four sections corresponding to the four objects already listed. In the first section we present the basics of *statements* and *variables*. Following that is the subject of *matrices*. The third section presents an introduction to *graphics*, and the fourth section is a discussion on the important topic of *scripts* and *M-files*.

F.2 STATEMENTS AND VARIABLES

Statements have the form shown in Fig. F.1. *Matlab* uses the assignment so that equals ("=") implies the assignment of the expression to the variable. The command prompt is two right arrows, ">>." A *typical* statement is shown in Fig. F.2, where we are entering a

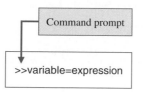

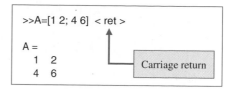

FIGURE F.1
Matlab statement form.

FIGURE F.2
Entering and displaying a matrix **A**.

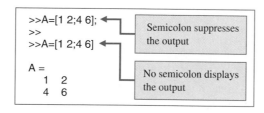

FIGURE F.3
Using semicolons to suppress the output.

FIGURE F.4
Using the calculator mode.

2×2 matrix to which we attach the variable name A. The statement is executed after the carriage return (or enter key) is pressed. The carriage return is not explicitly denoted in the remaining examples in this appendix and the chapter sections on *Matlab*.

The matrix **A** is automatically displayed after the statement is executed following the carriage return. If the statement is followed by a semicolon (;), the output matrix **A** is suppressed, as seen in Fig. F.3. The assignment of the variable **A** has been carried out even though the output is suppressed by the semicolon. It is often the case that your *Matlab* sessions will include intermediate calculations for which the output is of little interest. Use the semicolon whenever you have a need to reduce the amount of output. Output management has the added benefit of increasing the execution speed of the calculations, since displaying screen output takes time.

The usual mathematical operators can be used in expressions. The common operators are shown in Table F.1. The order of the arithmetic operations can be altered by using parentheses.

The example in Fig. F.4 illustrates that *Matlab* can be used in a "calculator" mode. When the variable name and "=" are omitted from an expression, the result is assigned to the generic variable *ans*. *Matlab* has available most of the trigonometric and elementary

TABLE F.1   Mathematical Operators

+	Addition
−	Subtraction
*	Multiplication
/	Division
^	Power

TABLE F.2   Common Mathematical Functions

sin(X)	Sine of the elements of X
cos(X)	Cosine of the elements of X
asin(X)	Arcsine of the elements of X
acos(X)	Arccosine of the elements of X
tan(X)	Tangent of the elements of X
atan(X)	Arctangent of the elements of X
atan2(X,Y)	Four quadrant arctangent of the real elements of X and Y
abs(X)	Absolute value of the elements of X
sqrt(X)	Square root of X
imag(X)	Imaginary part of X
real(X)	Real part of X
conj(X)	Complex conjugate of X
log(X)	Natural logarithm of the elements of X
log10(X)	Logarithm base 10 of the elements of X
exp(X)	Exponential of the elements of X

math functions of a common scientific calculator. The *User's Guide* has a complete list of available trigonometric and elementary math functions; the more common ones are summarized in Table F.2.

Variable names begin with a letter and are followed by any number of letters and numbers (including underscores). Keep the name length to 19 characters, since *Matlab* remembers only the first 19 characters. It is a good practice to use variable names that describe the quantity they represent. For example, we might use the variable name *vel* to represent the quantity *aircraft velocity*. Generally, we do not use extremely long variable names even though they may be legal *Matlab* names.

Since *Matlab* is *case sensitive*, the variables *M* and *m* are not the same. By *case* we mean upper- and lowercase, as illustrated in Fig. F.5. The variables *M* and *m* are recognized as different quantities.

Matlab has several predefined variables, including *pi, Inf, Nan, i,* and *j*. Three examples are shown in Fig. F.6. *Nan* stands for *Not-a-Number* and results from undefined operations. *Inf* represents $+\infty$, and *pi* represents π. The variable $i = \sqrt{-1}$ is used to represent complex numbers. The variable $j = \sqrt{-1}$ can be used for complex arithmetic by those who prefer it over *i*. These predefined variables can be inadvertently overwritten. Of course, they can also be purposely overwritten in order to free the variable name for other uses. For instance, one might want to use *i* as an integer and reserve *j* for complex arithmetic. Be safe and leave these predefined variables alone, as there are plenty of alternative names that can be used. Predefined variables can be reset to their default values by using clear *name* (e.g., clear *pi*).

```
>>z=3+4*i
z =
     3.0000 + 4.0000i

>>Inf
ans =
     ∞

>>0/0
Warning:  Divide by zero
ans =
     NaN
```

FIGURE F.6
Three predefined variables *i, Inf,*
and *Nan.*

```
>>M=[1 2];
>>m=[3 5 7];
```

FIGURE F.5
Variables are case
sensitive.

```
>>who
Your variables are:
A      M      ans      m      z

leaving 675516 bytes of memory free.
```

FIGURE F.7
Using the **who** function to display variables.

The matrix **A** and the variable *ans*, in Figs. F.3 and F.4, respectively, are stored in the *workspace*. Variables in the workspace are automatically saved for later use in your session. The who function gives a list of the variables in the workspace, as shown in Fig. F.7.

Matlab has a host of built-in functions. Refer to the *User's Guide* for a complete list. Each function will be described as the need arises.

The whos function lists the variables in the workspace and gives additional information regarding variable dimension, type, and memory allocation. Figure F.8 gives an example of the whos function. The memory allocation information given by the whos function can be interpreted as follows. Each element of the 2×2 matrix **A** requires 8 bytes of memory for a total of 32 bytes, the 1×1 variable *ans* requires 8 bytes, and so forth. All the variables in the workspace use a total of 96 bytes. The amount of remaining free

FIGURE F.8
Using the **whos**
function to display
variables.

```
>>whos
        Name      Size      Total    Complex

        A         2 by 2      4        No

        M         1 by 2      2        No

        ans       1 by 1      1        No

        m         1 by 3      3        No

        z         1 by 1      2        Yes

Grand total is (12 * 8) = 96 bytes,
leaving 664912 bytes of memory free.
```

```
>>clear A

>>who

Your variables are:

M       ans       m       z

leaving 663780 bytes of memory free.
```

FIGURE F.9
Removing the matrix **A** from the workspace.

memory depends on the total memory available in the system. Computers with virtual memory will not display the remaining free memory.

Variables can be removed from the workspace with the clear function. Using the function clear, by itself, removes all items (variables and functions) from the workspace; clear variables removes all variables from the workspace; clear *name1 name2 . . .* removes the variables *name1, name2,* and so forth. The procedure for removing the matrix **A** from the workspace is shown in Fig. F.9.

A simple calculation shows that clearing the matrix **A** from memory freed up more than 32 bytes. In some cases, clearing a variable may not change the value of the displayed free memory at all. The who function displays the amount of *contiguous* remaining free memory. So, depending upon the "location" of the variable in the workspace, clearing the variable may or may not increase the displayed amount of remaining free memory. Available free memory may be more than displayed with the who or whos functions.

All computations in *Matlab* are performed in *double precision.* However, the screen output can be displayed in several formats. The default output format contains four digits past the decimal point for nonintegers. This can be changed by using the format function shown in Fig. F.10. Once a particular format has been specified, it remains in effect until

FIGURE F.10
Output format control illustrates the four forms of output.

```
>>pi
ans =
   3.1416                    <-- 5-digit scaled fixed point

>>format long; pi
ans =
   3.14159265358979          <-- 15-digit scaled fixed point

>>format short e; pi
ans =
   3.1416e+00                <-- 5-digit scaled floating point

>>format long e; pi
ans =
   3.141592653589793e+00     <-- 15-digit scaled floating point
```

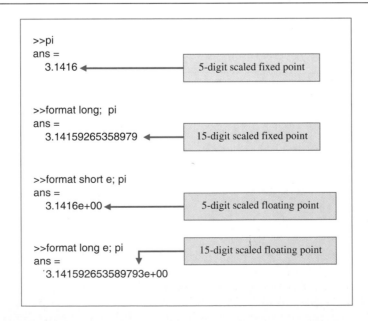

```
>>WHO
??? Undefined function or variable
Symbol in question ==> WHO

>>Who
??? Undefined function or variable
Symbol in question ==> Who
```

FIGURE F.11
Function names
are case sensitive.

altered by a different format input. Remember that the output format does not affect the *Matlab* computations—all computations are in double precision. On the other hand, the number of digits displayed does not necessarily reflect the number of significant digits of the number. This is problem dependent, and only the user can know the true accuracy of the numbers input and displayed by *Matlab*.

Since *Matlab* is case sensitive, the functions who and WHO are not the same functions. who is a built in function, and typing who lists the variables in the workspace. On the other hand, typing the uppercase WHO results in the error message shown in Fig. F.11. Case sensitivity applies to all functions.

F.3 MATRICES

Matlab is short for *matrix laboratory*. The *User's Guide* describes the program as a high-performance interactive software package designed to provide easy access to the *Linpack* and *Eispack* matrix software. Although we will not emphasize the matrix routines underlying our calculations, we will learn how to use the interactive capability to assist us in the control system design and analysis. We begin by introducing the basic concepts associated with manipulating matrices and vectors.

The basic computational unit is the matrix. Vectors and scalars can be viewed as special cases of matrices. A typical matrix expression is enclosed in square brackets, [·]. The column elements are separated by blanks or commas, and the rows are separated by semicolons or carriage returns. Suppose we want to input the matrix **A**, where

$$\mathbf{A} = \begin{bmatrix} 1 & -4j & \sqrt{2} \\ \log(-1) & \sin(\pi/2) & \cos(\pi/3) \\ \arcsin(0.5) & \arccos(0.8) & \exp(0.8) \end{bmatrix}.$$

One way to input **A** is shown in Fig. F.12. The input style in Fig. F.12 is not unique.

Matrices can be input across multiple lines by using a carriage return following the semicolon or in place of the semicolon. This practice is useful for entering large matrices. Different combinations of spaces and commas can be used to separate the columns, and different combinations of semicolons and carriage returns can be used to separate the rows, as illustrated in Fig. F.12.

No dimension statements or type statements are necessary when using matrices; memory is allocated automatically. Notice in the example in Fig. F.12 that the size of the matrix **A** is automatically adjusted when the input matrix is redefined. Also notice that the

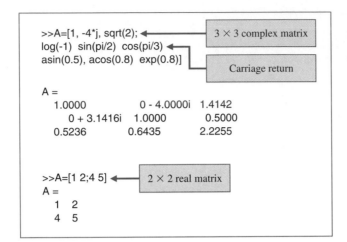

FIGURE F.12
Complex and
real matrix input
with automatic
dimension and
type adjustment.

matrix elements can contain trigonometric and elementary math functions, as well as complex numbers.

The important basic matrix operations are addition and subtraction, multiplication, transpose, powers, and the so-called array operations, which are element-to-element operations. The mathematical operators given in Table F.1 apply to matrices. We will not discuss matrix division, but be aware that *Matlab* has a left- and right-matrix division capability.

Matrix operations require that the matrix dimensions be compatible. For matrix addition and subtraction, this means that the matrices must have the same dimensions. If $\mathbf{A}$ is an $n \times m$ matrix and $\mathbf{B}$ is a $p \times r$ matrix, then $\mathbf{A} \pm \mathbf{B}$ is permitted only if $n = p$ and $m = r$. Matrix multiplication, given by $\mathbf{A} * \mathbf{B}$, is permitted only if $m = p$. Matrix-vector multiplication is a special case of matrix multiplication. Suppose $\mathbf{b}$ is a vector of length p. Multiplication of the vector $\mathbf{b}$ by the matrix $\mathbf{A}$, where $\mathbf{A}$ is an $n \times m$ matrix, is allowed if $m = p$. Thus $\mathbf{y} = \mathbf{A} * \mathbf{b}$ is the $n \times 1$ vector solution of $\mathbf{A} * \mathbf{b}$. Examples of three basic matrix-vector operations are given in Fig. F.13.

The matrix transpose is formed with the apostrophe ($'$). We can use the matrix transpose and multiplication operation to create a vector *inner product* in the following manner. Suppose $\mathbf{w}$ and $\mathbf{v}$ are $m \times 1$ vectors. Then the inner product (also known as the dot product) is given by $\mathbf{w}' * \mathbf{v}$. The inner product of two vectors is a scalar. The *outer product* of two vectors can similarly be computed as $\mathbf{w} * \mathbf{v}'$. The outer product of two $m \times 1$ vectors is an $m \times m$ matrix of rank 1. Examples of inner and outer products are given in Fig. F.14.

The basic matrix operations can be modified for element-by-element operations by preceding the operator with a period. The modified matrix operations are known as *array operations*. The commonly used array operators are given in Table F.3. Matrix addition and subtraction are already element-by-element operations and do not require the additional period preceding the operator. However, array multiplication, division, and power do require the preceding dot, as shown in Table F.3.

Consider $\mathbf{A}$ and $\mathbf{B}$ as 2×2 matrices given by

$$\mathbf{A} = \begin{bmatrix} a_{11} & a_{12} \\ a_{21} & a_{22} \end{bmatrix}, \qquad \mathbf{B} = \begin{bmatrix} b_{11} & b_{12} \\ b_{21} & b_{22} \end{bmatrix}.$$

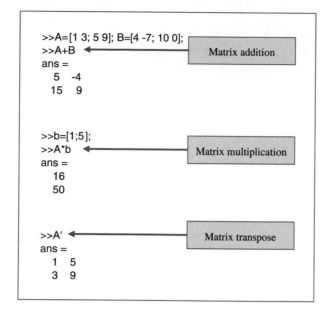

FIGURE F.13
Three basic matrix operations: addition, multiplication, and transpose.

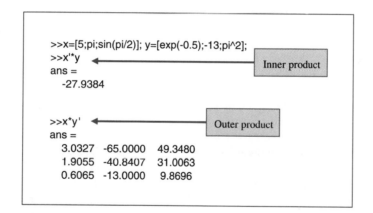

FIGURE F.14
Inner and outer products of two vectors.

TABLE F.3   Mathematical Array Operators

+	Addition
−	Subtraction
.*	Multiplication
./	Division
.^	Power

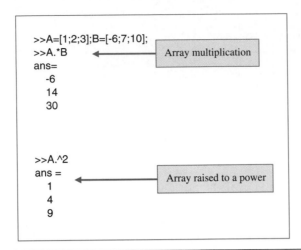

FIGURE F.15
Array operations.

Then, using the array multiplication operator, we have

$$
\mathbf{A} .* \mathbf{B} = \begin{bmatrix} a_{11}b_{11} & a_{12}b_{12} \\ a_{21}b_{21} & a_{22}\,b_{22} \end{bmatrix}.
$$

The elements of $\mathbf{A} .* \mathbf{B}$ are the products of the corresponding elements of $\mathbf{A}$ and $\mathbf{B}$. A numerical example of two array operations is given in Fig. F.15.

Before proceeding to the important topic of graphics, we need to introduce the notion of *subscripting using colon notation*. The colon notation, shown in Fig. F.16, allows us to generate a row vector containing the numbers from a given starting value, x_i, to a final value, x_f, with a specified increment, dx.

We can easily generate vectors using the colon notation, and as we shall soon see, this is quite useful for developing *x-y plots*. Suppose our objective is to generate a plot of $y = x \sin(x)$ versus x for $x = 0, 0.1, 0.2, \ldots, 1.0$. Our first step is to generate a table of x-y data. We can generate a vector containing the values of x at which the values of $y(x)$ are desired using the colon notation, as illustrated in Fig. F.17. Given the desired x vector, the vector $y(x)$ is computed using the multiplication array operation. Creating a plot of $y = x \sin(x)$ versus x is a simple step once the table of x-y data is generated.

FIGURE F.16
The colon notation.

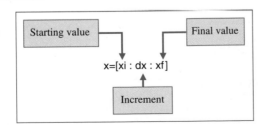

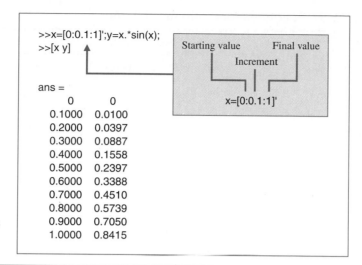

FIGURE F.17
Generating vectors using the colon notation.

F.4 GRAPHICS

Graphics plays an important role in both the design and analysis of control systems. An important component of an *interactive* control system design and analysis tool is an effective graphical capability. A complete solution to the control system design and analysis will eventually require a detailed look at a multitude of data types in many formats. The objective of this section is to acquaint the reader with the basic *x-y* plotting capability of *Matlab*. More advanced graphics topics are addressed in the chapter sections on *Matlab*.

Matlab uses a *graph display* to present plots. Some computer configurations allow both the command display and graph display to be viewed simultaneously. With computer configurations that allow only one to be viewed at a time, the command display will disappear when the graph display is activated. The graph display is activated automatically when a plot is generated using any function that generates a plot (e.g., the plot function). Switching from the graph display back to the command display is accomplished by pressing any key on the keyboard. The plot in the graph display is cleared by the clg function at the command prompt. The shg function is used to switch to the graph display from the command display.

TABLE F.4   Plot Formats

plot(x,y)	Plots the vector **x** versus the vector **y**.
semilogx(x,y)	Plots the vector **x** versus the vector **y**. The x-axis is $\log_{10}$; the y-axis is linear.
semilogy(x,y)	Plots the vector **x** versus the vector **y**. The x-axis is linear; the y-axis is $\log_{10}$.
loglog(x,y)	Plots the vector **x** versus the vector **y**. Creates a plot with $\log_{10}$ scales on both axes.

TABLE F.5   Functions for Customized Plots

title('text')	Puts 'text' at the top of the plot.
xlabel('text')	Labels the *x*-axis with 'text'.
ylabel('text')	Labels the *y*-axis with 'text'.
text(p1,p2,'text','sc')	Puts 'text' at (pl,p2) in screen coordinates where (0.0,0.0) is the lower left and (1.0,1.0) is the upper right of the screen.
subplot	Subdivides the graphics window.
grid	Draws grid lines on the current plot.

There are two basic groups of graphics functions. The first group, shown in Table F.4, specifies the type of plot. The list of available plot types includes the *x-y* plot, semilog plots, and log plots. The second group of functions, shown in Table F.5, allows us to customize the plots by adding titles, axis labels, and text to the plots and to change the scales and display multiple plots in subwindows.

The standard *x-y* plot is created using the plot function. The *x-y* data in Fig. F.17 are plotted using the plot function, as shown in Fig. F.18. The axis scales and line types are

FIGURE F.18
(a) *Matlab*
commands. (b) A
basic *x-y* plot of
x sin(*x*) versus *x*.

```
>>x=[0:0.1:1]';
>>y=x.*sin(x);
>>plot(x,y)
>>title('Plot of x sin(x) vs x ')
>>xlabel('x')
>>ylabel('y')
>>grid
```

(a)

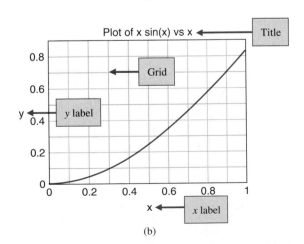

(b)

**TABLE F.6   Line Types for
Customized Plots**

-	Solid line
- -	Dashed line
:	Dotted line
-.	Dashdot line

automatically chosen. The axes are labeled with the xlabel and ylabel functions; the title is applied with the title function. A grid can be placed on the plot by using the grid function. A basic *x-y* plot is generated with the combination of functions plot, xlabel, ylabel, title, and grid.

Multiple lines can be placed on the graph by using the plot function with multiple arguments, as shown in Fig. F.19. The default line types can also be altered. The available line types are shown in Table F.6. The line types will be automatically chosen unless specified by the user. The use of the text function and the changing of line types are illustrated in Fig. F.19.

The other graphics functions—loglog, semilogx, and semilogy—are used in a fashion similar to that of plot. To obtain an *x-y* plot where the *x*-axis is a linear scale and the *y*-axis

FIGURE F.19
(a) *Matlab*
commands. (b) A
basic *x-y* plot with
multiple lines.

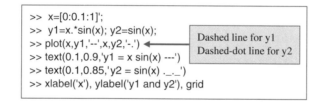

```
>> x=[0:0.1:1]';
>> y1=x.*sin(x); y2=sin(x);
>> plot(x,y1,'--',x,y2,'-.')      ◄——    Dashed line for y1
>> text(0.1,0.9,'y1 = x sin(x) ---')      Dashed-dot line for y2
>> text(0.1,0.85,'y2 = sin(x) ._._')
>> xlabel('x'), ylabel('y1 and y2'), grid
```

(a)

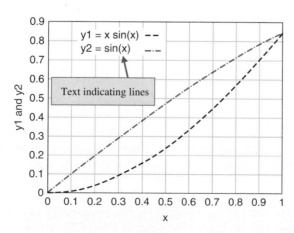

(b)

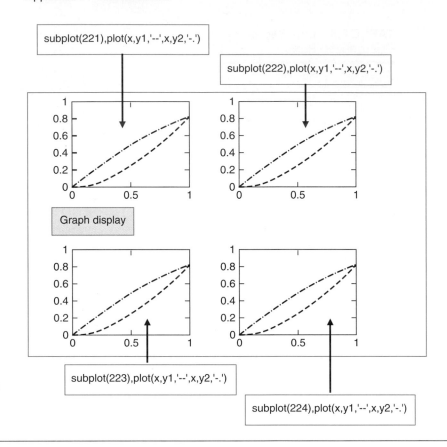

subplot(221),plot(x,y1,'--',x,y2,'-.')

subplot(222),plot(x,y1,'--',x,y2,'-.')

Graph display

subplot(223),plot(x,y1,'--',x,y2,'-.')

subplot(224),plot(x,y1,'--',x,y2,'-.')

FIGURE F.20
Using **subplot** to create a 2 × 2 partition of the graph display.

is a $\log_{10}$ scale, you would use the **semilogy** function in place of the **plot** function. The customizing features listed in Table F.5 can also be utilized with the **loglog**, **semilogx**, and **semilogy** functions.

The graph display can be subdivided into smaller subwindows. The function **subplot(mnp)** subdivides the graph display into an $m \times n$ grid of smaller subwindows, where $m \leq 2$ and $n \leq 2$. This means the graph display can be subdivided into two or four windows. The integer p specifies the window, numbering them left to right, top to bottom, as illustrated in Fig. F.20, where the graphics window is subdivided into four subwindows.

F.5 SCRIPTS

Up to this point, all of our interaction with *Matlab* has been at the command prompt. We entered statements and functions at the command prompt, and *Matlab* interpreted our input and took the appropriate action. This is the preferable mode of operation whenever the work sessions are short and nonrepetitive. However, the real power of *Matlab* for control system design and analysis derives from its ability to execute a long sequence of commands stored in a file. These files are called *M-files*, since the filename has the form *filename.m*. A *script* is one type of *M-file*. The *Control System Toolbox* is a collection of *M-files* designed specifically for control applications. In addition to the preexisting *M-files* delivered with *Matlab* and the toolboxes, we can develop scripts for our applications.

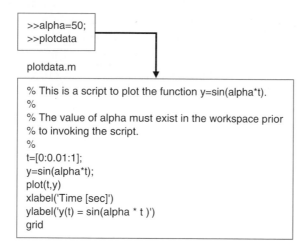

```
>>alpha=50;
>>plotdata
```

plotdata.m

```
% This is a script to plot the function y=sin(alpha*t).
%
% The value of alpha must exist in the workspace prior
% to invoking the script.
%
t=[0:0.01:1];
y=sin(alpha*t);
plot(t,y)
xlabel('Time [sec]')
ylabel('y(t) = sin(alpha * t )')
grid
```

FIGURE F.21
A simple script to
plot the function
$y(t) = \sin \alpha t$.

Scripts are ordinary ASCII text files and are created by using a text editor. Creating and storing scripts are computer platform–dependent topics, so contact the appropriate expert at your location for more information.

A script is just a sequence of ordinary statements and functions used at the command prompt level. A script is invoked at the command prompt level by simply typing in the filename. Scripts can also invoke other scripts. When the script is invoked, *Matlab* executes the statements and functions in the file without waiting for input at the command prompt. The script operates on variables in the workspace.

Suppose we want to plot the function $y(t) = \sin \alpha t$, where α is a variable that we want to vary. Using a text editor, we write a script that we call plotdata.m, as shown in Fig. F.21, then input a value of α at the command prompt, placing α in the workspace. Then we execute the script by typing in plotdata at the command prompt; the script plotdata.m will use the most recent value of α in the workspace. After executing the script, we can enter another value of α at the command prompt and execute the script again.

Your scripts should be well documented with *comments*, which begin with a %. Put a *header* in the script comprised of several descriptive comments regarding the function of the script, and then use the help function to display the header comments and describe the script to the user, illustrated in Fig. F.22.

Use plotdata.m to develop an interactive capability with α as a variable, as shown in Fig. F.23. At the command prompt, input a value of $\alpha = 10$ followed by the script filename, which in this case is plotdata. The graph of $y(t) = \sin \alpha t$ is automatically generated. One

FIGURE F.22
Using the **help**
function.

```
>>help plotdata

   This is a script to plot the function y=sin(alpha*t).

   The value of alpha must exist in the workspace prior
   to invoking the script.
```

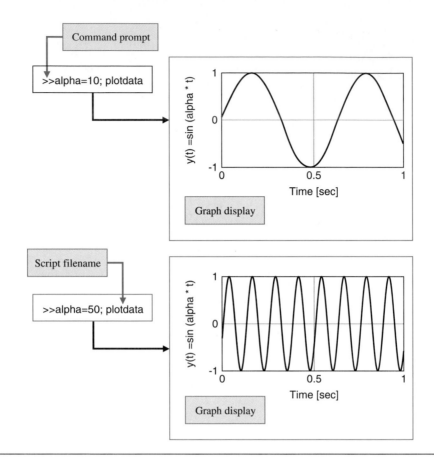

FIGURE F.23
An interactive
session using a
script to plot the
function $y(t) = \sin \alpha t$.

can now go back to the command prompt, enter a value of $\alpha = 50$, and run the script again to obtain the updated plot.

The graphics capability of *Matlab* extends beyond the introductory material presented here. We will investigate the issue of graphics in the chapter sections on *Matlab*. A table of *Matlab* functions used in this book is provided in Table F.7.

TABLE F.7   *Matlab* Functions

Function Name	Function Description
abs	Computes the absolute value
acos	Computes the arccosine
ans	Variable created for expressions
asin	Computes the arcsine
atan	Computes the arctangent (2 quadrant)
atan2	Computes the arctangent (4 quadrant)
axis	Specifies the manual axis scaling on plots
bode	Generates Bode frequency response plots

TABLE F.7 *Continued*

Function Name	Function Description
c2dm	Converts a continuous-time state variable system representation to a discrete-time system representation
clear	Clears the workspace
clg	Clears the graph window
cloop	Computes the closed-loop system with unity feedback
conj	Computes the complex conjugate
conv	Multiplies two polynomials (convolution)
cos	Computes the cosine
ctrb	Computes the controllability matrix
diary	Saves the session in a disk file
d2cm	Converts a discrete-time state variable system representation to a continuous-time system representation
dstep	Computes the unit step response of a discrete-time system
eig	Computes the eigenvalues and eigenvectors
end	Terminates control structures
exp	Computes the exponential with base e
expm	Computes the matrix exponential with base e
eye	Generates an identity matrix
feedback	Computes the feedback interconnection of two systems
for	Generates a loop
format	Sets the output display format
grid	Adds a grid to the current graph
help	Prints a list of HELP topics
hold	Holds the current graph on the screen
i	$\sqrt{-1}$
imag	Computes the imaginary part of a complex number
impulse	Computes the unit impulse response of a system
inf	Represents infinity
j	$\sqrt{-1}$
linspace	Generates linearly spaced vectors
load	Loads variables saved in a file
log	Computes the natural logarithm
log10	Computes the logarithm base 10
loglog	Generates log-log plots
logspace	Generates logarithmically spaced vectors
lsim	Computes the time response of a system to an arbitrary input and initial conditions
margin	Computes the gain margin, phase margin, and associated crossover frequencies from frequency response data
max	Determines the maximum value
mesh	Creates three-dimensional mesh surfaces
meshdom	Generates arrays for use with the mesh function
min	Determines the minimum value

(*continued*)

TABLE F.7   *Continued*

Function Name	Function Description
minreal	Transfer function pole–zero cancellation
NaN	Representation for Not-a-Number
ngrid	Draws grid lines on a Nichols chart
nichols	Computes a Nichols frequency response plot
num2str	Converts numbers to strings
nyquist	Calculates the Nyquist frequency response
obsv	Computes the observability matrix
ones	Generates a matrix of integers where all the integers are 1.
pade	Computes an nth order Padé approximation to a time delay
parallel	Computes a parallel system connection
plot	Generates a linear plot
poly	Computes a polynomial from roots
polyval	Evaluates a polynomial
printsys	Prints state variable and transfer function representations of linear systems in a readable form
pzmap	Plots the pole–zero map of a linear system
rank	Calculates the rank of a matrix
real	Computes the real part of a complex number
residue	Computes a partial fraction expansion
rlocfind	Finds the gain associated with a given set of roots on a root locus plot
rlocus	Computes the root locus
roots	Determines the roots of a polynomial
roots1	Same as the **roots** function, but gives more accurate answers when there are repeated roots
semilogx	Generates an *x*-*y* plot using semilog scales with the *x*-axis $\log_{10}$ and the *y*-axis linear
semilogy	Generates an *x*-*y* plot using semilog scales with the *y*-axis $\log_{10}$ and the *x*-axis linear
series	Computes a series system connection
shg	Shows graph window
sin	Computes the sine
sqrt	Computes the square root
ss2tf	Converts state variable form to transfer function form
step	Calculates the unit step response of a system
subplot	Splits the graph window into subwindows
tan	Computes the tangent
text	Adds text to the current graph
title	Adds a title to the current graph
tf2ss	Converts a transfer function to state variable form
who	Lists the variables currently in memory
whos	Lists the current variables and sizes
xlabel	Adds a label to the *x*-axis of the current graph
ylabel	Adds a label to the *y*-axis of the current graph
zeros	Generates a matrix of zeros

MATLAB BASICS: PROBLEMS

F.1   Consider the two matrices

$$A = \begin{bmatrix} 4 & 2\pi \\ 6j & 10 + \sqrt{2}j \end{bmatrix},$$

$$B = \begin{bmatrix} 6j & -13\pi \\ \pi & 16 \end{bmatrix}.$$

Using *Matlab,* compute the following:

(a) $A + B$ (e) B^{-1}

(b) AB (f) $B^T A^T$

(c) A^2 (g) $A^2 + B^2 - AB$

(d) A^T

F.2   Consider the following set of linear algebraic equations:

$$5x + 6y + 10z = 4,$$

$$-3x + 14z = 10,$$

$$-7y + 21z = 0.$$

Determine the values of x, y, and z such that the set of algebraic equations are satisfied. (*Hint:* Write the equations in matrix vector form.)

F.3   Generate a plot of

$$y(x) = e^{-0.5x} \sin \omega x,$$

where $\omega = 10$ rad/s and $0 \le x \le 10$. Utilize the colon notation to generate the x vector in increments of 0.1.

F.4   Develop a *Matlab* script to plot the function

$$y(x) = \frac{4}{\pi} \cos \omega x + \frac{4}{9\pi} \cos 3\omega x,$$

where ω is a variable input at the command prompt. Label the x-axis with *time(sec)* and the y-axis with $y(x) = (4/pi) * cos(wx) + (4/9pi) * cos(3wx)$. Include a descriptive header in the script, and verify that the help function will display the header. Choose $\omega = 1, 3, 10$ rad/s and test the script.

F.5   Consider the function

$$y(x) = 10 + 5e^{-x} \cos(\omega x + 0.5).$$

Develop a script to co-plot $y(x)$ for the three values of $\omega = 1, 3, 10$ rad/s with $0 \le x \le 5$ seconds. The final plot should have the following attributes:

Title	$y(x) = 10 + 5 \exp(-x)*\cos(wx+0.5)$
x-axis label	time (seconds)
y-axis label	$y(x)$
Line type	$\omega = 1$: solid line
	$\omega = 3$: dashed line
	$\omega = 10$: dotted line
Grid	

References

Chapter 1

1. O. Mayr, *The Origins of Feedback Control,* MIT Press, Cambridge, Mass., 1970.
2. O. Mayr, "The Origins of Feedback Control," *Scientific American,* 223, 4, October 1970, pp. 110–118.
3. O. Mayr, *Feedback Mechanisms in the Historical Collections of the National Museum of History and Technology,* Smithsonian Institution Press, Washington, D.C., 1971.
4. E. P. Popov, *The Dynamics of Automatic Control Systems,* Gostekhizdat, Moscow, 1956; Addison-Wesley, Reading, Mass., 1962.
5. J. C. Maxwell, "On Governors," *Proc. of the Royal Society of London,* 16, 1868; in *Selected Papers on Mathematical Trends in Control Theory.* Dover, New York, 1964, pp. 270–283.
6. I. A. Vyshnegradskii, "On Controllers of Direct Action," *Izv. SPB Tekhnolog. Inst.,* 1877.
7. H. W. Bode, "Feedback—The History of an Idea," in *Selected Papers on Mathematical Trends in Control Theory.* Dover, New York, 1964, pp. 106–123.
8. H. S. Black, "Inventing the Negative Feedback Amplifier," *IEEE Spectrum,* December 1977, pp. 55–60.
9. J. E. Brittain, *Turning Points in American Electrical History,* IEEE Press, New York, 1977, Sect. II-E.
10. G. J. Thaler, *Automatic Control Systems,* West Publishing, St. Paul, Minn., 1989.
11. G. Newton, L. Gould, and J. Kaiser, *Analytical Design of Linear Feedback Controls,* John Wiley & Sons, New York, 1957.
12. M. D. Fagen, *A History of Engineering and Science on the Bell Systems,* Bell Telephone Laboratories, 1978, Ch. 3.
13. G. Zorpette, "Parkinson's Gun Director," *IEEE Spectrum,* April 1989, p. 43.
14. R. C. Dorf and A. Kusiak, *Handbook of Automation and Manufacturing,* John Wiley & Sons, New York, 1994.
15. W. D. Rasmussen, "The Mechanization of Agriculture," *Scientific American,* September 1982, pp. 26–37.
16. R. Shoureshi, "Intelligent Control Systems," *Journal of Dynamic Systems,* ASME, June 1993, pp. 392–400.
17. A. G. Ulsoy, "Control of Machining Processes," *Journal of Dynamic Systems,* ASME, June 1993, pp. 301–307.
18. D. E. Hardt, "Modeling and Control of Manufacturing Processes," *Journal of Dynamic Systems,* ASME, June 1993, pp. 291–300.
19. P. Varaiya, "Smart Cars on Smart Roads," *IEEE Transactions on Automatic Control,* February 1993, pp. 195–207.
20. J. D. Powell, "Engine Control Using Cylinder Pressure," *Journal of Dynamic Systems,* ASME, June 1993, pp. 343–350.
21. P. M. Moretti and L. V. Divone, "Modern Windmills," *Scientific American,* June 1986, pp. 110–118.
22. B. Preising and T. C. Hsia, "Robots in Medicine," *IEEE Engineering in Medicine and Biology,* June 1991, pp. 13–22.
23. R. C. Dorf and J. Unmack, "A Time-Domain Model of the Heart Rate Control System," *Proceedings of the San Diego Symposium for Biomedical Engineering,* 1965, pp. 43–47.
24. R. C. Dorf, *Electrical Engineering Handbook,* CRC Press, Boca Raton, Fla., 1993, Section XI.
25. R. C. Dorf, *Introduction to Computers and Computer Science,* Boyd and Fraser, San Francisco, 3rd ed., 1982, Ch. 13, 14.
26. K. Sutton, "Productivity," in *Encyclopedia of Engineering,* McGraw-Hill, New York, pp. 947–948.
27. R. C. Dorf, *The Encyclopedia of Robotics,* John Wiley & Sons, New York, 1988.

28. R. C. Dorf, *Robotics and Automated Manufacturing,* Reston Publishing, Reston, Va., 1983.

29. S. S. Hacisalihzade, "Control Engineering and Therapeutic Drug Delivery," *IEEE Control Systems,* June 1989, pp. 44–46.

30. E. R. Carson and T. Deutsch, "A Spectrum of Approaches for Controlling Diabetes," *IEEE Control Systems,* December 1992, pp. 25–30.

31. J. R. Sankey and H. Kaufman, "Robust Considerations of a Drug Infusion System," *Proceedings of the American Control Conference,* San Francisco, Calif., June 1993, pp. 1689–1695.

32. K. Passino, "Bridging the Gap between Conventional and Intelligent Control," *IEEE Control Systems,* June 1993, pp. 12–18.

33. D. Auslander, "The Rise of Mechanical System Control, *Journal of Dynamic Systems,* ASME, June 1993, pp. 234–238.

34. T. S. Perry, "Improving Air Traffic Control System," *IEEE Spectrum,* February 1991, pp. 22–36.

35. P. J. Brancazio, "Science and the Game of Baseball," *Science Digest,* July 1984, pp. 66–70.

36. D. Scott, "Wingsail Trimaran," *Popular Science,* June 1990, pp. 116–117.

37. C. Klomp et al., "Development of an Autonomous Cow-Milking Robot Control System," *IEEE Control Systems,* October 1990, pp. 11–19.

38. M. Mittal et al., "Nonlinear Adaptive Control of a Twin Lift Helicopter System," *IEEE Control Systems,* April 1991, pp. 39–44.

39. S. Lee, "Intelligent Sensing and Control for Advanced Teleoperation," *IEEE Control Systems,* June 1993, pp. 19–28.

40. C. O'Malley, "Rapid Rails," *Popular Science,* June 1992, pp. 74–79.

41. G. B. Gordon and J. C. Roark, "ORCA: An Optimized Robot for Chemical Analysis," *Hewlett-Packard Journal,* June 1993, pp. 6–19.

42. K. Hollenback, "Destabilizing Effects of Muscular Contraction in Human-Machine Interaction," *Proceedings of the American Control Conference,* San Francisco, Calif., 1993, pp. 736–740.

43. H. Asade and C. C. Federspiel, "Human Centered Control in Robotics and Consumer Product Design," *Journal of Dynamic Systems,* ASME, June 1993, pp. 271–280.

44. O. Mayr, "Adam Smith and the Concept of the Feedback System," *Technology and Culture,* 12, 1, January 1971, pp. 1–22.

45. A. Goldsmith, "Autofocus Cameras," *Popular Science,* March 1988, pp. 70–72.

46. R. Johansson, *System Modeling and Identification,* Prentice-Hall, Englewood Cliffs, N.J., 1993.

47. C. Bradford, "Correcting Hubble's Vision," *Aerospace America,* July 1993, pp. 32–35.

48. K. Capek, *Rossum's Universal Robots,* English version by P. Selver and N. Playfair, Doubleday, Page, New York, 1923.

49. D. Hancock, "Prototyping the Hubble Fix," *IEEE Spectrum,* October 1993, pp. 34–39.

50. A. K. Naj, "Engineers Want New Buildings to Behave Like Human Beings," *Wall Street Journal,* January 20, 1994, p. B1.

51. K. Schmidt-Nielsen, "How Are Control Systems Controlled?" *American Scientist,* January 1994, pp. 38–44.

52. B. Rosewicz, "Hubble Photos Show Telescope in Good Shape," *Wall Street Journal,* January 14, 1994, p. B3.

Chapter 2

1. R. C. Dorf, *Electric Circuits,* 2nd ed., John Wiley & Sons, New York, 1993.

2. C. N. Dorny, *Understanding Dynamic Systems,* Prentice-Hall, Englewood Cliffs, N.J., 1993.

3. J. W. Nilsson, *Electric Circuits,* 4th ed., Addison-Wesley, Reading, Mass., 1993.

4. E. Kamen, *Introduction to Signals and Systems,* 2nd ed., Macmillan, New York, 1990.

5. F. Raven, *Automatic Control Engineering,* 2nd ed., McGraw-Hill, New York, 1990.

6. R. C. Dorf, *The Encyclopedia of Robotics,* John Wiley & Sons, New York, 1988.

7. R. R. Kadiyala, "A Toolbox for Approximate Linearization of Nonlinear Systems," *IEEE Control Systems,* April 1993, pp. 47–56.

8. R. Smith and R. Dorf, *Circuits, Devices and Systems,* 5th ed., John Wiley & Sons, New York, 1992.

9. Y. M. Pulyer, *Electromagnetic Devices for Motion Control,* Springer-Verlag, New York, 1992.

10. B. C. Kuo, *Automatic Control Systems,* 4th ed., Prentice-Hall, Englewood Cliffs, N.J., 1994.

11. J. L. Shearer, *Dynamic Modeling and Control of Engineering Systems,* Macmillan, New York, 1990.

12. R. C. Dorf, *Electrical Engineering Handbook,* CRC Press, Boca Raton, Fla., 1993.

13. H. Saadat, *Computational Aids in Control Systems Using Matlab,* McGraw-Hill, New York, 1993.

14. M. Jamshidi, *Computer Aided Analysis and Design of Linear Control Systems,* Prentice-Hall, Englewood Cliffs, N.J., 1992.
15. G. B. Gordon, "ORCA: Optimized Robot for Chemical Analysis," *Hewlett-Packard Journal,* June 1993, pp. 6–19.
16. B. W. Dickinson, *Mathematics of Controls, Signals and Systems,* Springer-Verlag, New York, 1993.
17. S. Bennett, "Nicholas Minorsky and the Automatic Steering of Ships," *IEEE Control Systems,* November 1984, pp. 10–15.
18. R. C. Rosenberg, "Reflections on Engineering Systems and Bond Graphs," *Journal of Dynamic Systems,* ASME, June 1993, pp. 242–250.
19. C. M. Close and D. K. Frederick, *Modeling and Analysis of Dynamic Systems,* 2nd ed., Houghton Mifflin, Boston, 1993.
20. H. S. Black, "Stabilized Feed-Back Amplifiers," *Electrical Engineering,* 53, January 1934, pp. 114–120. Also in *Turning Points in American History,* J. E. Brittain, ed., IEEE Press, New York, 1977, pp. 359–361.
21. M. W. Spong et al., *Robot Control Dynamics, Motion Planning and Analysis,* IEEE Press, New York, 1993.
22. W. J. Rugh, *Linear System Theory,* Prentice-Hall, Englewood Cliffs, N.J., 1993.
23. H. Kazerooni, "Human Extenders," *Journal of Dynamic Systems,* ASME, June 1993, pp. 281–290.
24. R. Johansson, *System Modeling and Identification,* Prentice-Hall, Englewood Cliffs, N.J., 1993.
25. S. P. Parker, *Encyclopedia of Engineering,* 2nd ed., McGraw-Hill, New York, 1993.
26. M. Vidyasagar, *Nonlinear System Analysis,* Prentice-Hall, Englewood Cliffs, N.J., 1993.
27. G. T. Pope, "Living-Room Levitation," *Discover,* June 1993, p. 24.
28. D. J. Bak, "Fast Trains Face Off," *Design News,* September 20, 1993, pp. 78–85.

Chapter 3

1. R. C. Dorf, *Electric Circuits,* 2nd ed., John Wiley & Sons, New York, 1993.
2. W. J. Rugh, *Linear System Theory,* Prentice-Hall, Englewood Cliffs, N.J., 1993.
3. C. E. Rohrs, J. L. Melsa, and D. Schultz, *Linear Control Systems,* McGraw-Hill, New York, 1993.

4. R. C. Dorf, *Encyclopedia of Robotics,* John Wiley & Sons, New York, 1988.
5. L. B. Jackson, *Signals, Systems, and Transforms,* Addison-Wesley, Reading, Mass., 1991.
6. J. L. Stein, "Modeling and State Estimator Design Issues for Model Based Monitoring Systems," *Journal of Dynamic Systems,* ASME, June 1993, pp. 318–326.
7. R. C. Rosenberg, "Reflections on Engineering Systems and Bond Graphs," *Journal of Dynamic Systems,* ASME, June 1993, pp. 242–250.
8. R. C. Dorf, *Electrical Engineering Handbook,* CRC Press, Boca Raton, Fla., 1993.
9. Y. M. Pulyer, *Electromagnetic Devices for Motion Control,* Springer-Verlag, New York, 1992.
10. C. M. Close and D. K. Frederick, *Modeling and Analysis of Dynamic Systems,* 2nd ed., Houghton Mifflin, Boston, 1993.
11. R. C. Durbeck, "Computer Output Printer Technologies," in *Electrical Engineering Handbook,* R. C. Dorf, Ed., CRC Press, Boca Raton, Fla., 1993, pp. 1958–1975.
12. B. Wie et al., "New Approach to Attitude/Momentum Control for the Space Station," *AIAA Journal of Guidance, Control, and Dynamics* 12, no. 5, 1989, pp. 714–722.
13. B. Etkin, *Dynamics of Atmospheric Flight,* John Wiley & Sons, New York, 1972.
14. P. C. Hughes, *Spacecraft Attitude Dynamics,* John Wiley & Sons, New York, 1986.
15. H. Ramirez, "Feedback Controlled Landing Maneuvers," *IEEE Transactions on Automatic Control,* April 1992, pp. 518–523.
16. M. W. Spong et al., *Robot Control Dynamics, Motion Planning and Analysis,* IEEE Press, New York, 1993.
17. R. R. Kadiyala, "A Toolbox for Approximate Linearization of Nonlinear Systems," *IEEE Control Systems,* April 1993, pp. 47–56.
18. K. E. Drexler, *Nanosystems,* John Wiley & Sons, New York, 1992.
19. E. Colle, "Localization of an Autonomous Vehicle with Beacons," *International Journal of Robotics and Automation* 8, no. 1, 1993, pp. 30–38.
20. W. Leventon, "Mountain Bike Suspension Allows Easy Adjustment," *Design News,* July 19, 1993, pp. 75–77.
21. H. Saadat, *Computational Aids in Control Systems Using MATLAB,* McGraw-Hill, New York, 1993.

22. M. Jamshidi, *Computer Aided Analysis and Design of Linear Control Systems,* Prentice-Hall, Englewood Cliffs, N.J., 1992.

23. R. Johansson, *System Modeling and Identification,* Prentice-Hall, Englewood Cliffs, N.J., 1993.

24. M. Fujita, "Synthesis of an Electromagnetic Suspension System," *Proceedings of the Conference on Decision and Control,* IEEE Press, New York, 1992.

25. D. Cho, "Magnetic Levitation Systems," *IEEE Control Systems,* February 1993, pp. 42–48.

26. E. C. Gwo, "Approximate Feedback Linearization," *Proceedings of the American Automatic Control Conference,* San Francisco, 1993, pp. 1495–1499.

27. H. Kazerooni, "Human Extenders," *Journal of Dynamic Systems,* ASME, June 1993, pp. 281–290.

28. C. N. Dorny, *Understanding Dynamic Systems,* Prentice-Hall, Englewood Cliffs, N.J., 1993.

29. Chen, Chi-Tsing, *Linear System Theory and Design,* Holt, Rinehart and Winston, New York, 1984.

30. G. E. Forsythe, M. A. Malcolm, and C. B. Moler, *Computer Methods for Mathematical Computations,* Prentice-Hall, Englewood Cliffs, N.J., 1977.

Chapter 4

1. R. C. Dorf, *Electrical Engineering Handbook,* CRC Press, Boca Raton, Fla., 1993.

2. R. C. Dorf, *Electric Circuits,* 2nd ed., John Wiley & Sons, New York, 1993.

3. C. E. Rohrs, J. L. Melsa, and D. Schultz, *Linear Control Systems,* McGraw-Hill, New York, 1993.

4. C. M. Close and D. K. Frederick, *Modeling and Analysis of Dynamic Systems,* 2nd ed., Houghton Mifflin, Boston, 1993.

5. B. K. Bose, *Modern Power Electronics,* IEEE Press, New York, 1992.

6. J. V. Wait and L. P. Huelsman, *Operational Amplifier Theory,* 2nd ed., McGraw-Hill, New York, 1992.

7. *Motomatic Speed Control,* Electro-Craft Corp., Hopkins, Minn., 1993.

8. M. W. Spong et al., *Robot Control Dynamics, Motion Planning and Analysis,* IEEE Press, New York, 1993.

9. R. C. Dorf, *Encyclopedia of Robotics,* John Wiley & Sons, New York, 1988.

10. D. J. Bak, "Dancer Arm Feedback Regulates Tension Control," *Design News,* April 6, 1987, pp. 132–133.

11. "The Smart Projector Demystified," *Science Digest,* May 1985, p. 76.

12. R. Meier, "Control of Blood Pressure During Anesthesia," *IEEE Control Systems,* December 1992, pp. 12–16.

13. G. N. Roberts, "Rudder Roll Stabilization for Ships," *Proceedings of the American Control Conference,* ASME, San Francisco, 1993, pp. 2403–2407.

14. C. N. Dorny, *Understanding Dynamic Systems,* Prentice-Hall, Englewood Cliffs, N.J., 1993.

15. C. W. DeSilva, *Control Sensors and Actuators,* Prentice-Hall, Englewood Cliffs, N.J., 1990.

16. S. P. Parker, *Encyclopedia of Engineering,* 2nd ed., McGraw-Hill, New York, 1993.

17. M. S. Markow, "An Automated Laser System for Eye Surgery," *IEEE Engineering in Medicine and Biology,* December 1989, pp. 24–29.

18. W. R. Perkins, "Sensitivity Function Methods in Control System Education," *Proceedings of IFAC Advances in Control Education,* June 1991, pp. 14–22.

19. Y. M. Pulyer, *Electromagnetic Devices for Motion Control,* Springer-Verlag, New York, 1992.

20. J. R. Layne, "Control for Cargo Ship Steering," *IEEE Control Systems,* December 1993, pp. 23–33.

Chapter 5

1. C. M. Close and D. K. Frederick, *Modeling and Analysis of Dynamic Systems,* 2nd ed., Houghton Mifflin, Boston, 1993.

2. R. C. Dorf, *Electric Circuits,* 2nd ed., John Wiley & Sons, New York, 1993.

3. B. K. Bose, *Modern Power Electronics,* IEEE Press, New York, 1992.

4. P. R. Clement, "A Note on Third-Order Linear Systems," *IRE Transactions on Automatic Control,* June 1960, p. 151.

5. R. N. Clark, *Introduction to Automatic Control Systems,* John Wiley & Sons, New York, 1962, pp. 115–124.

6. D. Graham and R. C. Lathrop, "The Synthesis of Optimum Response: Criteria and Standard Forms, Part 2," *Trans. of the AIEE* 72, November 1953, pp. 273–288.

7. J. V. Wait and L. P. Huelsman, *Operational Ampli-*

fier Theory, 2nd ed., McGraw-Hill, New York, 1992.

8. R. C. Dorf, *Encyclopedia of Robotics,* John Wiley & Sons, New York, 1988.

9. L. E. Ryan, "Control of an Impact Printer Hammer," *ASME Journal of Dynamic Systems,* March 1990, pp. 69–75.

10. T. C. Hsia, "On the Simplification of Linear Systems," *IEEE Transactions on Automatic Control,* June 1972, pp. 372–374.

11. E. J. Davison, "A Method for Simplifying Linear Dynamic Systems," *IEEE Transactions on Automatic Control,* January 1966, pp. 93–101.

12. R. C. Dorf, *Electrical Engineering Handbook,* CRC Press, Boca Raton, Fla., 1993.

13. A. G. Ulsoy, "Control of Machining Processes," *ASME Journal of Dynamic Systems,* June 1993, pp. 301–310.

14. C. N. Dorny, *Understanding Dynamic Systems,* Prentice-Hall, Englewood Cliffs, N.J., 1993.

15. W. J. Rugh, *Linear System Theory,* Prentice-Hall, Englewood Cliffs, N.J., 1993.

16. W. J. Book, "Controlled Motion in an Elastic World," *Journal of Dynamic Systems,* June 1993, pp. 252–260.

17. C. E. Rohrs, J. L. Melsa, D. Schultz, *Linear Control Systems,* McGraw-Hill, New York, 1993.

18. S. Lee, "Intelligent Sensing and Control for Advanced Teleoperation," *IEEE Control Systems,* June 1993, pp. 19–28.

19. R. Rosen and L. J. Williams, "The Rebirth of the Supersonic Transport," *Technology Review,* March 1993, pp. 22–29.

20. M. Alpert, "A Bullet Train for the U.S.," *Fortune,* May 1993, p. 32.

21. D. Hancock, "Prototyping the Hubble Fix," *IEEE Spectrum,* October 1993, pp. 34–39.

22. M. Hutton and M. Rabins, "Simplification of Higher-Order Mechanical Systems Using the Routh Approximation," *Journal of Dynamic Systems,* ASME, December 1975, pp. 383–392.

23. F. Taylor, *Principles of Signals and Systems,* McGraw-Hill, New York, 1994.

24. B. Rosewicz, "Hubble Photos Show Telescope in Good Shape," *Wall Street Journal,* January 14, 1994, p. B3.

Chapter 6

1. R. C. Dorf, *Electrical Engineering Handbook,* CRC Press, Boca Raton, Fla., 1993.

2. R. C. Dorf, *Electric Circuits,* 2nd ed., John Wiley & Sons, New York, 1993.

3. C. E. Rohrs, J. L. Melsa, and D. Schultz, *Linear Control Systems,* McGraw-Hill, New York, 1993.

4. W. J. Rugh, *Linear System Theory,* Prentice-Hall, Englewood Cliffs, N.J., 1993.

5. F. B. Farquharson, "Aerodynamic Stability of Suspension Bridges, with Special Reference to the Tacoma Narrows Bridge," *Bulletin 116, Part I,* The Engineering Experiment Station, University of Washington, 1950.

6. A. Hurwitz, "On the Conditions Under Which an Equation Has Only Roots with Negative Real Parts," *Mathematische Annalen* 46, 1895, pp. 273–284. Also in *Selected Papers on Mathematical Trends in Control Theory,* Dover, New York, 1964, pp. 70–82.

7. E. J. Routh, *Dynamics of a System of Rigid Bodies,* Macmillan, New York, 1892.

8. G. G. Wang, "Design of Turning Control for a Tracked Vehicle," *IEEE Control Systems,* April 1990, pp. 122–125.

9. B. K. Bose, "Modern Power Electronics," *IEEE Press,* New York, 1992.

10. R. C. Dorf, *Encyclopedia of Robotics,* John Wiley & Sons, New York, 1988.

11. R. C. Dorf and A. Kusiak, *Handbook of Manufacturing and Automation,* John Wiley & Sons, New York, 1994.

12. C. N. Dorny, *Understanding Dynamic Systems,* Prentice-Hall, Englewood Cliffs, N.J., 1993.

13. S. P. Parker, *Encyclopedia of Engineering,* 2nd ed., McGraw-Hill, New York, 1933.

14. M. Fujita, "Control of a Magnetic Bearing," *Proceedings of the American Automatic Control Conference,* San Francisco, 1993, pp. 8–12.

15. P. Varaiya, "Smart Cars on Smart Roads," *IEEE Transactions on Automatic Control,* February 1993, pp. 195–207.

16. G. E. Young and K. N. Reid, "Control of Moving Webs," *Journal of Dynamics,* ASME, June 1993, pp. 309–316.

17. D. W. Freeman, "Jump-Jet Airliner," *Popular Mechanics,* June 1993, pp. 38–40.

18. G. T. Pope, "America's Last and Best Shuttle," *Popular Mechanics,* June 1992, pp. 44–47.

19. S. Lee, "Intelligent Sensing and Control for Advanced Teleoperation," *IEEE Control Systems,* June 1993, pp. 19–28.

20. "Uplifting," *The Economist,* July 10, 1993, p. 79.

21. R. N. Clark, "The Routh–Hurwitz Stability Criterion, Revisited," *IEEE Control Systems,* June 1992, pp. 119–120.

22. S. G. Tzafestas, *Applied Control,* Marcel Dekker, New York, 1993, pp. 274–276.

Chapter 7

1. W. R. Evans, "Graphical Analysis of Control Systems," *Transactions of the AIEE,* 67, 1948, pp. 547–551. Also in G. J. Thaler, ed., *Automatic Control,* Dowden, Hutchinson, and Ross, Stroudsburg, Pa., 1974, pp. 417–421.

2. W. R. Evans, "Control System Synthesis by Root Locus Method," *Transactions of the AIEE,* 69, 1950, pp. 1–4. Also in *Automatic Control,* G. J. Thaler, ed., Dowden, Hutchinson, and Ross, Stroudsburg, Pa., 1974, pp. 423–425.

3. W. R. Evans, *Control System Dynamics,* McGraw-Hill, New York, 1954.

4. R. C. Dorf, *Electrical Engineering Handbook,* CRC Press, Boca Raton, Fla., 1993.

5. J. G. Goldberg, *Automatic Controls,* Allyn and Bacon, Boston, 1965.

6. R. C. Dorf, *The Encyclopedia of Robotics,* John Wiley & Sons, New York, 1988.

7. H. Ur, "Root Locus Properties and Sensitivity Relations in Control Systems," *I.R.E. Trans. on Automatic Control,* January 1960, pp. 57–65.

8. T. R. Kurfess and M. L. Nagurka, "Understanding the Root Locus Using Gain Plots," *IEEE Control Systems,* August 1991, pp. 37–40.

9. T. R. Kurfess and M. L. Nagurka, "Foundations of Classical Control Theory," *The Franklin Institute,* 330, no. 2, 1993, pp. 213–227.

10. R. C. Dorf and A. Kusiak, *Handbook of Manufacturing and Automation,* John Wiley & Sons, New York, 1994.

11. C. N. Dorny, *Understanding Dynamic Systems,* Prentice-Hall, Englewood Cliffs, N.J., 1993.

12. S. Ashley, "Putting a Suspension Through Its Paces," *Mechanical Engineering,* April 1993, pp. 56–57.

13. B. K. Bose, *Modern Power Electronics,* IEEE Press, New York, 1992.

14. P. Varaiya, "Smart Cars on Smart Roads," *IEEE Transactions on Automatic Control,* February 1993, pp. 195–207.

15. J. L. Jones and A. M. Flynn, *Mobile Robots,* A. K. Peters Publishing, New York, 1993.

16. R. Rosen and L. J. Williams, "The Rebirth of the Supersonic Transport," *Technology Review,* March 1993, pp. 22–29.

17. L. V. Merritt, "Tape Transport Head Positioning Servo Using Positive Feedback," *Motion,* April 1993, pp. 19–22.

18. G. E. Young and K. N. Reid, "Control of Moving Webs," *Journal of Dynamic Systems* ASME, June 1993, pp. 309–316.

19. S. P. Parker, *Encyclopedia of Engineering,* 2nd ed., McGraw-Hill, New York, 1993.

20. D. McLean, *Automatic Flight Control Systems,* Prentice-Hall, Englewood Cliffs, N.J., 1990.

21. T. B. Sheridan, *Telerobotics, Automation and Control,* MIT Press, Cambridge, Mass., 1992.

22. L. W. Couch, *Digital and Analog Communication Systems,* Macmillan, New York, 1993.

23. D. Hrovat, "Applications of Optimal Control to Automotive Suspension Design," *Journal of Dynamic Systems,* ASME, June 1993, pp. 328–342.

24. M. Alpert, "A Bullet Train for the U.S.," *Fortune,* May 1993, p. 32.

25. M. van de Panne, "A Controller for the Dynamic Walk of a Biped," *Proceedings of the Conference on Decision and Control,* IEEE, December 1992, pp. 2668–2673.

26. R. C. Dorf, *Electric Circuits,* 2nd ed., John Wiley & Sons, New York, 1993.

27. W. J. Cook, "The Invasion of Mars," *US News and World Report,* August 23, 1993, pp. 51–59.

Chapter 8

1. R. C. Dorf, *Electrical Engineering Handbook,* CRC Press, Boca Raton, Fla., 1993.

2. C. M. Close and D. K. Frederick, *Modeling and Analysis of Dynamic Systems,* 2nd ed., Houghton Mifflin, Boston, 1993.

3. R. C. Dorf, *Electric Circuits,* 2nd ed., John Wiley & Sons, New York, 1993.

4. H. W. Bode, "Relations Between Attenuation and Phase in Feedback Amplifier Design," *Bell System Tech. J.,* July 1940, pp. 421–454. Also in *Automatic Control: Classical Linear Theory,* G. J. Thaler, ed., Dowden, Hutchinson, and Ross, Stroudsburg, Pa., 1974, pp. 145–178.

5. M. D. Fagen, *A History of Engineering and Science in the Bell System,* Bell Telephone Laboratories, Murray Hill, N.J., 1978, Chapter 3.

6. C. N. Dorny, *Understanding Dynamic Systems,* Prentice-Hall, Englewood Cliffs, N.J., 1993.

7. R. C. Dorf and A. Kusiak, *Handbook of Manufacturing and Automation,* John Wiley & Sons, New York, 1994.

8. R. C. Dorf, *The Encyclopedia of Robotics,* John Wiley & Sons, New York, 1988.

9. T. B. Sheridan, *Telerobotics, Automation and Control,* MIT Press, Cambridge, Mass., 1992.

10. J. L. Jones and A. M. Flynn, *Mobile Robots,* A. K. Peters Publishing, New York, 1993.

11. D. McLean, *Automatic Flight Control Systems,* Prentice-Hall, Englewood Cliffs, N.J., 1990.

12. G. Leitman, "Aircraft Control Under Conditions of Windshear," *Proceedings of IEEE Conference on Decision and Control,* December 1990, pp. 747–749.

13. S. Lee, "Intelligent Sensing and Control for Advanced Teleoperation," *IEEE Control Systems,* June 1993, pp. 19–28.

14. R. A. Hess, "A Control Theoretic Model of Driver Steering Behavior," *IEEE Control Systems,* August 1990, pp. 3–8.

15. P. Varaiya, "Smart Cars on Smart Roads," *IEEE Transactions on Automatic Control,* February 1993, pp. 195–207.

16. J. Ackermann and W. Sienel, "Robust Yaw Damping of Cars with Front and Rear Wheel Steering," *IEEE Transactions on Control Systems Technology,* March 1993, pp. 15–20.

17. L. V. Merritt, "Differential Drive Film Transport," *Motion,* June 1993, pp. 12–21.

18. S. Ashley, "Putting a Suspension Through Its Paces," *Mechanical Engineering,* April 1993, pp. 56–57.

19. D. A. Linkens, "Anaesthesia Simulators," *Computing and Control Engineering Journal,* IEEE, April 1993, pp. 55–62.

20. J. R. Layne, "Control for Cargo Ship Steering," *IEEE Control Systems,* December 1993, pp. 58–64.

21. A. Titli, "Three Control Approaches for the Design of Car Semi-active Suspension," *IEEE Proceedings of Conference on Decision and Control,* December 1993, pp. 2962–2963.

Chapter 9

1. H. Nyquist, "Regeneration Theory," *Bell Systems Tech. J.,* January 1932, pp. 126–147. Also in *Automatic Control: Classical Linear Theory,* G. J. Thaler, ed., Dowden, Hutchinson, and Ross, Stroudsburg, Pa., 1932, pp. 105–126.

2. M. D. Fagen, *A History of Engineering and Science in the Bell System,* Bell Telephone Laboratories, Inc., Murray Hill, N.J., 1978, Chapter 5.

3. H. M. James, N. B. Nichols, and R. S. Phillips, *Theory of Servomechanisms,* McGraw-Hill, New York, 1947.

4. W. J. Rugh, *Linear System Theory,* Prentice-Hall, Englewood Cliffs, N.J., 1993.

5. D. A. Linkens, *CAD for Control Systems,* Marcel Dekker, New York, 1993.

6. H. Saadat, *Computational Aids in Control Systems Using MATLAB,* McGraw-Hill, New York, 1993.

7. R. C. Dorf, *Electrical Engineering Handbook,* CRC Press, Boca Raton, Fla., 1993.

8. D. Sbarbaro-Hofer, "Control of a Steel Rolling Mill," *IEEE Control Systems,* June 1993, pp. 69–75.

9. R. C. Dorf and A. Kusiak, *Handbook of Manufacturing and Automation,* John Wiley & Sons, New York, 1994.

10. J. J. Gribble, "Systems with Time Delay," *IEEE Control Systems,* February 1993, pp. 54–55.

11. C. N. Dorny, *Understanding Dynamic Systems,* Prentice-Hall, Englewood Cliffs, N.J., 1993.

12. R. C. Dorf, *Electric Circuits,* 2nd ed., John Wiley & Sons, New York, 1993.

13. T. B. Sheridan, *Telerobotics, Automation and Control,* MIT Press, Cambridge, Mass., 1992.

14. K. K. Chew, "Control of Errors in Disk Drive Systems," *IEEE Control Systems,* January 1990, pp. 16–19.

15. R. C. Dorf, *The Encyclopedia of Robotics,* John Wiley & Sons, New York, 1988.

16. D. W. Freeman, "Jump-Jet Airliner," *Popular Mechanics,* June 1993, pp. 38–40.

17. L. O'Connor, "Electrohydraulic Valves Take Control," *Mechanical Engineering,* June 1993, pp. 54–57.

18. B. K. Bose, *Modern Power Electronics,* IEEE Press, New York, 1992.

19. G. T. Pope, "America's Last and Best Shuttle," *Popular Mechanics,* June 1992, pp. 44–47.

20. A. T. Bahill and L. Stark, "The Trajectories of Sac-

cadic Eye Movements," *Scientific American,* January 1979, pp. 108–117.

21. J. V. Wait and L. P. Huelsman, *Operational Amplifier Theory,* 2nd ed., McGraw-Hill, New York, 1992.

22. A. G. Ulsoy, "Control of Machining Processes," ASME, *Journal of Dynamic Systems,* June 1993, pp. 301–310.

23. C. E. Rohrs, J. L. Melsa, and D. Schultz, *Linear Control Systems,* McGraw-Hill, New York, 1993.

24. J. L. Jones and A. M. Flynn, *Mobile Robots,* A. K. Peters Publishing, New York, 1993.

25. D. A. Linkens, "Adaptive and Intelligent Control in Anesthesia," *IEEE Control Systems,* December 1992, pp. 6–10.

26. R. H. Bishop, "Adaptive Control of Space Station with Control Moment Gyros," *IEEE Control Systems,* October 1992, pp. 23–27.

27. J. B. Song, "Application of Adaptive Control to Arc Welding Processes," *Proceedings of the American Control Conference,* IEEE, June 1993, pp. 1751–1755.

28. X. G. Wang, "Estimation in Paper Machine Control," *IEEE Control Systems,* August 1993, pp. 34–43.

29. R. Patton, "Mag Lift," *Scientific American,* October 1993, pp. 108–109.

Chapter 10

1. R. C. Dorf, *Electrical Engineering Handbook,* CRC Press, Boca Raton, Fla., 1993.

2. A. Saberi, *Loop Transfer Recovery,* Springer-Verlag, New York, 1993.

3. J. R. Mitchell and W. L. McDaniel, Jr., "A Computerized Compensator Design Algorithm with Launch Vehicle Applications," *IEEE Trans. on Automatic Control,* June 1976, pp. 366–371.

4. W. R. Wakeland, "Bode Compensator Design," *IEEE Transactions on Automatic Control,* October 1976, pp. 771–773.

5. J. R. Mitchell, "Comments on Bode Compensator Design," *IEEE Trans. on Automatic Control,* October 1977, pp. 869–870.

6. S. T. Van Voorhis, "Digital Control of Measurement Graphics," *Hewlett-Packard Journal,* January 1986, pp. 24–26.

7. R. H. Bishop, "Adaptive Control of Space Station with Control Moment Gyros," *IEEE Control Systems,* October 1992, pp. 23–27.

8. C. L. Phillips, "Analytical Bode Design of Controllers," *IEEE Transactions on Education,* February 1985, pp. 43–44.

9. W. Ali and J. H. Burghart, "Effects of Lag Controllers on Transient Response," *IEE Proceedings,* March 1991, pp. 119–122.

10. B. Sridhar, "Kalman Filters for Helicopter Flight," *IEEE Control Systems,* August 1993, pp. 26–33.

11. S. P. Parker, *Encyclopedia of Engineering,* 2nd ed., McGraw-Hill, New York, 1993.

12. T. B. Sheridan, *Telerobotics, Automation and Control,* MIT Press, Cambridge, Mass., 1992.

13. R. C. Dorf, *The Encyclopedia of Robotics,* John Wiley & Sons, New York, 1988.

14. R. L. Wells, "Control of a Flexible Robot Arm," *IEEE Control Systems,* January 1990, pp. 9–15.

15. H. Kazerooni, "Human Extenders," *Journal of Dynamic Systems,* ASME, June 1993, pp. 281–290.

16. R. C. Dorf and A. Kusiak, *Handbook of Manufacturing and Automation,* John Wiley & Sons, New York, 1994.

17. G. T. Pope, "America's Last and Best Shuttle," *Popular Mechanics,* June 1992, pp. 44–47.

18. K. Pfeiffer and R. Isermann, "Driver Simulation in Dynamical Engine Test Stands," *Proceedings of the American Control Conference,* IEEE, 1993, pp. 721–725.

19. M. S. Habib, "Vehicle-Driver Stability Domain Using Human Pilot Model," *Proceedings of the American Control Conference,* IEEE, 1993, pp. 726–730.

20. A. G. Ulsoy, "Control of Machining Processes," ASME, *Journal of Dynamic Systems,* June 1993, pp. 301–310.

21. B. K. Bose, *Modern Power Electronics,* IEEE Press, New York, 1992.

22. J. L. Jones and A. M. Flynn, *Mobile Robots,* A. K. Peters Publishing, New York, 1993.

23. J. M. Weiss, "The TGV Comes to Texas," *Europe,* March 1993, pp. 18–20.

24. P. Schmidt, "A Parameter Optimization Approach to Controller Partioning for Flight Control," *IEEE Transactions on Control Systems Technology,* March 1993, pp. 21–36.

25. H. Kazerooni, "A Controller Design Framework for Telerobotic Systems," *IEEE Transactions on Control Systems Technology,* March 1993, pp. 50–62.

26. F. N. Bailey, "High Performance Materials Testing," *IEEE Control Systems,* April 1992, pp. 63–70.

27. J. Ackermann, "Robust Yaw Damping of Cars with Front and Rear Wheel Steering," *IEEE Transactions on Control Systems Technology,* March 1993, pp. 15–20.

28. C. T. Chen, *Linear System Theory and Design,* Holt, Rinehart and Winston, New York, 1984.

29. C. T. Chen, *Analog and Digital Control System Design,* Holt Brace Jovanovich, Orlando, Fla., 1993.

30. J. H. Blakelock, *Automatic Control of Aircraft and Missiles,* 2nd ed., John Wiley & Sons, New York, 1991.

Chapter 11

1. R. C. Dorf, *Electrical Engineering Handbook,* CRC Press, Boca Raton, Fla., 1993.

2. W. J. Grantham and T. L. Vincent, *Modern Control Systems Analysis and Design,* Wiley and Sons, New York, 1993.

3. G. Vachtsevanos, "Fuzzy Logic Control of an Automotive Engine," *IEEE Control Systems,* June 1993, pp. 62–67.

4. J. Farrell, "Using Learning Techniques to Accommodate Unanticipated Faults," *IEEE Control Systems,* June 1993, pp. 40–48.

5. C. T. Chen, *Analog and Digital Control System Design,* Harcourt Brace Jovanovich, Orlando, Fla., 1993.

6. M. Bodson, "High Performance Control of a Permanent Magnet Stepper Motor," *IEEE Transactions on Control Systems Technology,* March 1993, pp. 5–14.

7. G. W. Van der Linden, "Control of an Inverted Pendulum," *IEEE Control Systems,* August 1993, pp. 44–50.

8. Y. Ishii, "Joint Connection Mechanism for Reconfigurable Manipulator," *IEEE Control Systems,* August 1993, pp. 73–78.

9. W. J. Book, "Controlled Motion in an Elastic World," *Journal of Dynamic Systems,* June 1993, pp. 252–260.

10. C. E. Rohrs, J. L. Melsa, and D. Schultz, *Linear Control Systems,* McGraw-Hill, New York, 1993.

11. C. N. Dorny, *Understanding Dynamic Systems,* Prentice-Hall, Englewood Cliffs, N.J., 1993.

12. W. J. Rugh, *Linear System Theory,* Prentice-Hall, Englewood Cliffs, N.J., 1993.

13. C. M. Close and D. K. Frederick, *Modeling and Analysis of Dynamic Systems,* 2nd ed., Houghton Mifflin, Boston, 1993.

14. B. W. Dickinson, *Mathematics of Controls, Signals and Systems,* Springer-Verlag, New York, 1993.

15. D. Hrovat, "Applications of Optimal Control to Automotive Suspension Design," *Journal of Dynamic Systems,* ASME, June 1993, pp. 328–342.

16. R. H. Bishop, "Adaptive Control of Space Station with Control Moment Gyros," *IEEE Control Systems,* October 1992, pp. 23–27.

17. R. C. Dorf, *Encyclopedia of Robotics,* John Wiley & Sons, New York, 1988.

18. T. B. Sheridan, *Telerobotics, Automation and Control,* MIT Press, Cambridge, Mass., 1992.

19. R. C. Dorf and A. Kusiak, *Handbook of Manufacturing and Automation,* John Wiley & Sons, New York, 1994.

20. C. T. Chen, *Linear System Theory and Design,* Holt, Rinehart and Winston, New York, 1984.

21. K. E. Drexler, *Nanosystems,* John Wiley & Sons, New York, 1993.

22. F. L. Chernousko, *State Estimation for Dynamic Systems,* CRC Press, Boca Raton, Fla., 1993.

23. M. A. Gottschalk, "Dino-Adventure Duels Jurassic Park," *Design News,* August 16, 1993, pp. 52–58.

24. Y. Z. Tsypkin, "Robust Internal Model Control," *Journal of Dynamic Systems,* ASME, June 1993, pp. 419–425.

25. M. Saif, "Robust Servo Design with Applications," *IEE Proceedings,* March 1993, pp. 87–92.

26. J. K. Pieper, "Control of a Coupled-Drive Apparatus," *IEE Proceedings,* March 1993, pp. 70–79.

27. Rama K. Yedavalli, "Robust Control Design for Aerospace Applications," *IEEE Transactions of Aerospace and Electronic Systems,* vol. 25, no. 3, 1989, pp. 314–324.

28. Bryan L. Jones and Robert H. Bishop, "H_2 Optimal Halo Orbit Guidance," *Journal of Guidance, Control, and Dynamics,* AIAA, 16, no. 6, 1993, pp. 1118–1124.

Chapter 12

1. R. C. Dorf, *The Encyclopedia of Robotics,* John Wiley & Sons, New York, 1988.

2. R. C. Dorf, *Electrical Engineering Handbook,* CRC Press, Boca Raton, Fla., 1993.

3. P. Dorato, *Robust Control,* IEEE Press, New York, 1987; B. Barmish, "The Robust Root Locus," *Automatica,* vol. 26, no. 2, 1990, pp. 283–292; M. Morari, *Robust Process Control,* Prentice-Hall, Englewood Cliffs, N.J., 1991.

4. P. Dorato, "Case Studies in Robust Control Design," *IEEE Proceedings of the Decision and Control Conference,* December 1990, pp. 2030–2031.

5. C. N. Dorny, *Understanding Dynamic Systems,* Prentice-Hall, Englewood Cliffs, N.J., 1993.

6. C. M. Close and D. K. Frederick, *Modeling and Analysis of Dynamic Systems,* 2nd ed., Houghton Mifflin, Boston, 1993.

7. G. T. Pope, "Living-Room Levitation," *Discover,* June 1993, p. 24.

8. P. Varaiya, "Smart Cars on Smart Roads," *IEEE Transactions on Automatic Control,* February 1993, pp. 195–207.

9. X. G. Wang, "Estimation in Paper Machine Control," *IEEE Control Systems,* August 1993, pp. 34–43.

10. D. Sbarbaro-Hofer, "Control of a Steel Rolling Mill," *IEEE Control Systems,* June 1993, pp. 69–75.

11. B. K. Bose, *Modern Power Electronics,* IEEE Press, New York, 1992.

12. J. M. Weiss, "The TGV Comes to Texas," *Europe,* March 1993, pp. 18–20.

13. S. Lee, "Intelligent Sensing and Control for Advanced Teleoperation," *IEEE Control Systems,* June 1993, pp. 19–28.

14. J. V. Wait and L. P. Huelsman, *Operational Amplifier Theory,* 2nd ed., McGraw-Hill, New York, 1992.

15. J. L. Jones and A. M. Flynn, *Mobile Robots,* A. K. Peters Publishing, New York, 1993.

16. R. Shoureshi, "Intelligent Control Systems," *Journal of Dynamic Systems,* June 1993, pp. 392–400.

17. M. Alpert, "A Bullet Train for the U.S.," *Fortune,* May 1993, p. 32.

18. L. W. Couch, *Digital and Analog Communication Systems,* Macmillan, New York, 1993.

19. A. G. Ulsoy, "Control of Machining Processes," ASME, *Journal of Dynamic Systems,* June 1993, pp. 301–310.

20. C. T. Chen, *Analog and Digital Control System Design,* Harcourt Brace Jovanovich, Orlando, Fla., 1993.

21. T. B. Sheridan, *Telerobotics, Automation and Control,* MIT Press, Cambridge, Mass., 1992.

22. W. J. Grantham and T. L. Vincent, *Modern Control Systems Analysis and Design,* John Wiley & Sons, New York, 1993.

23. K. Capek, *Rossum's Universal Robots,* English edition by P. Selver and N. Playfair, Doubleday, Page, New York, 1923.

24. L. L. Cone, "Skycam: An Aerial Robotic Camera System," *Byte,* October 1985, pp. 122–128.

25. H. Kazerooni, "Human Extenders," *Journal of Dynamic Systems,* ASME, June 1993, pp. 281–290.

26. C. Lapiska, "Flight Simulation," *Aerospace America,* August 1993, pp. 14–17.

27. D. E. Bossert, "A Root-Locus Analysis of Quantitative Feedback Theory," *Proceedings of the American Control Conference,* June 1993, pp. 1698–1705.

28. J. A. Gutierrez and M. Rabins, "A Computer Loopshaping Algorithm for Controllers," *Proceedings of the American Control Conference,* June 1993, pp. 1711–1715.

29. J. W. Song, "Synthesis of Compensators in Linear Uncertain Plants," *Proceedings of the Conference on Decision and Control,* December 1992, pp. 2882–2883.

30. M. Gottschalk, "Part Surgeon—Part Robot," *Design News,* June 7, 1993, pp. 68–75.

31. D. E. Whitney, "From Robots to Design," *Journal of Dynamic Systems,* ASME, June 1993, pp. 262–270.

32. S. Jayasuriya, "Frequency Domain Design for Robust Performance Under Uncertainties," *Journal of Dynamic Systems,* June 1993, pp. 439–450.

33. L. S. Shieh, "Control of Uncertain Systems," *IEE Proceedings,* March 1993, pp. 99–110.

34. M. van de Panne, "A Controller for the Dynamic Walk of a Biped," *Proceedings of the Conference on Decision and Control,* IEEE, December 1992, pp. 2668–2673.

35. S. Bennett, "The Development of the PID Controller," *IEEE Control Systems,* December 1993, pp. 58–64.

36. J. C. Doyle, *Feedback Control Theory,* Macmillan, New York, 1992.

Chapter 13

1. R. C. Dorf, *The Encyclopedia of Robotics,* John Wiley & Sons, New York, 1988.
2. C. T. Chen, *Analog and Digital Control System Design,* Harcourt Brace Jovanovich, Orlando, Fla., 1993.
3. G. F. Franklin, *Digital Control of Dynamic Systems,* 2nd ed., Addison-Wesley, Reading, Mass., 1990.
4. S. H. Zak, "Ripple-Free Deadbeat Control," *IEEE Control Systems,* August 1993, pp. 51–56.
5. C. Lapiska, "Flight Simulation," *Aerospace America,* August 1993, pp. 14–17.
6. T. B. Sheridan, *Telerobotics, Automation and Control,* MIT Press, Cambridge, Mass., 1992.
7. W. J. Grantham and T. L. Vincent, *Modern Control Systems Analysis and Design,* John Wiley & Sons, New York, 1993.
8. R. C. Dorf, *Electrical Engineering Handbook,* CRC Press, Boca Raton, Fla., 1993.
9. H. Kazerooni, "Human Extenders," *Journal of Dynamic Systems,* ASME, June 1993, pp. 281–290.
10. A. G. Ulsoy, "Control of Machining Processes," ASME, *Journal of Dynamic Systems,* June 1993, pp. 301–310.
11. C. E. Rohrs, J. L. Melsa, and D. Schultz, *Linear Control Systems,* McGraw-Hill, New York.
12. R. C. Dorf and A. Kusiak, *Handbook of Manufacturing and Automation,* John Wiley & Sons, New York, 1994.
13. L. W. Couch, *Digital and Analog Communication Systems,* Macmillan, New York, 1993.
14. M. van de Panne, "A Controller for the Dynamic Walk of a Biped," *Proceedings of the Conference on Decision and Control,* IEEE, December 1992, pp. 2668–2673.
15. K. S. Yeung and H. M. Lai, "A Reformation of the Nyquist Criterion for Discrete Systems," *IEEE Transactions on Education,* February 1988, pp. 32–34.
16. A. Tarczynski, "Simple Discrete-Time Regulator

Minimizing Sensitivity," *Automatica,* vol. 29, no. 3, 1993, pp. 793–796.
17. T. R. Kurfess, "Predictive Control of a Robotic Grinding System," *Journal of Engineering for Industry,* ASME, November 1992, pp. 412–420.
18. T. Studt, "Why You Really Need a DSP," *R&D Magazine,* June 1993, pp. 32–34.
19. D. M. Auslander, "The Computer as Liberator," *Journal of Dynamic Systems,* June 1993, pp. 234–238.
20. R. Shoureshi, "Intelligent Control Systems," *Journal of Dynamic Systems,* June 1993, pp. 392–400.
21. D. J. Leo, "Control of a Flexible Frame in Slewing," *Proceedings of American Control Conference,* 1992, pp. 2535–2540.
22. M. Alpert, "A Bullet Train for the U.S.," *Fortune,* May 1993, p. 32.
23. J. L. Jones and A. M. Flynn, *Mobile Robots,* A. K. Peters Publishing, New York, 1993.
24. V. Skormin, "On-Line Diagnostics of a Self-Contained Flight Actuator," *IEEE Transactions on Aerospace and Electronic Systems,* January 1994, pp. 130–141.

Appendix C

1. R. C. Dorf, *Matrix Algebra—A Programmed Introduction,* John Wiley & Sons, New York, 1969.
2. C. R. Wylie, Jr., *Advanced Engineering Mathematics,* 4th ed., McGraw-Hill, New York, 1975.

Appendix F

1. N. E. Leonard and W. S. Levine, *Using Matlab to Analyze and Design Control Systems,* Benjamin/Cummings, Redwood City, Calif., 1992.
2. H. Saadat, *Computational Aids in Control Systems Using Matlab,* McGraw-Hill, New York, 1993.
3. B. Shahian and M. Hassul, *Computer-Aided Control System Design Using Matlab,* Prentice-Hall, Englewood Cliffs, N.J., 1993.
4. *The Student Edition of Matlab,* Prentice-Hall, Englewood Cliffs, N.J., 1992.

Index

Acceleration error constant, 235
Acceleration input, 218, 235
Accelerometer, 56, 73
Ackermann's formula, 622
ACSL, 70
Actuator The device that causes the process to provide the output; the device that provides the motive power to the process, 49
Addition of matrices, 756
Adjoint of a matrix, 760
Aircraft auto pilot, 670
Airplane control, 212, 379, 596
All-pass network, 406
Amplidyne, 55, 99
Amplifier, feedback, 174
Analog-to-digital converter, 714
Analogous variables, 35
Analysis of robustness, 652
Angle of departure The angle at which a locus leaves a complex pole in the *s*-plane, 329
Artificial heart, 14
Asymptote The path the root locus follows as the parameter becomes very large and approaches infinity. The number of asymptotes is equal to the number of poles minus the number of zeros, 322
Asymptote centroid The center of the linear asymptotes, σ_A, 323
Asymptote of root locus, 322
Asymptotic approximation for a Bode diagram, 395
Automatic control, history of, 4
Automatic fluid dispenser, 160
Automatic test system, 627, 631
Automation The control of an in-

dustrial process by automatic means, 18
Automobile steering control system, 10, 439
Auxiliary polynomial The equation that immediately precedes the zero entry in the Routh array, 282

Backward difference rule, 735
Bandwidth The frequency at which the frequency response has declined 3 db from its low-frequency value, 412, 475
Biological control system, 14
Black, H. S., 7, 102
Block diagram Unidirectional, operational block that represents the transfer functions of the elements of the system, 59
reduction, 62, 83
Bobbin drive, 673
Bode, H. W., 394
Bode plot The logarithm of the magnitude of the transfer function is plotted versus the logarithm of ω, the frequency. The phase, ϕ, of the transfer function is separately plotted versus the logarithm of the frequency, 394
asymptotic approximation for, 395
Bode plot using *Matlab*, 419
Bounded response, 275
Break frequency The frequency at which the asymptotic approximation of the frequency response for a pole (or zero) changes slope, 395

Breakaway point The point on the real axis where the locus departs from the real axis of the *s*-plane, 325

Calculus of matrices, 762
Cascade compensation network A compensator network placed in cascade or series with the system process, 528
Cauchy's theorem If a contour encircles *Z* zero and *P* poles of *F(s)* traversing clockwise, the corresponding contour in the *F(s)*-plane encircles the origin of the *F(s)*-plane $N = Z - P$ times clockwise, 448
Characteristic equation The relation formed by equating to zero the denominator of a transfer function, 41
Characteristic roots of a matrix, 286, 761
Characteristic vector, 761
Circles, constant magnitude, 469
constant phase, 471
Closed-loop feedback control system A system that uses a measurement of the output and compares it with the desired output, 3, 170
Closed-loop frequency response The frequency response of the closed-loop transfer function $T(j\omega)$, 469
Closed-loop sampled-data system, 720
Closed-loop transfer function, 62
Column vector, 755
Compensation The alteration or adjustment of a control system

Standard Variables and Parameters

$C(s)$ output
$D(s)$ disturbance
$E(s)$ error
$G(s)$ plant or process
$G_c(s)$ compensator
$G_p(s)$ prefilter
$H(s)$ feedback block transfer function
K gain constant
$q(s)$ polynomial in s, the characteristic equation
$R(s)$ input or command
s complex variable of Laplace transform
S_G^T sensitivity function
T sampling period; also time delay
T_s settling time
$T(s)$ closed-loop transfer function
$T_d(s)$ torque disturbance
$x(t)$ state variable
ω_n natural frequency
z complex variable of z-transform
ζ, zeta damping ratio

Modern History of Control Systems

1769 James Watt's flyball governor for engine speed control
1868 James C. Maxwell's analysis of the flyball governor
1880 E. J. Routh's stability analysis
1910 Elmer A. Sperry develops the gyroscope and autopilot
1927 H. S. Black's feedback amplifier
1932 H. Nyquist's stability criterion
1938 H. W. Bode demonstrates the logarithmic frequency diagram
1947 N. B. Nichols's chart is provided for frequency analysis
1948 Walter R. Evans develops the root locus method
1958 J. Engelberger and G. Devol build the first modern industrial robot
1969 W. Hoff develops the microprocessor
1970 State variable feedback control widely used
1980 Robust control system design widely studied
1990 Export-oriented manufacturing companies emphasize automation
1994 Feedback control widely used in automobiles